Supported by the National Fund for Academic Publication in Science and Technology

Taxonomy and Systematics of the Genus *Macromitrium* (Orthotrichaceae, Moss) in the World

世界蓑藓属植物的分类和系统学研究

GUO Shuiliang　　LI Dandan　　YU Jing

内 容 简 介

本书是作者对世界蓑藓属植物 20 余年分类研究工作的总结。书中展示了 256 个种和 15 个变种及亚种，并编制了分种检索表，具体包括每个种的主要特征、引证的模式标本和普遍标本，及其异名、分布等信息，同时配有绝大部分种的墨线图，及主要形态类群的彩色图片。书中还基于分子系统发育分析，提出将该属分为 2 个亚属和 22 个组，并讨论了其生物地理格局、可能的起源区域和形态性状的进化方向。

本书内容丰富，资料翔实，可供植物分类、生物多样性与进化等领域的研究人员阅读，也可作为高等院校相关专业师生的参考资料。

图书在版编目（CIP）数据

世界蓑藓属植物的分类和系统学研究 = Taxonomy and Systematics of the Genus *Macromitrium* (Orthotrichaceae, Moss) in the World：英文 / 郭水良，李丹丹，于晶著. -- 北京：科学出版社，2025.3. -- ISBN 978-7-03-081319-0

Ⅰ. Q949.35

中国国家版本馆 CIP 数据核字第 2025EL4129 号

责任编辑：王海光 / 责任校对：胡小洁
责任印制：肖　兴 / 封面设计：张云峰

科学出版社 出版

北京东黄城根北街 16 号
邮政编码：100717
http://www.sciencep.com

北京建宏印刷有限公司印刷
科学出版社发行　各地新华书店经销

*

2025 年 3 月第 一 版　　开本：889×1194　1/16
2025 年 3 月第一次印刷　　印张：29 1/2
字数：951 000

定价：398.00 元
（如有印装质量问题，我社负责调换）

Collection information: China, Xizang Aut. Reg., Mêdog Co., alt. 1193 m, *Li D.D. 057*, 2024. 8. 3
Species information: *Macromitrium giraldii* Müll. Hal.

Foreword

Twenty years ago, Dr. Guo Shuiliang went to University of Helsinki as a visiting scholar for a year supported by the State Scholarship Fund of China. Before he left, I told him that Orthotrichaceae is one of the largest moss families with a lot of problems in taxonomy and systematic and worthwhile to be studied. Since then, Dr. Guo has persisted in research of the family Orthotrichaceae, focusing on taxonomic revision of the genus *Macromitrium*. The present monograph on taxonomy and systematics of the genus *Macromitrium* in the world is the main result of Guo and his cooperators working on the genus for more than 20 years. I highly recommend this monograph book to bryologists as well as botanists at home and abroad herewith.

Macromitrium Brid. (Orthotrichaceae, Bryopsida) is the third largest moss genus in the world, with the highest diversity around the Pacific Ocean and a typical pan-tropical distribution. It is also one of the most taxonomically difficult moss genera due to a high degree of morphological variation and a huge number of species. Since its inception by Bridel in 1819, a total of 959 species names have been placed under the genus. By the end of the 20th century, about 360 species of the genus were still recognized in the world. The present monograph is basically a systematic summary of type specimen-based taxonomic revision, together with the previous taxonomic revisions by other bryologists. Major sources of material used in this monograph are type and ordinary specimens. The type specimens were mainly obtained from thirty-one herbaria in the world (B, BM, E, FH, FI, FIELD, G, GACP, GOET, H, H-BR, HIRO, Hottori, HSNU, IBSC, IFP, JE, KRAM-B, KUN, L, MEL, MICH, MO, NICH, NY, PC, PE, S, SHM, SHTU, US). The first author went to H (2003. 12-2004.11), MO (2006.12- 2007.4. 2013.1-9), IBSC (2009.8), KUN (2014.8) and PE (2015.6) to check the *Macromitrium* specimens in detail. Field collections were conducted in China, including Fujian, Guangdong, Guangxi, Guizhou, Hainan, Henan, Hunan, Jiangxi, Shannxi, Sichuan, Xizang, Yunnan and Zhejiang. Finally, more than 4000 specimens, including 730 type specimens, were checked during the revisions. As a result of taxonomic revisions, 271 species (including 12 varieties and three subspecies) are confirmed, and with about 30 taxonomically doubtful species of the genus *Macromitrium* in the world in this book. The book provided information of the literature, description of morphology, geographic distribution, habitats and figures in detail for each species. The key to 271 species of the genus *Macromitrium* in the world is also given. In addition, a preliminary system and classification of *Macromitrium* in the world was proposed based on molecular phylogenic analysis. In conclusion, this book is a comprehensive monograph about the genus *Macromitrium* in the world. I believe that the book is a valuable and useful reference for the study of mosses in the world.

In view of the large number of species of the genus, high morphological variations, some type specimens being unavailable to the authors, the taxonomic status of some species, the division, evolution and origin of the genus are still open to study.

Taxonomy is hard and time-consuming work. Congratulations to the authors for the publication of this English book with more than 400 pages. I hope they will be successful in future bryological research.

CAO Tong
Professor of Shanghai Normal University
From Shanghai, China

Contents

CHAPTER 1
Introduction 1

CHAPTER 2
Taxonomic History 3
2.1 Circumscription of the genus 3
2.2 Preliminary taxonomic history 3
2.3 Division and systematics of the genus 9

CHAPTER 3
Material and Methods 13
3.1 Specimen examination 13
3.2 Phylogenetic analysis 13
    3.2.1 Molecular protocols 13
    3.2.2 Phylogenetic tree construction 14
3.3 Data analysis about geographical distribution 14

CHAPTER 4
Taxonomy 17
4.1 Morphological notes 17
4.2 Key to recognized species 48
4.3 Descriptions of recognized species 65
    1. *Macromitrium acuminatum* (Reinw. & Hornsch.) Müll. Hal. 65
    2. *Macromitrium acutirameum* Mitt. 65
    3. *Macromitrium amaniense* P. de la Varde 65
    4. *Macromitrium amboroicum* Herzog 69
    5. *Macromitrium angulatum* Mitt. 69
    6. *Macromitrium angulosum* Thwaites & Mitt. 69
    7. *Macromitrium angustifolium* Dozy & Molk. 70
    8. *Macromitrium antarcticum* C.H. Wright 70
    9. *Macromitrium archboldii* E.B. Bartram 76
    10. *Macromitrium archeri* Mitt. 76
    11. *Macromitrium argutum* Hampe 78
    12. *Macromitrium atratum* Herzog 80
    13. *Macromitrium atroviride* R.S. Williams 80
    14. *Macromitrium attenuatum* Hampe 80
    15. *Macromitrium aurantiacum* Paris & Broth. 84
    16. *Macromitrium aurescens* Hampe 84
    17. *Macromitrium aureum* Müll. Hal. 85
    18. *Macromitrium austrocirrosum* E.B. Bartram 85
    19. *Macromitrium bifasciculare* Müll. Hal ex Dusén 88

20. *Macromitrium bifasciculatum* Müll. Hal. 88
21. *Macromitrium binsteadii* Dixon 93
22. *Macromitrium bistratosum* E. B. Bartram 93
23. *Macromitrium blumei* Nees ex Schwägr. 93
24. *Macromitrium brachypodium* Müll. Hal. 94
25. *Macromitrium brevicaule* (Besch.) Broth. 98
26. *Macromitrium brevihamatum* Herzog 98
27. *Macromitrium brevisetum* Mitt. 101
28. *Macromitrium caldense* Ångstr. 101
29. *Macromitrium caloblastoides* Müll. Hal. 101
30. *Macromitrium calocalyx* Müll. Hal. 102
31. *Macromitrium calomicron* Broth. 102
32. *Macromitrium calymperoideum* Mitt. 108
33. *Macromitrium campoanum* Thér. 108
34. *Macromitrium cardotii* Thér. 108
35. *Macromitrium cataractarum* Müll. Hal. 112
36. *Macromitrium catharinense* Paris 112
37. *Macromitrium chloromitrium* (Besch.) Wilbraham 113
38. *Macromitrium cirrosum* (Hedw.) Brid. 113
39. *Macromitrium clastophyllum* Cardot 114
40. *Macromitrium comatum* Mitt. 119
41. *Macromitrium concinnum* Mitt. ex Bosch & Sande Lac 119
42. *Macromitrium constrictum* Hampe & Lorentz 119
43. *Macromitrium crassirameum* Müll. Hal. 123
44. *Macromitrium crassiusculum* Lorentz 123
45. *Macromitrium crinale* Broth. & Geh. 123
46. *Macromitrium crispatulum* Mitt. 124
47. *Macromitrium crosbyorum* B.H. Allen & Vitt 124
48. *Macromitrium cuspidatum* Hampe 130
49. *Macromitrium cylindricum* Mitt. 130
50. *Macromitrium densum* Mitt. 130
51. *Macromitrium diaphanum* Müll. Hal. 131
52. *Macromitrium dielsii* Broth. ex Vitt & H. P. Ramsay 136
53. *Macromitrium divaricatum* Mitt. 136
54. *Macromitrium diversifolium* Broth. 136
55. *Macromitrium dubium* Schimp. ex Müll. Hal. 139
56. *Macromitrium dusenii* Müll.Hal. ex Broth. 139
57. *Macromitrium echinatum* B.H. Allen 139
58. *Macromitrium ecrispatum* Dixon 143
59. *Macromitrium eddyi* B.C. Tan & Shevock 143
60. *Macromitrium ellipticum* Hampe 143
61. *Macromitrium emersulum* Müll. Hal. 146
62. *Macromitrium erubescens* E.B. Bartram 146
63. *Macromitrium erythrocomum* H.P. Ramsay, A. Cairns & Meagher 146
64. *Macromitrium evrardii* Thér. 147
65. *Macromitrium exsertum* Broth. 147
66. *Macromitrium falcatulum* Müll. Hal. 150
67. *Macromitrium fendleri* Müll. Hal. 150
68. *Macromitrium fernandezianum* Broth. 150
69. *Macromitrium ferriei* Cardot & Thér. 154
70. *Macromitrium fimbriatum* (P. Beauv.) Schwägr. 154
71. *Macromitrium flavopilosum* R.S. Williams 155
72. *Macromitrium flexuosum* Mitt. 155
73. *Macromitrium formosae* Cardot 160
74. *Macromitrium fortunatii* Cardot & Thér. 160
75. *Macromitrium fragilicuspis* Cardot 160
76. *Macromitrium francii* Thér. 161

77. *Macromitrium frondosum* Mitt. 161
78. *Macromitrium frustratum* B.H. Allen 167
79. *Macromitrium fulgescens* E.B. Bartram 167
80. *Macromitrium fulvum* Mitt. 167
81. *Macromitrium funicaule* Schimp. ex Besch. 171
82. *Macromitrium funiforme* Dixon 171
83. *Macromitrium fuscescens* Schwägr. 171
84. *Macromitrium fuscoaureum* E.B. Bartram 172
85. *Macromitrium galipense* Müll. Hal. 176
86. *Macromitrium gigasporum* Herzog 176
87. *Macromitrium giraldii* Müll. Hal. 176
88. *Macromitrium glabratum* Broth. 178
89. *Macromitrium glaziovii* Hampe 178
90. *Macromitrium gracile* (Hook.) Schwägr. 184
91. *Macromitrium greenmanii* Grout 184
92. *Macromitrium grossirete* Müll. Hal. 184
93. *Macromitrium guatemalense* Müll. Hal. 185
94. *Macromitrium gymnostomum* Sull. & Lesq. 186
95. *Macromitrium hainanense* S.L. Guo & S. He 191
96. *Macromitrium harrisii* Paris 191
97. *Macromitrium helmsii* Paris 191
98. *Macromitrium hemitrichodes* Schwägr. 194
99. *Macromitrium herzogii* Broth. 195
100. *Macromitrium hildebrandtii* Müll. Hal. 195
101. *Macromitrium holomitrioides* Nog. 199
102. *Macromitrium hortoniae* Vitt & H. P. Ramsay 199
103. *Macromitrium huigrense* R.S. Williams 201
104. *Macromitrium humboldtense* Thouvenot & Frank Müll. 203
105. *Macromitrium incurvifolium* (Hook. & Grev.) Schwägr. 203
106. *Macromitrium involutifolium* (Hook. & Grev.) Schwägr. 204
107. *Macromitrium japonicum* Dozy & Molk. 205
108. *Macromitrium krausei* Lorentz 206
109. *Macromitrium laevigatum* Thér. 210
110. *Macromitrium laevisetum* Mitt. 210
111. *Macromitrium lanceolatum* Broth. 210
112. *Macromitrium laosianum* Paris & Broth. 213
113. *Macromitrium larrainii* Thouvenot & K.T. Yong 215
114. *Macromitrium lauterbachii* Broth. ex Fleisch. 215
115. *Macromitrium lebomboense* van Rooy 215
116. *Macromitrium leprieurii* Mont. 215
117. *Macromitrium leratii* Broth. & Paris 216
118. *Macromitrium ligulaefolium* Broth. 221
119. *Macromitrium ligulare* Mitt. 221
120. *Macromitrium lomasense* H. Rob. 224
121. *Macromitrium longicaule* Müll. Hal. 224
122. *Macromitrium longifolium* (Hook.) Brid. 226
123. *Macromitrium longipapillosum* D.D. Li, J. Yu, T. Cao & S.L. Guo 227
124. *Macromitrium longipes* (Hook.) Schwägr. 228
125. *Macromitrium longipilum* A. Braun ex Müll. Hal. 228
126. *Macromitrium longirostre* (Hook.) Schwägr. 231
127. *Macromitrium lorifolium* Paris & Broth. 231
128. *Macromitrium macrocomoides* Müll. Hal. 232
129. *Macromitrium macrosporum* Broth. 235
130. *Macromitrium macrothele* Müll. Hal. 235
131. *Macromitrium maolanense* Ze Y. Zhang, D.D. Li, J. Yu & S.L. Guo 236
132. *Macromitrium masafuerae* Broth. 236
133. *Macromitrium mcphersonii* B.H. Allen 236

134. *Macromitrium megalocladon* M. Fleisch. 237
135. *Macromitrium melinii* Roiv. 237
136. *Macromitrium menziesii* Müll. Hal. 244
137. *Macromitrium microcarpum* Müll. Hal. 244
138. *Macromitrium microstomum* (Hook. & Grev.) Schwägr. 244
139. *Macromitrium minutum* Mitt. 246
140. *Macromitrium mittenianum* Steere 251
141. *Macromitrium moorcroftii* (Hook. & Grev.) Schwägr. 251
142. *Macromitrium mosenii* Broth. 251
143. *Macromitrium nanothecium* Müll. Hal. ex Cardot 252
144. *Macromitrium nematosum* E.B. Bartram 252
145. *Macromitrium nepalense* (Hook. & Grev.) Schwägr. 258
146. *Macromitrium nigricans* Mitt. 258
147. *Macromitrium noguchianum* W. Schultze-Motel 261
148. *Macromitrium norrisianum* Vitt 261
149. *Macromitrium nubigenum* Herzog 261
150. *Macromitrium oblongum* (Taylor) Spruce 262
151. *Macromitrium ochraceoides* Dixon 262
152. *Macromitrium ochraceum* (Dozy & Molk.) Müll. Hal. 266
153. *Macromitrium onraedtii* Bizot 266
154. *Macromitrium orthophyllum* Mitt. 266
155. *Macromitrium orthostichum* Nees ex Schwägr. 267
156. *Macromitrium osculatianum* De Not. 268
157. *Macromitrium ousiense* Broth. & Paris 268
158. *Macromitrium ovale* Mitt. 269
159. *Macromitrium pallidum* (P. Beauv.) Wijk & Margad. 269
160. *Macromitrium panduraefolium* Thouvenot 277
161. *Macromitrium paridis* Besch. 277
162. *Macromitrium parvifolium* Dixon 277
163. *Macromitrium parvirete* E.B. Bartram 278
164. *Macromitrium pellucidum* Mitt. 278
165. *Macromitrium peraristatum* Broth. 282
166. *Macromitrium perdensifolium* Dixon 282
167. *Macromitrium perfragile* E. B. Bartram 282
168. *Macromitrium perichaetiale* (Hook. & Grev.) Müll. Hal. 283
169. *Macromitrium perpusillum* Müll. Hal. 283
170. *Macromitrium perreflexum* Steere 288
171. *Macromitrium pertriste* Müll. Hal. 288
172. *Macromitrium petelotii* Tixier 288
173. *Macromitrium picobonitum* B.H. Allen 289
174. *Macromitrium pilicalyx* Dixon ex E. B. Bartram 289
175. *Macromitrium piliferum* Schwägr. 289
176. *Macromitrium pilosum* Thér. 290
177. *Macromitrium proliferum* Mitt. 290
178. *Macromitrium prolongatum* Mitt. 299
179. *Macromitrium prorepens* (Hook.) Schwägr. 299
180. *Macromitrium proximum* Thér. 301
181. *Macromitrium pseudofimbriatum* Hampe 301
182. *Macromitrium pseudoserrulatum* E.B. Bartram 303
183. *Macromitrium pulchrum* Besch. 304
184. *Macromitrium pullenii* Vitt 304
185. *Macromitrium punctatum* (Hook. & Grev.) Brid. 305
186. *Macromitrium pyriforme* Müll. Hal. 306
187. *Macromitrium quercicola* Broth. 306
188. *Macromitrium ramsayae* Vitt 313
189. *Macromitrium raphidophyllum* Müll. Hal. 314
190. *Macromitrium refractifolium* Müll. Hal. 314

191. *Macromitrium regnellii* Hampe	314
192. *Macromitrium renauldii* Thér.	315
193. *Macromitrium repandum* Müll. Hal.	315
194. *Macromitrium retusulum* Müll. Hal.	316
195. *Macromitrium retusum* Hook. f. & Wilson	316
196. *Macromitrium rhacomitrioides* Nog.	321
197. *Macromitrium rimbachii* Herzog	322
198. *Macromitrium rufipilum* Cardot	322
199. *Macromitrium ruginosum* Besch.	322
200. *Macromitrium rugulosum* Ångstr.	327
201. *Macromitrium rusbyanum* E. Britton	327
202. *Macromitrium saddleanum* Besch. ex Müll. Hal.	327
203. *Macromitrium salakanum* Müll. Hal.	328
204. *Macromitrium savatieri* Besch.	328
205. *Macromitrium schmidii* Müll. Hal.	329
206. *Macromitrium scoparium* Mitt.	329
207. *Macromitrium sejunctum* B.H. Allen	330
208. *Macromitrium semperi* Müll. Hal.	330
209. *Macromitrium serpens* (Burch. ex Hook. & Grev.) Brid.	340
210. *Macromitrium sharpii* H.A. Crum ex Vitt	340
211. *Macromitrium similirete* E.B. Bartram	341
212. *Macromitrium solitarium* Müll. Hal.	341
213. *Macromitrium soulae* Renauld & Cardot	345
214. *Macromitrium speirostichum* Müll. Hal.	345
215. *Macromitrium st-johnii* E.B. Bartram	345
216. *Macromitrium standleyi* E.B. Bartram	346
217. *Macromitrium stoneae* Vitt & H.P. Ramsay	346
218. *Macromitrium streimannii* Vitt	351
219. *Macromitrium subbrevihamatum* Broth.	351
220. *Macromitrium subcirrhosum* Müll. Hal.	351
221. *Macromitrium subcrenulatum* Broth.	352
222. *Macromitrium subdiscretum* R.S. Williams	352
223. *Macromitrium subhemitrichodes* Müll. Hal.	353
224. *Macromitrium subincurvum* Cardot & Thér.	353
225. *Macromitrium sublaeve* Mitt.	353
226. *Macromitrium sublongicaule* E.B. Bartram	354
227. *Macromitrium submucronifolium* Müll. Hal. & Hampe	363
228. *Macromitrium subperichaetiale* Thér.	363
229. *Macromitrium subscabrum* Mitt.	363
230. *Macromitrium subtortum* (Hook. & Grev.) Schwägr.	366
231. *Macromitrium subulatum* Mitt.	366
232. *Macromitrium sulcatum* (Hook.) Brid.	368
233. *Macromitrium swainsonii* (Hook.) Brid.	370
234. *Macromitrium taiheizanense* Nog.	371
235. *Macromitrium taiwanense* Nog.	377
236. *Macromitrium taoense* Thér.	377
237. *Macromitrium tenax* Müll. Hal.	377
238. *Macromitrium thwaitesii* Broth. ex M. Fleisch.	381
239. *Macromitrium tocaremae* Hampe	381
240. *Macromitrium tongense* Sull.	381
241. *Macromitrium tosae* Besch.	382
242. *Macromitrium trachypodium* Mitt.	383
243. *Macromitrium trichophyllum* Mitt.	383
244. *Macromitrium trinitense* R.S. Williams	384
245. *Macromitrium tuberculatum* Dixon	384
246. *Macromitrium turgidum* Dixon	385
247. *Macromitrium tylostomum* Mitt. ex Broth & Sande Lac.	395

248. *Macromitrium ulophyllum* Mitt. 395
249. *Macromitrium uraiense* Nog. 395
250. *Macromitrium urceolatum* (Hook) Brid. 396
251. *Macromitrium validum* Herzog 396
252. *Macromitrium vesiculosum* Tixier 396
253. *Macromitrium viticulosum* (Raddi) Brid. 397
254. *Macromitrium xenizon* B.H. Allen & W.R. Buck 403
255. *Macromitrium yuleanum* Broth. & Geh. 403
256. *Macromitrium zimmermannii* M. Fleisch. 405
4.4 Doubtful species 406

CHAPTER 5

Results and Discussion 413
5.1 Taxonomy 413
5.2 Habitats 413
5.3 Geographic distribution 414
5.4 Morphological variations at the intraspecific level 416
5.5 Systematics and classification of *Macromitrium* – a preliminary result 425
    5.5.1 Molecular phylogeny of *Macromitrium* 425
    5.5.2 Toward a new classification of *Macromitrium* 427
5.6 Preliminary thought about phylogeny and biogeography of *Macromitrium* 430
    5.6.1 About the morphology of *Macromitrium* 430
    5.6.2 About general evolutionary directions and geographic region of origin 430
    5.6.3 About the endemism, disjunction and speciation time 432

Acknowledgements 433
References 435
Index to Scientific Names 443
List of all names in the genus *Macromitrium* 449

Chapter 1

Introduction

Macromitrium Brid. (Orthotrichaceae, Bryopsida) is the third largest moss genus in the world, with the highest diversity around the Pacific Ocean and a typical pan-tropical distribution. It is also one of the most taxonomically difficult moss genera due to a high degree of morphological variation and a huge number of species. Since its inception by Bridel in 1819, a total of 959 species names have been placed under the genus. By the end of the 20th century, about 360 species of the genus were still recognized in the world.

    The regional taxonomical revisions of the genus have been made for New Zealand (Vitt, 1983), Australia (Vitt & Ramsay, 1985), Papua New Guinea (Vitt *et al.*, 1995), Mexico (Vitt, 1994), South Africa (van Rooy & van Wyk, 1992), Japan (Noguchi, 1967), New Caledonia (Thouvenot, 2019) and China (Guo *et al.*, 2013). Eddy (1996) preliminarily recorded *Macromitrium* species from the Malesiana region, which included Malaysia, Indonesia, the Philippines and Melanesia (New Caledonia, Fiji, Solomon Islands, Papua New Guinea). The taxonomic work on African *Macromitrium* is weak and inadequate. Sporadic taxonomical revisions have recently been conducted on this genus in Africa by Wilbraham and her cooperators (Wilbraham, 2007, 2008, 2015, 2016, 2018; Wilbraham & Ellis, 2010). Valente *et al.* (2020) made preliminary taxonomic treatments of the Brazilian species of *Macromitrium*, though they didn't give detailed information or rationality about their nomenclatural treatments. Despite the above work, there are a large number of *Macromitrium* species, which had been reported from South Asia, South America, Africa, and Oceania, are still poorly understood and are urgently in need of taxonomic revision.

    With the encouragement of Professor Cao Tong of Shanghai Normal University and the support by National Natural Science Foundation of China (Grants: 32100171, 32071643, 31370233, 30970184, 30570121) and the Shanghai Sailing Program (No. 20YF1435500), we have conducted a continuous taxonomic revision on the genus since 2003, first focusing on Chinese species, then on Asian, finally on worldwide species of the genus. So far, we have published 30 taxonomical papers about the genus, describing three new species and one new variety, proposing two new combinations and 38 new synonyms. The present monograph is a preliminary systematic summary based on previous regional revisions and our work on the genus in the past twenty years. This monography included fifteen new synonyms and excluded three species from the genus. Finally, we confirmed 256 species and 15 varieties and subspecies based on our examinations of their type and ordinary specimens. According to our preliminary molecular phylogenetic analysis, we suggested two subgenera and 22 groups, which were partially supported by our phylogenetic evidence. We also discussed biogeographic and intraspecific variation patterns, possible original regions and evolutionary directions in morphological traits of the genus.

    Despite worldwide taxonomical revisions on the genus, there are still thirty-six doubtful species because their type specimens were unavailable to us. Therefore, the classification system constructed here is preliminary, and ongoing work is needed to construct a more natural classification system for the genus.

Chapter 2

Taxonomic History

2.1 Circumscription of the genus

Macromitrium is a sizable genus with up to 271 presently confirmed species that are widely distributed in tropical and subtropical regions in the world, with a high diversity in tropical America and tropical Asia, the East Indies and Australasia.

Plants of the genus highly vary in size, dark-green, yellow-green, brown, rusty brown to olive-brown, often in dense mats on trees, rarely on rocks. Primary stems are creeping to various degrees, often with inconspicuous and caduceus leaves; secondary stems erect-ascending, irregularly branched. Leaves are often variously twisted and contorted, some appressed and spirally curved to twisted or coiled around the branches when dry, erect, erect-spreading, spreading to squarrose-spreading when moist, keeled or plane above, most lanceolate, oblong-lanceolate, ligulate-lanceolate, oblong-ligulate to ligulate, occasionally lingulate or long, narrowly linear lanceolate, even threadlike; apices are long and narrowly acuminate, acuminate, acuminate-acute, acute, obtuse, apiculate, mucronate, occasionally retuse; costae are single, subpercurrent, percurrent, or excurrent into an arista in various length; margins are sometimes entire throughout, but often crenulate, serrulate, serrate or dentate above and entire below; upper marginal cells differentiated from their ambient cells to form a conspicuous or inconspicuous border or completely undifferentiated; the cells at the outmost marginal row at the insertion occasionally enlarged and teeth-like for some species from tropical America; upper cells often rounded, quadrate, elliptic or elongate, smooth, bulging to conic-bulging as mammillose, uni- or pluri-papillose, cells gradually elongate toward the base; low and basal cells are often linear, rarely short-rectangular, smooth or tuberculate to varying degrees, porose or not. Most species are pseudoautoicous. Perichaetial leaves are often differentiated from branch leaves to various degrees. Setae are often smooth, occasionally papillose, erect to twisted to the left, and a few to the right. Capsules are often long-exserted, some emergent to short-exserted, rarely immersed, with superficial stomata; capsule urns ovoid, ellipsoid, oblong-cylindric to cylindric, smooth or furrowed, some constricted beneath the mouth, or with a four-angled mouth; opercula are conic-rostrate; peristomes are double, single or absent; exostome teeth 16, some 8 or 32, truncate or lanceolate, papillose or papillose-striate, endostome hyaline, slightly papillose, often consisting of a delicate, at times lacerated, membrane, segments present or absent. Spores are isosporous or anisosporous, smooth to papillose. Calyptrae are mitrate, mitrate-campanulate or cucullate, variously lacerate at the base, smooth or plicate, hairy or naked.

2.2 Preliminary taxonomic history

The world was divided into 21 moss geographical units (MGUs) in the *Index Muscorum* by Wijk *et al.* (1959). About 993 subordinate taxon names under the genus *Macromitrium* have been reported from all these geographical units except Ant, Eur, Am1, As1 and As5. These names included 806 taxonomically valid names and 187 invalid names, 30 subgenera or sections, subsections (including 23 valid names and 7 invalid names), 100 intraspecific taxa (including 84 valid names and 16 invalid names) and 862 species (including 698 valid names and 164 invalid names). Bridel (1819) first used the name *Macromitrium* and took *M. aciculare* Brid. (= *M. pallidum* (P. Beauv.) Wijk & Margad.) as the type species of the genus. However, the earliest species of the genus was *Macromitrium cirrosum* (Hedw.) Brid., which was described by Hedwigia (1801) as *Anictangium cirrosum* in his book '*Species Muscorum Frondosorum*'. The name *Anictangium cirrosum*, nomenclaturally dating from the pre-Hedwigian name *Hypnum cirrosum* Swartz (1788), which was described from Swartz's collection from Jamaica between 1783 and 1787.

In the early 19th century from 1801 to 1812, four *Macromitrium* species were described as the members of *Orthotrichum* (*M. fimbriatum* (P. Beauv.) Schwägr. and *M. pallidum* (P. Beauv.) Wijk & Margad.), *Weissia* (*M. uncinatum* (Brid.) Brid.) and *Schlotheimia* (*M. squarrosum* (Brid.) Müll. Hal.). Several years later, W. J. Hooker and R. K. Greville described 26 *Macromitrium* species (as numbers of *Orthotrichum*, Hooker, 1818; Hooker & Greville, 1824). Later these early names were transferred into *Macromitrium* by Schwägrichen (1823, 1824, 1826 and 1827), and by Bridel (1826), Hampe (1844) and Müller (1845). Later in the 1820s, Schwägrichen (1826, 1827) described ten *Macromitrium* species, among them four were placed at first in *Orthotrichum*. It is somewhat strange that only two species (*M. nigrescens* Kunze and *M. tenerum* Kunze) were described during the whole 1830's (Holl, 1830).

From the Mid-19th century (1840) to the Mid-20th century (1965), a total of 678 valid names of *Macromitrium* at specific and intraspecific levels were reported, accounting for 86.70% of the total valid names of the genus. During that period, most species names (487) were dependently published by Müller C (up to 149 species), Mitten W (75), Brotherus VF (57), Thériot I (37). Bartram EB (39), Dixon NH (31), Bescherelle É (28), Cardot J (22), Hampe GEL (20), Herzog T (17), Hornschuch CF (13), Bosch RB and Sande Lac CM (12). Additionally, 53 species names were reported by Thériot I (7), Cardot J (14), Brotherus VF (24), Mitten W (5) and Hampe GEL (3) as coauthors. Namely, nearly 80% species names of the genus from 1840 to 1965 were described by the bryologists mentioned above.

Since the 1960s, about fifty *Macromitrium* species have been described, mainly by Vitt and Ramsay (8) from Australia and New Zealand, by Allen, Vitt and Buck (7) from Central America, by Guo, He, Li *et al*. (4) from Japan and China, by Thouvenot, Frank Muell, Young (3) from New Caledonia, and by Tixier (2) from Vietnam.

Regional revisions of the genus have been conducted for Japan (Noguchi, 1967), New Zealand (Vitt, 1983), Australasia (Vitt & Ramsay, 1985), South Africa (van Rooy & Wyk, 1992), Mexico (Vitt, 1994), Central America (Allen, 2002), China (Guo *et al*., 2012), Brazil (Valente *et al*., 2020), New Caledonia (Thouvenot, 2019). Considering the huge number of species names for the genus, here we mainly reviewed the taxonomical history about the species names, which have been taxonomically still accepted so far.

As 2, mainly including most of China, Japan and the Korean peninsula, with about 30 presently-accepted species of the genus, is a relatively independent geographical region of *Macromitrium* flora in the world. Most *Macromitrium* species reported from As 2 are endemic to this region.

In China, Lindberg (1864) reported the first *Macromitrium* species, *M. japonicum* Dozy & Molk. (as *Dasymitrium incurvum* Lindb.). Later, Thériot and Cardot (Thériot, 1906, 1909) described *M. cavaleriei* Cardot & Thér., *M. fortunatii* Cardot & Thér. from Guizhou province; Brotherus described *M. ousiense* Broth. & Paris from Wuxi (Jiangsu, as Ou Si) (Paris, 1910), and Dixon (1933) described *M. tuberculatum* from Hong Kong. From Taiwan, Cardot (1905) described *M. formosae*, Noguchi described *M. taiheizanense*, *M. holomitrioides*, *M. taiwanense*, *M. uraiense*, and *M. rhacomitrioides* (Noguchi, 1936, 1938). Recently, Guo and his cooperators added three new *Macromitrium* taxa into the moss flora of China, including *M. schmidii* var. *macroperichaetialium* from Guangdong (Guo *et al*., 2007a), *M. hainanense* from Hainan province (Guo & He, 2008a) and *M. maolanense* from Guizhou and Hunan (Zhang *et al*., 2019).

Additionally, many species of the genus, which had been described as new taxa from China, later were synonymized. Representative examples are *M. sinense* E.B. Bartram, *M. syntrichophyllum* Thér. & P. de la Varde, *M. syntrichophyllum* var. *longisetum* Thér. & Reimers, *M. gebaueri* Broth., *M. giraldii* var. *acrophylloides* Müll. Hal., *M. rigbyanum* Dixon, *M. handelii* Broth., and *M. cancellatum* Y.X. Xiong, which were reduced to synonymy of *M. cavaleriei* Cardot & Thér. (Guo & He, 2008b; Lou *et al*., 2014). *Macromitrium melanostomum* Paris & Broth., *M. cylindrothecium* Nog., *M. chungkingense* P.C. Chen, and *M. courtoisii* Broth. & Paris were placed in synonymy with *M. tosae* Besch. (Yu *et al*., 2012); *Macromitrium incurvum* (Lindb.) Mitt. (as *Dasymitrium incurvum* Lindb.) and *M. giraldii* Müll. Hal. were placed in synonymy with *M. japonicum* Dozy & Molk. (Ignatov & Afonina, 1992; Wijk *et al*., 1964); and *Macromitrium heterodictyon* Dix (= *M. ousiense* Broth. & Par.) and *M. brevituberculatum* Dix. (= *M. gymnostomum* Sull. & Lesq.) were synonymized by Yu *et al*. (2013) and Guo *et al*. (2007b), respectively.

From Japan, thirteen *Macromitrium* species (including varieties) were described as new species. The following species names are still taxonomically recognizable: *M. comatum* Mitt. (Mitten, 1891), *M. ferriei* Cardot & Thér. (Thériot, 1908), *M. gymnostomum* Sull. & Lesq. (Sullivant & Lesquereux, 1859), *M. japonicum* Dozy & Molk. (Dozy & Molkenboer, 1844), *M. longipapillosum* D.D. Li, J. Yu, T. Cao & S.L. Guo (Li *et al*., 2017), *M. prolongatum* Mitt. (Mitten, 1891), *M. tosae* Besch. (Bescherelle, 1898a). *Macromitrium nipponicum* Nog. (= *M. comatum* Mitt.), *M. comatulum* Broth. ex Okamura *in* Matsumura (= *M. ferriei* Cardot & Thér.), *M. insularum* Sull. & Lesq. (=*M. japonicum* Dozy & Molk.), *M. japonicum* var. *makinoi* (Broth.) Nog. (as *Dasymitrium makinoi* Broth., = *M. japonicum* Dozy & Molk.), *M. brachycladulum* Broth. & Paris (= *M. prolongatum* Mitt.), and *M. prolongatum* var. *brevipes* Cardot (= *M. prolongatum* Mitt.) were synonymized later (Noguchi & Iwatsuki, 1989).

Macromitrium clastophyllum Cardot is the only species described from the Korean peninsula (Cardot, 1904) and is still taxonomically recognizable now.

As 3, mainly includes South Asia and Indochina. The region is also rich in *Macromitrium* species. From Indochina (including Vietnam, Thailand, Laos and Myanmar), nine *Macromitrium* species (taxonomically accepted up to now) were described. Among them, three species (*M. aurantiacum* Paris & Broth, *M. lorifolium* Paris & Broth. and *M. laosianum* Paris & Broth.) were described by Paris and Brotherus (1907a, 1907b, 1908) and two species (*M. petelotii* Tixier and *M. vesiculosum* Tixier, both from Vietnam) by Tixier (1966). Mitten (1856), Thériot (1931), and Dixon (1932) described *M. calymperoideum* Mitt., *M. evrardii* Thér., and *M. turgidum* Dixon, respectively.

Macromitrium flora in South Asia (including India, Nepal, Bangladesh and Sri Lanka) is regionally specific. From South Asia, a total of fifteen species, which have been accepted so far, were described as new species. Among them, *M. sulcatum* (Hook.) Brid. is the earliest species described from this region, which was described as *Schlotheimia sulcate* Hook. by W. J. Hooker from Nepal in his *Musci Exotici* in 1818-1819. Soon later in 1824, *M. moorcroftii* (Hook. & Grev.) Schwägr., *M. nepalense* (Hook. & Grev.) Schwägr. and *M. subtortum* (Hook. & Grev.) Schwägr. from India were described by Hooker and Greville as members of *Orthotrichum*. From Nepal, Mitten (1859a) described *M. densum* Mitt. Later Müller (1854), Mitten (1859a) and Brotherus (1899) described *M. schmidii* Müll. Hal., *M. nigricans* Mitt., and *M. leptocarpum* Broth., respectively. From Sri Lanka, Mitten (1859a), Müller (1870), Hampe (1872), Mitten (1873a), Fleischeler (1904) and Dixon (1930) described *M. fulvum* Mitt., *M. nietneri* Müll. Hal., *M. ellipticum* Hampe, *M. angulosum* Thwaites & Mitt., *M. thwaitesii* Broth. ex M. Fleisch. and *M. binsteadii* Dixon, respectively.

As 4, mainly including Indonesia, Papua New Guinea, New Caledonia, Malaysia and Philippines, is rather rich in *Macromitrium* species. About 50 species, which have been still taxonomically recognizable so far, were described as new species from this region. The first record of *Macromitrium* from As 4 is *M. incurvifolium* (Hook. & Grev.) Schwägr., which was described as new species (as a member of *Orthotrichum*) based on Dickson's materials collected from Indonesia (together with the material by other anonymous collectors from Australia) by Hooker and Greville.

Vitt *et al.* (1995) conducted a taxonomical revision of the genus in the Huon Peninsula and Papua New Guinea. They confirmed 29 species and described three new species (*M. norrisianum* Vitt, *M. pullenii* Vitt and *M. streimannii* Vitt), reduced 38 names as synonyms, and selected 32 lectotypes. Among the 29 confirmed species in this region, six were first described as new species from Papua New Guinea, which are *M. yuleanum* Broth. & Geh. (Brotherus, 1895a), *M. macrosporum* Broth. (Brotherus, 1898), *M. parvifolium* Dixon (Dixon 1942), *M. sublongicaule* E.B. Bartram and *M. similirete* E.B. Bartram (Bartram, 1945, 1953), and *M. noguchianum* W. Schultze-Motel (Schultze-Motel, 1962).

In *A Handbook of Malesian Mosses*, Eddy (1996) recorded 35 *Macromitrium species* from Melanesia region, which included Malaysia, Indonesia, the Philippines and Melanesia (New Caledonia, Fiji, Solomon, Papua New Guinea).

Thouvenot (2019) taxonomically revised the genus *Macromitrium* in New Caledonia. Among the 46 taxa of the genus previously recorded in this area, 24 taxa are taxonomically recognizable at specific or infraspecific rank. From 2015 to 2019, Thouvenot and his cooperators described three new species (*M. humboldtense* Thouvenot & Frank Müll., *M. panduraefolium* Thouvenot, *M. larrainii* Thouvenot & K.T. Yong), and proposed three new combinations (*M. pulchrum* var. *neocaledonicum* (Besch.) Thouvenot, *M. aurescens* var. *caledonicum* (Thér.) Thouvenot, and *M. hemitrichodes* var. *sarasinii* (Thér.) Thouvenot). Among the other eighteen accepted species of the genus in New Caledonia, eleven species were originally described from this region. These species include six described by Thériot in 1907 and 1910 (*M. cardotii* Thér., *M. francii* Thér., *M. laevigatum* Thér., *M. renauldii* Thér., *M. taoense* Thér., and *M. pilosum* Thér.), *M. brachypodium* Müll. Hal. (Müller, 1857a), three species by Bescherelle in 1873 (*M. ptychomitrioides* Besch., *M. pulchrum* Besch. and *M. brevicaule* (Besch.) Broth.) and *M. rufipilum* Cardot (1908).

From the Mid-19th century to the Mid-20th century, bryologists described 24 *Macromitrium* species from Indonesia, including *M. acuminatum* (Reinw. & Hornsch.) Müll. Hal. (as *Schlotheimia acuminata* Reinw. & Hornsch. 1829), *M. orthostichum* Nees ex Schwägr. (1842), *M. blumei* Nees ex Schwägr. by Schwägrichen (1842), *M. ochraceum* (Dozy & Molk.) Müll. Hal by Dozy as *Schlotheimia ochracea* Dozy & Molk. (1844), *M. angustifolium* Dozy & Molk. (1844), *M. cuspidatum* Hampe by Hampe (1844), *M. longicaule* Müll. Hal. (1849), *M. longipilum* A. Braun ex Müll. Hal. (1851), *M. salakanum* Müll. Hal. (1851), *M. concinnum* Mitt. ex Bosch & Sande Lac. (Dozy & Molkenboer, 1861), *M. tylostomum* Mitt. ex Broth & Sande Lac. (Dozy & Molkenboer, 1861), *M. minutum* Mitt. (1873a), *M. crinale* Broth. & Geh. (Geheeb, 1898), *M. zimmermannii* M. Fleisch. (1904), *M. complicatulum* Müll. Hal. (1900), *M. orthostichum* subsp. *micropoma* M. Fleisch., *M. megalocladon* M. Fleisch. by Fleischer, *M. lauterbachii* Broth. ex Fleisch. (1904, 1911), *M. orthostichum* var. *burgeffii* Herzog (Herzog, 1932), *M. perdensifolium* Dixon and *M. marginatum* Dixon (1934, 1935), *M. erubescens* E.B. Bartram, *M. austrocirrosum* E.B. Bartram, *M. archboldii* E.B. Bartram (Bartram, 1942). Later in 1962, Froehlich described *M. kinabaluense* J. Froehl. from North Borneo.

Only three *Macromitirum* species were originally described from the Philippines, which included two species described by Müller in 1874a (*M. semperi* Müll. Hal. and *M. falcatulum* Müll. Hal.), and a variety by Herzog (1932, *M. orthostichum* var. *siccosquarrosum* Herzog). The survey of *Macromitrium* in Malaysia seems inadequate. So far only three species, which are taxonomically acceptable nowadays, have been reported from this country. Based on the collection of Richard Eric Holttum from Malaysia, Dixon (1935) described *M. ochraceoides* Dixon. Based on material collected from Mt. Kinabalu, Malaysia, Bosch & Sande Lacoste described *M. striatum* in 1960, and Froehlich described *M. stephanodictyon* J. Froehl. in 1962 [1963].

Australasia includes two MGUs, Austr 1 and Austr 2. Vitt revised *Macromitrium* species in New Zealand in 1983. A total of 31 species names of *Macromitrium* have been proposed based on specimens collected at least partially from New Zealand since 1818. Among these names, twelve have been taxonomically acceptable so far. The earliest species of *Macromitrium* described based on the material from New Zealand are *M. gracile* (Hook.) Schwägr., *M. longipes* (Hook.) Schwägr., *M. longirostre* (Hook.) Schwägr., *M. prorepens* (Hook.) Schwägr., *M. involutifolium* (Hook. & Grev.) Schwägr. and *M. recurvifolium* (Hook. & Grev.) Brid. The former five species were first placed in the genus *Orthotrichum* by W. J. Hooker in his *Musci Exotici* in 1818-1819 based on Menzies's collection from Dusky Bay, New Zealand and later transferred into *Macromitrum* by Schwägrichen (1823, 1824, 1826, 1827). *Macromitrium recurvifolium* (Hook. & Grev.) Brid. was first described *Orthotrichum recurvifolium* by W. J. Hooker and R. K. Grevillie in 1824 and later placed in *Macromitrium* by Bridel in 1826. It was not until the 1850's, Mitten (1859b) described *M. orthophyllum* Mitt. and *M. ligulare* Mitt., Müller and Hampe (1853), Hooker (1855) described *M. submucronifolium* Müll. Hal. & Hampe and *M. retusum* Hook. f. & Wilson, respectively. Throughout the 19th century, only two *Macromitrium* species were described from New Zealand, which are *M. helmsii* Paris (Paris, 1900) and *M. ramsayae* Vitt (Vitt, 1983).

Two years later in 1985, Vitt & Ramsay revised Australian *Macromitrium*. In their revision, they described three new species (*M. hortoniae* Vitt & H.P. Ramsay, *M. stoneae* Vitt & H.P. Ramsay, and *M. dielsii* Broth. ex Vitt & H.P. Ramsay) and recognized 33 taxa of the genus. So far 59 names have been reported based on specimens collected at least partially from Australia, among them 20 taxa have been taxonomically acceptable so far. The earliest Australian species of the genus alia, which are still taxonomically accepted nowadays, are *M. microstomum* (Hook. & Grev.) Schwägr. and *M. involutifolium* (Hook. & Grev.) Schwägr. These two species were at first placed in *Orthotrichum* by Hooker and Greville (1824) and later transferred into *Macromitrium* by Schwägrichen (1827). The type material of *M. microstomum* was sent by Spence and Neill from Tasmania (Van Dieman's Land), while that of *M. involutifolium* was from New South Wales (Hooker & Greville, 1824). *Macromitrium incurvifolium* (Hook. & Grev.) Schwägr. was possibly described partially from Australia (Vitt & Ramsay, 1985). As well in 1827, Schwägrichen described *M. hemitrichodes* Schwägr. based on the material collected by Sieber from Nova Hollandia. In the second half of the 19th century, based on materials from Australia, Mitten added *M. archeri* Mitt. (Hooker, 1859) and *M. subulatum* Mitt. (1882), Müller described *M. diaphanum* Müll. Hal. and *M. weissioides* Müll. Hal. (1872), *M. repandum* Müll. Hal. (1883), *M. subhemitrichodes* Müll. Hal., *M. lonchomitrioides* Müll. Hal. and *M. caloblastoides* Müll. Hal. (1898a), Brotherus (1893, 1898) published *M. peraristatum* Broth., *M. exsertum* Broth. and *M. ligulaefolium* Broth., and Hampe (1860) added *M. aurescens* Hampe into Australian *Macromitrium* flora.

Am 2 includes Mexico, and seven countries in Central America (Guatemala, Nicaragua, El Salvador, Honduras, Panama, Costa Rica, Belize). Vitt (1994) recognized ten *Macromitrium* species in Mexico. Allen (2002) revised and confirmed 31 species in *Macromitrium* flora of Central America. So far thirty-four *Macromitrium* species are taxonomically recognized in this MGU. Among these species, eighteen were originally described based on the specimens collected from this region. In the Mid-19th century, Müller described three species of the genus based on the specimens from Guatemala (*M. guatemalense* Müll. Hal., 1851), Costa Rica (*M. subcirrhosum* Müll. Hal., 1862), and Mexico (*M. perpusillum* Müll. Hal., 1874b). Among these three species, *M. guatemalense* is the earliest *Macromitrium* species described as new from Am 2. Wilson (1857) described *M. patens* Wilson from Panama; Mitten (1869) described *M. ulophyllum* Mitt. based on the material from Guatemala. In first half of the 20th century, Bartram (1928, 1944) reported four new *Macromitrium* species based on the materials collected from Costa Rica, which are *M. parvirete* E.B. Bartram, *M. standleyi* E.B. Bartram, *M. fuscoaureum* E.B. Bartram and *M. fulgescens* E.B. Bartram; Cardot (1909), Williams (1911) and Grout (1944) described *M. fragilicuspis* Cardot from Mexico, *M. flavopilosum* R.S. Williams from Panama, *M. greenmanii* Grout from Costa Rica, respectively. Based on the collections of Aaron J. Sharp in 1944 from Durango, Mexico, Vitt (1979) described *M. sharpii* H.A. Crum ex Vitt, which was first recognized by Howard Crum but he didn't give a name to the species. At the turn of 21th century, Allen (1998) added five new species into the *Macromitrium* flora of Am 2, which are four species from Honduras (*M. frustratum* B.H. Allen, *M. picobonitum* B.H. Allen, *M. sejunctum* B.H. Allen, and *M. mcphersonii* B.H. Allen) (Allen, 1998, 2002), and one species from Panama (*M. echinatum* B.H. Allen).

Am 3 includes Caribbean island countries in Central America such as Cuba, Haiti, Dominica, Jamaica, the Guadeloupe Island, etc. Seven taxonomically acceptable *Macromitrium* species were described from Am 3. *Macromitrium cirrosum* (Hedw.) Brid. is certainly the earliest species described from Am 3. It is a bit confused about the types of the species. Wilbraham and Price (2013) designated a lectotype for the species from the material of a single herbarium sheet bearing the name '*Anictangium cirrosum*' in the Hedwig-Schwägrichen herbarium (G). The type materials of *M. cirrosum* were collected by Olof Swartz from Jamaica as well as material from Montserrat. Hooker and Greville (1824) described *Orthotrichum perichaetiale* based on the collection by A. Menzies from the island of St. Vincent, Müller (1845) transferred it into *Macromitrium*. From St. Kitts, Müller (1849) described *M. dubium* Schimp. ex Müll. Hal. Twenty years later, Mitten (1869) added *M. scoparium* Mitt. based on Crüger's material from Jamaica. In the first half of the 20th century, Paris (1900) and Thériot (1940) described *M. harrisii* Paris from Jamaica and *M. subperichaetiale* Thér. from Cuba, respectively; Grout (1944) reported *M. clavatum* Schimp. ex Grout from the Guadeloupe Island.

Am 4 covers Northwest South America, including Venezuela, Colombia, Ecuador (with Galapagos Islands), Peru and Bolivia. This region is rather rich in *Macromitrium* species in the south hemisphere. About 70 species which are taxonomically recognizable nowadays, were described from this region, with the richest in Bolivia. The earliest species of the genus, which was described from Am 4, is *M. longifolium* (Hook.) Brid. The species was first reported as *Orthotrichum longifolium* based on the collection of Bonpland from Caraccas, Venezuela by W. J. Hooker in 1818-1819, and later was transferred to *Macromitrium* by Bridle in 1826. Among eight *Macromitrium* species (taxonomically accepted nowadays) from Venezuela, six were described by Müller (1848, 1851, 1879), which are *M. fendleri* Müll. Hal. and *M. raphidophyllum* Müll. Hal. based on the collections by A. Fendler in 1879, *M. macrothele* Müll. Hal. and *M. pyriforme* Müll. Hal. based on the materials collected from Galipan, Venezuela (1848, 1851), *M. acutissimum* Müll. Hal. based on the material from Tovar and *M. stricticuspis* Müll. Hal. in 1897a. Paris (1897) described *M. venezuelense* Paris based on the material collected from Venezuela.

Considering its territorial area, Ecuador, a country in the Andes Mountains, is relatively rich in *Macromitrium* species. A total of seventeen *Macromitrium* species, which are taxonomically accepted so far, were described from Ecuador. Among them, the earliest species name is *M. osculatianum* De Not. described by De Notaris (1859). In 1868, Hampe and Lorentz reported *M. constrictum* Hampe & Lorentz based on H. Krause's collection from Ecuador. One year later, Mitten (1869) described twelve new *Macromitrium* species in his 'Musci Austro-Americani', including *M. crispatulum* Mitt., *M. cylindricum* Mitt., *M. divaricatum* Mitt., *M. frondosum* Mitt., *M. laevisetum* Mitt., *M. ovale* Mitt., *M. sublaeve* Mitt., *M. subscabrum* Mitt., *M. trachypodium* Mitt. and *M. trichophyllum* Mitt. In the same publication in 1869, Mitten transferred *Schlotheimia oblonga* into *Macromitrium*, which was described by Taylor based on the collection of William Jameson from Andes Quitense, Ecuador in 1846. Since 1869, only four *Macromitrium* species (taxonomically acceptable nowadays) have been reported, which are *M. huigrense* R.S. Williams in 1927, *M. mittenianum* Steere in 1948, *M. rimbachii* Herzog in 1952, and *M. perreflexum* Steere in 1982.

Peru is a medium-sized country with a territorial area of more than 1.28 million km^2, but only three species (taxonomically accepted nowadays) were described from this country, which are *M. brachycarpum* Mitt. described by Mitten based on the collection of Apruce in 1869, *M. melinii* Rovi. by Roivainen in 1936, and *M. lomasense* H. Rob. described by Robinson based on the collection of F. Ayala from Prov. Trujillo in 1971, Therefore, the investigation of *Macromitrium* in Peru was conducted relatively late, and possibly inadequate. Similar to Peru, Colombia has a land area of 1.14 million km^2 and diverse environmental conditions, but only three taxonomically acceptable *Macromitrium* species were described from that country. All these three species were described in the Mid-20th century, two by Hampe (*M. tocaremae* Hampe in 1862 and *M. attenuatum* Hampe in 1865) and one by Müller (*M. aureum* Müll. Hal. in 1857b).

Bolivia has a similar territorial area to Peru or Colombia, however, among 72 *Macromitrium* species acceptable in Am 4 so far, as many as seventeen *Macromitrium* species have been described from this country. All these species were described from 1896 to 1920. Based on the material collected by Rusby H. H. from Unduavi, E. Britton described *M. rusbyanum* in 1896. One year later, Müller (1897b) described four new species (*M. solitarium*, *M. cataractarum*, *M. refractifolium* and *M. crassirameum*). At the turn of the 20th century, Williams (1903) described two new species based on the materials from Apolo (*M. atroviride* and *M. subdiscretum*). Herzog made a great contribution to *Macromitrium* flora of Bolivia. By the beginning of the 20th century, based on his own specimens collected from Bolivia, Herzog described five new species, one in 1909 (*M. amboroicum*) and four in 1916 (*M. validum*, *M. nubigenum*, *M. gigasporum* and M. *brevihamatum*). In the same publication by Herzog in 1916, Brotherus added two new species (*M. subcrenulatum* and *M. herzogii*) and one variety (*M. solitarium* var. *brevipes*). Four years later, Brotherus (in Herzog 1921) described the other two species (*M. glabratum* and *M. subrevihamatum*). The above five taxa reported by Brotherus are all based on Herzog's collection from Bolivia.

Am 5 includes Brazil, Paraguay, Guiana, Trinidad and Tobago. According to Valente *et al*. (2020), 64 species have been cited for Brazil (Yano, 1981; Costa *et al*., 2011; Flora online http://floradobrasil.jbrj.gov.br/). Among these species, fourteen were taxonomically accepted and 21 were synonymized, and seven species remained unknown status. It should be noted that Valente didn't give detailed information and rationality about their nomenclatural treatments. For example, Valente *et al*. (2020) considered *M. nematosum* as a synonym of *M. argutum*. However, *M. nematosum* distinctly differs from *M. argutum* in having leaves with all cells smooth, numerous rhizoids at the base and abundant branched gemmae present on the adaxial surface. *M. perfragile* was synonymized with *M. longifolium* by Valente *et al*. 2020. Actually, *M. perfragile* differs from *M. longifolium* in having long lanceolate leaves with the fragile subulate upper portion, with cells at the outermost marginal row near insertion differentiated, hyaline enlarged, and those near costa larger and hyaline, forming a "cancellina region".

According to nomenclature and our revisionary work on the genus, among the species described from Am 5, twenty-six species are still taxonomically acceptable, most from Brazil. The earliest *Macromitrium* species described from Brazil included *Orthotrichum swainsonii* described by W. J. Hooker based on the collection of D. Swainson (1819), *Schlotheimia viticulosa* described by Raddi (1822) and *Orthotrichum punctatum* (Hooker & Greville, 1824). The above three species were transferred into *Macromitrium* by Bridel in 1826 and are still acceptable so far. By the 1840s, four more species were reported from this MGU, which are *M. leprieurii* reported by Montagne (1840) from Guiana, *M. argutum*, *M. fragile* and *M. regnellii* reported by Hampe (1847, 1849) from Brazil. In the second half of the 19th century, eleven *Macromitrium* species were described from Brazil, including *M. proliferum* and *M. pellucidum* by Mitten (1869), *M. glaziovii* and *M. pseudofimbriatum* by Hampe (1875), *M. rugulosum* and *M. caldense* by Ångström (1876), *M. diversifolium* and *M. mosenii* by Brotherus (1895b, 1895c), *M. strictfolium*, *M. eriomitrium* and *M. undatum* by Müller (1898b). In the first half of the 20th century, *M. catharinense* Paris (1900), *M. atratum* Herzog (1925), *M. nematosum* E. B. Bartram, *M. perfragile* E. B. Batram (1952) were described from Brazil, and *M. trinitense* R. S. Williams was described from Trinidad and Tobago (1922). *Macromitrium angulicaule* is the only species of the genus *Macromitrium* described from Paraguay (Müller, 1897c). *Macromitrium xenizon* is the latest species described from Am 5, which was described by Allen and Buck from French Guiana in 2003.

Am 6 includes Chile, Argentina, Uruguay and the Falkland Islands. A total of thirteen *Macromitrium* species (taxonomically acceptable nowadays) were reported as new species from this region. Among these species, *M. pseudoserrulatum* was described from Argentina by E. B. Bartram, and the other twelve species were described from Chile. In the second half of 19th century, from Chile Müller described six species (*M. microcarpum* 1849, *M. tenax* 1883, *M. saddleanum* 1885, *M. macrocomoides*, *M. pertriste* and *M. bifasciculatum* 1898a) and Lorentz described two in 1866 (*M. crassiusculum* and *M. krausei*). The other four species were reported in the first half of the 20th century, which include *M. fernandezianum* and *M. masafuerae* described by Brotherus in 1924, *M. bifasciculare* by Dusén in 1903 and *M. campoanum* by Thériot in 1939.

Africa, the second largest continent in the world, with its surrounding islands (Madagascar, Mauritius, Réunion, St. Helena, Madeira, Azores, Canary Islands and Kerguelen Islands) were divided into four MGUs. The taxonomical works on African *Macromitrium* is weak and inadequate. Taxonomical revisions have been sporadically conducted on this genus in Africa since 21th century (Wilbraham, 2007, 2008, 2015, 2016, 2018; Wilbraham & Ellis, 2010). Among the species which were first reported from these four MGUs (Afr 1-4), only 21 were recognized by our revision, most are from Madagascar (5 species), Mauritius (5) and St. Helena (4).

No species (taxonomically accepted nowadays) were described from Afr 1. Nine species (taxonomically accepted nowadays) were first described from Afr 2. Among these species, four were described from St. Helena Island with one of the earliest *Macromitrium* species for Africa (*M. fimbriatum* (P. Beauv.) Schwägr.). *Macromitrium fimbriatum* was first described as *Orthotrichum fimbriatum* P. Beauv. from St. Helena by P. Beauvois in 1805, later in 1823 Schwägrichen transferred the species into *Macromitrium*. In 1818, W. J. Hooker published *Orthotrichum urceolatum* from the St. Helena and Bridel placed it into *Macromitrium* in 1826. It was not until 1876 that Mitten described *M. acutirameum* from the island. Between the late 19th and the earlyt 20th century, three species were reported from Afr 2 (*M. liliputanum* from Kenya by Müller in 1890, *M. dusenii* from Cameroon by Müller in 1897 and *M. antarcticum* from St. Helena by Wright in 1905). The remaining three species are *M. megalosporum* (described by Thériot & Naveau from Gongo in 1927), *M. ecrispatum* (described by Dixon from Kenyan in Herzog 1935) and *M. amaniense* (reported by P. de la Varde from Tanzania in 1955).

Afr 3 includes islands in the Indian Ocean near Africa such as Comoros, Madagascar, Mauritius and Réunion. There are thirteen *Macromitrium* species which are taxonomically accepted nowadays. *Macromitrium pallidum* is the earliest species described from Afr 3, which was described as *Orthotrichum pallidum* by P. Beauvois based on the collection of Louis Marie Aubert du Petit Thouars from Mauritius in 1805 and later transferred into *Macromitrium* by Wijk and Wargad in 1960. *Macromitrium urceolatum* is also an early species name of the genus in Afr 3, which was described as a member of *Orthotrichum* by W. J. Hooker and R. K. Greville from Madagascar in 1824 and later transferred into *Macromitrium* by Bridel in 1826. Fifty years later, Müller (1876) described *M. hildebrandtii* based on the collection of J. M. Hildebrandt from Comoros. In the early 1880s, Bescherelle (1880) described *M. funicaule* and *M. chloromitrium* (as *M. fibriatum* var. *chloromitrium*) from Mauritius; Müller and Geheeb (1881) described *M. calocalyx* and *M. urceolatulum* from Madagascar. In the early 20th century, from Mauritius Brotherus (1908) described *M. lanceolatum* and *M. calomicron*, from Madagascar Müller (1915) reported *M. nanothecium* based on Hildebrand's specimens, Cardot (in Renauld 1915) reported *M. sclerodictyon* and Thériot (1925) published *M. proximum* based on the collection of Bathie de la P. The remaining species, *M. onraedtii*, was described from Madagascar by Bizot in 1974.

Afr 4 includes S. Africa and the Kerguelen Islands. Only two species (taxonomically accepted nowadays), *M. lebomboense* van Rooy and *M. serpens* (Burch. ex Hook. & Grev.) Brid., were described from this region. The latter was first described as a member of *Orthotrichum* by W. J. Hooker and R. K. Greville in 1824, and later was placed in *Macromitrium* by Bridel in 1826.

A total of seventeen species, which are taxonomically accepted nowadays, were first reported from Oc, most from Hawaii and French Polynesia. Among these species, twelve were described in the 19th century, three in the first half of the 20th century and one in the second half of the 20th century. Schwägrichen described *M. piliferum* from Hawaii in 1826 and *M. fuscescens* from Marianas in 1827, which are the two earliest *Macromitrium* species from Oc. In the second half of the 19th century, Sullivant (1859) described *M. tongense* from Tonga; Müller described four *Macromitrium* species from Oc, which are *M. menziesii* from Tahiti, French Polynesia (1862), *M. speirostichum* from Samoa (1874c), *M. altum* and *M. emersulum* from Hawaii (1896). At the same period, Mitten reported *M. angulatum* from Samoa (1868) and *M. brevisetum* from Hawaii (1873b), Bescherelle reported *M. savatieri* (1894), *M. paridis* and *M. ruginosum* (1898b) from French Polynesia. In the 1940s, five species were added into *Macromitrium* from Oc, which are *M. st-johnii* and *M. bistratosum* described by Bartram from French Polynesia in 1940, *M. okabei* by Sakurai from Ponapei in 1943, *M. pilicalyx* by Dixon from Fiji in 1948. Based on the collection of Nadeaud J from French Polynesia, Margadant (1972) described *M. tahitisecundum*.

Since 2004, with the support from the National Natural Science Foundation of China (32100171, 32071643, 31370233, 30970184, 30570121) and the Shanghai Sailing Program, China (No. 20YF1435500), the authors of this book have conducted continuous taxonomical and systematic studies of the genus in the world. During our taxonomic revision of the genus, we described four new *Macromitrium* taxa (*M. hainanense*, *M. longipapillosum*, *M. maolanense*, *M. schmidii* var. *macroperichaetialium*) (Guo & He, 2008a; Li *et al.*, 2017; Zhang *et al.*, 2019; Guo *et al.*, 2007a), proposed two new combinations, *M. sulcatum* var. *leptocarpum* (Yu *et al.*, 2018) and *M. blumei* var. *zollingeri* (Guo *et al.* 2006), and found 38 new synonyms (Guo *et al.*, 2006, 2007a, 2007b; Guo & He, 2008a, 2008b; Guo & He, 2014; Lou *et al.*, 2014; Guo *et al.*, 2012; Yu *et al.*, 2012, 2013, 2014, 2018; Li *et al.*, 2020, 2024).

2.3 Division and systematics of the genus

Since the global revision of the genus has not been completed and necessary molecular phylogenetic evidence is still lacking, there is currently no widely accepted classification system for the genus. Bryologists in different historical periods proposed different classification systems and divisions of the genus.

Early in 1845, Müller proposed a classification system for the genus. In his system, five sections were included, which were Sect. *Macrocoma* Hornsch. ex Müll. Hal., Sect. *Schlotheimia* Brid., Sect. *Chaenomitrium* Müll. Hal., Sect. *Cryptocarpon* (Dozy et Molk.) Müll. Hal. and Sect. *Macromitrium* Brid. When Müller proposed Sect. *Chaeomitrium*, he took *Macromitrium clavellatum* (Hook. & Grev.) Schwägr. as the type of the section. While *M. clavellatum* is actually a member of the genus *Drummondia* (Wijk *et al.*, 1964; Crum & Anderson, 1981; http://floranorthamerica.org/Category:Bryophyta). Based on *Cryptocarpon* Dozy & Molk., Müller (1845) estibalished Sect. *Cryptocarpon* (Dozy et Molk.) Müll. Hal. In the original publication of the section, Müller included two species, *Macromitrium brachiatum* Hook. & Wilson and *Cryptocarpon apiculatum* Dozy & Molk. However, these two species all became members of the genus *Desmotheca* (Vitt, 1990). Therefore, among the above five sections, the first four sections were excluded from *Macromitrium* (Müller, 1849; Cardot, 1897; Grout, 1944; Wijk *et al.*, 1964; Vitt, 1990).

Later in 1869, Mitten divided the species of the genus in South America into four sections: Sect. *Macrocoma* Hornsch. ex Müll. Hal., Sect. *Micromitrium* Mitt., Sect. *Goniostoma* Mitt. and Sect. *Leiostoma* Mitt. Among them, the first two sections were excluded from *Macromitrium* (Bescherelle, 1872; Grout, 1944; Crum & Steere, 1950; Florschütz, 1964). Three years later, Mitten (1873a, b) built two new sections under the genus, Sect. *Campylodictyon* Mitt. and Sect. *Cometium* Thwaites & Mitt.

Müller (1872, 1883, 1897a) proposed Sect. *Argyrothrix* Müll. Hal., Sect. *Ceratodontium* Müll. Hal., Sect. *Crispata* Müll. Hal. and Sect. *Longipila* Müll. Hal. The latter three sections, however, were invalid because of no description (www.tropicos.org).

Brotherus (1902) elevated Sections *Macrocoma* Hornsch. ex Müll. Hal., *Orthophyllina* Müll. Hal., *Cometium* Thwaites & Mitt., *Micromitrium* Mitt. and *Eumacromitrium* Brid. to the subgeneric level, further proposed a new subgenus *Trachyphyllum* Broth. with the type species *Macromitrium gracillimum* (Besch.) Broth. Additionally, Brotherus (1902) placed Sect. *Goniostoma* Mitt. and Sect. *Leiostoma* Mitt. under Subgenus *Macromitrium*. Based on molecular evidence, Goffinet & Vitt (1998) raised the subgenus *Trachyphyllum* Broth. to the generic level based on *Matteria gracillima* (Besch.) Goffinet (Basionym: *Macromitrium gracillima* Besch.). The Subg. *Micromitrium* (Mitt.) Broth., which included species with small calyptrae and branch leaves with short, smooth inner basal cells and a basal leaf limbidium of elongate cells, was later excluded from *Macromitrium* and raised to a new genus named *Groutiella* Steer. At present, only the Subg. *Eumacromitrium*, Subg. *Orthophyllina* and Subg. *Cometium* are taxonomically recognized and have been still kept within the genus *Macromitrium* (Bescherelle, 1872; Grout, 1944; Crum & Steere, 1950; Florschütz, 1964).

Cardot (1904) established Sect. *Dasymitrium* (Lindb.) Cardot after he synonymized *Dasymitrium* Lindb. with *Macromitrium* Brid. and took *Dasymitrium incurvum* Lindb. as the type of Sect. *Dasymitrium*. However, *D. incurvum* was synonymized with *M. japonicum* Dozy & Molk., while the latter is the representative species of Sect. *Leiostoma* Mitt under the Subg. *Macromitrium* (Chen, 1978). Therefore, the rationality of Sect. *Dasymitrium* Lindb. is questionable.

Fleischer (1904) divided *Macromitrium* in Java (Indonesia) and proposed a different classification system for this genus in Java. He introduced new hierarchical levels including subgenus, section, and subsection, even *Macromitrium Sensu stricto*. In his system, Fleischer proposed two new subgenera, Subg. *Diplohymenium* M. Fleisch. and *Haplodontiella* M. Fleisch., the former included two sections, Sect. *Epilimitrium* M. Fleisch. and Sect. *Cometium* Mitt. In contrast to Brotherus (1902), who advocated for elevating Sect. *Cometium* Thwaites & Mitt. to the level of subgenus, Fleischer maintained *Cometium* as section level and kept it under Subg. *Diplohymenium* M. Fleisch. Additionally, within Sect. *Cometium*, Fleischer further introduced two new subsections: Subsect. *Macrocometium* and Subsect. *Microcometium*.

Based on the collection of Dr. Rusby from Unduavi, Bolivia, Britton (1896) described *Macromitrium rusbyanum* E. Britton. The species was characterized by large plants in yellowish-brown tufts, long creeping stems up to 10 cm long, stems repeatedly branched, with caduceus leaves; leaves at branch tips spirally twisted when dry, long lanceolate-linear from a broader yellow or brown base; costa ending in the channeled apex; capsules almost globose, smooth and thick-walled, brown and shining; peristome double, spores rather large, about 80 μm. One year later, Müller (1897c) proposed a new genus *Teichodontium* based on this species (*Teichodontium rusbyanum* (E. Britton) Müll. Hal.). However, Herzog (1916) treated *Teichodontium* as a subgenus of *Macromitrium*, which included *Macromitrium gigasporum* Herzog, a species with smooth capsules. In "*A Dictionary of Mosses (third printing)*", Crosby & Magill (1981) also didn't adopt Müller's view and placed *Teichodontium* Müll. Hal. in synonymy with *Macromitrium* Brid.

Chen (1978) basically adopted Brotherus' *Macromitrium* classification system, using the name Subgen. *Macromitrium* Brid. instead of Subgen. *Eumacromitrium* Broth., and did not agree with Herzog (1916), who treated *Teichodontium* as Subgen. *Teichodontium*, not as a Section within Subg. *Macromitrium*. Chen (1978) also didn't place Sect. *Cometium* under Subgen. *Diplohymenium* M. Fleisch., which had been proposed by Fleischler (1904), but supported Brotherus' view and retained Subg. *Cometium* under the genus *Macromitrium*.

Buck (1990) conducted a study on the genus *Macromitrium* in Guyana and established a new section, Sect. *Reverberatum* W.R. Buck.

To date, five subgenera and eight sections have been taxonomically recognized as the subordinate classification units under the genus *Macromitrium*.

Five taxonomically recognized subgenera are as follows:

Subgen. *Cometium* (Thwaites & Mitt.) Broth. — Plants rather slender, in less rigid mats; stems with forked to tufted-branched branches; branch leaves sparsely-twisted when dry, spreading when moist; most leaf cells rounded except those near the outmost margins at base, thin-walled, papillose to varying degrees; setae papillose; capsules small, oval to almost spherical, smooth; peristomes double, exostome short and truncated, united together; endostome consisting of a membrane of the same height as the exostome teeth; calyptrae with long hairs (Brotherus, 1902).

Subgen. *Diplohymenium* M. Fleisch. — Most of the leaf cells of rounded, papillose, tuberculate towards the base of the leaf; setae mostly papillose; peristomes always double, formed from two low, usually adhesive membranes, without or with hints of rudimentary teeth; calyptrae glabrous, hairy to various degrees, lacerate (Fleischer, 1904).

Subgen. *Haplodontiella* M. Fleisch. — Leaf cells near the base smooth; peristome single, with 16 lanceolate teeth; setae always smooth (Fleischer, 1904).

Subgen. *Macromitrium* Brid. — Plants from small to large; stems with forked to tufted branches; cells of the lamina mostly rounded above and elongated at the base of the leaf, thickened with a narrow lumen; calyptrae large, mitrate, very rarely slit on one side (Brotherus, 1902).

Subgen. *Orthophyllina* (Müll. Hal.) Broth. — Plants large and strong, in wide, dark green to brownish mats, with branches erect short or up to 2 cm long, branches simple or forked; branch leaves appressed below, more or less twisted spirally around the branch when dry, lanceolate-lingulate, pointed; leaf cells thickened, with a small, rounded lumen and smooth upper, elongated towards the base of the leaf, thickened, with a linear lumen, smooth, at the edge a row of hyaline, rectangular, thin-walled cells; setae up to 5 mm long; capsules oval, folded at the mouth, exostome peristome absent, endostome membraneous and densely papillose at the top; calyptrae small and mitrate (Brotherus, 1902).

Eight taxonomically recognized sections are listed below:

Sect. *Argyrothrix* Müll. Hal. — Branch leaves irregularly and abruptly narrowed to a notched apex with flexuose, hyaline awn. e.g. *Macromitrium diaphanum* Müll. Hal. (Müller, 1872).

Sect. *Campylodicton* Mitt. — Low and basal cells of leaves curved. *e.g. Macromitrium incurvifolium* (Hook. & Grev.) Schwägr. (Mitten, 1873b).

Sect. *Cometium* Mitt. — Plants slender to strong, branches dichotomy tufted, stiff; leaf cells rounded, rounded-square to elliptical, thick-walled, smooth to coarsely papillose; setae papillose; capsule urns spherical to ovoid, with a small mouth and usually 4-6 stripes; calyptrae hairy to various degrees. *e.g. Macromitrium orthostichum* Nees ex Schwägr., *M. angulatum* Mitt., *M. minutum* Mitt. (Mitten, 1873a).

Subsect. 1. *Microcometium* M. Fleisch. — Plants small; leaf cells rounded to circular, large, smooth or finely papillose; setae short and papillose; capsule often long wrinkled beneath the mouth; calyptrae hairy. *e.g. M. minutum* Mitt., *M. orthostichum* Nees ex Schwägr. (Fleisher, 1904).

Subsect. 2: *Macrocometium* M. Fleisch. — Plants robust; leaf cells rounded square to elliptical, thick-walled, coarsely papillose; setae papillose. *e.g. M. ochraceum* (Dozy & Molk.) Müll. Hal. (Fleisher, 1904)).

Sect. *Epilimitrium* M. Fleisch. — Plants usually strong; leaf cells small, rounded-quadrate; basal leaf cells with large papillae; setae smooth or papillate, usually elongated; peristome double; calyptrae naked. *e.g. M. longipilum* A. Braun ex Müll. Hal., *M. lauterbachii* Broth. ex Fleisch., *M. concinnum* Mitt. ex Bosch & Sandel Lac., and *M. blumei* Nees ex Schwägr. (Fleisher, 1904).

Sect. *Goniostoma* Mitt. — Setae smooth; capsule usually small-mouthed, wrinkled, puckered or folded at the mouth; endostome peristome absent, exostome peristome developed to various degrees, discrete, sometimes caducous and inconspicuous. *e.g. M. orthophyllum* Mitt., *M. longipes* (Hook.) Schwägr., *M. microstomum* (Hook. & Grev.) Schwägr., M. *viticulosum* (Raddi) Brid., *M. longirostre* (Hook.) Schwägr. and *M. repandum* Müll. Hal. (Mitten, 1869; Brotherus, 1902).

Sect. *Leiostoma* Mitt. — Capsules with a wide or small mouth, but not wrinkled, puckered or plicated at the mouth; peristome usually double, sometimes single or absent; calyptrae elongate; leaves abruptly acuminate; low and basal leaf cells elongate. *e.g. M. ovale* Mitt., *M. microcarpum* Müll. Hal., *M. densum* Mitt., *M. swainsonii* (Hook.) Brid., M. *trichophyllum* Mitt., *M. ulophyllum* Mitt., *M. scoparium* Mitt., *M. regnellii* Hampe and *M. sublaeve* Mitt. (Mitten, 1869).

Sect. *Reverberatum* W.R. Buck — Branch leaves long linear-lanceolate; leaf cells thick-walled; calyptrae cucullate, smooth and naked. *e.g. M. dubium* Schimp. ex Müll. Hal. (Buck, 1990).

Sect. *Teichodontium* (Müll. Hal.) Chen — Species with smooth and globous capsules. e.g. *M. rusbyanum* E. Britton and *M. gigasporum* Herzog (Chen, 1978).

Problems still exist in the relationships and systematic status of taxonomic units within the genus *Macromitrium*, because Sect. *Goniostoma* Mitt., Sect. *Leiostoma* Mitt. and Sect. *Teichodontium* (Müll. Hal.) Chen were placed under Subg. *Macromitrium*, while the other five sections are under the genus *Macromitrium*. The current subdivisions of the genus *Macromitrium* were mostly established by different scholars at different times for species from different regions, and no bryologist has explored the systematic relationships between them so far.

Due to the chaotic state in the classification system of the genus *Macromitrium*, contemporary bryologists no longer followed the previously established systems (Vitt, 1994; Eddy, 1996; Allen, 2002; Ramsay *et al.*, 2006), proposed respective classification systems for species within their own research region, no longer using traditional classification ranks such as "subgenus", "section" and "subsection", but using "group" as their classification units under the genus. For example, Vitt (1994) divided ten *Macromitrium* species of Mexico into three groups; Allen (2002) divided 31 *Macromitrium* species recorded from Central America into five groups; and Vitt and Ramsay (1985) divided 32 *Macromitrium* species of Australia into five groups. From the accepted 24 *Macromitrium* species recorded from New Caledonia, Thouvenot (2019) identified five morphological groups. Among the above subdivisions of the genus, Thouvenot (2019) and Allen (2002) didn't follow the phylogenetic relationships to propose their morphological groups. Although Vitt (1983) and Vitt and Ramsay (1985) proposed groups according to phylogenetic relationships among species, they only used morphological characters to construct phylogenetic trees. Currently, bryologists no longer mention whether these "groups" of the genus are at the level of "subgenus" or "section", nor the relationship between these "groups" and the previous subordinate taxa (subgenus, section and subsection) of the genus.

Because of the lack of molecular data on global species of the genus for its phylogenetic analysis, up to now no classification system based on molecular phylogenetic relationships of the species could be proposed for *Macromitrium* in the world. As the third largest genus of mosses, so far no preliminary classification system of the genus *Macromitrium* has been widely recognized, which can not be said to be a big defect in the systematics of mosses.

Chapter 3

Material and Methods

3.1 Specimen examination

This monograph is basically a systematic summary of our type specimen-based taxonomic revision, together with the previous taxonomic revisions by other bryologists. Major sources of material used in this monograph are type and ordinary specimens. The type specimens were mainly obtained from thirty-one herbaria in the world (B, BM, E, FH, FI, FIELD, G, GACP, GOET, H, H-BR, HIRO, Hottori, HSNU, IBSC, IFP, JE, KRAM-B, KUN, L, MEL, MICH, MO, NICH, NY, PC, PE, S, SHM, SHTU, US). The first authors went to H (2003. 12-2004.11), MO (2006.12-2007.4. 2013.1-9), IBSC (2009.8), KUN (2014.8) and PE (2015.6) to check the *Macromitrium* specimens in detail. Field collections were conducted in China, including in Fujian, Guangdong, Guangxi, Guizhou, Hainan Island, Henan, Hunan, Jiangxi, Shannxi, Sichuan, Xizang, Yunnan and Zhejiang. Finally, more than 4000 specimens, including 730 type specimens, were checked during our revisions.

For type specimens, branches in dry and wet habits, sporophytes, perichaetial and vaginulae were observed and measured under stereo microscopes, a few leaves were taken from the middle of branches in order to describe their shape and areolation. The descriptions of some parts, which could not be dissected from the type specimen, e.g. perichaetium or capsule features, were referred from ordinary specimens for some species.

Microscopic examinations and measurements were taken with an Olympus-BX53 light microscope, while microphotographs were obtained for partial types with a DP74 camera mounted on the microscope or by a microcamera. Descriptions and illustrations of leaf cells were based on the leaves from the middle of the stems and branches. For the description of the perichaetial leaves, those most different from branch leaves were used, thus to show the degree of their differentiation from branch leaves.

The type specimens that have been seen were denoted by a (!), illegitimate names and later homonyms by a (*), and species names without description by a (**). Whenever possible, names without holotype have been lectotypified unless authentic material was not available or could not be located. The concepts of types followed the International Code of Plant Nomenclature (the Vienna Code). Type protologue in the original publication was given both for acceptable species and synonyms. If syntypes were assigned to a given species, then all syntypes were examined as could as possible. Considering space-saving for the book, the illustrations of the species from Papua New Guinea, New Caledonia, New Zealand and Australia were basically omitted because they were given in the taxonomic revision of the genus in these regions.

Not all specimens, which were examined in our taxonomic revision, were included as this practice is extremely space-consuming, especially for some widely distributed species. The distributions presented here are mainly based on our examination of the specimens from the above 31 herbaria, and from our field collection in China, and the records in authentic taxonomic revisions. A species was considered as doubtful if neither its type specimens were available to us nor its taxonomic revision had been conducted.

3.2 Phylogenetic analysis

3.2.1 Molecular protocols

A total of 122 samples were used for the analyses, representing 40 species of *Macromitrium*. *Schlotheimia macgregorii* Broth. & Geh., *S. grevilleana* Mitt., *Macrocoma tenuis* subsp. *sullivantii* (Müll. Hal.) Vitt, *Groutiella tomentosa* (Hornsch.) Wijk & Margad. were used as outgroups.

One nuclear and two plastid markers were chosen: the nuclear ribosomal internal transcribed spacer region ITS1-5.8S-ITS2 (hereafter, ITS2), the tRNA (Gly) (UCC) (hereafter, *trn*G), and the *trn*L-*trn*F intergenic spacer (*trn*L-F). These three regions have been widely used in phylogenetic analyses of pleurocarpous mosses (Draper & Hedenäs, 2009; Hedenäs, 2012).

DNA was extracted by using a commercially available kit (DNeasy Plant Mini Kit, Qiagen, Valencia, CA, USA) following the manufacturer's instructions. DNA was stored at -20°C. The primers in Table 1 were used to amplify the three makers. PCR amplification was carried out with the LA Taq PCR kit (TaKaRa Bio Inc., Shiga, Japan) by using 2 μL DNA template and 1.5 μL 10 μmol/L of each primer in a 30 μL total reaction volume according to the manufacturer's protocol, using a program of 35 cycles of 95°C for 30 s; 52-56°C for 30s; 72°C for 1 min and finishing with 72°C for 15 min. PCR products were purified and bidirectionally sequenced by BGI (www.genomics.cn) using the amplification primers.

Table 1. Primers of amplification and the related references

Genes	Primer name	Direction	Primer sequence (5'-3')	References
*trn*L-F	*trn*C	forward	CGAAATCGGTAGACGCTACG	Taberlet *et al.*, 1991
	*trn*F	reverse	ATTTGAACTGGTGACACGAG	
*trn*G	*trn*GF	forward	GGCTAAGGGTTATAGTCGGC	Werner *et al.*, 2009
	*trn*GR	reverse	CGGGTATAGTTTAGTGG	Pacak & Szweykowska-Kulińska, 2000
ITS2	5.8SF	forward	GACTCTCAGCAACGGATA	Hartmann *et al.*, 2006
	26SR	reverse	AGATTTTCAAGCTGGGCT	

3.2.2 Phylogenetic tree construction

Sequences were manually aligned and assembled using PhyDE version 0.997 (Müller *et al.*, 2010). Regions of partially incomplete data at the beginnings and ends of sequences were excluded from subsequent analyses. Gaps were treated as missing data. Maximum likelihood (ML) analysis was performed using RAxML-HPC v.8 (Stamatakis, 2006; Stamatakis, 2014) on the XSEDE Teragrid of the CIPRES Science Gateway (https://www.phylo.org) (Miller *et al.*, 2010) with rapid bootstrap analysis, followed by 1000 bootstrap replicates. The final tree was selected amongst suboptimal trees from each run by comparing likelihood scores under the GTRGAMMA substitution model. Trees were visualized and annotated in FigTree v1.4.4 (Rambaut, 2018).

After the deletion of regions incomplete for some samples at the beginnings and ends of the alignment, the total number of aligned sites from the three genes was 1913. Among them, 1072 sites were variable characters and 952 were parsimony-informative. The numbers of sites in each locus based on sequence length, parsimony variable sites and parsimony informative sites were given in Table 2.

Table 2. Number of sites in each locus based on sequence length (bp), parsimony variable (p.v.) sites and parsimony informative (p.i.) sites

Gene	length (bp)	p.v. sites	p.i. sites
*trn*L-F	469	176	123
*trn*G	577	210	178
ITS2	867	686	651
total	1913	1072	952

3.3 Data analysis about geographical distribution

Distribution of 271 confirmed species in the moss geographic units (referred as MGU hereafter, "Index Muscorum", Wijk *et al.*, 1959) was used for calculation and numerical analysis. A matrix with the distribution data of 271 species in MGUs (presence / absence, 1/0) were constructed based on our examination of available specimens and relevant literature. Based on the matrix, the floristic relationship of the genus *Macromitrium* in these MGUs were quantified with two approaches: Detrended Correspondence Analysis (Hill, 1979; Hill & Gauch, 1980) and Systematic Clustering Analysis. Jaccard's similarity index (S_{ij}) and Ward's linkage scheme were applied in the Clustering Analysis.

$$S_{ij} = \frac{C_{ij}}{A_i + A_j + C_{ij}}$$

Where C_{ij} is the number of species in common between MGU_i and MGU_j, A_i and A_j are the numbers of unique species in MGU_i and MGU_j, respectively.

The geographical elements of the 271 species were also quantified by using the clustering analysis (Guo *et al.*, 2017).

Chapter 4

Taxonomy

4.1 Morphological notes

Plants — Plants of *Macromitrium* are acrocarpous. The primary stem is creeping and prostrate, giving rise to erect or ascending secondary branches that terminate, when fertile, with archegonia. The plants highly vary in size, most are medium-sized with branches up to 10-20 mm tall, but some species are small in dense mats with creeping stems and short branches less than 5 mm tall, while others are large and robust in somewhat wefts with branches up to 50 mm long. The species of the genus are perennial and produce populations from several centimeters to as large as 40 cm in diameter.

Colour — Plants are also greatly diverse in colour, with light yellow upper portion and dark-yellow below portion, some plants are olive-greens, occasionally rather dull to slightly lustrous (*M. campoanum, M. ecrispatum, M. flavopilosum, M. fuscoaureum*), or rusty-yellow to reddish-brown (*M. macrosporum, M. perdensifolium, M. vesiculosum, M. yuleanum*).

Branches — Branches highly vary in length among species, from very short (*M. aurescens, M. brevicaule, M. brachypodium, M. densum, M. nepalense, M. hemitrichodes* var. *sarasinii, M. tongense*) to rather long (*M. aureum, M. echinatum, M. frondosum, M. gigasporum, M. glabratum, M. longicaule, M. longipes, M. mosenii, M. orthostichum, M. proliferum, M. saddleanum*), from rather thin (*M. orthostichum*) to thick (*M. aureum*).

Asexual reproductive structures — Most of *Macromitrium* species don't have specialized asexual reproductive structures, but a few can produce multicellular clavate gemmae. Representative species are *M. aurescens, M. brevicaule, M. gymnostomum, M. maolanense, M. nematosum* and *M. subhemitrichodes*. Many species of the genus are able to reproduce asexually by fragmentation, particularly for some species with fragile branch leaves or aristae.

Sets of branch leaves — Leaf sets are dramatically different between moist and dry conditions. When moist, the leaves of all species are straight and more or less wide-spread, without two much diagnostic values. However, in dry condition, the sets of branch leaves highly vary among different species, which could be roughly divided into the following 24 types.

Type 1 — Densely leaved, leaves regularly and spirally twisted-curved in rows, the upper portions curved toward one side to curly, forming a rope-like appearance (funiculate) to varying degrees (*M. funicaule, M. gracile, M. longipes, M. microstomum, M. onraedtii* and *M. taoense*) (Figure 1).

Type 2 — Densely leaved, leaves spirally twisted-curved in irregular rows, the upper portions flexuous, curved to curly, somewhat funiculate (*M. blumei, M. funiforme* and *M. microstomum*) (Figure 2).

Type 3 — Branches short, densely leaved, leaves obliquely appressed to spirally appressed-curved or spirally twisted (*M. acutirameum, M. brevicaule, M. crassiusculum, M. densum, M. diaphanum, M. pellucidum, M. tongense* and *M. urceolatum*) (Figure 3).

Type 4 — Leaves regularly to irregularly curved-erect, twisted, sometimes loosely arranged, with a curly-flexuose upper portion (*M. aurescens, M. brachypodium, M brevisetum, M. fernandezianum, M. fortunatii, M. lebomboense, M. lomasense, M. microcarpum, M. microstomum, M. nepalense, M. nepalense* as *M. incrustatifolium, M. repandum, M. serpens, M. swainsonii* and *M. tosae*) (Figures 4 & 5).

Type 5 — Leaves strongly contorted-twisted-crisped, the apex basically hidden in the inrolled cavity (*M. amboroicum, M. antarcticum, M. caloblastoides, M. fendleri, M. gymnostomum* var. *brevisetum, M. gymnostomum, M. holomitroides, M. involutifolium, M. japonicum, M. ligulaefolium, M. macrothele, M. microstomum* as *M. macropelma, M. nanothecium, M. ousiense, M. serpens, M. sharpii, M. stoneae, M. sulcatum* and *M. uraiense*) (Figures 6 & 7).

Type 6 — Leaves erect and individually twisted-flexuous, twisted-curved to curly above, apices curly and basically not hidden in an inrolled cavity (*M. exsertum, M. fernandezianum, M. fimbriatum, M. fimbriatum* var. *chloromitrium, M. ligulare, M. speirostichum, M. sublaeve, M. sulcatum* var. *leptocarpum* and *M. zimmermannii*) (Figure 8).

Type 7 — Branch leaves strongly keeled, erect below, spirally-twisted, variously flexuous-curly, or somewhat contorted-flexuous above (*M. fragilicuspis*) (Figure 9).

Type 8 — Leaves erecto-patent and twisted below, strongly and individually twisted-curved to twisted-crispate above, occasionally with apices hidden in the inrolled cavity, somewhat funiculate (*M. amaniense, M. austrocirrosum, M. dielsii, M. fuscescens, M. fuscescens* as *M. glaucum, M. japonicum, M. proximum, M. subtortum* and *M. viticulosum*) (Figure 10).

Type 9 — Leaves erect below, individually twisted-contorted-flexuous, strongly rugose and undulate above, apices often adaxially curved to inrolled (*M. atratum, M. calocalyx, M. crispatulum, M. glaziovii, M. guatemalense, M. moorcroftii, M. nematosum, M. paridis, M. parvirete, M. perpusillum, M. punctatum* and *M. sulcatum*) (Figure 11).

Type 10 — Leaves erect below, individually twisted, contorted-crisped-flexuose, apices twisted-flexuous or curved to inrolled (*M. acuminatum, M. angustifolium, M. constrictum, M. ferrriei, M. giraldii, M. hainanense, M. pallidum, M. quercicola, M. savatieri, M. schmidii, M. subtortum, M. subtortum* as *M. rhizomatosum* and *M. sulcatum*) (Figure 12).

Type 11 — Leaves long lanceolate, loosely arranged, erect below, irregularly and strongly twisted-contorted, contorted-crispate to contorted-flexuous above, apices twisted-flexuous or curved to inrolled (*M. archboldii, M. atroviride, M. crassirameum, M. fulgescens, M. glabratum, M. harrisii, M. macrosporum, M. ovale, M. pseudoserrulatum, M. pyriforme, M. regnellii* as *M. contextum, M. rimbachii, M. sejunctum* and *M. sulcatum*) (Figure 13).

Type 12 — Leaves stiffly erect and appressed below, erecto-patent and curved individually or to one side above (*M. orthostichum* and *M. salakanum*) (Figure 14).

Type 13 — Leaves stiffly appressed to erect below, spreading and flexuous-twisted above (*M. angulatum, M. brevihamatum, M. fimbriatum, M. lanceolatum, M. orthostichum* and *M. ruginosum*) (Figure 15).

Type 14 — Leave not very shriveled, erect below, loosely and individually spirally-curved to spirally-twisted, sometimes slightly flexuous above (*M. aureum, M. bistratosum, M. cardotii, M. crosbyorum, M. galipense, M. longifolium* as *M. crenulatum, M. perdensifolium, M. pseudofimbriatum, M. retusulum, M. subdiscretum* and *M. yuleanum*) (Figure 16).

Type 15 — Leave not very shriveled, erect below, loosely, spreading and divergent, sometimes slightly flexuous above (*M. cuspidatum, M. dubium, M. krausei, M. leprieurii* as *M. crumianum, M. ochraceum, M. subperichaetiale* and *M. ulophyllum*) (Figure 17).

Type 16 — Branches thick, leaves erect to erecto-patent, with spreading contorted-flexuous upper portions (*M. cirrosum* as *M. cirrosum* var. *stenophyllum, M. cuspidatum* var. *gracile* and *M. greenmanii*) (Figure 18).

Type 17 — Leaves stiffy, appressed below, erecto-patent, twisted to slightly twisted, weakly divergent, flexuous above (*M. bifasciculare, M. bifasciculatum, M. campoanum, M. macrocomoides, M. pertriste, M. saddleanum, M. tenax* as *M. coriaceum* and *M. trachypodium*) (Figure 19).

Type 18 — Leaves rather stiffy and straight, tightly appressed, or appressed curved to one side above (*M. orthophyllum*) (Figure 20).

Type 19 — Branch loosely leaved, leaves long-linear or threadlike, erecto-patent below, spreading and weakly (contorted- or twisted-) flexuous, loose and divergent above (*M. echinatum, M. frustratum, M. standleyi* as *M. standleyi* var. *subundulatum, M. standleyi* and *M. trinitense*) (Figure 21).

Type 20 — Leaves long lanceolate, clasping at base or erecto-patent below, squarrose to widely-spreading, flexuous and divergent above (*M. catharinense, M. cylindricum, M. divaricatum, M. fuscoaureum, M. hildebrandtii, M. trichophyllum* and *M. validum*) (Figure 22).

Type 21 — Leaves long lanceolate, loosely erect below, spreading, flexuous to weakly contorted-flexuous, contorted-twisted or contorted-crispate, curved above (*M. argutum, M. attenuatum, M. cataractarum, M. catharinense* as *M. drewii, M. cirrosum, M. constrictum, M. ecrispatum, M. flavopilosum, M. flexuosum, M. frondosum, M. gigasporum, M. guatemalense, M. herzogii, M. huigrense, M. longifolium, M. megalocladon, M. mosenii, M. nubigenum, M. osculatianum, M. ovale, M. perfragile, M. raphidophyllum, M. regnellii, M. rusbyanum, M. scoparium, M. solitarium* var. *brevipes, M. subbrevihamatum, M. subcirrhosum, M. subcrenulatum, M. subscabrum, M. sulcatum* and *M. tocaremae*) (Figures 23 & 24).

Type 22 — Leaves loosely, regularly and strongly contorted-twisted, twisted-flexuose, apices decurved to incurved (*M. semperi, M. sublongicaule, M. subtortum* as *M. subpungens* and *M. sulcatum* subsp. *ramentosum*) (Figure 25).

Type 23 — Leaves loosely and regularly arranged, regularly and strongly squarrose-recurved and often back bent (*M. perreflexum*) (Figure 26).

Type 24 — Leaves erect below, irregularly squarrose-recurved to bent back with contorted-flexuous upper portions (*M. megalocladon*) (Figure 27).

Fig. 1 Type 1 — Plants of *Macromitrium* 1: *M. microstomum* as *M. hornschuchii*, lectotype BM 000873100. 2: *M. funicaule*, isotype BM 000873889. 3: *M. microstomum*, MEL 2052518. 4: *M. gracile* as *M. mossmanianum*, isotype JE 04008719. 5: *M. onraedtii*, isotype PC 0137528. 6: *M. longipes*, MEL 2076463.

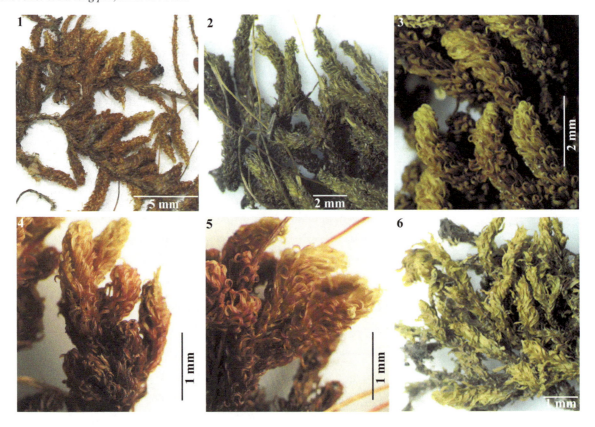

Fig. 2 Type 2 — Plants of *Macromitrium* 1: *M. microstomum* as *M. fasciculare* var. *angustifolium*, isotype PC 0105916. 2: *M. funiforme*, MEL 2241029. 3: *M. funiforme*, MEL 2226222B. 4, 5: *M. microstomum*, SHTU 20120913329. 6: *M. blumei*, MEL 239761.

Fig. 3 Type 3 — Plants of *Macromitrium* 1: *M. brevicaule*, BM 000982724. 2: *M. brevicaule*, MEL 2246639. 3: *M. densum*, isolectotype E 00165193. 4: *M. crassiusculum*, isotype NY 00518316. 5: *M. diaphanum*, MEL 2214757. 6: *M. acutirameum*, lectotype NY 00518221. 7: *M. tongense*, isotype US 00070281. 8: *M. pellucidum*, isotype NY 01086500. 9: *M. peraristatum*, MEL 1046448.

Fig. 4 Type 4 — Plants of *Macromitrium* 1: *M. nepalense* as *M. incrustatifolium*, holotype US. 2: *M. microstomum* as *M. fasciculare*, BM 000919519. 3: *M. microstomum* as *M. subnitidum*, isolectotype NY 01086589. 4: *M. brachypodium*, MEL 2096367. 5: *M. nitidum*, E 00002986. 6: *M. microstomum* as *M. pinnulatum*, syntype JE 04008712. 7: *M. urceolatum*, E 00348715. 8: *M. microcarpum*, isotype JE 04008720. 9: *M. repandum*, isotype JE 04008686.

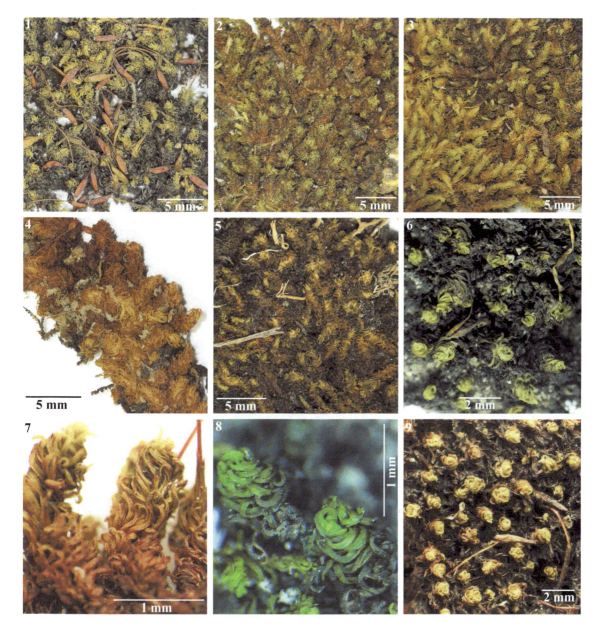

Fig. 5 Type 4 — **Plants of *Macromitrium*** 1: *M. tosae* as *M. chungkingense*, isotype JE 04008728. 2: *M. swainsonii* as *M. vesiculatum*, holotype JE 04008705. 3: *M. fernandezianum*, S-B 163396. 4: *M. serpens* as *M. anomodictyon*, isolectotype PC 0106570. 5: *M. lomasense*, holotype US 00070266. 6: *M. aurescens*, MEL 2264418. 7: *M. tosae*, SHTU 20180131065. 8: *M. nepalense*, SZG 15034. 9: *M. fortunatii*, SHTU 20181203002.

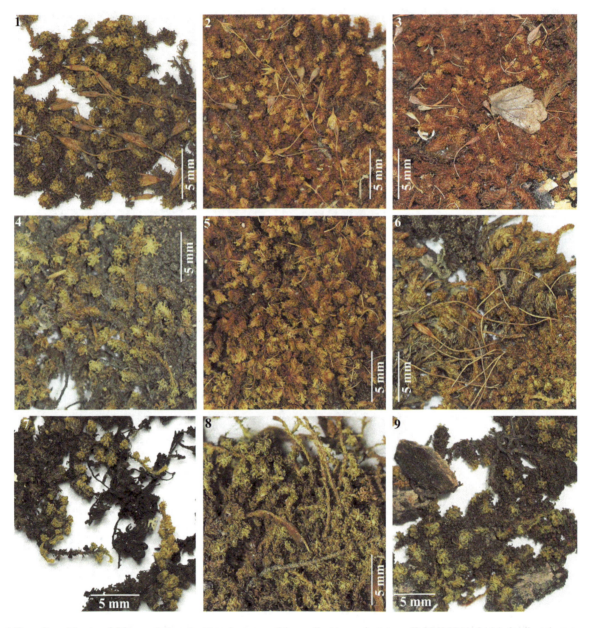

Fig. 6 Type 5 — Plants of *Macromitrium* 1: *M. sulcatum* as *M. muellerianum*, lectotype E 00625564. 2: *M. fendleri*, lectotype NY 01086590. 3: *M. macrothele*, isosyntype NY 01086495. 4: *M. involutifolium* as *M. daemelii*, isotype JE 04008747. 5: *M. sharpii*, isotype MICH 525909. 6: *M. microstomum* as *M. macropelma*, S-B 163955. 7: *M. serpens* as *M. astroideum*, E 00002458. 8: *M. amboroicum*, holotype JE 04008730. 9: *M. japonicum* as *M. polygonostomum*, isosyntype US 00070273.

Fig. 7 Type 5 — **Plants of *Macromitrium*** 1: *M. gymnostomum* var. *brevisetum*, F-C 0001228F. 2: *M. ligulaefolium* as *M. brevipilosum*, JE 04008700. 3: *M. antarcticum*, lectotype PC 0137579. 4: *M. nanothecium*, lectotype PC 0137518. 5: *M. antarcticum*, PC 0137580. 6: *M. caloblastoides*, MEL 1062981. 7: *M. involutifolium*, MEL 2264732. 8: *M. stoneae*, MEL 2261932. 9: *M. uraiense*, SHTU 20120916526. 10: *M. holomitrioides*, SHTU 20180131015. 11: *M. gymnostomum*, SHTU 20180131113. 12: *M. ousiense*, SZG CBLXH0253.

Fig. 8 Type 6 — **Plants of *Macromitrium*** 1: *M. zimmermannii*, isotype JE 04008698. 2: *M sulcatum* var. *leptocarpum*, isosyntype MICH 525887. 3: *M. sublaeve*, isotype MICH 525916. 4: *M. speirostichum*, JE 04008687. 5: *M. ligulare* as *M. luehmannianum*, isotype JE 04008749. 6: *M. exsertum*, MEL 2260913.

Fig. 9 Type 7 — **Plants of *Macromitrium*** 1, 2: *M. fragilicuspis*, isotype NY 00792493.

Taxonomy

Fig. 10 Type 8 — Plants of *Macromitrium* 1: *M. fuscescens* as *M. glaucum*, isotype F-C 0001109F. 2: *M. amaniense*, holotype PC 0105920. 3: *M. austrocirrosum*, isotype MICH 525866. 4: *M. viticulosum* as *M. intortifolium*, isotype BM 000873097. 5: *M. subtortum* as *M. mauritianum*, isolectotype BM 000873885. 6: *M. proximum*, lectotype PC 0073305. 7: *M. fuscescens*, MEL 2361948. 8: *M. dielsii*, MEL 2241074. 9: *M. japonicum*, SZG.

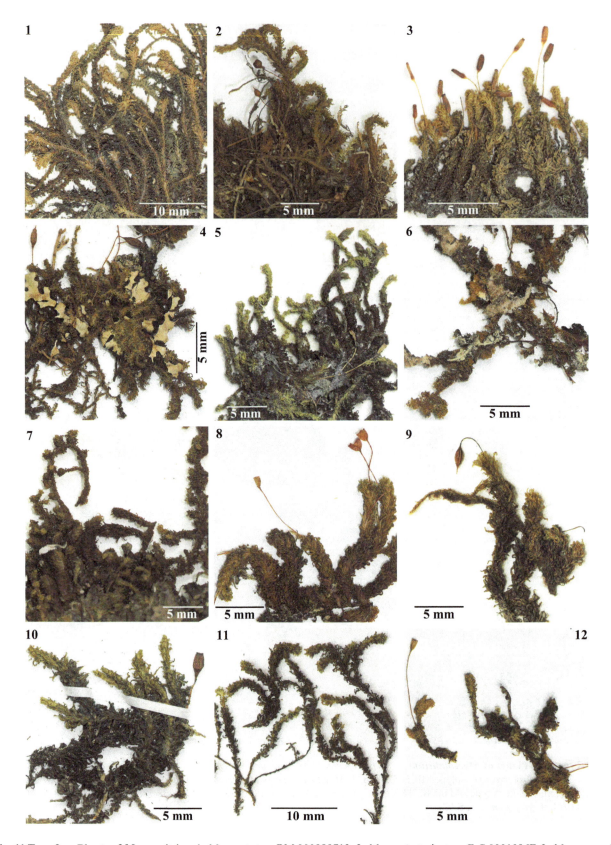

Fig. 11 Type 9 — Plants of *Macromitrium* 1: *M. punctatum*, BM 000989713. 2: *M. parvirete*, isotype F-C 0001096F. 3: *M. moorcroftii*, isotype E 00428995. 4: *M. crispatulum*, holotype NY 00518318. 5: *M. glaziovii*, isotype G 00050747. 6: *M. paridis*, isotype PC 0137864. 7: *M. guatemalense*, L 0060440. 8: *M. calocalyx* as *M. semipapillosum*, isolectotype PC 0106789. 9: *M. sulcatum*, PC 0721946. 10: *M. atratum*, holotype JE 04006253. 11: *M. nematosum*, holotype FH 00213653. 12: *M. perpusillum*, isotype PC 0137875.

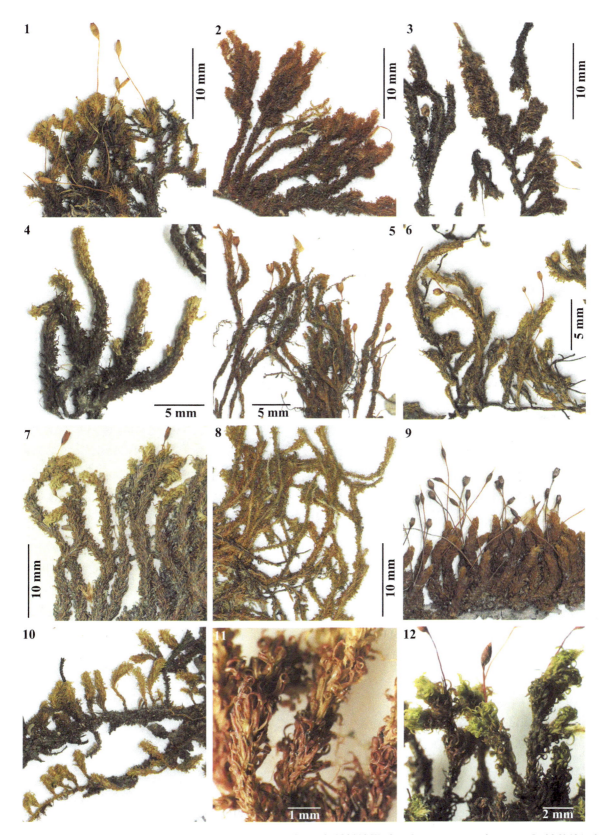

Fig. 12 Type 10 — Plants of *Macromitrium* 1: *M. quercicola*, F-C 0001125F. 2: *M. acuminatum*, lectotype L 0060421. 3: *M. angustifolium*, lectotype L 0060422. 4: *M. hainanense*, holotype MO. 5: *M. subtortum* as *M. rhizomatosum*, isotype PC 0137539. 6: *M. subtortum* as *M. sanctae-mariae*, isotype PC 0105699. 7: *M. schmidii*, JE 04008692. 8: *M. savatieri*, isotype PC 0148121. 9: *M. pallidum*, paratype BM 000873859. 10: *M. sulcatum* as *M. seriatum*, isolectotype PC 0106738. 11: *M. ferriei*, SHTU 20120910081. 12: *M. giraldii*, SHTU 20230614017.

Fig. 13 Type 11 — Plants of *Macromitrium* 1: *M. pseudoserrulatum*, holotype FH 00213665. 2: *M. harrisii*, isotype BM 000873105. 3: *M. rimbachii*, holotype JE 04008690. 4: *M. atroviride*, isotype F-C 0000994F. 5: *M. regnellii* as *M. contextum*, isotype PC 0137645. 6: *M. archboldii*, MICH 525864. 7: *M. fulgescens*, holotype FH 00213618. 8: *M. glabratum*, isosyntype JE 04008716. 9: *M. sejunctum*, holotype MO. 10: *M. crassirameum*, isolectotype JE 04008731. 11: *M. ovale*, isotype NY 01086488. 12: *M. pyriforme*, isotype NY 01086639.

Taxonomy

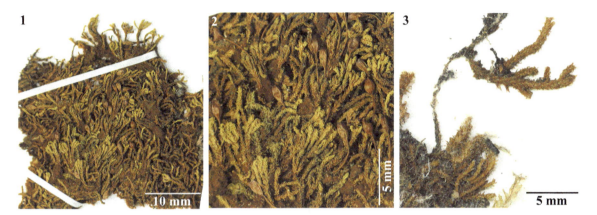

Fig. 14 Type 12— Plants of *Macromitrium* 1, 2: *M. orthostichum*, JE 04008737. 3: *M. salakanum*, isotype JE 04008691.

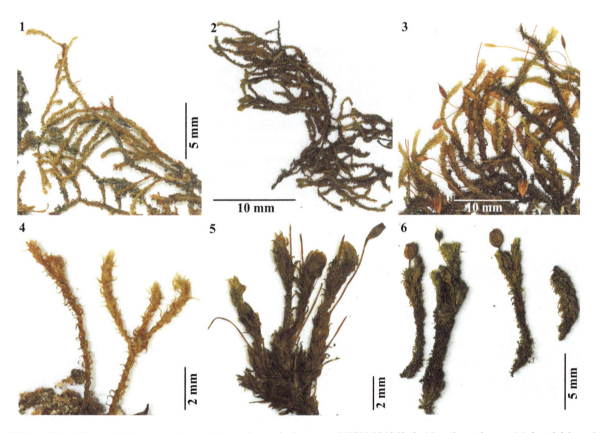

Fig. 15 Type 13 — Plants of *Macromitrium* 1: *M. angulatum*, isolectotype MICH 525863. 2: *M. orthostichum* as *M. fragilifolium*, MO. 3: *M. brevihamatum*, isosyntype S-B 162958. 4: *M. lanceolatum*, lectotype PC 0137511. 5: *M. fimbriatum*, PC 0137508. 6: *M. ruginosum*, S-B 164827.

Fig. 16 Type 14 — Plants of *Macromitrium* 1: *M. subdiscretum*, isotype JE 04008706. 2: *M. perdensifolium*, isotype JE 04008727. 3: *M. yuleanum*, MEL 2202626. 4: *M. bistratosum*, isotype BM 000982723. 5: *M. aureum*, isotype BM 000879970. 6: *M. crosbyorum*, holotype MO. 7: *M. retusulum*, lectotype FH 00213668. 8: *M. galipense*, isotype NY 01086536. 9: *M. cardotii*, isotype JE 04006255.

Taxonomy

Fig. 17 Type 15 — **Plants of *Macromitrium*** 1: *M. cuspidatum*, MEL 2356178. 2: *M. ochraceum*, MEL 2360480. 3: *M. subperichaetiale*, isotype PC 0137724. 4: *M. krausei*, lectotype NY 01202035. 5: *M. ulophyllum*, holotype NY 00322438. 6: *M. leprieurii* as *M. crumianum*, isotype MO 2846905.

Fig. 18 Type 16 — **Plants of *Macromitrium*** 1: *M. greenmanii*, isotype MO 406389. 2: *M. cuspidatum* var. *gracile*, isotype US 00070255. 3: *M. cirrosum* as *M. cirrosum* var. *stenophyllum*, E 00165151.

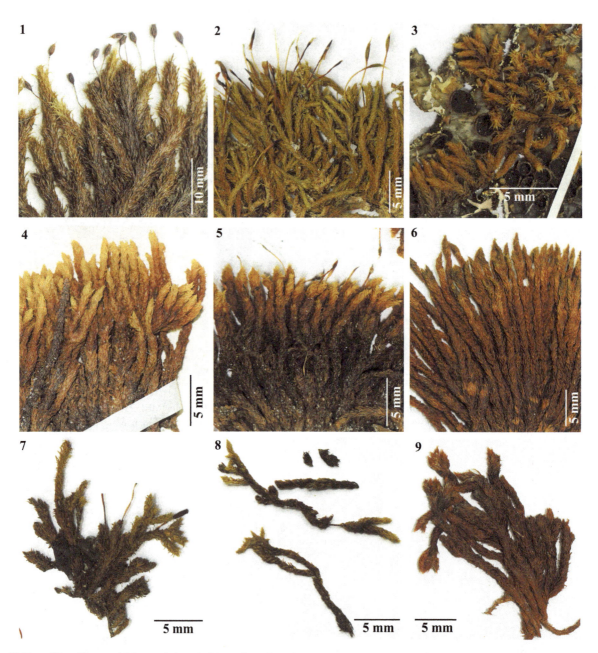

Fig. 19 Type 17 — Plants of *Macromitrium* 1: *M. trachypodium*, isosyntype E 00165165. 2: *M. tenax* as *M. coriaceum*, isolectotype F-C 0001238F. 3: *M. campoanum*, holotype JE 04000907. 4: *M. bifasciculatum*, BM 000989836. 5: *M. bifasciculare*, lectotype US 00070247. 6: *M. saddleanum*, isotype PC 0138019. 7: *M. pertriste*, isotype PC 0137885. 8: *M. macrocomoides*, isotype PC 0137796. 9: *M. bifasciculatum*, BM 000873313.

Taxonomy

Fig. 20 Type 18 — **Plants of *Macromitrium*** 1, 2: *M. orthophyllum*, MEL 1030346.

Fig. 21 Type 19 — **Plants of *Macromitrium*** 1: *M. standleyi*, isoparatype JE 04008695. 2: *M. trinitense*, BM 000873204. 3: *M. echinatum*, holotype MO. 4: *M. standleyi*, isotype NY 01243660. 5: *M. standleyi* as *M. standleyi* var. *subundulatum*, NY 01243662. 6: *M. frustratum*, holotype MO.

Fig. 22 Type 20 — Plants of *Macromitrium* 1: *M. catharinense*, isolectotype JE 04006251. 2: *M. cylindricum*, isotype E 00165152. 3: *M. trichophyllum*, NY 010866021. 4: *M. catharinense,* as *M. schiffneri* NY 01243611. 5: *M. hildebrandtii*, isotype PC 0137509. 6: *M. fuscoaureum*, US 00070259. 7: *M. divaricatum*, isotype E 00165153. 8: *M. fuscoaureum*, isotype NY 01086522. 9: *M. validum*, holotype JE 04008704.

Fig. 23 Type 21 — **Plants of *Macromitrium*** 1: *M. catharinense* as *M. drewii*, holotype US 00070256. 2: *M. gigasporum*, JE 04008679. 3: *M. huigrense*, US 00070264. 4: *M. frondosum*, isotype NY 01086533. 5: *M. herzogii*, isosyntype JE 04008714. 6: *M. longifolium*, BM 000720646. 7: *M. tocaremae*, isotype BM 000873210. 8: *M. frustratum*, F-C 0001080F. 9: *M. guatemalense* as *M. verrucosum*, isotype JE 04008693. 10: *M. cirrosum*, isosyntype E 00002459. 11: *M. attenuatum*, isotype PC 0137588. 12: *M. osculatianum*, lectotype NY 01202207.

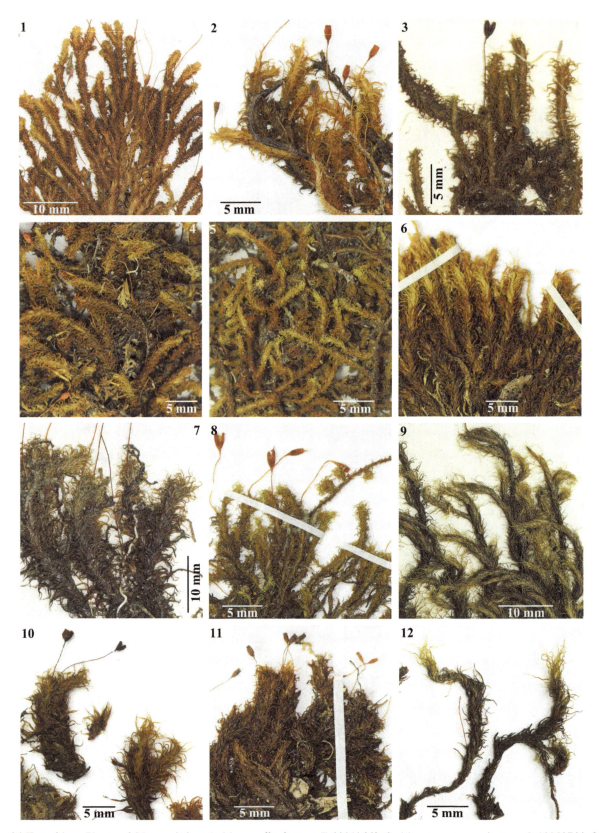

Fig. 24 Type 21 — Plants of *Macromitrium* 1: *M. regnellii*, isotype E 00011668. 2: *M. cataractarum*, isotype G 00050744. 3: *M. subscabrum*, lectotype NY 01086591. 4: *M. mosenii*, isosyntype FH 00213651. 5: *M. perfragile*, syntype FH 00290386. 6: *M. subcrenulatum*, isosyntype JE 04008703. 7: *M. ovale*, isotype E 00165158. 8: *M. subbrevihamatum*, isotype JE 04008732. 9: *M. frondosum*, isotype NY 01086534. 10: *M. flavopilosum*, isotype FH 00213612. 11: *M. nubigenum*, syntype JE 04008711. 12: *M. scoparium*, isotype S-B 165032.

Taxonomy

Fig. 25 Type 22 — Plants of *Macromitrium* 1: *M. sublongicaule,* isotype MICH 525917. 2, 3: *M. subtortum* as *M. subpungens*, isotype S-B 165699. 4: *M. sulcatum* subsp. *ramentosum*, lectotype MO 365341. 5: *M. semperi*, MEL 2202620.

Fig. 26 Type 23 — **Plants of *Macromitrium*** 1-3: *M. perreflexum*, isotype MICH 525897.

Fig. 27 Type 24 — **Plants of *Macromitrium*** 1, 2: *M. megalocladon*, MEL 2202654.

Shapes of branch leaves — In outline, branch leaves are often lanceolate, typically with a wider lower portion, oval to oblong, and a narrower and acuminate upper portion (*M. longipes* and *M. longirostre*), some are lingulate (*M. densum* and *M. swaisonii*), ligulate (*M. brevicaule, M. nepalense, M. serpens* and *M. subhemitrichodes*), or rather long and threadlike (*M. echinatum, M. standleyi* and *M. trinitense*). The leaf shapes of most species are intermediate transitional between the above shapes such as oblong-lingulate (*M. microcarpum*), ovate-lanceolate (*M. glaziovii, M. hemitrichodes* and *M. krausei*), ligulate-lanceolate (*M. hemitrichodes* and *M. lomasense*), oblong-lanceolate (*M. peraristatum*), lingulate-ligulate (*M. pellucidum*). Most ligulate to lanceolate leaves are strongly carinate, which could be seen with a common magnifying glass, but some soft, or long linear lanceolate, threadlike leaves are not carinate. The leaf shapes of some species are rather peculiar, lingulate, vesicular-inflated on both sides between the costa and the margin on the dorsal view, cucullate with arcuately rolled edges (*M. swaisonii*). Leaf shapes vary not only among different species, but also within the same species. For example, branch leaves of *M. microstomum* vary from shortly ovate-lanceolate with acute to acuminate apices, ligulate, oblong-ligulate, ligulate-lanceolate with acute or cuspidate apices (Yu *et al.*, 2018). In length, branch leaves highly vary among different species, 2.5-4.0 mm long for most species, but very long, up to 18 mm in *M. standleyi*, 10 mm in *M. echinatum, M. fulgescens* and *M. fuscoaureum*), and rather short, shorter than 1.5 mm in *M. densum*.

Textures of branch leaves — Most leaves are moderately soft, but some are rather stiffy (*M. bifasciculare, M. bifasciculatum, M. diaphanum, M. macrocomoides, M. pertriste* and *M. saddleanum*) due to their bi- to multi-stratose laminae. A few species have rather soft leaves (*M. crosbyorum, M. dubium, M. humboldtense* and *M. leprieurii*). When moist, the stiffy leaves are often widely spread out flat, while the others sometimes undulate, rugose, or twisted in the upper portions.

Apices of branch leaves — Apices of branch leaves in *Macromitrium* also highly vary, being obtuse (*M. japonicum*), acute (*M. microstomum, M. nepalense, M. serpens, M. sharpii, M. st-johnii* and *M. sublaeve,*), mucronate (*M. densum, M. subhemitrichodes* and *M. swainsonii*) to retuse (*M. retusum*) in lingulate to ligulate leaves, or acuminate (*M. longipes*), cuspidate (*M. cuspidatum*) to long acuminate (*M. petelotii*) in lanceolate to linear-lanceolate leaves. Leaf apices of some species are rather peculiar. For example, the branch leaves of *M. diaphanum* irregularly and abruptly narrowed to an irregular, notched tip, the awn flexuose, hyaline, broad below, sometimes short and nearly absent or broken off; the young leaves of *M. retusum* have very long, linear, acute, green, smooth, stiffly flexuose aristae, these aristae are fragile and easily broken off when mature; the apices of branch leaves are often asymmetrical in *M. blumei*, truncate with lamina extends on both sides of the costa, forming short asymmetric spiny protuberances in *M. perichaetiale*, rounded, mucronate and distinctively cucullate in *M. densum*, and fragile and easily broken off in *M. frustratum, M. helmsii, M. maolanense, M. perfragile, M. retusum, M. sejunctum* and *M. uraiense*.

Margins of branch leaves — Margins of branch leaves vary to an extent, many are entire throughout the margin (*M. japonicum, M. nepalense* and *M. serpens*), some are serrate (*M. caldense, M. frustratum* and *M. guatemalense*), serrulate (*M. osculatianum*), crenulated (*M. ruginosum* and *M. sharpii*), dentate (*M. ovale*), spinose-denticulate (*M. standleyi*) or notched (*M. brevihamatum* and *M. regnellii*) near the apex and entire below. The outmost marginal cells at the insertion are not differentiated in most species, or differentiated and enlarged (*M. subscabrum*) or enlarged and teeth-like in about 25 species from tropical America (*e.g. M. atratum, M. atroviride, M. amboroicum, M. atroviride, M. brevihamatum, M. guatemalense, M. melinii, M. nematosum, M. oblongum, M. regnellii, M. subbrevihamatum, M. swaisonii* and *M. xenizon*). In some species, the marginal cells near the insertion are longer and narrower than their ambient cells, forming a differentiated border of 5-10 rows (*M. brachypodium, M. brevicaule* and *M. norrisianum*), while in other species, their upper outmost marginal cells are smaller or narrower than their ambient interior cells, forming a distinctly differentiated border (*M. osculatianum, M. perfragile* and *M. perreflexum*).

Laminae of branch leaves — The laminae of branch leaves are unistratose for most species, but those of some species are partially and sporadically bistratose (*e.g. M. bifasciculare, M. bifasciculatum, M. crassiusculum, M. fragilicuspis, M. longipes, M. longirostre, M. maolanense* and *M. saddleanum*) or frequently bistratose (*M. bistratosum* and *M. funicaule*), or irregularly 1-(3)4-stratose (*M. diaphanum, M, nepalense, M. onraedtii, M. serpens* and *M. tongense*) (Figures 28 & 29).

Papillosity of leaf cells — Leaf cells highly vary in papillosity among species. Among the 271 accepted species (varieties and subspecies), about 40 species are characterized by their leaves with smooth cells throughout (*e.g. M. cuspidatum, M. cylindricum, M. dubium, M. echinatum, M. flavopilosum, M. frondosum, M. leprieurii, M. microstomum, M. orthophyllum, M. peraristatum, M. perichaetiale, M. subperichaetiale, M. trachypodium, M. trichophyllum* and *M. ulophyllum*). Leaf cells are papillose to varying degrees for the remaining species, papillose throughout the leaf (e.g. *M. aurantiacum, M. ligulaefolium, M. masafuerae, M. mittenanium* and *M. zimmermannii*), or papillose in upper (medial) portions and smooth in (low) basal portions (*e.g. M. acutirameum, M. brevicaule, M.*

clastophyllum, *M. fuscescens* and *M. rusbyanum*), or the other way around (namely, papillose in low and basal portions and smooth in upper and medial portions (*e.g. M. flexuosum, M. frustratum, M. taoense* and *M. thwaitesii*). Unipapillose, tuberculate, pluripapillose, or conic-bulging to mammillose leaf cells existed in different species. For *M. longipapillosum* and *M. taiwanense,* they are characterized by their branch leaf cells with a single large, occasionally forked tuberculate papilla up to 20 μm tall.

Cell shapes and arrangements of branch leaves — For most species, the upper cells of branch leaves are small, often isodiametric, more or less rounded, rounded-quadrate, subquadrate; cells elongate, rectangular-elongate, evenly thick-walled or incrassate, with straight, curved, curved-sigmoid or sigmoid lumens in low and basal portions. Some species are characterized by their branch leaves with cells all longer than wide and porose through the leaf (*e.g. M. cuspidatum, M. echinatum, M. fuscoaureum, M. trachypodium, M. trichophyllum* and *M. ulophyllum*). Upper cells of branch leaves are often arranged in longitudinal rows to varying degrees, but occasionally in diagonal rows from the costa (*e.g. M. subcirrhosum, M. sulcatum* and *M. ulophyllum*). In some species, the cells along costa thin-walled, smooth and pellucid, distinctly larger than their adjacent cells, appearing as a "cancellina region" (*M. giraldii, M. mosenii* and *M. ousiense*). A few species are characterized by their bulging and collenchymatous cells in upper and medial cells of branch leaves (e. g. *M. aureum, M. holomitrioides, M. macrothele, M. pyriforme* and *M. solitarium*).

Costae of branch leaves — The costae of branch leaves are subpercurrent, percurrent to excurrent, some vanishing several cells before the apex (*e.g. M. crassiusculum* and *M. macrocomoides*), or long excurrent to form aristae of different lengths. Costae sometimes concealed in adaxial view by overlapping folds of laminae (*e.g. M. mosenii, M. nanothecium, M. pertriste* and *M. pulchrum*).

Perichaetial leaves — Perichaetial leaves are often differentiated from branch leaves to varying degrees in size and shape for most species of the genus. The perichaetial leaves of partial species are distinctly differentiated from branch leaves in shape (*e.g. M. ecrispatum, M. lanceolate, M. peraristatum, M. perichaetiale, M. perreflexum, M. rugulosum, M. stoneae* and *M. subperichaetiale*), much shorter than branch leaves (*e.g. M. angulatum, M. incurvifolium, M. rusbyanum, M. sulcatum, M. sublaeve* and *M. validum*), or much longer than branch leaves (*e.g. M. exsertum, M. greenmanii, M. laevisetum, M. peraristatum* and *M. perichaetiale*). Even among different populations, perichaetial leaves highly vary for some species. For example, the perichaetial leaves of *M. microstomum* vary from very short (0.6 mm) to rather long (2.0 mm) and from oblong-ovate to oblong-lanceolate (Yu *et al.*, 2018).

Setae — Setae highly vary in length, from rather short (0.3-0.7 mm in *M. hainanense*, 0.5 mm in *M. brevisetum*, 1.0-1.2 mm in *M. taiheizanense*, 1.0-1.5 mm in *M. francii*, 1.3 mm in *M. brachypodium*, and 2.0 mm in *M. prolongnatum*) to very long (35 mm in *M. pulchrum*, 25 mm in *M. longipilum* and *M. renauldii*), mostly smooth, but some papillose (*M. angulatum, M. erubescens, M. greenmanii, M. huigrense, M. longifolium, M. longipilum, M. mcphersonii, M. noguchianum, M. norrisianum, M. orthostichum, M. ochraceum* and *M. perrefexum*), often twisted to the left, occasionally to the right or not twisted. In *M. longifolium, M. huigrense* and *M. nubigenum*, there is a whitish collar wrapping around the seta beneath the urn.

Capsules — Capsules are often long-exserted, rarely almost immersed, vary in size, shape and surface smoothness. The capsules are often 1.5-2.5 mm long, ovoid-ellipsoid, ellipsoid, ellipsoid-cylindric and cylindric; few are small (ca. 1.0 mm in *M. hortoniae* and *M. microcarpum*), globular, ovoid, cupulate, narrowly ellipsoid-oblong; smooth or furrowed (plicate), constricted or not beneath the mouth when dry; mouths wide or puckered; opercula conic-rostrate; Peristome are double (in most tropical American species), or single to lacking (mainly in the species from Asia, Australasia, Africa and Oc); exostome teeth 8, 16 or 32, lanceolate, or truncate and united forming a short membrane, papillose or papillose-striate, endostome hyaline, finely papillose, often consisting of a delicate and lacerated membrane. Spores smooth to papillose, isosporous or anisosporous (Figures 30, 31, 32).

Calyptrae — Calyptrae of *Macromitrium* species are often large, covering the whole capsule, but a few are small, only partially covering the capsule (e.g. *M. brevicaule*). Most calyptrae are mitrate or mitrate-campanulate, a few are cucullate, naked (in most species from Tropical America), or hairy (in most *Macromitrium* species from Asia), often lacerate or lobed below (Figure 33). A few species have variable calyptra hairiness, with some populations having naked calyptrae, and others with densely hairy calyptrae (*e.g. M. longifolium*). Partial species of the genus have calyptrae with sparse hairs (*e.g. M. gigasporum, M. hemitrichodes*). Most species, if have hairy calyptrae, the hairs are flexuose (*e.g.* most some species from East Asia), but some are straight (*e.g.* some species from Australia) (Figure 33).

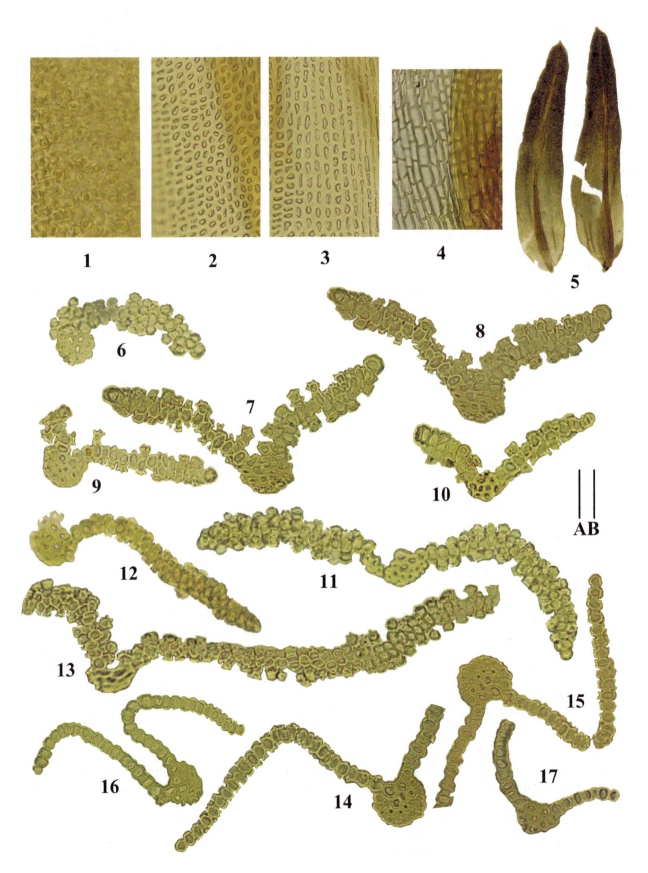

Fig. 28 *Macromitrium nepalense* (Hook. & Grev.) Schwägr. 1: Upper cells of branch leaf. 2: Medial cells of branch leaf. 3: Low cells of branch leaf. 4: Basal cells near costa of branch leaf. 5: Branch leaves. 6-8: Upper transverse sections of of branch leaves. 9-13: Medial transverse sections of of branch leaves. 14-15: Low transverse sections of of branch leaves. 16-17: Basal transverse sections of of branch leaves (all from lectotype of *M. nepalense*: Nepal. *Wallich s.n.*, BM 000982533). Line scales: A = 400 μm (5); B = 40 μm (1-4, 6-17).

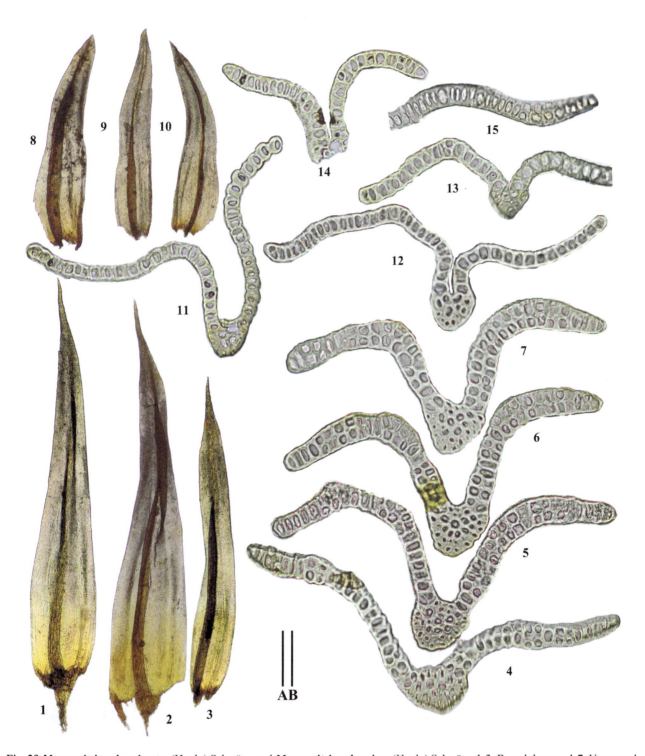

Fig. 29 *Macromitrium longirostre* (Hook.) Schwägr. and *Macromitrium longipes* (Hook.) Schwägr. 1-3: Branch leaves. 4-7: Upper and medial transverse sections of of branch leaves. 8-10: Branch leaves. 11-12: Low transverse sections of of branch leaves. 13-15: Upper and medial transverse sections of of branch leaves (1-8 from *M. longirostre*, *J. Milne 316*, MEL 2371528; 9-16 from *M. longipes*, *Ruth D. Svihla 5057*, H 3090510). Line scales: A = 400 µm (1-3, 8-10); B = 40 µm (4-7, 11-15).

Fig. 30 Capsules of *Macromitrium* 1: *M. herzogii*, isosyntype FH 00213628. 2: *M. lorifolium*, PC 0083690. 3: *M. mosenii*, syntype H-BR 2628005. 4: *M. subscabrum*, MO 6231629. 5: *M. taoense*, isotype H-BR 2618016. 6: *M. incurvifolium*, H-BR 2615001. 7: *M. parvirete*, isotype F-C 0001096F. 8: *M. lorifolium*, H-BR 2630007. 9: *M. trachypodium*, syntype NY 01086605. 10: *M. catharinense* as *M. drewii*, holotype US 00070256. 11: *M. gigasporum*, FH 00213619. 12: *M. bistratosum*, holotype FH 00213572.

Fig. 31 Capsules of *Macromitrium* 1: *M. greenmanii*, isotype MO 406389. 2: *M. subcrenulatum*, syntype H-BR 2625005. 3: *M. argutum*, H-BR 2633010. 4: *M. argutum*, BM 000879982. 5: *M. subscabrum*, MO 6231629. 6: *M. calymperoideum*, H-BR 2582004. 7: *M. concinnum*, H-BR 2602005. 8: *M. solitarium*, lectotype NY 01086658. 9, 17: *M. microstomum*, MO 2861162. 10: *M. sulcatum* as *M. tortifolium*, PC 0106745. 11: *M. quercicola*, syntype H-BR 2581011. 12: *M. soulae*, H-BR 2626007. 13: *M. soulae*, H-BR 2626008. 14, 15: *M. divaricatum*, MO 5913957. 16: *M. fuscoaureum*, MO 5239362. 18: *M. mosenii*, isosyntype FH 00213651.

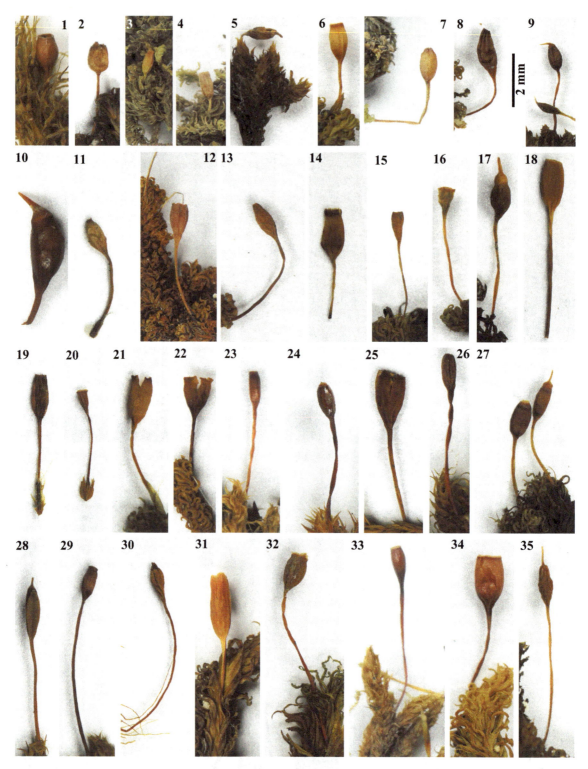

Fig. 32 Capsules of *Macromitrium* 1: *M. echinatum*, paratype H 3194059. 2: *M. acuminatum*, H-BR 2616007. 3, 4: *M. emersulum*, MO 60046931. 5: *M. crassiusculum*, isotype NY 00518316. 6: *M. lebomboense*, isotype MO 743664. 7: *M. microstomum*, MEL 2129409. 8: *M. gymnostomum*, isolectotype NY 00512839. 9: *M. microcarpum*, isotype JE 04008720. 10: *M. fuscoaureum*, MO 4462548. 11: *M. amboroicum*, holotype JE 04008730. 12: *M. macrothele*, NY 01086495. 13: *M. angulosum*, isotype H-BR 2535001. 14: *M. prolongatum*, isotype E 00165192. 15: *M. archeri*, MO 5210871. 16: *M. funicaule*, isotype BM 000873889. 17: *M. lomasense*, holotype US 00070266. 18: *M. densum*, BM 000745393. 19: *M. schmidii*, H-BR 2595002. 20: *M. macrothele*, H-BR 2603007. 21: *M. comatum*, H-BR 2575006. 22: *M. falcatulum* as *M. winkleri*, holotype H-BR 2580001. 23: *M. campoanum*, MO 5638417. 24: *M. longirostre*, H-BR 2528001. 25: *M. megalosporum*, PC 0098376. 26: *M. longirostre*, isotype H-BR 2528002. 27: *M. giraldii* as *M. cavaleriei*, PC 0083627. 28: *M. bistratosum*, holotype FH 00213572. 29: *M. blumei* var. *zollingeri*, H-BR 2601001. 30: *M. trichophyllum*, holotype NY 01086621. 31: *M. schmidii*, H-BR 2595006. 32: *M. nubigenum*, syntype H-BR 2625011. 33: *M. campoanum*, MO 5638408. 34: *M. mosenii*, syntype H-BR 2628005. 35: *M. hemitrichodes*, H-BR 2555001.

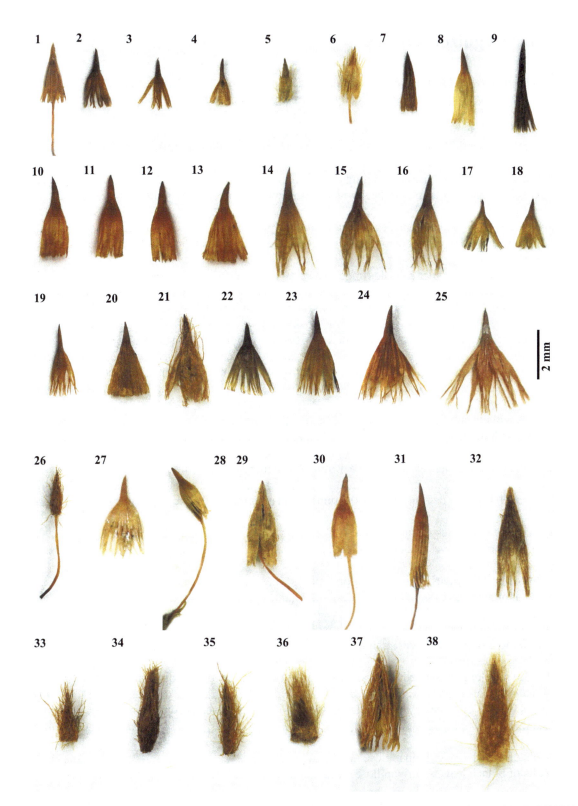

Fig. 33 Calyptrae of *Macromitrium* 1: *M. acuminatum*, H-BR 2616007. 2: *M. blumei* var. *zollingeri*, isotype H-BR 2601008. 3: *M. fendleri*, lectotype NY 01086590. 4: *M. gymnostomum*, isolectotype NY 00512839. 5: *M. emersulum*, MO 60046931. 6: *M. incurvifolium*, H-BR 2618006. 7: *M. microstomum*, BM 000873099. 8: *M. lebomboense*, isotype H 3090420. 9: *M. perdensifolium*, MO 1075838. 10-13, 30: *M. bistratosum*, holotype FH 00213572. 14-16: *M. greenmanii*, FH 00213619. 17-18: *M. fendleri*, isolectotype FH 00213610. 19: *M. subdiscretum*, holotype F-C 0001097F. 20: *M. glabratum*, isosyntype JE 04008716. 21: *M. argutum*, BM 000879982. 22: *M. lorifolium*, isolectotype H-BR 2630006. 23: *M. lorifolium*, PC 0083690. 24: *M. laevisetum*, isosyntype E 00165151. 25: *M. trachypodium*, syntype NY 01086605. 26, 33: *M. ferriei*, isolectotype PC 0083649. 27: *M. tocaremae*, BM 000873213. 28: *M. picobonitum*, paratype MO. 29: *M. aurescens*, H-BR 2586006. 31: *M. sulcatum* var. *leptocarpum*, isosyntype MICH 525887. 32: *M. densum*, BM 000745393. 34, 35: *M. fuscoaureum*, MO 4462548. 36: *M. comatum*, holotype NY 00518291. 37: *M. schmidii*, H-BR 2595005. 38: *M. calymperoideum*, H-BR 2582004.

4.2 Key to recognized species

1. Laminae of branch leaves unistratose..23
1. Laminae of branch leaves partially, sporadically or frequently bistratose, or irregularly 1-(3)4-stratose..................2

2. Branch leaves have irregularly 1-3(4)-stratose laminae..19
2. Branch leaves have regularly or sporadically bistratose, or with irregularly multistratose subulae..........................3

3. Branch leaves have regularly bistratose laminae..18
3. Branch leaves have partially and sporadically bistratose laminae or irregularly multistratose subulae.......................4

4. Stems conspicuously creeping, branches not dichotomous branching..7
4. Stems inconspicuously or weakly creeping, with agminated and somewhat dichotomous branches.........................5

5. Rectangular or short rectangular cells confined to a small area near the base in branch leaves......................................
.... 202. *M. saddleanum*
5. Rectangular, elongate cells covering the low and basal portions of branch leaves...6

6. Laminae of branch leaves often bistratose or partially bistratose, with a multistratose apex; calyptrae cucullate.....
.... 19. *M. bifasciculare*
6. Laminae of branch leaves occasionally partially bistratose, with a unistratose apex; calyptrae mitrate...................
.... 20. *M. bifasciculatum*

7. Branch leaves without subulae..10
7. Branch leaves with fragile multistratose subulae..8

8. Subulae of branch leaves slender and long (> 1 mm), basal marginal teeth-like large cells differentiated..............
.... 75. *M. fragilicuspis*
8. Subulae of branch leaves broad and short (< 0.5 mm), basal marginal teeth-like large cells not differentiated...........9

9. Most basal cells of branch leaves tuberculate.. 90. *M. gracile*
9. All basal cells of branch leaves smooth and clear.. 131. *M. maolanense*

10. Upper cells of branch leaves conic-bulging or papillose 13
10. Upper cells of branch leaves smooth 11

11. Basal cell lumens of branch leaves curved to sigmoid .... 124. *M. longipes*
11. Basal cells lumens of branch leaves straight, not curved to sigmoid...12

12. Setae 1.0-3.0 mm long; basal cells of branch leaves not porose, becoming quadrate to rectangular and larger at outmost marginal row.. 44. *M. crassiusculum*
12. Setae 3.5.8.0 mm long; basal inner cells of branch leaves porose, becoming narrower and longer toward margins... 126. *M. longirostre*

13. Upper cells of branch leaves unipapillose.. 171. *M. pertriste*
13. Upper cells of branch leaves pluripapillose..14

14. Capsules weakly ribbed, peristome narrowly oblong... 115. *M. lebomboense*
14. Capsules smooth when dry, peristome often caducous and rudimentary...15

15. Basal cells of branch leaves with straight lumens, not porose .... 216. *M. st-johnii*
15. Basal cells of branch leaves with curved to sigmoid lumens, porose 16

16. Upper margins of branch leaves completely entire; peristome absent........................... 159. *M. pallidum*
16. Upper margins of branch leaves entire to crenulate-papillose; peristome present and single....................17

17. Branch leaves sporadically bistratose at upper portions, or occasionally jagged............................ **3.** *M. amaniense*
17. Branch leaves partly bistratose at the apex... **117.** *M. leratii*

18. Basal cells with narrowly curved or somewhat sigmoid lumens; peristome single..................... **22.** *M. bistratosum*
18. Basal cells with straight lumens; peristome absent... **81.** *M. funicaule*

19. Branch leaves irregularly and abruptly narrowed to a notched, hyaline, fragile awn................... **51.** *M. diaphanum*
19. Apices of branch leaves acute, mucronate, acuminate-acute, without awn...20

20. Basal cells of branch leaves with straight lumens... **145.** *M. nepalense*
20. Basal cells of branch leaves with curved-sigmoid lumens...21

21. Branch leaves stiffly curved-erect, appressed, with apices curved outward or deflexed, spirally curved around branch to form rope-like appearance when dry... **240.** *M. tongense*
21. Branch leaves curved to twisted, with the apex hidden in the cavity when dry...22

22. Branch leaves without densely reddish brown rhizoids... **153.** *M. onraedtii*
22. Branch leaves densely with reddish brown rhizoids.. **209.** *M. serpens*

23. Enlarged teeth-like cells not differentiated at basal margins of branch leaves..48
23. Enlarged teeth-like cells differentiated at basal margins of branch leaves...24

24. Branch leaves with branched multicellular gemmae on adaxial surface.................................. **144.** *M. nematosum*
24. Branch leaves without multicellular, branched gemmae on adaxial surface..25

25. Plants without dense reddish or rusty rhizoids...37
25. Plants with dense reddish or rusty rhizoids..26

26. Branch leaves often broken off at upper and medial portions... **254.** *M. xenizon*
26. Branch leaves not broken off at upper and medial portions...27

27. Upper and medial cells of branch leaves larger, up to 15 μm wide....................................... **12.** *M. atratum*
27. Upper and medial cells of branch leaves small to medium-sized, often smaller than 10 μm wide..........................28

28. Basal cells near costa of branch leaves similar to their ambient cells, not forming a "cancellina region"...............32
28. Basal cells near costa of branch leaves enlarged and hyaline, forming a "cancellina region" 29

29. Upper and medial cells of branch leaves strongly bulging (to mammillose) to tuberculate...................................31
29. Upper and medial cells of branch leaves flat, clear and smooth...30

30. Perichaetial leaves oblong- to oval-lanceolate; upper cells rounded-quadrate to elliptic, papillose; basal cells elongate, thick-walled, smooth to tuberculate... **13.** *M. atroviride*
30. Perichaetial leaves broadly oblong-, triangular- to lingulate-lanceolate; all cells smooth and clear, longer than wide... **200.** *M. rugulosum*

31. Basal cells rectangular, occasionally tuberculate, marginal teeth-like cells distinctly differentiated, weakly porose... **4.** *M. amboroicum*
31. Basal cells elongate-rectangular, tuberculate, marginal teeth-like cells weak and inconspcious, tuberculate, porose... **181.** *M. pseudofimbriatum*

32. Upper cells in outmost marginal row not differentiated, without a differentiated border....................................34
32. Upper cells in outmost marginal row of branch leaves narrower than their ambient cells, forming a differentiated border..33

33. Upper and medial cells of branch leaves smooth and clear, flat.. **156.** *M. osculatianum*
33. Upper and medial cells of branch leaves conic-bulging, mammillose........................... **219.** *M. subbrevihamatum*

34. Plants small, rusty, setae 4-5 mm long; branch leaves with an apex hidden in the inrolled cavity.. **67. *M. fendleri***
34. Plants medium-sized to large, setae longer than 10 mm; apices of branch leaves not hidden in the inrolled cavity 35

35. Branch leaves broadly lanceolate, strongly and longitudinally undulate when moist; perichaetial leaves oblong- to ovate-lanceolate, capsules smooth.. **135. *M. melinii***
35. Branch leaves ligulate- to linear-lanceolate, not undulate when moist; perichaetial leaves long lanceolate, with a long arista; capsules furrowed when dry..36

36. Plants medium-sized; setae 10-15 mm long; branch leaves strongly squarrose-recurved spreading, often conduplicate when moist, upper margins plane.. **26. *M. brevihamatum***
36. Plants large; setae up to 30 mm long; branch leaves erecto-patent, some still twisted when moist, upper margins undulate... **150. *M. oblongum***

37. Branch leaves lingulate, with rounded to obtuse, emarginated to short mucronated apices......... **233. *M. swainsonii***
37. Branch leaves oblong-lanceolate, ligulate-lanceolate, lanceolate, narrowly lanceolate or ovate-lanceolate, with acuminate to acute apices...38

38. Upper cells of branch leaves moderately to strongly bulging, or unipapillose...44
38. Upper cells of branch leaves flat and smooth...39

39. Setae shorter, 3-4 mm long, upper marginal cells of branch leaves somewhat differentiated............ **99. *M. herzogii***
39. Setae longer, up to 7 mm, upper marginal cells of branch leaves not differentiated...............................40

40. Low and basal cells of branch leaves strongly tuberculate; perichaetial leaves without aristae............42
40. Low and basal cells sometimes unipapillose; perichaetial leaves acuminate, with or without aristae....41

41. Perichaetial leaves with a long arista.. **35. *M. cataractarum***
41. Perichaetial leaves without a long arista.. **43. *M. crassirameum***

42. Branch leaves squarrose-recurved spreading when moist, long lanceolate with an acuminate apex, upper cells in distinct longitudinal row... **100. *M. hildebrandtii***
42. Branch leaves spreading when moist, ligulate-lanceolate with a mucronate, acute to acuminate-acute apex, upper cells not in distinct longitudinal row...43

43. Capsules ovoid to ellipsoid-cylindric, distinctly furrowed when dry; margins of branch leaves almost entire near the apex, marginal enlarged teeth-like cells weakly differentiated... **14. *M. attenuatum***
43. Capsules ellipsoid-cylindric, smooth to plicate when dry; margins of branch leaves irregularly serrate near the apex, marginal enlarged teeth-like cells distinctly differentiated at insertion................................. **142. *M. mosenii***

44. Capsules ellipsoid-cylindric, strongly to moderately furrowed; upper margins of branch leaves entire or undulate, serrulate to serrate...46
44. Capsules ovoid or cupulate, smooth to weakly ribbed or strongly furrowed; upper margins of branch leaves irregularly and bluntly notched to serrate..45

45. Capsules strongly furrowed, margin of branch leaves distinctly serrate... **191. *M. regnellii***
45. Capsules almost smooth, margin of branch leaves weakly serrate or notched............................. **239. *M. tocaremae***

46. Upper margins of branch leaves undulate, serrulate to serrate... **93. *M. guatemalense***
46. Upper margins of branch leaves plane and entire..47

47. Plants small to medium-sized; branch leaves sometimes conduplicate and slightly curved adaxially above when moist.. **46. *M. crispatulum***
47. Plants robust; branch leaves weakly squarrose-recurved spreading when moist........................ **110. *M. laevisetum***

48. Cells of branch leaves papillose (mammillose or conic-bulging) to varying degrees.........................87
48. All cells of branch leaves clear and smooth...49

49. Branch leaves distinctly and strongly recurved from apex to base at one side, and along medial and low margins at the other side; all cells thick-walled, often collenchymatous... **128. *M. macrocomoides***
49. Branch leaves plane or recurved but not from apex to base; cells in upper and medial portions often not thick-walled; leaf cells not collenchymatous 50

50. Perichaetial leaves not or slightly to moderately differentiated, shorter than, as long as, or at most 1.2 times longer than branch leaves..54
50. Perichaetial leaves distinctively differentiated, at least 1.5 times longer than branch leaves...............................51

51. Apices of branch leaves sharply acuminate to long cuspidate, or acute...53
51. Apices of branch leaves truncate, with lamin extends on both sides of the costa, forming short asymmetric spiny protuberances...52

52. Branches to 20-25 mm long, 1.5 mm thick.. **168. *M. perichaetiale***
52. Branches up to 50 mm long and 2.5 mm thick.. **228. *M. subperichaetiale***

53. Apices of branch leaves acute... **154. *M. orthophyllum***
53. Apices of branch leaves sharply acuminate to long cuspidate...................................... **165. *M. peraristatum***

54. At least partial cells of branch leaves isodiametric, quadrate, subquadrate, rounded-quadrate, oblate...................70
54. All cells of branch leaves longer than wide...55

55. Upper and medial interior cells of branch leaves not in radiating diagonal rows from the costa............................57
55. Upper and medial interior cells of branch leaves in radiating diagonal rows from the costa................................56

56. Setae distinctly papillose, capsules cupulate, ovoid, ellipsoid to short cylindric, strongly and distinctly furrowed.....
.... **242. *M. trachypodium***
56. Setae smooth, capsules obovoid to ellipsoid, smooth... **248. *M. ulophyllum***

57. Branch leaves lingulate to ligulate, with apiculate to mucronate apices...................................... **164. *M. pellucidum***
57. Branch leaves long linear-lanceolate, linear-lanceolate, lanceolate to long ligulate-lanceolate, oblong- to ovate-lanceolate, apices not apiculate or mucronate..58

58. Apices of branch leaves asymmetric, sharply contracted to a long cuspidate or a long aristate point (up to 1.5 mm long) .... **48. *M. cuspidatum***
58. Apices of branch leaves symmetric, not sharply contracted to a long cuspidate..59

59. Branch leaves not clasping at base..66
59. Branch leaves distinctly clasping at base...60

60. Upper and medial marginal cells of branch leaves not much different from their ambient cells, without a differentiated broader.. **243. *M. trichophyllum***
60. Upper and medial marginal cells of branch leaves distinctly different from their ambient cells, forming a differentiated border..61

61. Branch leaves strongly squarrose-recurved spreading when moist 64
61. Branch leaves spreading, not or slightly squarrose-recurved when moist..62

62. Branch leaves up to 9 mm long, costae excurrent to a long awn, awns up to 1-2 mm long.......... **71. *M. flavopilosum***
62. Branch leaves up to 6 mm long, costae not excurrent to a long awn...63

63. Capsules ellipsoid, strongly furrowed .... **77. *M. frondosum***
63. Capsules ellipsoid, smooth or somewhat furrowed... **86. *M. gigasporum***

64. Capsules ellipsoid, ellipsoid-cylindric, smooth, weak to moderately furrowed........................... **57. *M. echinatum***
64. Capsule ovoid to cupulate, with a wide mouth, or cylindric, smooth to weakly furrowed.....................................65

65. Capsules ovoid to cupulate, smooth.. **49. *M. cylindricum***
65. Capsules cylindric, smooth to weakly furrowed.. **84. *M. fuscoaureum***

66. Branch leaves twisted-contorted, spirally twisted-contorted and flexuous when dry...69
66. Branch leaves appressed to erect below, slightly flexuous or somewhat spirally coiled above when dry................67

67. Perichaetial leaves distinctly different from branch leaves, broadly oblong, oblong-lanceolate with cuspidate apices.. **33. *M. campoanum***
67. Perichaetial leaves not much different from branch leaves, linear lanceolate..68

68. Branch leaves longer, 6-9 mm long, apices long acuminate and with a filament............................. **55. *M. dubium***
68. Branch leaves shorter, 3-5 mm long, apices short-acuminate, acute to apiculate........................... **116. *M. leprieurii***

69. Peristome single.. **9. *M. archboldii***
69. Peristome double.. **80. *M. fulvum***

70. Branch leaves without a long arista or long slender acuminate apex...74
70. Branch leaves gradually narrowed to a long, slender acuminate apex, or apices retuse with a long arista...............71

71. Branch leaves large, up to 10 mm long.. **79. *M. fulgescens***
71. Branch leaves shorter, at most 3 mm long...72

72. Branch leaves asymmetrical above... **113. *M. larrainii***
72. Branch leaves symmetrical above..73

73. Branch leaves gradually narrowed to a long, slender acuminate-subulate apex................................. **52. *M. dielsii***
73. Branch leaves with retuse apices, young leaves have a very long, linear, stiffly flexuous arista.. **195. *M. retusum***

74. Branch leaf cells smooth but not strongly bulging, outmost marginal cells of branch leaf similar to their ambient cells..76
74. Branch leaf cells strongly bulging but smooth, outmost marginal cells smaller than their ambient cells................75

75. Plants with dense reddish rhizoids; branch leaves lanceolate, acute; perichaetial leaves distinctly differentiated, wide ovate, abruptly constrict forming a narrowly linear upper portion..................................... **111. *M. lanceolatum***
75. Plants without dense rhizoids; branch leaves oblong-lanceolate to ligulate-lanceolate, acute to apiculate, acuminate-acute; perichaetial not differentiated or slightly shorter than branch leaves............... **120. *M. lomasense***

76. When dry, branch leaves erecto-patent below, moderately to strongly contorted-twisted and crisped, curved or inrolled with apices hidden in the cavity, or squarrose-recurved and flexuous above, not funiculate....................81
76. When dry, branch leaves appressed or obliquely appressed below, spirally or slightly spirally coiled, or curved around the branch, funiculate or somewhat funiculate...77

77. Branches densely with rusty brown rhizoids; capsules smaller... **137. *M. microcarpum***
77. Branches without dense rusty brown rhizoids; capsules larger..78

78. Plant large and robust; branch leaves keeled, appressed below, slightly twisted-contorted and apices somewhat incurved when dry.. **213. *M. soulae***
78. Plant small to medium-sized; branch leaves appressed or obliquely appressed below, flexuous-curved to erect-curved above, or curved around the branch, or flexuous-curved to erect-curved above..79

79. Plants small to medium-sized; setae longer, 15-25 mm long... **138. *M. microstomum***
79. Plants small; setae relatively shorter, 3-12 mm long..80

80. Low and basal cells of branch leaves not porose.. **108. *M. krausei***
80. Low and basal cells of branch leaves strongly porose... **237. *M. tenax***

81. Branch leaves long linear-lanceolate, gradually narrowed to a fragile subulate upper portion, upper portions easily broken..**167.** *M. perfragile*
81. Branch leaves oblong-ligulate, ligulate-lanceolate, lanceolate, apices of branch leaves obtuse, obtuse-acute, acute, acute to acuminate, upper portions often unbroken and intact...82

82. Calyptrae hairy..86
82. Calyptrae naked..83

83. Plants without dense rhizoids...85
83. Plants densely with rusty rhizoids..84

84. Peristome single.. **161.** *M. paridis*
84. Peristome double..**212.** *M. solitarium*

85. Branch leaves ligulate-lanceolate, margins entire throughout, with an acute apex........................ **8.** *M. antarcticum*
85. Branch leaves narrowly lanceolate, margins conspicuously dentate to bluntly notched, with an acuminate apex......
....**53.** *M. divaricatum*

86. Plants robust; branch leaves with an inrolled apex when dry; setae relatively longer, 4-6 mm long; peristome present... **30.** *M. calocalyx*
86. Plants small to medium-sized; branch leaves without an inrolled apex when dry; setae shorter, 2-3 mm long; peristome absent.. **157.** *M. ousiense*

87. Upper cells of branch leaves rather small, obscure and pluripapillose... **136.** *M. menziesii*
87. Upper cells of branch leaves moderately large, conic-bulging, mammillose, unipapillose or pluripapillose..........88

88. Setae completely smooth..109
88. Setae papillose to varying degrees...89

89. At least partial cells of branch leaves isodiametric, subquadrate, rounded, rounded-quadrate.............................94
89. All cells of branch leaves longer than wide..90

90. Branch leaves regularly and loosely arranged on the branch, regularly and strongly squarrose-recurved when dry, still strong squarrose-recurved spreading when moist; perichaetial leaves at least three times longer than branch leaves.. **170.** *M. perreflexum*
90. Branches densely leaved, not strong squarrose-recurved when dry or moist; perichaetial leaves at most 1.5 times longer than branch leaves..91

91. Branch leaves with an acuminate apex..93
91. Branch leaves acuminate to an arista...92

92. Setae minutely papillose, up to 25 mm long... **125.** *M. longipilum*
92. Setae strongly papillose, 10-15 mm long... **133.** *M. mcphersonii*

93. Branch leaves with a clasping base; calyptrae cucullate, with long brownish hairs................... **36.** *M. catharinense*
93. Branch leaves without a clasping base, calyptrae mitrate, naked or sparsely hairy................ **147.** *M. noguchianum*

94. Basal and low marginal cells of branch leaves becoming narrow and smooth, different from their ambient cells, forming a distinctive differentiated border... **148.** *M. norrisianum*
94. Basal and low marginal cells of branch leaves not much different from ambient cells, without a differentiated border...95

95. Setae without a collar 98
95. Some setae with a whitish collar beneath the urn..96

96. Calyptrae naked.. **149.** *M. nubigenum*
96. Calyptrae hairy to varying degrees...97

97. Branch leaves strong recurved-squarrose and twisted above when moist... **103.** *M. huigrense*
97. Branch leaves spreading when moist.. **122.** *M. longifolium*

98. Calyptrae hairy...102
98. Calyptrae naked..99

99. Branch leaves not spirally coiled, costae percurrent or subpercurrent, not forming a conspicuous apiculus or awn
101
99. Branch leaves densely arranged, spirally coiled, funiculate to varying degrees when dry, costae excurrent to form
 a conspicuous apiculus or awn 100

100. Awns of branch leaves shorter (40 -180 µm long), densely spirally around the stem, conspicuously funiculate...
 .... **23.** *M. blumei* var. *blumei*
100. Awns of branch leaves longer (140 -530 µm long), curled loosely around the stem, slightly funiculate.............
 .... **23a.** *M. blumei* var. *zollingeri*

101. Setae weakly prorate above, smooth below, capsules ellipsoid to obovoid, smooth, peristome single ....
 .... **18.** *M. austrocirrosum*
101. Setae conspicuously papillose, capsule ovoid, ovoid-ellipsoid to ellipsoid-cylindric, furrowed, peristome double
 .... **91.** *M. greenmanii*

102. Elongate, rectangular cells covering low and basal portions of branch leaves..107
102. Elongate, rectangular short-elliptic cells restricted to a small area near the base of branch leaves.....................103

103. Upper and medial cells of branch leaves strongly pluripapillose...105
103. Upper and medial cells of branch leaves strongly unipapillose...104

104. Plants small; setae 2-5 mm long.. **5.** *M. angulatum*
104. Plants small to medium-sized; setae 6-10 mm long.. **62.** *M. erubescens*

105. Capsule urns small, almost spherical, ca. 0.7 mm in diameter, wide-mouthed after dehiscence ....
 .... **139.** *M. minutum*
105. Capsule urns 0.9-1.5 × 0.5-0.7 mm, ovoid, obovoid to ellipsoid..106

106. Capsule urns obovoid to ellipsoid; basal cells of branch leaves short-elliptic....................... **155.** *M. orthostichum*
106. Capsule urns ovoid; basal cells of branch leaves elongate.. **199.** *M. ruginosum*

107. Upper cells of branch leaves with conspicuous intercellular spaces (appearing as 'dots' under a microscope),
 marginal cells near the apex without intercellular spaces in 2-4 rows, forming a shiny apical border....................
 .... **152.** *M. ochraceum*
107. Upper cells of branch leaves without intercellular space, upper marginal cells not differentiated.................... 108

108. Upper cells strongly unipapillose to high tuberculate or with a forked papilla; costa percurrent, not forming an
 arista.. **134.** *M. megaloclodon*
108. Upper cells flat and smooth; costa long excurrent, forming a long, flexuous, hyaline arista...................................
 .... **151.** *M. ochraceoides*

109. Cells of branch leaves papillose in upper and medial portions while smooth in low and basal portions, or
 papillose (tuberculate) in low and basal portions while smooth in upper and medial cells 189
109. Cells of branch leaves papillose (tuberculate) or conic-bulging (mammillose) to varying degrees from upper to
 low (some to basal) cells...110

110. Leaf cells isodiametric, pluripapillose from apex to base, similar to those of *Zygodon*.......... **140.** *M. mittenianum*
110. Leaf cells elongate toward the base, low and basal cells often tuberculate, papillosity different from those in
 Zygodon...111

Taxonomy

111. Medial and upper cells of branch leaves conic-bulging (mammillose), unipapillose or pluripapillose...............113
111. Medial and upper cells of branch leaves unipapillose to pluripapillose..112

112. Perichaetial leaves shorter than branch leaves, ovate-lanceolate to lanceolate..................... **118. *M. ligulaefolium***
112. Perichaetial leaves larger (longer and wider) than branch leaves, broadly lanceolate.............. **132. *M. masafuerae***

113. Upper cells strong bulging to conic-bulging (mammillose) or unipapillose..145
113. Upper cells pluripapillose to varying degrees..114

114. Branch leaves without aristae or with aristae shorter than 0.15 mm..117
114. Branch leaves with a long arista (0.4 -1.5 mm long) 115

115. Apices often asymmetric, aristae green, fragile, mostly breaking before mature............................. **97. *M. helmsii***
115. Apices often symmetric, aristae not easily broken, often present after mature..116

116. Branch leaves narrowly lanceolate to narrowly triangular, gradually tapering, long acuminate, ending in a piliform apex... **104. *M. humboldtense***
116. Branch leaves ligulate, apices obtuse to truncate, short acute with a long arista.......................... **198. *M. rufipilum***

117. All cells of branch leaves isodiametric, rounded-quadrate.. **15. *M. aurantiacum***
117. Low and basal cells of branch leaves longer than wide, elongate, long rectangular, rectangular, occasionally oval to irregular rectangular, rhomboid...118

118. Calyptrae hairy to varying degrees..125
118. Calyptrae smooth and naked..119

119. Branch leaves oblong to oblong-lanceolate, lanceolate to broadly lanceolate, ligulate to ligulate-lanceolate, or ligulate from a wider basal portion...121
119. Branch leaves long- to linear lanceolate...120

120. Setae rather long, up to 35 mm long... **34. *M. cardotii***
120. Setae relatively short, about 3 mm.. **59. *M. eddyi***

121. Setae up to 35 mm long; branch leaves lanceolate, ligulate from a wider basal portion............... **183. *M. pulchrum***
121. Setae shorter than 18 mm; branch leaves lanceolate, oblong to oblong-lanceolate, ligulate to ligulate-lanceolate... 122

122. Setae 14-18 mm long; capsules ovoid and 8-plicated... **256. *M. zimmermannii***
122. Setae shorter than 8 mm; capsules oblong, wrinkled at the mouth, or narrowly ovoid to ellipsoid-cylindric, or fusiform-ellipsoid to ellipsoid..123

123. Branch leaves strongly twisted-contorted and crisped, with inrolled apices when dry; perichaetial leaves longer than branch leaves... **89. *M. glaziovii***
123. Branch leaves appressed-curved or flexuous-twisted; perichaetial leaves shorter than branch leaves, with the upper portion strongly inrolled and the apices hidden in the cavity when dry...124

124. Branch leaves appressed-curved, upper cells strongly bulging, collenchymatous, clear and most smooth or slightly pluripapillose... **29. *M. caloblastoides***
124. Branch leaves flexuous-twisted, upper cells occasionally smooth, weakly pluripapillose, not collenchymatous .... **119. *M. ligulare***

125. Peristome present...127
125. Peristome absent..126

126. Capsules plicate or ribbed under the mouth when dry... **205. *M. schmidii***
126. Capsules 4-furrowed, or with a four-angled mouth when dry.. **245. *M. tuberculatum***

127. Plants often broken when old, forming many branches as separate individual "plants"...... **106. *M. involutifolium***
127. Plants not easily broken when old..128

128. Perichaetial leaves similar to or longer than branch leaves..130
128. Perichaetial leaves small, shorter than branch leaves..129

129. Plants without dense rusty rhizoids, vaginula covered by perichaetial leaves, branch leaves oblong-ligulate, with a cucullate, hooked mucronate apex.. **16. *M. aurescens***
129. Plants with dense rusty rhizoids, vaginula exposed; branch leaves ligulate to lanceolate-ligulate, broadly acute-apiculate .... **66. *M. falcatulum***

130. Basal cells near costa not much different from their ambient cells, not forming a "cancellina region"...............135
130. Basal cells near costa irregularly rectangular, thin-walled, pellucid, forming a "cancellina region"..................131

131. Upper cells of branch leaves bulging to varying degrees, hyaline or slightly obscure, weakly to moderately pluripapillose...133
131. Upper cells rather obscure, flat and densely pluripapillose..132

132. Branch leaves oblong lanceolate, ca. 2.5 mm long.. **69. *M. ferriei***
132. Branch leaves long and narrowly lanceolate, ca. 2.5 - 3.0 mm long .... **196. *M. rhacomitrioides***

133. Low cells of branch leaves ellipsoid-rhomboid, short oblong, distinctly unipapillose, upper cells strongly conic-bulging... **187. *M. quercicola***
133. Low cells of branch leaves rhomboid, rectangular to sublinear, weak to moderately unipapillose, upper cells flat to moderately bulging..134

134. Branch leaves strong keeled and crinkled, twisted-contorted and inrolled above with the apex hidden in the inrolled cavity when dry, oblong, oblong-lanceolate to ligulate-lanceolate; calyptrae with yellow-brownish, straight hairs... **32. *M. calymperoideum***
134. Branch leaves irregularly twisted-contorted and inrolled above but the apex not hidden in an inrolled cavity when dry, lanceolate, oblong- to ovate-lanceolate; calyptrae with numerous yellowish or brown-yellowish, flexuous hairs.. **87. *M. giraldii***

135. Upper cells of branch leaves oblate to rounded, most wider than long.................................... **218. *M. streimannii***
135. Upper cells of branch leaves rounded, rounded-quadrate..136

136. Branch leaves ligulate, ligulate-lanceolate, oblong, oblong-lanceolate, lanceolate, apices acute, mucronate, acuminate-apiculate...138
136. Branch leaves narrowly lanceolate to lanceolate from an ovate base, apices variably acuminate.....................137

137. Upper cells of branch leaves without forked papillae.. **121. *M. longicaule***
137. Upper cells of branch leaves with forked papillae .... **217. *M. stoneae***

138. Setae longer, longer than 2 mm; branch leaves lanceolate to ligulate-lanceolate, longer than 1.5 mm................140
138. Setae rather short; branch leaves ligulate to ligulate-lanceolate, shorter than 1.5 mm long..............................139

139. Upper marginal cells of branch leaves smaller than ambient cells, often oblate.................................. **76. *M. francii***
139. Upper marginal cells of branch leaves similar to ambient cells.. **176. *M. pilosum***

140. Upper cells of branch leaves with forked papillae...143
140. Upper cells of branch leaves without forked papillae..141

141. Setae slender, about 2 mm long; basal cells of branch leaves near margins not much differentiated from ambient cells; calyptrae mitrate, with numerous straight hairs... **174. *M. pilicalyx***
141. Setae longer, 4-11 mm long; basal cells near margins at one side often differentiated, rectangular, pellucid with thinner walls; calyptrae cucullate, with numerous flexuous hairs..142

142. Upper cells of perichaetial leaves longer than wide... **224.** *M. subincurvum*
142. Upper cells of perichaetial leaves isodiametric... **241.** *M. tosae*

143. Basal cells near margins of branch leaves thin-walled, wider than their ambient cells, form a short border..............
.... **179.** *M. prorepens*
143. Basal cells near margins of branch leaves not much different from ambient, without a border..........................144

144. Branch leaves ovate-lanceolate, lanceolate to ligulate-lanceolate, narrowed to an apiculus or stout mucro; capsules urns oblong-ovoid to narrowly ovate, smooth below, 8-plicate beneath a small mouth, exostome teeth broken off when old; calyptrae with sparse and straight hairs... **98.** *M. hemitrichodes*
144. Branch leaves lanceolate from an elliptic low portion to ligulate-lanceolate, acute to stoutly acuminate-apiculate; capsule urns ovoid-obloid, slightly 4-plicate beneath the mouth, exostome teeth well-developed; calyptrae sparsely to densely hairy.. **227.** *M. submucronifolium*

145. Upper cells of branch leaves often isodiametric, rounded to rounded-quadrate...150
145. All cells of branch leaves longer than wide...146

146. Upper marginal cells of branch leaves longer and narrower than their ambient cells, forming a distinct border
....148
146. Upper marginal cells of branch leaves not differentiated from their ambient cells...147

147. Branch leaves broadly lanceolate, oblong lanceolate, costae extending in a long, smooth and flexuous hair up to 2 mm long.. **125.** *M. longipilum*
147. Branch leaves long linear-lanceolate, apices acuminate to setaceous-acuminate........................ **251.** *M. validum*

148. Branch leaves rather long, up to 18 mm long... **215.** *M. standleyi*
148. Branch leaves shorter than 6 mm...149

149. Setae up to 22 mm long, branch leaves linear-lanceolate, with a long acuminate apex.............. **206.** *M. scoparium*
149. Setae short, 6-9 mm long, branch leaves linear-lanceolate, frequently oblong-lanceolate, long acuminate into a long and fine arista... **244.** *M. trinitense*

150. Upper marginal cells of branch leaves not differentiated, not forming a border..155
150. Upper marginal cells of branch leaves differentiated, forming a conspicuous or inconspicuous border.............151

151. Costae of branch leaves excurrent into a hair tip.. **173.** *M. picobonitum*
151. Costae percurrent...152

152. Upper cells of branch leaves collenchymatous, bulging and mammillose............................... **47.** *M. crosbyorum*
152. Upper cells of branch leaves not collenchymatous, bulging, occasionally unipapillose.....................153

153. Peristome absent or reduced to a basal membrane.. **63.** *M. erythrocomum*
153. Peristome double..154

154. Dry leaves shrived, gold-yellowish, clasping at base, individually twisted, narrowly lanceolate, apices acuminate
.... **190.** *M. refractifolium*
154. Dry leaves not very shriveled, dark brown, spirally-curved twisted, ligulate, oblong-lignulate to ligulate-lanceolate, apices acute... **222.** *M. subdiscretum*

155. Upper cells of branch leave without a linear central papilla...158
155. Upper cells of branch leaves with a single large linear central papilla up to 14-18 μm high.............................156

156. Costae of branch leaves excurrent to a long hyaline arista... **235.** *M. taiwanense*
156. Costae of branch leaves percurrent, not forming a long hyaline arista..157

157. Calyptrae mitrate, densely covered with delicate, yellowish-hyaline, long hairs; peristome absent... **92. *M. grossirete***
157. Calyptrae campanulate, with many long yellowish hairs; peristome single..................... **123. *M. longipapillosum***

158. Upper and medial cells of branch leaves not collenchymatous...162
158. Upper and medial cells of branch leaves collenchymatous..159

159. Peristome single; calyptrae hairy... **101. *M. holomitrioides***
159. Peristome double; calyptrae naked...160

160. Plants rather large; branch leaves undulate and contorted-flexuous above when dry, distinctly squarrose-recurved when moist; linear lanceolate, 3-5 mm long, apices long acuminate; setae 10-15 mm long... **17. *M. aureum***
160. Plants small; branch leaves flexuous, twisted-contorted above when dry, erect-spreading when moist; oblong-lanceolate to lanceolate, 1.5-2.5 mm long; perichaetial leaves broadly oblong-lanceolate, setae 3-5 mm..........161

161. Perichaetial leaves with the widest at the base.. **85. *M. galipense***
161. Perichaetial leaves lanceolate with a widely oblong low part... **130. *M. macrothele***

162. Branch leaves narrowed to a long, fragile, denticulate, twisted, yellowish or hyaline arista up to 2-3 mm or longer... **175. *M. piliferum***
162. Branch leaves without long aristae, costae subpercurrent, percurrent or short excurrent.......................163

163. Branch leaves relatively loosely arranged, not or somewhat spirally coiled, curved, circinate, contorted-twisted, not funiculate when dry, narrowly lanceolate, lanceolate, ligulate-lanceolate, oblong-lanceolate......................167
163. Branch leaves densely arranged, conspicuously spirally coiled around stem, funiculate when dry, lingulate, lingulate-lanceolate, oblong-lingulate to oblong-lanceolate..164

164. Calyptrae hairy...166
164. Calyptrae naked to scabrous above..165

165. Branch leaves with a rounded, mucronate and distinctively cucullate apex, margins strongly rugose above and plicate below.. **50. *M. densum***
165. Branch leaves acute to somewhat apiculate, or shortly cuspidate.. **194. *M. retusulum***

166. Plants without dense reddish rhizoids; branch leaves with a short awn, upper cells conic-bulging and mammillose .... **41. *M. concinnum***
166. Plants with dense reddish rhizoids; branch leaves without awns, upper cells strongly conic-bulging, distinctly unipapillose.. **74. *M. fortunatii***

167. Branch leaves not fragile, with or without dense reddish rhizoids...170
167. Branch leaves fragile, easily broken when mature, densely with reddish rhizoids.................................168

168. Upper margins of branch leaves distinctly and strongly erose-dentate, basal cells short-rectangular weakly tuberculate, marginal cells in the outmost row enlarged, short-rectangular to quadrate............ **207. *M. sejunctum***
168. Upper margins entire or weakly denticulate, basal cells elongate, moderately to strongly tuberculate, marginal cells at base not differentiated from their ambient cells..169

169. Branch leaves longer, 5.0-6.0 mm long, costae excurrent to a short awn.. **172. *M. petelotii***
169. Branch leaves shorter, 2.7-4.0 mm long, costae percurrent.. **249. *M. uraiense***

170. Basal cells of branch leaves with straight lumens..172
170. Basal cells of branch leaves with straight to curved-sigmoid lumens..171

171. Plants reddish to chestnut-brown, with numerous rusty rhizoids; basal cells of branch leaves strongly tuberculate.. **31. *M. calomicron***
171. Plants brown-green, without numerous rhizoids; basal cells weakly tuberculate or smooth................................
.... **129. *M. macrosporum***

172. Basal cells of branch leaves near costa large similar to ambient, not forming a "cancellina region"...................176
172. Basal cells of branch leaves near costa large, straight-walled, rectangular to irregularly rectangular, smooth, forming a "cancellina region"..173

173. Cells in 8-10 marginal rows at the base smooth, linear-rectangular, different from their ambient cells, forming a distinct border... **54. *M. diversifolium***
173. Marginal cells at the base not much different from their ambient cells, not forming a differentiated border.........
....174

174. Branch leaves undulate, strongly twisted-contorted and curly when dry, upper cells strongly bulging, obscure and unipapillose... **225. *M. sublaeve***
174. Branch leaves erect below, crisped to flexuous-contorted above when dry, upper cells conic-bulging (mammillose) 175

175. Plants small to medium-sized, without dense rhizoids; capsule urns cupulate, branch leaves oblong-lanceolate to ligulate, margins undulate and crenulate above.. **163. *M. parvirete***
175. Plants small, densely with reddish rhizoids; capsule urns ellipsoid-ovoid, ellipsoid-cylindric, branch leaves lanceolate, margins crenulate above.. **210. *M. sharpii***

176. Peristome single, absent or fragmentary..179
176. Peristome double..177

177. Basal marginal cells of branch leaves distinctly differentiated, wider and shorter than their ambient cells; capsules ellipsoid to ellipsoid-cylindric, weakly to moderately wrinkled or furrowed...... **221. *M. subcrenulatum***
177. Basal marginal cells of branch leaves not much different from ambient cells, capsules ellipsoid, ellipsoid-cylindric, cupulate to hemispheric, smooth or weakly furrowed at neck..178

178. Upper margin serrate, sometimes with a weak limbidium... **38. *M. cirrosum***
178. Upper margin notched, without a limbidium... **189. *M. raphidophyllum***

179. Branch leaves flexuous or incurved apices, the apex not hidden in a cavity when dry...181
179. Branch leaves inrolled above with their apices hidden in their inrolled cavities when dry..................................180

180. Calyptrae smooth and naked, oblong-lanceolate or ligulate-lanceolate, slightly squarrose-recurved spreading, conduplicate above when moist.. **6. *M. angulosum***
180. Calyptrae hairy, broadly oblong-lanceolate, lingulate-lanceolate, apices obtuse or obtuse-acute, spreading and slightly adaxially incurved upper when moist... **21. *M. binsteadii***

181. Plants not rust-brownish, creeping stems not covered with numerous rusty reddish rhizoids.............................183
181. Plants rust-brownish creeping stems covered with numerous rusty reddish rhizoids..182

182. Branch leaves long lanceolate, acuminate, upper and medial cells conic-bulging and mammillose........................
.... **252. *M. vesiculosum***
182. Branch leaves ovate-, oblong-, or ligulate-lanceolate, sharply acute to mucronate-acuminate, upper and medial cells strongly bulging, unipapillose, papillae occasionally 2-3 forked, becoming smaller and oblate toward margins in upper portion.. **255. *M. yuleanum***

183. Calyptrae naked..185
183. Calyptrae hairy...184

184. Branch leaves densely arranged, twisted-contorted and flexuose, cells not in regularly longitudinal rows............ .... **64.** *M. evrardii*
184. Branch leaves spirally coiled, curved or circinate when dry, all cells in regularly longitudinal rows, with a longitudinally striated appearance... **192.** *M. renauldii*

185. Branch leaves relatively longer, 3.0-4.0 mm long, oblong- to ligulate-lanceolate, abruptly narrowed to a long-cuspidate or mucronate apex... **166.** *M. perdensifolium*
185. Branch leaves relatively short, 1.3-2.2 mm long, ligulate to ligulate-lanceolate or lanceolate, abruptly narrowed to an obtuse or asymmetric short-cuspidate apex, or apices shortly cuspidate, mucronate-acuminate, or acute to apiculate-acuminate..186

186. Capsules narrowly to widely ellipsoid, smooth; peristome single................................ **211.** *M. similirete*
186. Capsules broadly ellipsoid, furrowed, or smooth in low portion and 8-plicate in upper half, with a small puckered mouth, or ovoid, 4-angled to slightly 8-plicate just below the mouth; peristome single, fragmentary, or absent..187

187. Setae relatively shorter, 3-8 mm; cells of branch leaves near the costa larger than those at the margins............... .... **10.** *M. archeri*
187. Setae relatively longer, 9-12 mm; cells of branch leaves near the costa similar to those at the margins..........188

188. Capsules with a small and puckered mouth... **162.** *M. parvifolium*
188. Capsules with a wide and not puckered mouth... **169.** *M. perpusillum*

189. Cells of branch leaves unipapillose or tuberculate to varying degrees in low (basal) portions while smooth in upper (medial) portions..229
189. Cells of branch leaves papillose or mammillose to varying degrees in upper (medial) portions while smooth in low (basal) portions..190

190. Branch leaves crisped, contorted or twisted to varying degrees...195
190. Branch leaves appressed and spirally curved or coiled around the branch, or squarrose-recurved and flexuose above when dry..191

191. Medial and upper cells of branch leaves mostly smooth, occasionally unipapillose, or strongly bulging and weakly papillose.. **2.** *M. acutirameum*
191. Medial and upper cells often pluripapillose..192

192. Costae of branch leaves excurrent to conspicuous hyaline aristae.....................................**160.** *M. panduraefolium*
192. Costae of branch leaves percurrent or excurrent in a mucro or ending just before the apex.................193

193. Stem leaves without clavate gemmae; branch leaves narrowly lanceolate; setae longer, ca. 15 mm long.................. .... **250.** *M. urceolatum*
193. Stem leaves occasionally with clavate gemmae; branch leaves ligulate; setae shorter, 3.5 mm long................194

194. Branch leaves with several rows of longer marginal cells, forming an indistinct border.............. **25.** *M. brevicaule*
194. Branch leaves without several rows of longer marginal cells, not forming a border....... **223.** *M. subhemitrichodes*

195. Branch leaves lanceolate, ligulate-lanceolate to ligulate, narrowly acuminate, obtuse or mucronate.................198
195. Branch leaves narrowly lanceolate, setaceous-acuminate, or gradually narrowed to a slender acumen, subula, or a long fragile arista..196

196. Plants small, branch leaves gradually narrowed to a long fragile arista................................. **39.** *M. clastophyllum*
196. Plants robust, branch leaves setaceous-acuminate, gradually narrowed to a slender acumen or subula...............197

197. Branch leaves with a clasping base, bulging and unipapillose in upper portions; peristome double..................... .... **201.** *M. rusbyanum*
197. Branch leaves without a clasping base, densely pluripapillose, the papillae forming a continuous covering, which is similar to those in *Zygodon*; peristome single.. **231.** *M. subulatum*

198. Upper cells of branch leaves rather obscure, densely pluripapillose, sharply (2-3 cells) grading to long, clear, hyaline low cells with curved-sigmoid lumens.. **83. *M. fuscescens***
198. Upper cells of branch leaves weakly to densely pluripapillose, clear or obscure, gradually elongate or gradually becoming hyaline towards base, low and basal cells with straight, or curved-sigmoid lumens..........................199

199. Low and basal cells without curved-sigmoid lumens..208
199. Low and basal cells with curved-sigmoid lumens...200

200. Setae relatively longer, often longer than 4 mm..202
200. Setae short, 2-3 mm or rather short, only 0.3-0.7 mm long..201

201. Setae rather shorter, only 0.3-0.7 mm long, with numerous smooth paraphyses at vaginula; calyptrae with numerous erect, slightly coarse hairs.. **95. *M. hainanense***
201. Setae 2-3 mm long, vaginulae naked; calyptrae naked.. **109. *M. laevigatum***

202. Peristome absent... **230. *M. subtortum***
202. Peristome single...203

203. Calyptrae naked..207
203. Calyptrae hairy...204

204. Perichaetial leaves much shorter than branch leaves, capsules ovoid to ellipsoid, smooth to weakly 8-plicate .... **105. *M. incurvifolium***
204. Perichaetial leaves longer than branch leaves; capsules ovoid to ellipsoid, smooth...205

205. Perichaetial long lanceolate, gradually acute, costae excurrent to an arista................................ **1. *M. acuminatum***
205. Perichaetial ligulate- to broadly long triangular-lanceolate..206

206. Plants with numerous rusty rhizoids.. **203. *M. salakanum***
206. Plants without too much rusty rhizoids.. **214. *M. speirostichum***

207. Plants small to medium-sized, branch leaves gradually narrowed to a slenderly acuminate.......................................
.... **7. *M. angustifolium***
207. Plants medium-sized to large, branch leaves acuminate to bluntly acuminate................................ **208. *M. semperi***

208. Calyptrae hairy to varying degrees...217
208. Calyptrae naked and smooth..209

209. Peristome present...211
209. Peristome absent...210

210. Upper cells of branch leaves strongly bulging and sparsely pluripapillose; calyptrae mitrate.................................
.... **70. *M. fimbriatum***
210. Upper cells of branch leaves rather obscure, densely pluripapillose, often look like blackish patch when moist; calyptrae cucullate.. **94. *M. gymnostomum***

211. Upper cells of branch leaves pluripapillose (occasionally unipapillose or smooth) 213
211. Upper cells of branch leaves unipapillose...212

212. Plants medium-sized; branch leaves not undulate and not rugose above, basal cells much larger than upper and medial cells.. **58. *M. ecrispatum***
212. Plants robust; branch leaves undulate and rugose above, basal cells elongate................................ **60. *M. ellipticum***

213. Capsules ellipsoid, smooth...215
213. Capsules long cylindric, conspicuously furrowed or short cylindric, longitudinally wrinkled..........................214

214. Capsules long cylindric, conspicuously furrowed.. **180. *M. proximum***
214. Capsules short cylindric, longitudinally wrinkled.. **182. *M. pseudoserrulatum***

215. Branch leaves lanceolate, apices acuminate, calyptrae conic-mitrate... **146. *M. nigricans***
215. Branch leaves ligulate-lanceolate, apices obtuse or mucronate, calyptrae mitrate..216

216. Upper and medial cells of branch leaves pluripapillose, some unipapillose to smooth; capsules small, urns ca. 1 mm long.. **102. *M. hortoniae***
216. Upper and medial cells pluripapillose; capsules larger, often longer than 1.5 mm................ **37. *M. chloromitrium***

217. Setae longer; capsules exserted..219
217. Setae rather short, ca. 0.5 mm long; capsules immersed..218

218. Branch leaves lingulate-oblong, upper outmost marginal cells differentiated, much smaller than their ambient cells.. **27. *M. brevisetum***
218. Branch leaves ligulate-lanceolate, upper outmost marginal cells not much differentiated......... **61. *M. emersulum***

219. Peristome present...221
219. Peristome absent..220

220. Plants small; perichaetial leaves not much differentiated from branch leaves; calyptrae with many long, brown-yellowish, flexuous hairs, weakly lacerate.. **73. *M. formosae***
220. Plants large; perichaetial leaves longer than branch leaves; calyptrae with a few straight hairs, deeply lacerate..
.... **204. *M. savatieri***

221. Upper and medial cells of branch leaves pluripapillose..223
221. Upper and medial cells of branch leaves bulging mammillose..222

222. Peristome double.. **185. *M. punctatum***
222. Peristome single.. **253. *M. viticulosum***

223. Branch leaves ligulate to oblong-ligulate, fairly broad, apices obtuse to mucronate, elongate cells restricted to insertion, longer near margins and forming an indistinct border of 5-10 rows..................... **24. *M. brachypodium***
223. Branch leaves long narrow lanceolate, lanceolate-ligulate, oblong-lanceolate, oblong-ligulate, elongate cells not restricted to insertion, without differentiated marginal cells...224

224. Branch leaves not or inconspicuously curly above, without apices hidden, spreading when moist....................226
224. Branch leaves circinate-curly above, with apices hidden in an inrolled cavity when dry, spreading but apices often adaxially incurved when moist..225

225. Perichaetial leaves ovate-oblong, acuminate; setae 3-4 mm long; calyptrae cucullate.............. **107. *M. japonicum***
225. Perichaetial leaves oblong, with acuminate-acute to long acuminate apices; setae 8-10 mm; calyptrae mitrate...
.... **112. *M. laosianum***

226. Branch leaves linear or long lanceolate.. **178. *M. prolongatum***
226. Branch leaves ligulate to ligulate-lanceolate...227

227. Setae shorter, ca. 3.5-4.0 mm long.. **40. *M. comatum***
227. Setae longer, up to 8 mm...228

228. Branch leaves strongly curly and keeled, crisped and contorted when dry; calyptrae densely with yellowish or brownish straight hairs.. **141. *M. moorcroftii***
228. Branch leaves individually twisted, carinate, their apices circinate to coiled when dry; calyptrae sparsely hairy.
.... **143. *M. nanothecium***

229. Upper and medial cells of branch leaves often rounded, subquadrat, rounded-quadrate, isodiametric............233
229. All cells of branch leaves longer than wide..230

230. Peristome single...**45. *M. crinale***
230. Peristome double...231

231. Upper cells of branch leaves not porose...**238. *M. thwaitesii***
231. Upper cells of branch leaves conspicuously porose..232

232. Branch leaves lanceolate from an oblong or broadly oblong low part, slightly sheathing at base, with acuminate apices, upper cells not in radiating diagonal rows from the costa, upper margins strongly serrate..**72. *M. flexuosum***
232. Branch leaves broadly lanceolate from a long oblong low part, with subulate-acuminate apices, upper cells occasionally in distinctly radiating diagonal rows from the costa, upper margins weakly serrate..**220. *M. subcirrhosum***

233. Branch leaves linear-lanceolate, lanceolate to oblong- or ligulate-lanceolate, crisped, or twisted-contorted, flexuous to varying degrees when dry..237
233. Branch leaves lingulate, oblong-lingulate to ligulate, appressed and spirally curved to twisted around the branch when dry..234

234. Branch leaves frequently retuse, ending in a long hyaline arista, upper portion often rugose.........**184. *M. pullenii***
234. Branch leaves with acute to obtuse, apiculus apices, without hyaline aristae, upper portions not rugose...........235

235. Calyptrae densely with yellow hairs..**56. *M. dusenii***
235. Calyptrae naked or occasionally with a few short hairs...236

236. Calyptrae naked or occasionally with a few short hairs, basal cells curved-sigmoid, occasionally and sparsely tuberculate..**193. *M. repandum***
236. Calyptrae naked, basal cells straight, densely tuberculate..**236. *M. taoense***

237. Branch leaves with a long fragile subula, upper marginal cells distinctly longer and narrower than ambient cells, forming a conspicuous differentiated border...**78. *M. frustratum***
237. Branch leaves without a fragile subula, upper marginal cells similar to their ambient cells, not or forming an inconspicuous border...238

238. Branch leaves not strong rugose above when moist...240
238. Branch leaves strong rugose above when moist...239

239. Perichaetial leaves similar to branch leaves, lanceolate and broadly acuminate....................**127. *M. lorifolium***
239. Perichaetial leaves different from branch leaves, lanceolate from a broadly oblong base, acuminate to a long and fine arista..**197. *M. rimbachii***

240. Branch leaves linear-lanceolate, lanceolate to oblong- to ligulate-lanceolate, margins entire or weakly serrate or crenulate above..242
240. Branch leaves oblong-lanceolate to long lanceolate, margins distinctly serrate above......................................241

241. Branch leaves oblong-lanceolate, apices acute...**28. *M. caldense***
241. Branch leaves long lanceolate, apices acuminate...**158. *M. ovale***

242. Peristome present; setae longer, often longer than 4 mm..244
242. Peristome absent; setae shorter than 3 mm...243

243. Calyptrae naked and smooth; plants small to medium-sized; branch leaves oblong- to ligulate-lanceolate, not longer than 4 mm, broadly acuminate at upper portions; setae about 3 mm long...............**68. *M. fernandezianum***
243. Calyptrae hairy; plants large; branch leaves long linear-lanceolate, up to 7 mm long, gradually narrowed to a slender acuminate acumen or subula from a long oblong low portion; setae about 1-1.2 mm long..**234. *M. taiheizanense***

244. Peristome double..248
244. Peristome single...245

245. Plants frequently broken to form separate branches; calyptrae densely hairy........................ **247. *M. tylostomum***
245. Plants not easily broken to form separate branches, calyptrae naked or with a few, stiff, thick erect hairs..........246

246. Branch leaves 4.0-6.0 mm long; setae shorter, 2.5-4.0 mm long.. **226. *M. sublongicaule***
246. Branch leaves shorter than 3.0 mm; setae longer, 8-11 mm long..247

247. Branch leaves 2.0-3.0 mm long, lanceolate to narrowly lanceolate, sharply and slenderly acute to acuminate....
.... **65. *M. exsertum***
247. Branch leaves 1.5-2.5 mm long, narrowly ovate-lanceolate to lanceolate, abruptly acuminate to narrowly acute, some slenderly cuspidate-acuminate.. **82. *M. funiforme***

248. Calyptrae naked..250
248. Calyptrae hairy...249

249. Cells of branch leaves moderately large, gradually elongate toward base from the medial portion; capsules ovoid, longitudinal wrinkled or distinctly plicate.. **11. *M. argutum***
249. Cells of branch leaves rather large from upper to base, basal cells short to long rectangular, confined to a small area of the leaf; capsules ellipsoid, ellipsoid-cylindric, strongly furrowed................................ **177. *M. proliferum***

250. Upper cells of branch leaves bulging, collenchymatous; capsules obovate or cupulate.............. **186. *M. pyriforme***
250. Upper cells of branch leaves flat to bulging, not collenchymatous; capsules ovate, ellipsoid, fusiform-cylindric to cylindric..251

251. Upper cells of branch leaves small, often in regularly oblique rows in broad leaves and in longitudinal rows in narrow leaves, basal cells near the costa distinctly larger than their ambient cells, forming a "cancellina region".. **232. *M. sulcatum***
251. Upper cells of branch leaves moderately large, not arranged in oblique rows; basal cells near the costa not distinctly larger than their ambient cells, not forming a "cancellina region"......................................252

252. Branch leaves lanceolate to long linear lanceolate, acuminate..254
252. Branch leaves broadly lanceolate, acute to shortly cuspidate, broadly acuminate-apiculate..............................253

253. Upper and medial cells of branch leaves rather small; setae longer, 10-13 mm....................... **114. *M. lauterbachii***
253. Upper and medial cells of branch leaves moderately large; setae short and stout, 2.5-4.0 mm....................................
.... **188. *M. ramsayae***

254. Plants in loose mats, stems rather slender, long creeping, sparsely with branches..
.... **232b. *M. sulcatum* var. *torulosum***
254. Plants in mats, stems creeping or weakly creeping, densely with branches..255

255. Plants without dense brown-reddish rhizoids...257
255. Plants with dense brown-reddish rhizoids..256

256. Upper margins of branch leaves entire... **96. *M. harrisii***
256. Upper margins of branch leaves slightly dentate.. **246. *M. turgidum***

257. Capsules cylindric, furrowed.. **88. *M. glabratum***
257. Capsules ellipsoid, smooth or furrowed..258

258. Capsules smooth; medial and upper cells of branch leaves rather small; perichaetial leaves similar to branch leaves, lanceolate to narrowly lanceolate, and acuminate... **42. *M. constrictum***
258. Capsules furrowed; medial and upper cells of branch leaves moderately sized; perichaetial leaves longer than branch leaves, triangular-lanceolate with the widest part at base, long acuminate to an arista..............................
.... **229. *M. subscabrum***

4.3 Descriptions of recognized species

1. *Macromitrium acuminatum* (Reinw. & Hornsch.) Müll. Hal., Bot. Zeitung (Berlin) 3: 544. 1845. (Figure 34)
Basionym: *Schlotheimia acuminata* Reinw. & Hornsch., Nova Acta Phys.-Med. Acad. Caes. Leop. -Carol. Nat. Cur. 14: 711. 1829. Type protologue: (Indonesia) in Java insula. Type citation: *Schlotheimia* nova spec. *no. 67* and Java-Reinhardt (lectotype: L 0060421!).
*Macromitrium elongatum Dozy & Molk. ex Bosch. & Sande Lac., Bryol. Jav. 1: 116. 93. 1859, *fide* Vitt *et al.*, 1995. Type protologue: (Indonesia) Habitat insulam Javae; ad arbores montis Salak *Zollinger coll. no. 1734* (holotype: L).

(1) Plants medium-sized, somewhat lustrous, in loose mats; stems not much differentiated from branches. (2) Branch leaves erect to patent below, individually twisted, loosely contorted-flexuose, curved-twisted above, with wide-spreading-curved to deflexed-curved-twisted upper portions when dry, squarrose-recurved when moist, 2.7-3.0 mm long, narrowly lanceolate from an oblong, sheathing base, keeled; margins plane above, reflexed to recurved below, entire; apices slenderly acute to acuminate; costae percurrent to excurrent; upper and medial cells rounded, quadrate, oblate, occasionally longer than wide, obscurely and densely pluripapillose, marginal cells smaller and oblate, slightly differentiated from their ambient cells; low and basal cells narrowly rectangular, thick-walled, with curved-sigmoid lumens, clear to yellow, smooth and flat. (3) Perichaetial leaves erect, loosely sheathing vaginula; 3-4 mm long, long lanceolate, gradually acute; all cells longer than wide; costa excurrent into an arista; vaginulae without hairs. (4) Setae 4.0-4.3 mm long, straight, smooth, twisted to the left. (5) Capsule urns ellipsoid-ovoid, smooth; peristome single, exostome of 16, well-developed, teeth finely papillose-striate, incurved; anisosporous. (6) Calyptrae mitrate, oblong-conic, plicate and hairy.

According to the type, *M. acuminatum* (Reinw. & Hornsch.) Müll. Hal. is rather similar to *M. angustifolium* Dozy & Molk. by having isodiametric, rounded pluripapillose upper cells, but the former can be distinguished from the latter by its hairy calyptrae. *Macromitrium acuminatum* is similar to *M. fuscescens* Schwägr., *M. salakanum* Müll. Hal. and *M. semperi* Müll. Hal. in their basal leaf cells with sigmoid-curved lumens. However, the plants of *M. semperi* are rather large, rusty-brown or reddish-brown; the upper cells of *M. fuscescens* are very small and rather obscure, sharply grading to long, clear and hyaline basal cells; the branch leaves of *M. salakanum* coiled around the branch forming "rope-like" shoots.

Distribution: Indonesia, Papua New Guinea (Vitt *et al.*, 1995), the Solomon Islands.
Specimen examined: **THE SOLOMON ISLANDS**. *D. H. Norris & G. L. Roberts 19584B* (H 3090086).

2. *Macromitrium acutirameum* Mitt., J. Linn. Soc., Bot. 15: 62. 1876. (Figure 35)
Type protologue: collection during Challenger Expedition, Tristan d'Acunha. Type citation: [South Atlantic Ocean] Tristan d' Acunha, *Challenger Exp. s.n.* (lectotype designated here: NY 00518221!).

(1) Plants small to medium-sized; stems long-creeping, densely with short and erect branches. (2) Stem leaves squarrose flexuose when dry, spreading or somewhat abaxially curved when moist, ovate-lanceolate, triangular-lanceolate; the apices acute, bluntly-acuminate, somewhat cuspidate; upper and medial cells round-quadrate to oblate, thick and smooth, basal cells elongate and thick-walled, smooth, the outmost marginal cells at the insertion enlarged, shorter and broader, quadrate to short quadrate. (3) Branch leaves keeled, obliquely appressed below, spirally twisted to coiled around the branch above when dry, spreading and the costa often concealed in adaxial view by overlapping folds of laminae under the apex when moist, oblong-lanceolate, some with the widest at the base, plicate below, frequently recurved above, keeled; costae subpercurrent; apices acute, bluntly acuminate, mucronate; margins entire throughout; apical cells smaller, rounded, rounded-quadrate, elliptic to oblong; upper and medial cells rounded, rounded-quadrate, obliquely oblate, occasionally unipapillose; basal cells elongate and thick-walled, smooth and not porose, marginal cells enlarged.

Distribution: Tristan d'Acunha.

3. *Macromitrium amaniense* P. de la Varde, Ark. Bot., n.s 3: 174. f. 26. 1955. (Figure 36)
Type protologue and citation: [Tanzania] Usambara Mts: near Amani agric, station, 1000 m, epiphytic, with *Leucoloma Holstii* Broth. (2046, type) (holotype: PC 0105920!).

(1) Plants medium-sized, in mats; primary stems prostrate and creeping, about 4-5 cm long. (2) Branch leaves erect-appressed below, twisted-crisped above, with curved to circinate apices when dry, spreading to slightly abaxially curved spreading when moist, long ligulate-lanceolate, 2.0-2.5 mm long, 0.25-0.30 mm wide; apices mucronate, acute, to acuminate-acute; margins plane, entire to crenulate-papillose above, entire below; costae subpercurrent to percurrent; leaf upper parts in cross sections sporadically bistratose or occasionally jagged; upper

and medial cells rounded-quadrate, rather obscure, pluripapillose; low and basal cells elongate, long rectangular, lumen curved to sigmoid, thick-walled and porose, clear and smooth. (3) Perichaetial leaves not much different from branch leaves. (4) Setae about 5 mm long, smooth, not twisted. (5) Capsule urns ovoid-ellipsoid, smooth; peristome single, exostome of 16, lanceolate with obtuse apices, papillose.

Macromitrium amaniense P. de la Varde is similar to *M. st-johnii* E. B. Bartram in their upper portions of branch leaves with partial, variable bistratose, papillose and jagged laminae. The latter could be separated from the former by its gymnostomous capsules.

Distribution: Tanzania.

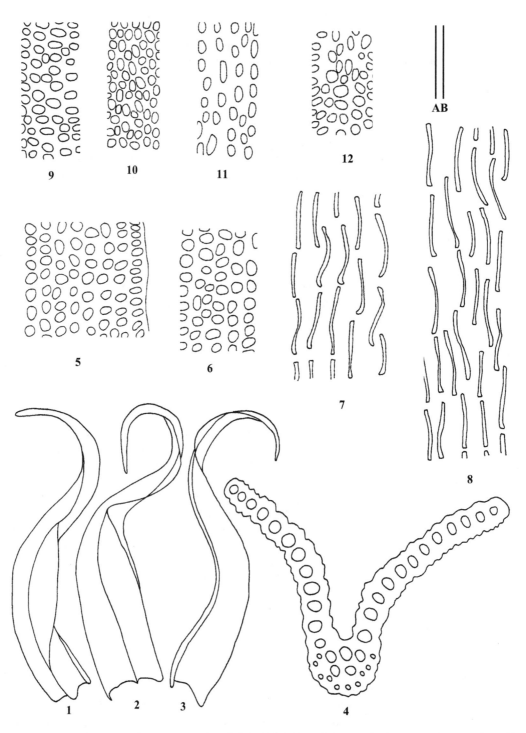

Fig. 34 *Macromitrium acuminatum* (Reinw. & Hornsch.) Müll. Hal. 1-3: Branch leaves. 4: Upper transverse section of branch leaf. 5, 9, 10: Upper cells of branch leaves. 6, 11, 12: Medial cells of branch leaves. 7: Low cells of branch leaf. 8: Basal cells of branch leaf (all from lectotype, L 0060421). Line scales: A = 1 mm (1-3); B = 67 μm (4-12).

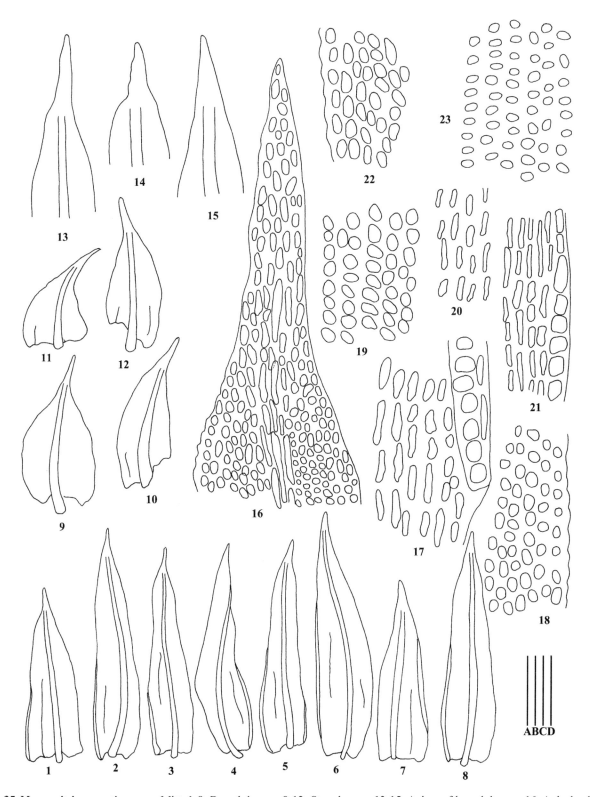

Fig. 35 *Macromitrium acutirameum* Mitt. 1-8: Branch leaves. 9-12: Stem leaves. 13-15: Apices of branch leaves. 16: Apical cells of branch leaf. 17: Basal cells of branch leaf. 18: Upper cells of stem leaf. 19: Medial cells of branch leaf. 20: Basal cells near costa of stem leaf. 21: Basal marginal cells of stem leaf. 22: Upper cells of branch leaf. 23: Medial cells of stem leaf (all from lectotype, NY 00518221). Line scales: A = 1 mm (1-12); B = 400 μm (13-15); C = 100 μm (16); D = 67 μm (17-23).

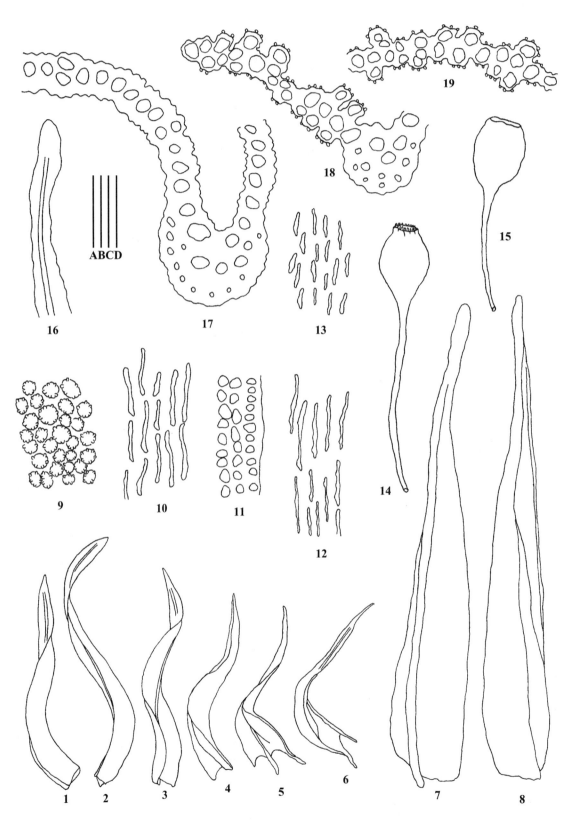

Fig. 36 *Macromitrium amaniense* P. de la Varde 1-3: Branch leaves. 4-6: Stem leaves. 7-8: Perichaetial leaves. 9: Upper cells of branch leaf. 10: Basal cells of branch leaf. 11: Upper cells of perichaetial leaf. 12: Low cells of Perichaetial leaf. 13: Medial cells of branch leaf. 14-15: Capsules. 16: Apex of branch leaf. 17: Medial transverse section of branch leaf. 18-19: Upper transverse sections of branch leaves (all from holotype, PC 0105920). Line scales: A = 2 mm (14-15); B = 1 mm (1-6); C = 400 μm (7-8, 16); D = 67 μm (9-13, 17-19).

4. *Macromitrium amboroicum* Herzog, Beih. Bot. Centralbl., Abt. 26(2): 68. 1909. (Figure 37)

Type protologue: Bolivia, Auf Bäumasten in der Cinchonaregion des Cerro Amboró (Ostcordillere), about 1300 m; Oktober, 07. Type citation: Bolivia, Santa Cruz, Auf Baumästen in der Cinchonaregion des Cerro Amboró (Ostcordillere), about 1300 m; *Herzog T.C.J.*, October 07 (holotype: JE 04008730!).

(1) Plants small to medium-sized; yellowish green above, dark brown below; stems long and creeping, densely with reddish or rusty rhizoids, branches up to 10 mm long. (2) Stem leaves erect-appressed below, flexuose above, costae percurrent; all cells clear and smooth. (3) Branch leaves erect below, keeled, strongly twisted-flexuose to contorted-crisped above, apices incurved to inrolled when dry, oblong-lanceolate to ligulate-lanceolate, recurved at one side below, with acute to apiculate the apices; margins entire throughout; costae stout and percurrent; upper and medial cells rounded, rounded-quadrate, rounded-hexagonal, in longitudinal rows, strongly bulging, smooth and clear; low and basal cells rectangular, thick-walled, not porose, occasionally tuberculate; basal cells near costa enlarged and hyaline, marginal enlarged teeth-like cells at insertion differentiated.

In shape of branch leaves, *M. amboroicum* Herzog is similar to *M. pseudofimbriatum* Hampe, *M subbrevihamatum* Broth., and *M. punctatum* (Hook. & Grev.) Brid. *Macromitrum amboroicum* differs from *M. punctatum* by its branch leaves with entire margins, basal cells not porose, and enlarged teeth-like cells at insertion. From *M. pseudofimbriatum*, *M. amboroicum* can be separated by its basal cells of branch leaves without porose, and stems densely with reddish rhizoids. *Macromitrium subbrevihamatum* differs from *M. amboroicum* by its branch leaves with smooth basal cells.

Distribution: Bolivia.

Specimen examined: **BOLIVIA**. Depto. Santa Cruz. Prov. Ichilo, *Cerro Amboró. Marko lewis 37834 d-3* (MO 3961419).

5. *Macromitrium angulatum* Mitt., J. Linn. Soc. Bot. 10: 167. 1868. (Figure 38)

Type protologue: (Samoa) Tutuila, on *Hibiscus* trees (1000 ft.), *no. 67.* Type citation: Samoa. Tutuila, on Hibiscus trees, 1000 ft., *Powell 67* (lectotype: NY; isolectotype: MICH 525863!).

Macromitrium papuanum Dixon in Forbes, J. Bot. 61 (Suppl.): 63. 1923, *fide* Vitt *et al.*, 1995. Type protologue: Papua New Guinea. Central, Sogere, 2000 ft., epidendric, *Forbes 515b* (isotype: BM).

Macromitrium wellingtonianum Vitt, J. Hattori Bot. Lab. 54: 177-178. 1983, *fide* Vitt & Ramsay, 1985. Type protologue: New Zealand. North Island, Tararua Range, 60 km North of Wellington, 800 meters elev., 28. XII. 1972 Balázs NZ-S/I/J.

(1) Plants rather small, slender, dull and stiff, in sparse to dense mats, olive-green to green-brown, rigidly branched; stems prostrate, densely with rusty rhizoids; branches slender and filiform, up to 2.0 cm long. (2) Stem and branch leaves similar, stiffly appressed to erect below, spreading and slightly curved toward one side to individually flexuose-twisted above when dry, spreading to wide-spreading-recurved when moist, 0.7-1.5 mm long, oblong-ovate to oblong- or ligulate-lanceolate; margins plane, crenulate to denticulate from bulging cells throughout leaf margin; apices acute-apiculate to broadly acuminate-acute; costa shortly excurrent to rarely percurrent; all cells rounded to rounded-quadrate; upper and medial cells 7-12 × 7-12 µm, strongly bulging, smooth or weak unipapillae, or 2-3 small, obscure papillae; marginal cells clear, smaller than their ambient cells; basal cells somewhat larger than medial cells, marginal cells usually with branched, papillose rhizoids. (3) Perichaetial leaves shorter than branch leaves, about 0.5 mm long, erect, ovate-lanceolate, lanceolate from an oblong low portion. (4) Setae 2-5 mm long, stout, prorate throughout, ridged, twisted to the left; vaginulae exposed, sparsely hairy. (5) Capsule urns ellipsoid to ovoid, smooth, 4-angled to round near the mouth, and smaller at the mouth; peristome single, exostome fused to a low, smooth membrane; spores distinctly anisosporous. (6) Calyptrae conic-mitrate, sometimes plicate, densely hairy.

Macromitrium angulatum Mitt., *M. erubescens* E.B. Bartram and *M. orthostichum* Nees ex Schwägr. are similar in their small plants, rounded to rounded-quadrate leaf cells, prorate setae, and 4-angled capsules. However, *M. orthostrichum* has pluripapillose upper leaf cells, *M. erubescens* has longer setae (6.0-10. mm long), costa of perichaetial leaves excurrent to form a short arista.

Distribution: Indonesia, New Zealand (Vitt, 1983), Papua New Guinea, Thailand, Samoa.

Specimens examined: **PAPUA NEW GUINEA**. Morobe, *H. Streimann 18911* (MO 4462193), *T. J. Koponen 29639* (MO 4428637); West Sepik, *A. Touw 15084* (MO 5371062), *A. Touw 15024* (MO 5375811), *A. Touw 17789* (MO 5375901), *A. Touw 17619* (MO 5375899). **THAILAND**. Peninsular, *T. Smitinand & E. Warncke 1487* (MO 3971199).

6. *Macromitrium angulosum* Thwaites & Mitt., J. Linn. Soc., Bot. 13: 300. 1873. (Figure 39)

Type protologue and citation: In Ceylon, *Dr. Thwaites* (isotype: H-BR 2535001!).

(1) Plants medium-sized, in dense mats, olive-green above, dark-brownish below; stems long creeping. (2) Branch leaves strongly crisped and contorted, with the apices hidden in the inrolled cavity when dry, slightly abaxially

curved spreading, conduplicate above when moist, oblong-lanceolate or ligulate-lanceolate, somewhat plicate below, with acuminate-acute the apices; margins entire, plane or narrowed recurved at one side; upper cells quadrate-round, or irregular quadrate, 4-5 μm wide, thin-walled, obscure, moderately bulging, unipapillose; medial cells short-rectangular, 5-6 × 6-8 μm wide, bulging to conic-bulging, clear, unipapillose; basal cells thin-walled, short-rectangular, hyaline, slightly bulging, smooth or unipapillose; costae subpercurrent. (3) Perichaetial leaves slightly shorter than branch leaves, lanceolate, erecto-patent when moist. (4) Setae smooth, 6-8 mm long, twisted to the left. (5) Capsule urns ovoid to ellipsoid, weakly plicate with a small and puckered mouth when dry; opercula conic-rostrate with a long beak; peristomes and spores not seen. (6) Calyptrae mitrate, smooth and naked.

Macromitrium angulosum Thwaites & Mitt. is slightly similar to *M. japonicum* Dozy & Molk., but differs from the latter in having leaves with smooth or unipapillose cells, calyptrae smooth and naked.

Distribution: Sri Lanka.

Specimen examined: **SRI LANKA**. *G. H. K. Thwaites 37* (MO 5278954).

7. *Macromitrium angustifolium* Dozy & Molk., Ann. Sci. Nat., Bot., sér. 3, 2(5): 311. 1844. (Figure 40)

Type protologue: (Indonesia) Sumatra, Java, Borneo. Type citation: Indonesia, Java (lectotype: L 0060422!); Indonesia, Sumatra (isosyntype: H-BR 2604007!).

Macromitrium fruhstorferi Cardot, Rev. Bryol. 28: 113. 1901, *fide* Wijk *et al.*, 1964.

(1) Plants small to medium-sized, in spreading mats, brown to reddish Brown below, yellowish-green above when young; stems freely branched, with branches 0.5-2.0 cm long, often bearing numerous branchlets. (2) Stem leaves inconspicuous, appressed when dry, erect-spreading when moist, oblong-ovate at base, sublinear above, covered with rhizoids. (3) Branch leaves individually twisted, upper portions contorted to deflexed curved when dry, abaxially curved spreading when moist, 2.0-3.5 mm long, narrowly lanceolate to lanceolate from an oblong low portion, gradually narrowed to a slenderly acuminate apex; margins broadly reflexed, serrulate near the apex; costae yellowish brown, extending to the apex; apical cells elongate and smooth; upper cells round to oblate, 4.0-6.5 μm, evenly thick-walled, moderately pluripapillose to almost smooth, in distinct longitudinal rows; medial cells hyaline, subquadrate, thick-walled, smooth, in distinct longitudinal rows; basal cells rectangular or sublinear, 17-30 μm long, thick-walled, smooth, lumen sigmoid-curved. (4) Perichaetial leaves erect-appressed, oblong-lanceolate, acute to short acuminate; costae percurrent to excurrent; all cells longer than wide, smooth, curved-sigmoid. (5) Setae 6-8 mm long, smooth, twisted to the left. (6) Capsule urns ellipsoid, constricted and slightly plicate beneath the mouth; peristome single, exostome 16, lanceolate; spores anisosporous. (7) Calyptrae mitrate, naked, plicate.

Macromitrium angustifolium Dozy & Molk. is similar to *M. salakanum* Müll. Hal., but the apices of the perichaetial leaves in the latter are obtuse and mucronate to sharply apiculate-acuminate, and those in the former are short acuminate to acute.

Distribution: Indonesia, Philippines, Vietnam, Papua New Guinea (Vitt *et al.*, 1995).

Specimens examined: **INDONESIA**. South Celebes, *W. Meijer 11523a* (MO 5361149); Java, *E. Nyman 166* (MO 2860677). **PHILIPPINES**. Mindanao, *A. D. E. Elmer 11792* (MO1950190); Luzon, *J. V. Pancho 3431* (MO 2505476), *B. Allen 3125* (MO 3987169), *B. F. Hernaez 1537* (MO 2555516), *G. E. Edaño 12834* (MO 4416218), *G. E. Edaño 1197* (MO 4416219), *R. S. Williams 20626* (MO 3365029). **VIETNAM**. Lam Dong, *S. He & K. Nguyen 42907* (MO 6239370).

8. *Macromitrium antarcticum* C.H. Wright, J. Linn. Soc., Bot. 37: 264. 1905. (Figure 41)

Type protologue: [St. Helena] Endemic in Gough Island. Type citation: Gough Island 22-4-1904, Coll. *R.N. R. Brown s.n.* (lectotype designated here: PC 0137579!).

(1) Plants small, in dense mats; stems long-creeping, densely with short and erect branches and rusty rhizoids. (2) Stem leaves erect-flexuose when dry, spreading but abaxially curved above when moist. (3) Branch leaves strongly keeled, erect and flexuose-twisted below, contorted-flexuose-crisped, occasionally the apices hidden in the inrolled cavity when dry, widely spreading when moist, ligulate-lanceolate, with an acute apex; costae subpercurrent; margins entire throughout; all cells clear and smooth and not porose; upper and medial cells isodiametric, quadrate, subquadrate, elongate towards the base; low and basal cell elongate-rectangular, thick-walled, with straight lumens, not porose. (4) Perichaetial leaves lanceolate from an ovate- or oblong base, with an acuminate or somewhat cuspidate apex; slightly plicate and sheathing at the base; all cells smooth and not porose, upper cells short-rectangular, quadrate, thick-walled; medial cells elongate-rectangular, thick-walled. (5) Setae about 3 mm long, smooth, twisted to the left. (6) Capsule urns ellipsoid, smooth, narrowed and plicate beneath the mouth when dry, conic-rostrate, with a straight beak; peristome single, teeth oblong with obtuse apices. (7) Calyptrae mitrate, plicate, smooth and naked, not lobed or cleft at one side.

Distribution: St. Helena.

Specimen examined: **ST. HELENA**. Gough or Diego Alvarez, Scottish Nat. Antarctic Expedition, 1904 (PC 0137580).

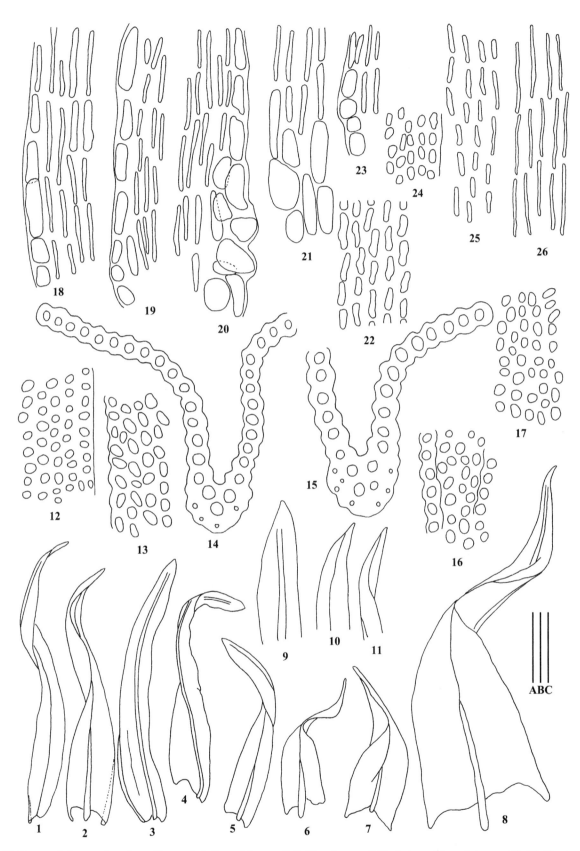

Fig. 37 *Macromitrium amboroicum* Herzog 1-5: Branch leaves. 6-8: Stem leaves. 9-11: Apices of branch leaves. 12, 16: Upper cells of branch leaves. 13: Upper cells of stem leaf. 14-15: Medial transverse sections of branch leaves. 17: Medial cells of branch leaf. 18-21, 23: Basal marginal cells of branch leaves. 22: Basal cells of stem leaf. 24: Medial cells of stem leaf. 25: Low ells of branch leaf. 26: Basal cells of branch leaf (all from holotype, JE 04008730). Line scales: A = 1 mm (1-7); B = 400 μm (8-11); C = 67 μm (12-26).

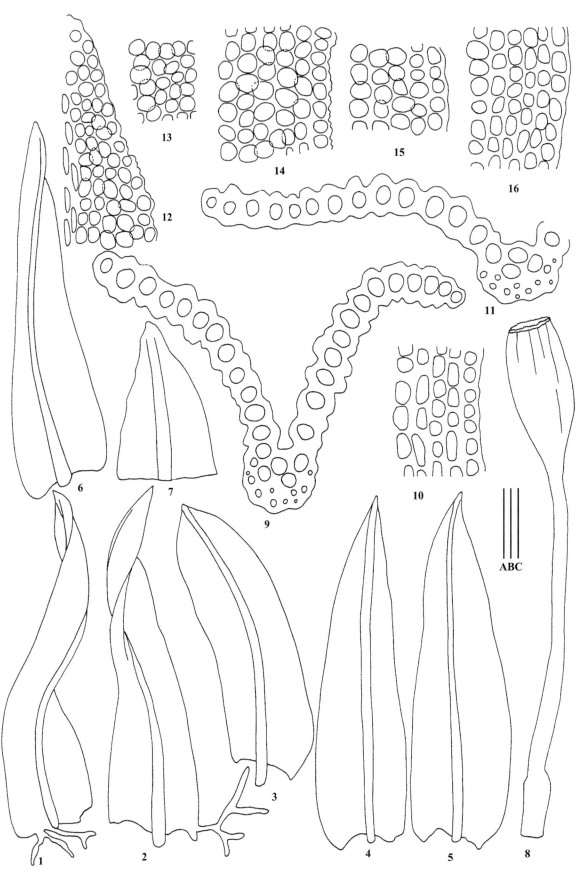

Fig. 38 *Macromitrium angulatum* Mitt. 1-6: Branch leaves. 7: Apex of branch leaf. 8: Capsule. 9: Medial transverse section of branch leaf. 10, 16: Basal cells of branch leaves. 11: Basal transverse section of branch leaf. 12, 13: Apical cells of branch leaves. 14: Medial cells of branch leaf. 15: Low cells of branch leaf (all from isolectotype, MICH 525863). Line scales: A = 2 mm (8); B = 400 μm (1-7); C = 67 μm (9-16).

Taxonomy

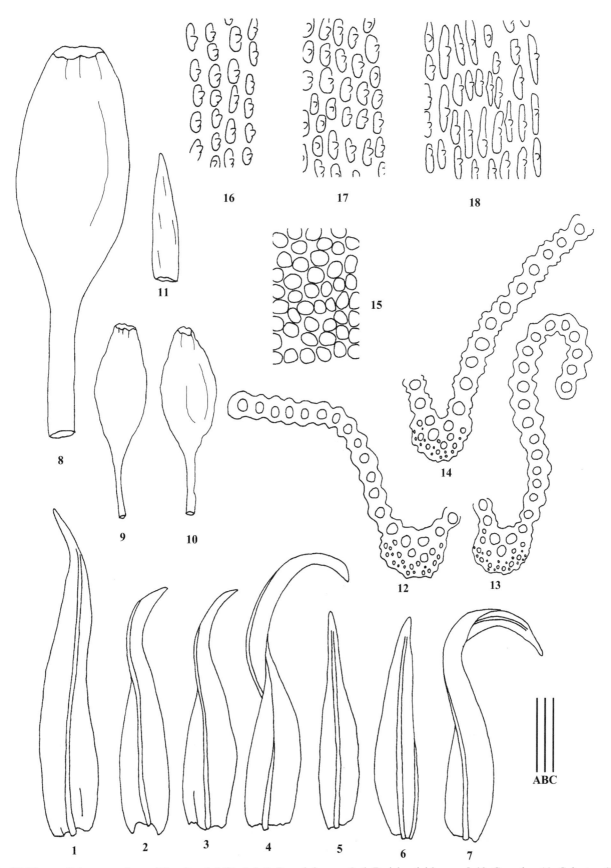

Fig. 39 *Macromitrium angulosum* Thwaites & Mitt. 1-4, 7: Branch leaves. 5, 6: Perichaetial leaves. 8-10: Capsules. 11: Calyptra. 12-14: Transverse sections of branch leaves. 15: Upper cells of branch leaf. 16, 17: Medial cells of branch leaves. 18: Basal cells of branch leaf (all from isotype, H-BR 2535001). Line scales: A = 0.88 mm (9-11); B = 0.44 mm (1-8); C = 44 μm (12-18).

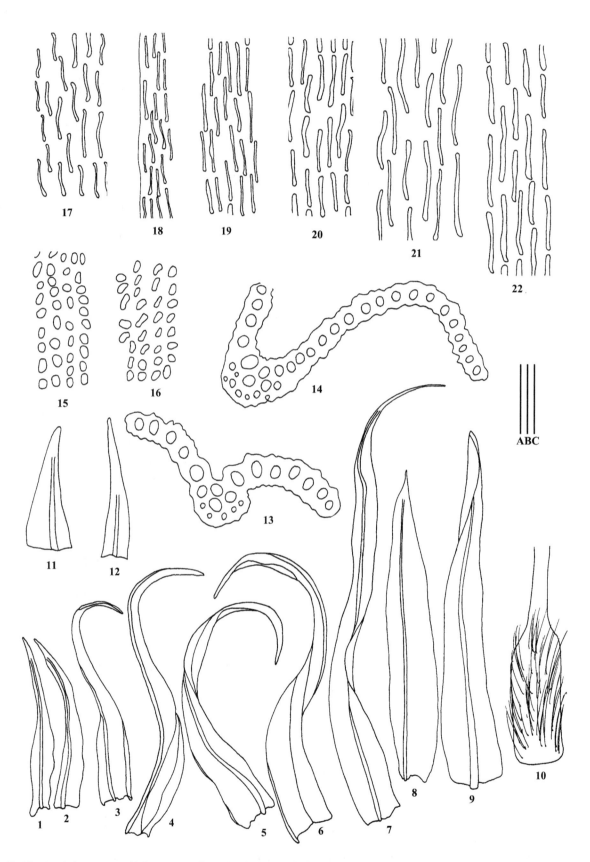

Fig. 40 *Macromitrium angustifolium* Dozy & Molk. 1-7: Branch leaves. 8-9: Perichaetial leaves. 10: Vaginula. 11-12: Apices of perichaetial leaves. 13: Upper transverse section of branch leaf. 14: Low transverse section of branch leaf. 15: Medial cells of branch leaf. 16: Upper cells of branch leaf. 17: Basal cells of branch leaf. 18: Apical cells of perichaetial leaf. 19: Medial cells of perichaetial leaf. 20-21: Low cells of perichaetial leaves. 22: Basal cells of perichaetial leaf (all from lectotype, L 0060422). Line scales: A = 1 mm (1-10); B = 400 μm (11, 12); C = 67 μm (13-22).

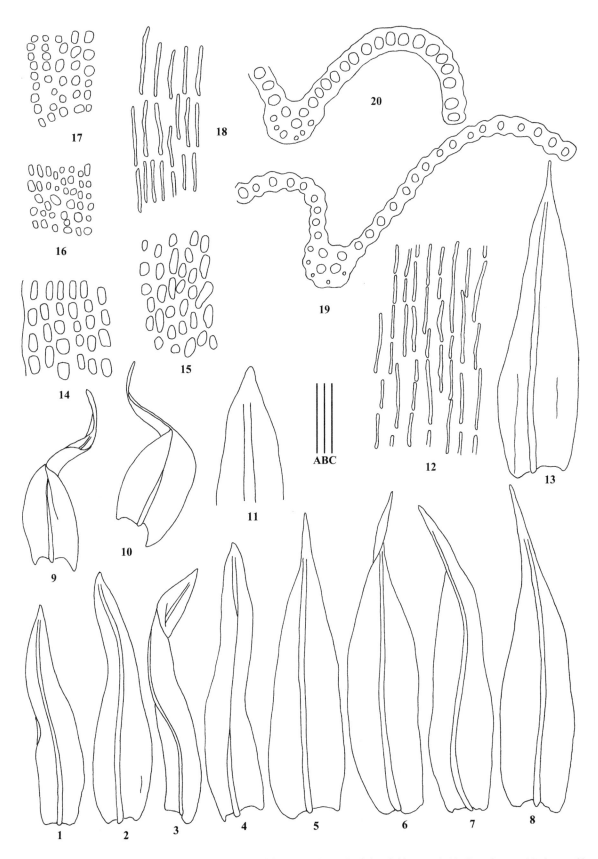

Fig. 41 *Macromitrium antarcticum* C.H. Wright 1-4: Branch leaves. 5-8, 13: Perichaetial leaves. 9-10: Stem leaves. 11: Apex of branch leaf. 12: Medial cells of perichaetial leaf. 14: Low cells of branch leaf. 15: Upper cells of perichaetial leaf. 16: Upper cells of branch leaf. 17: Medial cells of branch leaf. 18: Basal cells of branch leaf. 19: Basal transverse section of branch leaf. 20: Medial transverse section of branch leaf (all from lectotype, PC 0137579). Line scales: A = 1 mm (1-10, 13); B = 400 μm (11); C = 67 μm (12, 14-20).

9. *Macromitrium archboldii* E.B. Bartram, Lloydia 5: 272. 34. 1942. (Figure 42)

Type protologue: (Papua New Guinea) Mt. Wilhelmina: very abundant in subalpine forest, cushioned on trees and ground, 2 km. E. of Wilhelmina-top, 3800 m, *no. 10188* type; epiphyte in subalpine forest, E. of Wilhelmina-top, 3800 m, 10323. Type citation: N. Guinea, Mt. Wilhelmina; 3800 m, IX, 1938, collector: *Brass* & *MyerDaees 10188*, Det: Bartram (isotype: L 0060423!).

Macromitrium ruberrimum Dixon, Farlowia 1: 31. 1943, *fide* Vitt *et al.*, 1995. Type protologue: Indonesia. West Irian, Paniai. Meerendal, Mt. Carstensz, 4000-4100 m, *Wissel 8*, *Wissel 9*.

(1) Plants robust, dull to somewhat lustrous; stems and branches up to 18 cm, forming loose mats. (2) Stem and branch leaves similar, individually twisted, contorted-flexuous and loosely wide-spreading from an erect base when dry, loosely wide-spreading-flexuose to abaxially curved when moist, 4.0-6.5 mm long, narrowly lanceolate, lanceolate from an oblong base, long and slenderly acuminate in upper portion; margins plane to broadly reflexed, or reflexed-recurved on one side below, serrate to notched in the upper portion; costa percurrent or ending several cells beneath the apex; all cells clear and smooth (except some weakly unipapillose in upper portion), longer than wide, in longitudinal rows, thick-walled; upper cells elliptic to elliptic-rectangular, $10-12 \times 10-30$ μm, irregularly thick-walled; medial cells rounded-quadrate to long-rectangular-rhombic, $9-12 \times 8-32$ μm, almost flat to bulging, porose; basal cells $8-10 \times 34-70$ μm, elongate-rectangular, porose, the marginal cells longer and narrower than their ambient cells. (3) Perichaetial leaves stiffly erect, sheathing at base, narrowly lanceolate, gradually narrowed to a slender acuminate apex, 5-8 mm long; all cells longer than wide. (4) Setae 10-12 mm long, prorate with papillae, ridged, twisted to the left; vaginulae naked. (5) Capsule urns broadly ovoid to long-ellipsoid, smooth to lightly 8-ribbed in upper 1/3, contracted to the seta, slightly smaller at the mouth when old; peristome single, exostome of 16, short, well-developed; spores anisosporous. (6) Calyptrae mitrate, plicate, naked, lobed at base.

Macromitrium archboldii E.B. Bartram and *M. cuspidatum* Hampe are similar in their long and smooth cells of branch leaves. But *M. cuspidatum* has branch leaves with distinct cuspidate apices.

Distribution: Papua New Guinea.

Specimens examined: **PAPUA NEW GUINEA**. West Sepik, *A. Touw 16623* (MO 5371054), *A. Touw 16368* (MO 5371054).

10. *Macromitrium archeri* Mitt., Fl. Tasman. 2: 183. 173. f 6. 1859. (Figure 43)

Type protologue: (Australia) Hab, On trees: Chestnut, *Archer*. on dead branches of trees: Ker-mandie Rivulet, *Oldfield*.

Macromitrium asperulum Mitt., Fl. Tasman. 2: 376. 1859, *fide* Vitt & Ramsay, 1985.Type protologue: On trees: *Lawrence* and *Gunn*. Found also in New Zealand, near Wellington, *Stephenson*, and elsewhere, *Knight* and *Lyall*.

Macromitrium muelleri Hampe, Linnaea 30: 634. 1860, *fide* Vitt & Ramsay, 1985.Type protologue: Hab. Sealers Cove.

Macromitrium pusillum Mitt., Fl. Tasman. 2: 183. 1859, *fide* Vitt & Ramsay, 1985. Type protologue: On stones: Cataract Hill, *Archer*.

(1) Pants small, dull, orange-brown, ochre, in dense, spreading mats; stems long creeping, with short, slender, erect branches; branches shorter, about 8 mm high, regularly spaced. (2) Branch leaves small, spirally twisted around the branch, with the apices exposed, never hidden in the inrolled cavity of the inrolled leaf when dry, widely spreading when moist, 1.4-1.8 mm long, ligulate to ligulate-lanceolate, with a strong apiculus, acute to apiculate- acuminate apex; costae excurrent; upper cells rounded-quadrate, elliptical at the margin, strongly conic-bulging, conic-papillose, the cells near the costa larger than those at the margin; transitional cells strongly unipapillose, sharply becoming smooth basal cells. (3) Perichaetial leaves ovate, with a short-acuminate apex; laminal cells similar to those of branch leaves but less papillose and longer above. (4) Setae erect, 3-8 mm long, smooth, twisted to the left. (5) Capsules long-exserted, urns ovoid to long obloid, 4-angled to slightly 8-plicate just below the mouth, peristome absent or fragmentary. (6) Calyptrae narrowly conical, naked and smooth, evenly lacerated below, faintly plicate.

Macromimitrium archeri Mitt. is similar to *M. ligulaefolium* (Hook.) Brid., but the latter can be separated from the former by its inrolled leaves, bulging, hardly papillose upper leaf cells. Additionally, the laminal cells of *M. ligulaefolium* are uniform in size across the laminae.

Distribution: Australia, New Zealand.

Specimens examined: **AUSTRALIA**. Victoria, *H. Streimann 61601* (MO 5210871); New South Wales, *H. Streimann 61294* (MO 5210941); Tasmania, *R. D. Seppelt 28531* (KRAM-B-186981); *William R. Buck 58018* (MO 6367659), *D.H. Vitt 29218* (H 3090100), *D.H. Vitt 29225* (H 3090101), *J. B. Moor, 297* (H-BR 2546017), *W. A. Weymouth 1591* (H-BR 2546009).

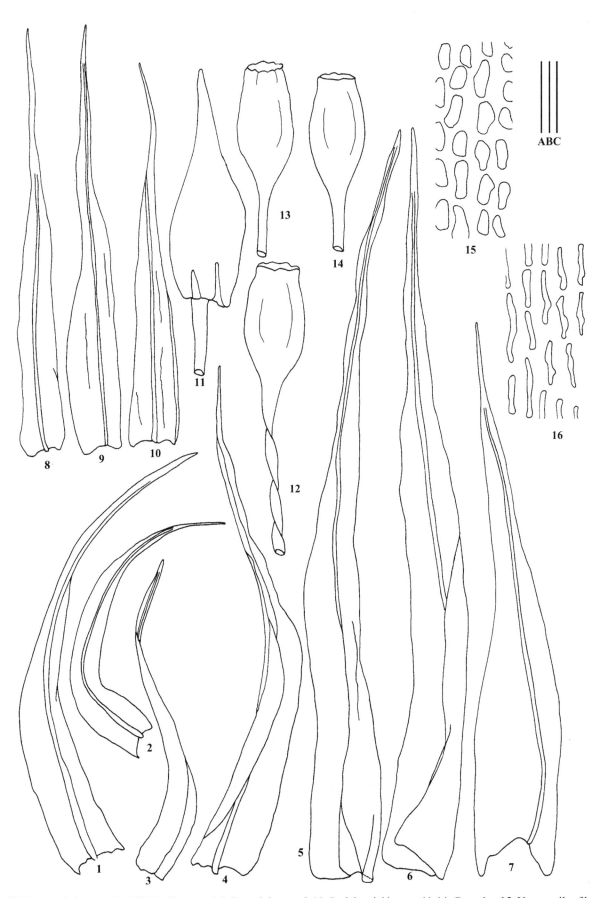

Fig. 42 *Macromitrium archboldii* E.B. Bartram 1-7: Branch leaves. 8-10: Perichaetial leaves. 11-14: Capsules. 15: Upper cells of branch leaf. 16: Medial cells of branch leaf (all from isotype, L 0060423). Line scales: A = 2 mm (1-4, 8-10, 11-14); B = 1 mm (5-7); C = 67 μm (15, 16).

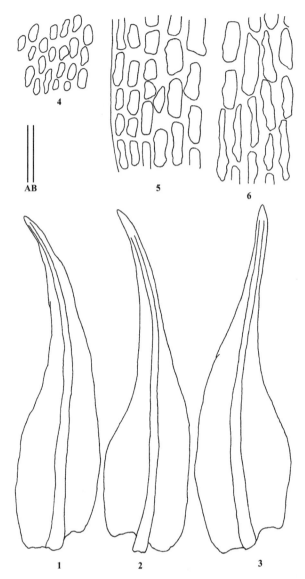

Fig. 43 *Macromitrium archeri* Mitt. 1-3: Branch leaves. 4: Medial cells of branch leaf. 5: Basal marginal cells of branch leaf. 6: Basal cells of branch leaf (all from H-BR 2539028). Line scales: A = 0.44 mm (1-3); B = 44 μm (4-6).

11. *Macromitrium argutum* Hampe, Linnaea 22: 581. 1849. (Figure 44)

Type protologue: no information about the type. Type citation: "Brasil, Rio Janeiro, *A. Glazion 9241*" (lectotype designated by Costa *et al*., 2016: BM 000879980!).

(1) Plants robust, brown-yellowish; stem creeping, branches long (up to 30 mm) and thick (1.5 mm). (2) Branch leaves appressed below, twisted-contorted and flexuose with divergent apices when dry, spreading and slightly abaxially curved, margin undulate above when moist, lanceolate from an oblong low portion, bluntly acuminate; margin weakly denticulate above, entire below; upper and medial cells large, thin-walled and bulging or conic-bulging, smooth, isodiametric except a few cells near the apices (longer than wide); low cells rounded- quadrate, short-elliptic rectangular, smooth and clear; basal cells elliptic-rectangular, 20 μm long, unipapillose, porose; marginal enlarged cells present at leaf insertion. (3) Perichaetial leaves similar to branch leaves in length, abruptly narrowed in the medial part from an ovate-oblong low portion to form an acuminate upper part or a long arista; all cells smooth, longer than wide. (4) Seta 7-10 mm long, smooth. (5) Capsule urns ovoid, longitudinal wrinkled or distinctively plicate; peristome double, exostome teeth hyaline and united, forming a short membrane, papillose, endostome segments hyaline and smooth. (6) Calyptrae mitrate, lacerate at base, hairy.

Distribution: Brazil.

Specimens examined: **BRAZIL**. Rio de Janeior, *M. C. Vaughan Bandeira s.n.* (MO); *A. Glaziou no. 7900* (BM 000879976, BM 000879982), *9421* (BM 000879979), *9900* (BM 00879981); São Paulo, *V. Schiffner s.n.* (BM 000989821), *J. J. Pirggair 428* (H-BR 2633011), *J. J. Pirggair 422* (H-BR 2633010); *Glazou 9097* (BM 000989820); Sant do Rio, *M. Bandeira 92* (H 3090112).

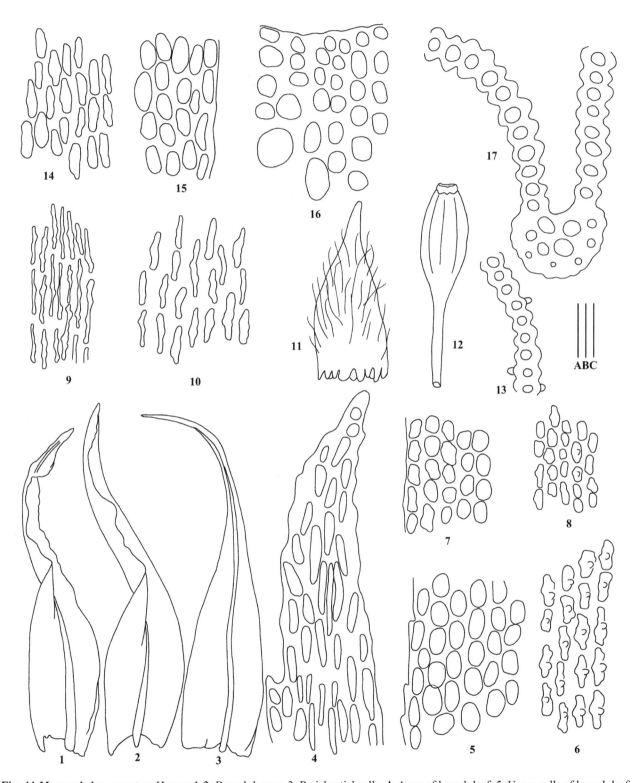

Fig. 44 *Macromitrium argutum* Hampe 1-2: Branch leaves. 3: Perichaetial cells. 4: Apex of branch leaf. 5: Upper cells of branch leaf. 6: Basal cells of branch leaf. 7: Medial cells of branch leaf. 8: Low cells of branch leaf. 9: Basal cells of perichaetial leaf. 10: Low cells of perichaetial leaf. 11: Calyptra. 12: Capsule. 13: Basal transverse section of branch leaf. 14: Medial cells of perichaetial leaf. 15: Upper cells of perichaetial leaf. 16: Cells of capsule mouth. 17: Medial transverse section of branch leaf (all from holotype, BM 000879982). Line scales: A = 2 mm (11-12); B = 1 mm (1-3); C = 67 μm (4-10, 13-17).

12. *Macromitrium atratum* Herzog, Repert. Spec. Nov. Regni Veg. 21: 30. 1925. (Figure 45)

Type protologue and citation: Brasilia: Prov. São Paulo: Campos do Jordao, Serra da mantilqueira, 1700 m, auf Podocarpus Lamberii, spärlich zwischen Pilotrichella flexills, leg. *F. C. hoehne*, 17. IX. 23 (herb. No. 637a) (holotype: JE 04006253!).

(1) Plants medium-sized to large, olive-green above, dark brown below; stems creeping, with branches up to 25 mm long. (2) Branch leaves strongly keeled and undulate, twisted-contorted and crisped, apices mostly incurved-twisted when dry, not widely open and flat but conduplicate and slightly abaxially curved when moist, oblong-lanceolate with a moderately sheathing base and dense rusty-reddish rhizoids at base, about 5 mm long; upper margins finely denticulate; apices acute or acuminate-acute, bluntly acuminate; apical cells elongate and smooth; medial and upper cells larger, 12-15 μm wide, round-quadrate, bulging, clear and smooth; low cells rectangular, frequently unipapillose; basal cells elongate, long rectangular, smooth, thick-walled and distinctively porose; basal marginal enlarged teeth-like cells at insertion differentiated. (3) Perichaetial leaves differentiated, shorter than branch leaves, lanceolate from an ovate-oblong low portion, bluntly acuminate, plicate below. (4) Setae smooth, up to 10 mm long, twisted to the left. (5) Capsule urns ovoid to ellipsoid-cylindrical, moderately to strongly furrowed; peristome double, exostome teeth united forming a short, light brown membrane, endostome segments hyaline and membranous at base; spores globular, finely papillose.

Distribution: Brazil, Bolivia, Colombia.

Specimens examined: **BOLIVIA**. Nor Yungas, *Marko Lewis 89-72* (MO). **COLOMBIA**. Tolima, Ibague, *G. Morales J. 041* (MO); Narino, Yacuanque, *Bernardo R. Ramirez Padilla 8534* (MO).

13. *Macromitrium atroviride* R.S. Williams, Bull. New York Bot. Gard. 3(9): 131. 1903. (Figure 46)

Type protologue and citation: [Bolivia] Apolo, 1500 meters, on tree trunks, April 17, 1902 (1822) (isotypes: JE 04008733!, NY 00518242!, H-BR 2625024!, F-C 0000994F!).

(1) Plants medium-sized, dark green; stems long-creeping, densely with erect branches and rusty-reddish rhizoids, branches up to 20 mm long. (2) Stem leaves small and inconspicuous, triangular-lanceolate, some abaxially curved when moist, upper and medial cells rounded-quadrate, elliptic-oblong, smooth; low and basal cells elongate, long rectangular, clear and smooth. (3) Branch leaves erect below, contorted-twisted-flexuose above when dry, spreading, often abaxially curved, somewhat flexuose and twisted when moist, ligulate to ligulate-lanceolate, occasionally narrowly lanceolate; apices broadly acute to bluntly acuminate-acute; upper margins distinctly and irregularly dentate; costa stout, subpercurrent, ending several cells beneath the apex; upper, medial and low cells rounded, rounded-quadrate, oblate, small, flat and smooth; cells gradually elongate towards base; Basal cells long rectangular to narrowly rectangular, thick-walled, porose, strongly tuberculate near the base; basal cells near costa enlarged and hyaline, marginal enlarged teeth-like cells at the insertion strongly differentiated. (4) Perichaetial leaves shorter, oblong- to oval-lanceolate, broadly acuminate; upper cells rounded-quadrate to elliptic, papillose; basal cells elongate, thick-walled, and smooth to tuberculate; cells near costa enlarged and hyaline, marginal enlarged teeth-like cells at the insertion differentiated; costa vanishing beneath the apex. (5) Setae smooth, to 15 mm, twisted to the left; vaginulae hairy. (6) Capsule urns ovoid, smooth to weakly furrowed; peristome double, exstome teeth lanceolate, the apices discrete and obtuse, papillose-striate. (7) Calyptrae mitrate, naked, deeply lacerate.

Distribution: Bolivia.

Specimens examined: **BOLIVIA**. La Paz: FranzTamayo, *Fuentes A. & Aldana C*, *6479* (MO 5914159); *Fuentes A. & Aldana C*, *6429* (MO 59140588).

14. *Macromitrium attenuatum* Hampe, Ann. Sci. Nat., Bot., sér. 5, 4: 329. 1865. (Figure 47)

Type protologue: [Colombia] Bogota, Pacho, altit. 2200 metr., in sylv. Ad arbores, Julio, 1863, leg. *A. Lindig*; Type citation: Nova Granada (= Kolumbien), Bogota Pacho, alt. 2200 m. Jul. 1863, leg. *Lindig s.n.* (isotypes: PC 0137587!, PC 0137588!, GOET 012310!).

(1) Plants medium-sized to large, brownish-yellow to yellowish-green; stems loosely creeping, branches up to 50 mm long. (2) Branch leaves erect below, individually twisted-contorted and flexuose above, with divergent apices when dry, widely spread or slightly abaxially curved when moist, ligulate-lanceolate, with aucte, bluntly acuminate or short cuspidate apices; margin weakly and sparsely denticulate upper, without a differentiated border; costae percurrent to excurrent into a short cuspidate point; upper and medial cells rather small (4-5 μm wide), thick-walled and flat, rounded-quadrate, oblong, oblate, smooth and clear; low and basal cells thick-walled, densely and strongly tuberculate; marginal enlarged teeth-like cells at insertion present; (3) Perichaetial leaves differentiated, ovate-oblong to oblong-lanceolate, clasping and plicate below; costae excurrent to a short arista. (4) Setae smooth, up to 22 mm long, twisted to the left. (5) Capsule urns ovoid-ellipsoid to ellipsoid-cylindric, distinctively 8-furrowed; peristome double, exostome teeth yellowish, bluntly lanceolate, endostome a white or colorless membrane; opercula conic-rostrate. (6) Calyptrae mitrate, smooth and naked.

Macromitrium attenuatum Hampe is similar to *M. oblongum* (Taylor) Spruce, but the former could be distinguished from the latter by its branch leaves with rather small, clear and smooth upper and medial cells.

Distribution: Colombia.

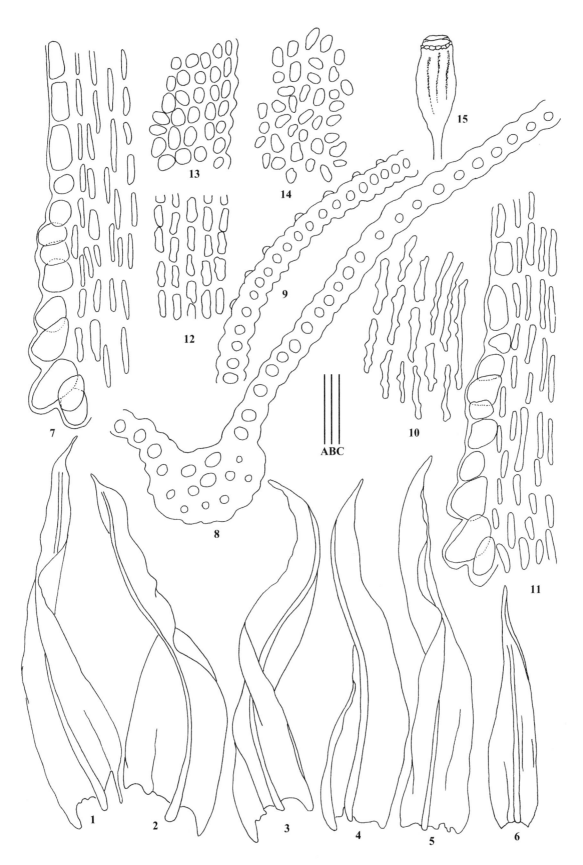

Fig. 45 *Macromitrium atratum* Herzog. 1-5: Branch leaves. 6: Perichaetial leaf. 7, 11: Basal marginal cells of branch leaves. 8: Basal transverse section of branch leaf. 9: Low transverse section of branch leaf. 10: Basal cells of branch leaf. 12: Low cells of branch leaf. 13: Apical cells of branch leaf. 14: Medial cells of branch leaf. 15: Capsule (all from holotype, JE 04006253). Line scales: A = 4 mm (15); B = 1 mm (1-6); C = 67 μm (7-14).

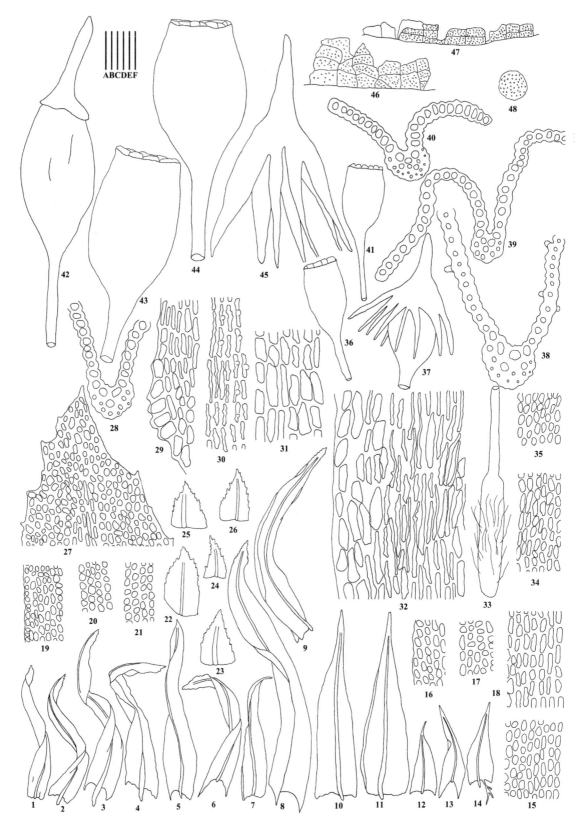

Fig. 46 *Macromitrium atroviride* R.S. Williams 1-9: Branch leaves. 10-11: Perichaetial leaves. 12-14: Stem leaves. 15, 17: Medial cells of stem leaf. 16: Upper cells of stem leaf. 18: Basal cells of stem leaf. 19: Medial marginal cells of branch leaf. 20: Low cells of branch leaf. 21: Medial cells of branch leaf. 22-26: Apices of branch leaves. 27: Apical cells of branch leaf. 28: Medial transverse section of branch leaf. 29: Basal marginal cells of branch leaf. 30: Basal cells of branch leaf. 31: Basal cells near costa of branch leaf. 32: Basal cells of perichaetial leaf. 33: Vaginula. 34: Medial cells of perichaetial leaf. 35: Upper cells of perichaetial leaf. 36, 41: Dry capsules. 37, 42-44: Wet capsules. 38, 39: Basal transverse sections of branch leaves. 40: Upper transverse section of branch leaf. 45: Calyptra. 46, 47: Peristome. 48: Spore (all from isotype, F-C 0000994C). Line scales: A = 2 mm (36, 37, 41); B = 1 mm (1-14, 33, 42-45); C = 400 μm (22-26, 47); D = 200 μm (46); E = 100 μm (48); F = 67 μm (15-21, 27-32, 34-35, 38-40).

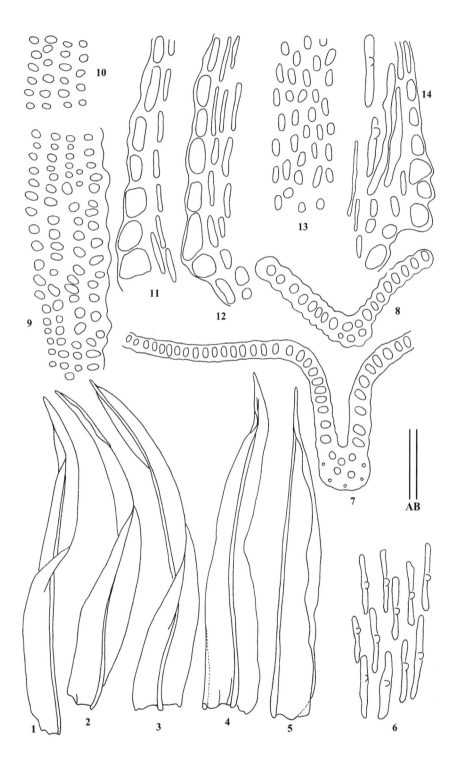

Fig. 47 *Macromitrium attenuatum* Hampe 1-4: Branch leaves. 5: Perichaetial leaves. 6: Basal cells of branch leaf. 7: Basal transverse section of branch leaf. 8: Upper transverse section of branch leaf. 9: Upper cells of branch leaf. 10: Medial cells of branch leaf. 11-12, 14: Basal marginal cells of branch leaves 13: Upper cells of perichaetial leaf (all from isotype, PC 0137588). Line scales: A = 1 mm (1-5); B = 67 μm (6-14).

15. *Macromitrium aurantiacum* Paris & Broth., Rev. Bryol., 34: 43. 1907. (Figure 48)

Type protologue: [Vietnam] Arbres dans la forêt sur la route de Djirin, 1500 m. Type citation: *Eberhardt, 506*, 1906, marked with "*Macromitrium aurantiacum* B. P. *n. sp*" (lectotype designated here: H-BR 2586001!).

(1) Plants small, in dense mats; stems creeping. (2) Branch leaves twisted crispate-contorted when dry, spreading but conduplicate, slightly abaxially curved and twisted when moist, broadly oblong-lanceolate, ligulate, 2.0 × 0.5 mm; margins entire and plane above, revolute below; apices long cuspidate; costae excurrent into an arista; all cells rounded-quadrate, strongly conic-bulging; upper cells obscure, pluripapillose; low and basal cells distinctly bulging and unipapillose. (3) Perichaetial leaves smaller than branch leaves, lanceolate, long-acuminate upper. (4) Setae 6-7 mm long, smooth, twisted to the left; vaginulae with numerous yellow and long paraphyses. (5) Capsule urns ellipsoid-cylindric; opercula long conic-rostrate. (6) Calyptrae campanulate, orange yellow, plicate, hairy.

Distribution: Vietnam.

16. *Macromitrium aurescens* Hampe, Linnaea 30: 633. 1860. (Figure 49)

Type protologue: [Australia] Ad. Delabechiam. Type citation: On the stem of Shomas mtchlis's bottle, Delabechia, Dawton River, Australia, *F. V. Műller s.n.* (isotype: NY 00518244!).

Macromitrium cylindromitrium Müll. Hal., Hedwigia 37: 146. 1898, *fide* Vitt & Ramsay, 1985. Type protologue: [Australia] Queensland, distr. Widebay pr. Gayndah: Dämel 1874 in Hb. Godeffroy Hamburgo; Enoggera: F. M. Bailey in Hb. Brotheri 1890. Type citation: Australia, Queensland, Widebay pr. Gayndah, Dämel (syntype: H-BR 2586004!); Queensland, Enoggera, *F. M. Bailey 795* (lectotype designated by Vitt & Ramsay, 1985: H-BR 2586008!).

Macromitrium sordidevirens Müll. Hal., Linnaea 37: 153. 1872, *fide* Vitt & Ramsay, 1985. Type protologue: Brisbane River orientalis: *Amalie Dietrich* 1864 legit cum M. diaphano consociatum. Type citation: Queensland, *Amalie Dietrich* (lectotype by designated Vitt & Ramsay, 1985: H-BR 2586006!).

(1) Pants medium-sized, in dense mats; stems creeping, densely with short, stout branches; branches 5-6 mm long, 2 mm thick. (2) Stem leaves triangular-lanceolate, moderately abaxially curved spreading when moist. (3) Branch leaves irregularly erect-curved to twisted-curved, but most spirally twisted to spiral-whorled around the branch when dry, erect-spreading and often conduplicate upper when moist, 2.0-2.6 mm long, oblong-ligulate, with a cucullate, hooked mucronate apex; costa excurrent in the mucro; upper cells 7-13 μm wide, bulging, rounded-quadrate, pluripapillose; medial cells 9-11 μm wide, rounded to elliptical, more elongated towards the base, with 1 or 2 conical to forked papillae per cell; transitional cells rhomboid to short-rectangular, irregularly thick-walled, distinctly porose, with a large, tall and conic papillae on most cells; basal cells 15-35 μm long, 10-17 μm wide, flat, short-rectangular, rhomboid, bulging, strongly papillose-tuberculate, thick-walled, porose, restricted to 3-7 tiers at the insertion, becoming long towards the margin. (4) Perichaetial leaves shorter than branch leaves, lanceolate-lingulate to ovate-lingulate; apices distinctly cuspidate; upper and medial cells quadrate, rounded-quadrate, bulging and smooth; basal cells short-rectangular, tuberculate. (5) Setae 3-7 mm long, smooth, twisted to the left. (6) Capsules short-exserted, urns long-ellipsoid to cylindrical, 2-3 mm long, smooth; peristome single, exostome of 16, well developed, lanceolate, papillose; spores indistinctly anisosporous. (7) Calyptrae long conic-mitrate, plicate, split 1/3-1/2, smooth, with dense long-flexuose hairs near the apex.

Macromitrium aurescens Hampe, *M. brevicaule* (Besch.) Broth. and *M. brachypodium* Müll. Hal. are similar in their broad, oblong-ligulate, conduplicate, mucronate leaves with conspicuous costa, densely papillose upper laminal cells and short basal cells. *Macromitrium brevicaule* differs from *M. aurescens* in having naked calyptrae, and brood-bodies on the abaxial leaf surface. *Macromitrium brachypodium* can be separated from *M. aurescens* by its short setae (< 1.3 mm), ellipsoid capsules, short conic calyptrae with densest hairs near the base, as well as its shorter branch leaves with smooth basal cells.

Distribution: Australia, Thailand, Vietnam.

Specimens examined: **AUSTRALIA**. New South Wales, *H. Streimann 43950* (KRAM-B-093940); Queensland, *H. Streimann 52302* (MO 4419014), *H. Streimann 52419* (MO 4417785); Queensland, Brisbane, J, K. Bailey (H-BR 2536007). **THAILAND**. North-eastern Phu Kradung, *Charoenphol, Larsen & Warncke 4662* (MO 3971201); Northern Doi Pha Dam, *Larsen, Santisuk & Warncke 2196* (MO 3971200), *Larsen, Santisuk & Warncke 2194* (MO 3971186); Northern Bo Luang, *Larsen, Santisuk & Warncke 1925* (MO 3971198); Northern S of mae Sariang, *Larsen, Santisuk & Warncke 2297* (MO 3971185); Phitsanulok, *Larsen, Smitinand & Warncke 910* (MO 3971159); Udawn, sandstone massive Phu (Mt.) Krading, *A. Touw 10936* (MO 2158138); Udawn, *A. Touw 10361* (MO 2154143). **VITENAM**. Ninh Binh, *S. He & K. Nguyen 42123* (MO 6165912).

Macromitrium aurescens* var. *caledonicum (Thér.) Thouvenot, Cryptog., Bryol. 40(16): 177. 2019. (Figure 50)

Basionym: *Macromitrium cylindromitrium* var. *caledonicum* Thér., Bull. Acad. Int. Géogr. Bot. 19: 21 (1909). Type protologue: [Caledonia] Hab. Bourail. Type citation: Nouvelle Calédoniem, Bourail, 1905, *Le Rat 35*, Herbier I. Thériot (lectotype designated by Thouvenot, 2019: PC 0083647!)

Macromitrium aurescens var. *aurescens* and var. *caledonicum* mainly differ in quantitative characters. The latter could be separated from the former by its relatively small plants, short branches and leaves with smooth basal cells.
Distribution: New Caledonia.

17. *Macromitrium aureum* Müll. Hal., Bot. Zeitung (Berlin) 15: 580. 1857. (Figure 51)
Type protologue: [Colombia] Nova Granada, Cerro Pelado: Schlim 1852 legit. Collect. Linden. Sine No. cum Zygodonte commixtum. Type citation: "Nova Granada, *Schlim s.n.* (lectotype: H-BR 2625062); Nova Granada, Cerro Pelado, *Schlim s.n*" (isotypes: BM 000879971!, BM 000879970!).
(1) Plants rather large, look like wefts, dark brown-reddish, yellowish green; stems slightly creeping, branches up to 50 mm long, 2.5 mm thick. (2) Branch leaves erect below, individually twisted, margins undulate, contorted-flexuose in upper portion when dry, distinctively abaxially curved when moist, 3-5 mm long, linear-lanceolate, with a long acuminate apex; margins serrate or dentate above, entire below; costae percurrent; all cells thick-walled; upper and medial cells rounded-quadrate, elliptic to rhombic in longitudinal rows, 20-25 µm long, conic-bulging to strongly mammillose, collenchymatous; gradually elongate towards the base; low and basal cells rectangular to long rectangular, porose, strongly tuberculate; enlarged teeth-like cells at basal margin not differentiated. (3) Perichaetial leaves lanceolate with a widely oblong low part, distinctively plicate; all cells longer than wide; upper cells smooth, gradually elongate from upper to base; low and basal cells tuberculate. (4) Setae 10-15mm long, smooth. (5) Capsule urns 1-2 mm long, often ovoid, furrowed; peristome double, exostome teeth membranous, reddish-organ, densely papillose, endostome segments hyaline, papillose; spores hyaline, anisosporous, 12-34 µm in diameter, papillose to different degrees. (6) Calyptrae large, 3-4 mm long, mitrate, naked, deeply lacerate.
Macromitrium aureum Müll. Hal. is similar to *M. cirrosum* (Hedw.) Brid. in seta length, leaf shape, and mammillose lamina cells. *Macromitrium cirrosum* can be separated from *M. aureum* by its smaller plants and smooth capsules. *Macromitrium aureum* is also similar to *M. longifolium* (Hook.) Brid. in their gametophytes, but the latter has papillose setae.
Distribution: Bolivia, Colombia, Costa Rica, Ecuador.
Specimens examined: **BOLIVIA**. La Paz: Franz Tamayo, Parque Nacional Madidi. *Fuentes, Aflredo F. & H. Huaylla 13110* (MO 6363710); Santa Cruz: Manuel M. Caballero, Serrania de Siberia. *S. Churchill, R. Ledezma & Muñoz 22367* (MO 5647768). **COLOMBIA**. Antioquia: Medellin Municipio, *R. Callejas 7374* (MO 4433127); Cauca, Macizo Colombiano, Valle de Las Papas, Terricola, *H. Bischler 1087* (MO 5137004), Cauca, Depto. Cauca, Macizo Colombiano, *H. Bischler 931* (MO 5137011). **COSTA RICA**. Cartago, Cordillera de Talamanca, *I. Holz & B. Allen CR 00-695* (KRAM-B-146827, MO 5282908). **ECUADOR**. Azuay, *L. Holm-Nielsen, S. Jeppesen, B. Lojtnant et al. 4960,* (MO 2561628); Prov. De Chimborazo, *Adrian juncosa 891* (MO 3081270).

18. *Macromitrium austrocirrosum* E.B. Bartram, Lloydia 5: 273. 35. 1942. (Figure 52)
Type protologue: [Indonesia] Lake Habbema: cushioned on low trees of open secondary forest on landslip, 9 km. NE. of Lake Habbema, 2800 m., *no. 10961, 10962*; epiphyte on open edges of a landslip, 9 km. NE. of Lake Habbema 2800 m., *no. 11012* type. Type citation: Dutch New Guinea, 9 km NE of Lake Habbema, 2800 m. camp, Oct. 1938, *L. J. Brass, 11012* (isotype: MICH 525866!).
(1) Plants medium-sized; stems inconspicuous, covered by rhizoids, easily breaking apart with branches appearing separated plants, branches up to 4 cm long. (2) Branch leaves individually twisted-contorted from an erect, often clasping base, upper portion distinctly inrolled to spreading-divergent when dry, spreading, wide-spreading, or squarrose when moist, 3-4 mm long, narrowly lanceolate, gradually narrowed (tapered) to an acuminate apex, channeled just below the apex; margins plane above, reflexed below, crenulate-papillose to irregularly denticulate near apex, entire below; costae ending just beneath the apex to shortly excurrent; upper cells rounded to rounded-quadrate, 12-14 µm, bulging, thin-walled, most smooth, some unipapillose, smaller at the margin (8-10 µm wide), longer near the apex, 6 × 20-25 µm, long elliptic; medial cells 10-12 ×16-24 µm, irregularly short-rectangular to short-elliptic, bulging, smooth, in longitudinal rows, transition to basal cells sharp; basal cells 9-11 µm × 35-90 µm long, elongate to elongate-rectangular, smooth, flat, clear, thick-walled, lumens straight to irregularly porose. (3) Perichaetial leaves 3.4-3.8 mm long, erect-flexuose, sheathing in low portion, lanceolate, long acuminate, with a long excurrent costa; vaginulae without hairs. (4) Setae 4-9 mm long, slender, lightly prorate above, smooth below, twisted to the left. (5) Capsules ellipsoid to obovoid, smooth; peristome single; exostome of 16, lanceolate, blunt, well-developed, finely papillose on outer surface, coarsely and irregularly striate on inner surface, erect-inflexed, yellow-brown; spores anisosporous. (6) Calyptrae plicate, mitrate, naked, deeply lacerate.
The species is similar to *M. archboldii* E.B. Bartram, but the plants of the latter are larger, with upper cells longer than wide.
Distribution: Indonesia.

Specimens examined: **PAPUA NEW GUINEA.** Morobe Prov., *T. Koponen 29546* (H 3195782); Southern Highlands, *H. Streimann 26949* (MO 5143196); West Sepik, *A. Touw 17639* (MO 5375900, H 3205114), *A. Touw 17496* (MO 5371040), *A. Touw 17878* (MO 5375902).

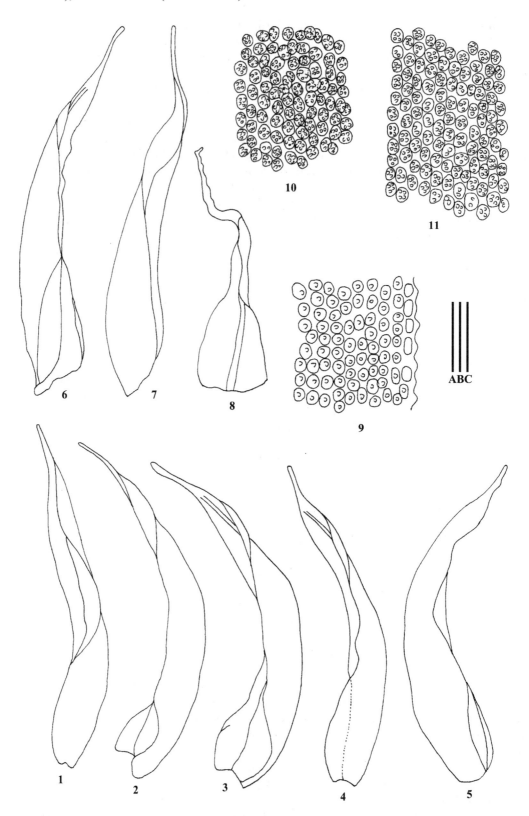

Fig. 48 *Macromitrium aurantiacum* Paris & Broth. 1-7: Branch leaves. 8: Apex of branch leaf. 9: Basal cells of branch leaf. 10: Upper cells of branch leaf. 11: Medial cells of branch leaf (all from lectotype designated here, H-BR 2586001). Line scales: A = 500 μm (1-7); B = 200 μm (8); C = 50 μm (9-11).

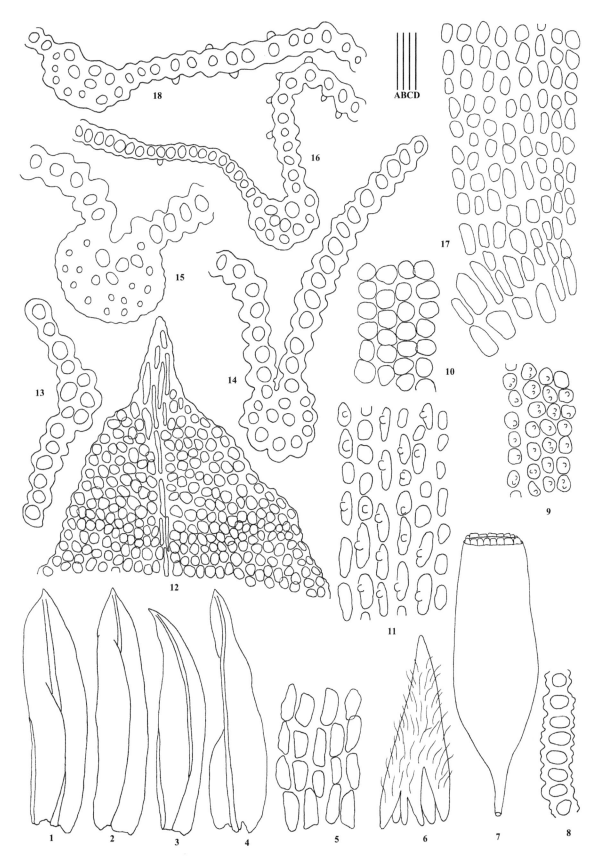

Fig. 49 *Macromitrium aurescens* Hampe 1-4: Branch leaves. 5, 17: Basal cells of branch leaf. 6: Calyptra. 7: Wet capsule. 8, 13: Upper transverse sections of branch leaves. 9: Upper cells of branch leaf. 10: Medial cells of branch leaf. 11: Low cells of branch leaf. 12: Apical cells of branch leaf. 14: Medial transverse section of branch leaf. 15: Low transverse section of branch leaf. 16, 18: Basal transverse sections of branch leaves (all from isotype, NY 00518244). Line scales: A = 2 mm (6); B = 1 mm (1-4, 7); C = 100 μm (12); D = 67 μm (5, 8-11, 13-18).

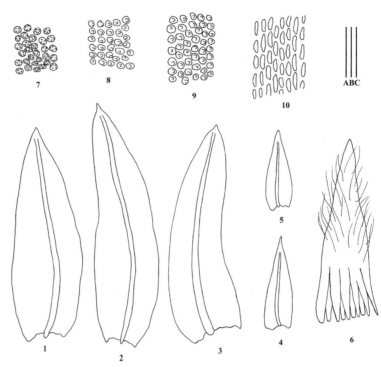

Fig. 50 *Macromitrium aurescens* var. *caledonicum* (Thér.) Thouvenot 1-3: Branch leaves. 4-5: Perichaetial leaves. 6: Calyptra. 7: Upper cells of branch leaf. 8: Medial cells of branch leaf. 9: Low cells of branch leaf. 10: Basal cells of branch leaf (all from lectotype, PC 0083647). Line scales: A = 1 mm (4-6); B = 400 μm (1-3); C = 67 μm (7-10).

19. *Macromitrium bifasciculare* Müll. Hal ex Dusén, Rep. Princeton Univ. Exp. Patagonia, Bot. 8(3): 83. 12, 9 f. 5. 1903. (Figure 53)

Type protologue: [Chile & Argentina] Hab. Fuegia australis ad Villarino in rupibus litoreis (?). Type citation: [Argentina] Patagonia, Villarino, in rupibus...., 3 V. 1897. *P. Dusén, 612* (lectotype designated here: US 00070247!).

(1) Plants medium-sized, in dense mats, brown-yellowish above, dark brown below; stems weakly creeping, with erect simple or forked branches; branches densely clustered, up to 20 mm long and about 1.0 mm thick. (2) Stem leaves inconspicuous and caducous. (3) Branch leaves individually to weakly spirally appressed-twisted when dry, spreading and flat when moist, lanceolate, ligulate- to oblong-lanceolate, rarely and slightly plicate below; apices broadly acuminate, acute to mucronate, with multistratose cells on transverse section; costae strong, keeled, ending several cells beneath the apex; margins entire throughout, often narrowly recurved on one side; laminae often bistratose or partially bistratose above; all cells thick-walled, smooth and not porose; upper cells quadrate, subquadrate, rounded, oblong or oblate; medial cells subquadrate, short-rectangular, basal cells elongate rectangular, thick-walled. (4) Perichaetial leaves differentiated, broadly oblong to ovate-oblong, with an acute or mucronate apex; costa ending several cells beneath the apex. (5) Setae about 3 mm long, smooth, straight to slightly twisted to the right. (6) Capsule urns ellipsoid, furrowed when dry; opercula conic-rostrate, some with an oblique beak. (7) Calyptrae cucullate, naked and smooth.

Distribution: Argentina.

20. *Macromitrium bifasciculatum* Müll. Hal., Hedwigia 37: 150. 1898. (Figure 54)

Type protologue: [Chile] Habitatio. Fuegia, fretum Magellanicum occidentale: Prof. *Pirotta* misit 1885 ex Hb. Horti Romani. Type citation: Fuürland: Caracciolo-Bay, mist. Prof. *Priotta* 1885. Typus! (isotype: PC 0137594!).

(1) Plants large, in dense tufts; stems inconspicuous, branches densely clustered, upright, up to 25 mm long, thick, dark rusty-reddish, with short branchlet above and obtuse terminal. (2) Stem leaves inconspicuous and caducous. (3) Branch leaves individually to weakly spirally appressed-twisted when dry, spreading and flat when moist, ligulate- to oblong-lanceolate, lanceolate, with an acute to acuminate apex; costae brown-purple, channeled in adaxial view, vanishing far from the apex; margins entire throughout, narrowly recurved below; laminae sporadically and partially bistratose above; all cells thick-walled, smooth and clear, not porose; upper and medial cells quadrate, rounded, a few oblong; low cells ovate, elliptic, oblong-rectangular; basal cells slightly elongate. (4) Perichaetial leaves differentiated, broadly oblong to ovate-oblong, abruptly narrowed above, with an obtuse or mucronate apex, recurved at one side; costae ending several cells beneath the apex. (5) Setae about 3 mm long, smooth, slightly twisted to the right. (6) Capsule urns ellipsoid, smooth or furrowed when dry; opercula conic-rostrate, some with an oblique beak. (7) Calyptrae mitrate, naked and smooth.

Taxonomy

Distribution: Chile.
Specimens examined: **CHILE**. Patagonia, *Dusén s. n.* (BM 000989835); *Dusén 612* (BM 000989836).

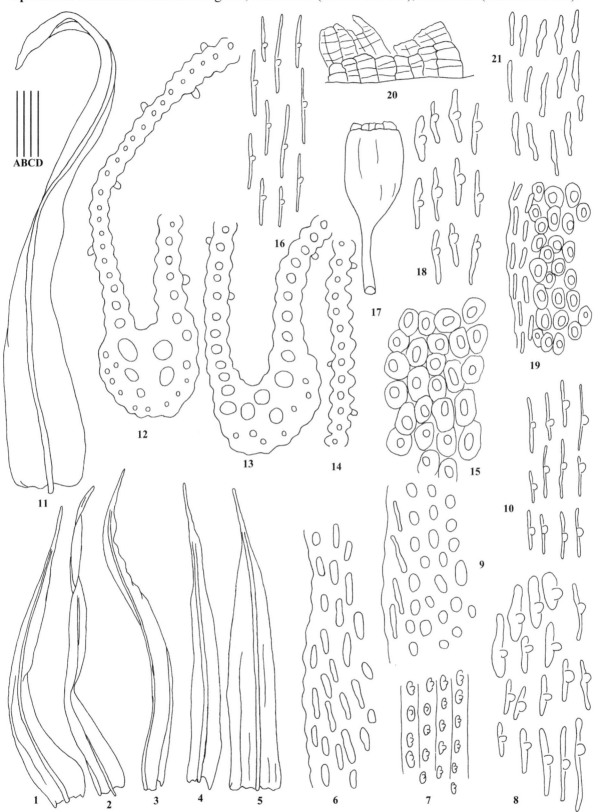

Fig. 51 *Macromitrium aureum* Müll. Hal. 1-4, 11: Branch leaves. 5: Perichaetial leaf. 6: Apical cells of branch leaf. 7: Medial cells of branch leaf. 8: Low cells of branch leaf. 9, 15: Upper cells of branch leaf. 10: Basal cells of branch leaf. 12-14: Basal transverse sections of branch leaves. 16: Basal cells of perichaetial leaf. 17: Capsule. 18: Low cells of branch leaf. 19: Upper marginal cells of branch leaf. 20: Peristome. 21: Upper cells of perichaetial leaf (1-5, 7, 10, 15-21 from isotype, BM 000879971; 6, 8-9, 11-14 from isotype, MICH 525865). Line scales: A = 2 mm (1-5, 17); B = 1 mm (11); C = 400 μm (20); D = 67 μm (6-10, 12-16, 18-19, 21).

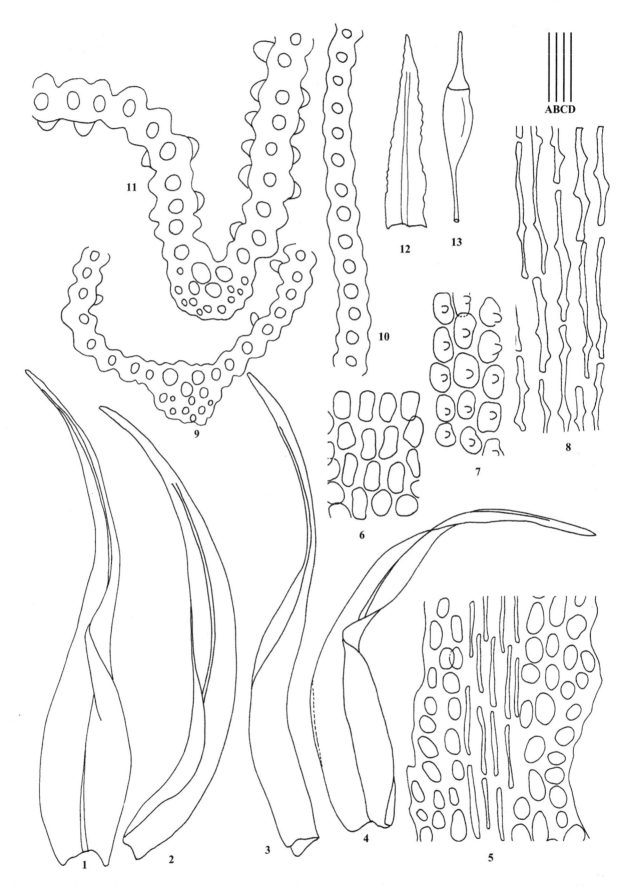

Fig. 52 *Macromitrium austrocirrosum* E.B. Bartram 1-4: Branch leaves. 5: Upper marginal cells of branch leaf. 6: Medial cells of branch leaf. 7: Upper cells of branch leaf. 8: Basal cells of branch leaf. 9, 11: Upper transverse sections of branch leaves. 10: Medial transverse section of branch leaf. 12: Apex of branch leaf. 13: Capsule (all from isotype, MICH 525866). Line scales: A = 2 mm (13); B = 1 mm (1-4); C = 400 μm (12); D = 67 μm (5-11).

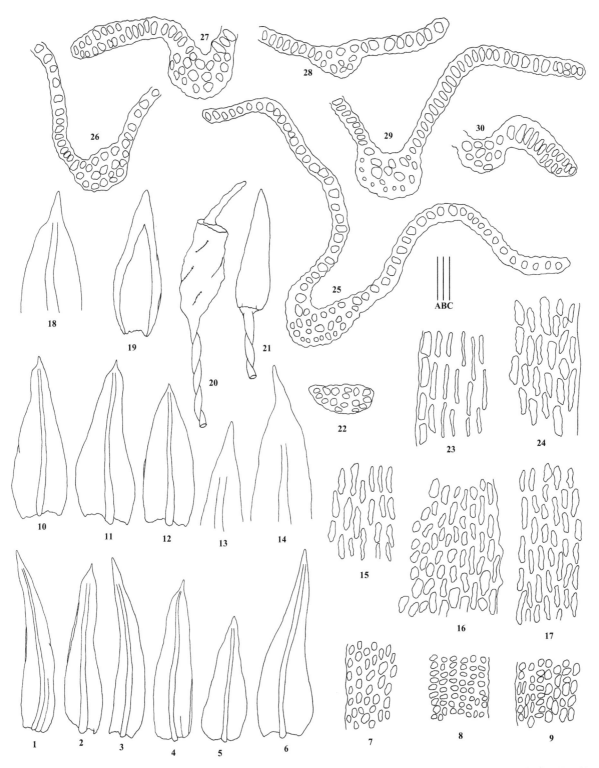

Fig. 53 *Macromitrium bifasciculare* Müll. Hal ex Dusén 1-4: Branch leaves. 5-6, 10-12: Perichaetial leaves. 7: Apical cells of branch leaf. 8: Upper cells of branch leaf. 9: Medial cells of branch leaf. 13-14, 18: Apices of branch leaves. 15: Basal cells of perichaetial leaf. 16: Apical cells of perichaetial leaf. 17: Medial cells of perichaetial leaf. 19: Calyptra. 20-21: Capsules. 22: Apical transverse section of branch leaf. 23: Basal cells of branch leaf. 24: Basal cells near costa of perichaetial leaf. 25-26: Basal transverse sections of branch leaves. 27-28, 30: Upper transverse sections of branch leaves. 29: Medial transverse section of branch leaf (all from lectotype designated here, US 00070247). Line scales: A = 1 mm (1-6, 10-12, 19-21); B = 400 μm (13-14, 18); C = 67 μm (7-9, 15-17, 22-30).

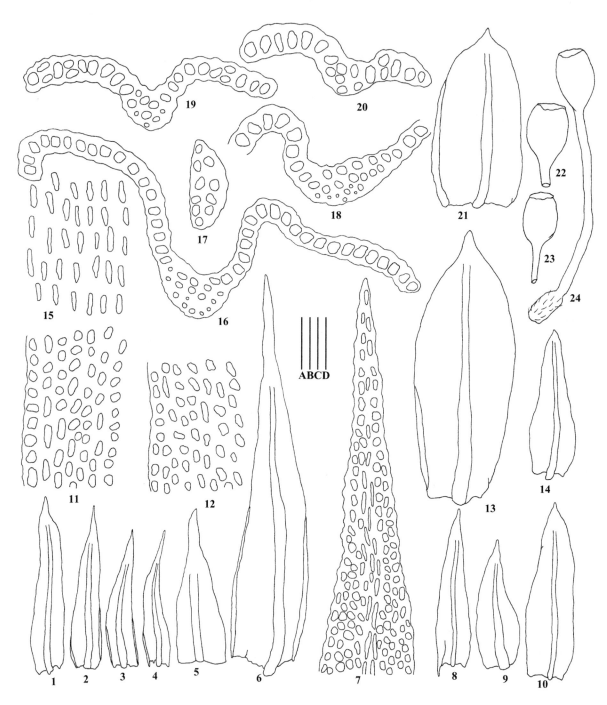

Fig. 54 *Macromitrium bifasciculatum* Müll. Hal. 1-4, 6, 8-10, 14: Branch leaves. 5: Apex of branch leaf. 7: Apical cells of branch leaf. 11: Upper cells of branch leaf. 12: Medial cells of branch leaf. 13, 21: Perichaetial leaves. 15: Low cells of branch leaf. 16, 18: Basal transverse sections of branch leaves. 17: Transverse section of stem. 19-20: Upper transverse section of branch leaf. 22-24: Capsules (1-7, 11-12, 15-20 from isotype, BM 000873313; 8-10, 13-14, 21-24 from holotype, BM 000989835). Line scales: A = 2 mm (22-24); B = 1 mm (1-4, 8-10, 14); C = 400 μm (5-6, 13, 21); D = 67 μm (7, 11-12, 15-20).

21. *Macromitrium binsteadii* Dixon, Ann. Cryptog. Exot. 3: 189. 1930. (Figure 55)

Macromitrium assimile Broth. et Dix., J. Bot. 53: 264. 540. f. 4. 1915 hom. Illeg, *fide* Wijk *et al.*, 1964. Type protologue: [Sri Lanka] Sunny rock, Udapassellawa, *405*. Type citation: Ceylon, Sunny rock, Udapassellawa, leg. *Binstead, 405*, 1913 (holotype: H-BR 2572022!).

(1) Plants medium-sized, in dense mats; stems long-creeping, with short branches, branches up to 8 mm long, densely leaved. (2) Stem leaves deflexed to suberect when moist, ovate-lanceolate to triangular-lanceolate. (3) Branch leaves keeled, appressed below, and strongly twisted-contorted, with apices hidden in the inrolled cavity when dry, widely spreading and slightly adaxially incurved upper when moist, to 2 mm long, broadly oblong-lanceolate, lingulate-lanceolate; apices obtuse or obtuse-acute; costa slender, ending beneath the apex; margins entire throughout; upper and medial cells irregularly quadrate or rounded-quadrate, about 8 µm wide, clear, in longitudinal row, frequently unipapillose; low and basal cells slightly elongate to oblong-rectangular, unipapillose, not porose. (4) Perichaetial leaves ovate-lanceolate to triangular-lanceolate, bluntly acuminate; upper cells oblong-rectangular, ovate-oblong, unipapillose; medial and low cells elongate-rectangular, smooth. (5) Setae erect, smooth, up to 10 mm long. (6) Capsule urns ellipsoid-cylindrical, slightly plicated when dry; peristome single, exostome of 16, papillose, short lanceolate, densely papillose, with obtuse the apices. (7) Calyptrae mitrate, hairy.

Macromitrium binsteadii Dixon is similar to *M. tosae* Besch., differing from the latter by its branch leaves with unipapillose cells in medial and upper portions.

Distribution: India, Sri Lanka.

Specimen examined: **INDIA**. Madras State, *G. Foreau 1929* (H 3090121).

22. *Macromitrium bistratosum* E. B. Bartram, Occas. Pap. Bernice Pauahi Bishop Mus., 15(27): 339. 8. 1940. (Figure 56)

Type protologue: [French Polynesia] Rapa: Mangaoa, on tree trunks, moist open woods, alt. 400 m., July 18, 1934, Fosberg 11586 (type); other specimens cited: Taratika, east side of Mt. Perahu, tree trunk, alt. 450 m., St. John and Maireau 15574; east side of Mt. Perahu, tree trunk, alt. 615 m., St. John and Maireau 15662. Type citation: Mangaoa, on tree trunks, moist open woods, alt. 400 m, coll. *F. R. Fosberg no. 11586*, July 18. 1934 (holotype: FH 00213572!; isotype: BM 000982723!); Taratika, east side of Mt. Perahu, on tree trunk, alt. 615 m, 21, July 1934, coll. H. St. John, E. R. Fosberg & Jean Maireau, 15662 (isoparatype: MICH 525867!).

(1) Plants robust, light-yellowish above and brownish below, lustrous; stems long and creeping, densely with erect branches, branches up to 20 mm long, simple or forked above, densely leaved. (2) Branch leaves appressed to erect below, curved or individually twisted and flexuose above, with unevenly divergent apices when dry, spreading when moist, narrowly lanceolate, acuminate, about 3 mm long; keeled-concave; margins entire, costa subpercurrent; upper and medial cells rounded, incrassate, frequently bistratose, smooth to weakly pluripapillose; basal cells elongate, clear and smooth, incrassate, with narrow curved or somewhat sigmoid lumens. (3) Perichaetial leaves differentiated, larger than branch leaves, lanceolate from a broadly oblong low portion; all cells longer than wide, clear and smooth, thick-walled. (4) Setae about 7 mm long, erect or slightly curved, smooth. (5) Capsule urns ovoid to ellipsoid, strongly ridged when young, becoming smooth with age, not wrinkled beneath the mouth; opercula rostrate, with a long beak; peristome single, exostome lanceolate and pale, densely papillose; spores anisosporous, papillose. (6) Calyptrae about 2.5 mm long, conic-mitrate, smooth and naked, somewhat scabrous above, not or weakly lobed at base.

Distribution: French Polynesia.

Specimen examined: **FRENCH POLYNESIA**. Austral Islands, Rapa: *Perau, K. R. Wood 9499* (KRAM-B-179082).

23. *Macromitrium blumei* Nees ex Schwägr., Sp. Musc. Frond. Suppl. 4 316B. 1842. (Figure 57: 1-17)

Type protologue and citation: Indonesia, Java, Blume, legit, misit Al. Braun ad Muellerum, Herb. Muell., comm. Schliephacke (isotype: H-BR 2600011!).

Macromitrium assimile Broth., Leafl. Philipp. Bot. 6: 1978. 1913, *fide* Wijk *et al.*, 1964. Type protologue and citation: [Philippines] Todaya (Mt. Apo), District of Davao, Mindanao. Sep 1909, *A. D. E. Slmer 11663* (holotype: H-BR 2602009!).

Macromitrium copelandii Broth., Philipp. J. Sci. 3: 16. 1908, *fide* Wijk *et al.*, 1964. Type protologue: [Philippines] Luzon, Province of Bataan, Mount Mariveles, on trees (Copeland) (holotype: H-BR 2599001!).

Macromitrium horridum Dixon, Ann. Bryol. 5: 30. 1932, *fide* Eddy, 1996.

Macromitrium teres (Dozy & Molk.) Müll. Hal., Bot. Zeitung (Berlin), 3: 544. 1845, *fide* Wijk *et al.*, 1964.

(1) Plants small to medium-sized, stems long creeping, yellow-green at shoot tips; branches about 20 mm long, and often with branchlets. (2) Branch leaves densely arranged, regularly and spirally twisted-curved in rows, the upper portions curved toward one side, funiculate to varying degrees when dry, widely erect-spreading when moist,

1.1-1.9 × 0.3-0.4 mm, oblong to lingulate, plicate below, keeled; margins entire or slightly crenulate above, plane or recurved below; apices obtuse or rounded, often asymmetrical; costae strong, excurrent forming a conspicuous apiculus or awn (0.14-0.53 mm long); upper cells isodiametric with a round lumen, thick-walled, bulging to strongly mammillose, somewhat obscure, in longitudinal rows; low cells elongate, thick-walled, strongly tuberculate, cells at insertion smooth. (3) Perichaetial leaves about as long as branch leaves but broader, sheathing at base. (4) Setae 12-18 mm long, smooth or sometimes papillose or only roughened beneath the capsule urn. (5) Capsules erect, urns urceolate, smooth, narrowed to a shallowly cylindrical mouth; peristome double; exostome rudimentary; endostome a white or colorless membrane, finely striate-papillose. (6) Calyptrae naked and plicate.

Distribution: China, Indonesia, Malaysia, Philippines, Vietnam.

Specimens examined (var. *blumei*): **CHINA**. Hainan, *P. J. Lin & L. Zhang 978* (IBSC). **INDONESIA**. Java, Tjibodas, *M. Fleischer 1594* (H 3090128); *Hjalmar Möller 51* (H-BR 2600001); *Hjalmar Möller 52* (H-BR 2600002); *Hjalmar Möller 103* (H-BR 2600003); Tjikorai, *E. Nyman 150* (H-BR 2600004); Tjikorai, E. Nyman 159 (H-BR 2600005); Sumatra, Westkust, *Sjamsoeddin 4225* (H 3090127). **MALAYSIA**. Pahang, *M. G. Monte 2468* (MO 4449244). **PHILIPPINES**. *H. M. Curran & M. L. Merritt 8188* (MO). **VIETNAM**. Lam Dong, *S. He & K. Nguyen 42968* (MO 6239374); *S. He & K. Nguyen 42935* (MO 6239371); *S. He & K. Nguyen 42753* (MO 6239373).

Macromitrium blumei var. *zollingeri* (Mitt. ex Bosch & Sande Lac.) S.L. Guo, B.C. Tan & Virtanen, Nova Hedwigia 82: 476. 2006. (Figure 57: 18-26)

Basionym: *Macromitrium zollingeri* Mitt. ex Bosch & Sande Lac., Bryol. Jav., 1: 113. 90. 1859. Type protologue: [Indonesia] Habitat insulam Javae Zollinger itineris Jav. secundi coll. no. 3716. Type citation: Java 1852, leg. *Zollinger 3716* (isotype: H-BR 2601008!).

Macromitrium contortum Thwaites & Mitt., J. Linn. Soc., Bot., 13: 301. 1873, *fide* Wijk et al., 1964. Type protologue: Hab. Ins. Ceylon. Dr. Thwaites. Type citation: Ceylon, Central Prov. *J. Thwaites 36* (isotypes: H-BR 2602008!, MO!).

Macromitrium magnirete Dixon, Bull. Torrey Bot. Club, 51: 234. 1924, *fide* Eddy, 1996. Type protologue: HAB. Gunong Tahan, Pahang, 7000 ft., 1922; Mohammed Haniff and M. Nur (7907) (isotype: H-BR 2602001!)

Macromitrium annamense Brotherus & Paris in Brotherus, Nat. Pfl. Ed. 2, 11: 39. 1925, *fide* Guo et al., 2006. Type protologue: Annam, Lang Bian, alt. 1500m, 25 Nov. 1903, leg. *W. Micholitz 290* (isotype: JE 04006258!).

This variety differs from the type variety in having longer leaves, longer leaf awn, and ligulate leave outline with a higher leaf length/width ratio. In appearance, the leaves of *M. blumei* var. *zollingeri* curled loosely around the stem because of its longer branch leaves and longer leaf awn, rather than coiling tightly around the stems to give the shoots a rope-like appearance as in the type variety.

Distribution: China, Indonesia, Malaysia, Philippines, Vietnam.

Specimens examined: **INDONESIA**. Java, *E. Nyman 165* (H-BR 260006); Java, Avdjoeno, *C. Lauterbach 452* (H-BR 2601007); Sumatra, *A. Ernst* (H-BR 2600017), Bukit Basar, *K. Giesenhagen* (H-BR 2600018); Celebes, Mangkasse, Bojong, *O. Warburg* (H-BR 2600019). **MALAYSIA**. Borneo, *J. Clemens 3557* (H-BR 3090124). **PHILIPPINES**. Luzon, *M. Jacobs B641* (H 3090123), Mindoro, *E.D. Merrill 5505* (H-BR 2600022).

24. *Macromitrium brachypodium* Müll. Hal., Bot. Zeitung (Berlin) 15: 778. 1857.

Type protologue: Isle of pines Novae Caledoniae: Collect. *Cuming*. Type citation: Isle of Pine, leg *Cuming s.n.* (lectotype designated by Vitt & Ramsay 1985: H-BR 2572015!).

Macromitrium brevisetaceum Hampe, Linnaea 38: 663. 1874, *fide* Vitt & Ramsay, 1985. Type protologue: Lord Howe's Island. ad arbores.

(1) Plants small, in dense spreading mats; stems creeping, densely with short branches, branches about 5 mm long. (2) Stem leaves about 1.0 mm long, irregularly erect-flexuose when dry, recurved-flexuose when moist, lanceolate, gradually narrowed and acuminate; all cells uniform and rounded. (3) Branch leaves irregularly to spirally twisted or twisted-contorted, with curved upper portions when dry, erecto-patent when moist, 1.5-2.5 mm long, ligulate to oblong-ligulate, fairly broad; apices obtuse to mucronate; costae percurrent; upper cells 7-9 µm wide, subquadrate-rounded to hexagonal-rounded, slightly bulging, pluripapillose; medial cells similar to upper cells, becoming more clear and smooth; basal cells 10-30 µm long, 5-7 µm wide, restricted to the insertion, short-rectangular to short-elongate, smooth, flat, longer near the margin and forming an indistinct border of 5-10 rows. (4) Perichaetial leaves not much differentiated. (5) Setae rather short, only 1.3 mm long, smooth, straight or slightly twisted to the left. (6) Capsules emergent to short-exserted, urns ellipsoid, smooth, distinctly darker at rim, wide-mouthed; peristome single, exostome of 16, linear-lanceolate. (7) Calyptrae mitrate, short-conic, fimbriate-lacerate, smooth, plicate, with sparse to dense, straight and thick hairs.

Distribution: Australia, New Caledonia.

Specimens examined: **AUSTRALIA**. New South Wales, Lord Howe Island, *D. H. Vitt 28343* (H 3090132), *D. H. Vitt 28564* (H 3090133), *D. H. Vitt 28621* (H 3090134), *W. W. Watts no. 41* (H-BR 2588002), *William A. Weber & M. Colson B-77019* (MO 3973331), *H. Streimann 55958* (KRAM-B-114512).

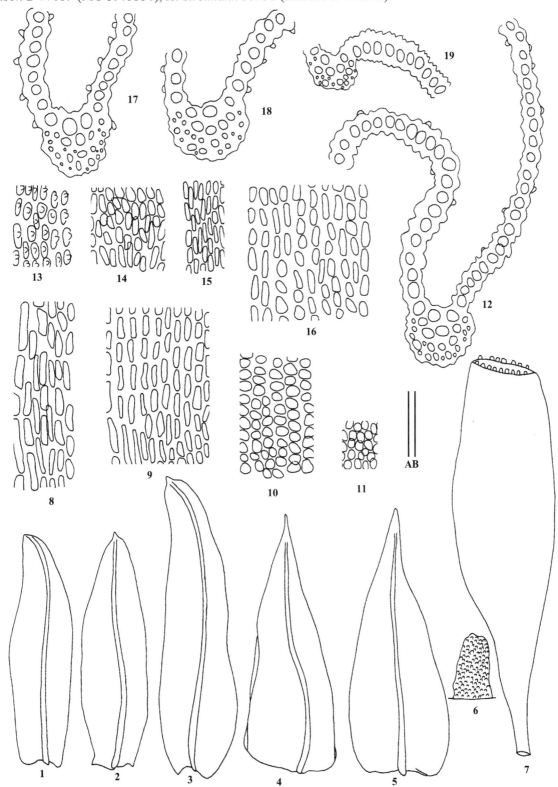

Fig. 55 *Macromitrium binsteadii* Dixon 1-3: Branch leaves. 4, 5: Perichaetial leaves. 6: Peristome. 7: Capsule. 8: Basal cells of branch leaf. 9: Low cells of branch leaf. 10: Medial cells of branch leaf. 11: Upper cells of branch leaf. 12: Upper transverse section of branch leaf. 13: Upper cells of perichaetial leaf. 14: Medial cells of perichaetial leaf. 15: Basal cells of perichaetial leaf. 16: Low cells of perichaetial leaf. 17, 18: Medial transverse sections of branch leaves. 19: Upper transverse section of branch leaf (all from holotype, H-BR 2572022). Line scales: A = 0.44 mm (1-5, 7); B = 44 µm (6, 8-19).

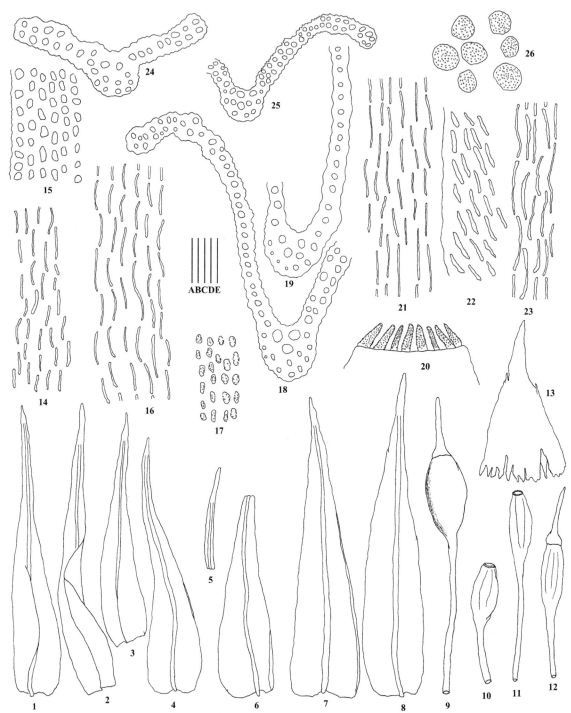

Fig. 56 *Macromitrium bistratosum* E. B. Bartram 1-4, 6: Branch leaves. 5: Apex of branch leaf. 7-8: Perichaetial leaves. 9-12: Capsules. 13: Calyptra. 14: Low cells of branch leaf. 15: Upper cells of branch leaf. 16: Basal cells of branch leaf. 17: Medial cells of branch leaf. 18: Low transverse section of branch leaf. 19: Basal transverse section of branch leaf. 20: Peristome. 21: Medial cells of perichaetial leaf. 22: Upper cells of perichaetial leaf. 23: Basal cells of perichaetial leaf. 24: Upper transverse section of branch leaf. 25: Medial transverse section of branch leaf. 26: Spores (all from holotype, FH 00213572). Line scales: A = 2 mm (9-13); B = 1 mm (1-8); C = 400 μm (20); D = 100 μm (26); E = 67 μm (14-19, 21-25).

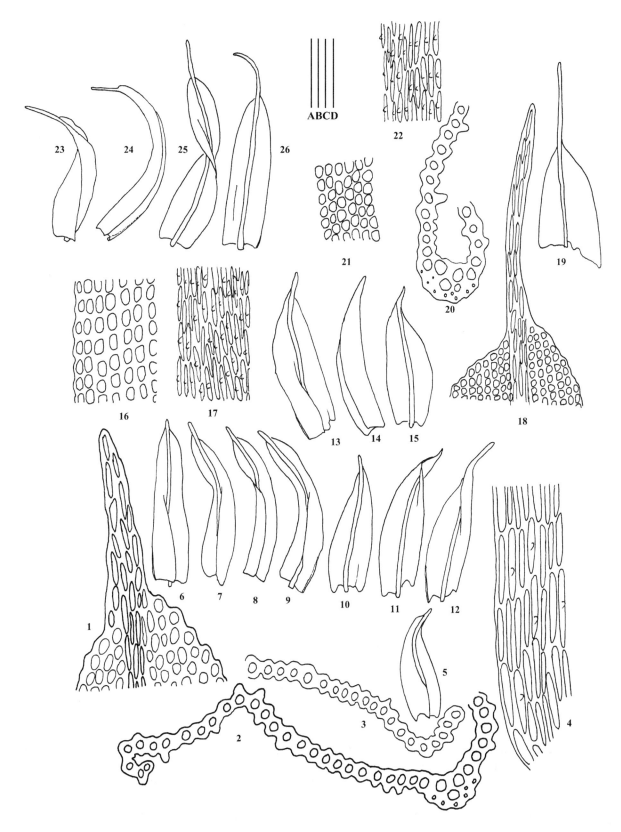

Fig. 57 *Macromitrium blumei* var. *blumei* Nees ex Schwägr. & *M. blumei* var. *zollingeri* 1: Apical cells of branch leaf. 2-3, 20: Basal transverse sections of branch leaves. 4: Basal marginal cells of branch leaf. 5-13: Branch leaves. 14-15: Perichaetial leaves. 16: Medial cells of branch leaf. 17: Basal cells near costa of branch leaf. 18: Apical cells of branch leaf. 19: Apex of branch leaf. 21: Medial cells of branch leaf. 22: Basal cells of branch leaf. 23-26: Branch leaves (1-17 from isotype of *M. blumei* in H-BR; 18-26 from the plant of *M. blumei* var. *zollingeri* mixed in the isotype packet of *M. blumei* preserved at H-BR). Line scales: A = 0.4 mm (5-15, 23-26); B = 160 μm (19); C = 64 μm (18); D = 40 μm (1-4, 16-17, 21-22). (Guo *et al.*, 2006).

25. Macromitrium brevicaule (Besch.) Broth., Die Nat. Pflanzenfam. I(3): 486. 1903. (Figure 58)

Basionym: *Micromitrium brevicaule* Besch., Ann. Sci. Nat., Bot., sér. 5, 18: 211. 1873, Type protologue and citation: [New Caledonia] Balade (*Viellard, no. 1734*) (syntype: BM 000982724!), ad truncus prope Noumea (*Balansa, no. 2563*) (syntype: BM 000982725!).

Macromitrium mucronulatum Müll. Hal., Hedwigia 37: 146. 1898, *fide* Vitt & Ramsay, 1985. Type protologue: [Australia] Queensland, Burpangary, Majo 1888: C. Wild in Hb. Brotheri (holotype: H-BR!).

Macromitrium subbrevicaule Broth. & Watts, Proc. Linn. Soc. New South Wales 40: 371. 1915. Type protologue: [Australia] Lord Howe Island, Growing mostly on cliffs at North head and on the Northern Hills *no. 236, 239, 478, 504, 507*). Type citation: [Australia] Lord Howe Island, North head, leg. *Watts 507*, Aug, 1911 (syntype: H-BR 2563005!).

Macromitrium subfragile Dixon & Sainsbury, Trans. & Proc. Roy. Soc. New 75: 178 1945, *fide* Vitt & Ramsay, 1985. Type protologue: [New Zealand] Top and inner side of rock against cliff on seashore, Stony Bay, Cape Colville; coll. *L.B. Moore, 19/5/1933, no. 770*. On cliffs, North Cape; coll. *L.W. Millener, 10/12/1934, no. 781*, type.

Macromitrium wattsii Broth., Öfvers. Finska Vetensk.-Soc. Förh., 40: 169, 1898, *fide* Vitt & Ramsay, 1985. Type protologue and citation: [Australia] New South Wales, Richmond River, E. Balina, ad arbores, *W. W. Watts, n. 329* (syntype: H-BR 2565002!), *719, 1041* (syntype: H-BR 2565004!), North Creek *(n. 1108)* (syntype: H-BR 2565005!), Wollongong, ad rupes (*n. 109*) (syntype: H-BR 2565001!).

(1) Plants slender, in dense, compact, spreading mats; stems prostrate, long creeping, with very short, erect branches, branches up to 7 mm long. (2) Stem leaves broadly lanceolate to ovate-lanceolate; the apices bluntly acute to narrowly obtuse, erect-twisted when dry, spreading when moist, occasionally with brown, multicellular clavate gemmaes. (3) Branch leaves densely and tightly leaved, keeled, leaves obliquely appressed to spirally appressed-curved when dry, erect-spreading and straight with weakly flexuose apices when moist, 1.0-1.8 mm long, ligulate, oblong-ligulate to broadly oblong; apices mucronate; margins entire or slightly crenulate above; costae very strong and conspicuous, excurrent in a mucro or ending just below the apex; upper cells 8-12 μm wide, subquadrate to irregular-rounded, obscure by low, dense pluripapillae; medial cells similar to upper cells, rounded, 9-10 μm wide, pluripapillose; basal cells shorter, rounded to short elliptic-rectangular, thick-walled, clear, flat and smooth, with several rows of longer marginal cells to form an indistinct border. (4) Perichaetial leaves similar to branch leaves. (5) Setae 3-5 mm long, straight or flexuose, smooth, twisted to the left. (6) Capsule urns ovoid-ellipsoid when moist, narrowly ovate to ovate-cylindrical when dry, smooth to slightly wrinkled, not constricted below the mouth; peristome single, exostome of 16, irregular, blunt, finely papillose to papillose-striate teeth; spores indistinctly anisosporous. (7) Calyptrae short, conic-mitrate, smooth and naked, lightly plicate, covering half of the capsule.

Distribution: Australia, New Caledonia, New Zealand (Vitt & Ramsay, 1985).

Specimens examined: **AUSTRALIA**. Capt. Cook Monument. *H. Streimann 32013* (H 3194602); New South Wales, *H. Streimann 49906* (KRAM-B-105193), *H. Streimann 55913* (KRAM-B-114557); Norflok Island, *H. Streimann 31770* (H 3194601), *H. Streimann 49589* (KRAM-B-103042), *H. Streimann 53786* (KRAM-B-110844); Queensland, *H. Streimann 52387* (KRAM-B-107179).

26. Macromitrium brevihamatum Herzog, Bibliotheca Botanica 87: 68. 25. 1916. (Figure 59)

Type protologue: [Bolivia] An Bäumen im unteren Coranital, about 1800 m, *No. 4714* u. *5064*; im Bergwald des Rio Tocorani *No. 4065*. Type citation: Bolivia, Im unteren Coranital, about 1800 m, *T. C. J. Herzog 4714* (syntype: JE 04008709!; isosyntypes: S-B 162958!, S-B 162959!); Bolivia, Im Berg- wald des Rio Tocorani, about 2200 m, *T. C. J. Herzog 4065* (syntypes: JE 04008707!, JE 04008708!; isosyntype: H-BR 2625009!).

(1) Plants medium-sized, in loose mats, brown to rusty-brown above, dark brown below; stems prostrate-creeping, loose branches up to 40 mm long. (2) Branch leaves erect below, twisted-contorted and flexuose, undulate above when dry, strongly abaxially curved spreading, often conduplicate when moist, lanceolate to ligulate-lanceolate, acute or acute-acuminate, occasionally decurrent at base, basal margin often have reddish rhizoids; margins serrulate, notched near the apex, entire below; costae single and stout, subpercurrent to percurrent; apical cells elliptic-rhombic; upper and medial cells rounded, rounded-quadrate, ovate, bulging and smooth, those near the costa elongate; low and basal cells elongate, long rectangular, thick-walled, weakly porose, strongly tuberculate, enlarged teeth-like cells distinctly differentiated at the insertion. (3) Perichaetial leaves oval- to oblong-lanceolate with the widest portion at the base, plicate below, sharply narrowed in the upper part to an arista; margins plane and entire throughout; costae single, stout, percurrent; all cells longer than wide, low cells tuberculate. (4) Setae smooth, to 30 mm long, twisted to the left. (5) Capsule urns large, red-brown, ellipsoid to ellipsoid cylindrical, constricted beneath the mouth, distinctly furrowed; peristome double, exostome teeth yellowish brown, endostome hyaline; spores globular, finely papillose. (6) Calyptrae mitrate, smooth and naked, deeply lacerate.

Macromitrium brevihamatum Herzog is similar to *M. guatemalense* Müll. Hal., differing from the latter in having smooth upper and medial leaf cells and rather long setae. The former is likely a synonym of the latter.

Distribution: Bolivia.

Specimens examined: **BOLIVIA**. Wdldgrewze, Tocoraui, Zapawillas, about 2900-3000 m Jul. 1911, *Th. Herzog. 3845/s* (H-BR 2625028); Bäumen im unteren Coranital, *Th. Herzog 5065* (H-BR 2625030).

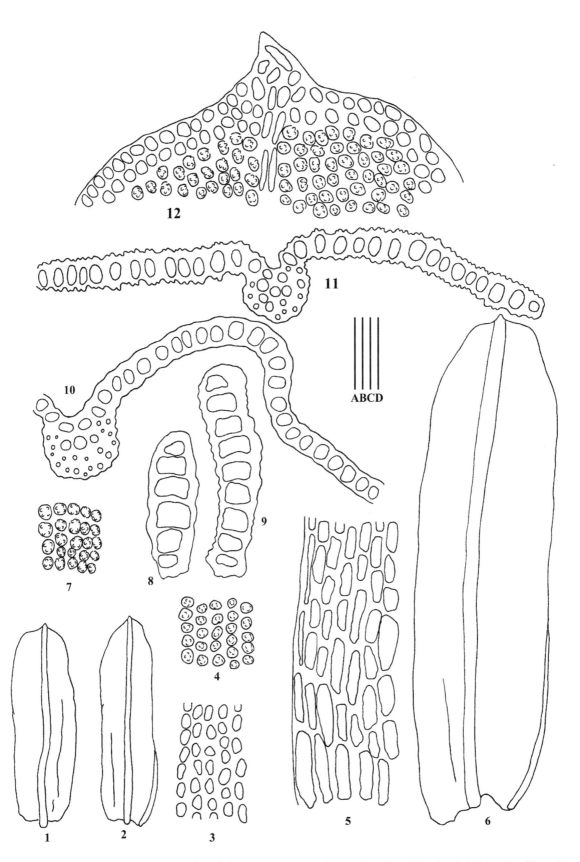

Fig. 58 *Macromitrium brevicaule* (Besch.) Broth. 1-2, 6: Branch leaves. 3: Low cells of branch leaf. 4: Medial cells of branch leaf. 5: Basal cells of branch leaf. 7: Upper cells of branch leaf. 8-9: Gemmae. 10: Medial transverse section of branch leaf. 11: Basal transverse section of branch leaf. 12: Apical cells of branch leaf (all from syntype, BM 000982724). Line scales: A = 1 mm (1-2); B = 400 μm (6); C = 100 μm (8-9); D = 67 μm (3-5, 7, 10-12).

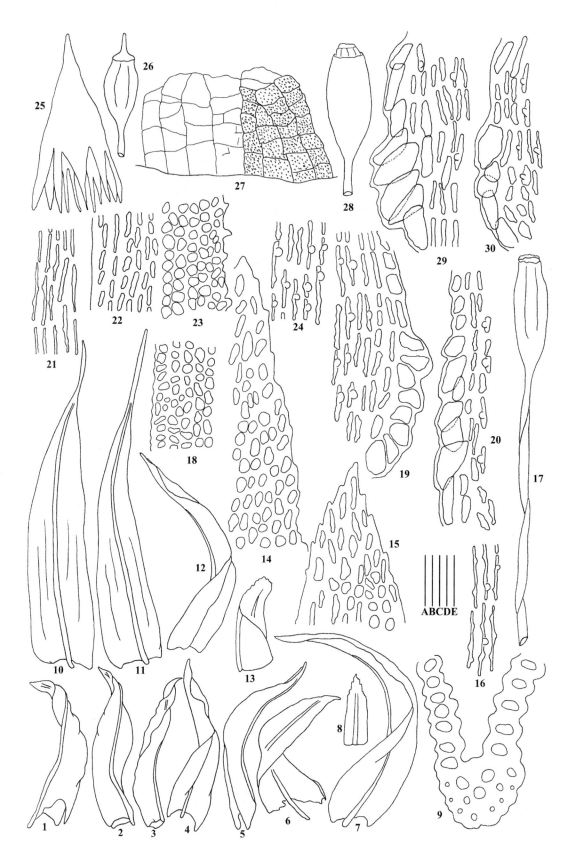

Fig. 59 *Macromitrium brevihamatum* Herzog 1-7, 12: Branch leaves. 8, 13: Apices of branch leaves. 9: Medial transverse section of branch leaf. 10-11: Perichaetial leaves. 14-15: Apical cells of branch leaves. 16: Low cells of perichaetial leaf. 17, 26: Dry capsules. 18: Medial cells of branch leaf. 19, 20, 29, 30: Basal marginal cells of branch leaves. 21: Medial cells of perichaetial leaf. 22: Upper cells of perichaetial leaf. 23: Upper cells of branch leaf. 24: Low cells of branch leaf. 25: Calyptra. 27: Peristome. 28: Wet capsule (all from syntype, JE 04008709). Line scales: A = 2 mm (17, 25, 26, 28); B = 1 mm (1-7, 10-12); C = 400 μm (8, 13); D = 100 μm (15, 27); E = 67 μm (9, 14, 16, 18-24, 29-30).

Taxonomy

27. *Macromitrium brevisetum* Mitt. in Seemann, Fl. Vit. 379. 1873. (Figure 60)
Type protologue and citation: [Hawaii] Oahu (Beechey) (lectotype: NY 00518258!; isolectotype: FH 00213573!).
Macromitrium aristocalyx Müll. Hal., Flora, 82: 454. 1896, *fide* Wijk *et al*., 1964. Type protologue: Insulae Hawaiieae, sine loco speciali: Dr. W. Hillebrand. Type citation: Hawaii, leg. *Hillebrand s.n.* (lectotype: H-BR 255802!).

(1) Plants medium-sized; stems creeping, with short and erect branches. (2) Branch leaves keeled, contorted and slightly crisped, with an incurved upper portion when dry, widely spreading, open and flat when moist, 2.5-3.0 × 0.4 mm, lingulate, lingulate-lanceolate, with obtuse acute to mucronate the apices; costa percurrent to shortly excurrent; margins crenulate with papillae except at base; upper cells rounded-hexagonal, 8-10 μm in diameter, obscure and papillose; low cells oblong, narrowly rectangular towards to the margin, shorter and broader towards the costa, smooth and incrassate. (3) Perichaetial leaves 4 mm long, slightly longer than branch leaves, costae excurrent in a long denticulate arista. (4) Setae rather short, about 0.5 mm long, erect, reddish. (5) Capsules immersed, bright reddish brown, ellipsoid, about 1.5 mm long, wide-mouthed, weakly plicate or longitudinally wrinkled in the upper half when dry; peristome absent; opercula conic-rostrate; spores papillose, large, up to 50 μm in diameter. (6) Calyptrae campanulate, lobed at base, densely pilose with coarse, strict, minutely scabrous hairs.

Distribution: Hawaii.
Specimens examined: **HAWAII**. Mauna, Hsoop, *J. Roca s.n.* (H 3090156); Oahu, South Koolau Mountains, *D. H. Vitt 8188* (H 3090158); S. Slope of Hualalai, *Carl Skottsberg 163* (H-BR 2557001); Hawaii Ins. Dry forest Mauna Loa, *J. Rock s. n.* (H-BR 2557002).

28. *Macromitrium caldense* Ångstr., Öfvers. Förh. Kongl. Svenska Vetensk.-Akad. 33(4): 12, 1876. (Figure 61)
Type protologue: [Brazil] S. Henschen cum macromitrio Regnellii retulit. Type citation: Brassilia, Minas Gerais Caldes, leg. *S. Henschen* (lectotype designated here: H-BR 2628014!).

(1) Plants medium-sized, in sparse mats; stems creeping, irregularly branched, dark-brown below, tomentose on the low portion of leaves. (2) Branch leaves appressed below, strongly twisted-contorted and crisped, keeled above when dry, erect-spreading and somewhat twisted and incurved above when moist, slightly inrolled at the apex, lanceolate; apices short acute, obtuse, blunt acute; margins serrate above, entire below, recurved on one side below, plicate; costae brownish, percurrent; upper cells quadrate, rounded-quadrate to rounded, small and smooth; medial cells round to oblong, arranged in longitudinal rows; low linear or long rectangle, porose, strongly tuberculate. (3) Perichaetial oval to oblong, acuminate, cells smooth and clear, hyaline towards margin. (4) Setae about 6 mm long, smooth, twisted to the right. (5) Capsule urns ovoid-ellipsoid to ellipsoid-cylindrical, smooth, constricted and wrinkled beneath the mouth, brown. (6) Calyptrae dark yellow below, brown above, smooth and naked, not lacerated.

Distribution: Brazil.
Specimens examined: **BRAZIL**. São Paulo. Legit *Schiffner 1329*, 15, VIII. 1901 (H 3204881, KRAM-B-077053).

29. *Macromitrium caloblastoides* Müll. Hal., Hedwigia 37: 151. 1898. (Figure 62)
Type protologue and citation: Australia tropica, Queensland, sine loco speciali: F. M. Bailey in Hb. Brotheri (holotype: H-BR 2535004!).
Macromitrium dimorphum Müll. Hal., Hedwigia 37: 152. 1898, *fide* Vitt & Ramsay, 1985. Type protologue and citation: Australia tropica, Queensland, sine loco speciali: *F. M. Bailey* in Hb. Brotheri (holotype: H-BR 2535003!).

(1) Plants slender to medium-sized, in dense, compact, spreading mats; stems prostrate, creeping, branches to 10 mm long, regularly spaced on the stems. (2) Stem leaves 1.0-1.2 mm long, ovate-lanceolate to broadly lanceolate, bluntly acute, erect-flexuose when dry, widely spreading when moist; all cells uniform throughout, rounded to elliptic. (3) Branch leaves strongly keeled, irregularly and strongly twisted-flexuose-contorted above, apices hidden in the inrolled cavity, not funicuate when dry, flexuose-spreading with conduplicate above, and cucullate, incurved apices when moist,1.3-2.0 mm long, broadly ligulate to oblong; apices bluntly and broadly acute, occasionally somewhat apiculate; margins broadly reflexed, entire below, weakly crenulate near the apex; costae ending a few cells beneath the apex or percurrent; upper cells 9-12 μm wide, quadrate-rounded to hexagonal-rounded, thin-walled, strongly bulging, collenchymatous, clear and most smooth or slightly pluripapillose; transitional cells short, clear, strongly bulging and smooth, in longitudinal rows, rounded to elliptic; basal cells 14-35 × 70-10 μm, evenly thin-walled, short-rectangular to short-elongate, slightly bulging, smooth or weakly unipapillose. (4) Perichaetial leaves 1.2-1.5 mm long, shorter than branch leaves, ovate-lanceolate to oblong; apices acuminate to acute. (5) Setae 5-7 mm long, stout and smooth, twisted to the left. (6) Capsule urns fusiform-ellipsoid to ellipsoid-ovoid, smooth and 8-plicate below the narrow, puckered mouth; peristome single and greatly reduced or absent; spores isosporous, finely papillose.

(7) Calyptrae conic-mitrate, smooth and naked, slightly plicate and lobed in the low portion, irregularly lobed at the base.

Macromitrium caloblastoides Müll. Hal. is similar to *M. ligulaefolium* (Hook.) Brid., but differs from the latter in having 1) branch leaves broad ligulate, tightly and regularly inrolled, with cucullate apices hidden in the inrolled cavity; 2) capsules puckered at the mouth and distinctly 8-plicate; 3) peristome single and reduced to a low membrane, and 4) calyptrae conic-mitrate, and naked. *Macromitrium caloblastoides* is also similar to *M. involutifolium* (Hook. & Grev.) Schwägr., but differs from the latter in having longer setae (up to 5-7 mm long), and naked calyptrae.

Distribution: Australia.

Specimens examined: **AUSTRALIA**. New South Wales, *W. W. Watts 3403* (H 3 090160), *N. W. W. Watts 1601* (H-BR 2535014), *H. Streimann 80686* (KRAM-B-137394).

30. *Macromitrium calocalyx* Müll. Hal., Abh. Naturwiss. Vereins Bremen 7: 208. 1881. (Figure 63)

Type protologue: (Madagascar) Wald von Ambatondrazaka, (Rutenberg), 6 Decbr. 1877. Type citation: Madagascar, leg. *Rutenberg* (lectotype designated by Wilbraham, 2018: H-BR 2545005!).

Macromitrium semipapillosum Thér. & P. de la Varde, Recueil Publ. Soc. Havraise Études Diverses, 91: 89. 1924, *fide* Wilbraham, 2018. Type: Madagascar, leg. *Carrougeau s.n.* (lectotype designated by Wilbraham 2018: PC 0106790!; isolectotype: PC 0106789!).

(1) Plants robust, in compact mat; stems strongly creeping, densely with erect branches up to 15 mm long. (2) Branch leaves keeled, individually twisted and crisped, undulate, apices curved to inrolled when dry, spreading with an incurved and conduplicate upper portion when moist, 2.0-3.3 mm long, lanceolate from an oblong base, narrowly acuminate above, occasionally fragile; upper cells oblate, subquadrate, about 5 µm wide, bulging, obscure and pluripapillose, arranged in longitudinal rows; low and basal cells narrowly rectangular, 20-40 µm long, thick-walled, strongly tuberculate; basal cells near costa hyaline and large, forming a distinct "cancellina region". (3) Perichaetial leaves erect, 2.5-4.0 mm long, longer than branch leaves, narrowly oblong-lanceolate, plicate below, with a widest portion at base, with a subulate upper portion, gradually narrowed to an acuminate-acute apex; all cells longer than wide, thick-walled and smooth. (4) Setae 4-6 mm long, smooth. (5) Capsule urns ovoid, smooth, wide-mouthed; peristome single, exostome of 16 teeth, short-lanceolate, reduced and fragmentary when old. (6) Calyptrae large, lacerate, sparsely hairy.

Distribution: Madagascar.

Specimen examined: **MADAGASCAR**. Mt. Tsaratanana, *Ferriei* (H-BR 2626005).

31. *Macromitrium calomicron* Broth. in Voeltzk., Reise Ostafr., Syst. Arbeit. 3: 55. 8 f. 8. 1908. (Figure 64)

Type protologue: Fundnetiz: Mauritius, an Baumrinde. Type citation: Mauritius, 1904, leg. *A. Voeltzkow s.n.* (lectotype designated here: H 3090161!).

(1) Plants small, in mats, brown-green, dull; stems long and slender, prostrate and forked, covered with rusty rhizoids; branches erect, shorter than 5 mm, densely leaved. (2) Branch leaves incurved when dry, erect spreading when moist, lanceolate from an oblong base, apices acute; margins plane and entire; costae light reddish, percurrent or excurrent to form a short point; upper and medial cells rounded, oblate, subquadrate, smooth to weakly papillose; low and basal cells linear rectangular, strongly thick-walled, lumens curved-sigmoid, frequently strongly tuberculate. (3) Inner perichaetial leaves differentiated, shorter and wider than branch leaves, gradually becoming narrow, acuminate; costae excurrent to a short arista; all cells strongly thick-walled, clear and smooth; upper cells rounded; basal cells elongate. (4) Setae about 5-7 mm long, smooth. (5) Capsules erect, ellipsoid, distinctly wrinkled beneath the mouth; peristome single, exstome teeth lanceolate, short, densely papillose.

Distribution: Mauritius.

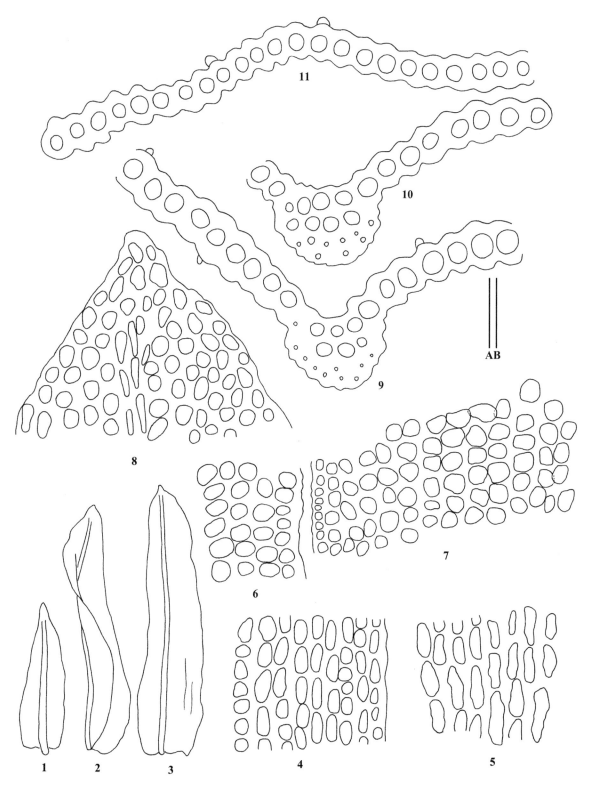

Fig. 60 *Macromitrium brevisetum* Mitt. 1-3: Branch leaves. 4: Low cells of branch leaf. 5: Basal cells of branch leaf. 6: Medial cells of branch leaf. 7: Upper cells of branch leaf. 8: Apical cells of branch leaf. 9: Medial transverse section of branch leaf. 10: Upper transverse section of branch leaf. 11: Basal transverse section of branch leaf (all from isolectotype, FH 00213573). Line scales: A = 1 mm (1-3); B = 67 μm (4-11).

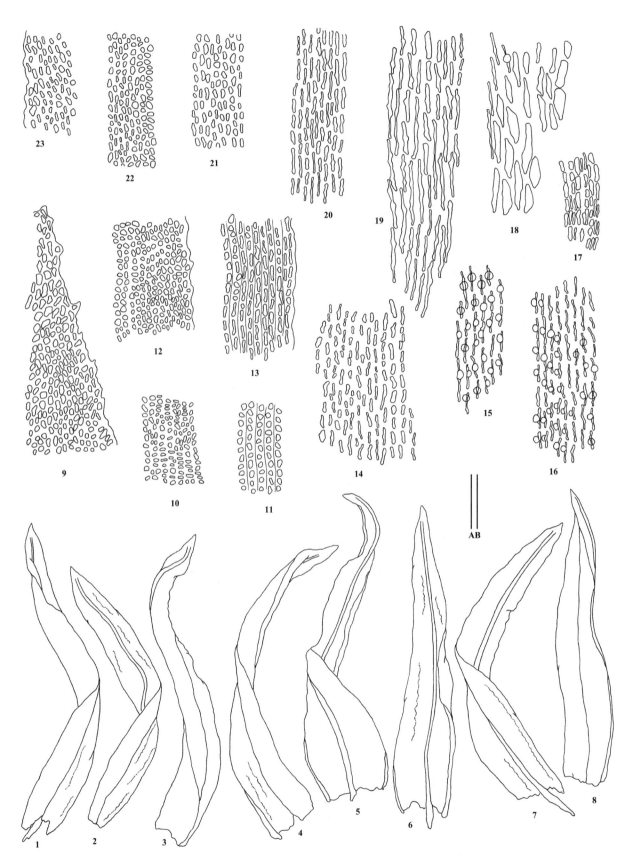

Fig. 61 *Macromitrium caldense* Ångström 1-5, 7: Branch leaves. 6, 8: Perichaetial leaves. 9: Apical cells of branch leaf. 10-11: Medial cells of branch leaves. 12: Upper cells of branch leaf. 13-14: Low cells of branch leaves. 15-16: Basal cells of branch leaves. 17: Basal marginal cells of branch leaf. 18: Basal cells near costa of branch leaf. 19: Basal cells of perichaetial leaf. 20: Low cells of perichaetial leaf. 21: Medial cells of perichaetial leaf. 22: Upper cells of perichaetial leaf. 23: Apical cells of perichaetial leaf (all from H-BR 2628017). Line scales: A = 0.5 mm (1-8); B = 50 μm (9-23).

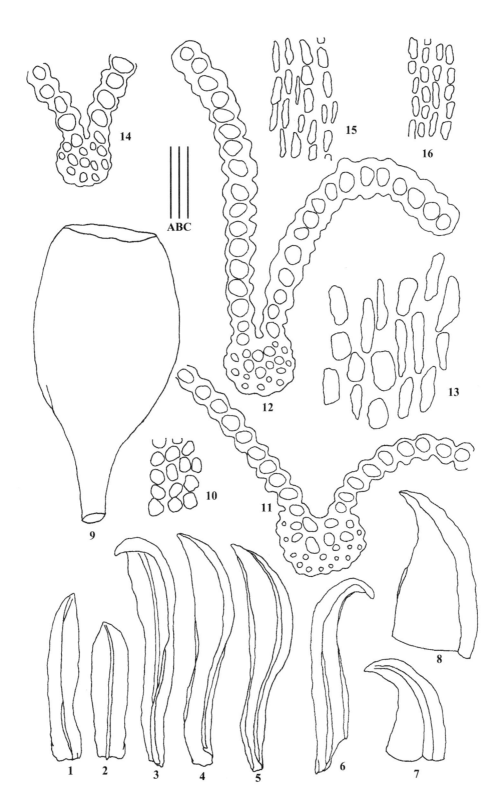

Fig. 62 *Macromitrium caloblastoides* Müll. Hal. 1-6: Branch leaves. 7-8: Apices of branch leaves. 9: Calyptra. 10: Upper cells of branch leaf. 11: Basal transverse section of branch leaf. 12: Medial transverse section of branch leaf. 13: Exothecial cells of capsule. 14: Upper transverse section of branch leaf. 15: Basal cells of branch leaf. 16: Low cells of branch leaf (1-2 from holotype of *M. dimorphum*, H-BR 2535003; 3-16 from holotype of *M. caloblastoides*, H-BR 2535004). Line scales: A = 0.44 mm (1-6, 9); B = 176 μm (7-8); C = 44 μm (10-16).

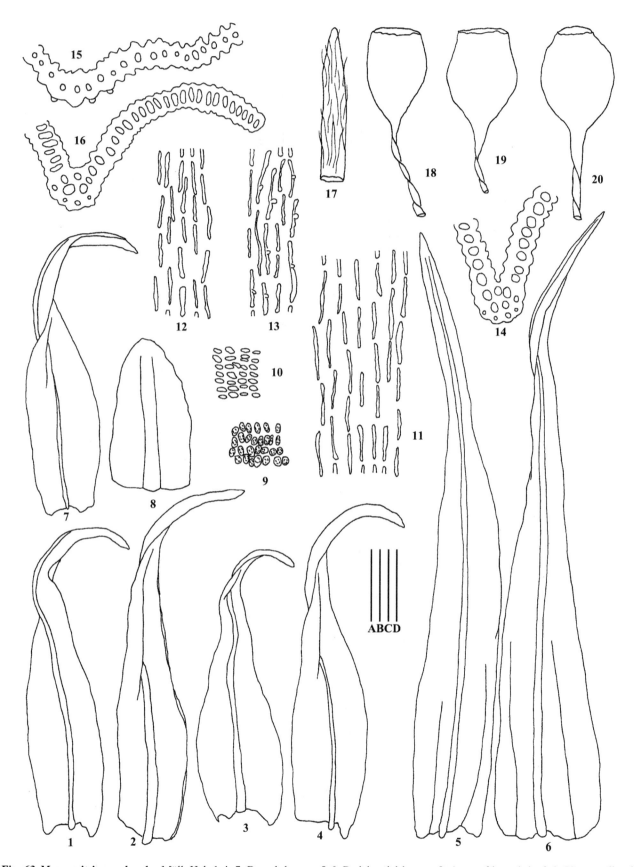

Fig. 63 *Macromitrium calocalyx* Müll. Hal. 1-4, 7: Branch leaves. 5-6: Perichaetial leaves. 8: Apex of branch leaf. 9: Upper cells of branch leaf. 10: Medial cells of branch leaf. 11: Basal cells of branch leaf. 12: Medial cells of perichaetial leaf. 13: Basal cells of perichaetial leaf. 14: Upper transverse section of branch leaf. 15: Basal transverse section of branch leaf. 16: Medial transverse section of branch leaf. 17: Calyptra. 18-20: Capsules (all from H-BR 2626005). Line scales: A = 0.88 mm (17-20); B = 0.44 mm (1-7); C = 70 μm (8); D = 44 μm (9-16).

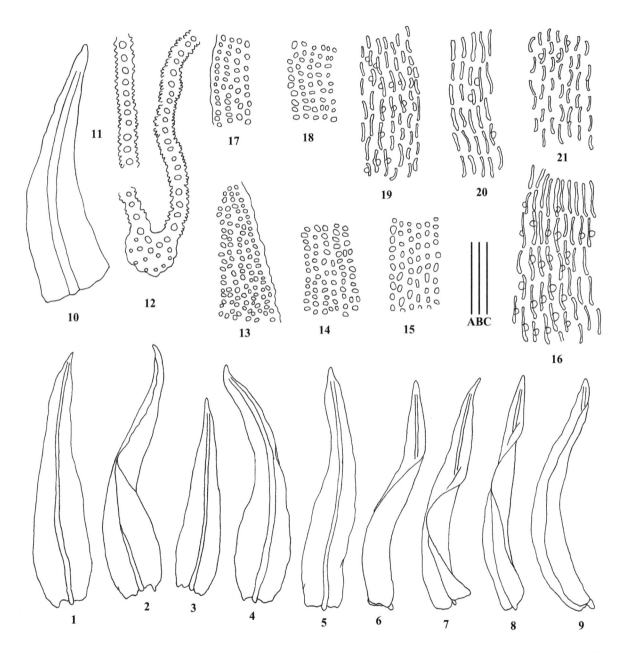

Fig. 64 *Macromitrium calomicron* Broth. 1-9: Branch leaves. 10: Apex of branch leaf. 11-12: Medial transverse sections of branch leaves. 13, 17: Apical cells of branch leaves. 14, 18: Upper cells of branch leaves. 15: Medial cells of branch leaf. 16, 20: Basal cells of branch leaves. 19, 21: Low cells of branch leaves (all from lectotype, H 3090161). Line scales: A = 0.5 mm (1-9); B = 200 μm (10); C = 50 μm (11-21).

32. *Macromitrium calymperoideum* Mitt., Hooker's J. Bot. Kew Gard. Misc. 8: 354. 1856. (Figure 65)

Type protologue and citation: [Myanmar] on a tree near Yavoy (*No. 90*) (isotypes: NY 00518321!, NY 00518322!).

(1) Plants medium-size, stems strongly creeping; branches short, densely leaved. (2) Branch leaves keeled, strongly keeled and crinkled, twisted-contorted and inrolled above with apices hidden in the inrolled cavity when dry, spreading but distinctly rugose or somewhat undulate, and still keeled and adaxially curved above when moist, plicate below, oblong, oblong-lanceolate to ligulate- lanceolate; apices acute; margins flat and entire, narrowly recurved below, occasionally undulate at the upper portion; costae somewhat reddish, percurrent; upper and medial cells rounded-quadrate, densely pluripapillose, rather obscure; low and basal cells elongate-rectangular, smooth or unipapillose; the outmost marginal cells at one side often differentiated, regularly rectangular, pellucid with thinner walls, those near costa sometimes enlarged, thin-walled and pellucid, forming a "cancellina region". (3) Perichaetial leaves similar to branch leaves. (4) Setae 8-12 mm long, smooth, twisted to the left. (5) Capsule urns ellipsoid-cylindric, smooth to weakly plicate when dry; peristome single, exostome of 16, narrowly lanceolate and papillose; spores anisosporous. (6) Calyptrae large, long mitrate, covering the whole capsule, with straight and brown-yellowish hairs.

Macromitrium calymperoideum Mitt. is similar to *M. tosae* Besch., but differing from the latter in having branch leaves with smooth or weak unipapillae in low and basal cells, and with a "cancellina region" near costa at insertion, as well as adaxially curved upper portion when moist. *Macromitrium calymperoideum* is also similar to *M. japonicum* Dozy & Molk., but differing from the latter in having large calyptrae with straight and brown-yellowish hairs, and leaves with acute apices and papillose low and basal cells. Mitten (1856) thought that *M. calymperodieum* was similar to *M. nepalense* (Hook. & Grev.) Schwägr. in size and habit. But the branch leaves of *M. nepalense* spirally coiled with hook-like apices, distinctly funiculate when dry, widely spreading when moist, irregular 1-3-stratose proliferation with pluripapillose cells on both dorsal and ventral laminal surfaces (Guo & He, 2014).

Distribution: India, Myanmar.

Specimens examined: **MYANMAR**. *C. S. P. Parish 7* (NY 00780046); [Mon kingdom] Pegu, *Kwiz 2924* (H-BR 2582002).

33. *Macromitrium campoanum* Thér., Arch. Esc. Fárm. Fac. Ci. Méd. Córdoba 7: 49. 1939. (Figure 66)

Type protologue and citation: Chile, Prov. Valdivia, Dep. Corral, Quitaluto, in der Küstenkordillere, Sphagnum-"alerce"-Sumpf, 430 m, auf altem Holz, *C.C. Hosseus 687 A*, 2, II. 1935 (holotype: JE 04000907!; isotype: KRAM-B-203339!).

(1) Plants small, brown and lustrous; stems long creeping with short erect branches, branches 5 mm long and 0.8 mm thick. (2) Branch leaves obliquely appressed below, spirally twisted and slightly flexuose above when dry, erecto-patent when moist, lanceolate and acuminate; margins entire throughout; costae stout and brown-reddish, subpercurrent to percurrent; all cells clear, flat, smooth, in distinctly longitudinal rows, elongate from upper to base; cells near the apex oblong, round-rectangular; upper cells rounded-quadrate, oblong-rectangular, rhombic; medial cells quadrate, oblong, oblong-rectangular; basal cells rectangular to long rectangular, weakly porose. (3) Perichaetial leaves differ from branch leaves, broadly oblong, oblong-lanceolate with cuspidate apices, entire; all cells clear, smooth; upper and medial cells rounded-quadrate, elliptic-rectangular, rhomboidal; low cells near costa oblong, becoming narrower and rectangular near margin; basal cells long, smooth, slightly porose.

Macromitrium campoanum Thér. is similar to *M. tenax* Müll. Hal., differing from the latter in having broadly oval-oblong perichaetial leaves with cuspidate apices. The perichaetial leaves of *M. tenax* are oblong-lanceolate with acute apices. *Macromitrium campoanum* is also similar to *M. microstomum* (Hook. & Grev.) Schwägr., but differing from the latter in having broadly lanceolate branch leaves with rounded-quadrate, isodiametric medial lamina cells. The medial cells of branch leaves in *M. microstomum* are rectangular to long rectangular.

Distribution: Chile.

34. *Macromitrium cardotii* Thér., Diagn. Esp. Var. Nouv. Mouss. 8: 5. 1910. (Figure 67)

Type protologue and citation: New Caledonia. Tao, forêt, sur les écorces, alt. 600 à 800 m, 1910, *Franc s.n.* (isotypes: JE 04006255!, H-BR 2551009!).

(1) Plants robust, rusty-brown; stems creeping and short, branches up to 8-20 mm long, 2 mm thick; red-tinged to brown. (2) Branch leaves not very shriveled, individually to spirally twisted and flexuose above, apices curved to circinate when dry, erect to patent when moist, 3-4.5 × 0.35-0.75 mm, long lanceolate to linear lanceolate, gradually narrowed to an acuminate, mucronate apex or forming an arista; margins papillose-crenulate, slightly recurved in one side near base; upper and medial cells relatively large, 10-20 × 7-12 μm, obscure, rounded-quadrate, oblong, oblate, bulging, thick-walled, pluripapillose or forked papillae; transitional cells short rectangular, thick-walled, unipapillose; low cells elongate, rectangular to linear, translucent, thick-walled, somewhat porose, unipapillose to tuberculate;

Taxonomy

costae excurrent. (3) Perichaetial leaves similar to branch leaves, with a longer arista. (4) Setae smooth, rather long, up to 35 mm long; vaginulae without hairs, but with short paraphyses. (5) Capsule urns narrowly ellipsoid, smooth, plicate beneath the constricted mouth; peristome single and caducous. (6) Calyptrae plicate and naked.

Distribution: New Caledonia.

Fig. 65 *Macromitrium calymperoideum* Mitt. 1: Apex of branch leaf. 2-6: Branch leaves. 7: Capsules. 8: Spores. 9: Peristome. 10, 11: Upper cells of branch leaves. 12: Basal transverse section of branch leaf. 13: Medial cells of branch leaf. 14, 15: Basal cells of branch leaves (all from isotype, NY 00518321). Line scales: A = 0.44 mm (2, 3); B = 176 μm (1, 4-7); C = 70 μm (8, 9); D = 44 μm (10-15).

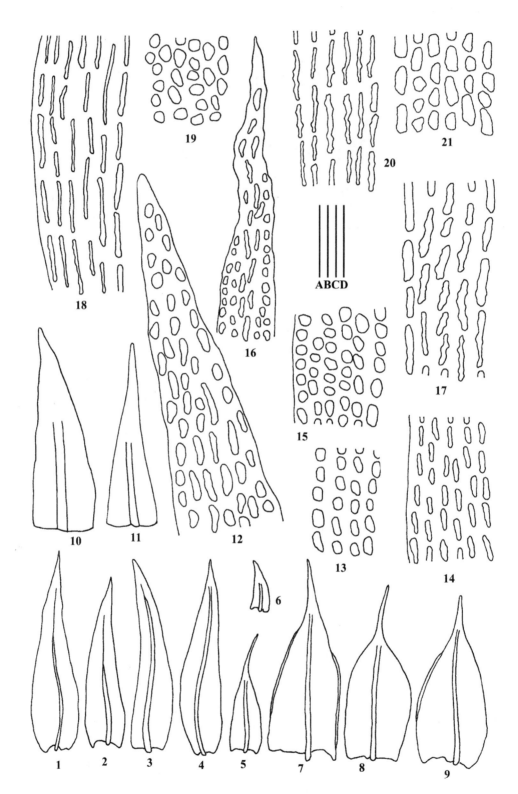

Fig. 66 *Macromitrium campoanum* Thér. 1-5: Branch leaves. 6-9: Perichaetial leaves. 10-11: Apices of branch leaves. 12, 16: Apical cells of branch leaves. 13: Medial cells of branch leaf. 14: Low cells of branch leaf. 15: Upper cells of branch leaf. 17: Low cells of stem leaf. 18: Basal cells of branch leaf. 19: Upper cells of perichaetial leaf. 20: Basal cells of perichaetial leaf. 21: Medial cells of perichaetial leaf (all from holotype, JE 04000907). Line scales: A = 1 mm (1-9); B = 400 µm (10-11); C = 100 µm (16); D = 67 µm (12-15, 17-21).

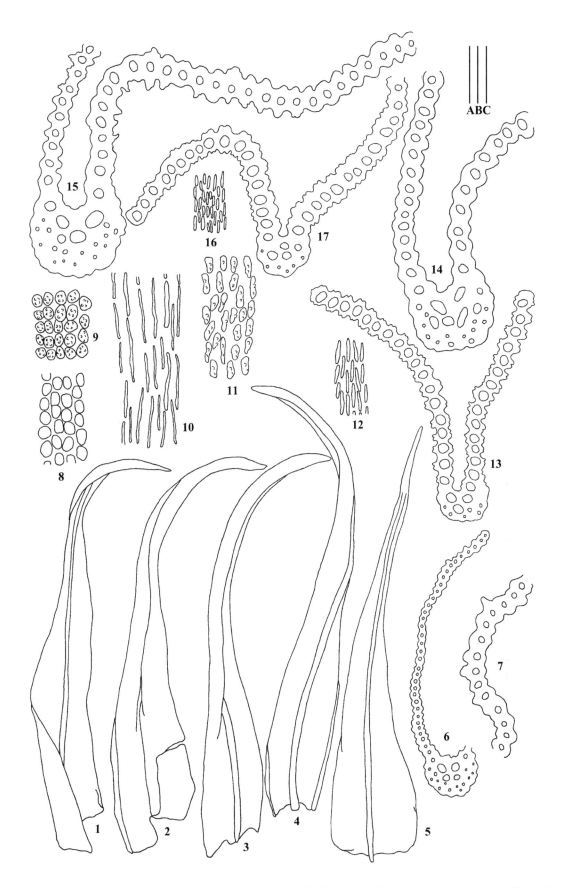

Fig. 67 *Macromitrium cardotii* Thér. 1-4: Branch leaves. 5: Perichaetial leaf. 6-7, 14-15: Basal transverse sections of branch leaves. 8: Medial cells of branch leaf. 9: Upper cells of branch leaf. 10: Basal cells of branch leaf. 11: Low cells of perichaetial leaf. 12: Medial cells of perichaetial leaf. 13, 17: Upper transverse sections of branch leaves. 16: Upper cells of perichaetial leaf (all from isotype, H-BR 2551009). Line scales: A = 0.44 mm (1-5); B = 70 μm (6, 16); C = 44 μm (7-15, 17).

35. *Macromitrium cataractarum* Müll. Hal., Nuovo Giorn. Bot. Ital., n.s. 4: 124. 1897. (Figure 68)

Type protologue and citation: Habitatio. Bolivia, prov. Larecaja, Soratu, ad margines cataractarum, 3400 m. alt. Aug. 1858: *G. Mandon. Coll. No. 1655* (isotype: G 00050744!).

(1) Plants medium-sized, in prostrate mates, stems creeping, loosely branched; branches simple or with branchlets above, brown-yellow. (2) Stem leaves inconspicuous and caducous. (3) Branch leaves contorted-twisted and flexuose, upper portions sometimes incurved to curled when dry, spreading and flat, some still twisted, conduplicate and keeled above when moist, lanceolate, narrowly lanceolate or ovate-lanceolate, plicate and narrowly recurved below, somewhat sheathing at the base, with an acuminate apex; costae vanishing several cells before the apex; margins minutely notched above and entire below; apical cells slightly elongate, oblong to rounded-quadrate; upper and medial cells isodiametric, rounded-quadrate, flat and smooth, medial cells in longitudinal rows; low cellse longate-rectangular, thick-walled, not porose, frequently unipapillose; basal cells elongate, thick-walled, sometimes unipapillose and slightly porose, marginal enlarged teeth-like cells differentiated at the insertion. (4) Perichaetial leaves differentiated, larger and longer than branch leaves, oblong-lanceolate to triangular-lanceolate, with the widest portion at or near the base, somewhat sheathing, incurved and plicate below, acuminate with a long arista; costae vanishing far from the apex; all cells elongate, longer than wide, thick-walled and smooth, occasionally somewhat porose; the outmost marginal cells at the insertion swollen, larger than their ambient cells. (5) Setae about 10 mm long, smooth, not or slightly twisted to the left. (6) Capsules ellipsoid-cylindric, short-cylindric, smooth to furrowed, with a wide mouth; peristome double, both membraneous. (7) Calyptrae mitrate, naked.

Distribution: Bolivia, Brazil.

Specimen examined: **BRAZIL**. São Paulo, *V. Schiffner s.n.* (H-BR 2649002).

36. *Macromitrium catharinense* Paris, Index Bryol. Suppl. 237. 1900. (Figure 69)

Basionym: *Macromitrium prolongatum* Müll. Hal., Bull. Herb. Boissier 6: 99. 1898. Illegitimate, later homonym. Type protologue: Brasilia, Sa. Catharina, Serra Geral, in ramis arborum, Januario 1890, cum fructibus junioribus No 847 e; ad ramjos arborum marginis Serrae ejusdem, Aprili 1891 cum fructu vetusta et ramis aureis No 1017; Serra Itatiaia, 2000 m. alt., 2 Martio 1894 sterile, No 1835. Type citation: Brazil, Prov. S. Catharina, Serra Geral, ad ramos arborum, m. Apr. 1891. leg. *E. Ule. 1017*, with "E. Ule. Bryotheca brasiliensis. 134. *Macromitrium prolongatum* C. Müll. *n. sp.*" (lectotype designated by Li *et al.*, 2019: H-BR 2649005!; isolectotypes: JE 04006251!, JE 04006252!, GOET 012311!, FI!); Estado de Sta. Catharina, Serra Geral, Januario 1890. leg. *E. Ule 847* (syntype: FI!).

Macromitrium catharinense var. *gracilius* (Müll. Hal.) Paris, Index Bryol. Suppl. 237. 1900. ≡ *Macromitrium prolongatum* var. *gracilius* Müll. Hal., Bull. Herb. Boissier 6: 100. 1898, *fide* Li *et al.*, 2019. Type protologue: Brasilia, Rio de Janeiro, in arboribus Serræ dos Orgaus, Dee 1891, c. fr. Reluslis et junioribus, Macrom. Profusum mihi, *no. 1242*. Type citation: 1242. *Macromitrium profusum* Müll. Hal. = *Teichodontium catharinense* var. *gracilior*, Brasilia, Minas Geraës, Serra dos Orgãos, auf Bauman in der Serra, Dec. 1891, leg *E. Ule* (lectotype: H-BR 2649007!).

Macromitrium drewii H. Rob., Bryologist 70: 320. f. 21-24, 1967, *fide* Li *et al.*, 2019. Type protologue and citation: Ecuador. Imbabra: Cordillera Oriental, Camp Sphagnum, 1 km W of Llanacocha, E of Volcan de Cayambe, 11,400 feet. *Drew E-399* (holotype: US 00070256!).

Macromitrium schiffneri Broth., Ergebn. Bot. Exp. Südbras., Musci, 290. 1924, *fide* Valente *et al.*, 2020. Type protologue and citation: [Brazil] São Paulo, Campo Grande, *V. Schiffner 558* (holotype: H-BR 2628016!; isotypes: NY 01243611!, BM 000873236!).

(1) Plants large and robust, in loose mats, dark brown below, yellowish brown above; stems creeping, primary branches up to 40 mm long, with several branchlets. (2) Branch leaves clasping at base, squarrose to widely-spreading, flexuous and divergent above when dry, squarrose-spreading to squarrose-recurved when moist, 3.2-4.0 × 0.6-0.8 mm, abruptly narrowed to a long lanceolate upper part from an ovate-oblong clasping base, upper gradually narrow to an acuminate apex; margins plane, denticulate to crenulate from projecting cells above, entire below, marginal cells not differentiated; costae ending in a few cells beneath the apex; all cells longer than wide; upper cells irregularly rounded-oblong, rhombic, strongly bulging, 12-16 × 5-6 μm, unipapillose; cells near the apex elongate, long-oblong, oblong-rectangular to rectangular; cells gradually elongate from upper to base; medial cells short-rectangular, rounded-oblong, moderately thick-walled, unipapillose; low cells linear-oblong, 35-55 × 2-3 μm, thick-walled, porose, tuberculate. (3) Perichaetial leaves differentiated from and longer than branch leaves, 5.5-7.8 mm long, oblong-lanceolate, upper part abruptly narrowed to form a long setaceous-acuminate apex; costa long-excurrent into a naked awn; all cells longer than wide, thick-walled; upper cells narrowly oblong, 30-35 μm × 3-4 μm, distinctively unipapillose; low and basal cells 55-75 × 2-3 μm, smooth, inconspicuously porose. (4) Capsules erect, ovoid, ovoid-ellipsoid to ellipsoid, smooth; peristome double, exostome lanceolate, yellowish, endostome a short membrane, smooth and hyaline; spores globular, 40-70 μm in diameter. (5) Setae 6-15 mm, smooth or somewhat papillose, twisted to the left. (6) Calyptrae cucullate, with long brownish hairs.

Distribution: Brazil, Colombia, Ecuador.

Specimens examined: **BRAZIL**. São Paulo. *Daniel M. Vital & William R. Buck 20506* (KRAM-B-178792). **COLOMBIA**. Cauca, Inza, *S. P. Churchill & Julio C. Betancur B. 18031* (MO); Nariño, Pasto, *Bernardo R. Ramírez Padilla 7677, 10902, 10586, 10452*; *Bernardo R. Ramírez Padilla & J.A. Cuayal 3868, 4527, 4436* (MO); *S. P. Churchill & Wilson Rengifo M. 17426, 17528-B* (MO); Nariño, Tuquerres, *Bernardo R. Ramírez Padilla & Max Weigend 6859* (MO); Nariño, Sapuyes, *Bernardo R. Ramírez Padilla 9026, 2943, 4746* (MO). **ECUADOR**. Carchi, *Simon Lægaard 101679H, 101679D* (MO); *Holm-Nielsen, Jeppesen, LØjtnant & Øllgaard, 5395* (MO).

37. *Macromitrium chloromitrium* (Besch.) Wilbraham, Cryptog., Bryol. 31: 52. 2010.

Basionym: *Macromitrium fimbriatum* var. *chloromitrium* Besch., Ann. Sci. Nat., Bot., sér. 6, 9: 359. 1880. Type protologue: Maurice: montagne du Pavillon, BOIVIN (Herb. Mus. Par.). Type citation: Mauritius, leg. Boivin, ex hb. Bescherelle (isotype: PC 0137506!); Mauritius, Montagne du Pavillon (holotype: PC 0137505!).

(1) Plants small, in dense mats; stems long creeping, densely with branches. (2) Branch leaves erect below, individually twisted and crisped above, some apices hidden in the inrolled cavity when dry, widely spreading when moist, shorter than 2 mm, oblong lanceolate, lingulate lanceolate; apices blunt acute to obtuse; costae vanishing several cells beneath the apex; upper cells small and bulging, irregularly rounded, partially obscure and pluripapillose; basal cells differentiated, long rectangular, thick-walled, not porose, clear and smooth, confined to a relatively small area. (3) Perichaetial leaves broadly and short lanceolate, with acute apices. (4) Setae shorter, about 2 mm long, smooth. (5) Capsule urns ellipsoid, smooth, slightly constricted beneath the mouth, rim dark brown or brown-red; peristome single; spores anisosporous. (6) Calyptrae mitrate, slightly plicate, naked and lacerated below.

Distribution: Mauritius, Réunion Island, Tanzania (Wilbraham & Ellis, 2010).

38. *Macromitrium cirrosum* (Hedw.) Brid., Bryol. Univ., 1: 316 1826. (Figure 70)

Basionym: *Anictangium cirrosum* Hedw. Sp. Musc. Frond. 42. 1801. Type protologue: Locus, Jamaica, Montserrat. Type citation: Jamaica (lectotype designated by Wilbraham & Price, 2013: G00040187).

Anoectangium cirrosum (Hedw.) Schwägr., Sp. Musc. Frond., Suppl. 1: 38. 1811.

Hypnum cirrhatum Brid., Muscol. Recent. 2(2): 185. 1801, *fide* Grout, 1944.

Macromitrium barbense Renauld & Cardot, Bull. Soc. Roy. Bot. Belgique 31(1): 157. 1893, *fide* Bowers, 1974. Type protologue: Costa Rica, Forêts du Barba, mélange à l'espece suivante *Pittier no 5541*.

Macromitrium costaricense E.B. Bartram, Contr. U.S. Natl. Herb., 26: 88. f. 27. 1928, *fide* Florschütz, 1964. Type protologue: Costa Rica, on tree, la Palma, Province of San Jose, Costa Rica, *Paul C. Standley*, March 17, 1924, *no. 38023*.

Macromitrium cirrosum var. *stenophyllum* (Mitt.) Grout, Bryologist 47: 9. 1944, *fide* Churchill, 2016.

Macromitrium cubensicirrhosum Müll. Hal., Hedwigia 37: 236. 1898, *fide* Wijk *et al.*, 1964. Type protologue: Cuba, ad arbores in montibus altioribus: *Charles Wright, Coll. No. 51*. Sub *Macromitrio cirrhoso* Sulliv. Nec. Sw.

Macromitrium erectopatulum Müll. Hal., Nuovo Giorn. Bot. Ital., n.s. 4: 124. 1897, *fide* Grout, 1944. Type protologue: (Boliva) no detail information.

Macromitrium hoehnei Herzog, Arch. Bot. São Paulo 1(2): 63. 1924, *fide* Valente *et al.*, 2020. Type protologue: [Brazil], São Paulo: Estação Biológica do Alto da Serra, s.col. hb n. 4340-hb T. Herzog 7500 (syntypes: SP 060004, PH 00003401).

Macromitrium mammillosum E.B. Bartram, Contr. U. S. Natl. Herb., 26: 87. f. 26. 1928, *fide* Churchill & Linares, 1995. Type protologue: Costa Rica, In tree, Cerros de Zurqui, northeast of San Isidro, Province of Heredia, Costa Rica, altitude 2000 to 2400 meters, *Paul C. Standley and Juvenal Valerio*, March 3, 1926, *no. 50384*.

Macromitrium microtheca Mitt., J. Linn. Soc., Bot. 12: 208. 1869, *fide* Grout, 1944. Type protologue: Hab. Andes Quitenses, in sylva Canelos (4000 ped.), *Spruce, n. 85*.

Macromitrium praelongum Mitt., J. Linn. Soc., Bot., 12: 207. 1869, *fide* Wijk *et al*. 1964. Type: Hab. Ins. Sti. Vincentis, in devexis montis S. Andrew, Guilding; Jamaica, Wilson; S. Christopheris, Breutel; Dominica, Herb. Hooker.

Macromitrium pseudocirrosum Müll. Hal., Hedwigia 37: 237. 1898, *fide* Wijk *et al.*, 1964. Type protologue: Portorico, sine loco speciali, ad truncus filicum arborescentium: *P. Sintenis* in Hb. Mönkemeyeri, qui misit 1886.

Macromitrium schwaneckeanum Hampe, Linnaea 25: 360. 1852[1853], *fide* Grout 1944. Type protologue: Ad arbor. trunc.

Macromitrium stenophyllum Mitt., J. Linn. Soc., Bot. 12: 215. 1869, *fide* Churchill & Linares, 1995. Type protologue: Hab Ins. Jamaica, *Wilson, n. 607, in. Herb.Hookers*; etiam ex India occidentali, an Jamaica?, *Herb. R. Brown*.

Macromitrium substrictifolium Müll. Hal., Bull.Herb. Boissier 6: 98. 1898, *fide* Valente *et al.*, 2020. Type protologue: Brasilia, Rio de Janeiro, Tijuca, *E. Ule 1672* (syntypes: US 02482517, R 00014297).

Macromitrium werckleanum Thér., Recueil Publ. Soc. Havraise Études Diverses, 1921: 307. 1. 1921, *fide* Allen,

2002. Type protologue: Costa Rica, la Palam, prés San josé, alt. 1500 m, C. Wercklé, arvil 1910.

(1) Plants medium-sized, yellowish green; stems moderately to strongly creeping, branches 20-50 mm long. (2) Branch leaves erect below, flexuose spreading, spirally contorted-twisted when dry, spreading when moist, narrowly long lanceolate, lanceolate to oblong-lanceolate, 2-4 mm long; apices acute, shortly acuminate to acuminate; upper margins serrate, sometimes with a weak limbidium; costas shortly excurrent or percurrent; the cells near the apex irregularly oblong or rectangular; upper cells subquadrate, quadrate, rounded-quadrate, somewhat longer than wide, thick-walled, often in longitudinal rows, clear, smooth to weakly bulging or mammillose, gradually elongate from medial part to the base; basal cells linear, thick-walled, porose, weakly to strongly tuberculate. (3) Perichaetial leaves shorter than branch leaves, lanceolate from a widely oblong low part; all cells longer than wide, smooth, strongly porose at low and basal portions. (4) Setae 5-15 mm long, smooth, not twisted or slightly twisted to the left beneath the capsule. (5) Capsule urns cupulate to hemispheric, smooth or weakly furrowed at neck; peristome double, exostome teeth united to form a short membrane, yellow and distinctively papillose, endostome segments 16, hyaline, papillose; spores anisosporous, 14-30 µm, papillose. (6) Calyptrae mitrate and naked, 3 mm long, deeply lacerate.

Mitten (1869) described *M. jamaicense* (Arn.) Müll. Hal. based on Wilson's material collected from Jamica (type protologue: Jamaica, *Wilson 832*; ex insulis Indiae occidentalis). Grout (1944) treated it as *M. cirrosum* var. *jamaicense* because the species was essentially the same as *M. cirrosum*, only the upper leaf cells in *M. jamaicense* were arranged in rows, irregular and small in diameter, and its basal marginal leaf cells narrower than the medial cells. Allen (2002) remarked that *M. cirrosum* is a common species in Central America, extremely varying in nearly all its features. Additionally, *Macromitrium werckleanum* Thér. was placed in synonymy with *M. cirrosum* var. *jamaicense* in 1944 by Grout, but also synonymized with *M. cirrosum* by Allen (2002). Considering a high morphological variation of *M. cirrosum*, *M. jamaicense* is rather likely conspecific with *M. cirrosum*.

Based on the specimens collected from Bolivia, Müll (1897) described *Macromitrium erectopatulum* Müll. Hal., and a variety, var. *grossirete*. The variety differed from var. *erectopatulum* by its more mammillose leaf cells. Considering *M. erectopatulum* had been synonymized with *M. cirrosum* by Grout (1944), here we suggest placing *M. erectopatulum* var. *grossirete* in synonymy with *M. cirrosum*.

Distribution: Belize, Bolivia, Brazil, Costa Rica, Honduras, Jamaica, St. Vincent, Suriname, United States, Venezuela, Dominica (Allen, 2002).

Specimens examined: **BELIZE**. Cayo, *B. Allen 15536* (MO 3981467), Toledo, *B. Allen 18767* (MO 4415502); Toledo, Bladen Nature Reserve, *A.T. Whittmore et al 6441* (MO 5910525). **BOLIVIA**. Tunari, *James A. Duke s.n.* (MO 6001077); Santa cruz, *A. Fuentes 1253* (MO 5141413); Depto. La Paz, Prov. Inquisivi. *Marko Lewis 89-830 d-3* (MO 3962958). **BRAZIL**. São Paulo, *Schiffner 314* (H 3204889). **COSTA RICA**. Puntarenas, Monteverde. *M. J. Lyon 128* (MO 4443509); Cerro Zurquí, *Paul C. Standley 50515* (as *M. werckleanum*) (H-BR 2632037). **HONDURAS**. Cortes, Cusuco National Park. *B. Allen 14136* (MO 3972044). **JAMAICA**. John Crow Peak, *W. Harris s.n.* (H 3090174); Sir John Peak, *W. Harris s. n.* (H 3090177); Maccasncker Bump, St. Thomas, *William R. Maxon 9544* (as *M. jamaicense*) (H-BR 2632014). **ST. VINCENT**. Anon. Nr.6910a (H-BR 2634007). **SURINAME**. Sipaliwini, Tafclberg, *B. Allen 23159* (MO 5648868), *B. Allen 23480* (MO 5644737), *B. Allen 23161* (MO 5644734), *B. Allen 20704* (MO 5123493), *B. Allen 23160* (MO 5644659), *B. Allen 23735* (MO 5644657), *B. Allen 23605* (MO 5644658), *B. Allen 20506* (MO 5123510) , *B. Allen 20703* (MO 5123492), *B. Allen 20684* (MO 5123227), *B. Allen 20667* (MO 5123523), *B. Allen 23438* (MO 5644736), *B. Allen 20681* (MO 5123525), *B. Allen 23106* (MO 5644662), *B. Allen 20718a* (MO 5123497), *B. Allen 20670a* (MO 5123524), *Casado 484* (MO 5629981), *Casado 566* (MO 5629982). **UNITED STATES**. Puerto Rico: Rio Grande Co. *M. J. Price 915* (MO 5138740). **VENEZUELA**. Territorio Federal Amazonas, Sierra Parima, *Julian A. Steyermark 105985* (MO 2554330, MO 2554217), Monagas, *R.A. Pursell 8980a* (MO 4453060).

39. *Macromitrium clastophyllum* Cardot, Beih. Bot. Centralbl. 17: 12. 7. 1904. (Figure 71)

Type protologue: [Korea] Seoul (*no. 94*, ster.). Type citation: Coree: Seuol, leg. *Fauriei, no. 094*, Juin, 1901 (isotype: H-BR 2572020!).

(1) Plants small, forming dense mats, brown-yellowish above, brown-blackish below; stems long creeping, with erect to ascending branches, usually up to 4-8 mm high. (2) Branch leaves strongly obliquely appressed, spirally coiled around the branch when dry, 2.6-4.0 × 0.25-0.35 mm, long and narrowly lanceolate, gradually narrow to form a long fragile aristate or subula, aristae of old leaves easily broken; upper and medial cells small, about 4.0-5.5 µm wide, quadrate-rounded, rather obscure, pluripapillose; low cells rounded-quadrate, clear and smooth; basal cells 9.0-18.0 × 4.0-6.0 µm, elongate-rectangular, smooth; costa end in or just beneath the apex, often nearly filling the acumen, sometime excurrent.

Morphologically, *M. clastophyllumn* Cardot species is similar to *M. ferriei* Cardot & Thér., but differing from the latter in having rather narrow and long leaves with easily broken upper parts and basal smooth lamina cells.

Distribution: China, Indonesia, Papua New Guinea, South Korea. Habitat: on rocks.

Taxonomy

Specimens examined: **CHINA**. Liaoning, *Y.C. Zhu 940 (006445)* (IFP orig., ALTA dup. as *Groutiella tomentosa*). **INDONESIA**. Batudulang, *Lostermans 18316* (H 3068936). **PAPUA NEW GUINEA**. New Britain, *H. Streimann 41618* (H 3207145).

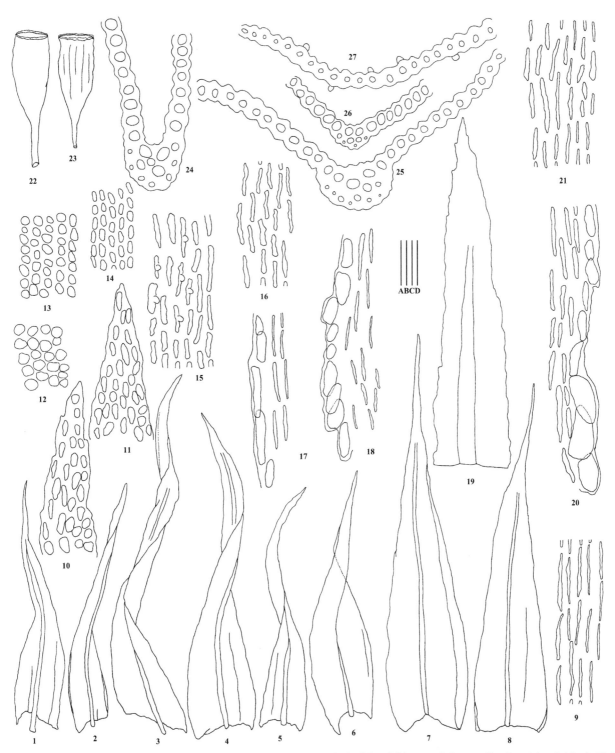

Fig. 68 *Macromitrium cataractarum* Müll. Hal. 1-6: Branch leaves. 7-8: Perichaetial leaves. 9: Low cells of perichaetial leaf. 10-11: Apical cells of branch leaves. 12: Upper cells of branch leaf. 13-14: Medial cells of branch leaves. 15-16: Low cells of branch leaves. 17-18: Basal marginal cells of branch leaves. 19: Apex of branch leaf. 20: Basal cells of perichaetial leaf. 21: Upper cells of perichaetial leaf. 22-23: Capsules. 24: Medial transverse section of branch leaf. 25, 27: Basal transverse sections of branch leaves. 26: Upper transverse section of branch leaf (all from isotype, G 00050744). Line scales: A = 2 mm (22-23); B = 1 mm (1-8); C = 100 μm (19); D = 67 μm (9-18, 20-21, 24-27).

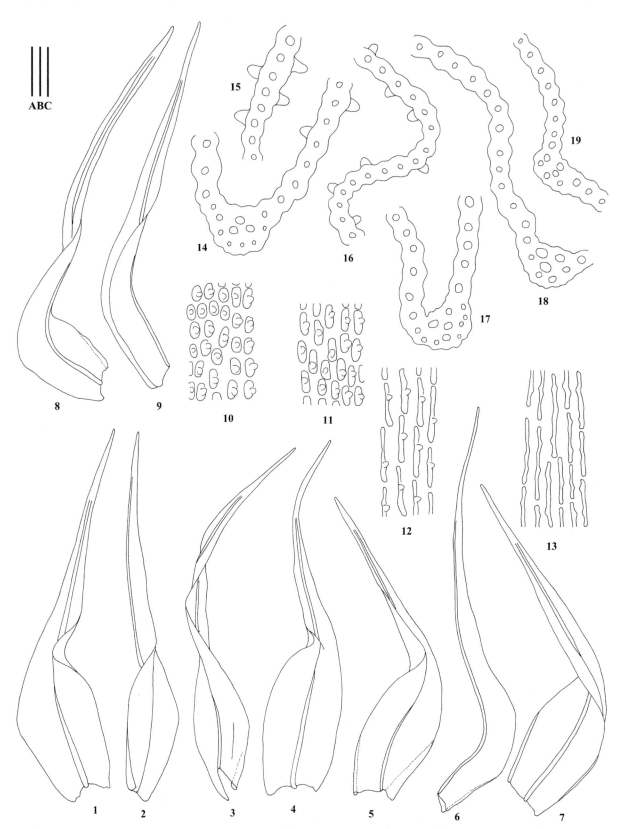

Fig. 69 *Macromitrium catharinense* Paris 1-9: Branch leaves. 10: Upper cells of branch leaf. 11: Medial cells of branch leaf. 12: Low cells of branch leaf. 13: Basal cells of branch leaf. 14, 16: Low transverse sections of branch leaves. 15: Medial transverse section of stem leaf. 17-19: Basal transverse sections of branch leaves (1-3, 6, 8 from the lectotype of *M. catharinense*, H-BR 2649005; 4, 5, 7, 9-19 from the holotype of *M. drewii*, US 00070256). Line Scales: A = 0.5 mm (1-9); B = 50 μm (10-13); C = 33 μm (14-19). (Li *et al.*, 2019).

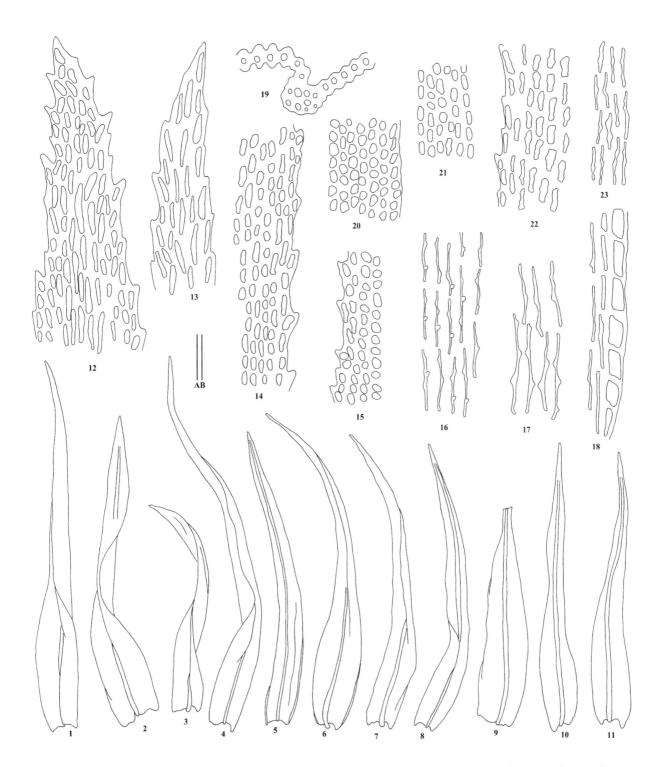

Fig. 70 *Macromitrium cirrosum* (Hedw.) Brid. 1-8: Branch leaves. 9-11: Perichaetial leaves. 12, 13: Apical cells of branch leaves. 14, 15: Upper cells of branch leaves. 16: Basal cells of branch leaf. 17: Basal cells of perichaetial leaf. 18: Basal marginal cells of branch leaf. 19: Upper transverse section of branch leaf. 20, 21: Medial cells of branch leaves. 21: Medial marginal of branch leaf. 22: Low marginal cells of branch leaf. 23: Upper cells of perichaetial leaf (all from isosyntype, E 00002459) Line scales: A = 1 mm (1-11); B = 67 μm (12-23).

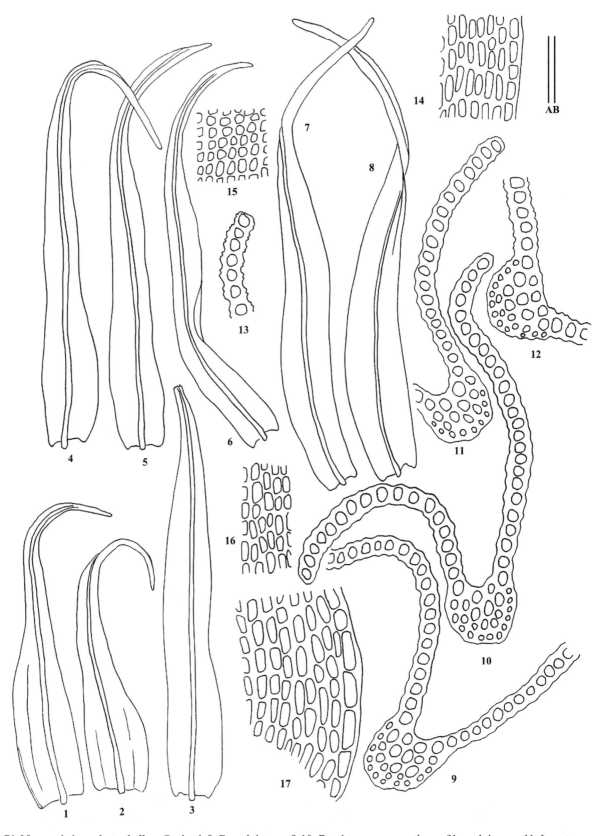

Fig. 71 *Macromitrium clastophyllum* Cardot 1-8: Branch leaves. 9-10: Basal transverse sections of branch leaves. 11: Low transverse section of branch leaf. 12-13: Upper transverse sections of branch leaves. 14, 17: Basal marginal cells of branch leaves. 15: Low cells of branch leaf. 16: Basal cells of branch leaf (1-3, 9-10, 15, 17 from isotype, H-BR 2572020; 4-8, 11-14, 16 from Liaoning, *Y. C. Zhu 940* (006445), ALTA). Line scales: A = 0.50 mm (1-8); B = 50 μm (9-17). (Guo *et al.*, 2007).

40. *Macromitrium comatum* Mitt., Trans. Linn. Soc. London, Bot. 3: 163. 1891. (Figure 72)

Type protologue: [Japan] Umagayeshi to Chiusenji, in loose tufts with Lasia fruticella and Anomdon tritis; fruit immature in September, *Bisset*. Type citation: Japan, Umagayeshi to Chiusenjj, Sept. 1886, *J. Bisset* (holotype: NY 00518291!; isotype: NY 00518290!).

Macromitrium nipponicum Nog., J. Hattori Bot. Lab. 20: 281. 2. 1958, *fide* Noguchi & Iwatsuki, 1989. Type protologue and citation: [Japan] Nagano Pref.: Mt. Ontake, Nog. (holotype).

(1) Plants medium-sized, forming dense yellowish-green mats, dark-brown below; stems long creeping, with erect branches; branches densely leaved, up to 15 mm long, with a few branchlets. (2) Stem leaves entire, 1.0-1.4 × 0.25-0.45 mm, spreading when moist, ovate-lanceolate; apices acute, incurved, keeled; costae percurrent. (3) Branch leaves keeled, curly above when dry, widely spreading and slightly incurved above when moist, 2.0-3.0 × 0.4-0.5 mm, oblong, lanceolate-lingulate, slightly plicate below; apices acute, obtuse-acute or broadly acuminate; margins plane or recurved on one side; costa percurrent; upper and medial cells pellucid, rounded-quadrate, hexagonal, 9-12 μm wide, thin-walled, bulging, smooth or slightly pluripapillose; low cells rectangular to sublinear, 22-30 × 6.5-8.5 μm, with thick yellowish brown walls, smooth, not porose. (4) Perichaetial leaves shorter than branch leaves, ovate-oblong, abruptly narrow to an acuminate upper part; costae vanishing far below the leaf apex; cells oblong, thick-walled, smooth or papillose. (5) Setae smooth, 3.5-4.0 mm, vaginulae with numerous papaphyses. (6) Capsule urns ellipsoid-cylindric, smooth; peristome single, exostome teeth linear-lanceolate, translucent, obtuse, densely papillose. (7) Calyptrae large, campanulate, with many long, brown-yellowish hairs.

Macromitrium comatum Mitt. is similar to *M. tosae* Besch., but differing from the latter in having branch leaves with smooth low and basal cells. *Macromitrium comatum* is sometimes confused with *M. japonicum* Dozy & Molk., but *M. japonicum* differs from *M. comatum* in having strongly incurved to contorted dry leaves with apices hidden in the inrolled cavity, as well as wet leaves with adaxially incurved apices.

Distribution: China, Japan.

Specimens examined: **CHINA**. Fujian, *H. H. Chung m4*, NY! (ex Farlow Herbarium); Hubei, *Sino-Amer. Exped. 822* (MO 2844283). **JAPAN**. Honshu, Hiroshima, *H. Ando* (H 3090196, H 3090197); Yamanashi, *T. Osada & N. Suzuki* (H 3090198), *T. Osada & N. Suzuki 1530* (MO 2862896); Iwate, *Z. Iwatsuki* (H 3090199); Kinki, *K. Yamada 77-100* (MO 5146694); Shizuoka, *Z. Iwatsuki s. n.* (MO 4460719).

41. *Macromitrium concinnum* Mitt. ex Bosch & Sande Lac., Bryol. Jav. 1: 132, 110. 1860. (Figure 73)

Type protologue: [Indonesia] Habitat insulam Java Zollinger itin. Jav. cecundi coll. *n° 3716*? Type citation: Zollinger itin Javanicum Secundum *3176 ?* Herb Dozy et Molk, holotype: NY 00518292!).

(1) Plants small; stems long-creeping; branches 5-10 mm long. (2) Branch leaves densely arranged, spirally coiled around the branch, somewhat funiculate when dry, erect-spreading when moist, oblong to lingulate, plicate below; keeled, margins distinctly crenulate above and entire below; upper cells isodiametric, conic-bulging to strongly mammillose; cells slightly elongate towards the base; basal cells short elliptic, strongly tuberculate. (3) Setae erect and smooth, 8-20 mm long. (4) Capsule urns ellipsoid-cylindric. (5) Calyptrae hairy.

Macromitrium concinnum Mitt. ex Bosch & Sande Lac. is similar to *M. blumei* Nees ex Schwägr., only the former has completely smooth setae, strongly papillose upper lamina cells.

Distribution: Indonesia.

Specimen examined: **INDONESIA**. Lombok, *Ecbart 1537* (H-BR 2602005).

42. *Macromitrium constrictum* Hampe & Lorentz, Bot. Zeitung (Berlin) 26: 798. 1868. (Figure 74)

Type protologue: [Ecuador] Hab. In einer sumpfigen Stelleder Cordinllere, au Zweigen herabhängend, 5000-6000' (Standortsverwechselung?). Type citation: Ecuador. einer Stelleder Cordinllere, von Zweigen herabhängend, 5-6000, *Lorentz*, leg *Krause* (holotype: BM 000879966!; isotype: NY 00518301!).

(1) Plants robust, yellowish green above, dark brown below; stems weakly creeping, branches 25-30 mm long, 2 mm thick. (2) Branch leaves appressed below, individually twisted, contorted-flexuose and loosely inrolled above when dry, erectly spreading and occasionally abaxially curved or slightly twisted when moist, lanceolate to narrowly lanceolate, not sheathing at the base; margins entire, with an inconspicuous limbidium; upper and medial cells rounded-quadrate, subquadrate, elliptic, with various sizes, 11-15 μm, bulging, clear and smooth, gradually elongate from medial to base; low cells rectangular, thick-walled; basal cells long to liner-rectangular, 41-81 μm long, thick-walled and porose, strongly tuberculate, marginal enlarged teeth-like cells at insertion not differentiated. (3) Setae smooth, 14-18 mm long, twisted to the left. (4) Capsule urns ovoid to ellipsoid-cylindric, smooth to furrowed; peristome double, exostome teeth united to a membrane, brown yellow, endostome segments hyaline, membranous. (5) Calyptrae mitrate and naked.

Macromitrium constrictum Hampe & Lorentz is similar to *M. cirrosum* (Hedw.) Brid., only differing from the latter in having larger plants, and branch leaves with weak papillae in most cells. *Macromitrium constrictum* is also similar to *M. subscabrum* Mitt., but the latter could be separated from the former by its branch leaves with notched or crnulate upper margins and differentiated border with enlarged outmost marginal cells at the insertion.

Distribution: Bolivia, Ecuador.

Specimens examined: **BOLIVIA**. La Paz, *Marko Lewis 89-830* (MO); *Alfredo F. Fuentes & H. Huaylla 13110* (MO). **ECUADOR**. Azuay, *L.B. Holm-Nielsen & et al. 4960* (MO); Zamora-Chinchipe, *Simon Lægaard 18731A* (MO).

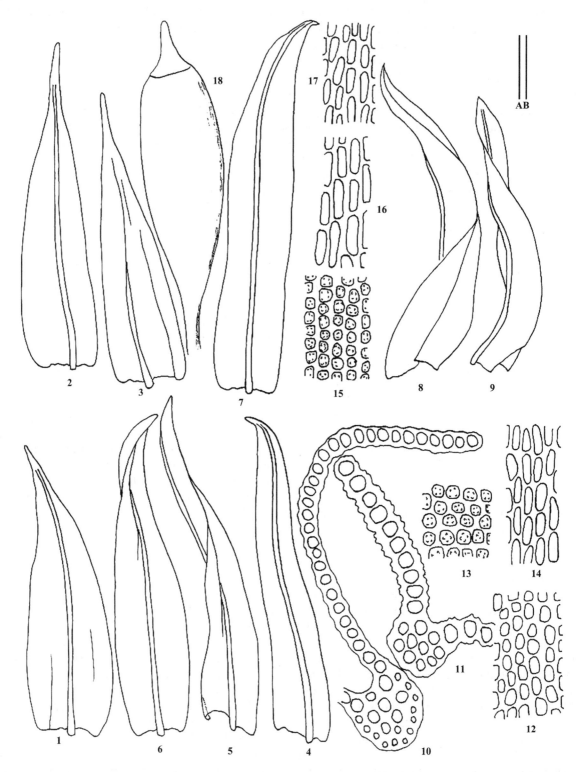

Fig. 72 *Macromitrium comatum* Mitt. 1-3: Perichaetial leaves. 4-9: Branch leaves. 10: Basal transverse section of branch leaf. 11: Upper transverse section of branch leaf. 12: Medial cells of branch leaf. 13, 15: Upper cells of branch leaf. 14, 16,17: Basal cells of branch leaves. 18: Capsule (2-3, 7-9, 16-18 from holotype of *M. comatum* (NY); 1, 4-6, 7-17 from Fukian, *Chung no. m 4* (NY)). Line scales: A = 0.50 mm (1-9, 18); B = 50 μm (10-17).

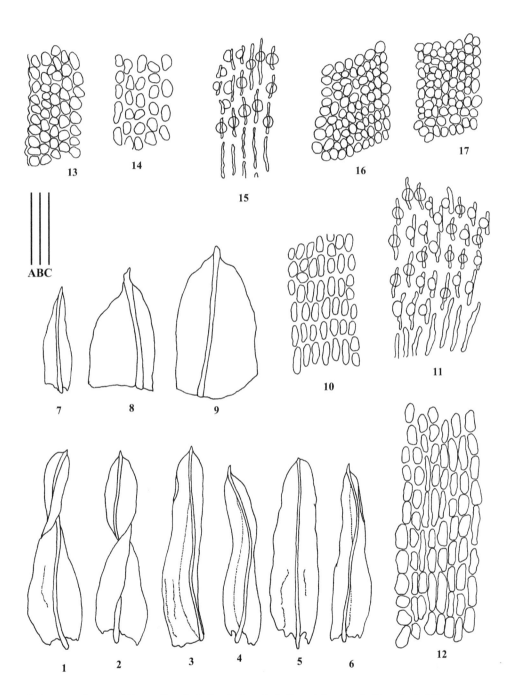

Fig. 73 *Macromitrium concinnum* Mitt. ex Bosch & Sande Lac. 1-6: Branch leaves. 7: Stem leaf. 8-9: Apices of branch leaves. 10: Low cells of branch leaf. 11: Basal cells of branch leaf. 12: Basal marginal cells of branch leaf. 13: Upper cells of stem leaf. 14: Medial cells of stem leaf. 15: Basal cells of stem leaf. 16: Upper cells of branch leaf. 17: Medial cells of branch leaf (all from holotype, NY 00518292). Line scales: A = 1 mm (1-7); B = 200 μm (8-9); C = 50 μm (10-17).

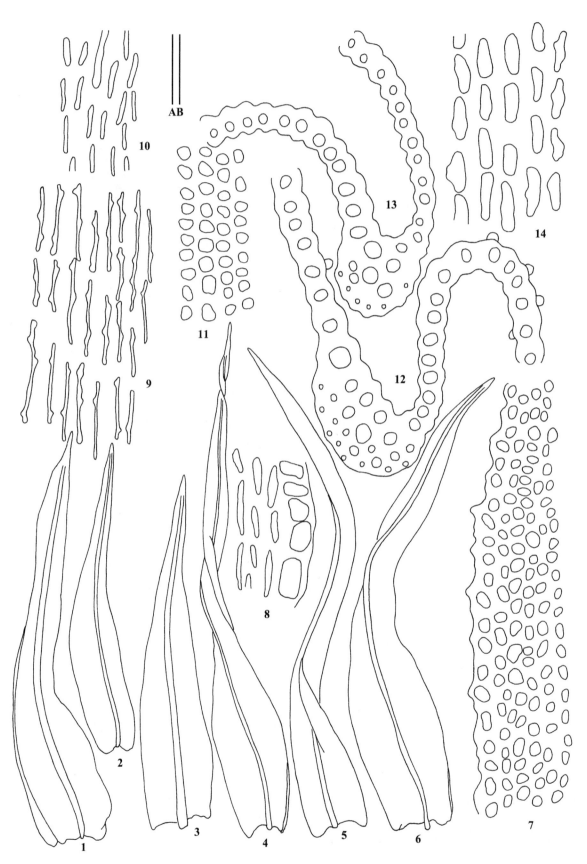

Fig. 74 *Macromitrium constrictum* Hampe & Lorentz. 1-6: Branch leaves. 7: Apical cells of branch leaf. 8: Basal marginal cells of branch leaf. 9: Basal cells of branch leaf. 10, 14: Low cells of branch leaves. 11: Medial cells of branch leaf. 12-13: Low transverse sections of branch leaves (all from isotype, NY 00518301). Line scales: A = 1 mm (1-6); B = 67 μm (7-14).

43. *Macromitrium crassirameum* Müll. Hal., Nuovo Giorn. Bot. Ital., n.s. 4: 125. 1897. (Figure 75)

Type protologue: [Bolivia]... prima linea in species 136 Bryothecae Levierianae Florentinae, a cl. Germain pro Dom. Obertituer (Rennes Galliae) 1889 prope Choquecamata provinciae Cochabamba Bolivianae. Type citation: Bolivia, provincia Cochabamba prope Choquecamata, Jun 1889 Germain, determ. C. Müller *n° 1233* (lectotype designated here: G 00050746!; isolectotypes: H-BR 2625018!, G 00050745!, JE 04008731!).

(1) Plants medium-sized; stems creeping, with erect, long and thick branches, branches simple or occasionally forked. (2) Stem leaves caducous and inconspicuous. (3) Branch leaves loosely spaced, individually twisted, strong contorted-twisted-crisped when dry, spreading or sometimes twisted and conduplicate above when moist, long lanceolate with an acuminate apex, occasionally plicate, and recurved at one side below; costa light rusty red, vanishing beneath the apex; upper and medial cells small, subquadrate, quadrate, or rounded, smooth and clear; low and basal cells elongate, thicken-walled, moderately porose, unipapillose; marginal enlarged teeth-like cells at the insertion somewhat differentiated. (4) Perichaetial leaves differentiated, erect, oblong, long-acuminate and sheathing at base, all cells smooth. (5) Setae smooth, twisted to the right, erect, about 20 mm long. (6) Capsule urns ellipsoid to ovoid, smooth to light furrowed; peristome double, exostome teeth with an obtuse apex, yellowish, endostome teeth linear. (7) Calyptrae mitrate, deeply lacerate, smooth and naked.

Distribution: Bolivia.

44. *Macromitrium crassiusculum* Lorentz, Bot. Zeitung (Berlin) 24: 187. 1866. (Figure 76)

Type protologue and citation: [Chile] Auf sonnigen Klippen am Meeresstrande. Type citation: Ad Mopulos apricos ad mare, prope Corral Puerto de Valdivia, Leg. *Krause s.n.*, det. *P. G. Lorentz*, marked with '*Macromitrium crassiusculum* Lortentz. *n. sp*' (isotypes: NY 00518313!, NY 00518316!); Auf sonnigen Klippen am Meeresstrande, Corral, Puerto de Valdivia. Leg. *Krause, P. G. Lorentz* (isotype: NY 00518314!); Auf sonnigen Klippen am Meeresstrande, *Lorentz* (isotype: NY 01797585!).

(1) Plants small to medium-sized, lustrous and brown, dark brown below; stems weakly creeping, with erect branches; branches up to 15 mm long and 1.5 mm thick, stiff and thick, simple or occasionally forked. (2) Stem leaves caducous and inconspicuous. (3) Branch leaves densely and tightly appressed, obliquely appressed to spirally appressed-curved when dry, erecto-patent when moist, oblong- to lingulate-lanceolate, sporadically bistratose in medial and upper portions; apices acute to apiculate; margins entire, occasionally recurved on one or both sides; costae stout, vanishing several cells beneath the apex; all cells flat, clear and smooth; upper and medial cells rounded-quadrate, oblate; low cells elliptic-rhombic, short oblong; basal cells near costa elongate, oblong-rectangular to narrow rectangular, some cells wider than long, thick-walled, not porose; marginal cells at outmost row large, quadrate to rectangular. (4) Setae smooth, 2-3 mm long, twisted to the left. (5) Capsule urns ellipsoid, furrowed beneath a puckered mouth; peristome single, membranous, light yellow, truncate and lacerated above.

Distribution: Chile.
Specimens examined: **CHILE**. Valdivia, *Dr. H. Krause s.n.* (NY 00518313, NY 00518316)

45. *Macromitrium crinale* Broth. & Geh., Biblioth. Bot. 44: 11. 10. 1898. (Figure 77)

Type protologue and citation: Indonesia. West Irian. Manokwari, Mt. Arfak ad Hatam, 5 000-7 000 ft, VII. 1875, *O. Beccari, no. 186* (lectotype by Vitt *et al*., 1995, H-BR!, isotype: L 0060434!).

Macromitrium mindorense Broth., Philipp. J. Sci. 2: 340. 1907, *fide* Eddy, 1996. Type protologue: [Philippines] On trees, 2000 to 2350 m. alt. (Nos. 5559, 6165). Type citation: (Philippines) MT. Halcon, on trees 6500 ft. Minodor, *no. 5559, Elmer D. Merrill*, Nov. 1906 (syntype: H-BR 2622001!; isosyntype: MO!); Mt. Halcon, on trees, 7800 ft., Minodoro, *Elmer D. Merrill, no. 6165,* Nov. 1906 (syntype: H-BR 2622002!).

(1) Plants rather robust, lustrous, chestnut-brown, darker and reddish below; stems creeping, with stiff, erect branches up to 4 cm long. (2) Stem leaves lanceolate, appressed. (3) Branch leaves erect-flexuose to flexuose-curved, individually flexuose-twisted when dry, erect-spreading to wide-spreading-flexuose when moist, 4.0-6.0 × 1.0-1.5 mm, broadly lanceolate to ovate-oblong, shortly aristate to long filiform-acuminate, sometimes rugose above, plicate below; margins plane above, entire and reflexed below, sometimes crenulate above; costae ending in or at base of the acumen; all cells longer than wide, distinctly porose; upper and medial cells 6-12 × 10-30 µm, rounded-elliptic, short-rhombic, rhombic-rectangular, flat to bulging, smooth, irregularly thick-walled, lumens sigmoid to curved, in longitudinal rows, becoming elongate at margin and forming an indistinct border; basal cells 12-18 × 35-70 µm long, long-rectangular, lumens 2-4 µm wide, curved to sigmoid, very irregularly thick-walled, flat, most cells strongly tuberculate, longer and narrower at margin; cells at the insertion mostly smooth. (4) Perichaetial leaves erect, 3.7-4.0 mm long, ovate-oblong to lanceolate-oblong, long-acuminate; all cells smooth. (5) Setae 22-32 mm long, slender, erect-flexuose, smooth, twisted to the left; vaginulae not hairy. (6) Capsule urns 2.0-3.5 mm long, fusiform to narrowly ovoid-ellipsoid, smooth to lightly 8-plicate at rim, gradually narrowed at the mouth and to the seta; peristome single; exostome membranous, low and coarsely papillose. (7) Calyptrae plicate, mitrate, naked (Vitt *et al.,*

1995).

Macromitrium crinale Broth. & Geh. is similar to *M. ochraceum* (Dozy & Molk.) Müll. Hal., but the medial cells of the latter are tuberculate.

Distribution: Indonesia, Papua New Guinea, Philippines.

Specimens examined: **PAPUA NEW GUINEA**. Morobe, *D. H. Norris 62033* (H 3195892), *D. H. Norris 62026* (MO 4435573), *T. J. Koponen 30238* (H 3195906), *T. J. Koponen 30499* (H 3195911), *T. J. Koponen 30240* (MO 4435574); Eastern Highlands, *W. A. Weber B-34131* (MO 5263300); West Sepik, *A. Touw 18035* (MO 3994821).

46. *Macromitrium crispatulum* Mitt., J. Linn. Soc., Bot. 12: 210. 1869. (Figure 78)

Type protologue: [Ecuador] Andes Quitenses, Antombos ad pedem montis Tunguragua (5000 ped.), *Spruce 86*. Type citation: Antombos, *Spruce 86* (holotype: NY 00518318!).

(1) Plants small to medium-sized, brown-yellow, dark brown when old; stems weakly creeping, branches up to 3.5 cm tall. (2) Branch leaves keeled, individually twisted, contorted-crisped-flexuose, undulate above when dry, spreading and sometimes conduplicate and slightly adaxially curved above when moist, oblong-lanceolate, ligulate-lanceolate; apices acute, apiculate; margins entire; costae stout, percurrent; upper and medial cells rounded-quadrate, subquadrate, moderately bulging, somewhat obscure, smooth; marginal cells not differentiated; low and basal cells rectangular, elongated-rectangular, distinctly unipapillose; marginal enlarged teeth-like cells at the insertion strongly differentiated. (3) Perichaetial leaves differentiated from branch leaves, broadly oblong-lanceolate with the widest at the base, sheathing and plicate below; the apices acuminate-acute with an aristate point; costa stout, disappearing several cells beneath the apex; all cells longer than wide, long- and linear-rectangular, thick-walled, smooth. (4) Setae smooth, 7-18 mm long, twisted to the left, vaginulae hairy. (5) Capsule urns ellipsoid-cylindric, cylindrical, rounded cupped, wide-mouthed, weakly to strongly furrowed; opercula conic-subulate; peristome double, both membranous, exostome teeth papillose, endostome smooth and hyaline. (6) Calyptrae mitrate, deeply lacerate, smooth and naked.

Macromitrium crispatulum Mitt. is similar to *M. regnellii* Hampe and *M. atroviride* R.S. Williams, differing from the latter two species in having branch leaves with almost entire margins, smooth upper and medial cells, perichaetial leaves with cells all longer than wide, and also having ellipsoid-cylindrical, cylindrical capsules.

Distribution: Bolivia, Ecuador.

Specimens examined: **BOLIVIA**. Depto. Santa Cruz. Prov. Ichilo: Cerro Amboro, *Marko Lewis 37854 d-3* (MO 3961439), *Marko Lewis 37853 d-3* (MO 3961431).

47. *Macromitrium crosbyorum* B.H. Allen & Vitt, Novon 8: 113. f. 1. 1998. (Figure 79)

Type protologue and citation: Costa Rica San josé: along Inter American Highway, about 10 km NW of summit at La Ascension, 9°37'N, 83°48'W, *Crosby & Crosby 6089* (holotype: MO!; isotype: US 00603983!); Costa Rica San José: summit of Pan American highway at Cerro de la Muerte, 9°30'N, 83°45'W, *Crosby 3906* (paratype: MO!). PANAMA. Bocas del Toro: Cordillera de Talamanca, 2 airline km SW of the main peak of Cerro Fabrega along the NW ridge of the massif, 9°08'N, 82°53'W, *Davidse et al. 25327* (paratype: MO!).

(1) Plants robust, rusty-brown, reddish brown; stems long creeping, branches to 2-3 cm long, densely with reddish rhizoids below. (2) Branch leaves not very shriveled, undulate, spirally twisted and contorted-flexuose when dry, slightly abaxially curved and erect-spreading when moist, lanceolate to long lanceolate, 4-6×1.0 mm, apices acuminate; margins undulate and serrate above, frequently serrulate near the base, recurved below, plane above; costae percurrent; upper and medial interior cells rounded-quadrate, round hexagonal, rhombic, collenchymatous, bulging and mammillose; upper marginal cells narrower and longer than their ambient cells, forming a weak border; basal cells long rectangular, thick-walled and porose, strongly tuberculate, 26-44 μm long. (3) Perichaetial leaves long lanceolate, recurved at one side below, all cells longer than wide, smooth and clear, basal cells thick-walled and porose. (4) Setae smooth, 7-12 mm long, upright below and twisted to the left above. (5) Capsule urns brown-ochre, ovoid to ellipsoid-cylindric, slightly constricted beneath the mouth, strongly furrow; opercula long-rostrate, 1-1.5 mm long; peristome double, exostome teeth truncate, yellow, densely papillose-striate, united to form a membrane, endostome hyaline, weakly papillose; spores anisosporous, small spores smooth to weakly papillose, large spores densely papillose. (6) Calyptrae large, up to 5 mm long, mitrate, naked, deeply lacerate.*Macromitrium crosbyorum* B.H. Allen & Vitt is similar to *M. aureum* Müll. Hal. in their robust plants, flexuose to spirally contorted and undulate leaves when dry, upper interior cells isodiametric, rounded-quadrate to hexagonal, bulging to mammillose, low and basal cells rectangular and strongly tuberculate, smooth setae, furrowed ovoid capsules. However, *M. crosbyorum* can be separated from *M. aureum* by its branch leaves having collenchymatous cells in the upper interior portion and a differentiated marginal border with narrow and long cells.

Distribution: Colombia, Costa Rica, Panama.

Specimens examined: **COLOMBIA**. Cauca, *S. Churchill & W. Rengifo M. 17306* (MO 4463485, H 3194589).

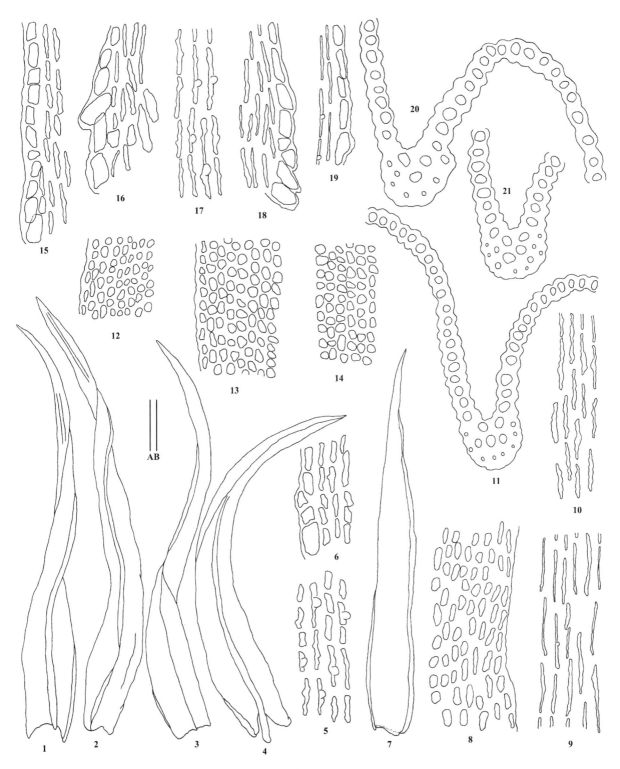

Fig. 75 *Macromitrium crassirameum* Müll. Hal. 1-4: Branch leaves. 5: Low cells of branch leaf. 6, 15-16, 18: Basal marginal cells of branch leaves. 7: Perichaetial leaf. 8: Upper cells of perichaetial leaf. 9: Low cells of perichaetial leaf. 10: Medial cells of perichaetial leaf. 11, 21: Basal transverse sections of branch leaves. 12: Upper leaf cells. 13: Medial marginal cells of branch leaf. 14: Medial cells of branch leaf. 17: Basal cells of branch leaf. 19: Basal cells of perichaetial leaf. 20: Medial transverse section of branch leaf (all from isolectotype, JE 04008731). Line scales: A = 1 mm (1-4, 7); B = 67 μm (5-6, 8-21).

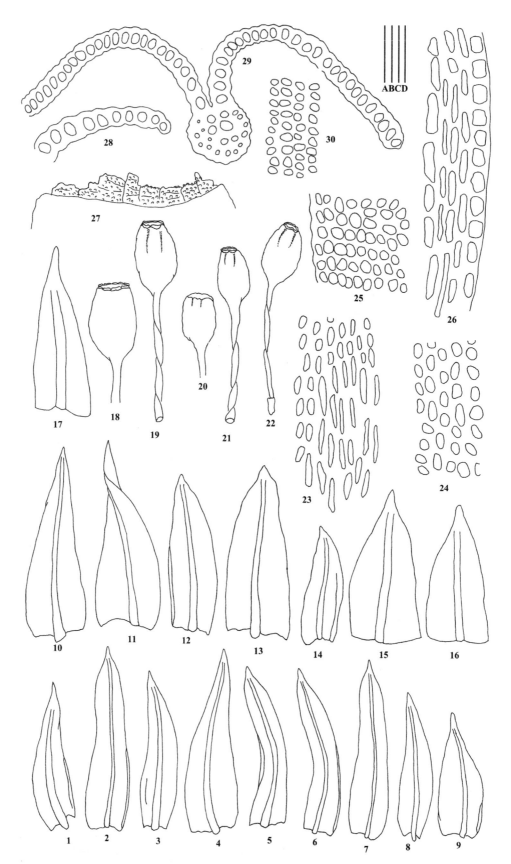

Fig. 76 *Macromitrium crassiusculum* Lorentz 1-11, 14: Branch leaves. 12-13: Perichaetial leaves. 15-17: Apices of branch leaves. 18-22: Capsules. 23: Basal marginal cells of branch leaf. 24: Low cells of branch leaf. 25: Upper cells of branch leaf. 26: Basal cells of branch leaf. 27: Peristome. 28: Upper transverse section of branch leaf. 29 Medial transverse section of branch leaf. 30: Medial cells of branch leaf (all from isotype, NY 00518316). Line scales: A = 2 mm (18-22); B = 1 mm (1-14); C = 400 μm (15-17, 27); D = 67 μm (23-26, 28-30).

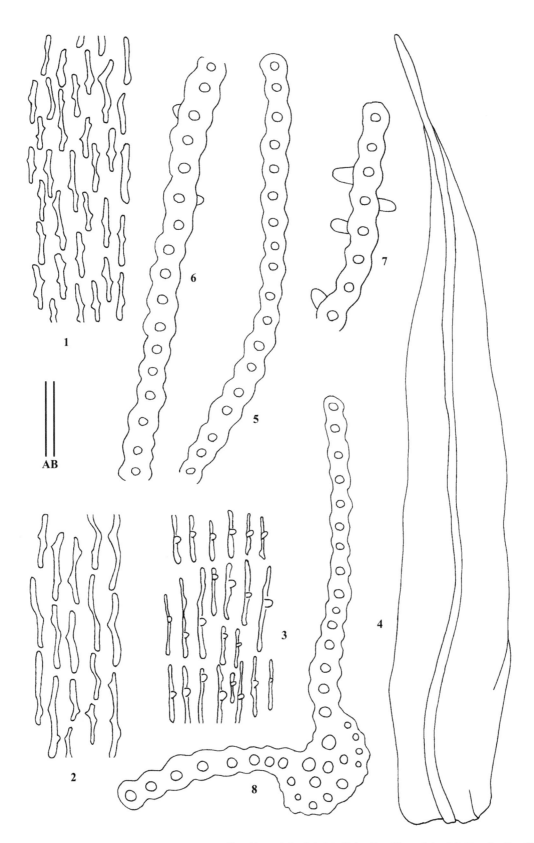

Fig. 77 *Macromitrium crinale* Broth. & Geh. 1: Upper cells of branch leaf. 2: Medial cells of branch leaf. 3: Basal cells of branch leaf. 4: Branch leaf. 5: Medial transverse section of branch leaf. 6: Low transverse section of branch leaf. 7: Basal transverse section of branch leaf. 8: Upper transverse section of branch leaf (all from isotype, L 0060434). Line scales: A = 1 mm (4); B = 67 μm (1-3, 5-8).

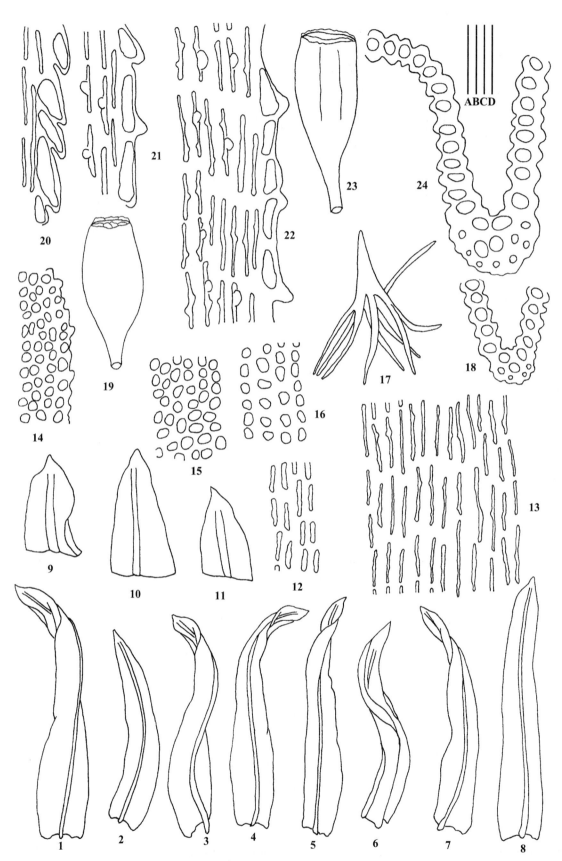

Fig. 78 *Macromitrium crispatulum* Mitt. 1-8: Branch leaves. 9-11: Apices of branch leaves. 12: Low cells of branch leaf. 13: Basal cells of branch leaf. 14: Apical cells of branch leaf. 15: Upper cells of branch leaf. 16: Medial cells of branch leaf. 17: Calyptra. 18: Upper transverse section of branch leaf. 19, 23: Capsules. 20-22: Basal marginal cells of branch leaves. 24: Medial transverse section of branch leaf (all from holotype, NY 00518318). Line scales: A = 2 mm (17, 19, 23); B = 1 mm (1-8); C = 400 μm (9-11); D = 67 μm (12-16, 18, 20-22, 24).

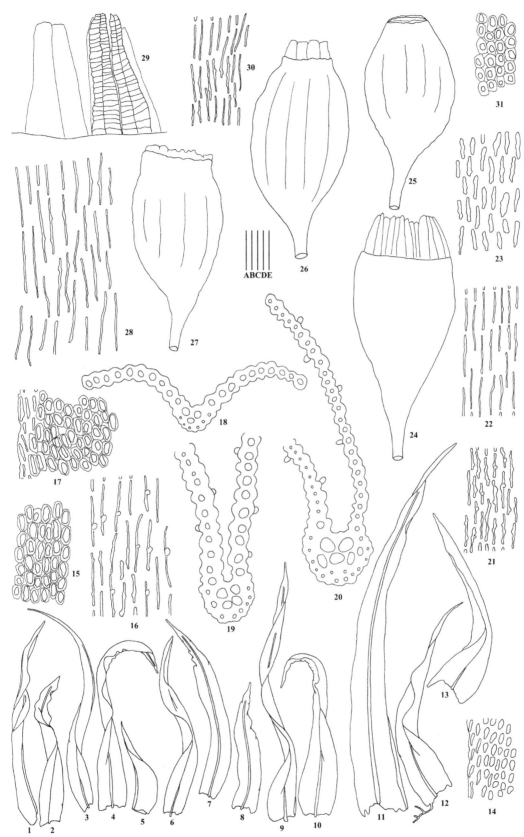

Fig. 79 *Macromitrium crosbyorum* B.H. Allen & Vitt 1-10: Branch leaves. 11: Perichaetial leaves. 12-13: Stem leaves. 14: Upper cells of stem leaf. 15: Medial cells of branch leaf. 16: Basal cells of branch leaf. 17: Upper cells of branch leaf. 18: Upper transverse section of branch leaf. 19: Medial transverse section of branch leaf. 20: Basal transverse section of branch leaf. 21: Basal cells of stem leaf. 22: Medial cells of perichaetial leaf. 23: Upper cells of perichaetial leaf. 24-27: Capsules. 28, 30: Basal cells of perichaetial leaves. 29: Peristome. 31: Medial cells of stem leaf (1-9, 11-31 from holotype, MO 406382; 10 from paratype, MO 406381). Line scales: A = 2 mm (1-10); B = 1 mm (11-13, 24-27); C = 400 μm (29); D = 100 μm (22, 30); E = 67 μm (14-21, 23, 28, 31).

48. *Macromitrium cuspidatum* Hampe, Icon. Musc. (Hampe) pl. 20. 1844. (Figure 80)
Type protologue and citation: Indonesia, In insula Iava legi diligentissimus *Dr. Junguhu*; ab amicissimo Dr. *Gottsche* Communicatum (lectotype: BM!).
Macromitrium elongatum Dozy & Molk., Ann. Sci. Nat., Bot., sér. 3, 2: 311. 1844, *fide* Wijk *et al.*, 1964. Type protologue: Borneo, Sumatra (isotype: H-BR 2617017!).

(1) Plants robust, large, brown-yellowish, weakly glossy; stems prostrate, with blackish rhizoids; branches up to 30 mm long, 1.5 mm thick. (2) Stem leaves small, inconspicuous, lanceolate-acuminate. (3) Branch leaves not very shriveled, keeled, loosely erect below, spreading and divergent, sometimes with deflexed apices (abaxially curved) when dry, widely spreading-recurved when moist, 3.0-4.8 mm long, oblong-lanceolate, ovate-lanceolate to elliptic-lanceolate, apical area mostly asymmetric, sharply contracted to a long cuspidate or a long aristate point (up to 1.5 mm long), sometimes undulate above; margins subentire near apex, plane above and reflexed to recurved and plicate below; costae long excurrent; all cells longer than wide, flat and smooth; upper cells elliptic-rounded, long-oval, 20-30 × 5-7 μm; medial cells short to long rectangular, 18-35 × 5-7 μm, with straight to somewhat curved lumens; basal cells 25-60 × 8-12 μm, long- rectangular, thick-walled, lumens curved to sigmoid, 2-3 μm wide, hyaline, distinctly porose, yellow at the insertion. (4) Perichaetial leaves erect, not much differentiated from branch leaves, lanceolate to acuminate; all cells longer than wide, smooth. (5) Setae 4-5 mm long, smooth, twisted to the left. (6) Capsule urns erect, ovoid, smooth, with a slightly small mouth; peristome single; exostome of 16, well developed, pale, blunt teeth, coarsely papillose; spores anisosporous, coarsely papillose. (7) Calyptrae campanulate, hairy, covering the whole capsule, with smooth and erect hairs, lacerate below.

Distribution: China, Indonesia, Malaysia, Papua New Guinea, Philippines.
Specimens examined: **CHINA**. Hainan, *P. L. Jr. Redfearn 36210* (MO 3974405). **INDONESIA**. Java, Salak, *S. Kurz 845* (H-BR 2620014). **MALAYSIA**. Pahang, *A. Schäfer-Verwimp 18535* (MO 5367862). **PAPUA NEW GUINEA**. West Sepik, *T. J. Koponen 34863* (MO 4435576), *T. J. Koponen 35068* (H 3195918), *T. J. Koponen 34863* (H 3195919). **PHILIPPINES**. Palawan island, *H. M. Curran 3888* (H-BR 2620002); Central Luzon, Bataan, *R. S. Williams 822* (H-BR 2620005).

Macromitrium cuspidatum var. *gracile* Dixon ex E.B. Bartram, Philippines J. Sci. 68: 180. 1939.
Type protologue and citation: [Philippines] Panay, Antique Province, *McGregor 32652* (isotype: US 00070255!).
The variety differs from the type variety in having smaller plants, slender branches, and branch leaves with a strongly crispate hair-point.
Distribution: China, Indonesia, Philippines, Malaysia, New Guinea, Papua New Guinea; Kampuchea (Tan & Iwatsuki, 1993).

49. *Macromitrium cylindricum* Mitt., J. Linn. Soc., Bot. 12: 210. 1869. (Figure 81)
Type protologue: [Ecuador] Andes Quitenses, in Cordillera occidentali in monte Chimborazo, in sylvis Cinchonce succirubrce (4000 ped.), *Spruce, n.82*. Type citation: Ecuador, Mt. Chimborazo, 4000 ft *Spruce* (isotype: E 00165152!).

(1) Plants large, light yellowish brown; stems weakly creeping, branches prostrate and loosely leaved, to 55 mm long. (2) Branch leaves clasping below, strongly abaxially curved, flexuose and divergent above when dry, abaxially curved spreading when moist, long linear-lanceolate, clasping at base; apices long-subulate, narrowly acuminate; margins distinctly serrulate above; costae excurrent into the arista; all cells longer than wide, clear and smooth and clear, slightly bulging; upper marginal cells narrower and longer, differentiated from their ambient cells; upper and apical cells long-oblong, rhombic; basal cells long-rectangular, weakly porose. (3) Perichaetial leaves shorter than branch leaves, acuminate-acute. (4) Setae smooth, 10-15 mm long. (5) Capsule urns ovoid to cupulate, smooth, wide-mouthed; peristome double, exostome teeth brown-yellow, endostome hyaline, membranous at base, segments lanceolate. (6) Calyptrae mitrate, smooth and naked.

Macromitrium cylindricum Mitt. is similar to *M. fuscoaureum* E.B. Bartram, only differs from the latter in having ovoid to cupulate capsules. These two species are likely conspecific. *Macromitrium cylindricum* is also similar to *M. echinatum* B.H. Allen, but differs from the latter in having naked calyptrae and smooth basal cells.
Distribution: Ecuador.

50. *Macromitrium densum* Mitt., J. Proc. Linn. Soc., Bot., Suppl. 1: 51. 1859. (Figure 82)
Type protologue and citation: Nepal, *Wallich s.n.* (lectotype: NY!; isolectotypes: BM 000852422 (H666b)!, BM 000852423(H666b)!, BM 000852425(H666b)!, E 00165193(H666b)!).
Macromitrium brevissimum Dix., J. Siam Soc. Nat. Hist. Suppl. 9: 19. 1932, *fide* Yu *et al.*, 2014. Type protologue and citation: (Thailand) Hab. Udawn. Nakawn Panom, Muk Dāhān, Siam, circa 200 m. alt., on rock in open deciduous

forest, Feb. 1924; *A. F. G. kerr* (74) (holotype: Hb Dixon, BM 000845010!), isotype: Hb Kerr, BM 000825423!); Lôi, Wang Sapung, circa 300 m. alt., on rock in deciduous forest, Mar., 1924; *A. F. G. Kerr* (*79*) (isoparatypes: BM 000825420!, BM 000825424!).

(1) Plants small, in large, dense, compact, spreading mats; stems covered by numerous rhizoids, with short prostrate shoots giving rise to densely erect branches; branches up to 4-6 mm tall. (2) Branch rather short, densely leaved, leaves spirally twisted around the branch and weakly flexuose above when dry, erect-spreading to wide-spreading when moist, lanceolate-linguate to short-lingulate, 1.0-1.5 × 0.3-0.4 mm, with a rounded, mucronate and distinctly cucullate apex, strongly rugose and plicate; margins entire; costae short-excurrent with 1-2 cells extending into the mucro, keeled at leaf base; upper and medial cells rounded- quadrate, 5-10 μm, strongly bulging to mammillose; basal cells short-rhombic to elliptic-rhombic, 10-32 × 6-10 μm, unipapillose to tuberculate. (3) Perichaetial leaves erect to erecto-patent, similar to branch leaves. (4) Setae smooth, 6-9 mm long, twisted to the left. (5) Capsule urns ovoid, ellipsoid-ovoid to short-cylindric, contracted beneath the mouth; peristome single, exostome teeth fused into a continuous membrane, erect to recurved at the top in both dry and wet conditions; teeth reticulate-papillose on both dorsal and ventral sides; spores anisosporous. (6) Calyptrae mitrate, plicate, lacerated at the base, naked or scabrous above, almost cover the whole capsule.

Macromitrium densum Mitt. is somewhat similar to *M. swainsonii* (Hook.) Brid.. from Central and Southern America, but differing from the latter in the absence of marginal enlarged teeth-like cells at the insertion, and in having cucullate leaf apices. *Macromitrium densum* has often been mistaken for *M. concinnum* Mitt. ex Bosch & Lac. for their similarity in plant size, branch leaf shape and the bulging and smooth upper leaf cell features. By its rather rugose branch leaves with cucullated apices, and shorter and smooth setae, *M. densum* could be easily separated from *M. concinnum* (Yu *et al.*, 2014).

Distribution: India (Gangulee, 1976), Myanmar (Tan & Iwatsuki, 1993), Nepal, Peninsular Malaysia, Thailand (Northern, Northeastern, Southeastern), Vietnam (Yu *et al.*, 2014).

Habitat: on tree trunks and branches, or on granitic rocks or sandstones, usually in rather open habitats.

Specimens examined: **INDIA**. Tamil Nadu, Tirunveiveli, Narakkad, *A. E. D. Daniels 3537* (MO 6368663). **NEPAL**. *Wallich, s. n., H.6666.b* (BM 000852425 as *M. densifolium*); *N. Wallich, s. n., H.6666.b* (BM 000745391). **THAILAND**. Pu Wieng, Kaunkan, *A. F. G. Kerr 461* (BM 00825421); Aran Pratet, *A. F. G. Kerr 546* (BM 00825422); Tumpawa, *A. F. G. Kerr 463* (BM 00825427); Sitan, Loi, Udawn, Siam, *A. F. G. Kerr 563* (BM 00825426); Bo Luang, *K. Larsen et al. 2224* (MO 3971188, H 3090208); *K. Larsen et al. 2756a* (MO 3971189); Mae Sariang, *K. Larsen et al. 2266a* (MO 3971190); Phitsanulok, Tung Salaeng Luang, *Larsen, Smitinand & Warncke 860* (KRAM-B-100382, H 3090207); Northern, *Larsen, Smitinand & Warncke 2224* (H 3090208).

51. *Macromitrium diaphanum* Müll. Hal., Linnaea 37: 151, 1872. (Figure 83)

Type protologue: [Australia] Brishane River Nova Hollandia orient. Ubi Amalie Dietrich 1861 specimina pauca nter *macromitria* alia collegit.

Macromitrium circinicladum Müll. Hal., Hedwigia 37: 145, 1898, *fide* Vitt & Ramsay, 1985. Type protologue: New South Wales, Richmond River: Miss Hodgkinson in Hb. Melbourne 1880.

(1) Plants slender, very dull, in dense, compact, spreading mats; stems creeping, with short, erect, simple branches, branches only 5 mm long. (2) Branch leaves stiffy, appressed and erect-curved to twisted-curved when dry, widely spreading when moist, 2.0-2.5 mm long, broadly oblong-ovate to oblong-lanceolate, irregularly and abruptly narrowed to an irregular, notched awn, the awn flexuose, hyaline, broad below, sometimes short and nearly absent or broken off; margins entire and strongly reflexed; costae excurrent, very strong; upper cells rounded to quadrate-rounded, very obscure, thick-walled, bulging, irregularly 1- to 4-stratose, the layers unevenly thickened, papillae simple or irregularly forked; transitional cells 8-10 μm wide, more clear, uni- to bistratose, strongly unipapillose, irregularly thick-walled, porose; basal cells rhombic-elongate to rectangular, up to 35 μm long, thick-walled, smooth or with occasional scattered tall spinulose papillae; outmost marginal cells thin-walled, shorter and broader than their ambient cells. (3) Perichaetial leaves erect, ovate-lanceolate to broadly oblong-ovate, 2.3-2.5 mm long, ending in a slender hyaline awn; upper cells rhombic, very thick-walled, smooth; basal cells elongate. (4) Setae erect, 4-8 mm long, smooth; vaginulae densely hairy. (5) Capsules exserted, urns fusiform-elliptical, 8-plicate in the upper third; rim darker and narrow; peristome absent; spores anisosporous. (6) Calyptrae conic-mitrate and smooth, with numerous slits, and dense, slender and flexuose hairs.

Distribution: Australia.

Specimens examined: **AUSTRALIA**. German Creek, *W. W. Watts* (H-BR 2559003); N S Wales, *W. Turner 336* (H-BR 2559009); Queensland, *D. H. Norris 38435* (H 3090209, MO 2855670).

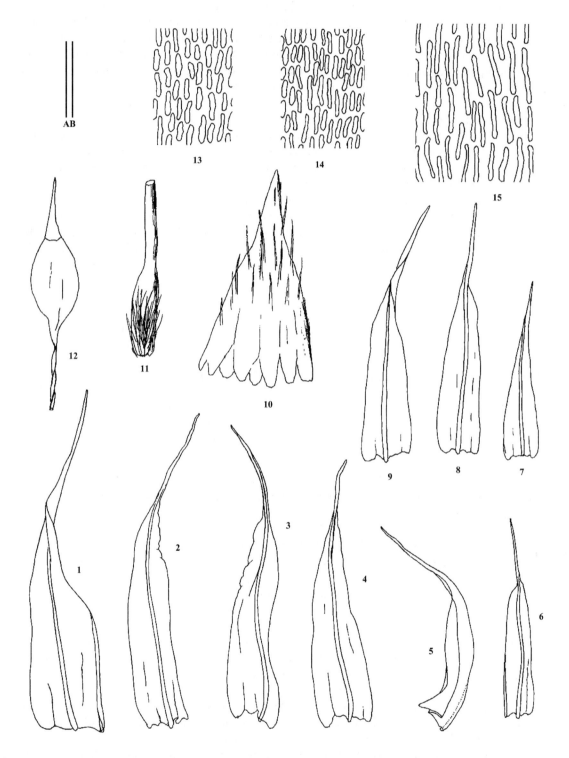

Fig. 80 *Macromitrium cuspidatum* Hampe 1-6: Branch leaves. 7-9: Perichaetial leaves. 10: Calyptra. 11: Vaginula. 12: Dry capsule. 13: Upper cells of branch leaf. 14: Medial cells of branch leaf. 15: Low cells of branch leaf (all from Ling shui, *P. L. Redfearn, Jr.* with *W. D. Reese et al., no. 36210,* MO). Line scales: A = 1 mm (1-12); B = 65 μm (13-15). (Guo *et al.*, 2012).

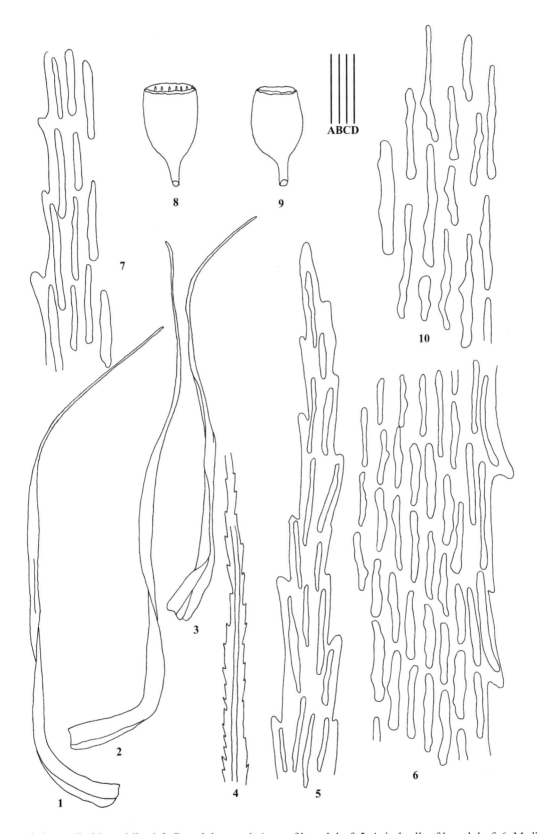

Fig. 81 *Macromitrium cylindricum* Mitt. 1-3: Branch leaves. 4: Apex of branch leaf. 5: Apical cells of branch leaf. 6: Medial cells of branch leaf. 7: Upper cells of branch leaf. 8-9: Capsules. 10: Basal cells of branch leaf (all from isotype, E 00165152). Line scales: A = 4 mm (8-9); B = 2 mm (1-3); C = 400 μm (4); D = 67 μm (5-7, 10).

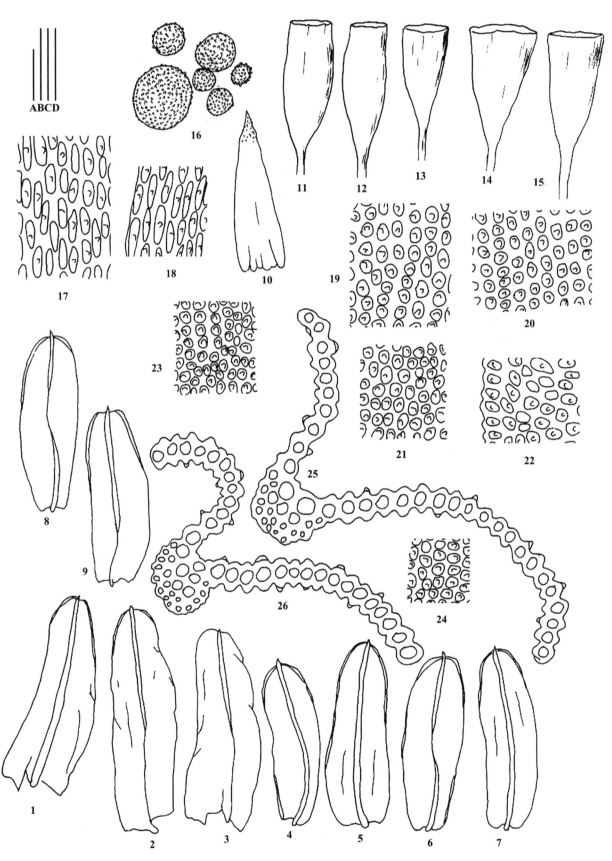

Fig. 82 *Macromitrium densum* Mitt. 1-9: Branch leaves. 10: Calyptra. 11-15: Dry capsules. 16: Spores. 17-18: Low marginal cells of branch leaves. 19-21: Medial cells of branch leaves. 22: Medial marginal cells of branch leaf. 23-24: Upper cells of branch leaves. 25: Medial transverse section of branch leaf. 26: Low transverse section of branch leaf (1-5, 15-17, 20, 22, 24 from BM 000745390; 6-14, 18-19, 21, 23, 25-26 from isolectotype, BM 00825423). Line scales: A = 0.66 mm (11-15); B = 0.44 mm (1-9); C = 44 μm (16-26); D = 0.44 mm (10). (Yu *et al.*, 2014).

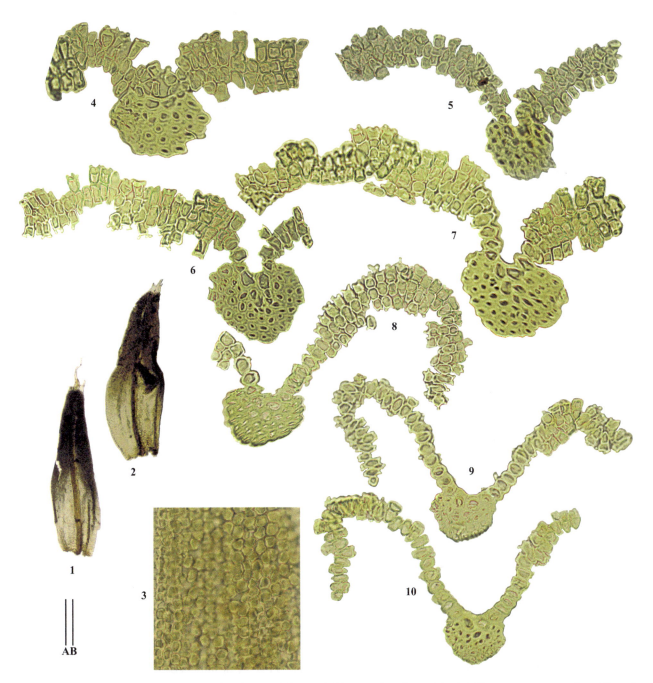

Fig. 83 *Macromitrium diaphanum* Müll. Hal. 1-2: Branch leaves. 3: Upper cells of branch leaf. 4: Upper transverse section of branch leaf. 5-7: Medial transverse sections of branch leaves. 8: Medial transverse section of branch leaf. 9-10: Low transverse sections of branch leaves (all from *D. H. Norris 38435*, H 3090209). Line scales: A = 400 μm (1-2), B = 400 μm (3-10).

52. *Macromitrium dielsii* Broth. ex Vitt & H. P. Ramsay, J. Hattori Bot. Lab. 59: 339. f. 4-21. 1985.

Type protologue: Australia, Queensland. Bellenden ker Range, Centre peak area around Telecom cable car summit area, 17°16'S, 145°51'E, Elfin rain forest of *Dracophyllum sayeri, Leptospermum wooroonooran, Alyxia ruscifolia,* and *Austromyrtus metrosideros, Vitt 27941.*

(1) Plants small, delicate, very lustrous; stems creeping. (2) Stem leaves narrowly ovate-lanceolate and gradually acuminate; costa excurrent and forming an entire subula. (3) Branch leaves regularly twisted with curved to circinate upper portions, somewhat funiculate when dry, narrowly lanceolate, very gradually narrowed to a long, slender, acuminate-subulate apex; costa strong, long excurrent and filling the subulate apex; all cells flat and smooth; upper marginal cells smaller, rounded-quadrate, elliptic-quadrate to short rectangular near costa; medial cells rectangular-elliptic; basal cells elongate to rectangular-elongate, evenly thick-walled and not porose; basal marginal cells broader. (4) Perichaetial leaves erect and loosely sheathing, oblong-lanceolate, attenuate-subulate, very slender above, gradually narrowed to the apex, all cells elongate and smooth. (5) Setae smooth, 12-16 mm long, twisted to the left. (6) Capsule urns ellipsoid to broadly ovoid, strongly 8-plicate and puckered beneath a small mouth; peristome single, exostome of 16, well-developed. (7) Calyptrae mitrate to cucullate, evenly fimbriate below, splitting by slits, naked and smooth (Vitt & Ramsay, 1985).

Macromitrium dielsii Broth. ex Vitt & H.P. Ramsay is similar to *M. microstomum* (Hook. & Grev.) Schwägr. in their ellipsoid to broadly ovoid, strongly 8-plicate capsules with single peristome, smooth and naked calyptrae, flat and smooth lamina cells. However, the branch leaves of *M. dielsii* narrowly lanceolate and very gradually narrowed to a long, slender subulate apex.

Distribution: Australia.

Specimens examined: **AUSTRALIA**. Queensland, I. G. Stone (MEL 2359725, MEL 2241090, MEL 2241074, MEL 2212879).

53. *Macromitrium divaricatum* Mitt., J. Linn. Soc. Bot. 12: 217. 1869. (Figure 84)

Type Protologue and citation: [Ecuador] Andes Quitenses, in Cordillera occidentali loco Lucmas (5000 ped.), *Spruce 87* (isotype: E 00165153!).

(1) Plants medium-sized, yellowish-green; stems weakly creeping, branches up to 20 mm long. (2) Stem leaves caducous and inconspicuous. (3) Branch leaves appressed below, abaxially curved and flexuose above when dry, widely-spreading or moderately abaxially curved spreading when moist, narrowly lanceolate, 2.5-3.5× 0.25-0.35 mm, with an acuminate apex; margins conspicuously dentate to bluntly notched; all cells smooth and clear; upper and medial cells rather small, isodiametric, 3-5 μm wide, limbidium absent, gradually elongate from the upper part to the base; low cells linear-rectangular, 20-35 × 3-4 μm, thick-walled, distinctively porose; basal cells near the costa enlarged, 12-25 × 4-7 μm, marginal cells at outmost row enlarged and hyaline, but not teeth-like. (4) Perichaetial leaves shorter than branch leaves, 2.0-2.5 × 0.25-0.35 mm, ovate-lanceolate, plicate below, gradually and narrowly acuminate to a fine arista; all cells longer than wide, smooth and clear. (5) Setae smooth, 8-12 mm long, twisted to the left. (6) Capsule urns ovoid-ellipsoid to short ellipsoid-cylindric, distinctively furrowed; peristome double, exostome teeth united to form a high membrane, light yellow, separated above, endostome segments lanceolate, discrete, hyaline. (7) Calyptrae mitrate, naked.

Macromitrium divaricatum Mitt. is rather similar to *M. pseudoserrulatum* E. B. Bartram. The latter could be distinguished from the former by its cylindric capsules, pluripapillose upper lamina cells. *Macromitrium divaricatum* is also similar to *M. glabratum* Broth., but the capsules of the latter are crinkled when dry.

Distribution: Colombia, Ecuador.

Specimens examined: **COLOMBIA**. Antioquia Urrao, *John James Pipoly, III no. 17317* (MO); Nariño, Tuquerres, *Bernardo R. Ramírez Padilla 6855* (MO).

54. *Macromitrium diversifolium* Broth., Hedwigia, 34: 126. 1895. (Figure 85)

Type protologue and citation: [Brazil] Goyaz: an Baumstammen des Corumbagebietes. *E. Ule n. 1562.* (holotype: H-BR 2628010!).

Macromitrium divortiarum Sehneum, Pesquisas Bot. 32: 24. pl. 7: b (pp. 68-69). 1978, *fide* Valente *et al.*, 2020. Type protologue: Brazil. Goias: Reserva das Aguas-Emendadas, 800-900 m, Collector and Number: *Sehnem 8605.*

(1) Plants small to medium-sized, reddish-rusty, stems long creeping, densely with brown rhizoids; branches erect, shorter than 10 mm, densely leaved, simple or branched. (2) Branch leaves erect below, keeled, contorted-twisted above, with a curved apex when dry, widely spreading when moist, ligulate to lanceolate-ligulate; apices somewhat cuspidate to mucronate; margins distinctly serrate near the apex, entire below, often recurved at one side below; costae percurrent into a short point, sometimes canceled by overlaping lamine from an adaxial view; upper and medial cells isodiametric, rounded-quadrate, or oblate, flat and clear, unipapillose, arranged in longitudinal rows, the outmost marginal cells somewhat different from their ambient cells; low cells rectangular and smooth; basal cells

elongate, long rectangular with a narrow lumen, thick-walled and slightly porose and strongly tuberculate; cells near the costa at the insertion smooth, enlarge and thin-walled, distinctly different from their ambient cells, forming a "cancellian region"; cells in 8-10 marginal rows at the base smooth, linear-rectangular and not porose, forming an distinct border. (3) Perchaetial leaves not much differentiated from branch leaves. (4) Setae erect, smooth, up to 8 mm long, twisted to the right. (5) Capsules erect, ellipsoid, reddish-brown, smooth. (6) Calyptrae mitrate, naked.

Distribution: Brazil.

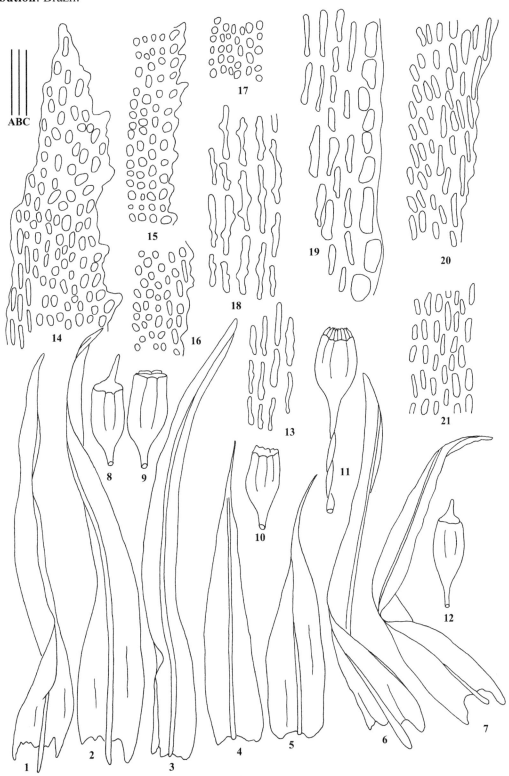

Fig. 84 *Macromitrium divaricatum* Mitt. 1-3, 6-7: Branch leaves. 4-5: Perichaetial leaves. 8-12: Capsules. 13: Basal cells of perichaetial leaves. 14: Apical cells of branch leaf. 15: Upper cells of branch leaf. 16: Medial marginal cells of branch leaf. 17: Medial cells of branch leaf. 18: Low cells of branch leaf. 19: Basal cells of branch leaf. 20: Upper marginal cells of perichaetial leaf. 21: Upper cells of perichaetial leaf (all from isotype, E 00165153). Line scales: A = 4 mm (8-12); B = 1 mm (1-7); C = 67 μm (13-21).

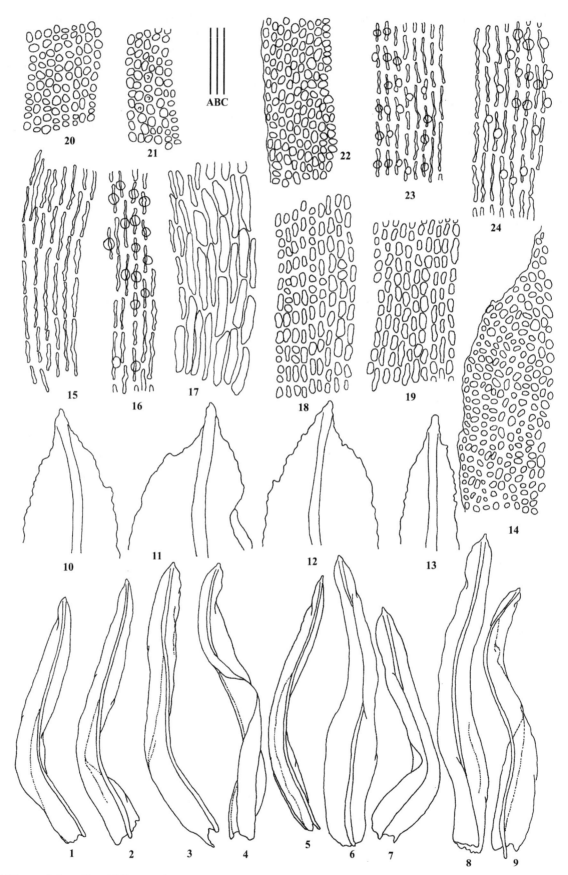

Fig. 85 *Macromitrium diversifolium* Broth. 1-9: Branch leaves. 10-13: Apices of branch leaves. 14: Apical cells of branch leaf. 15-16, 23-24: Basal cells of branch leaves. 17: Basal cells near costa of branch leaf. 18-19: Low cells of branch leaves. 20: Medial cells of branch leaf. 21-22: Upper cells of branch leaves (all from holotype, H-BR 2628010). Line Scales: A = 500 μm (1-9); B = 100 μm (10-13); C = 50 μm (14-24).

55. *Macromitrium dubium* Schimp. ex Müll. Hal., Syn. Musc. Frond. 1: 723. 1849. (Figure 86)

Type protologue: St. Kitts. In truncis arborum montium insulae, Hb. Schimper. Type citation: St. Kitts, 1841, *Anon. s.n.* (isotypes: NY 00667013!, NY 00667014!, NY 00667016!, MICH 525874!).

Macromitrium vernicosum Schimp. Ann. Sci. Nat., Bot., sér. 6, 3: 202, 1876. *fide* Grout, 1944. Type protologue: Guadeloupe (L'Herminier, in herb. Mus. Par.).

(1) Plants medium-sized, reddish brown; stems weakly creeping, branches up to 2.0 cm long. (2) Branch leaves not very shriveled, weakly keeled, appressed to erect below, erect-spreading, slightly flexuose when dry, erect-spreading when moist, 6-9 × 0.5-0.7 mm, linear-lanceolate; the apices long acuminate with a subula; margins plane, entire to weakly serrulate above; costae subpercurrent, vanishing far beneath the apex; all cells longer than wide, rectangular to linear rectangular, thick-walled, porose, flat, clear and smooth; upper cells 35 -50 × 10 µm, long to linear rectangular, gradually elongate towards the base; basal cells 60-70 ×10 µm, linear-rectangular. (3) Setae 15-20 mm long, smooth, somewhat twisted to the left. (4) Capsule urns 2.0-2.5 mm long, short cylindric, ellipsoid- cylindric, smooth, with a wide mouth; peristome absent; spores isosporous, papillose. (5) Calyptrae cucullate, naked, entire at base.

Macromitrium dubium Schimp. ex Müll. Hal. is rather similar to *M. leprieurii* Mont. in their linear, weakly keeled leaves with cells all smooth and clear, as well as longer than wide, together with their cucullate calyptrae. The latter differs from the former by its shorter leaves, with acute to short-acuminate apices, and capsules with rudimentary peristome.

Distribution: Costa Rica, Dominica, Guadeloupe, Panama, St. Kitts and Nevis (Allen, 2002).

Specimens examined: **COSTA RICA**. Guanacaste, la Cruz Canton, *MICH ael H. Grayum & Alexander F. Rojas A. 12847* (MO 6360629). **DOMINICA**. Parish of St. David, *H. A. Imshaug 33226* (H 3090213). **PANAMA**. Province Panama, summit of Cerro jeffe, *B. H. Allen 4915* (H 3090212, MO).

56. *Macromitrium dusenii* Müll.Hal. ex Broth., Bot. Jahrb. Syst. 24: 241. 1897. (Figure 87)

Type protologue: Kamerun (Cameroon): an Baumästen bei Ekundu Etitti (*Dusén n. 73*). Type citation: Kamerun, Ekundu Etitti, 26, 1890, *P. D s.n* (lectotype designated here: H-BR 2627010!).

(1) Plants small, in widely spreading mats, densely with short branches, branches with an obtuse terminal. (2) Branch leaves erect and appressed below, indistinctly spirally twisted above when dry, widely spreading when moist, oblong-lingulate, costa excurrent; apices mucronate; margins recurved at one side at base; upper lamina cellls large and rounded, becoming elongate towards the base; basal cells narrowly rectangular, strongly thick-walled, tuberculate. (3) Perichaetial leaves not much differentiated from branch leaves. (4) Setae about 5 mm long, twisted to the right, smooth. (5) Capsules small and erect, ovoid and smooth. (6) Calyptrae densely with yellow airs.

Distribution: Cameroon.

57. *Macromitrium echinatum* B.H. Allen, Novon 8: 115. f. 2. 1998. (Figure 88)

Type protologue and citation: Panama, Bocas del Toro: vicinity of Fortuna Dam, 2.8 road-miles along pipeline road leaving Chiriqui Grande road at Continental Divide, 8.55' N, 82.08' W, B. *Allen 5655* (holotype: MO!); *B. Allen 5688* (paratypes: MO!, H 3194059!); Prov. Bocas del Toro: Fortuna Dam region, along pipeline service road, c. 8°45' N, 82°15' W, B. *Allen 7846G* (paratype: MO!).

(1) Plants rather robust, yellowish-green above, brown below, lustrous; stems weakly creeping, branches up to 50 mm long. (2) Branch leaves keeled, erecto-patent below, undulate, flexuose, and weakly twisted above when dry, distinctly squarrose-spreading to abaxially curved when moist, 7-10 × 0.8-1.0 mm, long linear-lanceolate; apices setaceous-acuminate; margins revolute at base, plane above, serrate in upper part; all cells longer than wide, smooth and clear, thick-walled, distinctly porose; upper cells linear rhomboidal, upper marginal cells longer and narrower than their ambient cells, forming a differentiated border; basal cells linear-rectangular to linear-rhomboidal, up to 66 µm long; marginal enlarged teeth-like cells at insertion not differentiated; costae long-excurrent. (3) Perichaetial leaves broadly lanceolate, abruptly narrowed in the upper part, with an acuminate apex. (4) Setae smooth, 5-6 mm long. (5) Capsule urns ovoid, ellipsoid, ellipsoid-cylindric, 2 mm long, smooth or weakly to moderately furrowed when dry, wrinkled at neck; peristome double, exostome teeth triangular lanceolate, united at base to form a short membrane, discrete above, densely papillose, endostome segments hyaline, light yellow, weakly papillose; spores anisosporous. (6) Calyptrae 3-4 mm long, mitrate, densely hairy, deeply lacerate.

Macromitrium echinatum B.H. Allen is somewhat similar to *M. fuscoaureum* E.B. Bartram, but differing from the latter by its leaves not clasping at base, densely hairy calyptrae, and shorter setae. *Macromitrium echinatum* and *M. dubium* Schimp. ex Müll. Hal. are similar in their smooth cells all longer than wide, the latter could be separated from the former by its gymnostomous capsules and its branch leaves with costae vanishing far before the apex.

Distribution: Panama, Peru.

Specimens examined: **PANAMA**. Veraguas, *Robbin C. Moran 4016A* (MO 5244202); Bocas del Toro, *B. H. Allen 57168* (MO 3059582). **PERU**. Cajamarca: San Ignacio, *Jasmin Opisso et al. 628* (MO 5275520).

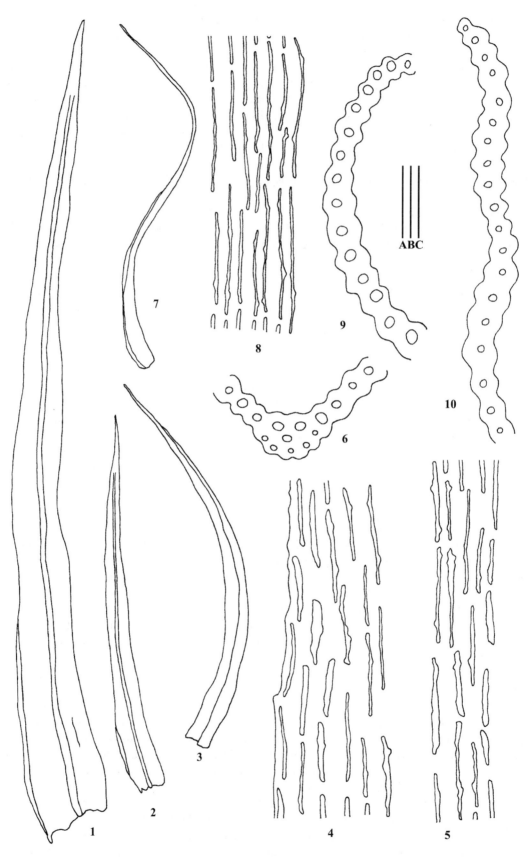

Fig. 86 *Macromitrium dubium* Schimp. ex Müll. Hal. 1-3, 7: Branch leaves. 4: Upper cells of branch leaf. 5: Medial cells of branch leaf. 6, 9, 10: Basal transverse sections of branch leaves. 8: Basal cells of branch leaf (1-2, 4-5, 8 from isotype, MICH 525874; 3, 6-7, 9-20 from isotype, NY 00667016). Line scales: A = 2 mm (2, 3, 7); B = 1 mm (1); C = 67 μm (4-6, 8-10).

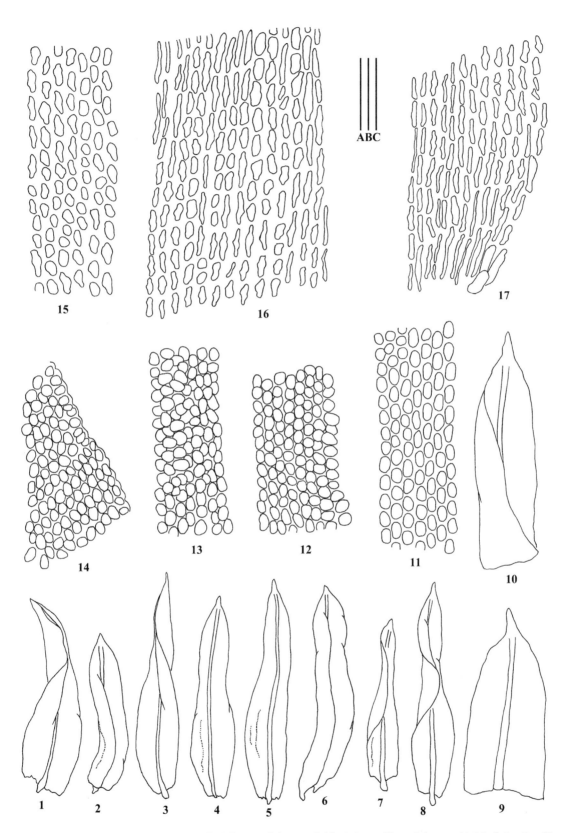

Fig. 87 *Macromitrium dusenii* Müll. Hal. ex Broth. 1-8: Branch leaves. 9-10: Apices of branch leaves. 11: Medial cells of branch leaf. 12-13: Upper cells of branch leaves. 14: Apical cells of branch leaf. 15-16: Low cells of branch leaves. 17: Basal cells of branch leaf (all from lectotype designated here, H-BR 2627010). Line scales: A = 1 mm (1-8); B = 200 μm (9-10); C = 50 μm (11-17).

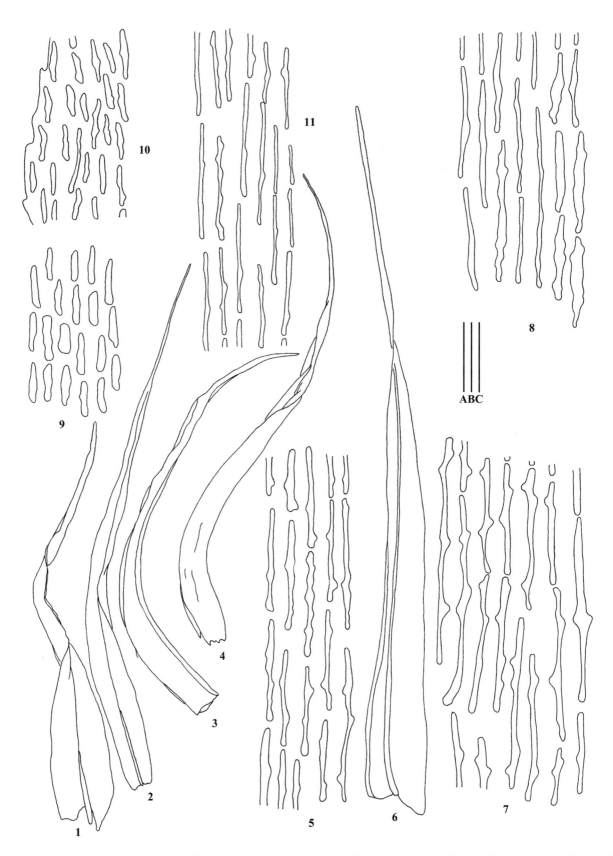

Fig. 88 *Macromitrium echinatum* B.H. Allen 1-4: Branch leaves. 5: Basal ells of perichaetial leaf. 6: Perichaetial leaf. 7: Basal cells of branch leaf. 8: Upper cells of perichaetial leaf. 9: Medial cells of branch leaf. 10: Upper cells of branch leaf. 11: Medial cells of perichaetial leaf (all from holotype, MO 406385). Line scales: A = 2 mm (1-4); B = 1 mm (6); C = 67 μm (5, 7-11).

58. *Macromitrium ecrispatum* Dixon in Herzog, Rep. Spec. Nov. Regn. Veg. 38: 104. 1935. (Figure 89)

Type protologue: [Kenya] Deutsch-Ost-Afrika: Meru, Erica arborea-Gürtel an Stämmen und Zweigen, 3300 m (leg. *C. Troll, n. 5800*). Type citation: D OstAfrika, Meru, Westflauke, der Erica arborea-Geholk, 3300 m, Dec 4 1934, *C. Troll, nr. 5800* (isotype: JE 04008750!).

(1) Plants medium-sized; stems creeping, brown-yellowish above, dark brown below, lustrous. (2) Branch leaves erect below, spreading, slightly contorted-crisped and flexuose or curved above when dry, widely spreading when moist, lanceolate, oblong-lanceolate; apices acuminate, broadly acuminate, or acute; cells near the apex round, subquadrate, smooth and clear; upper and medial cells rounded quadrate, rounded, smooth to pluripapillose; low cells elliptic to short rectangular, unipapillose; basal cells irregularly rectangular, clear and smooth, much larger than medial and upper cells. (3) Perichaetial leaves distinctly differentiated, longer than branch leaves, lanceolate from an oblong low portion, with an acuminate upper portion; all cells longer than wide, clear and smooth. (4) Setae smooth, up to 8 mm. (5) Capsules short ellipsoid-cylindric, ovoid. (6) Calyptrae almost smooth and naked.

Macromitrium ecrispatum Dixon is essentially similar to *M. subtortum* (Hook. & Grev.) Schwägr. However, the former could be separated from the latter by its branch leaves small and broader, not much crisped or contorted, only slightly curved when dry; setae longer, up to 8 mm; and short ellipsoid-cylindric or ovoid capsules.

Distribution: Kenya.

59. *Macromitrium eddyi* B.C. Tan & Shevock, Proc. Calif. Acad. Sci., ser. 4, 62(3): 542. 2015.

Type protologue: Papua New Guinea: Milne Bay District, Raba-Raba Sub-district, bottom of scarp of Tantam Plateau, Mt. Suckling, 1645 m., in shaded forest, common on wood, 20 Jul 1972, coll. *P.F. Stevens* [LAE 55716] (BM!; isotypes: CANB, E, L, LAE). Paratype: Philippines: Mindanao Island: Bukidnon Province, on access dirt road to trail less than 0.5 km above Lantapan Village toward Mt. Dulang-Dulang in Kitanglad Range Natural Park, on trunk of Gmelina in disturbed forest near cultivated field, 20 Apr 2014, *Shevock 44672* (CAS; isoparatypes: CMUH, NY, UC).

(1) Plants slender, dull, yellowish-green above, brown below; stems weakly creeping, with inconspicuous and caduceus leaves; branch prostrate-ascending, thin and up to 50 mm tall. (2) Branch leaves erect below, regularly crisped and occasionally flexuose above when dry, squarrose-recurved and arranged in five straight, longitudinal rows when moist; long lanceolate, gradually acuminate, apices pellucid; costae percurrent to short excurrent; upper cells small and isodiametric, 6-9 μm in diameter, densely pluripapillose and obscure; low and basal cells elongate, strongly thicken-walled, porose, with straight to sigmoid lumina, distinctly tuberculate. (3) Perichaetial leaves differentiated from branch leaves, narrowly triangular-lanceolate, plicate, very finely acuminate and filiform-pointed. (4) Setae short, about 3 mm long, smooth. (5) Capsules small, ca. 1.0 × 0.8 mm, ovoid and smooth; peristome single, exostome triangular, caduceus. (6) Calyptrae naked (Tan & Shevock, 2015).

Macromitrium eddyi B.C. Tan & Shevock is rather similar to *M. savatieri* Besch., but different from the latter by its low and basal cells with tuberculate papillae. *Macromitrium eddyi* is also similar to *M. cardotii* Thér., but the latter has very long setae up to 35 mm long.

Distribution: Papua New Guinea, Philippines.

60. *Macromitrium ellipticum* Hampe, Nuovo Giorn. Bot. Ital. 4: 274. 1872. (Figure 90)

Type protologue: no detail information, with the title: Muscri frondosi in insults Ceylon et Borneo Odoardo Beccari lecti. Type citation: Sri Lanka (Ceylon), *O. Beccari, 186* (lectotype designated here: BM 000852419!; isolectotypes: BM 000852418!, GOTT!).

(1) Plants robust, in dense mat, dark brownish below; stems creeping, densely with branches. (2) Branches erect, 20-25 mm long, often with several short branchlets, densely with leaves. (3) Branch leaves appressed below and twisted-crisped and inrolled above, widely spreading and somewhat undulate above when moist, ovate-lanceolate, oblong-lanceolate to ligulate-lanceolate, undulate and rugose above; apices obtuse or acute; margins minutely denticulate above, and entire below; upper cells small and unipapillose; medial cells irregular short-rectangular, subquadrate, unipapillose; basal cells elongate, bulging, pellucid, almost smooth, thin-walled. (4) Setae smooth, slightly twisted, 8-12 mm long. (5) Capsules ellipsoid, smooth, with a small mouth; peristome double, exostome lanceolate, papillose, broken above, endostome a short membrane, smooth or finely papillose; (6) Calyptrae mitrate, large, covering the whole capsule, deeply plicate, smooth and naked, deeply lacerate below.

Macromitrium ellipticum Hampe is similar to *M. lorifolium* Paris & Broth., but differing from the latter in having ovate- or ligulate-lanceolate branch leaves with acute or obtuse-acute apices, and having smooth capsules.

Distribution: Sri Lanka.
Specimen examined: **SRI LANKA**. Ceylon Am Hunasgiriaspik, *M. Fleischer* (H-BR 2630026).

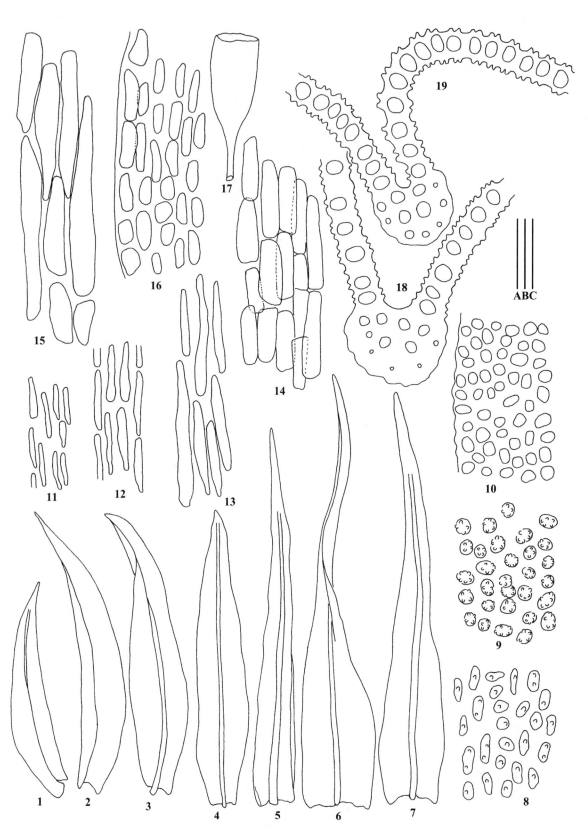

Fig. 89 *Macromitrium ecrispatum* Dixon 1-4: Branch leaves. 5: Inner perichaetial cells. 6-7: Perichaetial leaves. 8: Low cells of branch leaf. 9: Medial cells of branch leaf. 10: Apical cells of branch leaf. 11: Upper cells of perichaetial leaf. 12: Medial cells of perichaetial leaf. 13: Low cells of perichaetial leaf. 14: Basal cells near costa of branch leaf. 15: Basal cells of perichaetial leaf. 16: Basal marginal cells of branch leaf. 17: Capsule. 18-19: Medial transverse sections of branch leaves (all from isotype, JE 04008750). Line scales: A = 2 mm (17); B = 1 mm (1-7); C = 67 μm (8-16, 18-19).

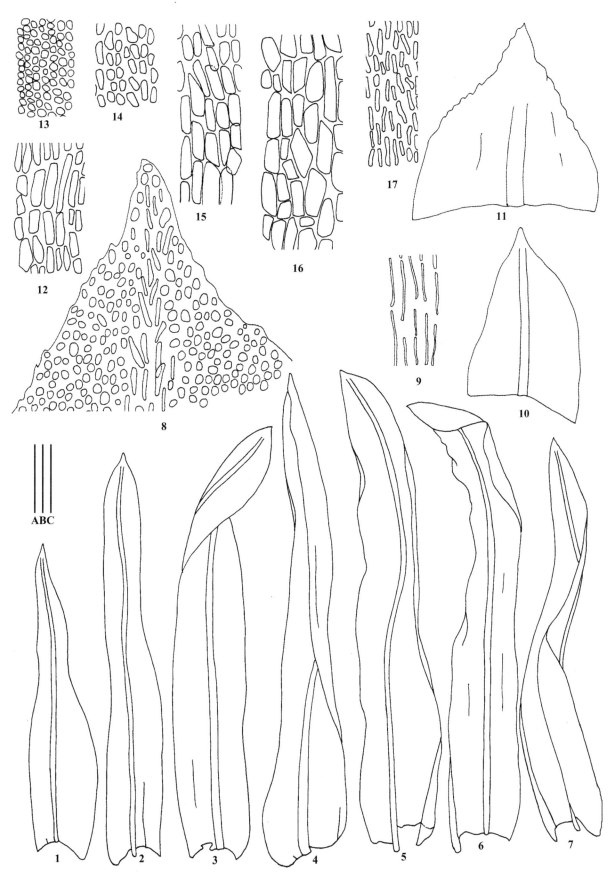

Fig. 90 *Macromitrium ellipticum* Hampe. 1-7: Branch leaves. 8: Apical cells of branch leaf. 9: Basal cells of branch leaf. 10, 11: Apices of branch leaves. 12: Low cells of branch leaf. 13: Upper cells of branch leaf. 14, 17: Medial cells of branch leaves. 15, 16: Basal marginal cells of branch leaves (all from lectotype, BM 000852419). Line scales: A = 0.44 mm (1-7); B = 176 μm (10, 11); C = 44 μm (8, 9, 12-17).

61. *Macromitrium emersulum* Müll. Hal., Flora 82: 452. 1896. (Figure 91)

Type protologue and citation: Hawaiieae, sine loco natali: *Dr. Hillebrand* (isotype: H-BR 2558026!).

(1) Plants small to medium-sized, brown-yellow above, dark brown below, in dense mats; stems strong creeping; branches short, up to 10 mm long. (2) Branch leaves twisted-contorted when dry, widely spreading when moist, oblong-ligulate; apices apiculate; margins weakly crenulate above, entire below; costae subpercurrent, disappearing several cells beneath the apex; upper cells rounded-quadrate, moderately bulging, clear and unipapillose to low and weakly pluripapillose; medial cells subquadrate, irregularly rounded-quadrate, short-elliptic, smooth; basal cells elongate-rectangular, straight-walled, smooth, those at the outmost row enlarged, short-rectangular, different from their ambient cells. (3) Perichaetial leaves narrower lanceolate, acuminate-aristate, easily broken off in the upper portion. (4) Setae shorter, capsules immersed among the perichaetial leaves. (5) Capsule urns ellipsoid, reddish brown, distinctly furrowed when dry. (6) Calyptrae small and mitrate, deeply lacerate, with straight brown-yellow hairs.

Distribution: Hawaii.

Specimens examined: **HAWAII**. W. Maui, *E. B. Bartram 410* (H 3090216); Oahu, *Degener 28.244* (H 3090217); Kauai, *Chani Joseph 3* (MO 6004688).

62. *Macromitrium erubescens* E.B. Bartram, Lloydia 5: 269, 28. 1942.

Type protologue and citation: Indonesia. West Irian, Javawijaya, Mt. Wilhelmina, subalpine forest, W slope, 7 km NE of Wilhelmina-top, 3560 m, *Brass 9634* (holotype: FH-Bartram); *Brass 9610, 9633, 9635, 9891, 9892* (paratypes: FH-Bartram); Lake Habbema: 9 km NE of Lake Habbema, 2800 m, *Brass 10963, 10965* (paratypes: FH-Bartram).

(1) Plants small to medium-sized, rusty-reddish, in dense mats; stems inconspicuous, densely covered with rusty rhizoidal tomentum, with erect, simple to forked branches, branches longer, up to 3-4 cm. (2) Stem leaves inconspicuous, smaller, densely covered with rusty rhizoids. (3) Branch leaves irregularly and individually flexuose-twisted when dry, erect-spreading to spreading when moist, lanceolate, oblong-lanceolate; apices short-acuminate-mucronate to acuminate-acute; costae shortly excurrent, often dark brown; all cells isodiametric except those very near the base; upper and medial cells round, rounded-quadrate to short-elliptic, bulging-conic, some strongly unipapillose, a few papillae apically forked; basal cells rounded-quadrate to short-elliptic, conic-bulging, sometimes extended as low papilla or tuberculum. (4) Perichaetial leaves ovate-lanceolate, acuminate; costae excurrent to form a short arista; basal cells elongate-rectangular. (5) Setae 6-10 mm long, ridged and erect, densely prorate, twisted to the left. (6) Capsule urns ovoid-ellipsoid to narrowly ellipsoid, with a 4-angled mouth; peristome single; exostome consisting of a pale membrane; anisosporous. (7) Calyptrate mitrate, deeply plicate, lacerate and densely hairy at base.

Distribution: Papua New Guinea.

Specimens examined: **PAPUA NEW GUINEA**. West Sepik, *A. Touw 16561* (H 3205270); *A. Touw 15918* (H 3205123); *A. Touw 15532* (MO 5371069).

63. *Macromitrium erythrocomum* H.P. Ramsay, A. Cairns & Meagher, Telopea 20: 262. 2017.

Type protologue: Australia, Queensland, Bellenden Ker Range, on horizontal branch of Leptospermum wooroonooran in microphyll fern forest, 17°15'52"S, 145°51'13"E, 1543 m asl, D. Meagher WT-1022 & A. Cairns, 15 Aug 2016 (holotype: BRI-AQ858157; isotype: CANB).

(1) Plants medium-sized to robust, somewhat lustrous, light-green above and rusty red to darker red below, in loose to dense spreading mats. (2) Primary stems creeping, stem leaves lanceolate-triangular, costa strong, shortly excurrent, leaf angles long-decurrent, often densely with red rhizoids. (3) Branches ascending, erect or wealy curved, 30-50(70) mm long, simple or sparsely branched; branch leaves loosely spreading and individually twisted-contorted from an erect adherent base, with squarrose leaf apices when dry, flexuose spreading to wide-spreading when moist; narrowly lanceolate with an ovate base, gradually tapering towards an acute to acuminate apex, unistratose throughout; margins plane, entire to minutely crenulate; costae red to rusty-red, percurrent to slightly excurrent at the apex; upper and medial cells strongly bulging-mammillose, isodiametric to shortly elliptic, often with a single low rounded papilla; arranged in longitudinal lows; the outmost marginal cells short-rectangular in the upper two-thirds of the leaf, with very thick outer walls curved strongly inwards producing a crescent-shaped lumen; basal cells gradually delineated from upper cells, thick-walled, porose, linear, lumens straight, sometimes with a single low papilla; basal juxtacostal cells irregularly rectangular, thin-walled, pellucid, forming a "cancellina region". (4) Perichaetial leaves triangular-lanceolate, shorter than branch leaves; costae excurrent. (5) Vaginulae slightly hairy; setae short, 3.5–4.0 (–6.0) mm stout, smooth, erect, not twisted. (6) Capsule urns ovoid-ellipsoid, with a narrow 4–6 plicate, puckered mouth; opercula conic-rostrate; peristome lacking or reduced to a basal membrane; spores anisomorphic, surface lightly papillose. (7) Calyptra mitrate, about 3 mm long, plicate, covering but not enclosing the capsule, with erect stiff rusty

red hairs, arising from base to top of calyptra; base lacerate into 8 segments to top of urn, sometimes with one split longer and the calyptra thus appearing cucullate (Ramsay *et al.*, 2017).

Macromitrium erythrocomum H.P. Ramsay is similar to *M. involutifolium* (Hook. & Grev.) Schwägr, particularly *M. involutifolium* subsp. *ptychomitrioides*. The former differs from the latter by its deeply lacerate calyptras with rigid red hairs, rusty red plants, leaves with an excurrent red costa that fills the apex, leaves at branch apices curved-squarrose with leaf apices exposed, not usually inrolled when dry (Ramsay *et al.*, 2017).

Distribution: Australia.

64. *Macromitrium evrardii* Thér., Rev. Bryol., n.s. 3: 181. 1. 1931. (Figure 92)

Type protologue: [Vietnam] Dalat, chemin de la montagne de l'Eléphant, ruisseau de Prenh (*no. 1926*). Type citation: Dalat, chemin de la montagne de l'Eléphant, ruisseau de Prenh (*no. 1926*), leg. *F. Evrard*, 29. Nov. 1924 (holotype: PC 0083648!).

(1) Plants medium-sized, in dense, rusty brownish mats; stems long creeping, with caducous leaves and dense, reddish tomentose below; branches erect, 5-7 mm long, with several short branchlets at upper part, densely leaved. (2) Branch leaves densely arranged, keeled, twisted-contorted and flexuose, with incurved apices, moderately funiculate when dry, widely spreading when moist, oblong-lanceolate to oblong-ligulate, sometimes plicated below, with an acute, acuminate-acute apex; margins entire throughout, not revoluted; upper and medial cells strongly conic-bulging to mammillose, rounded-quadrate, quadrate to subquadrate, 8-10 μm wide; low and basal cells elongate, oblong-rectangular, 16-18 × 6-8 μm, bulging, distinctly unipapillose to tuberculate; costae yellowish brown, percurrent, reaching to the apex to form a short point. (3) Perichaetial leaves lanceolate, oblong-lanceolate, to oval-lanceolate; apices acuminate, acute or apiculate; costae excurrent; all cells longer than wide, oblong, elliptic to oblong-rectangular, not porose, bulging, thick-walled, most unipapillose. (4) Setae smooth, about 20 mm long, twisted to the left. (5) Capsule urns erect, ellipsoid to ellipsoid-cylindric, not contracted beneath the mouth when old; opercula conic-subulate; peristome single, exostome teeth short- lanceolate, with obtuse the apices, densely papillose throughout; spores distinctly anisosporous. (6) Calyptra large, mitrate-cucullate, with yellow-brownish hairs.

Macromitrium evrardii Thér. is similar to *M. fortunatii* Cardot & Thér. in habits, coloration, leaf papillosity, seta length, capsule and calyptrae. However, the branch leaves of the former often oblong- to ligulate-lanceolate, twisted-contorted and flexuose when dry, and conduplicate above when moist, while those of the latter are ligulate, lanceolate-ligulate, spirally coiled around the branch and distinctly funiculate when dry.

Distribution: Vietnam.

65. *Macromitrium exsertum* Broth., Öfvers. Finska Vetensk.-Soc. Förh. 35: 46. 1893.

Type protologue: [Australia] Nova Hollandia austro-orientalis, Clyde Mountains alt. 3000 p.s.m. (*W. Bäuerlenn. 120b*); et ad sugar Loaf (*W. Bäuerlen n. 120a*). Type citation: Australia, orientalis-australia, Clyde, Mountains, 3000', by *W. Bäuerlenn. 120b,* 1888 (lectotype designated by Vitt & Ramsay, 1985: H-BR 2533002!); "Australia, orientalis-australia, Sugar Loaf, *W. Bäuerlen n. 120a, 1888*" (syntype: H-BR 2533003!).

(1) Plants roubst, rusty-brown, somewhat lustrous, in spreading mats; stems creeping, with branches up to 3 cm high. (2) Stem leaves broadly ovate to ovate-lanceolate, gradually acuminate. (3) Branch leaves irregularly funiculate, twisted-contorted, flexuose outwardly and in a slightly spiral arrangement when dry, erect-spreading to spreading-inflexed when moist, 2-3 mm long, lanceolate to narrowly lanceolate; apicessharply and slenderly acute to acuminate cuspidate; costae subpercurrent, percurrent or excurrent; upper cells 5-7 μm wide, subquadrate-rounded to rounded-quadrate, smooth and flat; medial cells long rhombic-elliptic to short elongate, smooth and flat; basal cells elongate, thick-walled, porose, tuberculate, with a differentiated marginal border. (4) Perichaetial leaves longer than branch leaves, subsheathing and plicate below, oblong-ligulate-lanceolate, long and gradually acuminate; elongate low cells continuing into the apex. (5) Setae 8-11 mm long, smooth, twisted to the left. (6) Capsule urns narrowly ovoid to ellipsoid-ovoid, smooth, narrowed to an 8-plicate and puckered, darker mouth; peristome single, exostome of 16, ligulate-lanceolate, well-developed. (7) Calyptrae mitrate, smooth and naked, lobed near base.

Macromitrium exsertum Broth., *M. dielsii* Broth. ex Vitt & H.P. Ramsay and *M. microstomum* (Hook. & Grev.) Schwägr. are three similar species with funiculate branch leaves, but *M. exsertum* can be separated from the latter two species by its larger, rusty-brown plants, longer leaves, basal tuberculate and porose cells.

Distribution: Australia.

Specimens examined: **AUSTRALIA**. New South Wales, *D. H. Vitt 28923* (H 3090233), *D. H. Vitt 28309* (KRAM-B-059212); Queensland, *D. H. Norris 34361* (H 3090237), *H. Streimann 52157* (KRAM-B-106884).

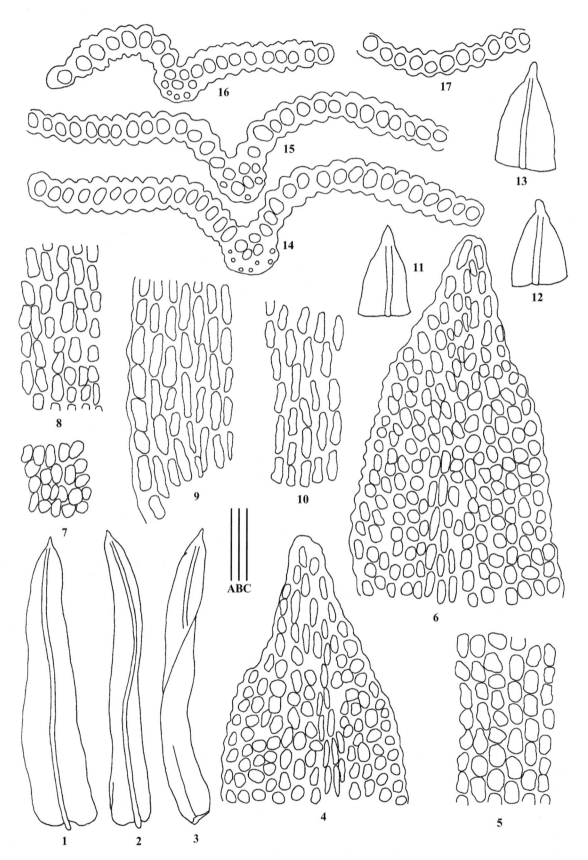

Fig. 91 *Macromitrium emersulum* Müll. Hal. 1-3: Branch leaves. 4, 6: Apical cells of branch leaf. 5: Medial cells of branch leaf. 7: Upper cells of branch leaf. 8, 10: Basal cells of branch leaf. 9: Basal marginal cells of branch leaf. 11-13: Apices of branch leaves. 14: Medial transverse section of branch leaf. 15, 17: Basal transverse sections of branch leaves. 16: Upper transverse section of branch leaf (all from isotype, H-BR 2558026). Line scales: A = 0.44 mm (1-3); B = 176 μm (11-13); C = 44 μm (4-10, 14-17).

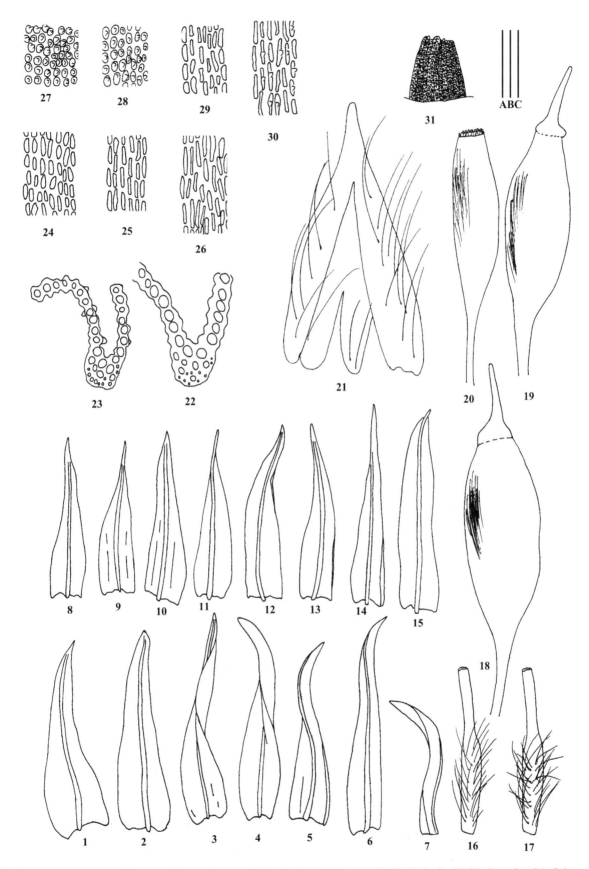

Fig. 92 *Macromitrium evrardii* Thér. 1-7: Branch leaves. 8-15: Perichaetial leaves. 16-17: Vaginula. 18-20: Capsules. 21: Calyptra. 22: Low transverse section of branch leaf. 23: Basal transverse section of branch leaf. 24: Upper cells of perichaetial leaf. 25: Medial cells of perichaetial leaf. 26: Basal cells of perichaetial leaf. 27: Upper cells of branch leaf. 28: Medial cells of branch leaf. 29: Low cells of branch leaf. 30: Basal cells of branch leaf. 31: Peristome (all from holotype, PC 0083648). Line scales: A = 1 mm (1-21); B = 200 μm (31); C = 67 μm (22-30).

66. *Macromitrium falcatulum* Müll. Hal., Linnaea 38: 558. 1874. (Figure 93)
Type protologue and citation: Philippines, *Cuming no. 2212* (isotype: S-B 88592!).
Macromitrium merrillii Broth., Öefvers. Förh. Finska Vetensk.-Soc. 47(14): 4. 1905.("*Merrillii*"), *fide* Vitt *et al.*, 1995. Type protologue and citation: Philippines. Luzon, Prov. of Tarlac, Concepcion, on branches of *Palaquim latifolium, Merrill 3590* (holotype: H-BR 2580003!).
Macromitrium winkleri Broth. in Winkler, Bot. Jahrb. 49: 350. 1913 ("*Winkleri*"), *fide* Vitt *et al.*, 1995. Type protologue and citation: Indonesia. S. O. Borneo: Hayup (*Winkler n. 2117*) (holotype: H-BR 2580001!).

(1) Small and dull plants in spreading mats; stem long-creeping and prostrate, with short, erect branches, branches up to 5-6 mm long and 1.2 mm thick. (2) Stem leaves densely covered with rusty rhizoidial tomentum, lanceolate-acuminate. (3) Branch leaves twisted-contorted and tightly inrolled to involute, most the apices hidden in the inrolled cavity, erect-incurved when moist, ligulate to lanceolate-ligulate, broadly acute-apiculate, cucullate; margins patent and crenulate; costae strong and percurrent; upper cells isodiametric, rounded, rounded-quadrate, bulging, pluripapillose; medial cells rounded-quadrate to short-elliptic, bulging, pluripapillose, marginal cells becoming smaller and oblate; basal cells shortly oval to rectangular, thick-walled, larger at insertion, clear, tuberculate to bulging-papillose. (4) Perichaetial leaves small, leaving the vaginula exposed, ovate-lanceolate to ovate-ligulate, acute; upper cells conic-bulging to tuberculate; low cells elliptic, tuberculate. (5) Setae short, 2.5-5 mm long, smooth, twisted to the right. (6) Capsule urns narrowly ellipsoid, smooth; peristome single; exostome of 16, well-developed, coarsely papillose-striate teeth. (7) Calyptrae mitrate, densely hairy.

Distribution: Indonesia, Malaysia, Papua New Guinea, Philippines (Vitt *et al.*, 1995).
Specimen examined: PHILIPPINES. Ins Philippines, Cumming *s.n.* (H-BR 258004).

67. *Macromitrium fendleri* Müll. Hal., Linnaea 42: 486. 1879. (Figure 94)
Type protologue: [Venezuela] with Nr 55 in the paper of Musci Fendleriani Venezuelenses. Type citation: No. 55, *Macromitrium fendleri n. sp.* Hab. Prope Coloniam Tocar legit *A. Fendler*, 1854-5. (lectotype designated here: NY 01086590!; isolectotype: FH 00213610!).

(1) Plants small, rusty reddish, in dense mats; stems creeping, densely branched; branches prostrately ascending, up to 10 mm long; stem leaves caducous and inconspicuous. (2) Branch leaves twisted-contorted to contorted-crisped, the apices often hidden in the inrolled cavity when dry, spreading, and often with an adaxially incurved and conduplicate upper portion when moist, oblong- to ligulate-lanceolate, acuminate at the apex; margins occasionally and partially recurved at one or both sides, entire below and notched, irregularly serrate near the apex; upper and medial cells subquadrate, rounded-quadrate, strongly unipapillose or mammillose, obscure and dark, low and basal cells elongate, thick-walled, strongly tuberculate; enlarged marginal teeth-like cells at insertion differentiated. (3) Perichaetial leaves differentiated, oblong with a short acuminate-acute apex. (4) Setae shorter, about 4-5 mm long, vaginulae with short paraphyses. (5) Capsules ellipsoid to ellipsoid-cylindric, furrowed when dry, peristome single, with short lanceolate teeth. (6) Calyptrae mitrate, smooth and naked, plicate and lacerate below.

Distribution: Venezuela.

68. *Macromitrium fernandezianum* Broth., Nat. Hist. Juan Fernandez, 2: 422. pl. 26: f. 25-28. 1924. (Figure 95)
Type protologue: Masatierra [Chile, Juan Fernandez Islands]: Puerto Frances, Loma Incienso; ad truncum Myrcengeniae (172). In jugo inter valles Piedra agujereada et Laura; 650 m. s. m. (171). In declivi septemtrionali montis Yunque; ad truncos arborum silvae; 4-500 m. s. m. (166). Valle Colonial, Quebrada seca; ad truncos arborum; 435 m. s. m. (167). Portezuelo de Villagra; ad Berberidem; c. 600 m. s. m. (170). Salsipuedes; ad Myrceugeniam (169). Puerto Ingles; in silva sicca jugi lapidosi; c. 550 m. s. m. (168). Type citation: Juna Fernandez, Masatierra, Valle Colonial, Quebrada seca; ad truncos arborum; 435 m. s. m. 20/12 1916, Carl o Inga Skottsberg (167) (isosyntype: NY 01201995!); Masatierra (Chile, Juan Fernandez Islands): Puerto Frances, Loma Incienso; ad truncum Myrcengeniae (172), M.S.M. 13/12 1916, Carl o. Inga Skottsberg (isosyntype: S-B 163394!).

(1) Plants small to medium-sized, grey yellow, not lustrous; stems long creeping, with erect and short branches up to 6 mm long, and 1-1.2 mm thick. (2) Stem leaves triangular-lanceolate, appressed below, recurved-squarrosed, spreading when moist; upper cells subquadrate to round-quadrate, bulging. (3) Branch leaves keeled, spirally curved to twisted around the branch, the upper portions twisted-curly or flexuose to curly when dry, erect to slightly patent when moist, sometimes fragile and broken off above when old; oblong- to ligulate-lanceolate, occasionally linear lanceolate; apices acuminate to narrowly acuminate; upper cells subquadrate to round-quadrate, or round, in distinctly longitudinal rows, smooth, bulging and somewhat dark; low cells elongate, rectangular and bulging, thick-walled, unipapillose; enlarged marginal teeth-like cells at insertion not differentiated. (4) Perichaetial leaves ovate-oblong to oblong-lanceolate, with an acuminate apex, plicate below, recurved at one side; costae vanishing beneath the apex, sometimes concealed in adaxial view by overlapping folds of laminae below, slightly sheathing at base; all cells longer than wide, tuberculate at low and basal portions. (5) Setae short, about 3 mm long, smooth and slightly twisted to left. (6) Capsule immature, somewhat ellipsoid. (7) Calyptrae mitrate, smooth and naked, lacerate below.

Distribution: Chile (Juan Fernandez Islands).

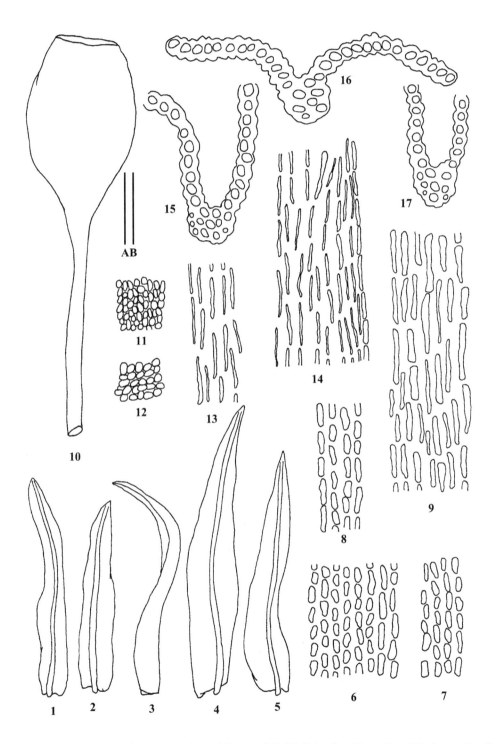

Fig. 93 *Macromitrium falcatulum* Müll. Hal. 1-5: Branch leaves. 6-7: Medial cells of branch leaf. 8: Low cells of branch leaf. 9, 13: Basal cells of branch leaf. 10: Capsule. 11-12: Upper cells of branch leaf. 14: Basal cells near costa of branch leaf. 15, 17: Basal transverse sections of branch leaves. 16: Medial transverse section of branch leaf (all from isotype, S-B 88592). Line scales: A = 0.44 mm (1-5, 10); B = 44 μm (6-9, 11-17).

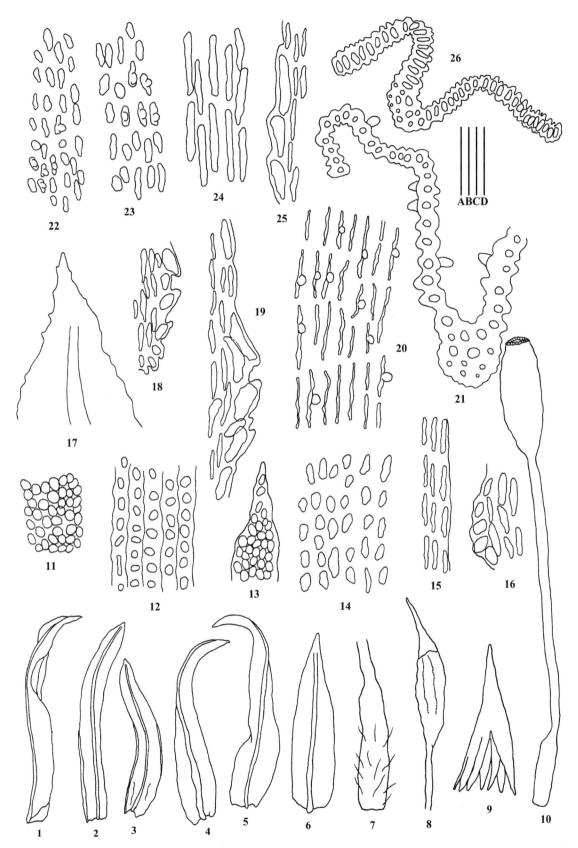

Fig. 94 *Macromitrium fendleri* Müll. Hal. 1-5: Branch leaves. 6: Perichaetial leaf. 7: Vaginual. 8: Capsule immatured. 9: Calyptra. 10: Capsule. 11: Upper cells of branch leaf. 12: Medial cells of branch leaf. 13: Apical cells of branch leaf. 14: Medial cells near costa of branch leaf. 15: Medial marginal cells of branch leaf. 16, 18-19: Basal marginal cells of branch leaves. 17: Apex of branch leaf. 20: Low cells of branch leaf. 21: Basal transverse section of branch leaf. 22: Upper cells of perichaetial leaf. 23: Medial cells of perichaetial leaf. 24: Basal cells of perichaetial leaf. 25: Basal marginal cells of perichaetial leaf. 26: Upper transverse section of branch leaf (all from isolectotype, FH 00213610). Line scales: A = 2 mm (8-10); B = 1 mm (1-7); C = 100 μm (17-18); D = 67 μm (11-16, 19-26).

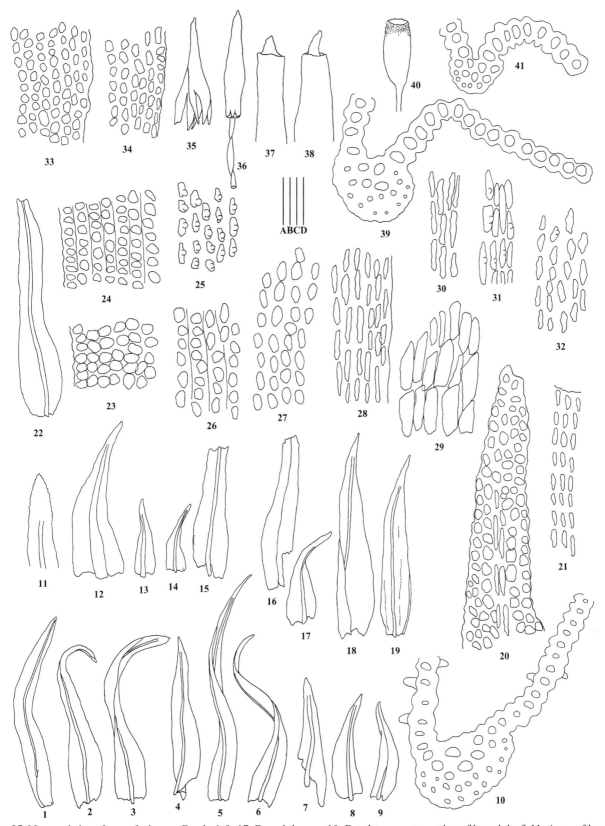

Fig. 95 *Macromitrium fernandezianum* Broth. 1-9, 17: Branch leaves. 10: Basal transverse section of branch leaf. 11: Apex of branch leaf. 12-14: Stem leaves. 15-16, 18-19, 22: Perichaetial leaves. 20: Apical cells of stem leaf. 21: Cells of capsule mouth. 23: Upper cells of branch leaf. 24: Medial cells of branch leaf. 25: Basal cells of branch leaf. 26: Low cells of branch leaf. 27: Upper cells of perichaetial leaf. 28, 32: Medial cells of perichaetial leaf. 29: Basal cells near costa of perichaetial leaf. 30-31: Basal cells of perichaetial leaf. 33: Medial cells of stem leaf. 34: Basal cells of stem leaf. 35: Calyptra. 36, 40: Capsules. 37-38: Apices of calyptrae. 39: Medial transverse section of branch leaf. 41: Upper transverse section of branch leaf (all from isosyntype, NY 01201995). Line scales: A = 2 mm (35-36, 40); B = 1 mm (1-9, 12-19, 22); C = 400 μm (11, 37-38); D = 67 μm (10, 20-21, 23-34, 39, 41).

69. Macromitrium ferriei Cardot & Thér., Bull. Acad. Int. Géogr. Bot. 18: 250. 1908 ("*Ferriei*"). (Figure 96)
Type protologue: [Japan] Oshima, Yowan-Dake. Type citation: Japan, Oshima, Yowan Dake, leg. *J. B. Ferrie*, 18, 10, 99 *J. B. Ferrié* (lectotype designated here: PC 0083650!; isolectotypes: PC 0083651!, PC 0083649!).

Macromitrium comatulum Broth. *ex* Okamura *in* Matsumura, Icon. Pl. Koisikavenses 3: 41, pl. 166. 1916. Lectotype citation: Japan. Prov. Tojima; Kimosaki, *G. Kono 225* (lectotype: NICH; isolectotype: H-BR 2575005!). Syntypes: Shikoku: Prov. Tosa; Yanase, *Okamura 167* (H-BR 2575007!)), *fide* Noguchi & Iwatsuki, 1989.

*Macromitrium comatulum Broth., Öfvers. Finska Vetensk.-Soc. Förh. 62A(9): 14. 1921, hom. illeg, *fide* Noguchi & Iwatsuki, 1989.

Macromitrium inflexifolium Dixon, J. Siam Soc., Nat. Hist. Suppl. 9: 20. 1932, *fide* Noguchi & Iwatsuki, 1989. Type protologue: (Thailand) Hab. Rachaburi, Kanburi, Siawat, Siam, cirea 800 m. alt., on Dendrobium, 17 Jan., 1928, coll. *A. F. G. Kerr (148)*, type. (Sri Lanka) On rocks in the jungle, Peradeniya, Ceylon, Sept., 1925; coll. *A. H. G. Alston (1593)*. On stone, Sugi, Prov. Tosa, Sikotu, Japan, Mar., 1928; coll. H. Sasaoka (4477).

Macromitrium nipponicum Nog., J. Hattori Bot. Lab. 20: 281. 1958, *fide* Noguchi & Iwatsuki, 1989.

(1) Plants medium-sized, in dense mats; stems creeping, densely branched; branches erect, short and simple, up to 1.5 cm long, densely foliate, obtuse at the apex. (2) Stem leaves suberect to squarrose when dry, oblong-lanceolate, 1.0-1.5×0.2-0.4 mm, distinctly keeled, acuminate at the apex; margin reflexed; costa subpercurrent; medial leaf cells hexagonal or shortly rectangular, with 1 to several large papillae; basal cells sublinear. (3) Branch leaves contorted when dry, oblong-lanceolate to linear-lanceolate, about 2.5 × 0.5 mm, keeled, more or less plicate at base, acuminate or obtuse at the apex; margins subentire or papillosely crenulate, recurved, especially in the low half; costa percurrent; upper and medial leaf cells hexagonal, 6.5-10 μm wide, inflated, thin-walled, densely pluripapillose, rather obscure. (4) Perichaetial leaves oblong-lanceolate, to 2.5 mm long, keeled, plicate below, narrowly acuminate at the apex; medial leaf cells rectangular, pellucid, unipapillose, thick-walled. (5) Setae erect, (3-) 7-10 mm long, smooth; vaginulae paraphyses numerous, 0.4-0.6 mm long. (6) Capsule urns erect, ellipsoid-ovoid or ellipsoid-cylindric, 1.2-1.6 × 0.5-0.9 mm, slightly constricted at the mouth when dry; opercula conic-rostrate, 0.7-0.8 mm long; peristome single, teeth linear to lanceolate, up to 0.2 mm long, obtuse at the apex, densely and finely papillose. (7) Calyptrae cucullate, 2-3 mm long, usually split or lobed at base, yellowish hairy. Spores spherical, 15-32 μm in diameter, finely papillose.

Macromitrium ferriei Cardot & Thér. is similar to *M. giraldii* Müll. Hal., but they are two distinctive species. For *M. giraldii*, the branch leaves are longer, up to 3.2 mm in length, narrowly acuminate to acuminate, occasionally broadly acuminate, the upper and medial leaf cells of branch leaves are pellucid, collenchymatous, bulging, moderately to slightly pluripapillose, or sometimes smooth, and the inner perichaetial leaves up to 2.5 mm long, abruptly constricted from an oblong base to a long acuminate or aristate apex, while for *M. ferriei*, the branch leaves shorter, up to 2 mm in length, broadly acuminate to narrowly acute, and the upper and medial cells of branch leaves are rather obscure, strongly pluripapillose, and the inner perichaetial leaves only 1.5 mm long, ovate- or triangular-lanceolate, gradually narrowed to a shortly acuminate apex.

Distribution: China, Japan, Korea.

Specimens examined: **CHINA**. Shandong, *leg. Schwabe-Beha* (JE); Sichuan, *B. Allen 7249* (MO); *B. Allen 6722* (MO); *P. L. Jr. Redfearn 34843* (MO); Xizang, *Y. G. Su 2337* (MO); Zhejiang, *P. C. Wu 198* (MO). **JAPAN**. Kyushu, *Z. Iwatsuki 1628* (MO); Kinki, *T. Nakajima, 1080* (MO).

70. Macromitrium fimbriatum (P. Beauv.) Schwägr., Sp. Musc. Frond., Suppl. 2 1(1): 37, 111. 1823. (Figure 97)
Basionym: *Orthotrichum fimbriatum* P. Beauv., Prodr. Aethéogam. 80. 1805. Type protologue: Cette espèce a été rapportée de l'Ile de Tristan d'Acunha, par M. Dupetit-Thouars, quime l'a communiquée. Type citation: [Sint-Helena] Tristan d A'Cunha (Petit Thouars) (isotypes: BM 000873857!, BM 000873898!); *Thouars s.n.* (isotypes: BM 000982406!, BM 000982407!).

Macromitrium nanothecium var. *sublaeve* Thér., Recueil Publ. Soc. Havraise Études Diverses, 96: 113, 1929, *fide* Wilbraham, 2018. Type protologue: Madagascar: Fort Dauphin, sur écorce, *R. Decary*. Type citation: Madagascar, Fort Dauphin, 30 June 1926, *Decary 36* (holotype: PC 0137524; isotypes: PC 0137525, PC 0137526, PC 0137527, MO!).

Macromitrium uncinatum (Brid.) Brid., Bryol. Univ. 1: 308, 1826, *fide* Crosby *et al.*, 1983.

Orthotrichum uncinatum (Brid.) Arn., Disp. Méth. Mousses, 17, 1825, *fide* O'Shea, 2006.

Weissia uncinata Brid., Muscol. Recent. Suppl. 1: 113, 1806, *fide* Crosby *et al.*, 1983. Type protologue: Germ. Hakenförmige Weiffie, Gall Verdule crochue.

(1) Plants small, in dense mats; stems creeping, with short branches. (2) Branch leaves individually to spirally twisted, twisted-flexuose-curved when dry, spreading below and conduplicate or keeled above when moist, 1-2.5 mm long, ligulate-lanceolate, with a rounded-obtuse, acute apex; upper cells larger, 7-13 μm wide, quadrate, rounded, oblate, strongly bulging and sparsely pluripapillose; medial cells ovate-elliptic, pluripapillose; elongated basal cells

restricting to the low 1/4 to 1/3 of the lamina, smooth, thick-walled, porose; costae vanishing several cells beneath the apex. (3) Perichaetial leaves differentiated from branch leaves, triangular-, oblong- to ovate-lanceolate, sometimes shorter than branch leaves. (4) Setae smooth, about 4 mm long. (5) Capsule urns ovoid to ellipsoid, smooth, constricted at the mouth; peristome absent. (6) Calyptrae mitrate, smooth and plicate, naked, lobed at base.

Distribution: Madagascar, Mauritius, Réunion Island (O'Shea, 2006).

71. *Macromitrium flavopilosum* R.S. Williams, Bull. Torrey Bot. Club 38: 34. 1911. (Figure 98)

Type protologue and citation: Panama, Cana, Prov. Dairen, *Williams 1063* (isotypes: FH 00213612!, NY 00456556!, MICH 525878!).

(1) Plants medium-sized to large, reddish-green to reddish-yellow, lustrous, in loose mats; stems weakly creeping, branches up to 4 cm long. (2) Branch leaves keeled, erect below, spreading, flexuose to spirally contorted and undulate above when dry, slightly abaxially curved and erect-spreading when moist, up to 8-9 mm long, often 5.5-7.0 × 1.0 mm, lanceolate to long ligulate-lanceolate, occasionally plicate below; apices acute to a long, setaceous; margins undulate or plane, slightly serrulate by protruding cells below, serrate near apex, recurved below, recurved to plane above; costae long- excurrent into a naked awn, awns up to 2 mm long; all cells longer than wide, thick-walled, flat, clear and smooth; upper and medial cells in rather straight longitudinal rows, distinctly porose; upper interior cells long-rhombic to fusiform, 12-16 μm long, upper marginal cells linear, forming a distinct differentiated border; basal cells narrowly long-rectangular, 45-55 μm long, enlarged marginal teeth-like cells at insertion not differentiated. (3) Perichaetial leaves not much different from branch leaves. (4) Setae 10-15 mm long, smooth, twisted to the left. (5) Capsule urns obovoid, ellipsoid to ovoid-ellipsoid, ellipsoid-cylindric, smooth, slightly constricted beneath the mouth; opercula long conic-rostrate, with a long, slender and straight beak; peristome double, both membraneous, papillose; spores anisosporous. (6) Calyptrae mitrate, 3 mm long, naked, deeply lacerate (Allen, 2002).

Macromitrium subcirrosum Müll. Hal. and *M. ulophyllum* Mitt. are similar to *M. flavopilosum* R.S. Williams in their leaves with long, setaceous, excurrent costa, all cells longer than wide. However, *M. ulophyllum* differs from *M. flavopilosum* by its broader, strongly undulate leaves with upper cells in diagonal rows from the costa, and with tuberculate basal cells. *M. subcirrosum* can be separated from *M. flavopilosum* by its leaves with tuberculate basal cells, and upper margin slightly or not differentiated with narrower cells.

Distribution: Guatemala, Honduras, Mexico, Nicaragua, Panama (Allen, 2002).

Specimen examined: **PANAMA**. Cerro Jefe, *Marshall R. Crosby 10907* (MO 2408351).

72. *Macromitrium flexuosum* Mitt., J. Linn. Soc., Bot. 12: 208. 1869. (Figure 99)

Type protologue: "Hab. Andes Quitenses, in sylva Canelos (3000 ped.), *Spruce 98*"; Type citation: "Mont de Canelos, *Spruce s.n.* (isotypes: NY 01086518!, NY 01086519!)"; " Ecuador, Andes Quitenses, Mont & Canelos, leg. *Spruce s.n.* (isotype: S-B 163499!)"; "Andes Quitens, in sylva Canelos, *Ric Spruce 98* (isotype: BM 000720530!)".

(1) Plants robust, large, upper brown-yellowish, dark-brown below; stems weakly creeping, branches up to 20 mm long, 2 mm wide, with branchlets. (2) Branch leaves erecto-appressed below, individually twisted, flexuose, slightly twisted-contorted above when dry, widely spreading to slightly squarrose–recurred when moist; lanceolate, acuminate upper part from an oblong or broadly oblong low part, slightly sheathing at base; margins serrate above, entire below, with a conspicuous limbidium; all cells longer than wide, thick-walled, and most porose from medial part to base; upper cells short rectangular to elliptic rectangular, smooth; leaf cells gradually elongate from medial part to base; lower and basal cells long to linear rectangular, tuberculate, strongly porose; marginal cells at basal insertion enlarged, subquadrate, quadrate to short rectangular, but not teeth-like. (3) Inner perchaetial leaves short, lanceolate from a broadly oblong low part, acuminate, involute below or with a sheathing base; outer perichaetial leaves similar to vegetative leaves but with a sheathing base. (4) Setae smooth, up to 15 mm long, twisted to the left. (5) Capsules oval-cylindric or short cylindric, smooth, peristome double, exostome teeth yellow-brownish, united to form a short membrane, smooth, endostome segments hyaline.

Macromitrium flexuosum Mitt. is similar to *M. mcphersonii* B.H. Allen, but the former could be distinguished from the latter by its smooth setae, branch leaves with cells longer than wide, thick-walled and porose throughout, and with a sheathing base, as well as serrate upper and medial margins. *Macromitrium flexuosum* also resembles *M. fuscoaureum* E.B. Bartram, but the latter can be separated from the former by its branch leaves with a clasping base and smooth lower and basal cells.

Distribution: Bolivia, Brazil, Colombia, Ecuador.

Specimens examined: **COLOMBIA**. Antioquia, *Carl A. Luer 10077A* (MO); Cauca, *Bernardo R. Ramírez Padilla 7246* (MO); Nariño, *Bernardo R. Ramírez Padilla 9057*(MO). **ECUADOR**. Zamora-Chinchipe, *Alberto Ortega U. 547a* (MO 2555699).

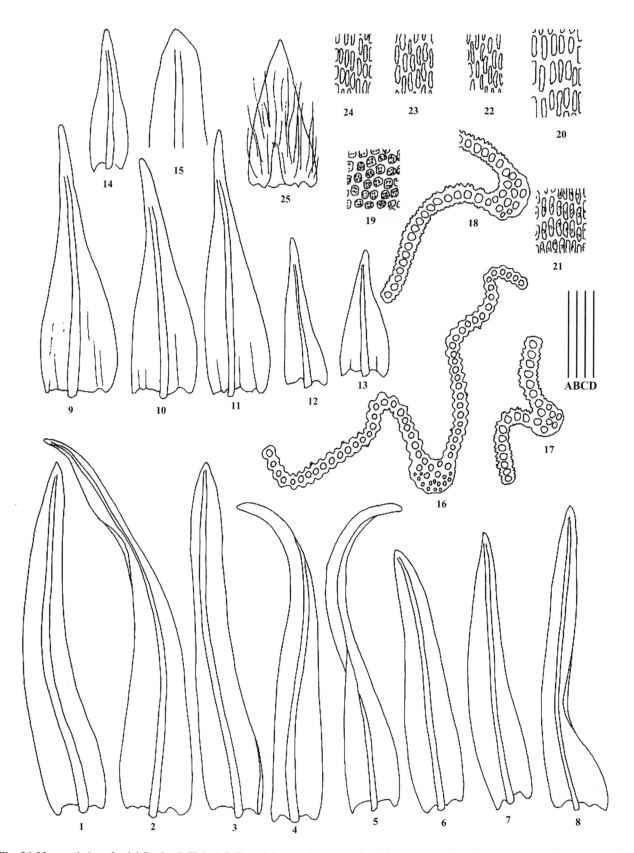

Fig. 96 *Macromitrium ferriei* Cardot & Thér. 1-8: Branch leaves. 9-13: Perichaetial leaves. 14: Stem leaf. 15: Apex of branch leaf. 16: Basal transverse section of branch leaf. 17: Upper transverse section of branch leaf. 18: Medial transverse section of branch leaf. 19: Upper cells of branch leaf. 20: Basal cells of branch leaf. 21: Low cells of branch leaf. 22: Upper cells of perichaetial leaf. 23: Medial cells of perichaetial leaf. 24: Basal cells of perichaetial leaf. 25: Calyptra (all from lectotype, PC 0083650). Line scales: A = 0.40 mm (1-14); B = 1 mm (25); C = 68 μm (16-24); D = 0.2 mm (15).

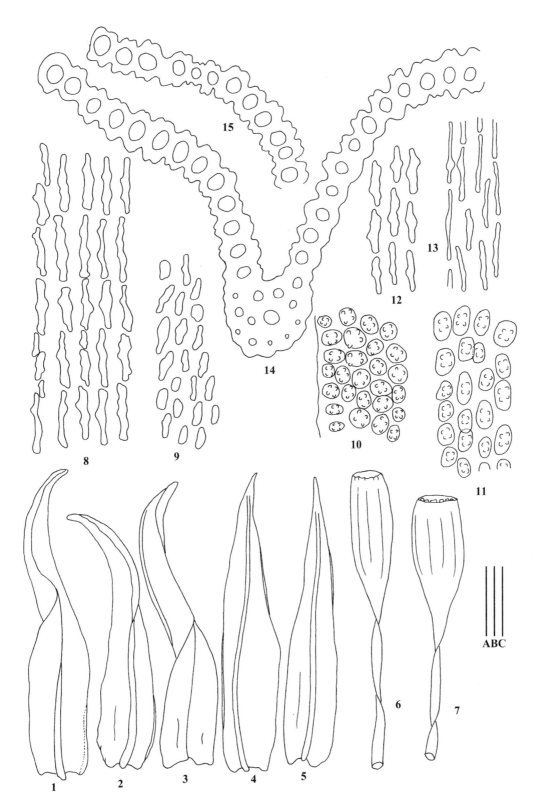

Fig. 97 *Macromitrium fimbriatum* (P. Beauv.) Schwägr., 1-3: Branch leaves. 4-5: Perichaetial cells. 6-7: Capsules. 8: Basal cells of branch leaf. 9: Upper cells of perichaetial leaf. 10: Upper cells of branch leaf. 11: Medial cells of branch leaf. 12: Medial cells of perichaetial leaf. 13: Basal cells of perichaetial leaf. 14: Medial transverse section of branch leaf. 15: Upper transverse section of branch leaf (all from isotype, BM 000982407). Line scales: A = 2 mm (6-7); B = 1 mm (1-5); C = 67 μm (8-15).

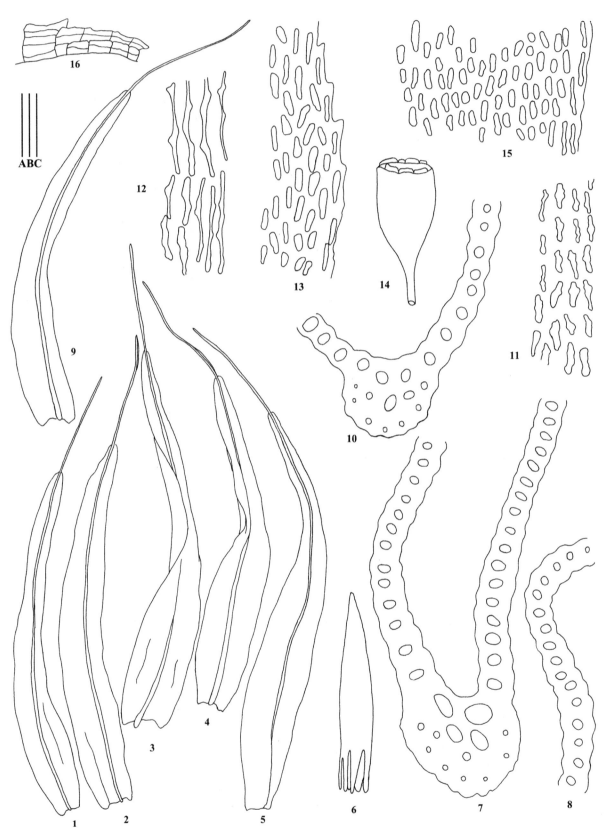

Fig. 98 *Macromitrium flavopilosum* R.S. Williams 1-5, 9: Branch leaves. 6: Calyptra. 7, 8: Basal transverse sections of branch leaves. 10: Low transverse section of branch leaf. 11: Medial cells of branch leaf. 12: Basal cells of branch leaf. 13: Apical cells of branch leaf. 14: Capsule. 15: Upper cells of branch leaf. 16: Peristome (4-6, 9 from isotype, NY 00456556; 1-3, 7-8, 10-15 from isotype, MICH 525876). Line scales: A = 2 mm (1-6, 9, 14); B = 400 μm (16); C = 67 μm (7-8, 10-13, 15).

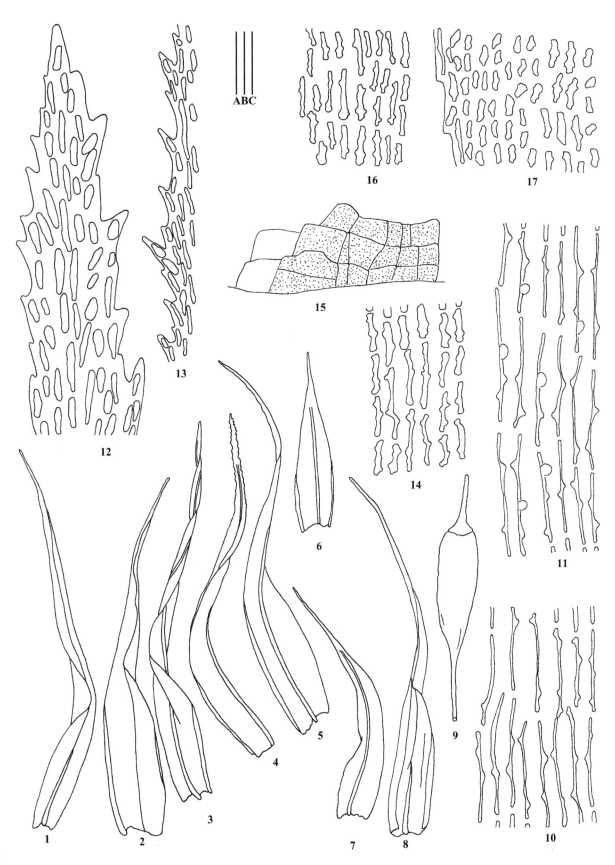

Fig. 99 *Macromitrium flexuosum* Mitt. 1-5: Branch leaves. 6: Inner perichaetial leaves. 7-8: Perichaetial leaves. 9: Capsule. 10: Basal cells of branch leaf. 11: Low cells of branch leaf. 12, 13: Apical cells of branch leaves. 14: Upper cells of branch leaf. 15: Peristome. 16: Medial cells of branch leaf. 17: Upper marginal cells of branch leaf (all from isotype, NY 01086519). Line scales: A = 2 mm (1-9); B = 100 μm (15); C = 67 μm (10-14, 16-17).

73. *Macromitrium formosae* Cardot, Beih. Bot. Centralbl., Abt. 2 19(2): 104. 8. 1905. (Figure 100)

Type protologue: [China] Kelung ([*M. I. Faruie*] *no. 181*; c. fruct); Type citation: [China] Taiwan: Kelung, leg. *Faurie, 1903, no. 181* (isotype: H 3090260!).

(1) Plants medial-sized, yellowish brown, in mats; stems creeping, with erect branches, branches up to 10 mm high. (2) Branch leaves contorted-curly when dry, flexuose-spreading when moist, 2.3-2.7 × 0.25-0.40 mm, lanceolate to linear-lanceolate, acuminate, revolute below on one side, and somewhat plicate at base; margins entire throughout; costae ending beneath the apex; cells unistratose from the apex to the base; upper and medial cells irregularly quadrate, rounded-quadrate, 7-8 μm wide, rather obscure, densely pluripapillose, the outmost marginal cells smooth and somewhat pellucid; basal cells elongate, 12-20 × 3-4 μm, thick-walled, clear and smooth, those along the basal margin regularly rectangular and smooth. (3) Setae 5-6 mm long, smooth. (4) Perichaetial leaves similar to branch leaves. (5) Capsule urns ovoid to ellipsoid, 1.2-1.4 × 0.65-0.80 mm, brown, with a small mouth, slightly ribbed below the mouth when dry; peristome absent. (6) Calyptrae campanulate, with many long, brown-yellowish hairs, not lacerate.

Distribution: China, Philippines (Tan & Iwatsuki, 1991).

74. *Macromitrium fortunatii* Cardot & Thér., Bull. Acad. Int. Géogr. Bot. 19: 19. 1909 ["i"]. (Figure 101)

Type protologue: [China] Pin-fa, sur rochers; leg. *Fortunat*; Type citation: Chine, Kouy Tcheou, Pin-fa, sur rochers, *Leg. Fortunat, no. 1749,* 5, April 1904 (lectotype designated by Guo & Yu, 2013: PC 0083654!; isolectotypes: PC 0083657!, PC 0719719!).

Macromitrium fortunatii var. *nigrescens* Tixier, Rev. Bryol. Lichénol. 34: 140. f. 8. 1966, *fide* Li *et al.*, 2018. Type protologue and citation: [Vietnam] Chapa, sur arbre et arbuste, lisière de foret, février 1929 (*Pételot, P.A. no. 141*) (lectotype: PC 0083660!, isolecotypes: PC 0721001!, PC 0719722!, PC 0137695!, S-B 115583!).

(1) Plants medium-sized, forming dense, rusty-brownish mats; stems long creeping, with densely rusty-reddish rhizoidal tomentose below, and dense erect branches, branches about 5-7 mm long, with several short branchlets at the upper part, densely leaved. (2) Stem leaves moderately deflexed to suberect when moist, triangular-lanceolate, basal cells rectangular, slightly sinuous, inflate and unipapillose; medial and upper cells quadrate to subquadrate, strong conic-bulging and unipapillose. (3) Branch leaves obliquely appressed, spirally twisted to coiled around the branch, densely arranged in spiral ranks, funiculate to varying degrees, widely spreading when wet, 1.7-2.2 × 0.35-0.55 mm, oblong-lanceolate to oblong-lingulate, keeled, sometimes plicated below; apices acute, acuminate-mucronate, shortly cuspidate to broadly acuminate; margins entire throughout; upper cells strongly conic-bulging, rounded-quadrate, clear, 8-10 × 8-10 μm, distinctly unipapillose; medial and low cells subquadrate to sub-rectangular, 16-18 × 6-8 μm, rather clear, bulging, distinctly unipapillose; basal cells rectangular, not much different from the upper cells, often unipapillose; costae yellowish brown, percurrent or excurrent to form a short point. (4) Perichaetial leaves 2.0-2.4 mm long, oblong to oblong-lanceolate, acuminate; upper cells 12 -16 × 2-4 μm, bulging, unipapillose, with thick walls; medial and low cells elongate, 25-30 × 2-4 μm, bulging, unipapillose. (5) Setae 6-20 mm long, smooth, twisted to the left; vaginulae with numerous hairs. (6) Capsules erect, urns ellipsoid to ellipsoid-cylindric, 1.4-1.8 × 0.7-0.9 mm, the mouth not contracted when old; opercula conic-rostrate, 0.7-0.8 mm long; peristome single, exostome teeth short- lanceolate, with an obtuse apex, densely papillose throughout; spores distinctly anisosporous, 20-53 μm in diameter. (7) Calyptrae large, 2-2.5 mm long, cucullate, with moderately pale-brownish hairs.

Distribution: China, Vietnam.

Specimens examined: **CHINA**. Guizhou, Pin-fa: sur rochers, *Cavalerie & Fortunati, s.n* (PC 0083652); *Fortunati & Cavalerie, no. 1552* (PC 0083653); *P. Cavarlerie* (PC 0083655); *P. Cavarlerie, no. 1992* (PC 0083656); *J. h. Erquirol, no. 3141* (PC 0083658-59) ; *Fortunat 1552* (PC0719719); *Fortunat s.n.* (S-B 163497, B163498); Gan Chouen Fou: *R. P. Cavalerie,* 1910 (S-B 115581); *R. P. Cavalerie,* 1912 (PC0719720); Kouy-Tcheou, Tong Tcheou, *Fortunati,* Oct. 1904 (H-BR 2581004), Libo, *Anonym* LB20151102046 (GACP), Guo & Li 20171109045, Libo country (SHTU); *Anonym* LL2014112912, DJ2016061017 (GACP). Guangdong: Jiaolin Co., *L. Deng no. 4834, 09517*(IBSC). **VIETNAM**. He & Nguyen *41699, 41853, 42205, 42320* (MO); Tonkin, Chapa, *Pételot, P.A. no. 138* (PC); Ha Giang, Si He & Khang Nguyen *41683* (MO 6164476), *41809* (MO 6165916); Ha Tay, Si He & Khang Nguyen *42320* (MO 6165718).

75. *Macromitrium fragilicuspis* Cardot, Rev. Bryol. 36: 109. 1909. (Figure 102)

Type protologue and citation: Mexico, Etat de Vera Cruz: prés de Jalapa, 1908 n. *Pringle 15168* (isotypes: NY 00792493!, H-BR 2632031!).

(1) Plants medium-sized, yellowish-brown above, dark-brown to rusty-brown below; stems creeping, with inconspicuous and caducous leaves, branches up to 2.5 cm long. (2) Branch leaves strongly keeled, erect below, spirally-twisted, flexuose-inrolled above when dry, erect-spreading when moist, 3.0 × 0.5 mm, lanceolate; apices abruptly rounded to long, fragile and multistratose subulate (up to 1 mm long); margins entire to slightly crenulate; costa percurrent; upper and medial interior cells quadrate, small, 4-6 μm, smooth to weakly bulging; basal cells

Taxonomy

rectangular, 15-25 μm long, thick-walled, porose, tuberculate; enlarged marginal teeth-like cells at insertion differentiated. (3) Setae smooth, 5-6 mm long, twisted to the left. (4) Capsule urns ellipsoid-cylindric, furrowed or wrinkled, somewhat constricted beneath the mouth; peristome double, exostome of 16, truncate, united forming a papillose, reddish-orange membrane, endostome hyaline; spores anisosporous. (5) Calyptrae mitrate, naked, deeply lacerate.

Macromitrium fragilicuspis Cardot is similar to *M. frustratum* B.H. Allen and *M. sejunctum* B.H. Allen in having fragile leaf tips, but the leaves of *M. frustratum* are much larger than those of *M. fragilicupis*, and the leaves of *M. sejunctum* has sharply dentate margin and short, smooth basal cells.

Distribution: Honduras, Mexico.

Specimens examined: **HONDURAS**. Lempira Department, *Bruce Allen 11803* (MO 3965334). **MEXICO**. Queretaro, *Garrie landry 15.4* (MO 5269359); Dosque de Juniperus con Dodonaea, *A. Cárdenas 129* (H 3090262).

76. *Macromitrium francii* Thér., Bull. Acad. Int. Géogr. Bot. 17: 308 17: 308. 1907. (Figure 103)

Type protologue: New Caledonia. 1906, *Franc s.n.* (Renauld missit).

Macromitrium contractum Thér., Bull. Acad. Int. Géogr. Bot. 18: 253. 1908, *fide* Thouvenot, 2019. Type protologue: New Caledonia, Mt Dzumac, 700 m, avril 1905, *Le Rat s.n.*

(1) Plants small to medium-sized, in dense mats; stems long creeping, densely with short branches, branches 5-10 mm long, simple or furcate. (2) Branch leaves often individually twisted flexuose, keeled, with incurved to circinate the apices when dry, erecto-patent with slightly incurved apices when moist, 1-1.5 × 0.3-0.4 mm, oblong, ligulate to lanceolate from a wider base, with acute to obtuse the apices; upper cells rounded to oval, thin-walled, bulging, pluripapillose, marginal cells smaller and usually oblate; transitional cells oval to short rectangular, thick-walled; low cells rectangular, thick-walled, occasionally unipapillose. (3) Perichaetial leaves similar to branch leaves. (4) Setae rather short, 1-1.5 mm long, vaginulae with long hairs reaching the capsule. (5) Capsules short exserted, urns ovoid to ellipsoid, smooth, plicate below the mouth; peristome single; spores anisosporous. (6) Calyptrae mitrate, with dense and erect hairs.

Macromitrium francii Thér. is similar to *M. involutifolium* (Hook. & Grev.) Schwägr., *M. ligulaefolium* Broth., and *M. hermitrichodes* var. *sarasinii* (Thér.) Thouvenot, but the branch leaves of the latter three species are longer, the stems of *M. involutifolium* often broken when old, forming many branches as separate individual "plants"; the setae of *M. ligulaefolium* longer, 5-6 mm long; and the basal cells of branch leaves in *M. hermitrichodes* var. *sarasinii* elongate, up to 50 μm.

Distribution: New Caledonia.

Specimens examined: **NEW CALEDONIA**. Reserve Montagne des Source, *Marchall R. Crosby 14167* (MO 4451776); West of last ridge before descent to Yaté on Nouméa-Yaté Road. *Marshall R. Crosby 14248* (MO 4451775); Aur de noruiea, *Franci s.n.* 1907 (H-BR 2561001, marked with "*sp. nov*").

77. *Macromitrium frondosum* Mitt., J. Linn. Soc. Bot. 12: 217. 1869. (Figure 104)

Type protologue: [Ecuador] Andes Quitenses, Paila-urcu in Cordillera occidentali ad Pangor (9000 ped.), *Spruce, no. 102*. Type citation: [Ecuador] Pangor, ad arb, *Spruce s.n.* (isotype: NY 01086534!); Paila urcu in rambis, *Spruce s.n.* (isotype: NY 01086533!); Ecuador, Paila-urcu in Cordillera at Pangor (9000 ft), *Spruce 102* (isotype: E 00165154!).

(1) Plants large and robust, yellowish-brown above, dark-brown below; stems creeping with caducous leaves, loosely branched, branches up to 50 mm long, with branchlets. (2) Branch leaves appressed below, spreading and slightly flexuose above when dry, spreading and slightly abaxially curved when moist, linear-lanceolate from an oblong and sheathing base, gradually narrowed to form a setaceous-acuminate apex with a row of cells; margins almost entire, revolute below, with a differentiated border with cells smaller than their ambient cells; all cells longer than wide, smooth and clear; upper and medial cells short-rectangular, 15-30 × 5 μm, gradually elongate towards base; low and basal cells elongate, long to linear-rectangular, thick-walled and porose; marginal enlarged teeth-like cells at insertion not differentiated. (3) Perichaetial leaves long-lanceolate, gradually narrowed to a setaceous-acuminate apex, all cells longer than wide, smooth and clear, rectangular, long-rectangular to linear-rectangular, porose at low and basal parts. (4) Setae smooth, about 10 mm long, twisted to the left. (5) Capsules ellipsoid, strongly furrowed, reddish brown; opercula conic-rostrate, beaks erect or oblique to various degrees; peristome double, both united, membraneous. (6) Calyptrae cucullate-mitrate, hairy.

The type specimens of *M. gigasporum* Herzog are rather similar to those of *M. frondosum* Mitt., only the capsules of the former smooth or furrowed to different degrees.

Macromitrium frondosum is characterized by its long lanceolate branch leaves with a clasping base, all cells smooth and longer than wide, without differentitated teeth-like large cells at basal margins, and by its furrowed capsules and hairy calyptrae.

Distribution: Ecuador.

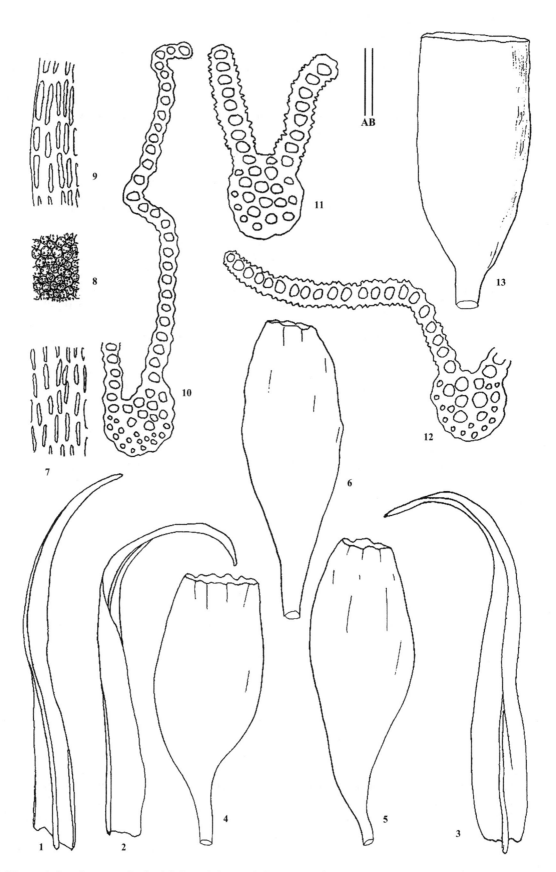

Fig. 100 *Macromitrium formosae* Cardot 1-3: Branch leaves. 4-6: Dry capsules. 7: Basal cells near costa of branch leaf. 8: Upper cells of branch leaf. 9: Basal marginal cells of branch leaf. 10: Basal transverse section of branch leaf. 11: Upper transverse section of branch leaf. 12: Medial transverse section of branch leaf. 13: Wet capsule (all from isotype, H-BR). Line scales: A = 0.50 mm (1-6, 13); B = 50 μm (7-12). (Guo *et al.*, 2007).

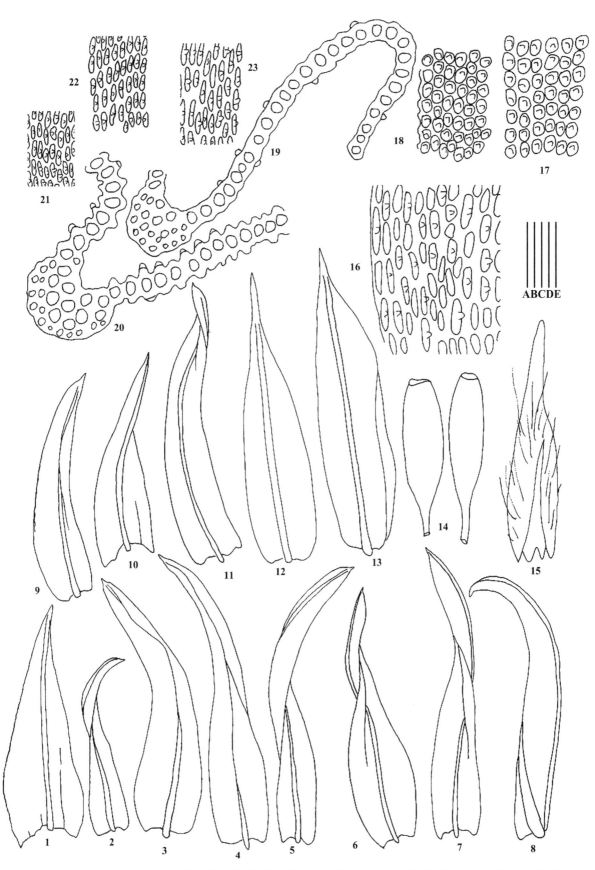

Fig. 101 *Macromitrium fortunatii* Cardot & Thér. 1-9, 11: Branch leaves. 10: Stem leaf. 12-13: Perichaetial leaves. 14: Capsule. 15: Calyptra. 16: Basal marginal cells of branch leaf. 17: Medial cells of branch leaf. 18: Upper cells of branch leaf. 19: Low transverse section of branch leaf. 20: Upper transverse section of branch leaf. 21: Upper cells of perichaetial leaf. 22: Medial cells of perichaetial leaf. 23: Low cells of perichaetial leaf (all from lectotype, PC 0083654). Line scales: A = 0.88 mm (14-15), B = 0.5 mm (1-2, 9-10); C = 0.44 mm (3-8, 11-13); D = 67 μm (21-23); E = 44 μm (16-20).

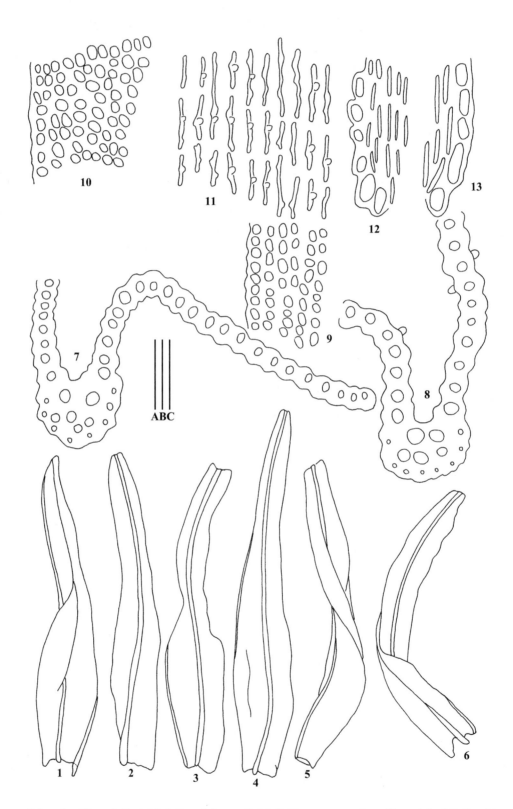

Fig. 102 *Macromitrium fragilicuspis* Cardot 1-6: Branch leaves. 7: Medial transverse section of branch leaf. 8: Basal transverse section of branch leaf. 9: Medial cells of branch leaf. 10: Upper cells of branch leaf. 11: Basal cells of branch leaf. 12-13: Basal marginal cells of branch leaves (all from isotype, NY 00792493). Line scales: A = 1 mm (1-6); B = 100 μm (12-13); C = 67 μm (7-11).

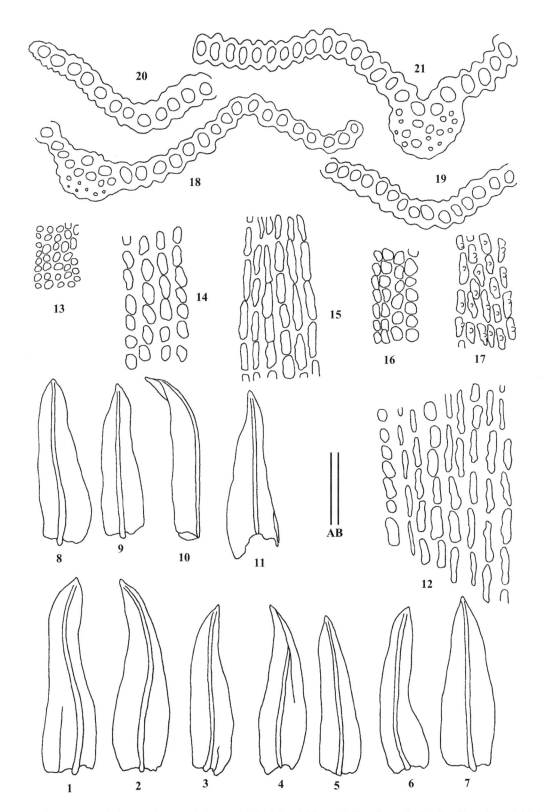

Fig. 103 *Macromitrium francii* Thér. 1-10: Branch leaves. 11: Perichaetial leaf. 12: Basal marginal cells of branch leaf. 13: Upper cells of branch leaf. 14: Medial cells of branch leaf. 15: Basal cells of branch leaf. 16: Medial cells of perichaetial leaf. 17: Basal cells of perichaetial leaf. 18: Basal transverse section of branch leaf. 19-20: Basal transverse sections of perichaetial leaves. 21: Medial transverse section of branch leaf (all from H-BR 2561001). Line scales: A = 0.44 mm (1-11); B = 44 μm (12-21).

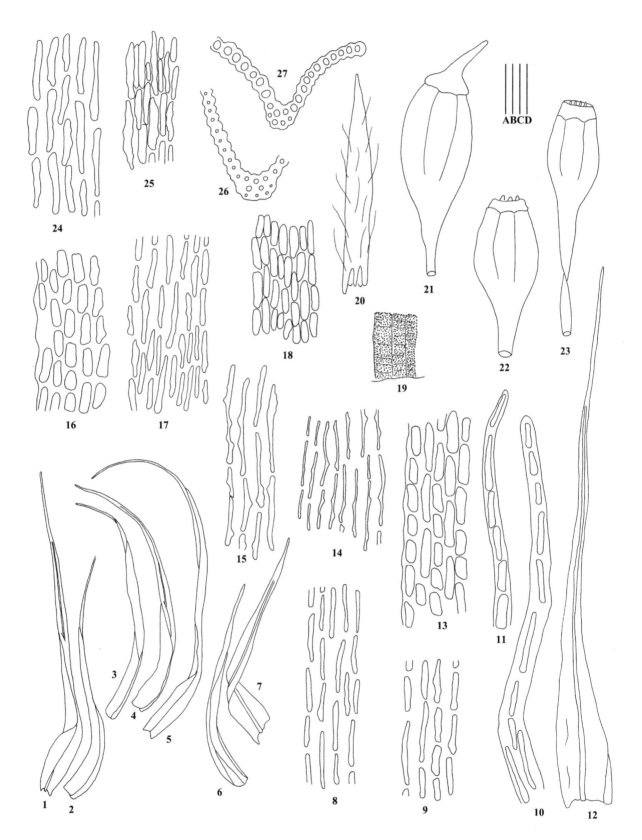

Fig. 104 *Macromitrium frondosum* Mitt. 1-7: Branch leaves. 8-9: Upper cells of branch leaves. 10: Apex of branch leaf. 11: Apex of perichaetial leaf. 12: Perichaetial leaf. 13: Upper marginal cells of branch leaf. 14: Basal cells of branch leaf. 15, 17: Low cells of branch leaves. 16: Medial cells of branch leaf. 18: Upper cells of branch leaf. 19: Peristome. 20: Calyptra. 21-23: Capsules. 24: Upper cells of perichaetial leaf. 25: Basal cells of perichaetial leaf. 26: Basal transverse section of branch leaf. 27: Upper transverse section of branch leaf (all from isotype, NY 01086534). Line scales: A = 2 mm (1-7, 12, 20-23); B = 400 μm (19); C = 100 μm (11, 26, 27); D = 67 μm (8-10, 13-18, 24-25).

78. *Macromitrium frustratum* B.H. Allen, Novon 8: 115. f. 3. 1998. (Figure 105)

Type protologue and citation: Honduras, Lempira: Montana de Celaque. Summit of Cerro la Castilla, ca 12 km SW of Gracias, 14°33′N, 88°41′W. On trunk of Pinus, *Allen 11542* (holotype: MO!; isotype: H 3194058!).

(1) Plants robust, yellowish--green above, dark rusty-brown below; stems slightly creeping; branches to 4 cm long. (2) Branch leaves strongly keeled, erect-clasping at base, flexuose to spirally twisted-contorted and undulate above, occasionally with the apices curved to circinate when dry, abruptly squarrose-spreading from an erect clasping base when moist, 6-9 × 0.5 mm, linear-lanceolate; apices fragile and long-subulate, acuminate; margins often undulate, serrate; upper and medial interior cells rounded-quadrate, subquadrate, smooth, thick-walled, in longitudinal rows, marginal cells longer and narrower than their ambient cells, forming a differentiated border; basal cells long-rectangular, 40-70 μm, thick-walled and porose, strongly tuberculate. (3) Perichaetial leaves not much differentiated from branch leaves; upper cells quadrate or subquadrate; medial cells rectangular or short-rectangular; basal cells elongate, smooth, with straight lumens. (4) Setae 20-30 mm long, smooth, not twisted or twisted to the left. (5) Capsule urns ovoid, ovoid-ellipsoid, ellipsoid to ellipsoid-cylindric, slightly constricted beneath the mouth, distinctly furrowed or wrinkled; opercula long conic-rostrate to conic-subulate; peristome double, exostome teeth truncate, yellow and densely papillose, fused at base forming an erect membrane, more or less discrete above, endostome hyaline, with basal membrane and upper segments; spores anisosporous. (6) Calyptrae mitrate, lacerate, naked.

Distribution: Guatemala, Honduras.

Specimens examined: **GUATEMALA**. Dept. Chiquimula, *Julian A. Steyermark, 31448* (F-C 0001080F). **HONDURAS**. Cortes, Cusuco National Park. *Bruce Allen 14213* (MO 3979197, H 3090263); Lempira, Montana de Celaque, *Bruce Allen 12078* (MO 3972911); lempira, Montana de Celaque, *Bruce Allen 12090* (H 3090264).

79. *Macromitrium fulgescens* E.B. Bartram in Grout, Bryologist 47: 12. 1944. (Figure 106)

Type protologue and citation: On tree, la Palma, province of San José, Costa Rica, altitude about 1800 meters, *Paul C. Standley*, march 17, 1924, *no. 38011* (holotype: FH 00213618!).

Macromitrium standleyi var. *subundulatum* E.B. Bartram, Contr. U.S. Natl. Herb., 26: 86. f. 24. 1928. *fide* Goffinet, 1993. Type protologue and citation: On tree, Cerro de las Caricias, north of San Isidro, Province of Heredia, Costa Rica, altitude 2000 t 2400 meters, *Paul C. Standley* and *juvenal Valerio*, March 11, 1926, *no. 52096*. (holotype: FH!; isotypes: NY 01243663!, US 00070279!).

**Macromitrium fuscescens* E.B. Bartram, Contr. U.S. Natl. Herb., 26: 89. f. 28. 1928. *fide* Bowers, 1974.

(1) Plants medium-sized to robust, lustrous, rusty to yellowish brown; stems creeping, branches to 6 cm long. (2) Branch leaves clasping at base, strongly contorted-flexuose when dry, abaxially curved spreading and flexuose when moist, 7-10 × 0.5 mm, long linear-lanceolate, threadlike; apices long-filiform acuminate; margins sinuate-denticulate below, sharply denticulate above, recurved below, plane above; costae excurrent; all cells smooth and clear; upper cells round-quadrate, elliptic, 8-18 × 8 μm, slightly bulging, in longitudinal rows; upper marginal cells elongate, forming a distinct border; basal cells narrowly rectangular, 40-48 μm, thick-walled and porose.

Macromitrium scoparium Mitt. and *M. fulgescens* E.B. Bartram are similar in their linear lanceolate with limbidium in upper margin, but the former differs from the latter by its leaves with tuberculate basal cells.

Distribution: Costa Rica.

80. *Macromitrium fulvum* Mitt., J. Proc. Linn. Soc., Bot., Suppl. 1: 52. 1859. (Figure 107)

Type protologue: In Ceylon, ad Rambodde, *Gardner (no. 233)*. Type citation: Ceylon, [Gardner] *233*, 39a (holotype: NY 00845352!; isotype: BM 000919515!).

(1) Plants medium-sized, brown or yellowish brown; stems creeping, in loose mats, with long and thick branches. (2) Branch leaves twisted-contorted and flexuous when dry, spreading when moist, linear-lanceolate, with a narrowly acuminate apex; all cells longer than wide, smooth, moderately conic-bulging, in distinctly longitudinal rows; costae percurrent or ending beneath the apex. (3) Perichaetal leaves lanceolate from an oblong base, acuminate, shorter than branch leaves, cells long rectangular, in regularly longitudinal rows. (4) Setae 4-10 mm long, smooth, vaginulae hairy. (5) Capsule urns ovoid, swollen, sulcate; peristome double, exostome yellowish membrane, endostome pale. (6) Calyptrae large, plicate, naked, lacerate to the medial portion.

Macromitrium fulvum Mitt. is similar to *M. sulcatum* (Hook.) Brid., but the latter can be distinguished from the former by its branch leaves with small (2.5-4.0 μm wide), irregularly quadrate, rounded-quadrate upper cells in regularly oblique rows and thick-walled and strongly tuberculate basak cells.

Habitats: on rocks or rotten trunks.

Distribution: Sri Lanka.

Specimens examined: **SRI LANKA**. Nuwara Eliya, *A. H. M. Jayasuriya 21/13-2* (MO 2859743), *20/47* (MO 2859756), *23/132-2* (MO 2859913); Nuwara Eliya, *W. Meijer 1857* (MO 5361159), *1780* (MO 5361157), *1872 a* (MO 5361158); Nallathanniya, *C. Ruinard & A. H. M. Jayasuriya 22/104* (MO 2859929).

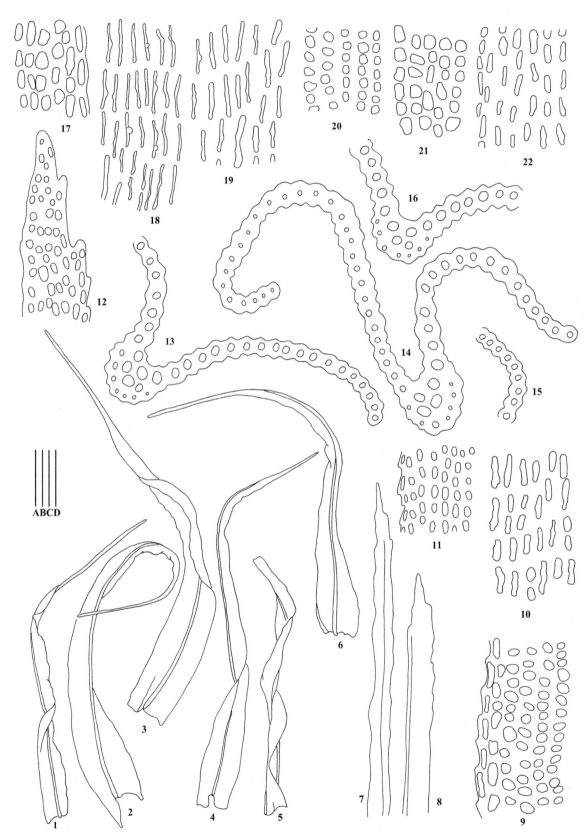

Fig. 105 *Macromitrium frustratum* B.H. Allen 1-6: Branch leaves. 7-8: Apices of branch leaves. 9: Upper cells of perichaetial leaf. 10: Low cells of perichaetial leaf. 11: Upper cells of branch leaf. 12: Apical cells of branch leaf. 13, 16: Medial transverse sections of branch leaves. 14, 15: Basal transverse sections of branch leaves. 17: Upper marginal cells of branch leaf. 18: Basal cells of branch leaf. 19, 22: Low cells of branch leaves. 20: Medial cells of branch leaf. 21: Medial cells of perichaetial leaf (all from holotype, MO 406388). Line scales: A = 2 mm (1-6); B = 400 μm (7-8); C = 100 μm (12); D = 67 μm (9-11, 13-22).

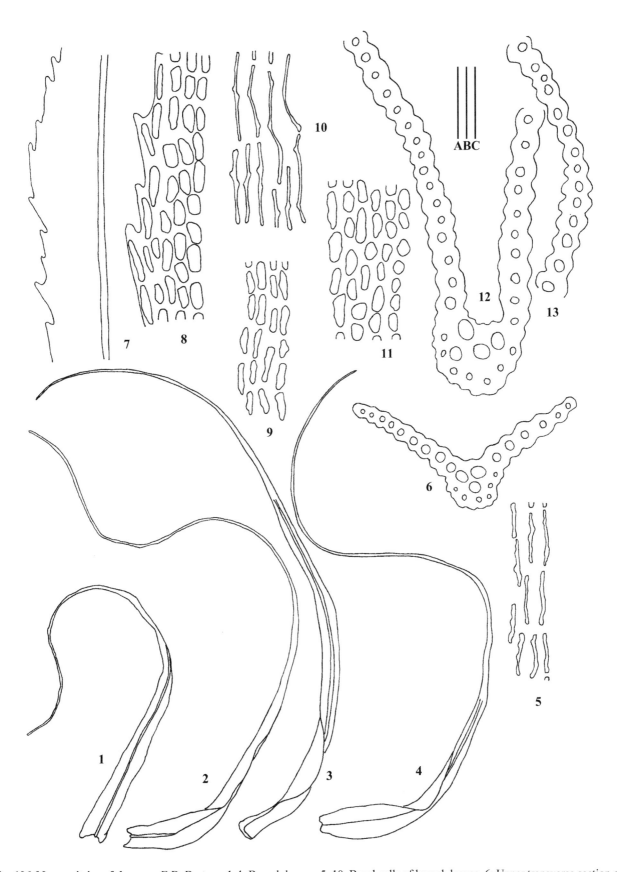

Fig. 106 *Macromitrium fulgescens* E.B. Bartram 1-4: Branch leaves. 5, 10: Basal cells of branch leaves. 6: Upper transverse section of branch leaf. 7: Margin of branch leaf. 8: Upper cells of branch leaf. 9: Low cells of branch leaf. 11: Medial cells of branch leaf. 12: Low transverse section of branch leaf. 13: Basal transverse section of branch leaf (all from holotype, FH 00213618). Line scales: A = 2 mm (1-4); B = 100 μm (7); C = 67 μm (5-6, 8-13).

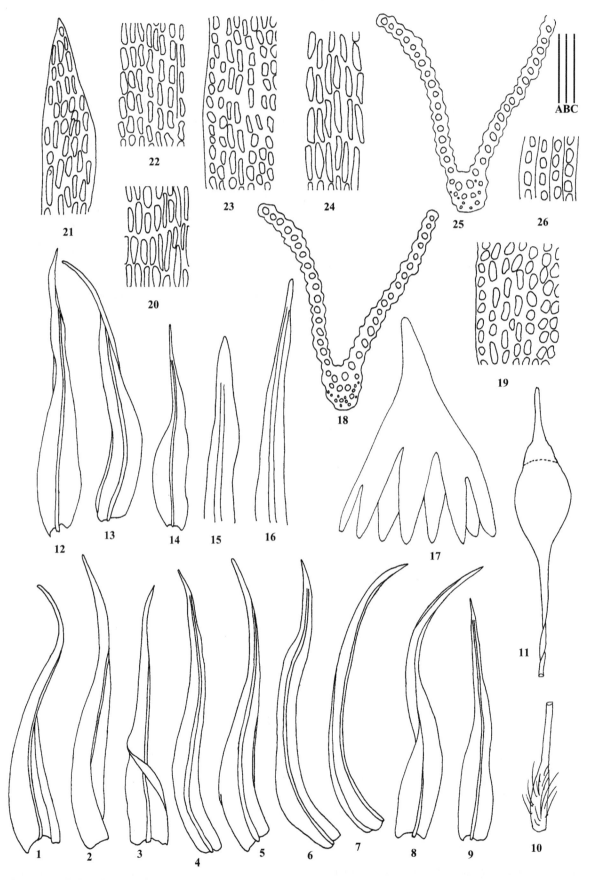

Fig. 107 *Macromitrium fulvum* Mitt. 1-8, 12-14: Branch leaves. 9: Perichaetial leaf. 10: Vaginula. 11: Capsule. 15, 16: Apices of branch leaves. 17: Calyptra. 18, 25: Basal transverse sections of branch leaves. 19, 22, 23, 26: Medial cells of branch leaves. 20, 24: Basal cells of branch leaves. 21: Apical cells of branch leaf (all from holotype, NY 00845352). Line scales: A =1 mm (1-14, 17); B = 200 μm (15, 16); C = 67 μm (18-26).

81. *Macromitrium funicaule* Schimp. ex Besch., Ann. Sci. Nat., Bot., sér. 6, 9: 359. 1880. (Figure 108)

Type protologue: [Mauritius] Maurice: sur les arbrre de la montagne Rouge, aux Trois-Ilets, *A. Darnty*, Juin 1874 (in herb Schmper). Type citation: Maurice: sur les arbrre de la montagne, 27 Juin 1874, *A. Darnty*, Herb. Bescherelle (isotype: BM 000873889!).

(1) Plants medium-sized, brown-yellowish above, dark brown below; stems weakly creeping, with erect and thick branches. (2) Branch leaves erect and appressed below, regularly and spirally twisted-curved in rows, the upper portions curved toward one side or curly when dry, funiculate when dry, spreading, open and flat with plane margins when moist, oblong- to ligulate-lanceolate, with an acute to mucronate apex; costae subpercurrent to percurrent, sometimes concealed in adaxial view by overlapping folds of laminae; margins entire and plane throughout; upper and medial lamina frequently bistratose, cells rounded to subquadrate, pluripapillose; low cells rectangular, clear and smooth; basal cells narrowly rectanguler, occasionally porose, thick-walled, smooth. (3) Perichaetial leaves broader than branch leaves. (4) Setae about 8 mm long, smooth. (5) Capsule urns ovoid to ellipsoid, wrinkled when dry, with a wide mouth; opercula conic-rostrate to conic-subulate; peristome absent. (6) Calyptrae mitrate, naked and smooth, lobed at base.

Distribution: Mauritius.

82. *Macromitrium funiforme* Dixon, Proc. Roy. Soc. Queensland 53: 30. 1941.

Type protologue: [Australia] Rain forest, eastern slope of Mount Bartle Frere, 28 Oct., 1939; coll. *H. Flecker (6411)*.

(1) Plants robust, lustrous, red-brown to golden-green above, dark chestnut-brown below, in spreading mats; stems creeping, with erect branches, branches up to 18 mm long, very regularly spaced. (2) Stem leaves flexuose-twisted when dry, widely recurved-spreading when moist, 1.5-2.0 mm long, ovate to ovate-lanceolate, gradually contracted to an acuminate-subulate upper portion; elongate low cells continuing nearly to apex. (3) Branch leaves funiculate, spirally twisted-curved in inconspicuous rows, the upper portions flexuose-curved when dry, spreading incurved when moist, 1.5-2.5 mm long, narrowly ovate-lanceolate to lanceolate, abruptly acuminate to narrowly acute, some slenderly cuspidate-acuminate; margins plane, entire; costae filling the acumen and excurrent; all cells flat, clear and smooth except basal cells occasionally papillose; upper cells 5-8 µm wide, subquadrate-rounded to irregularly-rounded, mostly thick-walled, chlorophyllose; medial cells short- to long-rectangular, lumens straight to curved, moderately and mostly evenly thick-walled, quickly grading into basal cells; basal cells 19-48 × 7-8 µm, with 3-5 µm wide lumens, elongate to rectangular-elongate, evenly thick-walled. (4) Perichaetial leaves 2.0-2.5 mm long, erect, straight or nearly so, with a subsheathing low portion, oblong-lanceolate to ovate-lanceolate, grading to a rather abrupt, long acumen; smooth, elongate low cells continuing to the apex or ending just below. (5) Setae 8-10 mm long, flexuose, smooth, twisted to the left; vaginulae naked. (6) Capsule urns ovoid to ellipsoid, smooth, narrowed to the 8-plicate and puckered, darker mouth; peristome single, exostome of 16, well-developed teeth; spores anisosporous. (7) Calyptrae mitrate, naked, plicate.

Macromitrium funiforme Dixon is somewhat similar to *M. microstomum* (Hook. & Grev.) Schwägr. in having funiculate branches, lanceolate branch leaves with smooth cells in upper and medial portions, smooth ovoid to ellipsoid capsules with a puckered mouth, and naked calyptrae. However, *M. funiforme* can be separated from *M. microstomum* by its larger plants and branch leaves occasionally with papillose basal cells.

Distribution: Australia.
Specimens examined: **AUSTRALIA**. Queensland, *D. H. Vitt 27925* (H 3090266), *H. Streimann 57129* (H 2090265).

83. *Macromitrium fuscescens* Schwägr., Sp. Musc. Frond., Suppl. 2 2(2): 129. pl. 191. 1827. (Figure 109)

Type protologue: (Mariana island) In insulis Marianis lectum misit cl. *Gaudichaud*. Type citation: Mariana *Gaudichaud s.n.*, lectotype designated by Vitt *et al.*, 1995: G 00042105!; isotype: G 00042106!).

Macromitrium calvescens Bosch & Sande Lac., Bryol. Jav. 1: 125. 103. 1860, *fide* Vitt *et al.*, 1995. Type protologue: Indonesia. Habitat insulam *Javae*, prop Tjiburrum legit Hasskarl. Type citation: Bryol. Java, Tjiburrum, leg. *Hasskarl* (isotype: H-BR 2609001!).

Macromitrium semipellucidum Dozy & Molk., Ann. Sci. Nat. Bot. sér. 3, 2: 311. 1844, *fide* Vitt *et al.*, 1995. Type protologue: (Indonesia) Borneo, Java. Type citation: Borneo (isosyntypes: H-BR 2613002!, H-BR 2612006!).

Macromitrium calvescens Bosch & Sande Lac., Bryol. Jav. 1: 125. 103. 1860, *fide* Vitt *et al.*, 1995. Type protologue and citation: (Indonesia) Habitat insulam *Javae*; prope Tjiburrum legit *Hasskarl*. (lectotype: L-Lacoste); Java, Prope Tjiburrum, Let. *Hasskarl* (isotype: H-BR 2609001!).

Macromitrium miquelii Mitt. ex Bosch & Sande Lac., Bryol. Jav. 1: 130. 108. 1860, *fide* Vitt *et al.*, 1995. Type protologue: (Indonesia) Habitat insulae *Javae* prov. *Buitenzorg*, anno 1843 lectum herb. Miquel (lectotype: L-Lacoste).

Macromitrium glaucum Mitt., J. Linn. Soc., Bot. 10: 167. 1868, *fide* Vitt *et al.*, 1995. Type protologue: (Samoa) Tutuila, forming extensive mats on Bread-fruit trees nearly at the sea-level, *T. Powell 109*. Type citation: Samoa, 170° Long.W, 14° Lat S., Rev.*T. Powell 109* (isotype: F-C 0001109F!).

Macromitrium eurymitrium Besch., Bull. Soc. Bot. France 45: 64. 1898, *fide* Vitt *et al.*, 1995. Type protologue: Society Islands. Tahiti. vallée de Puaa (1ʳᵉ herbor., no 262; 2ᵉ herbor., no 263; 3ᵉ herb., no 264; 6ᵉ herb., no 265). Type citation: Tahiti, 1896, legit *Dr. Nadeud s.n.* determ. Bescherelle (isosyntype: H-BR 2618021!); Tahiti, Legit *Dr. Nadearu, no. 263* (isosyntype: H-BR 2618023!).

Macromitrium subsemipellucidum Broth., Bernice P. Bishop Mus. Bull. 40: 19. 1927, *fide* Vitt *et al.*, 1995. Type protologue: Hawaii. West Maui: crater of Olowalu, on trunks, elevation 240 meters. Type citation: Insulae Sandwich, in crater Olowalu, ins Maui occid, in truncis, arborum, 800 ft. 1875 legit, *D. D. Baldwin 522* (lectotype: H-BR 2618017!; isolectotype: H-BR 2618018!).

(1) Plants small, dull; stems long-creeping, often covered with rhizoids; ascending branches about 1.5 cm long. (2) Stem leaves appressed, lanceolate. (3) Branch leaves erect-appressed below, curly, twisted-decurved in inconspicuous rows, sometimes strongly crisped above, funiculate to varying degrees when dry, spreading to wide-spreading, squarrose-spreading, with erect to incurved upper portion when moist, 1.3-2.5 mm long, lanceolate to broadly lanceolate, shortly acuminate-mucronate to acute-apiculate, channeled below apex; margins plane above, plane to broadly reflexed below, entire; costae excurrent and filling a short mucro or apiculus; upper and medial cells 4-6 × 3-6 μm, quadrate-rounded to subquadrate, densely pluripapillose, rather obscure, sharply (2-3 cells) grading to long, clear, hyaline, curved-sigmoid basal cells (6-10 × 16-30 μm). (4) Perichaetial leaves 1.5-2.0 mm long, erect, ovate-lanceolate, acute; costae short-excurrent, excurrent, percurrent, or ending just beneath the apex; all cells smooth and clear. (5) Setae 4-9 mm long, smooth, twisted to the left; vaginulae with a few hairs. (6) Capsule urns 1.0-1.4 × 0.6-0.8 mm, narrowly ellipsoid to oblong-ellipsoid, smooth or slightly ribbed at the mouth, sharply contracted to the seta; peristome single, exostome of 16, blunt, well-developed teeth, often deciduous, hyaline; endostome absent; spores 10-55 μm, anisosporous. (7) Calyptrae mitrate and hairy, lacerate at base.

Distribution: Indonesia, Malaysia, Papua New Guinea, Philippines, Thailand.

Specimens examined: **INDONESIA**. Borneo, *Native collector, 2738* (H-BR 2612004), *Native collector, 2-241* (H-BR 2612005); West Java, *Veldhuis 9449* (as *M. miquelii*, BM 000852405), *Buitenzorg* (BM 000852406). **MALAYSIA**. Pensiangan, *Klazenga N. s.n.* (MEL 2361948). **PAPUA NEW GUINEA**. Milne Bay, *Kumei 120* (MO 4435095); West Sepik, *T. J. Koponen 34864* (MO 4435572); Morobe, *A. Touw 14728* (H 3205162, MO 5375830); Morobe, *H. Streimann 25700* (MO 5138315, KRAM-B-137444); Northern, *S. He 44302* (MO). **PHILIPPINES**. Butuan, *C. M. Weber 1316* (H-BR 2613004); Luzon, *A. D. E. Elmer 22288* (MO 1036886); Mindanao, *C. M. Weber 1301* (MO 757070), *C. M. Weber 1316* (H-BR 2673001). **THAILAND**. Nakornsrithamarat, *A. Touw 12018* (MO 2148998), *A. Touw 11499* (MO 2145294), *A. Touw 11533* (MO 2147854); Peninsular, *C. Charoenphol, K. Larsen & E. Warncke 3910* (MO 3971187).

84. ***Macromitrium fuscoaureum*** E.B. Bartram, Contr. U.S. Natl. Herb. 26: 83. f. 20. 1928. (Figure 110)

Type protologue & Type citation: On tree, Cerros de Zurqui, northeast of San Isidro, Province of Heredia, Costa Rica, altitude 2000 to 2400 meters, March 3, 1926, *Paul C. Standley* and *Juvenal Valerio, no. 50403*. (holotype: FH!; isotypes: H-BR 2625037!, NY 01086521!, NY 01086522!).

(1) Plants large and robust, in loose wefts, reddish brown to dark-brown, lustrous; stems weakly creeping, branches to 40 mm long. (2) Branch leaves strongly keeled, erect-clasping at base, widely to squarrosely spreading, flexuose and spirally contorted above when dry, abruptly squarrose- recurved spreading from an erect clasping base when moist, 7-10×0.5 mm, narrow, long linear-lanceolate; apices acuminate to long-subulate; upper margins occasionally undulate and serrulate, low margins recurved; all cells longer than wide, thick-walled and porose, smooth and clear; upper inner cells 30-40 μm × 12 μm, fusiform to long-rhomboidal, upper marginal cells becoming longer and narrower, forming a conspicuous differentiated border; low and basal cells 40-70 μm, long-rectangular; basal marginal enlarged teeth-like cells at insertion not differentiated; costae percurrent. (3) Perichaetial leaves similar to branch leaves. (4) Setae smooth, up to 30 mm long, reddish. (5) Capsule urns ellipsoid-cylindric, smooth to weakly furrowed. (6) Calyptrae mitrate, smooth and naked, lacerate.

Macromitrium fuscoaureum E.B. Bartram is similar to *M. trichophyllum* Mitt., but differs from the latter in having shorter branch leaves with a strongly clasping base. *Macromitrium echinatum* B.H. Allen and *M. scoparium* Mitt. are also similar to *M. fuscoaureum* in leaf shape and upper differentiated border. However, *M. echinatum* differs from *M. fuscoaureum* in having branch leaves without clasping at base and widely spreading when moist, and with serrate upper margins, shorter setae and hairy calyptrae. *Macromitrium scoparium* differs from *M. fuscoaureum* in having shorter branch leaves (4-6 mm long) with a weakly tuberculate base.

Distribution: Colombia, Costa Rica, Ecuador, Panama, Peru; Nicaragua (Allen, 2002).

Specimens examined: **COLOMBIA**. Narino, Pasto Municipio, *S. Churchill & W. Rengifo M. 17512* (MO 4433026), *Robbin C. Moran 4016A* (MO 5244202), Bocas del Toro Province, *B. Allen, 5716B* (MO 3059582). **COSTA RICA**. Puntarenas, *M. J. Lyon 472A* (MO 4445311); Puntarenas *M. J. Lyon 275* (MO 4443519). **ECUADOR**. MoronaSantiago: Limòn indanza, *E. Toapanta 1649*. **PANAMA**. Veraguas, Barry, *E. Hammel 8576* (MO). **PERU**. Cajamarca: San Ignacio, *Jasmin Opisso et al. 628* (MO 5275520).

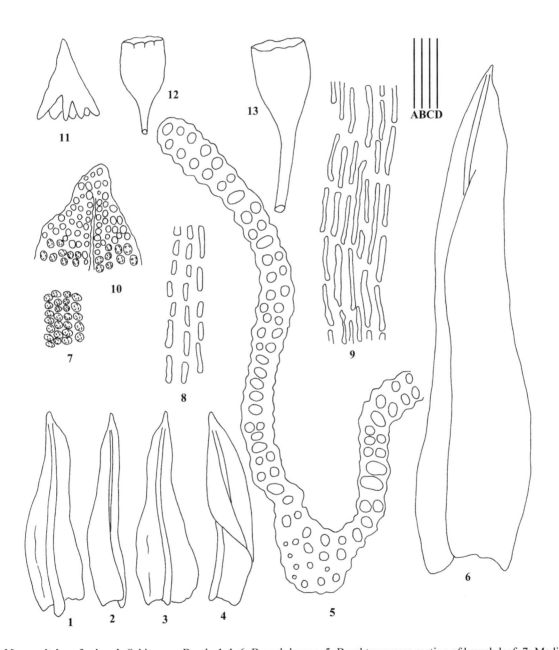

Fig. 108 *Macromitrium funicaule* Schimp. ex Besch. 1-4, 6: Branch leaves. 5: Basal transverse section of branch leaf. 7: Medial cells of branch leaf. 8: Low cells of branch leaf. 9: Basal cells of branch leaf. 10: Apical cells of branch leaf. 11: Calyptra. 12-13: Capsules (all from isotype, BM 000873889). Line scales: A = 2 mm (11-13); B = 1 mm (1-4); C = 400 μm (5); D = 67 μm (6-10).

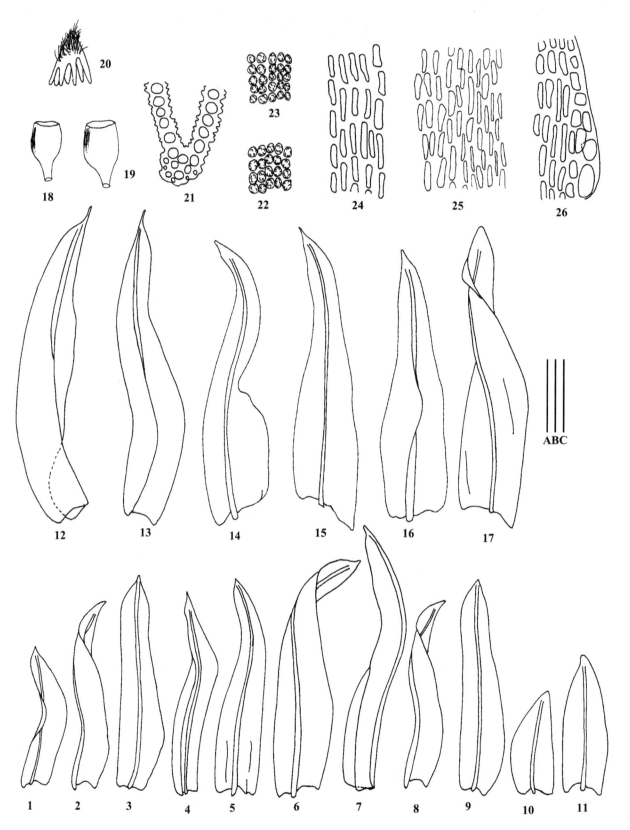

Fig. 109 *Macromitrium fuscescens* Schwägr. 1-17: Branch leaves. 18,19: Capsules. 20: Calyptra. 21: Upper transverse section of branch leaf. 22: Medial cells of branch leaf. 23: Upper cells of branch leaf. 24: Basal cells near costa of branch leaf. 25: Basal cells of branch leaf. 26: Basal marginal cells of branch leaf (all from lectotype of *Macromitrium miquekii*, L). Line scales: A = 1.76 mm (18-20); B = 0.44 mm (1-17); C = 44 µm (21-26).

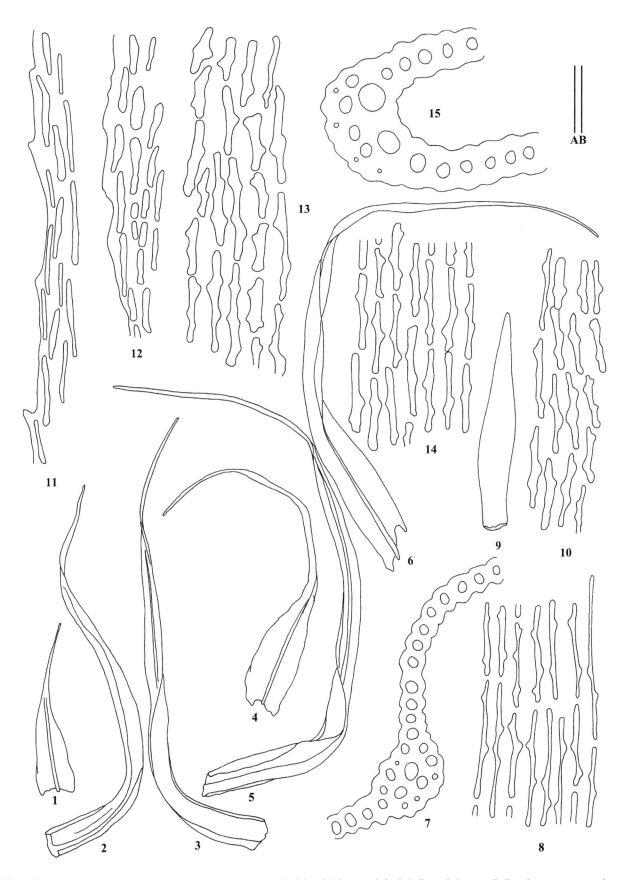

Fig. 110 *Macromitrium fuscoaureum* E.B. Bartram 1, 4: Perichaetial leaves. 2-3, 5-6: Branch leaves. 7: Basal transverse section of branch leaf. 8: Basal cells of branch leaf. 9: Calyptra. 10: Basal cells of perichaetial leaf. 11: Apical cells of branch leaf. 12: Apical cells of perichaetial leaf. 13: Basal cells of perichaetial leaf. 14: Medial cells of branch leaf. 15: Medial transverse section of branch leaf (all from holotype, US 00070259). Line scales: A = 2 mm (1-6, 9); B = 67 μm (7-8, 10-15).

85. *Macromitrium galipense* Müll. Hal., Syn. Musc. Frond. 2: 643. 1851. (Figure 111)
Type protologue: Venezuela, Galipan, ad arbores alt. 6000 pedum: Wagner m. Septbr. 1849 c. Type citation: Venezuela, Galipan, leg. *Wagner s.n.* (isotypes: S-B 163577!, NY 01086536!).

(1) Plants small, rusty brown; stems weakly creeping, branches 3-8 mm long, 1.5 mm thick, densely with rusty reddish rhizoids. (2) Branch leaves not very shriveled, keeled, appressed below, spirally twisted to flexuose when dry, somewhat funiculate, erect-spreading when moist, lanceolate, acuminate to acuminate-acute; margins serrulate near the apex, entire below; upper and medial cells rounded, rounded-quadrate, rounded-hexagonal, strongly bulging, smooth and collenchymatous; low and basal cells elongate, narrow rectangular, tuberculate. (3) Perichaetial leaves broader than vegetative, broadly oblong-lanceolate, with the widest part at base, acute; all cells longer than wide, elongate-rectangular, collenchymatous and smooth. (4) Setae 3-5 mm long, smooth, twisted to the left; vaginulae sparsely hairy. (5) Capsule urns ovoid to cupulate, moderately furrowed. (6) Calyptrae longer, lacerate at base, smooth and naked.

Macromitrium galipense Müll. Hal. is similar to *M. podcarpi* Müll. Hal. in having lanceolate leaves with bulging to mammillose upper cells, strongly tuberculate basal cells, furrowed capsules. However, *M. galipense* differs from *M. podcarpi* in having collenchymatous cells in the medial and upper portions of branch leaves, and marginal enlarged teeth-like cells at the insertion not differentiated. *Macromitrium galipense* is also similar to *M. crosbyorum* B.H. Allen & Vitt, but could be separated from the latter in having shorter branch leaves (1.5-2.2 mm), short setae, marginal cells at the outmost row not differentiated.

Distribution: Venezuela.

86. *Macromitrium gigasporum* Herzog, Biblioth. Bot. 87: 333. 1916. (Figure 112)
Type protologue: [Bolivia] Auf Bäumen im Nebelwald über Comarapa, about 2600 m, No. 3932, 3784; im oberen Coranital about 2600 m, No. 3396; an der Waldgrenez bei Tablas, about 3400 m, no. 2806. Type citation: [Bolivia] Auf Bäumen im Nebelwald über Comarapa, about 2600 m, leg. *Th. Herzog 3932*, April 1911 (syntype: JE 04008680!); *Th. Herzog 3396* (syntype: JE 04008681!); *Th. Herzog 2806* (syntype: JE 04008682!).

(1) Plants large and robust, yellowish-brown above, dark-brown below; stems creeping with caducous leaves, loosely branched, branches up to 50 mm long, with branchlets. (2) Branch leaves appressed below, spreading and flexuose above when dry, spreading and abaxially curved when moist, linear-lanceolate from an oblong and sheathing base, gradually narrowed to form a setaceous-acuminate apex with a row of cells; margins almost entire, revolute below, the cells at the outmost margin smaller than their ambient cells, forming a differentiated border; all cells longer than wide, smooth and clear; upper and medial cells short-rectangular, 15-30 × 5 μm, gradually elongate towards base; low and basal cells elongate, long to linear-rectangular, thick-walled and porose; marginal enlarged teeth-like cells at insertion not differentiated. (3) Perichaetial leaves long-lanceolate, with the widest part at the base, gradually narrowed to a setaceous-acuminate apex, plicate below, with a sheathing base; all cells longer than wide, smooth and clear, rectangular, long-rectangular to linear-rectangular, porose at low and basal parts. (4) Setae smooth, 10-15 mm long, twisted to the left; vaginulae with short straight hairs. (5) Capsule urns ellipsoid, smooth to plicate to varying degrees, reddish brown; opercula conic-rostrate, beaks erect or oblique to various degrees; peristome double, both united, membraneous. (6) Calyptrae cucullate-mitrate, sparsely hairy.

Distribution: Bolivia.
Specimens examined: **BOLIVIA**. im Nebelwald über Comarapa, *Th. Herzog, 57341* (BM 000873110); *Th. Herzog 3734* (JE 04008679!), *Th. Herzog s. n.* (FH 00213619); im Waldgnemse des Rio Saujana, h. Hezog, 3242 (JE 04008678).

87. *Macromitrium giraldii* Müll. Hal., Nuov. Giorn. Bot. Ital. n.s. 3: 106. 1896. (Figure 113)
Type protologue: [China, Shannxi] Inmonte Tui-kio-san, [*Patr Giraldi*] Aug. 1894. Type citation: China, Prov. Schan-Si, Mt. Tui-Kio-San, *Patr Giraldi* 1894, Hb. C. Müller (isotype: PC 0083663!).

Macromitrium cavaleriei Card. & Thér. in Thér., Bull. Acad. Int. Géogr. Bot. 16: 40. 1906, **syn. nov.** Type protologue: China, Kouy Techéou, J. Cavalerie n°. *833*, Nov, 1902. Type citation: China, prov. Kouy Techeou, Pin fa, Lou mong. Fouan, Coll. *R. J. Cavalerie*, nov. 1902, *no 833* (lectotype designated by Guo & He, 2008: PC 0083631!; isolectotypes: PC 0083629!, BM 000919466!).

Macromitrium sinense E.B. Bartram, Ann. Bryol. 8: 13, f. 7. 1936, *fide* Guo & He, 2008b. Type protologue and citation: [China] Kweichow Prov., Fan Ching Shan, alt. 2250m, on bark of tree, *S.Y. CHEO no. 430* (Type); Kwangsi Prov., Lin Wang Shan, San King Shien, *S. Y. CHEO 2754*. Type citation: China, Kweichow Prov. Fan Ching Shan, 2250 m., *S. Y. CHEO 430*, 1931 (isotype: PC 0083730!); China, Kweichow, Fan Ching Shan, Chiang K'ou Hsien, Collected in cooperation between the Farlow Herbarium of Harvard University and University of Nanking, by *S. Y. Cheo No. 430*, Det E. B. Bartram, IX/7/1931 (isotype: F!).

Macromitrium syntrichophyllum Thér. & P. de la Varde., Rev. Bot. Bull. Mens. 30: 347, 1. 1918, *fide* Guo & He,

2008b. Type protologue and citation: China or. prov. de Ngan Hoei, Leoufang (900 m. alt.,), Oct. 1910, coll. *F. Courtois no. 332* (syntype: PC 0083731; isosyntype: BM 000919468!); Tchan kia po (prov. Nagan hoi), Oct. 1910, *no. 335* (syntype).

Macromitrium syntrichophyllum var. *longisetum* Thér. et Reim., Hedwigia 71: 55. 1931, *fide* Guo & He, 2008b. Type protolgoue: China, Prov. Kwangsi: Yao-shan, 2000 m., 5, I. 1929. *S. S. Sin, no. 2075.* (holotype: PC 0083734!).

Macromitrium gebaueri Broth., Symb. Sin. 4: 72. 1 f. 10. 1929, *fide* Guo & He, 2008b. Type protologue: W-Y.: Im birm. Mons. an Bäumen in der wtp. St. des Schweili-Salwin-Scheidegebirges, 25°45', 2000-2800 m. 1914 (Gebauer). Type citation: China Yunnan, An Bäumen in der wtp. St. des Schweli- Salween- Scheidegebirge, 25°45', 2000-2800 m. 1914 (*A. K. GEBAUER*) (marked as n. sp.) (holotype: H-BR 2572001!).

Macromitrium giraldii var. *acrophylloides* Müll. Hal., Nouv. Giorn. Bot. Ital. n. s. 5: 187. 1898, *fide* Lou *et al.*, 2014. Type protologue: Habittio, China interior, Prov. Schensi meridionalis, in loco haud notato: *J. Giraldi* accepit 1896. Type citation: Type citation: China interior, Provincia Schen-si mer. Sept. 1896. Rev. *J. Giraldt, sub. n 2136*, Determ. Prof. C. Muller (lectotype: H-BR 2576003!).

Macromitrium rigbyanum Dixon, J. Bombay Nat. Hist. Soc. 777, 1937, *fide* Lou *et al.*, 2014. Type protologue: Kurseong, Himalayas, 1926; coll. Fr. Rigby, comm. Rev. G. Foreau (592), Type. Assam, Him. Parbat, 2000 m, 21 March 1934; coll. N. L. Bor (89). Ibidem, Peak, Charduar, 1850 m; 1934 (167). Pankim La, Abor Hills 300-2900 m; 1934 (148). On rock, Dzulake, Naga Hlls. 1850 m; 31 August 1934 (280 bis). Type citation: Kurseong, Himalayas, 1926, coll: Fr. Rigby, comm. rev. g. Foreau (592) (holotype: BM 000825431!) ; Peak, 6000', Charduar, Assam, Coll. N. L. Bor (167) (paratype: BM 000825433!).

Macromitrium handelii Broth., Sitzungsber. Ak. Wiss. Wien, Math. Nat. Kl. Abt. 1, 131: 212. 1922 ("*Handelii*"), *fide* Lou *et al.*, 2014. Type protologue: Prov. Yunnan: Prope vicum Jöschuitang ad septentr. urbis Yunnanfu, 25°26' lat., in regione calide temperata, ad arbores frondosas, about 1800 m (Nr. 450). Prov. Hunan austro-occid..: In monte Yün-schan prope urbem Wukang, in silva elata frondosa umbrosa ad lignum putridum; about 1150 m (Nr. 12.192). Type citatoin: An Laubbäumen in der wtp. St., c.sp. Y.: Bei Jöschuitang n von Yunnanfu, 1800m. 9. III. 1914 (450) (lectotype: H-BR 2572011!; isosyntypes: PC 0083672!, JE, S-B 115565!); SWH.: Im Walde des Yünschan bei Wukang,. 1150m. 21. VI. 1918 (12.192)" (syntype: H-BR; isosyntypes: S-B 115580, JE, BM 000919481!).

Macromitrium cancellatum Y. X. Xiong, Acta Bot. Yunnan. 22: 405. f. 1. 2000, *fide* Lou *et al.*, 2014. Type protologue and citation: *Xiong Yuan-xin SY96011*; July 26, 1996; China. Guizou [sic]. Suiyang, Wangcao, Guanghua, 206'N. 1012'E. Altitude about 1150 m. On surface of rocks under forest (holotype: GACP!).

(1) Plants small to medium-sized, in dense mats; stems long-creeping, with erect branches, 5-25 mm long, densely foliate, often with several short branchlets at upper part of branches. (2) Stem leaves suberect to strongly reflexed when moist, oblong-lanceolate to triangular-lanceolate, 0.6-1.4 × 0.2-0.3 mm, acuminate at the apex; upper and medial subquadrate or rounded-quadrate, bulging, smooth or slightly papillose; low cells slightly elongate; basal cells rectangular. (3) Branch leaves individually twisted, strongly contorted-crisped and flexuose or curved above when dry, spreading, open and flat, occasionally abaxially curved when moist, (1.5-) 2.5-4.2 × 0.25-0.50 mm, linear-lanceolate, lanceolate, ovate-lanceolate to oblong-lanceolate, plane or plicate below; apices mostly acuminate or narrowly acuminate, sometimes broadly acuminate, obtuse-acute, incurved; margins revolute at one side near base, entire, sometimes indistinct serrulate above due to protruding papillae; costae ending a few cells beneath the apex, occasionally percurrent or shortly excurrent; medial and upper cells quadrate to subquadrate, 8-12 × 8-10 μm, slightly to moderately bulging, thin-walled, hyaline or slightly obscure, often moderately pluripapillose, occasionally almost smooth and collenchymatous; low cells rhombic, rectangular to sublinear, 15-30 × 4-6 μm, nearly hyaline, somewhat thick-walled, rarely smooth, often weak to moderately unipapillose; basal cells near margin at one side often differentiated, regularly rectangular, pellucid with thinner walls, occasionally basal juxtacostal cells irregularly rectangular, thin-walled, pellucid, forming a "cancellina region". (4) Perichaetial leaves oblong-lanceolate, often plicate at base, up to 2.5 mm long, narrowly acuminate to long acuminate or aristate at the apex; costae excurrent; cells mostly longer than wide, hyaline, smooth, sometimes unipapillose in medial and upper portion. (5) Setae 4-10 mm long, smooth. (6) Capsules erect, ovoid, ellipsoid-ovoid to shortly cylindric, slightly constricted at the mouth when old; opercula conic-rostrate; peristome single, exostome of 16, lanceolate, obtuse at the apex, densely papillose throughout. (7) Calyptrae campanulate, medium-sized, with numerous yellowish or yellowish-brown hairs.

Macromitrium giraldii Müll. Hal. was once considered as a synonym of *M. japonicum* Dozy & Molk. *Macromitrium cavaleriei* Cardot & Thér. has been widely recorded from East Asia. Our morphological and phylogenetic evidence based on ITS2, *trn*G, *trn*L-F revealed that *M. giraldii* was more similar to *M. cavaleriei* than to *M. japonicum*. Having compared the types of *M. cavaleriei* and its former synonyms with that of *M. giraldii*, we found that *M. cavaleriei* was actually conspecific with *M. giraldii*. In view of nomenclature, *M. giraldii* was acceptable. *Macromitrium cavaleriei* as well as its previous eight former synonyms were placed in synonymy with *M. giraldii*. The synonymization extends the distribution of *M. giraldi* from Shaanxi province, China to East, Southeast, Central and Southwest China, and to Japan, Vietnam and India.

Macromitrium giraldii is similar to *M. ferriei*, but the latter could be distinguished from the former by its obscure upper and medial laminal cells with distinct pluripapillae, its shorter and oblong-lanceolate inner perichaetial leaves, and its branch leaves with an obtuse, subacute or broadly acuminate apex. *Macromitrium giraldii* is also similar to *M. quercicola* Broth. and *M. rhacomitrioides* Nog. in their gametophytes. From *M. giraldii*, *M. rhacomitrioides* Nog. could be separated by its long and narrowly lanceolate branch leaves with rather obscure, papillose upper cells, while *M. quercicola* could be distinguished by its ligulate-lanceolate branch leaves with oblong-rhomboid, short oblong and distinctly unipapillose basal cells.

Distribution: China, India, Japan (Lou *et al.*, 2014).

Specimens examined: **CHINA**. Anhui, *Z. L. Wan & J. X. Luo 9026, 9125, 9349, 9361* (MO, PE); Fujian, *H. H. Chung B351* (PC), *H. H. Chung B90-b* (PC 0083732); Guangdong, *S. He 40329* (MO); Guangxi, *S. Y. Cheo 2754* (MO), *Z. Y. Li 89125* (MO), *X. L. He 00546* (MO); Guizhou, *M. Bourell 4775* (MO); *C. Gao 32530* (MO); *P. C. Wu 23922-b* (MO); *P. C. Wu 667* (PC 0083627); *P. C. Wu 1931* (PC 0083632); *J. Cavalerie 667* (PC 0083633); *J. Cavalerie & Fortunati 1696* (PC 0083630); Henan, *J. X. Luo 233* (MO); Hubei, *P. C. Wu 393* (MO); Hunan, *H. Handel-Mazzetti 11308, 11451, 11502* (H-BR). Jiangsu, *P. C. Wu & Y. X. Jin 2143* (MO); Jiangxi, *S. Y. Cheo 1000* (FH); *P. C. Chen 157 a, 273* (MO, PE), *P. C. Wu 002, 003* (MO, PE), *S. Y. Cheo 1000, 2754* (FH), *H. H. Chung 4182, 4302* (FH); Sichuan, *M. Z. Wang 860390, 860801* (FH, MO), *B. Allen 6722, 7194* (MO), *P. C. Chen 5315* (MO); Taiwan, *J. R. Shevock 18153* (MO); *C. C. Chuang 2024* (NY); Xizang, *Y. G. Su 2573, 4241, 4782, 4759, 2324, 4037, 4436* (MO); Yunnan, *P. L. Jr. Redfearn, S. He, Y. G. Su 299, 528, 693, 704* (MO); *D. G. Long 32603, 32622, 33857* (MO); Zhejiang, *P. C. Wu 7761* (MO, PE). **JAPAN**. Shizouka, *Z. Iwatsuki s. n.* (MO).

88. *Macromitrium glabratum* Broth., Biblioth. Bot. 88: 12. 1920. (Figure 114)

Type protologue: [Bolivia] In Nebelwald über Comarapq, about 2600 m (3785, 4243/a). Type citation: Bolivia, Santa Cruz: Auf Baumästen im Nebelwald über Comarapa, about 2600 m, *Herzog, T. C. J., 4243/a*, (syntype: H-BR 2625029!; isosyntype: JE 04008716!). Bolivia, Nebelwald uber Comarapa, ca 2600m, [*Herzog T. C. J.*] *3785* (syntype: H-BR 2625027!).

(1) Plants robust, lustrous, yellowish-green; stems weakly creeping, branches up to 5 cm long. (2) Branch leaves appressed and erect below, individually twisted, contorted-flexuose to curly above when dry, spreading and abaxially curved, often conduplicate above when moist, narrowly long lanceolate, with an acuminate apex; margins entire or serrate to serrulate near the apex; costae percurrent, stout; all cells clear, bulging, in longitudinal rows, smooth or occasionally weakly unipapillose near the base; cells near the apex oblong; upper cells rounded to rounded-quadrate; medial and low cells slightly elongate, rounded to elliptical; basal cells elongate, narrowly rectangular, weakly porose, becoming broader and hyaline near the costa, marginal cells in the outmost row broadly rectangular or quadrate. (3) Perichaetial leaves larger and longer than branch leaves, oval- to oblong-lanceolate, tapered to a setaceous apex, weakly sheathing at the base, plicate below, entire; costae vanished far beneath the apex; all cells longer than wide, gradually elongate from the apex to the base, some weakly porose. (4) Setae smooth, 8-15 mm long, twisted to the left. (5) Capsule urns ellipsoid-cylindric, furrowed; peristome double, both membranous, exostome teeth papillose-striate, endostome smooth. (6) Calyptrae mitrate, deeply lacerate, naked.

Macromitrium glabratum Broth. is similar to *M. pseudoserrulatum* E.B. Bartram, the latter differs from the former in having pluripapillose medial and upper cells of branch leaves.

Distribution: Bolivia.

89. *Macromitrium glaziovii* Hampe, Vidensk. Meddel. Dansk Naturhist. Foren. Kjøbenhavn, 3, 6: 143. 1874. (Figure 115)

Type protologue: [Brazil] Sub [*Glaziov, A. F. M.*] *Nr. 6385* parce lectum. Type citation: Brazil, Rio-Janeiro, *Glaziov, A. F. M., 6385,* (isotypes: G 00050747!, H-BR 2534004!, S-B 163572!).

(1) Plants small-sized, in loose mats, olive-green above, brownish below; stems densely with reddish tomentose below; branches up to 20 mm long. (2) Branch leaves strongly twisted-contorted and crisped, with inrolled the apices when dry, widely-spreading when moist, oblong- to ligulate-lanceolate, plicate, with an acute to acuminate apex; costae single and strong, ending in the apex; the cells near the apex oval, oblong, or elliptic, smooth; upper and medial cells rounded-quadrate, oval, bulging and pluripapillose, the outmost marginal cells smooth; low and basal cells rhombic, rectangular, thick-walled and strongly porose, occasionally unipapillose. (3) Perichaetial leaves longer than branch leaves, narrowly lanceolate, with acuminate the apices, yellowish; all cells longer than wide, from rectangular near the apex to linear-rectangular at the basal. (4) Setae smooth, 6-8 mm long. (5) Capsule urns ellipsoid, longitudinal plicate, wrinkled at the mouth; opercula erect, conic-rostrate; peristome single, exstome short lanceolate. (6) Calyptrae campanulate-mitrate, smooth and naked.

Distribution: Brazil.

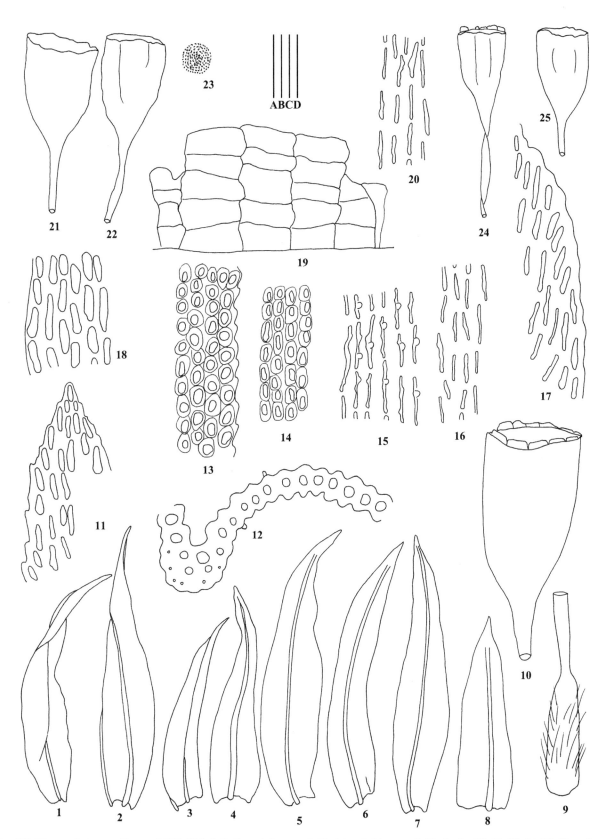

Fig. 111 *Macromitrium galipense* Müll. Hal. 1-7: Branch leaves. 8: Perichaetial leaves. 9: Vaginula. 10, 21-22, 24-25: Capsules. 11: Apical cells of branch leaf. 13: Upper cells of branch leaf. 14: Medial cells of branch leaf. 15: Basal cells of branch leaf. 16: Medial cells of perichaetial leaf. 17: Apical cells of perichaetial leaf. 18: Basal marginal cells of branch leaf. 18: Peristome. 20: Low cells of perichaetial leaf. 23: Spore (all from isotype, NY 01086536). Line scales: A = 2 mm (9-10, 21-22, 24-25); B = 1 mm (1-8); C = 100 μm (19, 23); D = 67 μm (11-18, 20).

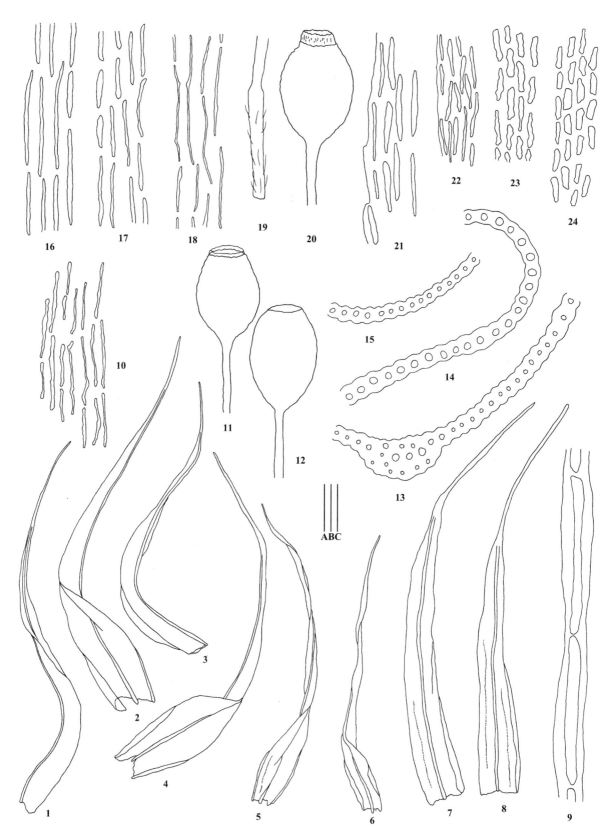

Fig. 112 *Macromitrium gigasporum* Herzog 1-6: Branch leaves. 7-8: Perichaetial leaves. 9: Apical cells of perichaetial leaf. 10: Basal cells of branch leaf. 11-12, 20: Capsules. 13-15: Basal transverse sections of branch leaves. 16: Basal cells of perichaetial leaf. 17: Upper cells of perichaetial leaf. 18: Low cells of perichaetial leaf. 19: Vaginula. 21: Apical cells of branch leaf. 22: Upper cells of branch leaf. 23-24: Medial cells of branch leaves (1-4, 7-24 from FH 00213619; 5-6 from JE 04008678). Line scales: A = 2 mm (5-8, 11-12, 19-20); B = 1 mm (1-4); C = 67 μm (9-10, 13-18, 21-24).

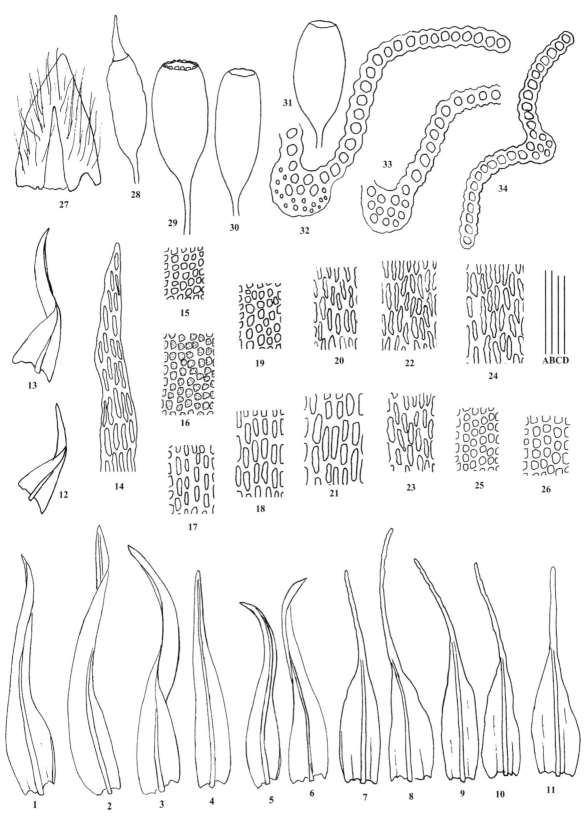

Fig. 113 *Macromitrium giraldii* Müll. Hal. 1-6: Branch leaves. 7-11: Perichaetial leaves. 12-13: Stem leaves. 14: Apical of pericahaetial leaf. 15: Upper cells of branch leaf. 16, 18: Low cells of branch leaves. 17: Medial cells of branch leaf. 19: Upper cells of perichaetial leaf. 20: Basal cells of branch leaf. 21: Low cells of perichaetial leaf. 22: Medial cells of perichaetial leaf. 23: Basal cells of perichaetial leaf. 24: Medial cells of stem leaf. 25: Basal cells of stem leaf. 26: Calyptra. 27: Dry capsule. 28-31: Wet capsules. 32: Low transverse section of branch leaf. 33: Medial transverse section of branch leaf. 34: Upper transverse section of branch leaf (1-3, 7-10, 14-15, 17-24, 27-34 from lectotype of *M. cavaleriei, J. Cavalerie 833*, PC 0083631; 4-6, 11-13, 25-26 from lectotype of *M. syntrichophyllum, Courtois 332*, PC 0083731; 16 holotype of *M. gebaueri*, H). Line scales: A = 1.00 mm (1-11, 27-31); B = 0.55 mm (12-13); C = 100 μm (17-19, 21); D = 68 μm (14-16, 20, 22-26, 32-34). (Guo & He, 2008b).

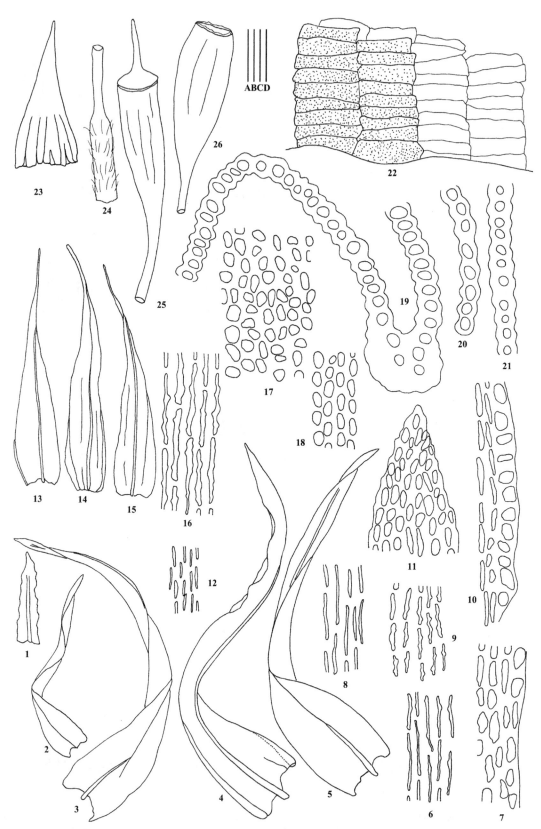

Fig. 114 *Macromitrium glabratum* Broth. 1: Apex of branch leaf. 2-5: Branch leaves. 6: Low cells of perichaetial leaf. 7: Basal cells of branch leaf. 8: Medial cells of perichaetial leaf. 9: Low cells of branch leaf. 10: Basal marginal cells of branch leaf. 11: Apical cells of branch leaf. 12: Upper cells of perichaetial leaf. 13-15: Perichaetial leaves. 16: Basal cells of perichaetial leaf. 17: Medial cells of branch leaf. 18: Upper cells of branch leaf. 19, 21: Low transverse sections of branch leaves. 20: Upper transverse section of branch leaf. 22: Peristome. 23: Calyptra. 24: Vaginula. 25, 26: Capsules (all from isosyntype, JE 04008716). Line scales: A = 2 mm (2, 13-15, 23-26); B = 1 mm (3-5); C = 100 μm (22); D = 67 μm (6-12, 16-21).

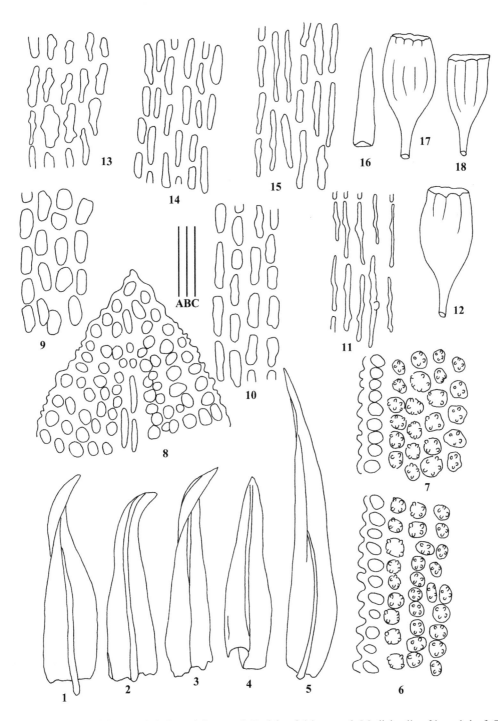

Fig. 115 *Macromitrium glaziovii* Hampe 1-4: Branch leaves. 5: Perichaetial leaves. 6: Medial cells of branch leaf. 7: Upper cells of branch leaf. 8: Apical cells of branch leaf. 9: Upper cells of perichaetial cells. 10: Low cells of branch leaf. 11: Basal cells of perichaetial leaf. 12, 17-18: Capsules. 13: Basal cells of branch leaf. 14: Medial cells of perichaetial leaf. 15: Low cells of perichaetial leaf. 16: Calyptra (all from isotype, G 00050747). Line scales: A = 4 mm (12, 16-18); B = 1 mm (1-5); C = 67 μm (6-11, 13-15).

90. *Macromitrium gracile* (Hook.) Schwägr., Sp. Musc. Frond., Suppl. 2, 1: 39. pl. 112. 1823. (Figure 116)
Basionym: *Orthotrichum gracile* Hook., Musci Exot 1: 27. 1818. Type protologue: In sinu Dusky Bay dicto, in Nova Zeelandia. *D. Menizies,* 1791. Type citation: New Zealand, leg. *D. Menzies* 1791 (isotypes: E 00002402!, E 00002403!).

*Macromitrium appendiculatum Müll. Hal., Hedwigia 37: 156. 1898, *fide* Wijk *et al.*, 1964.

Macromitrium mossmanianum Müll. Hal., Bot. Zeitung (Berlin) 9: 561. 1851, *fide* Wijk *et al.*, 1964. Type protologue: Nova Seelandia, ad rupes et truncos arborum humidos sylvarum ad Wairoa river prope Kaipara. Coll. 735. (Mossman, 1850). Type citation: New Zealand leg: *Mossman no. 755,* 1850 (isotype: JE 04008719!)

Macromitrium gracile var. *proboscideum* Dixon, New Zealand Inst. Bull. 3(6): 366. 1929, *fide* Vitt, 1983. Type protologue: On tree. *L. Waikaremoana, Hawkes Bay, G. O. K. Sainsbury,* Jan. 1924, *No. 48* (Type).

(1) Plants medium-sized to large, in stiff, loose and spreading mats; stems creeping, inconspicuous, with erect or ascending branches, branches up to 8 cm long. (2) Stem leaves ovate-lanceolate, gradually acuminate, upper cells rounded to elliptic, weakly pluripapillose, low cells rectangular-elongate, most tuberculate. (3) Branch leaves precisely and regularly twisted-curved, each leaf incurved or inrolled, with the upper portions curved to one side, somewhat funiculate when dry, spreading when moist; lanceolate from a broadly oblong or ovate base, gradually narrowed to a variable, broadly subula, the subula and upper portion often broken off; costae ending obscurely in or just below the apex of the subula; upper and medial cells rounded-quadrate to subquadrate, obscure, densely pluripapillose, flat, in distinct longitudinal rows, partially bistratose in subula, moderately thick-walled; transitional cells pluripapillose, rounded to elliptic-rectangular, gradually grading into low cells; basal cells elongate, thick-walled, mostly evenly so, most cells tuberculate but some smooth. (4) Perichaetial leaves stiffly erect, sheathing, broadly lanceolate to oblong-lanceolate, acuminate and shortly and stoutly subulate; subulae sometimes broken off; all cells longer than wide; upper cells elongate-rhombic and smooth; low cells tuberculate and elongate. (5) Setae 2-4 mm long, rather thick, smooth, twisted to the left. (6) Capsule urns narrowly ovoid, smooth or lightly 8-plicate; peristome single; exostome of 16, well-developed, easily broken off and often totally absent when old. (7) Calyptrae conic-mitrate, strongly plicate, naked or sparsely hairy, lacerate.

Distribution: New Zealand, Australia (Vitt & Ramsay, 1985).

Specimens examined: **NEW ZEALAND**. Auckland, *T.W.N. Beckett* (MEL 1030326); South Island, Central Otago, *D. H. Vitt 10537* (H 3090277), *W. Bell* (H-BR 2531017, H-BR 2531002); Track to Mt Stokes, *H. Streimann 51523* (KRAM-B-106784); North island: Mt. Egmont Area, *D. H. Vitt 29879* (H 3090282); Wellington Prov.: ex herb. Buchanan (H-BR 2531003).

91. *Macromitrium greenmanii* Grout, Bryologist 47: 11. 1944. (Figure 117)
Type protologue: Costa Rica, *Greenman* [sic], *5574* in Herb. Bartram. Type citation: Costa Rica, Feb, 13, 1922, *J. M. Greenman & M. T. Greenman 5574* (holotype: FH 00213622!; isotype: MO 406389!).

(1) Plants large and robust, reddish brown; stems weakly creeping, branches up to 3 cm high, with rusty-reddish rhizoids. (2) Branch leaves strongly keeled, erect below, flexuose to spirally contorted and undulate above when dry, erecto-patent when moist, 3-5 × 0.5-1.0 mm, lanceolate; apices shortly acuminate; margins entire and recurved below, serrulate and erectly plane above; costae percurrent; upper interior cells rounded, shortly rectangular to quadrate, smooth, bulging to mammillose, upper marginal cells not differentiated; medial cells elliptical and bulging; low and basal cells 24-40 µm long, rectangular, thick-walled and porose, strongly tuberculate near base. (3) Perichaetial leaves longer than branch leaves, lanceolate, gradually narrowed to a slender acumen or subula from a long oblong low portion; upper cells oval, rounded-quadrate; medial and low cells elongate, long-rectangular, thick-walled and porose. (4) Setae papillose, 10-15 mm long. (5) Capsule urns 2-3 mm long, ovoid, ovoid-ellipsoid, to ellipsoid-cylindric, furrowed; peristome double, exostome of 16, discrete and lanceolate, yellow to hyaline, papillose, endostome hyaline, smooth, basal membrane 140 µm high, segments 80 µm high; spores anisosporous. (6) Calyptrae 3-4 mm long, mitrate, naked, lacerate.

Macromitrium greenmanii Grout is almost identical to *M. longifolium* (Hook.) Brid., only the exostome teeth of *M. greenmanii* discrete and narrowly triangular while those of *M. longifolium* united to form a truncate, erect membrane.

Distribution: Costa Rica, Peru.

Specimens examined: **COSTA RICA**. Cartago, *I. Holz CR 00-0732* (MO 5262652). **PERU**. Amazonas, Chachapoyas, *J.-P. Frahm et al. 1103* (H 3090297).

92. *Macromitrium grossirete* Müll. Hal., Hedwigia 37: 153. 1898.
Type protologue: Nova Seelandia, insula sept., N. Canterbury, pattersons Creek, Weimacariri Gorge, ad truncus arborum: *T. W. Naylor Beckett* 1892 leg. et misit.

Macromitrium papillifolium Müll. Hal., Hedwigia 37: 154. 1898, *fide* Wijk *et al.*, 1964.

Macromitrium rigescens Broth. & Dix., *fide* Vitt 1983, Jour. Linn. Soc., Bot., 40: 446. 20-24. 1912. Type protologue and citation: Hab. Mt. Cook district (*Murray, no. 69*), c. fr. (lectotype: H-BR 2545015!).

(1) Plants medium-sized to robust; stems long-creeping, densely with erect branches, branches shorter and thicker, mostly simple, about 1.0 cm high. (2) Stem leaves erect-flexuose when dry, wide-spreading-recurved when moist, broadly lanceolate, gradually acuminate. (3) Branch leaves individually and irregularly twisted when dry, flexuose-spreading when moist, 2.5-4.5 mm long, long-ligulate, lanceolate-ligulate, abruptly acuminate-cuspidate to broadly and shortly acuminate; margins broadly reflexed, subentire to crenulate; upper cells rounded, thin-walled and thickened corners, chlorophyllose, strongly bulging, each surface with a strong, central, long-tuberculate papilla (up to 14 μm high); medial cells rounded to elliptic, others similar to upper cells; transitional cells elliptic-rectangular, distinctly porose, unipapillose to tuberculate, in longitudinal rows; basal cells long elongate, thick-walled, distinctly porose, smooth or occasionally tuberculate. (4) Perichaetial leaves oblong-lanceolate, acuminate, cells similar to those of branch leaves. (5) Setae 2.5-4.5 mm long, smooth, twisted to the right. (6) Capsule urns narrowly obovoid to narrowly ovoid, with a small mouth, 8-plicate beneath mouth, darker near the mouth; peristome absent. (7) Calyptrae large, mitrate, densely covered with delicate, smooth, yellowish-hyaline, long hairs.

Macromitriumg grossirete Müll. Hal. and *M. longipapillosum* D.D. Li, J. Yu, T. Cao & S.L. Guo are similar in their papillosity of branch leaves and lanceolate-ligulate leaf shape, but the latter can be separated from the former by its capsules with single peristome, forked tuberculae (up to 18 μm high) in the upper cells of the branch leaves.

Distribution: New Zealand.

Specimens examined: **NEW ZEALAND**. Otago prov., *W. Bell* (H-BR 2509001, H-BR 2509007); South Island: lake Rotoiti, *H. Streimann 58191* (H 3090301); Canterbury, *Irwin M. Brodo 24298* (MO 3066368, CANL), *Allan J. Fife 6254* (MO 3987279); Süd-Insel, Mt. Cook national park, *Schafer-Verwimp & Verwimp* (KRAM-B-101514).

93. *Macromitrium guatemalense* Müll. Hal., Syn. Musc. Frond. 2: 644. 1851. (Figure 118)
Type protologue: Guatemala, vulgatissimum et mensi Martii 1848. fruct. Matures supramaturis: Friedrichesthal? Ex horto dom. Van Houtte copiosissime communicavit Kegel, nunc hortulanus bot. Halensis. Type citation: Guatemala, Friedrichesthal (isotypes: NY 01202013!, FH 00213623!).

Macromitrium liberum Mitt., J. Linn. Soc., Bot. 12: 214. 1869, *fide* Vitt, 1994. Type protologue: Andes Novo-Granatenses, in arboribus ad viam inter pacho et Veragua (6500 pde.), *Weir 270.*

Macromitrium negrense Mitt., J. Linn. Soc., Bot. 12: 208. 1869, *fide* Valente *et al.*, 2020. Type protologue and citation: Fl. Negro, secus ostia, ad arbores rivuli Igarapé da Cachoeira, *Spruce 106* (isotype: E 00165156!).

Macromitrium paucidens Müll. Hal., Linnaea 42: 487. 1879, *fide* Grout, 1944.

Macromitrium penicillatum Mitt., J. Linn. Soc., Bot. 12: 217. 1869, *fide* Grout, 1944. Type protologue: Andes Quitenses, in Cordillera occidental ad Lucmas (5000 ped.). *Spruce 89*. Type citation: Ecuader, Andes Quite, ad Lumas in W. Cordilleres, @ 5000 ft, *Ric Spruce 89* (isotype: E00165155!).

Macromitrium rhystophylllum Müll. Hal., Bull. Herb. Boissier 5: 198. 1897, *fide* Grout, 1944. Type protologue: Habitatio, Inter Torsy et S. jeronimo, Augusto 1871. Coll. *No. 46.*

Macromitrium serrulatum Mitt., J. Linn. Soc., Bot. 12: 215. 1869, *fide* Vitt, 1994. Type protologue: Andes Quitenses, *Jameson s.n.*

Macromitrium subreflexum Müll. Hal., Bull. Herb. Boissier 5: 198. 1897, *fide* Grout, 1944. Type protologue: Habitatio, *Sn. Cristohal. Coll. No. 46.*

Macromitrium tortuosum Schimp. ex Besch., (Mém. Soc. Sci. Nat. Math. Cherbourg 16: 189. 1872, *fide* Grout, 1944. Type protologue: mejico, San Miguel (Liebmann in herb. Mus. Par.); Orizaba (F. Müller).

Macromitrium trianae Müll. Hal., Bot. Zeitung (Berlin) 15: 580. 1857, *fide* Grout, 1944. Type protologue: Patria. Nova Granada: J. Triana in Collect. *Linden. No.22 et No.12.*

Macromitrium verrucosum E.B. Bartram, Contr. U.S. Natl. Herb 26: 82. f. 19. 1928, *fide* Vitt, 1994. Type protologue: Costa Rica Prov. San Jose: near Quebradillas, 7 km N Sta. Maria de Dota, 1800 m, *Standley 43023* (isotype: JE 04008693!).

(1) Plants medium-sized; stems creeping, branches 1-2 cm long. (2) Branch leaves erect below, individually twisted, contorted-flexuose, and undulate above when dry, erect-spreading and somewhat rugose above when moist, 2-3 × 0.5-0.7 mm, oblong-lanceolate to lanceolate, with an acute apex; margins undulate above, sharply and regularly serrulate to serrate, recurved below, plane above; costa ending beneath the apex; upper cells rounded-quadrate, 5-7 μm, mammillose, upper marginal cells not differentiated; low and basal cells rectangular, thick-walled and porose, strongly tuberculate; marginal enlarged teeth-like cells at insertion distinctly differentiated. (3) Perichaetial leaves lanceolate, longer than branch leaves, all cells longer than wide. (4) Setae smooth, 4-15 mm long, twisted to the left. (5) Capsule urns ellipsoid-cylindric to cylindric, furrowed; peristome double, exostome teeth yellowish brown, densely papillose, united to form a truncate membrane, endostome hyaline, basal membrane present, segments rudimentary or absent; spores anisosporous. (6) Calyptrae mitrate, lacerate, naked or sparsely hairy.

Macromitrium guatemalense Müll. Hal. is similar to *M. parvirete* E.B. Bartram in leaf shape, rugose upper part of leaf, costa vanishing beneath the apex, strongly tuberculate basal cells, but the former differs from the latter in having marginal enlarged teeth-like cells at insertion differentiated.

Distribution: Bolivia, Colombia, Ecuador, Guatemala, Honduras, Mexico, Suriname.

Specimens examined: **BOLIVIA**. Depto. La Paz. *Marko Lewis 89-938 d-3* (MO 3963015). **COLOMBIA**. Cund inamarca, *D. Griffin 6176* (MO 3675244), *D. Griffin 6153* (MO 3674865); Antioquia Belmira, *Albert de Escobar 8716* (MO), *Escobar 101*(MO); **ECUADOR**. Prov. Loja. *William R. Buck 39469* (MO 6367775). **GUATEMALA**. Hb. Kegel 10014 (L 0060440), Hb. Kegel 10017 (L 0060441); **HONDURAS**. Olancho, Sierra de Algalta, *Allen12820* (MO 3965664). **MEXICO**. Chiapas. *S.R. Gradstein 8276* (KRAM-B-130746, MO 5135663). **SURINAME**. Sipaliwini. *B. Allen 19265A* (MO 4431784).

94. *Macromitrium gymnostomum* Sull. & Lesq., Proc. Am. Ac. Arts Sc. 4: 278. 1859. (Figure 119)

Type protologue: On rocks and trees, Simoda, Japan; and Ousima, one of the northern Loo Choo Islands. Type citation: Japan. Shizuoka Pref., Simoda, *Wright* ((lectotype designated by Noguchi, 1967, FH, not seen; isolectotype: NY 00512839!); Japan. Ousima, one of the northern Loo Choo Islands (syntype: NY 00512840!).

Macromitrium brevituberculatum Dixon, Hong Kong Natural. Suppl. 2: 14, 5. 1933, *fide* Guo *et al.*, 2007. Type protologue and citation: [China] Hab. Granite rock, 1,000-1,500 ft. alt., Shaukiwan, Hong Kong I., 5th October, 1931; coll. *Ah Nin (H.8)*. (holotype: BM 000576135!); Near sea level, Stanley, Hong Kong, 14th May, 1931; coll. *Ah Nin (H. 43 A)* (paratype: BM 000576134!).

Macromitrium robinsonii R. S. Williams, Bull. New York Bot. Gard. 8: 343. 1914, *fide* Guo *et al.*, 2007a. Type protologue and citation: [Philippines, in Central Luzon, Bataan] the Upper Lamao river, on tree, 1000 meters, Jan. 1904 (1760) (isotype: H-BR 2617006!).

Macromitrium rupestre Mitt., J. Linn. Soc., Bot. 8: 150. 1865[1864], *fide* Wijk *et al.*, 1962. Type protologue: Hab. Rocks at Nagasaki, Japan, Oldham.

Macromitrium gymnostomum var. *brevisetum* Thér., Ann. Cryptog. Exot., 5: 176. 1932, **syn. nov**. Type protologue: [China] Amoy (n° 364). Type citation: China, Fukien, Amoy Island, *H. H. Chung B. 364*, 5-12-1926 (isotype: F-C 0001228F!).

Macromitrium gymnostomum var. *robustum* Broth., Öfvers. Finska Vetensk.-Soc. Förh., 62A(9): 13. 1921, **syn. nov**. Type protologue and citation: [Japan] Shikoku: Prov. Tosa Takakawa (Sh. Okamur a 387) (holotype: H-BR 2542031!).

(1) Plants small to medium-sized, in dense mats; stems long creeping, densely covered with reddish rhizoids, frequently with brown, multicellular clavate gemmaes; branches erect, up to 5 mm long, simple, or with several short branchlets, obtuse at apex. (2) Branch leaves strongly contorted-twisted-crisped, with an apex adaxially incurved to circinate, often hidden in the inrolled cavity when dry, widely spreading when moist, 1.0-2.5 × 0.15-0.20 mm, linear or linear-lanceolate, keeled; apices obtuse-acute, acute or acuminate; low half of leaves yellowish and hyaline, upper half lamina rather obscure, often look like with blackish patch when moist; margins sometimes slightly recurved; costa yellowish brown, percurrent; medial and upper cells rather obscure, rounded or rounded-hexagonal, 4-5 µm, thin-walled, densely pluripapillose; low cells longer, rectangular, with or without papillae; basal cells linear, 8-25 µm long, with thickened walls, smooth. (3) Perichaetial leaves ovate-lanceolate, acuminate; costae percurrent; all cells hyaline and smooth. (4) Setae brown, smooth, usually 5-8 mm long; paraphyses numerous, slightly exserted beyond leaves. (5) Capsule urns ellipsoid-cylindric, brown, deeply plicate, constricted at mouth when dry, peristome absent; opercula conic-subulate; spores finely papillose, 20-25 µm in diameter. (6) Calyptrae cucullate, 1.7-2.0 mm long, somewhat lobed and plicate, naked.

Macromitrium gymnostomum Sull. & Lesq. is relatively stable in its morphological characters, but its setae varys to varying degrees among populations. The two varieties, var. *brevisetum* with short setae (shorter than 1.5 mm) and var. *robustum* with long setae (longer than 10 mm), mainly differ from var. *gymnostomum* in their seta length. However, the seta length of these two varieties falls into the range of seta variation of *M. gymnostomum*.

Distribution: China, Japan, Philippines.

Specimens examined: **CHINA**. Ning Koua, *R.P. Courtois no. 1120* (H-BR 2650013); Zhejiang, Tianmu Mountain, *W. R. Buck, no. 23916* (H); Kwangtung (Guangdong), *Y.W. Taam no. 402A* (H); Hongkong, Loh Fau Mountain (Lofaushan), *E. D. Merrill, no. 11159* (H-BR); Fujiang, Mt. Wuyi, *Guo S. L., 030001-030008* (SHTU); Omoy, *Chung s.n* (NY); Hunan, Daweishan, *Virtanen 61994* (H); Shunhuangshan, *Enroth 70674, 70760* (H); Yunshan. *Enroth 70130* (H). **JAPAN**. Sumiyo-mura, Island. Amami-ohshima, Kagoshima Pref., K. Saito, (H); Mt. Maya Settsu, Shutai Okamura, *no. 684* (H); Mt. Kuishi, Tosa, Shutai Okamura, *no. 202* (H); Kine Sushi Lili, *B. J. Ferrie* (H).

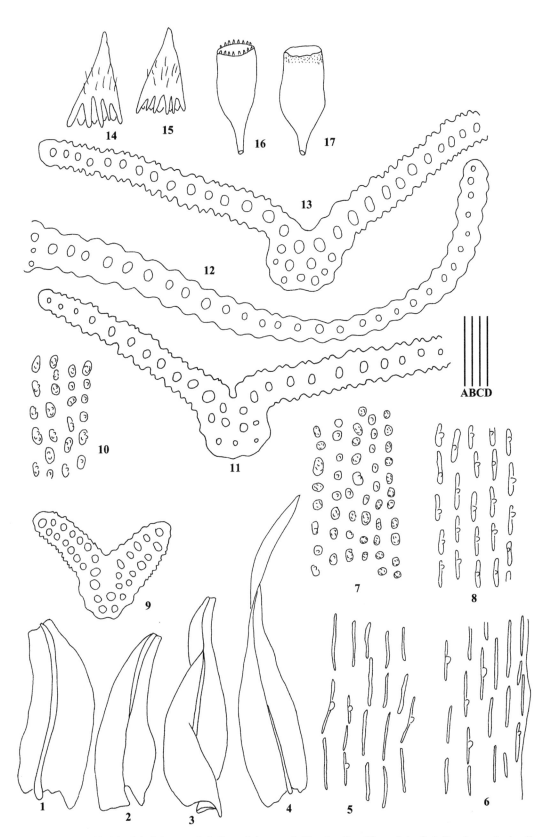

Fig. 116 *Macromitrium gracile* (Hook.) Schwägr. 1-4: Branch leaves. 5: Basal cells of branch leaf. 6: Basal marginal cells of branch leaf. 7: Upper cells of branch leaf. 8: Low cells of branch leaf. 9: Apical transverse section of branch leaf. 10: Medial cells of branch leaf. 11, 13: Upper transverse sections of branch leaves. 12: Basal transverse section of branch leaf. 14-15: Calyptrae. 16-17: Capsules (all from isotype, E 00002402). Line scales: A = 4 mm (14-17); B = 1 mm (1-4); C = 100 μm (9); D = 67 μm (5-8, 10-13).

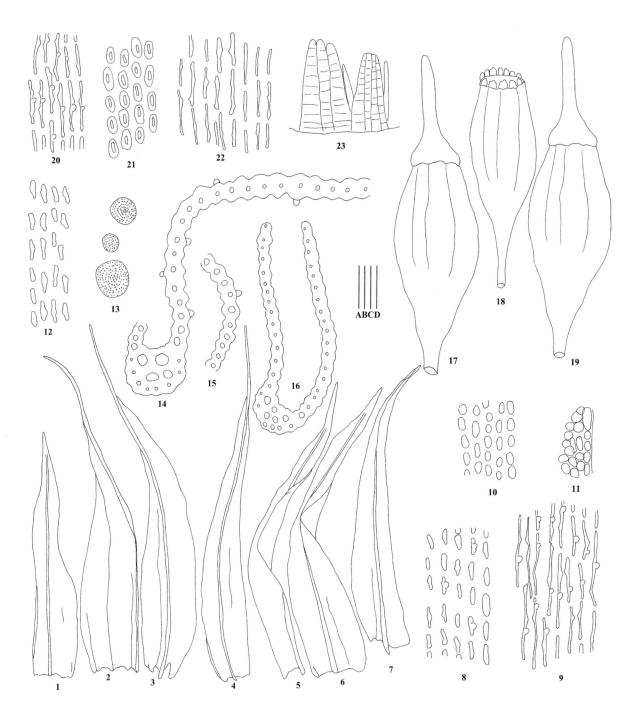

Fig. 117 *Macromitrium greenmanii* Grout 1-2, 5-7: Branch leaves. 3-4: Perichaetial leaves. 8: Medial cells of branch leaf. 9: Basal cells of branch leaf. 10: Upper cells of branch leaf. 11: Cells of capsule mouth. 12, 21: Upper cells of perichaetial leaves. 13: Spores. 14-15: Basal transverse sections of branch leaves. 16: Low transverse section of branch leaf. 17-19: Capsules. 20: Basal cells of perichaetial leaf. 22: Medial cells of perichaetial leaf. 23: Peristome (all from isotype, MO 406389). Line scales: A = 1 mm (1-7, 17-19); B = 400 μm (23); C = 100 μm (13); D = 67 μm (8-12, 14-16, 20-22).

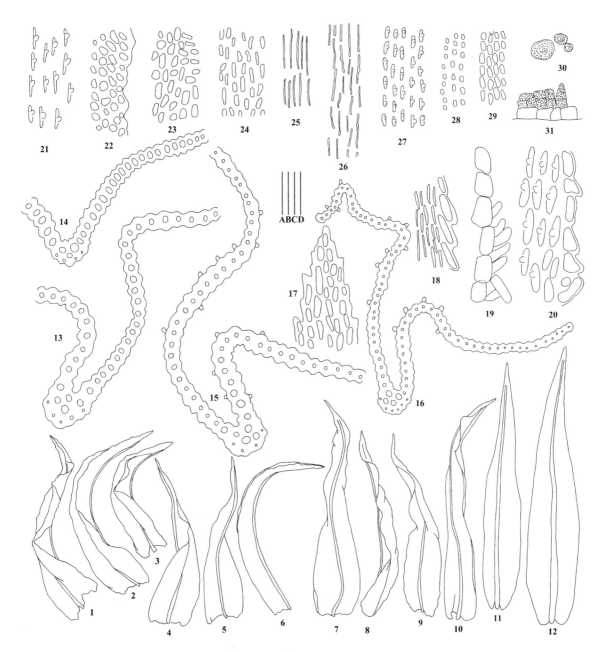

Fig. 118 *Macromitrium guatemalense* Müll. Hal. 1-10: Branch leaves. 11-12: Perichaetial leaves. 13: Low transverse section of branch leaf. 14: Medial transverse section of branch leaf. 15-16: Basal transverse sections of branch leaves. 17: Apical cells of branch leaf. 18-20: Basal marginal cells of branch leaves. 21, 27: Low cells of branch leaves. 22: Upper marginal cells of branch leaf. 23-24: Upper cells of perichaetial leaves. 25: Basal cells of perichaetial leaf. 26: Medial cells of perichaetial leaf. 28-29: Medial cells of branch leaves. 30: Spores. 31: Peristome (1-6, 13-17, 19-22, 27-32 from syntype, *10014*, L 0060440; 7-12, 18, 23-26 from syntype, *10017*, L 0040441). Line scales: A = 1 mm (1-12); B = 400 μm (31); C = 100 μm (30); D = 67 μm (13-29).

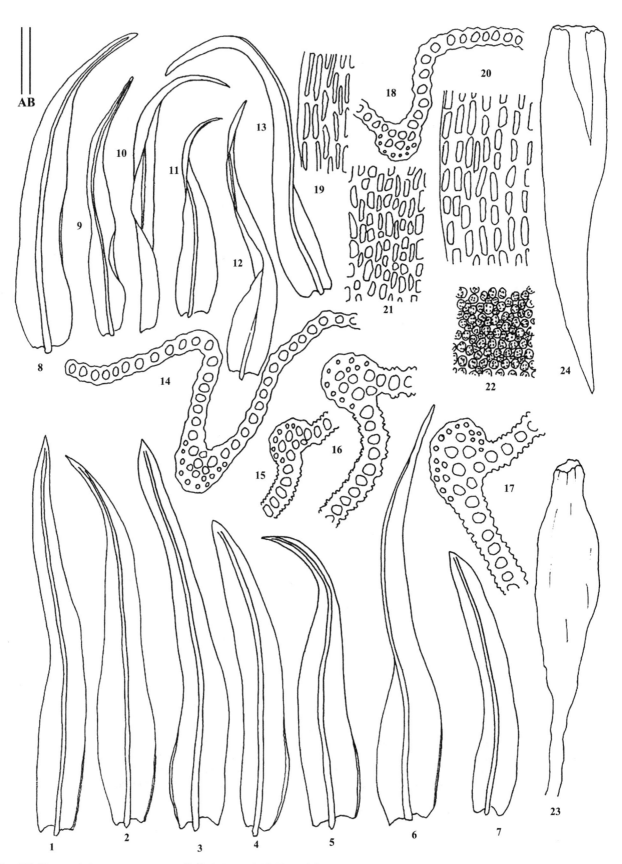

Fig. 119 *Macromitrium gymnostomum* Sull. & Lesq. 1-13: Branch leaves. 14, 18: Basal transverse sections of branch leaves. 16-17: Upper transverse sections of branch leaves. 19-20: Basal marginal cells of branch leaves. 21: Low cells of branch leaf. 23: Dry capsule. 24: Calyptra (1-7, 14-17, 20 from holotype of *M. brevituberculatum* Dixon in BM; 8-13, 18-19, 21-24 from isotype of *M. robinsonii* R. S. Williams in H-BR). Line scales: A = 0.50 mm (1-13, 23-24); B = 50 μm (14-22). (Guo *et al.*, 2007).

95. *Macromitrium hainanense* S.L. Guo & S. He, Bryologist 111: 505. 2008. (Figure 120)

Type protologue and citation: China, Hainan, Ling Shui County, vic. Diao Lu Shan Forestry Station. Ca 11 km NE of the station of flank of Mt. SanJiaoShan, on fallen tree over stream, 18°40'N, 109°55'E, alt. 800 m., *W. D. Reese 17956* (holotype: MO!).

(1) Plants robust, forming loose mates; primary and secondary stems creeping, up to 4.5 cm long, irregularly branched; branches erect, up to 20 mm long. (2) Primary stem leaves inconspicuous and caducous; secondary stem leaves individually twisted, flexuose to slightly contorted when dry, usually with tips broken, spreading, often deflexed when moist, 1.8-2.2 × 0.3-0.5 mm, broadly oblong-lanceolate to lanceolate, long acuminate towards the apex; margins entire, plane or slightly recurved below; costae percurrent to shortly excurrent; upper cells small, rounded-quadrate to oblate, pluripapillose; medial cells short-rectangular, moderately pluripapillose, becoming elongate to curved-sigmoid towards base, in distinct rows, less papillose to almost smooth; basal cells elongate-rectangular, thick-walled, lumens sigmoid-curved, hyaline, smooth. (3) Branch leaves strongly twisted and contorted when dry, unevenly spreading and deflexed when moist, 2.0-2.8 × 0.3-0.5 mm, ovate-lanceolate to broadly oblong-lanceolate, gradually narrowed to an acuminate apex, slightly keeled, plicated or slightly plicated below, somewhat sheathing at base; margins minutely roughened to entire above, entire below, more or less recurved on one or two sides; costa usually shortly excurrent; cells similar to those of secondary stem leaves. (4) Sexual condition unknown, only female plants seen; perichaetial leaves not much different from branch leaves, with slightly long acuminate apices, up to 2.5 mm long. (5) Setae very short, 0.3-0.7 mm long, with numerous smooth paraphyses at the vaginula. (6) Capsules almost immersed, erect, urns ovoid, 1.3-1.5 × 0.9-1.2 mm, moderately plicate; opercula long conic-rostrate, with beaks 0.4-0.5 mm long; peristome absent; spores isosporous, 25-40 µm in diameter, coarsely papillose. (7) Calyptrae campanulate-mitrate, deeply plicate, with numerous erect, slightly coarse hairs.

Distribution: China.

96. *Macromitrium harrisii* Paris, Index Bryol. Suppl. 238. 1900. (Figure 121)

Basionym: *Macromitrium peraristatum* Müll. Hal., Bulletin de l'Herbier Boissier 5: 560. 1897. Illegitimate, later homonym. Type protologue: Jamaica, prope plantationes Cinchonae, 4900 ped. alt., ad rupes: *W. Harris,* 24 Aprili 1896. Hb. Jamaicense 10,033. Type citation: W. Anchona, Jamaica, *W. Harris,* 24, Apr. 1896 as *Macromitrium peraristatum* C. M. *n. sp.*" (isotype: BM 000873105!).

(1) Plant medium-sized, brown above, dark brown below, lustrous; stems weakly creeping, densely covered with rusty reddish rhizoids; branches to 10-20 mm long, 1.5-18 mm thick. (2) Branch leaves erect below, individually twisted, contorted-flexuose above when dry, erect-spreading and occasionally twisted when moist, lanceolate to long narrowly lanceolate, narrowly acuminate; margins entire, slightly plicate below, irregularly recurved on both sides; upper and medial cell isodiametric, 5-8 µm wide, moderately bulging, smooth; upper marginal cells not differentiated, gradually elongate from upper part to base; low and basal cells long rectangular, 36-45µm long, thick-walled and porose, tuberculate, marginal enlarged teeth-like cells not differentiated at insertion; costae excurrent to form a hair point. (3) Perichaetial leaves oblong-lanceolate, all cells longer than wide, thick-walled, porose, smooth, plicate and slightly sheathing below. (4) Setae smooth, 8 mm long. (5) Capsule urns ovoid, distinctly furrowed; peristome double, exostome teeth united a short membrane, endostome longer, membraneous. (6) Calyptrae mitrate, naked, deeply lacerate.

Macromitrium harrisii Paris is similar to *M. cirrosum* (Hedw.) Brid. and *M. picobotinum* B.H. Allen, but the capsules of *M. harrisii* conspicuously furrowed.

Distribution: Dominica, Jamaica.

Specimens examined: **DOMINICA**. La Vega. *William Campbell Steere 23108* (H 3090326). **JAMAICA**. St. Thomas, *William R. Maxon 8915* (H-BR 2632011).

97. *Macromitrium helmsii* Paris, Ind. Bryol. Suppl. 238. 1900. (Figure 122)

Replaced synonym: *Macromitrium appendiculatum* Müll. Hal., Hedwigia 37: 156. 1898. *hom. illeg.*, *fide* Wijk *et al.*, 1964. Type protologue and citation: Nova Seelandia, insula australis, prope Greymouth, ubi Richard Helms specimen unicum incompletum 1885 legit (lectotype: H-BR!).

(1) Plants slender to moderately robust, dull; stems creeping with penicillate branches up to 2.5 cm high. (2) Branch leaves individually incurved-twisted with upper portions curved outward and down when dry, wide-spreading to spreading from an erect basal portion when moist, ligulate, narrowly ligulate to oblong-lanceolate, obtuse to retuse, or abruptly narrowed to the excurrent costa; apices often asymmetric, ending in a very long arista in young leaves (linear, green, sparsely papillose, stiffly flexuose, penicillate, fragile, mostly breaking before mature); upper cells irregularly rounded-quadrate, obscure, densely pluripapillose, in longitudinal rows; medial cells elliptic-rounded to rectangular-elliptic, papillose, in longitudinal rows; basal cells elongate, long-rectangular, irregularly thick-walled, smooth or weakly unipapillose, becoming longer and narrower near the margin, shorter near the costa. (3) Perichaetial

leaves similar to branch leaves, oblong, more gradually narrowed to an arista (Vitt, 1983).

Macromitrium helmsii Paris and *M. retusum* Hook. f. & Wilson are similar in their branch leaves with fragile, penicillate aristae, but the former differs from the latter in having more lanceolate branch leaves with papillose upper and medial cells.

Distribution: New Zealand.

Specimens examined: **NEW ZEALAND**. South Island, Buller, *Allan J. Fife 6046* (CHR 103483), *Allan J. Fife 6688* (MO 3656525), *Allan J. Fife 7294* (MO 3656496); North Island, Cattle ridge track, *H. Streimann 58102* (H 3090331).

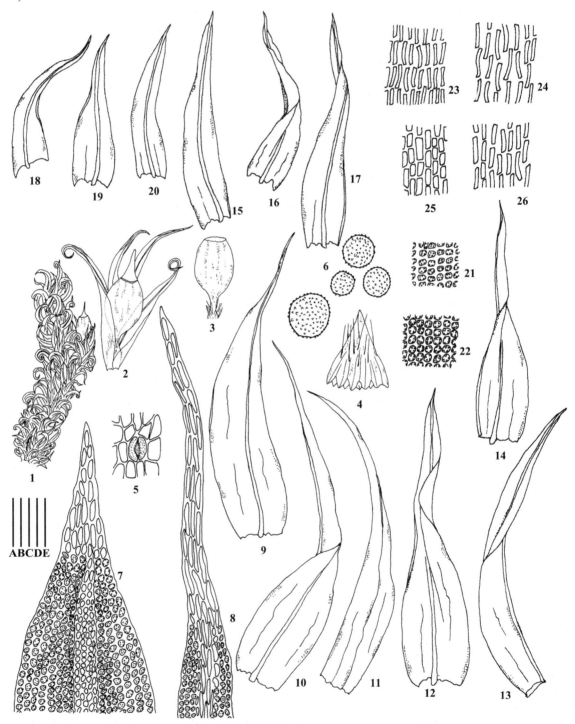

Fig. 120 *Macromitrium hainanense* S.L. Guo & S. He 1: Plant. 2: Capsule with perichaetial leaves. 3: Capsule. 4: Calyptra. 5: Stoma. 6: Spores. 7: Apex of branch leaf. 8: Apex of perichaetial leaf. 9-11: Perichaetial leaves. 12-17: Branch leaves. 18-20: Secondary stem leaves. 21: Upper cells of branch leaf. 22: Medial cells of branch leaf. 23: Low cells of branch leaf. 24: Basal cells of branch leaf. 25: Medial cells of perichaetial leaf. 26: Basal cells of perichaetial leaf (all from holotype, MO). Line scales: A = 2.3 mm (1); B = 1.3 mm (2-4); C = 0.8 mm (12-20); D = 0.6 mm (9-11); E = 42 μm (5-8, 21-26). (Guo *et al.*, 2008).

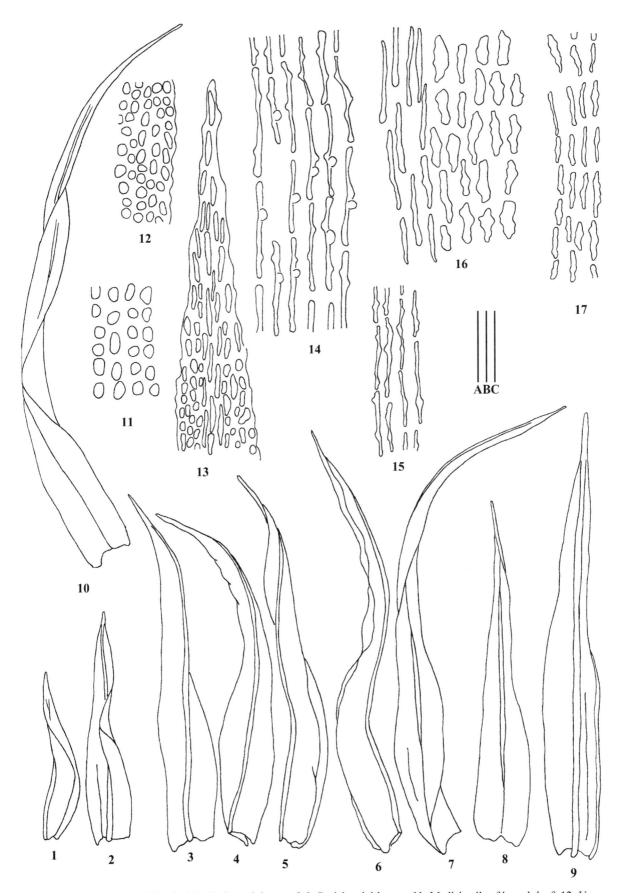

Fig. 121 *Macromitrium harrisii* Paris 1-7, 10: Branch leaves. 8-9: Perichaetial leaves. 11: Medial cells of branch leaf. 12: Upper cells of branch leaf. 13: Apical cells of branch leaf. 14: Low cells of branch leaf. 15: Low cells of perichaetial leaf. 16: Upper cells of perichaetial leaf. 17: Medial cells of perichaetial leaf (all from isotype, BM 000873105). Line scales: A = 1 mm (1-10); B = 100 μm (13); C = 67 μm (11-12, 14-17).

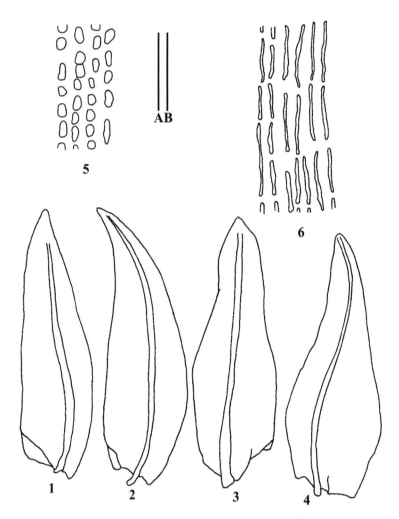

Fig. 122 *Macromitrium helmsii* Paris 1-4: Branch leaves. 5: Medial cells of branch leaf. 6: Basal cells of branch leaf (all from lectotype, H-BR 2529007). Line scales: A = 0.44 mm (1-4); B = 44 μm (5-6).

98. *Macromitrium hemitrichodes* Schwägr., Sp. Musc. Frond., Suppl. 2 2(2): 136. pl. 193. 1827. (Figure 123)
Type protologue: In nova Hollandia lectum debit Sieber, peregrinator intrejsidus [sic].
Macromitrium amoenum Hornsch. ex Müll. Hal., Syn. Musc. Frond. 740. 1849, *fide* Vitt & Ramsay, 1985. Type protologue: Nova Hollandia: Sieber.
Macromitrium baileyi Mitt., Trans. Proc. Roy. Soc. Victoria 19: 63. 1882, *fide* Vitt & Ramsay, 1985. Type protologue: Brisbane River, Bailey.
Macromitrium intermedium Mitt., Trans. Proc. Roy. Soc. Victoria 19: 63. 1882, *fide* Vitt & Ramsay, 1985. Type protologue: Brisbane River, Bailey.

(1) Plants small to medium-sized, in dense mats; stems long creeping, with short, erect branches, branches up to 15 mm high. (2) Branch leaves loosely and irregularly twisted-contorted, the upper portion inrolled and sometimes hidden in the inrolled cavity when dry, loosely erect-spreading when moist, 1.7-2.2 mm long, ovate-lanceolate, lanceolate to ligulate-lanceolate, narrowed to a strong apiculus or stout mucro; costae excurrent to form an apiculus; upper cells rounded to round-quadrate or irregularly rounded, 7-10 μm wide, obscure, bulging, pluripapillose, papillae irregularly branched; medial cells elliptical-rectangular, pluripapillose above, unipapillose below; basal cells elongate with straight lumens, 25-33 μm long, thick-walled, mostly smooth or some unipapillose. (3) Perichaetial leaves erect, broadly lanceolate to ovate-lanceolate; apices acute to acuminate; upper cells pluripapillose. (4) Setae 4-10 mm long, smooth, twisted to the left. (5) Capsule urns ovoid-elipsoid to narrowly ovoid, smooth below, 8-plicate beneath the small mouth; peristome single, exostome of 16, broken off when old; spores anisosporous. (6) Calyptrae conic, with numerous slits, sparse and straight hairs.

Distribution: Australia, New Caledonia.

Taxonomy

Specimens examined: **AUSTRALIA**. New South Wale, *H. Streimann 4954* (H 3090339); Larrys mtn, Moruya, *H. Streimann* (H 3090338); Hacking River below causeway, *R. G. Coveny 15862b & D. W. Hardin* (MO 5911721); Gloucester Bucketts, North coast, *R. G. Coveny 16148 & P. D. Hind* (MO 5634663); Queensland, Brisbane, *T. M. Bailey 352* (H-BR 2536005); Spring, on sandstone (H-BR 2552020); Tully Falls Road, Ravenshoe, *H. Streimann 30165* (H 3194600); Mt. Norman, Wallangarra, *H. Fagg* (H 3090347); Eukey-Wyberba Road, *H. Streimann 52916* (KRAM-B-134033); Paling Yard Creek, *H. Streimann 52953* (KRAM-B-130974); Hurdle Gully, Coominghah State Forest, *H. Streimann 52817* (KRAM-B-137558).

Macromitrium hemitrichodes Schwägr. var. *sarasinii* (Thér) Thouvenot. Crypt. Bryol. 40(16): 184. 2019.
Basionym: *Macromitrium sarasinii* Thér., *fide* Thouvenot 2019. Type protologue: New Caledonia, Mt Canala, alt. 650 m, 1911, *Sarasin 334*.

The differences between *Macromitrium hemitrichodes* var. *sarasinii* and *var. hemitrichodes* are mainly in their quantitative characters, with the former having longer lanceolate branch leaves (1.6 -2.7 mm) with a wider oblong base, an acuminate to acute apex, longer basal cells up to 50 μm.

Distribution: New Caledonia.

99. *Macromitrium herzogii* Broth., Biblioth. Bot. 87: 66. 1916. (Figure 124)
Type protologue: Im Nebelwald der Laguna verde bei Comarapa, alt. 2600 m, No. 4308, 4316. Type citation: Bolivia, Santa Cruz: Im Nebelwald über Comarapa, about 2600 m, *Herzog, T. C. J., 4308* (isosyntypes: JE 04008714!, FH 00213628!); Bolivia, Im Nebelwald über Comarapa, *T. C. J. Herzog, 4316* (syntype: H-BR 2625012!; isosyntype: NY 01086493!).

(1) Plants medium-sized, yellow-green; stems weakly creeping, branches frequently forked, up to 40 mm long. (2) Branch leaves erect below, widely spreading, flexuose above, with an incurved apex when dry, abaxially curved spreading and keeled above when moist, lanceolate from an oblong or oval-oblong base; apices acuminate; margins serrulate above, entire below; costae light reddish, percurrent; marginal cells sometimes differentiated from and smaller than their ambient cells; upper and medial cells oval, elliptic-rhombic, rounded-quadrate, clear, weakly bulging and smooth; low and basal cells elongate, linear-rectangular, porose, flat and occasionally tuberculate; the outmost marginal cells at the insertion enlarged, occasionally teeth-like. (3) Perichaetial leaves oblong-lanceolate, somewhat sheathing at the base, plicate below, acuminate-subulate upper; all cells longer than wider, clear and smooth; upper cells narrowly elongate-rectangular; basal cells enlarged, irregularly rectangular. (4) Setae smooth, 4-5 mm, twisted to the left; vaginulae sparsely hairy. (5) Capsules ellipsoid to ellipsoid-cylindric, slightly constricted beneath the mouth, strongly furrowed; opercula conic-subulate; peristome double, exostome yellowish, shortly membraneous, endostome hyaline. (6) Calyptrae campanulate-mitrate, naked, lobed at base.

Distribution: Bolivia.
Specimens examined: **BOLIVIA**. Santa Cruz: Im Nebelwald über Comarapa, about 2600 m, *T. C. J. Herzog, no. 4273* (JE 04008715).

100. *Macromitrium hildebrandtii* Müll. Hal., Linnaea 40: 248. 1876. (Figure 125)
Type protologue: Patria. Comoro-insula Johanna, 1000 met. supra mare, in lingo putrido sylvestri, cum Macromitrio subpungente Hpe. Associatum: *J. M. Hildebrandt Coll. no. 1814*, Junio-Aug. 1875. Type citation: Comoro-insul Johanna, Aug. 1875, *J. M. Hildebrandt, no. 1814* (isotypes: PC 0137509!, JE 04008744!, JE 04008745!).

(1) Plants larges, stems rather long, sparsely forked; branches in stem terminal up to 10 mm long, somewhat prostrate, simple or with short branchlets. (2) Stem and branch leaves similar, erect below, squarrose to widely-spreading, flexuous above when dry, erect-spreading above when moist, narrowly lanceolate, plicate below; apices linearly acuminate, acuminate, bluntly acuminate to acuminate-acute; margins irregularly serrate, or finely serrulate above, entire below, distinctly recurved at base; upper cells small, rounded- quadrate to elliptic, arranged in longitudinal rows, clear and smooth, thick-walled; basal cells long rectangular, strongly thick-walled, porose, strongly tuberculate, the outermost marginal cells distinctly differentiated at the insertion, enlarged and extraneous; costae rusty, percurrent or shortly excurrent, forming a sharp point. (3) Perichaetial leaves erect, plicate, and others similar to those of branch leaves. (4) Setae long and erect, smooth. (5) Capsules ovoid or globose, constricted towards the mouth, weakly furrowed; opercula conic-rostrate; peristome double, exostome yellowish, papillose, endostome membranous, hyaline. (6) Calyptrae smooth and naked, lobed at base.

Macromitrium hildebrandtii Müll. Hal. is somewhat similar to *M. sulcatum* (Hook.) Brid., but the former could be distinguished from the latter by its branch leaves strongly squarrose-recured when moist, with distinctly differentiated enlarged and extraneous cells at the outermost margin at the insertion.

Distribution: Comoros.

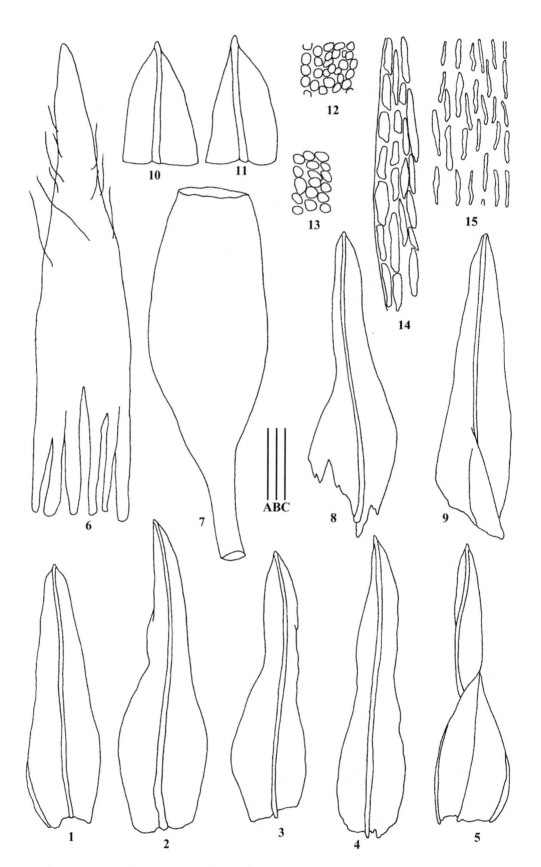

Fig. 123 *Macromitrium hemitrichodes* Schwägr. 1-5, 8-9: Branch leaves. 6: Calyptra. 7: Capsule. 10: Medial cells of branch leaf. 11: Upper cells of branch leaf. 12-13: Apices of branch leaves. 14: Basal marginal cells of branch leaf. 15: Basal cells of branch leaf (all from H 3090347). Line scales: A = 0.44 mm (1-9); B = 176 μm (10-11); C = 44 μm (12-15).

Taxonomy 197

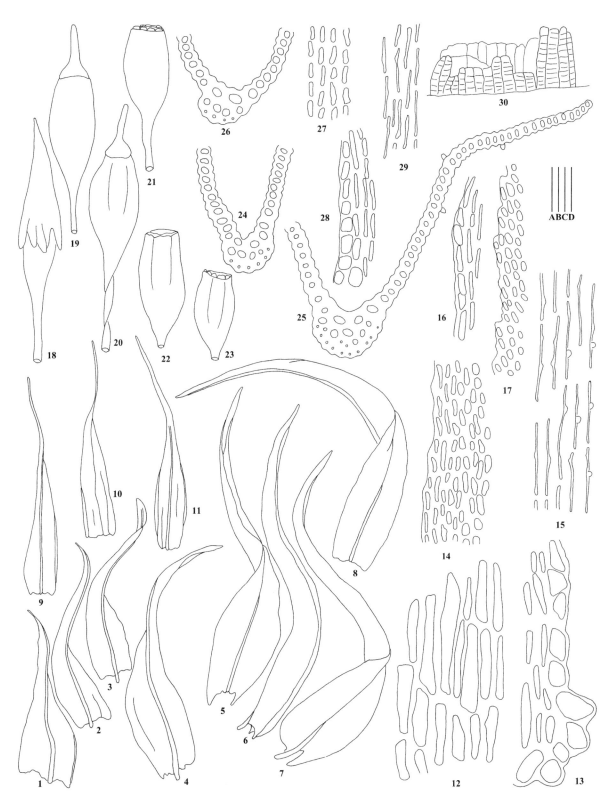

Fig. 124 *Macromitrium herzogii* Broth. 1-8: Branch leaves. 9-11: Perichaetial leaves. 12: Basal cells of perichaetial leaf. 13, 16, 28: Basal marginal cells of branch leaf. 14: Upper cells of branch leaf. 15: Basal cells of branch leaf. 17: Apical cells of branch leaf. 18-23: Capsules. 24: Medial transverse section of branch leaf. 25-26: Basal transverse sections of branch leaves. 27: Medial cells of branch leaf. 29: Medial cells of perichaetial leaf. 30: Peristome (all from isosyntype, FH 00213628). Line scales: A = 2 mm (9-11); B = 1 mm (1-8, 18-23); C = 400 μm (30); D = 67 μm (12-17, 24-29).

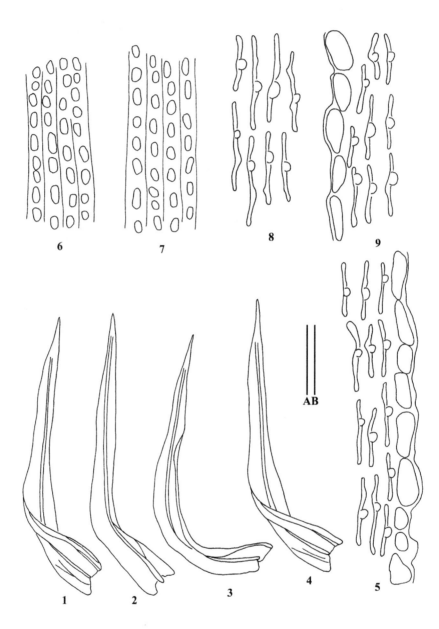

Fig. 125 *Macromitrium hildebrandtii* Müll. Hal. 1-4: Branch leaves. 5, 9: Basal marginal cells of branch leaves. 6: Upper cells of branch leaf. 7: Medial cells of branch leaf. 8: Low cells of branch leaf (all from isotype, PC 0137509). Line scales: A = 1 mm (1-4); B = 67 μm (5-9).

Taxonomy

101. *Macromitrium holomitrioides* Nog., J. Sci. Hiroshima Univ., Ser. B, Div. 2, Bot. 3: 135. 13 f. 1-9. 1938. (Figure 126)

Type protologue and citation: [China] Taiwan: Prov. Taihoku, Heisin, on Bark. Noguchi. Nov. 5, 1927, coll. *S. Suzuki*) ex Herb. Dr. *A. Noguchi, no. 9574* (holotype: NICH!; isotype: HIRO!).

(1) Plants in yellowish-green mats; stems creeping, with simple and short branches, branches only 4-6 mm tall. (2) Stem leaves ovate-lanceolate or oblong-lanceolate. (3) Branch leaves erect below, twisted-contorted-crisped above, the apices basically hidden in the inrolled cavity when dry, erect-spreading when moist, 2.5-3.5 × 0.50-0.8 mm, ovate-oblong, oblong-lanceolate to ligulate-lanceolate, keeled, weakly plicate below; apices acute and incurved; costa percurrent; medial and upper cells large, 10-15 µm wide, rounded-hexagonal, bulging and mammillose, collenchymatous, not obscure; low cells rectangular, often with a large papilla; basal cells sublinear to subrectangular, 30-60 µm long, thick-walled, somewhat porose. (4) Perichaetial leaves similar to the branch leaves but shorter, to 2 mm long, paraphyses numerous. (5) Setae up to 3 mm long, smooth. (6) Capsule urns 1.5-2.0 × 0.8-1.0 mm, ellipsoid, plicate, furrowed at the mouth when dry; opercula long-conic subulate; peristome absent. (7) Calyptrae large, campanulate, deeply cleft along the ridge, plicate, with sparse, brown hairs.

Macromitrium holomitrioides Nog. is distinguished from the Asian allies of the genus by its broad branch leaves with large, mammillose, collenchymatous upper and medial cells and gymnostomous capsules.

Distribution: China (Taiwan, Hainan), Japan (Kyushu). **Habitats**: on tree trunks.

Specimens examined: **CHINA**. Hainan, *P. C. Chen et al., 789b* (H), *P. C. Wu, W1185016* (H), *Paul L. Redfearn, Jr., 35766* (MO 3974398), *Paul L. Redfearn, Jr 35890* (MO 3974400), *W. D. Reese, 17612* (MO 5643693), *W. D. Reese, 17771* (MO 5643684); Hainan, Le Dong, *Paul L. Redfearn, Jr., no. 36010b*, 21. Mar, 1990 (MO 3974401).

102. *Macromitrium hortoniae* Vitt & H. P. Ramsay, J. Hattori Bot. Lab. 59: 367. f. 100-117. 1985. (Figure 127)

Type protologue and citation: Australia. Queensland: Brisbane area, Lamington National Park, 25 km S of Canungra on road to O'Reilly's Guesthouse, 28°17′S 153°00E. Open shrubby montane rain forest, *Vitt 28150* (isotype: H 3090363!).

(1) Plants rather small, in dense, compact, spreading mats; stems prostrate, creeping, with short, erect branches, branches about 6 mm long. (2) Stem leaves about 1.0 mm long, broadly lanceolate to narrowly ovate-lanceolate, gradually acute, erect-flexuose when dry, wide-spreading-flexuose when moist, cells similar to those of branch leaves. (3) Branch leaves curved to flexuose-decurved, with the upper portion strongly inrolled and the apex hidden in the inrolled cavity, not funiculate when dry, flexuose-spreading with inflexed apices when moist, 1.6-2.5 mm long, ligulate, lanceolate-ligulate; the apices obtuse to broadly acute; margins plane to reflexed below, entire below, minutely crenulate above; costae ending several cells beneath the apex; upper and medial cells 7-10 µm wide, quadrate-rounded to rounded, chlorophyllose, thin-walled, strongly bulging, clear to rather obscure, smooth to unipapillose or pluripapillose, small and oblate at marginal protion; basal cells near costa short, quadrate-rounded to short-rectangular, marginal cells elongate, rectangular, all cells firm-walled, evenly thickened and smooth. (4) Perichetial leaves 1.2-1.5 mm long, much shorter than branch leaves, lanceolate, acute, erect, stiff, not sheathing, cells mostly similar to those of branch leaves. (5) Setae short, 2-4 mm long, rather stout, smooth, twisted to the left; vaginulae naked. (6) Capsules small, urns about 1.0 mm long, ovoid-cylindric to narrowly ellipsoid, completely smooth, somewhat narrowed to a small mouth, or the rim firm and almost always wide-mouthed, with neither plications nor collapsed portion; capsules shortly exserted, peristome single, exostome of 16, erect to inflexed, sometimes irregularly lanceolate; spores anisosporous. (7) Calyptrae mitrate, smooth and naked, deeply plicate, deeply and uniformly lacerate, 10-15 slits split to mid-length (Vitt & Ramsay, 1985).

Distribution: Australia.

Specimens examined: **AUSTRALIA**. New South Wales, *W. W. Watts 1235* (H-BR 2535013); Queensland, *I. G. Stone* (MEL 2264750, MEL 2260917, MEL 2217892).

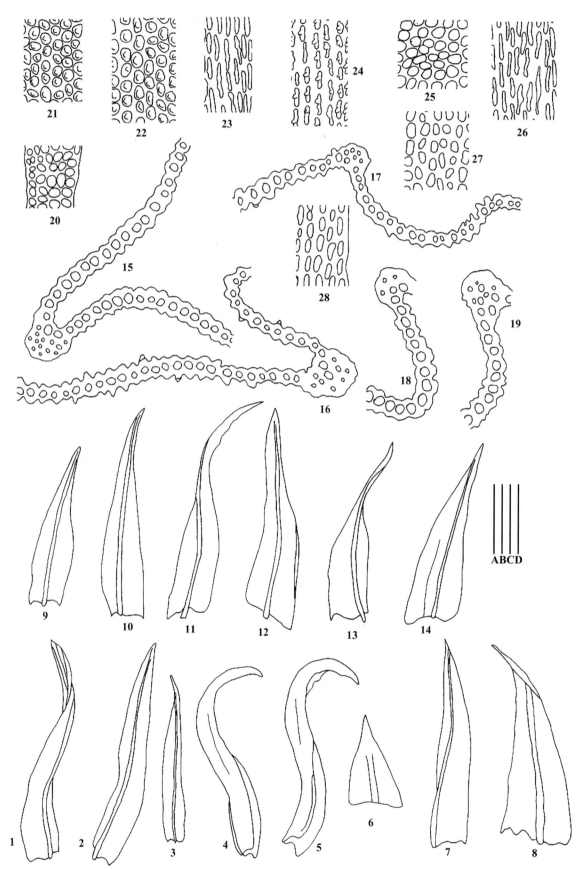

Fig. 126 *Macromitrium holomitrioides* Nog. 1-5: Branch leaf. 6: Apex of branch leaf. 9-14: Stem leaves. 15, 18-19: Medial transverse sections of branch leaves. 16-17: Low transverse sections of branch leaves. 20: Apical cells of branch leaf. 21: Upper cells of branch leaf. 22: Medial cells of branch leaf. 23: Basal cells of branch leaf. 24: Low cells of branch leaf. 25: Medial cells of stem leaf. 26: Basal cells of perichaetial leaf. 27: Upper cells of perichaetial leaf. 28: Medial cells of perichaetial leaf (all from isotype, HIRO). Line scales: A = 1 mm (1-5); B = 0.4 mm (7-14); C = 0.2 mm (6); D = 67 μm (15-28).

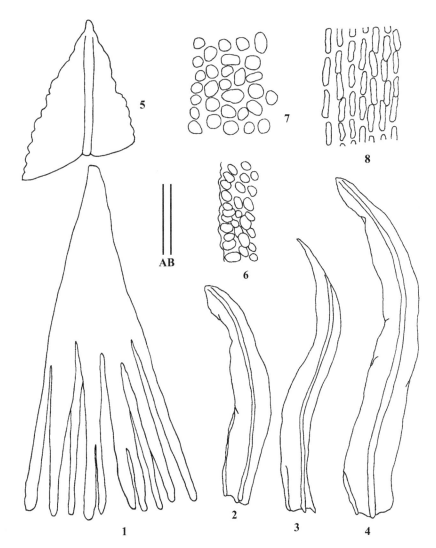

Fig. 127 *Macromitrium hortoniae* Vitt & H.P. Ramsay 1: Calyptra. 2-4: Branch leaves. 5: Apex of branch leaf. 6: Upper cells of branch leaf. 7: Medial cells of branch leaf. 8: Basal cells of branch leaf (all from H-BR 2535013). Line scales: A = 0.44 mm (1-4); B = 44 μm (5-8).

103. *Macromitrium huigrense* R.S. Williams, J. Washington Acad. Sci. 17: 492. f. a. 1927. (Figure 128)

Type protologue and citation: Ecuador, Vicinity of Huigra, at 9500 ft. elevation, Sept. 3, 1918, *J. N. & George Rose 23645*, type (holotype: NY 1086485!).

(1) Plants large and robust; stems inconspicuous and weakly creeping, with erect branches, branches up to 25-30 mm long and 1.5 mm thick. (2) Stem leaves caducous and inconspicuous. (3) Branch leaves erect below, loosely spiraled twisted and slightly contorted, flexuose above when dry, strongly abaxially curved spreading, sometimes twisted and undulate above when moist, long ligulate-lanceolate to lanceolate; margins entire below, irregularly serrulate above; cells near the apex longer than wide, narrowly oblong; upper and medial cells subquadrate, round-quadrate, oval, oblique rhombic, becoming smaller towards the margin, smooth and clear; low cells rhombic, strongly porose and thick-walled, elongate towards the base, basal cells rectangular to linear, strongly porose and thick-walled, tuberculate; marginal cells near the inertion enlarged, wider than their ambient inner cells, but not teeth-like. (4) Perichaetial leaves not much differentiated from branch leaves. (5) Setae erect, about 12-14 mm long, twisted to the left, strongly papillose, frequently with a whitish collar beneath the urn. (6) Capsules ovoid, ovoid-ellipsoid, strongly furrowed or plicate; opercula conic-subulate; peristome double, exostome teeth well developed, united to an erect membrane, endostome hyaline, united below and discrete above. (7) Calyptrae campanulate, lacerate below, scabrous and spinose, with brown sparse hairs, covering the whole capsule.

Macromitrium huigrense R.S. Williams is rather smiliar to *M. longifolium* (Hook.) Brid., only the branch leaves of the former are strong recurved-squarrose and twisted above when moist, while those of the latter widely spreading when moist.

Distribution: Ecuador.

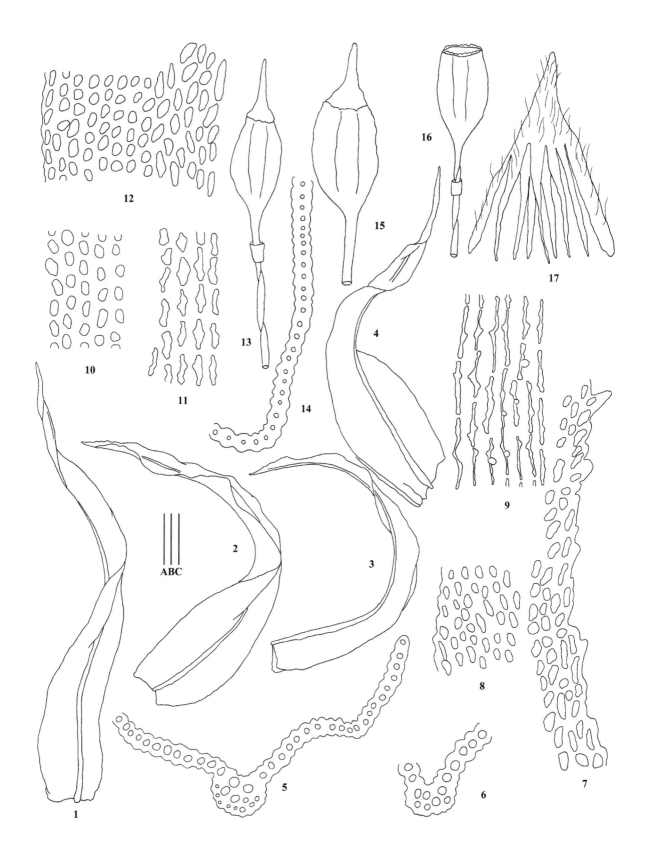

Fig. 128 *Macromitrium huigrense* R.S. Williams 1-4: Branch leaves. 5: Low transverse section of branch leaf. 6: Medial transverse section of branch leaf. 7: Apical cells of branch leaf. 8: Upper marginal cells of branch leaf. 9: Basal cells of branch leaf. 10: Medial cells of branch leaf. 11: Low cells of branch leaf. 12: Upper cells of branch leaf. 13, 15-16: Capsules. 14: Basal transverse section of branch leaf. 17: Calyptra (all from isotype, US 00070264). Line scales: A = 2 mm (13, 15-17); B = 1 mm (1-4); C = 67 μm (5-12, 14).

104. *Macromitrium humboldtense* Thouvenot & Frank Müll., Cryptog., Bryol. 37(3): 296, 1-6, 8-16. 2016.

Type protologue: New Caledonia, South Province, Mt Humboldt, ca 1600 m, 31 Aug. 2003, *F. Müller NC763* (holotype: DR; isotype: PC 0723602).

(1) Plants medium-sized, stems creeping, with crowded erect branches. (2) Branch leaves erecto-patent, twisted, loosely spirally with the leaf the apices turned in all directions, spreading when moist, upper lamina often undulate, 3.6-4.6 × 0.5-0.8 mm, narrowly lanceolate to narrowly triangular, somewhat keeled and plicate below, gradually tapering, long acuminate, ending in a piliform apex, hyaline or red-brown at the tip; margins crenulated-papillose to sharply toothed near the apex; costae percurrent to excurrent into a rough hyaline or pale red-brown arista, up to 250 μm long; upper and medial cells rounded quadrate, pluripapillose, occupying more than 3/4 the leaf length; transitional cells few and unipapillose; low and basal cells long rectangular to linear, with straight lumens, confined to a small area near the base, smooth or tuberculate. (3) Perichaetial leaves not differentiated, basal cells thin-walled and with wider lumens. (4) Setae long and flexuose, up to 15 mm, smooth; vaginulae glabrous with many paraphyses. (5) Immature capsule long exserted, cylindric; opercula with a long straight rostrum. (6) Calyptrae naked (Thouvenot & Müller, 2016).

Distribution: New Caledonia.

105. *Macromitrium incurvifolium* (Hook. & Grev.) Schwägr., Sp. Musc. Frond., Suppl. 2(2): 144. 1827.

Basionym: *Orthotrichum incurvifolium* Hook. & Grev., Edinb. J. Sci. 1: 177. 4. 1824. Type protologue: Indonesia. "Island of Ternate, and (Australia) in King George's Sound. Received from Mr. Dickson." (lectotype by Vitt & Ramsay, 1985: E-Greville; syntype: BM-Hooker).

Macromitrium beecheyanum Mitt., J. Linn. Soc., Bot. 10: 167. 1868, *fide* Whittier, 1976 and Staples *et al.*, 2004. Type protologue: Hab. Tutuila, on living trees, mostly Hibiscus tiliaceus (20-500ft.). No. 1.

Macromitrium cumingii Müll. Hal., Flora 82: 452, 1896, *fide* Staples *et al.*, 2004. Type protologue: Insulae Hawaiicae, sine loco speciali: *Cuming coll. no. 2214*. Type citation: Hawaii, sine loco spec. let. *Cuming s.n.* (isotype: FH 00213599!)

Macromitrium javanicum Bosch & Sande Lac., Bryol. Jav. 1: 123. 101. 1860. Type protologue: (Indonesia) Habitat insulam Javae; Herb. reg. Lugd. Bat. (lectotype designated by Vitt *et al.*, 1995: L-Lacoste).

Macromitrium kaernbachii Broth., in Schumann & Lauterb., Fl. Schutzgeb. Südsee 86. 1900. ('Kaernbachii'), *fide* Vitt *et al.*, 1995. Type protologue: (Papua New Guinea, Morobe) Kaiser Wilhelmsland: Sattelberg, Nuselang, am Wege nach Lukube, im Busch bei 850 m ü. M., 11.XII. 1893 *Kaernbach 62*. Type citation: New Guinea, Nuselang, Sattelberg, Wag nach Lukube, 2550 i, 11/12/1893, Leg. *Kaernback 62* (holotype: H-BR 2618006!).

Macromitrium leucoblastum Broth., in Schumann & Lauterb., Fl. Deutsch. Schutzgeb. Südsee 87. 1900, *fide* Vitt *et al.*, 1995. Type protologue: Papua New Guinea. Madang. Gogolexpedition, XI. 1890 *L. Kaernbach*.

Macromitrium planocespitosum Müll. Hal., Linnaea 38: 560. 1874, *fide* Vitt *et al.*, 1995. Type protologue and citation: (Philippines) Insulae Philippines, Luzon, Mahahai, reg. montosa: *G. Wallis 1871* (lectotype designated by Vitt *et al.*, 1995: H-BR!).

Macromitrium powellii Mitt., J. Linn. Soc., Bot. 10: 168. 1868, *fide* Vitt *et al.*, 1995. Type protologue: Samoa, Tutuila, on Cocoa-nut-trees near the sea-level. *Rev. T. Powell, no. 110*. Type citation: Samoa, 170° Long, W., 14 ° Lat., S., *Rev. T. Powell* (isotypes: MO 3365004!, MO 3365005!).

Macromitrium subtile Schwägr., Sp. Musc. Frond., Suppl. 2(2): 140. 192. 1827, *fide* Vitt & Ramsay, 1985. Type protologue: Society Islands, Tahiti, In insula Otaheite legit et dedit *D. Menzies*.

Macromitrium subtile subsp. *subuligerum* (Bosch & Sande Lac.) M. Fleisch., Musci Buitenzorg, 2: 439. 1904, *fide* Vitt & Ramsay, 1985.

Macromitrium subuligerum Bosch & Sande Lac., Bryol. Jav. 1: 124. 102. 1860, *fide* Vitt *et al.*, 1995. Type protologue: (Indonesia) Habitat insulam Javae; in montibus Gedé et Salak Teysmann. Type citation: Java, Gedé et Salak, Legit. Teysmann (isotypes: H-BR 2615002!, H-BR 2615003!).

Macromitrium zippelii Bosch & Sande Lac., Bryol. Jav. 1: 123. 100. 1860 ("*Zippelii*"). Type protologue: (Indonesia) Habitat Amboinam. Oct. Zippelius in herb. reg. Lugd. Bat. Type citation: Amboinam, Zippelii (isotype: H-BR 2617009!).

Orthotrichum undulatum Hook. & Grev., Edinburgh J. Sci. 1: 117. pl. 4: *O. undulatum*. 1824, *fide* Reese, 1997. Type protologue: Island of Ternate, communicated by Mr. Dickson.

(1) Plants medium-sized, in loose spreading mats; stems creeping, inconspicuous, with erect branches, branches up to 1.5 cm long. (2) Stem leaves irregularly twisted-contorted when dry, narrowly lanceolate to lanceolate, with slenderly acute apices. (3) Branch leaves individually irregularly twisted and contorted when dry, spreading with erect-incurved upper portion when moist, 1.8-2.2 mm long, ligulate-lanceolate, lanceolate to narrowly lanceolate, acuminate, acuminate-mucronate to acute; costae ending obscurely in the apex or shortly excurrent; upper cells

rounded-quadrate to subquadrate, densely pluripapillose; medial cells quadrate to short-rectangular, smooth, forming a gradual transition to basal cells; basal cells short- to long-rectangular, curved to sometimes shallow-sigmoid, smooth. (4) Perichaetial leaves much shorter than branch leaves, ligulate to ovate, abruptly narrowed to a stout cusp; all cells longer than wide, becoming elongate towards the base, all with curved lumens. (5) Setae 5-8 mm long, straight, smooth, twisted to the left. (6) Capsule urns oblong-ovoid to broadly ovoid, occasionally narrowly ellipsoid-ovoid, smooth to lightly 8-plicate; peristome single; exostome of 16; spores distinctly anisosporous. (7) Calyptrae conic mitrate, sparsely to densely hairy.

Macromitrium incurvifolium (Hook. & Grev.) Schwägr. is similar to *M. angustifolium* Dozy & Molk., *M. fuscescens* E.B. Bartram, *M. gracile* (Hook.) Schwägr., and *M. salakanum* Müll. Hal. From *M. incurvifolium*, *M. angustifolium* could be separated by its abaxially curved leaves when moist, *M. fuscescens* could be distinguished by its rather obscure, pluripapillose upper and medial cells, sharply grading to long, clear hyaline low and basal cells, *M. gracile* could be separated by its branch leaves with a fragile subula, *M. salakanum* differed by its fragile branch leaves (Vitt *et al*., 1995).

Distribution: Australia, Hawaii, Indonesia, Papua New Guinea, Philippines, Samoa, Society Islands, Tahiti.

Specimens examined: **AUSTRALIA**. Queenaland, *H. Streimann & T. Pócs 64815* (Orig. in H) (MO 5261264; KRAM-B-147831). **INDONESIA**. Amboinam (H-BR 2617007). **PAPUA NEW GUINEA**. New Britain, East Sepik, *H. Streimann 41600* (MO 4435200, KRAM-B-122633); Morobe, *A. Touw 18626* (MO 5371014), *A. Touw 14660* (MO 5371189), *H. Streimann 19137* (MO 4448116); Morobe, Yinimba, *H. Streimann 19137* (KARM-B- 107266). **SAMOA**. *R. T. Powell 18841* (MO 3365004 as *M. powellii*); *R. T. Powell 18465* (MO 3365005 as *M. powellii*). **TAHITI**. Insel Raiatea, *E. H. & P. D. Hegewald 11456* (MO 4432972); Windward, *E. H. & P. D. Hegewald 11473* (MO 945252), *D. Nadeaud*, 2/6, 1896 (H-BR 2614013).

106. *Macromitrium involutifolium* (Hook. & Grev.) Schwägr., Sp. Musc. Frond., Suppl. 2(2): 144. 1827. (Figure 129)

Basionym: *Orthotrichum involutifolium* Hook. & Grev., Edinb. J. Sci. 1: 117 5. 1824. Type protologue: Hab. King George's Sound; received from Mr. Dickson: from New Zealand and paramata, New South Wales, by Mr. Hobson. Type citation: (Australia, New SouthWales) *E. Hobson*, *H.655* (lectotype: BM 000852412!); King George' Sound, *Dickso*n (isosyntypes: BM 000852416!, E 00165196!); Parramatta, Hobson (syntype: E 00165195!; isosyntypes: BM 000852412!, BM 000852415!, NY-Mitten!).

Macromitrium daemelii Müll. Hal., Hedwigia 37: 153. 1898, *fide* Vitt & Ramsay, 1985. Type protologue: Australia tropica, Queensland, Distr. Wideboy, Gayndah: Damel primus legit 1874. Type citation: Australia, Queesland, *n. sp,* (isotype: JE 04008747!).

Macromitrium incurvulum Müll. Hal., Hedwigia 37: 155. 1898, *fide* Vitt & Ramsay, 1985. Type protologue: Australia tropica, Queensland, sine loco speciali: Rev. B. Scortechini in Hb. Saharampur. Dr. E. Levier 1893 misit. Type citation: Queensland, leg. *Scortechini* (isotype: H-BR 259002!).

Macromitrium malacoblastum Müll. Hal., Hedwigia 37: 150. 1898, *fide* Vitt & Ramsay 1985. Type protologue: Australia subtropica, New South Wales, Walcha: *A.R. Crawford* 1884 in Hb. Melbourne; Cambewarra: Thorpe Octobri 1884 begit fertile et mis. Hb. Melbourne; primus omnium legit *F. M. Reader* 1880 prope Tilba: Melbourne mis. 1881. Type citation: New South Wales, 1884, ex hb. C. Müller (isosyntype: H-BR 259004!).

Macromitrium noumeanum Besch., Ann. Sci. Nat. sér. 5, 18: 208. 1873, *fide* Vitt & Ramsay, 1985. Type protologue and citation: Ad truncus, in sylvis prope Noumea (*Balansa, 2535*). Canala, supra cataractum (*Balansa, 2539*, cum Rhacopilo cuspidigero mixtum). In Nova-Caledonia (Krieger, Schimper comm.). Type citation: (New Caledonia), Forêts situées au-dessus de la cascade, 1869-10, *Balansa no 2539* (lectotype: BM 000919524!; isosyntypes: PC 0083698!, PC 0083699!); Bosquets, écorces des arbres,1868, *Balansa no 2535* (isosyntypes: PC 0083696!, PC 0083697!).

(1) Plants medium-sized to large; stems prostrate to ascending, with ascending to erect branches, often broken when old, forming many branches as separate individual "plants". (2) Branch leaves loosely erect below, upper portion twisted-crisped, with apices curly to circinate, hidden in the inrolled cavity when dry, erect-incurved when moist, 2.0-2.8 mm long, narrowly lanceolate to ligulate; apices acute to acute-apiculate; margins plane or slightly reflexed, entire; costae slender, often ending a few cells beneath the apex, occasionally percurrent or excurrent; upper cells strongly bulging, rounded, 12-14 µm wide, smooth or rarely weakly pluripapillose; medial cells similar to upper cells, 12-20 × 12-15 µm; basal cells confined to a small area, rectangular-hexagonal, smooth or sometimes unipapillose. (3) Perichaetial leaves narrowly ovate-, oblong-lanceolate, 1.9-2.1 mm long, with a gradually acuminate apex. (4) Setae short and smooth, 2-5 mm long, slightly twisted to the left. (5) Capsules short- exserted, urns large and cylindrical, 1.8-2.5 mm long, with a wrinkled neck and a non-plicate smooth darkened rim, rarely collapsed; peristome single; exostome of 16, erect, lanceolate, short, reduced or absent in some populations; spores anisosporous,

finely papillose. (6) Calyptrae large, conic-mitrate, lacerate, with dense and straight hairs.

Distribution: Australia, Vietnam, New Caledonia (Thouvenot, 2019).

Specimens examined: **AUSTRALIA**. Queensland, Ranges, *Dr. Muller* (BM 000852411), Queensland, *D.H. Norris 34222* (H 3090401); New South Wales, *Whitelegge 350* (H-BR 2587003), *Whitelegge 245* (H-BR 2587005), *W. W. Watts 3475* (H 3090407), *W. W. Watts 3408* (H-BR 2593007), *W. W. Watts 4277* (H 3090403), *Richwar R, W.W. Watts s.n* (H-BR 2548003, H-BR 2548007). **VIETNAM**. Ha Giang, *D. K. Harder, L. V. Averyanov, T. H. Nguyên & S. A. Bodine 6238* (MO 5377212).

Macromitrium involutifolium subsp. *ptychomitrioides* (Besch.) Vitt & H.P. Ramsay, J. Hattori Bot. Lab., 59: 378. 1985.

Basionym: *Macromitrium ptychomitrioides* Besch., Ann. Sci. Nat., Bot., sér. 5, 5, 18: 208. 1873. Type protologue: (New Caledonia) Canala, in sylvis supra cataractam (*Balansa, no. 2540*).

Macromitrium carinatum Mitt., Trans. & Proc. Roy. Soc. Victoria, 19: 64. 1882, *fide* Vitt & Ramsay, 1985.

Macromitrium platyphyllaceum Müll. Hal., Hedwigia, 37: 154. 1898, *fide* Vitt & Ramsay, 1985. Type protologue: Australia tropica, Queensland, property Brisbane: F. M. Bailey 1888 in Hb. Broth.

Macromitrium plicatum Thér., Bull. Acad. Int. Géogr. Bot. 17: 307. 1907, *fide* Thouvenot, 2019. Type protologue: no detail information. Type citation: Nouvelle Calédonie, "environs, de Nouméa", 1907, leg. *Franc s.n.* (lectotype: PC 0093710!).

Macromitrium plicatum var. *aristatum* Thér., Bull. Acad. Int. Géogr. Bot. 20: 99. 1910, *fide* Thouvenot, 2019. Type protologue: Hab. Environs de Nouméa: leg. Le Rat, Comm. Museum Paris. Type citation: Nouvelle Calédonie, environs, de Nouméa, 1909, Le Rat (isotype: PC 0083716!).

Macromitrium plicatum var. *obtusifolium* Thér., Diagn. Esp. Var. Nouv. Mouss. 8: 4. 1910, *fide* Thouvenot, 2019. Type protologue: New Caledonia, "Monts Koghis, forêt, troncs, d´arbre, alt. 300 m" Franc *s. n.* Type citation: Nouvelle Calédonie, Mar 1907, Koghis, forêt, leg. *Franc* (lectotype: PC 0083719!).

Macromitrium suberosulum E.B. Bartram, Occas. Pap. Bernice Pauahi Bishop Mus. 15(27): 336. 6. 1940, *fide* Vitt & Ramsay, 1985. Type protologue: Austral Islands: Tubuai, south slope of Pance, on tree branches in mist upper forest, alt. 350 m., Aug. 23, 1934, St. John 1630a (holotype: FH-Bartram).

Macromitrium viridissimum Mitt., Trans. & Proc. Roy. Soc. Victoria 19: 64. 1882, *fide* Vitt & Ramsay, 1985. Type protologue: Ranges between the Burnett and Brishbane Rivers, F.v.M.; Toowoomba, Hartmann.

Macromitrium involutifolium has been divided into two subspecies: subsp. *involutifolium* and subsp. *ptychomitrioides*, the capsules of the former having single peristome, exostome of 16 erect teeth, with erect rims, exothecial cells many much longer than broad, elongate-rectangular to almost quadrate, while those of the latter without peristome, with collapsed rims, exothecial cells most about as long as broad or somewhat longer.

Distribution: Australia, New Caledonia.

Specimens examined: **AUSTRALIA**. Norfolk Island, *H. Streimann 53750* (KRAM-B-110766), *H. Streimann 53887* (KRAM-B-122607); Queensland, *D.H. Vitt 28083* (H 3090418).

107. *Macromitrium japonicum* Dozy & Molk., Ann. Sc. Nat. Bot. sér. 3, 2(5): 311. 1844. (Figure 130)

Type protologue and citation: Japonia, Siebold, lectotype L (910, 138-1128), lectotype designated by Noguchi, 1967.

Dasymitrium japonicum (Dozy & Molk) Lindb. ex Par., Ind. Bryol. Ed. 2, 5: 149, 1906.

Macromitrium japonicum var. *makinoi* (Broth.) Nog., Ill. Moss Fl. Japan 3: 606. 1989, **syn. nov.** - *Dasymitrium makinoi* Broth., Hedwigia 38: 215. 1899. Type protologue: [Japan] Shikoku: Setton, ad corticem arboris (Makino). Type citation: Japonia, Setton, leg *T. Makino*, 1889 (holotype: H-BR 2572002!).

Dasymitrium incurvum Lindb., J. Bot. 2: 385, 1864. - *Macromitrium incurvum* (Lindb.) Mitt., Trans. Linn. Soc. London Bot. ser. 2, 3: 163, 1891, *fide* Ignatov & Afonina 1992. Type protologue: In szxis insulæ, Tschea-schan (30°lat. Bor.) imperii chinensis, hedwigia albicanti, associatum, anno 1862, legit navarchus succicus *L. Ahlström*. Type citaton: China, ins Tschea-Schan (30°lat. Bor.), 1862, leg. *L. Ahlström* (isotypes: BM 000558422!, BM 000558422!).

Macromitrium bathyodontum Cardot, Beih. Bot. Centralbl. 17: 13. 8. 1904, *fide* Noguchi & Iwatsuki, 1989. Type protologue: Ouen-San (no. 32. c. fruet. vet. et juven). Type citation: Corēe: Quen-San, Aug. 1901, Leg. *Faurei, no. 32* (isotype: H-BR 2572017!).

Macromitrium dickasonii E.B. Bartram, Farlowia 1: 178. 1943, **syn. nov.** Type protologue: [Myanmar] Mogok, Apr. 1934, [*F. G. Dickason*] *no. 102*. Type citation: [Myanmar] Mogok, Apr. 1934, *F. G. Dickason 102* (holotype: FH 00213603!).

Macromitrium polygonostomum Dixon & P. de la Varde, Arch. Bot. Bull. Mens. 1(8-9): 181. 6. 1927, **syn. nov.** Type protologue: Hab.: Sirumalai, 4000 ft. alt., avril 1924, sterile (No 439)' Ibid., crête rocailleuse surplombant sentier d'Emmakkalapuram, mars 1927 (NOB 844A, 844B et 845); Type citation : India, Sirumalai, Mars 1927, leg. *G. Foreau 844B* (isosyntypes: US 00070273!, H-BR 2572016!; syntypes: PC, not seen).

Macromitrium insulanum Sull. & Lesq., Proc. Am. Ac. Arts Sc. 4: 278, 1859, *fide* Ignatov & Afonina, 1992. Type protologue: On trees, Loo Choo Islands, Ousima; Simoda, C. Japan.

(1) Plants small to medium size, in dense mats; stems long creeping, with dense, erect and short branches, branches often shorter than 6 mm, simple or with a few short branchlets, densely leaved. (2) Stem leaves recurved, from a triangular-ovate base gradually tapering to an oblong-lanceolate apex. (3) Branch leaves strongly keeled, erect-twisted below, twisted-crisped or twisted-contorted-crisped above, with apices often hidden in the inrolled cavity when dry, spreading but the apex often adaxially incurved when moist, often plicate at base, 1.3-2.5 × 0.25-0.40 mm, ligulate, lanceolate-ligulate; margin entire throughout; apices subacute, acute to obtuse; costae subpercurrent, ending just before the leaf apex; medial and upper cells obscure or slightly obscure, quadrate to hexagonal, 8-12 μm wide, thin-walled, with 3-5 small papillae; basal cells colorless, rectangular, 20-40 × 6-8 μm, with slightly thickened walls, smooth. (4) Perichaetial leaves shorter than branch leaves, 1.5 mm long, ovate-lanceolate or ovate-oblong, acuminate. (5) Setae smooth, 2-3 mm long. (6) Capsule urns ovoid, ovoid-ellipsoid, ellipsoid or subglobose, smooth, constricted at the mouth when dry; peristome single, exostome of 16, lanceolate, obtuse, irregular in outline, densely papillose on the entire surface, often caducous in old capsules; spores densely papillose, 20-30 μm in diameter. (7) Calyptrae cucullate, somewhat lobed at base, plicate, with numerous, long, yellowish hairs.

Distribution: China, Japan, Korea, Laos, Myanmar, Sri Lanka, Vietnam.

Specimens examined: **CHINA**. Anhui, *Wan ZL & Luo JX 9327.* Fujian, *Chung HH B4* (MO), *Chung HH B3* (PC 0083682, H 3090393, NY, FH, MO); *Chung HH B36/ad, B31* (FH); *Hance HB s.n.* (NY, as *M. spathylare*); *Chung HH B361* (S-B 115577). Zhejiang, *Guo SL N 13-1* (SHTU). Guangdong, *Wu Han B 95* (MO); *Redfearn PL 34384* (MO); *Magill R, Redfearn P & Crosby M 8127* (MO). Hainan, *Lau & Tsang 18295* (FH); *Reese WD 17498* (MO); Hebei, *Wang QW 976, 864* (MO). Heilongjiang, Manchuria *I. Kozlov 7.* (H 3090539). Henan, *Luo JX 128* (MO), *Boufford DE et al. 26349-B* (FH); *Luo JX 176, 196* (NY, MO); *Luo JX 321, 330a, 330, 352* (MO); *He ZZ 82* (MO); Hongkong. *Herklots GAC, 363* (BM); *Tsang WT 29731* (FH); Hubei, *Sino-Americ Exped. 822b* (NY). Hunan, *Handel-Mazzetti 11064* (H-BR); *Enroth 63564, 64691, 64712, 70982, Virtanen 61646, 61647, 61659, 61931* (H). Jiangxi, *Patterson WB s.n.* (as *M. incurvum*) (FH); *Chung HH 4045* (FH). Sichuan, *Chen PC 5869, Redfearn PL 35575, 34817, Bruce A 6688, 7219, 7373* (MO); *Koponen T 45900, 45958, 43425* (H). Shaanxi. *Giraldi J s.n.* (S-B 88595); *Giraldi J 2131* (H-BR 2576004, PC 0083662, JE); *Giraldi J 1496* (PC 0083661); *Giraldi J 1564* (H-BR 2576007); *Giraldi J 2133* (H-BR 2576007), *Giraldi J 1496b* (NY), *Girald J 1496d* (H- BR 2576002); *Giraldi J 2132* (H-BR 2576005); *Giraldi J 1565* (H-BR 2576008). Taiwan, *Schwabe-Beha. 9,10, 52, 56* (JE); *Sasarka 2490* (H-BR 2578031); *Sasaoka 2525* (H-BR 2578032); *Suzwk 2823* (H-BR 2578037); *Suzwk no. 2828* (H-BR 2578038); *Suzwk 2850* (H-BR 2578050); *Wichua 1648* (H-BR 2579013); *Faurie 95* (H-BR 2579014). Yunnan, *Wu PC 22152* (MO); *Magill R, Redfearn P & Crosby M 7987* (MO); *Redfearn RL et al. 1730, 1844* (MO). Zhejiang, *Wu PC 1231, 1258, 24910, Hu RL 92* (MO). **JAPAN**. Kiushiu, *S. Hattori 20* (MO); Honshiu, *K. Sakurai 141*; *Jhsiba 8278* (MO); Honshiu, *K. Sakurai 16707, M. Higuchi 12147* (MO); Yakushima, *Iwatsuki & Sharp 1029* (MO); Yoshio, *K. Mayebara 431* (MO); Kozushima, *T. Nakamura, 690* (MO, JE); Kanagawa, *P. L. Redfearn 757*. **KOREA**. Tongduch'on-Shi, *J. R. Shevock 16286* (MO).

108. *Macromitrium krausei* Lorentz, Bot. Zeitung (Berlin) 24: 187. 1866. (Figure 131)

Type protologue: [Chile prop Valdiviam et prope Corral lecti per Dr. Krause] An dünnen Aesten und Zweigen. Type citation: Chile, Corral Valdiviae, *P. G. Lorentz*, leg. *Krause s.n.* (lectotype designated here: NY 01202035!; isolectotypes: BM 000873306!, BM 000873307!).

(1) Plants small, brownish above, dark-brown below; stems creeping, branches 5 mm long. (2) Branch leaves not very shriveled, erect-appressed below, spreading and slightly flexuose above when dry, erecto-patent when moist, oblong-lanceolate, with an acute apex; margins entire; costa stout, vanishing far beneath the apex; all cells bulging, clear and smooth, arranged in longitudinal rows; upper and medial cells isodiametric and rounded, rounded-quadrate; low and basal cells rectangular, straight-walled, not porose. (3) Setae smooth, 5-7 mm long, twisted to the left. (4) Capsule urns ellipsoid-cylindric, smooth, weakly or moderately furrowed or longitudinally wrinkled. (5) Calyptrae mitrate, smooth and naked.

Macromitrium krausei Lorentz is somewhat similar to *M. microstomum* (Hook. & Grev.) Schwägr., but its medial cells are isodiametric, all cells moderate bulging, and some basal cells porose, as well as its branches not funiculate when dry.

Distribution: Chile.

Specimens examined: **CHILE**. Fuegia occidentalis, *Roivoinen H s.n.* (H 3090545); Insulis Guaitecas, in ramuli, *Dusen P s.n.* (MO 3987799).

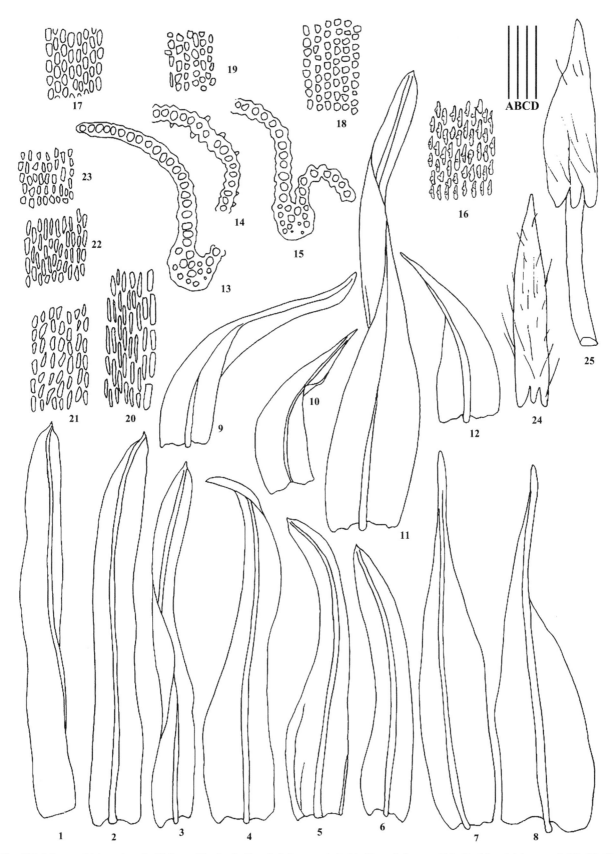

Fig. 129 *Macromitirium involutifolium* (Hook. & Grev.) Schwägr. 1-6, 11: Branch leaves. 7-8: Perichaetial leaves. 9-10, 12: Stem leaves. 13,15: Upper transverse sections of branch leaves. 14: Basal transverse section of branch leaf. 16: Basal cells of branch leaf. 17: Low cells of branch leaf. 18,19: Upper cells of perichaetial leaves. 20: Basal cells of stem leaf. 21: Low cells of stem leaf. 22: Medial cells of stem leaf. 23: Upper cells of stem leaf. 24: Calyptra. 25: Capsule (all from S-B 115522). Line scales: A = 2 mm (24-25); B = 1 mm (9-10, 12); C = 0.4 mm (1-8, 11); D = 67 μm (13-23).

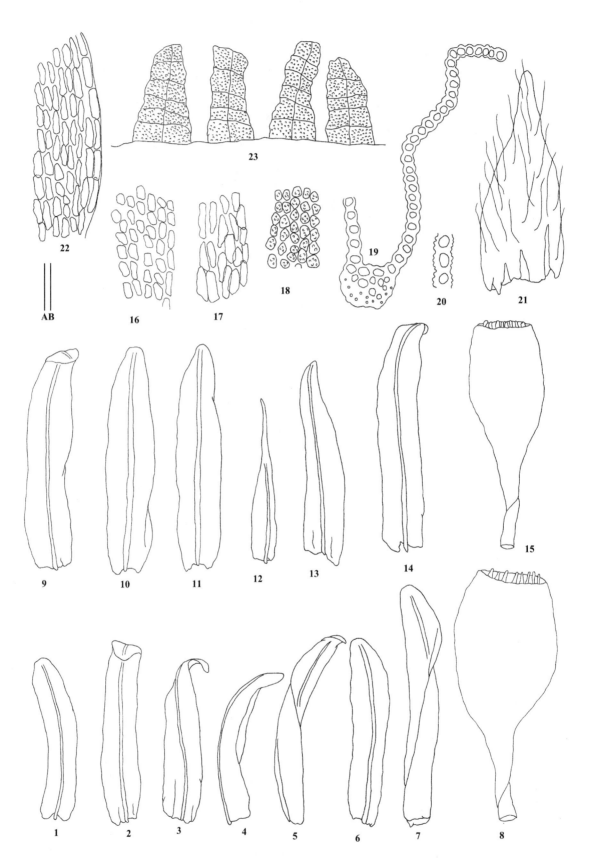

Fig. 130 *Macromitrium japonicum* Dozy & Molk. 1-7, 9-11, 13-14: Branch leaves. 12: Perichaetial leaves. 8, 15: Capsules. 16-17: Basal cells of branch leaves. 18: Medial cells of branch leaf. 19: Basal transverse section of branch leaf. 20: Upper transverse section of branch leaf. 21: Calyptra. 22: Basal marginal cells of branch leaf. 23: Peristome (all from syntype of *M. polygonostomum*, H-BR 2572016). Line scales: A = 0.44 mm (1-15, 21); B = 44 μm (16-20, 22-23). (Li *et al.*, 2024).

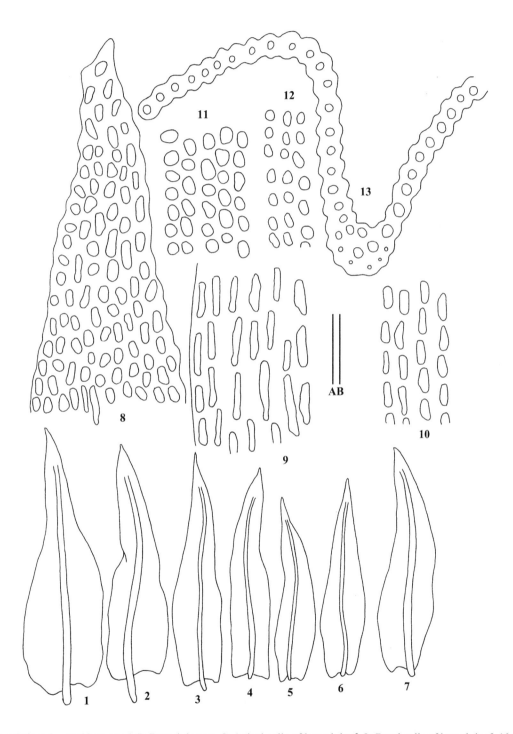

Fig. 131 *Macromitrium krausei* Lorentz 1-7: Branch leaves. 8: Apical cells of branch leaf. 9: Basal cells of branch leaf. 10: Low cells of branch leaf. 11: Upper cells of branch leaf. 12: Medial cells of branch leaf. 13: Basal transverse section of branch leaf (all from lectotype, NY 01202035). Line scales: A = 1 mm (1-7); B = 67 μm (8-13).

109. *Macromitrium laevigatum* Thér., Diagn. Esp. Var. Nouv. Mouss. 8: 5. 1910. (Figure 132)

Type protologue: New Caledonia. Mt Koghis, troncs d'arbre, *Franc s.n.* Type citation: (New Caledonia) Mont Koghis, troncs d'arbres, Leg. *Franc,* 1 Nov. 1909. (lectotype designated by Thouvenot, 2019: PC 0083685!; isolectotypes: PC 0083686!, PC 0083687!)

(1) Plants medium-sized; dark-brown, stems creeping, densely with branches; branches 10-20 mm long, straight, simple or furcate. (2) Stem leaves triangular-lanceolate. (3) Branch leaves strongly crisped and individually twisted, strongly keeled, with apices hidden in the inrolled cavity when dry, widely spreading when moist, lanceolate to ligulate from wide oblong basal portion, 1.3-1.7 × 0.3-0.4 mm; apices obtuse to shortly acute or apiculate; costae percurrent to excurrent; all cells thick-walled; upper cells small and obscure, rounded-quadrate, 7-8 μm wide, in conspicuous longitudinal rows, strongly pluripapillose; transitional parts shorter; low and basal cells rectangular elongated, 20-30 × 8-10 μm, smooth, lumens curved sigmoid. (4) Perichaetial leaves hyaline, oblong-ligulate to sub-triangular, with an obtuse apex. (5) Setae shorter, 2-3 mm long, smooth and straight, vaginulae naked. (6) Capsule urns ellipsoid, smooth, slightly plicate under the mouth; peristome single, exostome of 16. (7) Calyptrae mitrate, plicate, lobed at base, naked.

Distribution: Madagascar, New Caledonia.

Specimen examined: **MADAGASCAR**. Mahajanga, *Lala Roger Andriamiarisoa s.n.* (MO 5926326).

110. *Macromitrium laevisetum* Mitt., J. Linn. Soc., Bot. 12: 214. 1869. (Figure 133)

Type protologue: [Ecuador] Andes Quitenses, in montibus Tunguragua et Mulinúl (8000-10000 ped.), *Spruce, n. 95*; Brasilia, *Klotsch in Herb. Hooker*. Type citation: Musci Amazonici et Andini, leg. *R. Spruce 95*, deterinavit W. Mitten (isosyntype: E 00165151!).

(1) Plants robust, lustrous, light yellow above, dark brown below; stems weakly creeping, branches up to 35-40 mm long. (2) Branch leaves erect-appressed below, twisted-contorted and flexuose above when dry, weakly abaxially curved spreading, and keeled above when moist, narrowly lanceolate from an oblong base or broadly oblong base, acuminate, plicate below, occasionally decurrent; margins entire throughout; costae subpercurrent; upper cells quadrate, rhombic, oblong, bulging to conic-bulging; medial cells short to long rectangular, strongly porose; low and basal cells elongate, linear-rectangular, strong porose, tuberculate; basal cells near the costa enlarged and broader, marginal enlarged teeth-like cells distinctly differentiated. (3) Perichaetial leaves longer than branch leaves, broadly oblong-lanceolate, sheathing at base, sharply narrowed (cuspidate) in the upper portion, plicate below, with an acuminate-subulate apex and entire margins. (4) Setae smooth, up to 20 mm long, twisted to the left. (5) Capsule urns ellipsoid-cylindric, strongly furrowed, slightly constricted beneath the mouth; peristome double, both membranous, exostome teeth yellowish brown, endostome hyaline. (6) Calyptrae large, mitrate, deeply lacerate, smooth and naked.

Grout (1944) once treated *M. laveisetum* Mitt. as a synonym of *M. cirrosum* var. *stenophyllum* (Mitt.) Grout., and Wijk *et al.* (1964) treated it as a synonym of *M. cirrosum* (Hedw.) Brid. However, *M. laveisetum* distinctly differs from *M. cirrosum* in having leaves with marginal differentiated enlarged teeth-like cells at the insertions, and having strongly furrowed ellipsoid-cylindric capsules. *Macromitrium laveisetum* is similar to *M. brevihamatum* Herzog, but differs in the degrees of cell papillosity at low and basal portions of branch leaves, as well as differentiation of enlarged teeth-like cells at the insertion.

Distribution: Ecuador.

Specimens examined: **ECUADOR**. Pichincha, *Crosby M. R. 75-68* (MO 2330106), *Holm-Nielsen L & Brandbyge J 25050* (MO 3655808), *Holm-Nielsen H & Brandbyge J 25134* (MO 3655807), *Holm-Nielsen L 18199* (MO 3659319); Loja, *Harriet G. Barclay & Pedro Juajibioy 8503* (MO 2853718); Azuay, *LØjtnant B & Molau U 14426* (MO 4419647), *U 14711* (MO 4419627); Carchi, *LØjtnant B, Molau U & Madison M 12539* (MO 4419150); Cotopaxi, *Holm-Nielsen L 18325* (MO 3689041), *Holm-Nielsen L, Jeppesen S, LØjtnant B & Øllgaard B, 3355* (MO 2561515); Loja, *William C. Steere & Henrik Balslev 25863A* (KRAM-B-058370).

111. *Macromitrium lanceolatum* Broth. in Voeltzkow, Reise Ostafr., Syst. Arbeit. 3: 54. 8 f. 7. 1908. (Figure 134)

Type protologue: Fundnotiz: Mauritius. Type citation: Mauritius, 1904. leg. *Voeltzkow s.n.* com. Brotherus (lectotype designated here: PC 0137511!).

(1) Plants small, dark green, dull; stems long and creeping, densely with rusty rhizoids, with prostrate branches, branches up to 20 mm long, forked at obtuse terminal, moderately leaved. (2) Branch leaves erect-appressed below, spreading and individually twisted-flexuose above when dry, spreading and occasionally conduplicate when moist, lanceolate, with an acute apex; margins plane and entire; costae light reddish, subpercurrent; cells near the apex round and smooth, those in the topmost portion elongate; upper and medial cells rounded, subrounded, chlorophyllose, thin-walled, strongly bulging but almost smooth, the outmost marginal cells smaller than their ambient cells; basal cells smooth, near the costa elongate, marginal cells at the insertion rounded and smaller than their ambient cells. (3) Perichaetial leaves distinctly differentiated, wide ovate, abruptly constrict forming a narrowly linear upper portion; apices acuminate.

Distribution: Mauritius.

Taxonomy

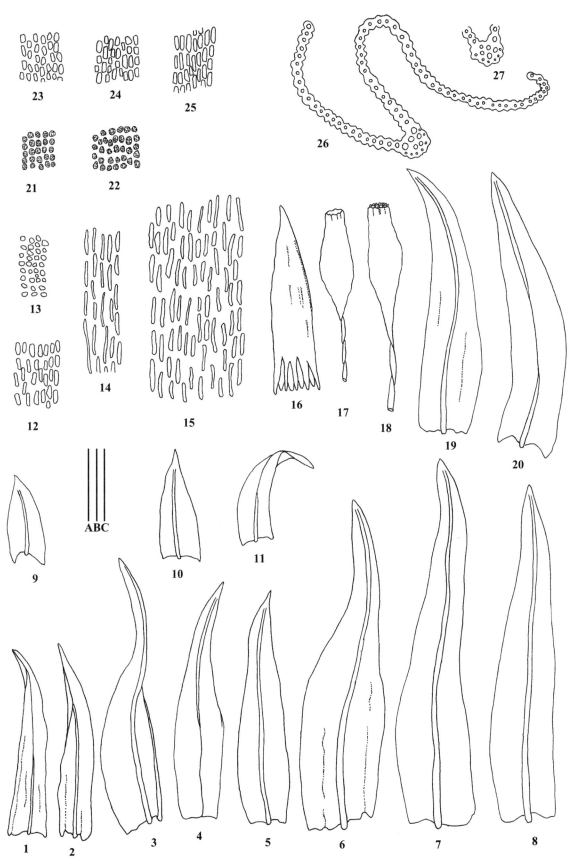

Fig. 132 *Macromitrium laevigatum* Thér. 1-2: Perichaetial leaves. 3-8, 19-20: Branch leaves. 9-11: Stem leaves. 12: Low cells of branch leaf. 13: Medial cells of branch leaf. 14-15: Basal cells of branch leaves. 16: Calyptra. 17-18: Capsules. 21-22: Upper cells of branch leaves. 23: Upper cells of stem leaf. 24: Medial cells of stem leaf. 25: Basal cells of stem leaf. 26-27: Basal transverse sections of branch leaves (all from lectotype, PC 0083685). Line scales: A = 1 mm (1-2, 16-18); B = 400 μm (3-11, 19-20); C = 67 μm (12-15, 21-27).

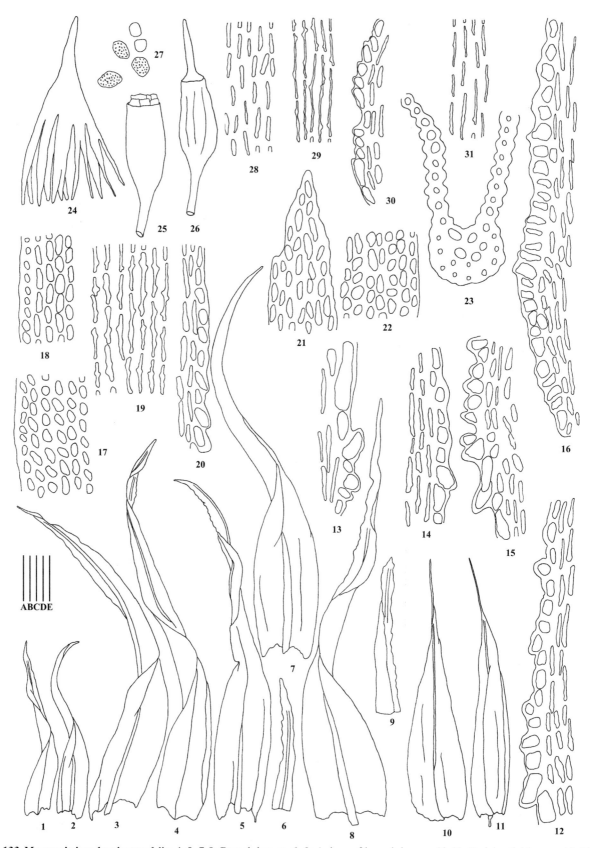

Fig. 133 *Macromitrium laevisetum* Mitt. 1-5, 7-8: Branch leaves. 6, 9: Apices of branch leaves. 10-11: Perichaetial leaves. 12-15, 20, 30: Basal marginal cells of branch leaves. 16: Basal marginal cells of perichaetial leaf. 17: Upper cells of branch leaf. 18: Low cells of branch leaf. 19: Basal cells of branch leaf. 21: Apical cells of branch leaf. 22: Medial cells of branch leaf. 23: Medial transverse section of branch leaf. 24: Calyptra. 25-26: Capsules. 27: Spores. 28: Upper cells of perichaetial leaf. 29: Basal cells of perichaetial leaf. 31: Medial cells of perichaetial leaf (all from isosyntype, E 00165151). Line scales: A = 2 mm (1-2, 10-11, 24-26); B = 1 mm (3-5, 7-8); C = 400 μm (6, 9); D = 100 μm (27, 30); E = 67 μm (12-23, 28-29, 31).

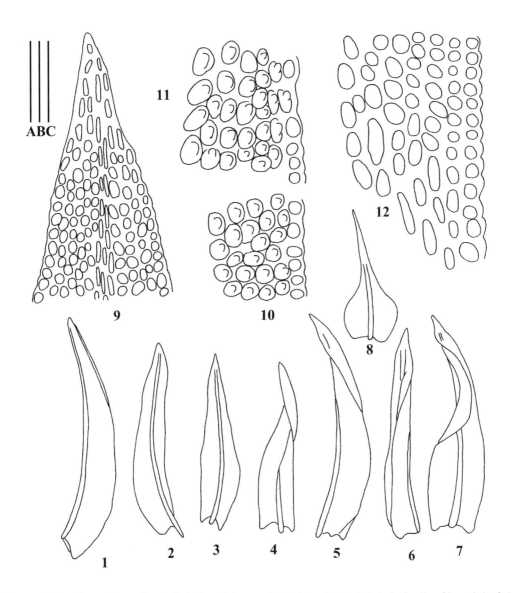

Fig. 134 *Macromitrium lanceolatum* Broth. 1-7: Branch leaves. 8: Perichaetial leaf. 9: Apical cells of branch leaf. 10: Upper cells of branch leaf. 11: Medial cells of branch leaf. 12: Basal cells of branch leaf (all from lectotype, PC 0137511). Line scales: A = 1 mm (1-8); B = 100 μm (9); C = 67 μm (10-12).

112. *Macromitrium laosianum* Paris & Broth., Rev. Bryol. 35: 50. 1908. (Figure 135)

Type protologue: (Laos) Muong Ho, 24.12. 06. Type citation: Laos, Muong Ho, 24.12. 06, leg. mission scientifique, d' exploraton de l'Indo-Chine, 737, *Macromitrium laosianum* n. sp. ex Herb. E. G. Paris (holotype: H-BR 2582001!).

(1) Plants small, in dense mat; stems long creeping, densely with short branches, branches 2-4 mm long. (2) Branch leaves erect-appressed below, spirally twisted-contorted with an inrolled or incurved apex when dry, erect-spreading, with adaxially incurved upper portion when moist, lanceolate-oblong to oblong-ligulate; apices rounded, obtuse to obtuse-acute; margins recurved below, plane above, entire from the apex to the base; upper and medial cells subquadrate, rounded-quadrate, obscure, puripapillose; elongate from medial to basal portion; basal cells rectangular, straight-walled, smooth. (3) Perichaetial ovate-lanceolate to oblong-lanceolate, acuminate. (4) Setae 8-10 mm long, smooth. (5) Capsule urns ellipsoid, smooth, peristome single, exostome lanceolate. (6) Calyptrae hairy, light yellowish.

Macromitrium laosianum Paris & Broth. is similar to *M. japonicum* Dozy & Molk., but the peristome of the former is not seen.

Distribution: Laos.

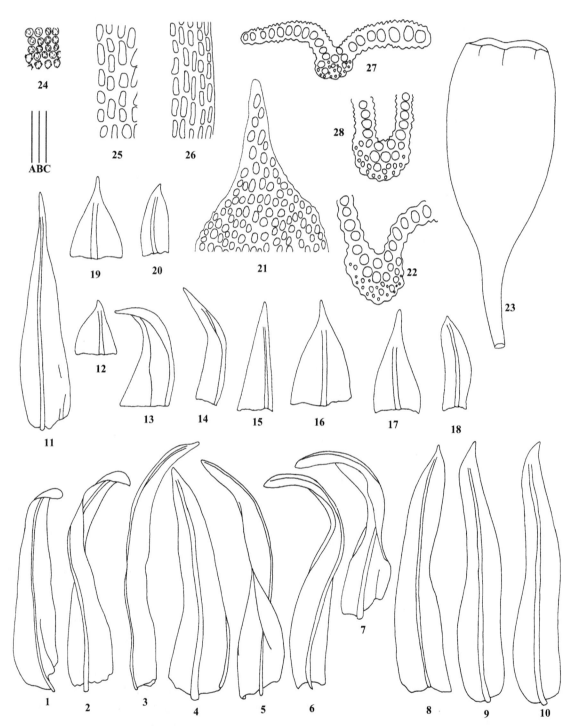

Fig. 135 *Macromitrium laosianum* Paris & Broth. 1-10: Branch leaves. 11: Perichaetial leaves. 12-20: Apices of branch leaves. 21: Apical cells of branch leaf. 22: Medial transverse section of branch leaf. 23: Capsule. 24: Upper cells of branch leaf. 25: Basal cells of branch leaf. 26: Basal marginal cells of branch leaf. 27: Upper transverse section of branch leaf. 28: Basal transverse section of branch leaf (all from holotype, H-BR 2582001). Line scales: A = 0.44 mm (1-11, 23); B = 176 μm (12-20); C = 44 μm (21-22, 24-27).

Taxonomy

113. *Macromitrium larrainii* Thouvenot & K.T. Yong, Cryptog., Bryol. 36(4): 344. 2015.

Type protologue: New Caledonia, North Province, Hienghène, summit of Mt Panié, ca 1640 m, 9 Oct. 2012, *Larraín 35846* (holotype: PC 0167650; isotypes: F, CONC, KLU, NOU, NY, S).

(1) Plants medium-sized, stem creeping, sparsely pinnate branching, branches to 25 mm, with a very shaggy habit due to the long, thin aristae of irregularly arranged branch leaves. (2) Branch leaves loosely erect to flexuose-twisted, irregularly but not spirally arranged around the branch when dry, erect to patent, incurved when moist, 1.2-2.7 × 0.2-0.5 mm, lanceolate, keeled below, distinctly slender, widest near the base, very narrow and asymmetrical above; thin aristae up to 1 mm long; costae excurrent; all cells smooth and thick-walled; upper cells rounded-quadrate to oblong, marginal cells smaller and isodiametric to oblate; medial cells oblong; basal cells linear, with straight lumens. (3) Perichaetial leaves not much differentiated. (4) Setae smooth, 8-10 mm long. (5) Capsule urns ovoid, orn 1.1-1.5 mm long, 0.6-0.8 mm wide, weakly plicate when dry, not constricted beneath the mouth, abruptly contracted to the seta; opercula conic, long-rostrate; peristome single, exostome of 16, densely papillose, erect. (6) Calyptrae conic-mitrate, covering the whole capsule, indistinctly lobed at base, strongly plicate, smooth (Thouvenot & Yong, 2015; Thouvenot & Müller, 2016; Thouvenot, 2019).

Macromitrium larrainii Thouvenot & K.T. Yong is similar to *M. piliferum* Schwägr., *M. dielsii* Broth. ex Vitt & H.P. Ramsay, *M. funiforme* Dixon, *M. peraristatum* Broth., *M. ochraceoides* Dixon, *M. longipilum* A. Braun ex Müll. Hal., *M. crinale* Broth. & Geh. and *M. cuspidatum* Hampe by piliferous branch leaves, but *M. larrainii* could be distinguished from the other piliferous species by its smooth basal laminal cells with straight lumens.

Distribution: New Caledonia.

114. *Macromitrium lauterbachii* Broth. ex Fleisch., Musci Buitenzorg 2: 420. 1904. (Figure 136)

Type protologue: [Indonesia] An Rinde? Ost-Java, am Ardjoenogebirge, 700 m., von Lauterbach entdeckt. Type citation: Java, Ardjoena, 15/2, 1890, leg *Lauterbach entdeckt* (lectotype designated here: H-BR 2602007!).

(1) Plants in dense mats; stem creeping and short, with dense branches. (2) Branch leaves irregularly crisped and twisted when dry, moderately spreading when moist, plane on both sides, keeled, broadly lanceolate with an acuminate apex, 3-4 mm long; margins serrulate near the apex; upper and medial cells rather small, rounded-quadrate, 3-5 µm wide, smooth; low cells elongate, rectangular to sublinear, thin-walled, occasionally mammillose to conic-papillose on one surface, strongly porose; costae ending beneath the apex. (3) Inner perchiaetial leaves appressed, slightly different from branch leaves, broadly lingulate; upper cells rounded, becoming oblong to sublinear towards the base. (4) Setae twisted to the right, smooth, 10-13 mm. (5) Capsule urns ellipsoid, ovoid, slightly furrowed, constricted below the mouth; peristome double, consisting of a low double membrane, exostome fragmentary, endostome finely papillose. (6) Calyptrae campanulate, naked, covering the entire capsule, plicate and lacerate at base.

Distribution: Indonesia.

115. *Macromitrium lebomboense* van Rooy, J. Bryol.16: 209. f. 1. 1990. (Figure 137)

Type protologue and citation: South Africa Natal: Lebombo Mountains, near Nambulugwana, along road from Mkuze to Jonzini, 340 m, *van Rooy 227* (isotypes: MO 743664!, H 3090420!).

(1) Plants medium-sized, forming mats; primary stem tomentose below; rhizoids reddish brown to brown; secondary stems numerous, 2-8 mm tall. (2) Stem leaves secund, lanceolate. (3) Branch leaves crowded, keeled, spirally coiled around the branch, twisted and tightly inrolled above when dry, erect-spreading when wet, rugulose below, unistratose or frequently with bistratose patches; lanceolate-ligulate, 1.6-2.3 mm long; ventral surface keeled; apices rounded-obtuse, acute; margins plane, frequently recurved on one side; costas ending before tha apex to percurrent; upper cells small, rounded-hexagonal to rounded-quadrate, incrassate, bulging, 7.5-10 µm, pluripapillose; basal cells smooth, rarely papillose below, rectangular, longitudinal walls incrassate, frequently sinuate. (4) Perichaetial leaves erect, longer than branch leaves, broadly lanceolate or oblong-lanceolate, bluntly acuminate or acuminate-acute, yellowish brown or reddish brown. (5) Capsule urns ovoid-cylindric, 1.5-1.8 mm long, weakly ribbed dry, mouth erect, neck differentiated; peristome single, teeth of 16, occasionally in pairs, narrowly oblong, blunt, 155-250 µm long, yellowish, papillose-striate; spores anisosporous, 12-30 µm in diameter, minutely papillose. (6) Calyptrae 2.5-3.5 mm long, mitrate, deeply plicate, lacerate below, naked or sparsely hairy.

Distribution: Natal, South Africa, Zululand (van Rooy, 1990).

Specimens examined: **NATAL**. Western escarpment of Lebombo Mountains, *J. van Rooy, 232* (MO 3986797); Mtunzini, *L. Smook 1524* (MO 3986801). **SOUTH AFRICA**. Ttranskei, *J. van Rooy 1714* (MO 3986802).

116. *Macromitrium leprieurii* Mont., Ann. Sci. Nat., Bot., sér. 2, 14: 347. 20 f. 5. 1840. (Figure 138)

Type protologue: (French Guiana) ad truncos arborum circà Cayennam.-*Lepr. Coll. n. 334.* Type citation: Guiana, leg *Leprieur* (isotype: S-B 163766!); Guiana, Montagne in Hb. R d Resch. (isotype: NY 01086486!).

Macromitrium crumianum Steere & Buck. Brittonia 31: 395, 1979. Type protologue and citation: Guatemala: on forest tree 10 m above ground, vicinity of Exmibal Camp 2 (La Gloria), NW of Lake Izabal, elev. 400-500 m. 7 May 1966, *Gayle c. Jones & Lynden Facey 3296* (isotype: MO 2846905!).

Macromitrium dussii Broth., Symb. Antill. 3: 424. 1903, *fide* Florschütz, 1964. Type protologue: Martinique in Montagne Pelée, ad terram: *Duss n. 351* (in Herb. Berol.).

(1) Plants small to medium-sized, in loose mats, distinctly lustrous; stems long creeping; branches to 0.5-1.0 cm long, sparsely covered with rusty rhizoids. (2) Stem leaves inconspicuous. (3) Branch leaves not very shriveled, weakly keeled, appressed to erect below, erect-spreading, slightly flexuose and undulate when dry, erecto-patent when moist, 3-5 mm × 0.5 mm, linear-lanceolate, plicate below; apices acute, apiculate to short-acuminate; margins plane or slightly revolute below, erose-denticulate near apex, subentire below; costae subpercurrent, percurrent to short excurrent; all cells smooth and clear, thick-walled, frequently porose, longer than wide, long-rectangular to linear-rectangular; gradually elongate from top to base; upper inner cells 17-45 × 2.5-8 µm, upper marginal cells not or occasionally smaller than inner cells; basal cells 32-70 µm long, lumens linear-rectangular, basal marginal cells enlarged, shorter and wider, but not teeth-like and not protruding. (4) Setae 10-25 mm long, smooth, slightly twisted to the left. (5) Capsule urns globular to long ovoid-cylindric, 1.5-2 mm long, smooth or with weak longitudinal wrinkles; peristome double, exostome teeth rudimentary, papillose; spores isosporous, papillose. (6) Calyptrae cucullate and naked, not lacerate.

Macromitrium leprieurii Mont. is similar to *M. dubium* Schimp. ex Müll. Hal., but differing from the latter in having shorter branch leaves with short-acuminate, acute to apiculate the apices, and relatively longer costa.

Distribution: Belize, Costa Rica, Dominica, French Guiana, Guiana, Guatemala, Nicaragua, Panama, Suriname, Venezuela (Allen, 2002).

Specimens examined: **DOMINICA**. St. Paul Parish. *Bernard Goffinet 2714* (MO 5370030). **FRENCH GUIANA**. Canton de Approuague-Kaw, *William R. Buck 37928* (MO 6002977), *William R. Buck 37995* (MO 6001520), *William R. Buck 37870* (MO 6001484). **GUIANA**. Pakaraima Mts., *P. J. M. Maas & L.Y. Th. Westra 4195* (MO 2843016); Pakaraima Mts, *P.J.M. Maas & L.Y. ATh. Westra 4195* (H 3090546). **SURINAME**. Brokopondo, *R.A. Pursell 11887* (MO 6002351), *R.A. Pursell 11888* (MO 6002352); Sipaliwini, Tafelberg. *B. Allen 23410* (MO 5648872), *B. Allen 23459* (MO 5648870), *B. Allen 23257* (MO 5644905), *B. Allen 23479* (KRAM-B-162724), *B. Allen 20616* (H 3194587). **VENEZUELA**. Estado Bolivar, *Julian A. et al. 106560* (MO 2559450).

117. *Macromitrium leratii* Broth. & Paris, Öfvers. Finska Vetensk.-Soc. Förh. 48(15): 12. 1906. (Figure 139)

Type protologue: Nouvelle Caledonie … In silvaticis Montis Dzumac, alt. 1100 m et in Monte Ouin, ad arbores, alt. 1000 m (*Le Rat*); Prony (*Etesse*). Type citation: Nova Caledonia, Mt. Dzumac, 1100m *Le Rat*, 1906 (lectotype: H-BR 2607009!); New Caledonia, Quin ad arbore alt. 1000 m. *Le Ra*t (syntype: H-BR 2607008!).

Macromitrium leratii var. *erectifolium* Thér., Rev. Bryol. 48: 16. 1921, *fide* Thouvenot, 2019. Type protologue and ctiation: New Caledonia, Île des Pins, 1874-76 (lectotype: PC 0083688!).

Macromitrium salakanum var. *majus* Besch., Ann. Sci. Nat. Bot., sér. 5, 18: 210. 1873, *fide* Thouvenot, 2019. Type protologue: New Caledonia, Ad arborum truncos montis Mou, 1200 m, *Balansa 2978, 2981*; "in monte Mi", *Balansa 916*.

(1) Plants robust, lustrous, rusty brown to chestnut-brown. (2) Branch leaves loosely, regularly and strongly twisted-flexuose, most apices decurved to incurved when dry, wide-spreading when moist, narrowly lanceolate from a broader basal area, 2-3.5 mm long, partly bistratose near the apex; apices acuminate; upper and medial cells quadrate with rounded to oval lumens, thick-walled, in distinct longitudinal rows, bulging, densely pluripapillose; cell shape abruptly changes in the transitional area, the transitional cells flat and smooth, rectangular, with curved lumens; basal cells flat and smooth, rectangular, with strongly curved to sigmoidal lumens, narrowly elongate, thick-walled and with narrow lumens near margin. (3) Perichaetial leaves erect, longer than branch leaves, forming a stout plicate sheath, oblong-lanceolate, gradually narrowed to an acuminate or stoutly cuspidate apex; elongate cells sigmoidal, continuing into apex, all cells flat and smooth. (4) Setae straight, slender, 5-7 mm long, smooth. (5) Capsules exserted, urns ovoid to obloid, slightly 8- plicate; rim firm, not plicate; peristome single; exostome of 16, large and well developed, often broken off and absent when old. (6) Calyptrae broadly conic-mitrate, strongly plicate, glossy, naked and smooth.

Distribution: Australia, New Caledonia.

Specimens examined: **AUSTRALIA**. New South Wales, *D. H. Vitt 27464* (H 3090423), *D. H. Vitt 27519* (H 3090424), *H. Streimann 60892* (KRAM-B-140175). **NEW CALEDONIA**. Along road to Mont Dzumac above Dombea Valley, *Marshall R. Crosby 14138* (H 3194609); Nova Caledonia, Mt. Dzumac, *Le Rat, 1010* (H-BR 2607005); Nova Caledonia, Prony, Etesse, 1906 (H-BR 2607010).

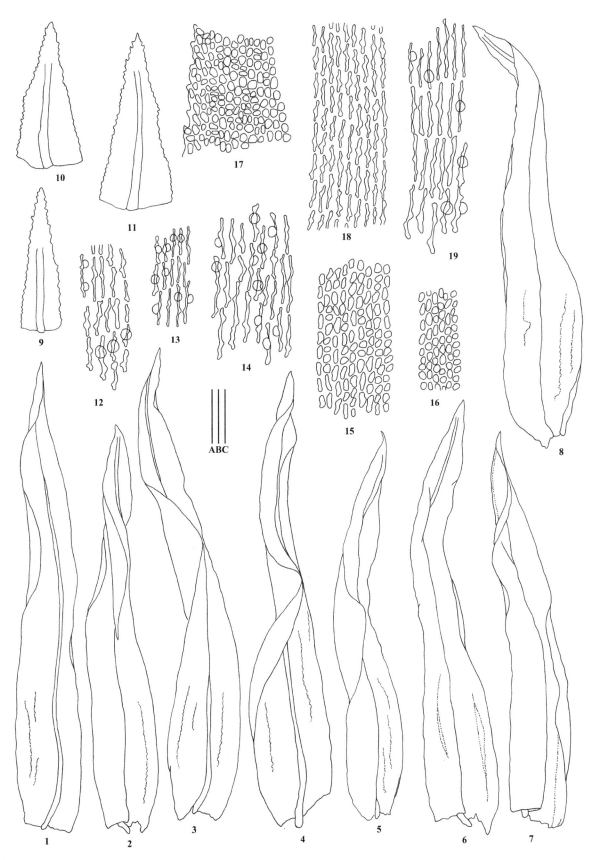

Fig. 136 *Macromitrium lauterbachii* Broth. ex Fleisch. 1-8: Branch leaves. 9-11: Apices of branch leaves. 12-14, 19: Basal cells of branch leaves. 15: Medial cells of branch leaf. 16: Upper cells of branch leaf. 17: Upper marginal cells of branch leaf. 18: Low cells of branch leaf (all from lectotype, H-BR 2602007). Line scales: A = 0.5 mm (1-8); B = 200 μm (9-11); C = 50 μm (12-19).

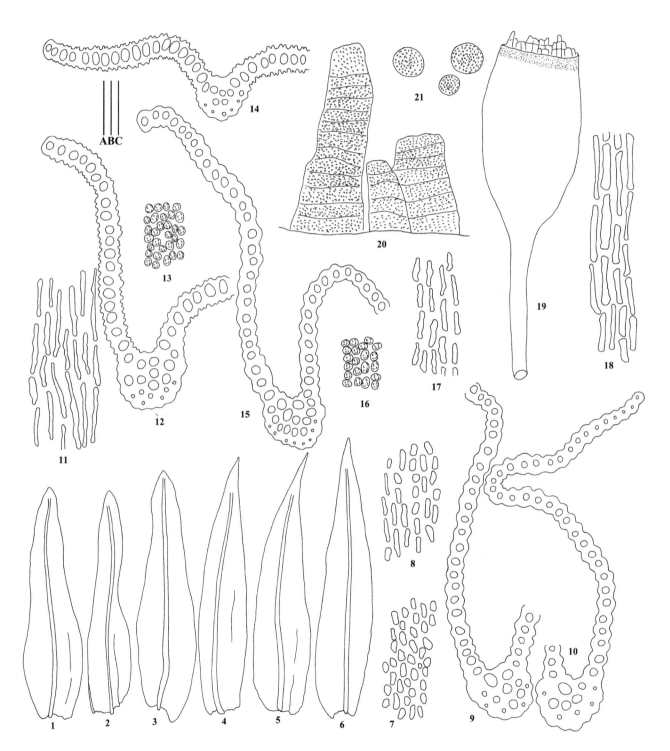

Fig. 137 *Macromitrium lebomboense* van Rooy. 1-3: Branch leaves. 4-6: Perichaetial leaf. 7: Upper cells of perichaetial leaf. 8: Medial cells of perichaetial leaf. 9-10: Basal transverse sections of branch leaves. 11: Basal cells of branch leaf. 12: Medial transverse section of branch leaf. 13: Upper cells of branch leaf. 14: Upper transverse section of branch leaf. 15: Low transverse section of branch leaf. 16: Medial cells of branch leaf. 17: Low cells of branch leaf. 18: Basal cells of perichaetial leaf. 19: Capsule. 20: Peristome. 21: Spores (all from isotype, MO 4420561). Line scales: A = 1 mm (1-6, 19); B = 100 μm (20-21); C = 67 μm (7-18).

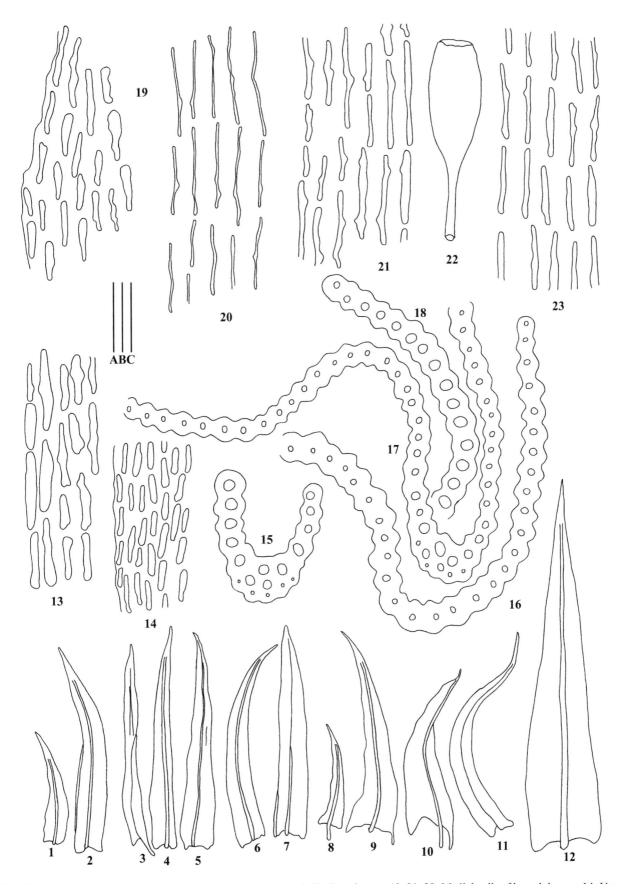

Fig. 138 *Macromitrium leprieuri* Mont. 1-7, 12: Branch leaves. 8-11: Stem leaves. 13, 21, 23: Medial cells of branch leaves. 14: Upper cells of branch leaf. 15: Apical transverse section of branch leaf. 16-18: Basal transverse sections of branch leaves. 19: Basal marginal cells of branch leaf. 20: Basal cells of branch leaf. 22: Capsule (all from isotype, NY 01086486). Line scales: A = 2 mm (1-11, 22); B = 1 mm (12); C = 67 μm (13-21, 23).

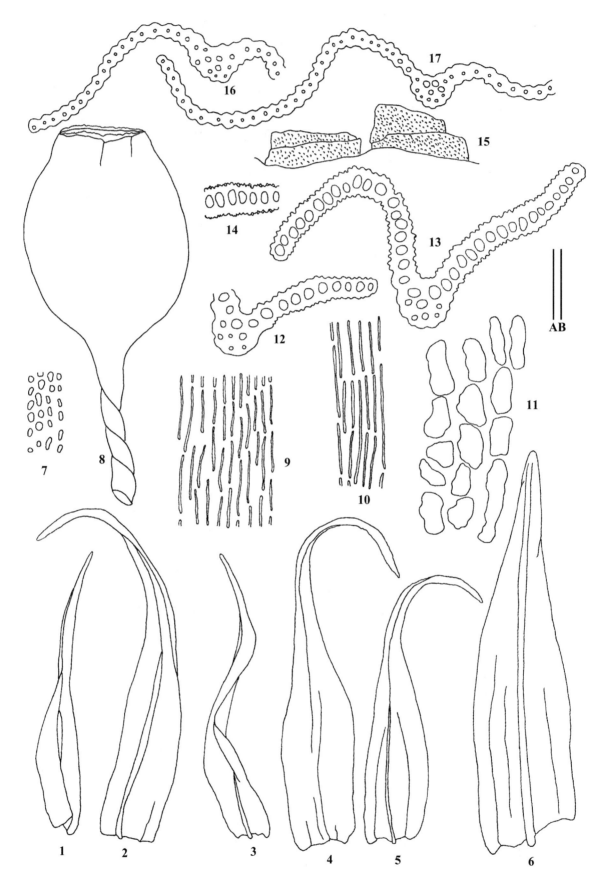

Fig. 139 *Macromitrium leratii* Broth. & Paris 1-5: Branch leaves. 6: Perichaetial leaf. 7: Upper cells of branch leaf. 8: Capsule. 9: Low cells of branch leaf. 10: Basal cells of branch leaf. 11: Exothecial cells of capsule. 12, 14: Upper transverse sections of branch leaves. 13: Medial transverse section of branch leaf. 15: Peristome. 16: Low transverse section of branch leaf. 17: Basal transverse section of branch leaf (all from H-BR 2607010). Line scales: A = 0.44 mm (1-6, 8); B = 44 μm (7, 9-17).

118. *Macromitrium ligulaefolium* Broth., Öfvers. Finska Vetensk.-Soc. Förh. 40: 170. 1898. (Figure 140)

Type protologue and citation: [Australia] New South Wales, Roseville Gully prope Sydney, ad rupes (*W. W. Watts n. 178*) (holotype: H-BR!).

Macromitrium brevipilosum Thér., Bull. Acad. Int. Géogr. Bot. 18: 253. 1908, *fide* Vitt & Ramsay, 1985. Type protologue: Nouvelle Calédonie: versant oust du Mont Koghis, 300m. *Franc*. Type citation: Neuvelle Calédonie, Mt Koghis, 300m, leg. *France*, 20. 10. 1907 (lectotype: PC 0137607; isotype: H-BR 2561003!).

Macromitrium cucullatum Thér., Bull. Acad. Int. Géog. Bot. 17(217): 307. 1907, *fide* Thouvenot, 2019. Type protologue and citation: environs ce Noumea Nouvelle-Calédonie, en 1906, par *M. Franc* (isotype: PC 0083639!).

Macromitrium ligulatulum Müll. Hal., Hedwigia 37: 151. 1898, *fide* Vitt, 1983. Type protologue: Australia subtropica, New South Wales, Richmond River: Miss Hodgkinson in Hb. Melbourne 1881) (lectotype designated by Vitt, 1983: H-BR).

Macromitrium perminutum Broth. & Paris, Öfvers. Finska Vetensk.-Soc. Förh. 51A(17): 15. 1909, *fide* Vitt & Ramsay, 1985. Type protologue: Nouvelle Caledonie L'Hermitage pr. Noumea, ad corticem arborum (*A. Le Rat*). Type citation: Nouvelle Caledonie L'Hermitage, leg. *Le Rat,* 1906, comm. Brotherus (holotype: H-BR 2563014!; isotype: PC 0083637!).

Macromitrium rapaense E.B. Bartram, Occas. Pap. Bernice Pauahi Bishop Mus. 15(27): 335. 5. 1940, *fide* Vitt & Ramsay, 1985. Type protologue: Rapa: Tapui islet, surface of rocks, alt. 4 mi., July 9 1934, *Fosberg 11441*. (holotype: FH-Bartram).

Macromitrium woollsianum Müll. Hal., Hedwigia 37: 156, 1898, *fide* Vitt, 1983. Type protologue: Australia subtropica, New South Wales, sine loco speciali: harriott et Wools lg.; Rever. Dr. W. Woolls in hb. Melbourne 1881. Type citation: N. S. Wales, ex. hb. C. Müller. (syntype: H-BR 2590003!).

Macromitrium woollsianum var. *chlorophyllosum* Müll. Hal., Hedwigia 37: 156. 1898, **syn. nov.** Type protologue: Australia, New South Wales, Richmond River: Miss Hodgkinson in Hb. Melbourne1880.

(1) Plants small, in dense, compact mats; stems prostrate and creeping, with short, erect branches, branches short, often about 5 mm high. (2) Stem leaves 1.3-1.5 mm long, broadly lanceolate to narrowly ovate-lanceolate; upper cells strongly conic-bulging, unipapillose, low cells slightly longer. (3) Branch leaves twisted-contorted and contorted-crisped, with upper portion strongly inrolled and apices hidden in the inrolled cavity when dry, spreading or slightly incurved above when moist, 2-3 mm long, linear-lanceolate, ligulate-lanceolate to ligulate; apices acute to broadly acute-apiculate; costae often subpercurrent, or percurrent, rarely excurrent; upper and medial cells rounded to hexagonal-rounded, rather thin-walled, strongly conic-bulging, clear to slightly obscure, smooth to pluripapillose or unipapillose; low cells short-rectangular, unipapillose; basal cells rectangular to short elongate-rectangular, tuberculate, some slightly bulging and smooth, or smooth in all basal cells. (4) Perichaetial leaves shorter than branch leaves, ovate-lanceolate to lanceolate. (5) Setae slender, 4-5 mm, smooth, twisted to the left; vaginulae sparsely hairy. (6) Capsule urns narrowly ovoid to ellipsoid, smooth, 8-plicate below the narrow, puckered mouth, darker beneath or at rim; peristome single and greatly reduced or absent. (7) Calyptrae conic mitrate or cucullate, lobed in low portion, divided by 1-5 slits, naked or sparsely hairy.

Macromitrium ligulaefolium Broth. is similar to *M. ligulare* Mitt., but the latter differs from the former by its collapsed capsule rims, a well-developed exostome of 16 teeth.

Distribution: Australia, New Caledonia, New Zealand (Vitt, 1983).

Specimens examined: **AUSTRALIA**. New South Wales, *H. Streimann 60050* (KRAM-B-137429), *H. Streimann 49113* (KRAM-B-102946), *Whitelegge s.n.* (H-BR 2539005); Queensland, *I.G. Stone.* (MEL 2264465, MEL 2264721); Tasmania, *J. Jarman* (MEL 2300337); Victoria, *H. Streimann* (MEL 2053775). **NEW CALEDONIA**. Mt. Koghis, *France s.n.* 1907 (JE 04008700); Yahoué, *France s.n.* (PC 0083641, PC 0083642, PC 0083643); Hermitage, *France s.n.* (PC 0083640), Hermitge, *Le Rat, 556* (PC 0083637).

119. *Macromitrium ligulare* Mitt., J. Proc. Linn. Soc., Bot. 4: 78. 1859. (Figure 141)

Type protologue: New Zealand, kerr; Waikeki, *Dr. Sinclair*. Type citation: Waikeki, N. Zealand, *Dr. Sinclair* (lectotype designated by Vitt & Ramsay, 1985: NY 01202041!; isosyntype: NY 01202038!).

Macromitrium luehmannianum Müll. Hal., Hedwigia 37: 152. 1898, *fide* Vitt & Ramsay, 1985. Type protologue: Australia, Victoria, Gippsland, prope dem Moe River; V. Luehmann 1881 in Hb. Melbourne. Type citation: Australia, Victoria, Gippsland, near Moe River, Luehmann 1881(isotypes: JE 04008748!, JE 04008749!).

(1) Plants slender, in dense, spreading mats; stems creeping, with erect and short branches, branches up to 15 mm high. (2) Stem leaves 1.0-1.3 mm long, lanceolate to broadly lanceolate, acuminate to narrowly obtuse or acute. (3) Branch leaves flexuose-twisted, with upper portion strongly contorted-crisped and apices hidden in the inrolled cavity when dry, flexuose-spreading with inflexed apices when moist, 1.5-2.5 mm long, ligulate to lanceolate-ligulate, rounded-obtuse to broadly acute, often with a short, one-celled apiculus; costae subpercurrent; upper and medial cells 9-15 µm, rounded-quadrate to rounded, strongly bulging and rather thin-walled, weakly pluripapillose, occasional

smooth; transitional cell 10-13 × 10-12 µm, rounded-elliptic; basal cells short-rectangular to rectangular, porose, some cells tuberculate; basal cells 14-28 µm long, lumens 4-6 µm wide, shortly rectangular to rectangular, restricted to the low 1/3 to 1/4 of the leaf. (4) Perichaetial leaves shorter than branch leaves, ovate-lanceolate, gradually acuminate or acute; upper cells elliptic-rounded and bulging-papillose, elongate below. (5) Setae 3-8 mm long, slender and smooth, twisted to the left. (6) Capsule urns narrowly ovoid to ellipsoid-cylindric, not constricted or ribbed at mouth, the mouth sometimes half-collapsed when old; peristome single, exostome of 16, pale and blunt. (7) Calyptrae conic, mitrate-cucullate, smooth and naked, split by one or two long slits, not lacerate, entire or only lobed at the base.

Distribution: Australia, New Zealand (Vitt & Ramsay, 1985).

Specimens examined: **NEW ZEALAND**. *W Bell s.n.* (H-BR 2538005); N. I. Valley, *W. Bell* (H-BR 2538001); South Island, *H. Streimann 58130* (KRAM-B-131063); Otago prov., *W. Bell 486* (H-BR 2538009); Otago prov., *W. Bell 291* (H-BR 2509017), *W. Bell s.n.* (H-BR 2509009).

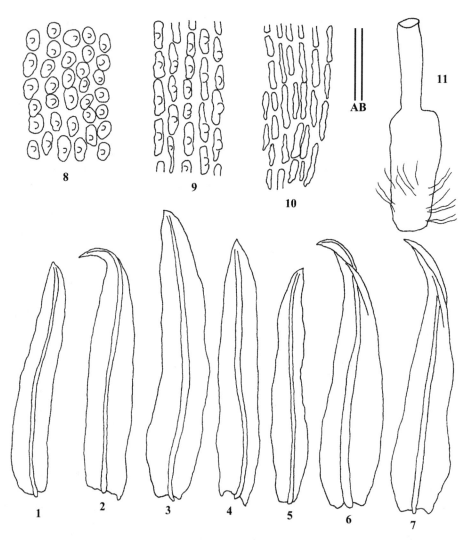

Fig. 140 *Macromitrium ligulaefolium* Broth. 1-5: Branch leaves. 6-7: Perichaetial leaves. 8: Upper cells of branch leaf. 9: Low cells of branch leaf. 10: Basal cells of branch leaf. 11: Vaginula (all from isotype of *M. brevipilosum*, H-BR 2561003). Line scales: A = 0.44 mm (1-7, 11); B = 44 µm (8-10).

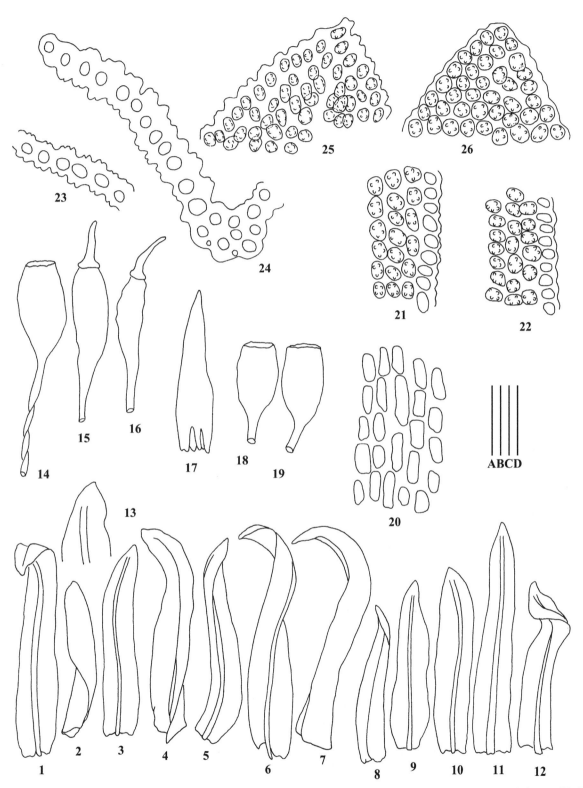

Fig. 141 *Macromitrium ligulare* Mitt. 1-12: Branch leaves. 13: Apex of branch leaf. 14-16, 18-19: Capsules. 17: Calyptra. 20: Basal cells of branch leaf. 21: Low cells of branch leaf. 22: Medial cells of branch leaf. 23, 24: Upper transverse sections of branch leaves. 25, 26: Upper cells of branch leaves (1-8, 15-17, 25 from isosyntype, NY 01202038; 9-12, 14, 18-24, 26 from lectotype, NY 01202041). Line scales: A = 2 mm (14-19); B = 1 mm (1-12); C = 400 μm (13); D = 67 μm (20-26).

120. *Macromitrium lomasense* H. Rob., Phytologia, 21: 392. f. 9-12. 1971. (Figure 142)

Type protologue and citation: Peru. La Libertad: Prov. Trujillo, Cerro Chiputur, 650 m, *Ayala 7124* (holotype: US 00070266!).

(1) Plants small, yellow-green above, dark brown below; stems weakly creeping, branches up to 5-6 mm long, about 1.0 mm thick. (2) Branch leaves strongly keeled, spirally twisted-coiled, with flexuose to twisted-flexuose upper portions when dry, moderately to conspicuously funiculate, erecto-patent when moist, oblong-lanceolate to ligulate-lanceolate; apices acute, apiculate to acuminate-acute; margins entire from the apex to the base; costa subpercurrent to percurrent; all cells smooth and clear; cells in upper 5/6 to 6/7 of the leaf length rounded, rounded-quadrate to rounded-hexgonal, conic-bulging to bulging, not porose; low and basal oblong-rectangular, oval, elliptic cells confined to a small area, slightly elongate, rectangular, straight-walled and smooth, occasionally weakly porose; the outmost marginal cells in medal and upper portion slightly differentiated, smaller and more oblate. (3) Perichaetial leaves not differentiated or slightly shorter than branch leaves. (4) Setae smooth, 3-6 mm long; vaginulae hairy. (5) Capsule urns ovoid to ovoid-ellipsoid, almost smooth or weakly furrowed; opercula conic-rostrate to conic-subulate; peristome double, exostome teeth well developed, yellowish brown, united to an erect membrane; endostome hyaline, membranous, truncate above; spores anisosporous, densely papillose, globular. (6) Calyptrae mitrate, lacerate below, rather scabrous and spinose, without hairs.

In the original description of M. *lomasense* H. Rob., the basal leaf cells are sometimes papillose, but those of the type specimen are completely smooth. *Macromitrium lomasense* is rather similar to *M. punctatum* (Hook. & Grev.) Brid., but the upper margins of leaves in the latter are serrulate to irregularly serrate, basal leaf cells occasionally tuberculate. *Macromitrium lomasense* also differs from *M. punctatum* in having funiculate branches with spirally coiled leaves when dry, as well as scabrous and spinous calyptrae.

Distribution: Peru.

Specimens examined: **PERU**. Dpto. La Libertad, Prov. Trujillo, *P. E. Hegewald 7458* (MO 3670999; MO 5134044, H 3090490); Lima, Lomas de Lachay, Peruvian coast, *B. Lowy PBR 187* (MO 5349142).

121. *Macromitrium longicaule* Müll. Hal., Syn. Musc. Froud. 1: 742. 1849.

Type protologue: (Indonesia) Java: Miquel Hb.

Macromitrium brachystele Dixon, J. Bot. 61 (Suppl.): 62. 1923, *fide* Vitt *et al.*, 1995. Type protologue: Papua New Guinea. Central, Sogere, 2 000 ft., epidendric, *Forbes 515 a*.

(1) Plants medium-sized to robust, dull and stiff, in loose to dense mats; stems creeping, covered with rusty-reddish rhizoids; branches simple to forked, stiff, up to 5 cm long. (2) Stem leaves small and inconspicuous. (3) Branch leaves individually twisted-contorted from an erect, clasping base, with widely divergent apices when dry, flexuose-spreading to wide-spreading when moist, 3-5 mm long, narrowly lanceolate to lanceolate from an ovate base, variably acuminate; margins plane to broadly reflexed below, erect near apex, entire to irregularly notched above; costa percurrent; upper cells quadrate-rounded to rounded, 10-14 µm wide, irregularly and densely pluripapillose; medial cells 11-15 µm wide, rounded to short-elliptic, clear, most papillose, bulging; basal cells 10-12 × 30-80 µm, elongate-rectangular, lumens straight to curved, thick-walled and strongly tuberculate. (4) Perichaetial leaves 4-5 mm long, narrowly lanceolate, long-acuminate from a slender base, cells longer than wide. (5) Setae 2.5-4.5 mm long, smooth, twisted to the left; vaginulae sparsely hairy. (6) Capsule urns 1.1-1.7 × 0.5-0.8 mm, ellipsoid-ovoid to ovoid, lightly 8-ribbed in upper 1/3, wide-mouthed, sharply contracted to seta, occasionally papillose in low portion; peristome single, exostome of 16, well-developed; spores anisosporous. (7) Calyptrae plicate, mitrate, naked or with a few, stiff hairs, lacerate at the base (Vitt *et al.*, 1995).

Macromitrium longicaule Müll. Hal. is similar to *M. sublongicaule* E.B. Bartram, but differing from the former by its upper cells of branch leaves almost smooth or with scattered low papillae, smooth medial lamina cells, as well as slightly longer marginal cells in the upper portion.

Distribution: Indonesia, Papua New Guinea, Philippines.

Specimens examined: **INDONESIA**. Java, *Wichma*, 2442 (H-BR 2604015). **PAPUA NEW GUINEA**. Morobe, *D. H. Norris 65019* (MO 4435585, H 3195800), *D.H. Norris 65393* (H 3195926), *Timo Koponen 34524* (KRAM-B-114324); West Sepik, *A. Touw. 18543* (MO 5371039). **PHILIPPINES**. Mindanao, *E. J. Reynoso, 2123b* (MO 4417625).

Taxonomy

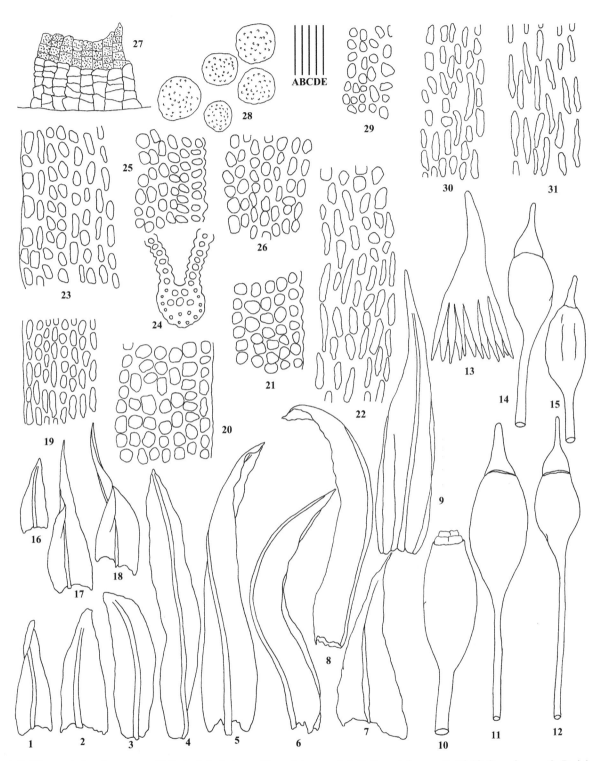

Fig. 142 *Macromitrium lomasense* H. Rob. 1-3: Apices of branch leaves. 4-6, 8: Branch leaves. 7, 16-18: Stem leaves. 9: Perichaetial leaf. 10-12, 14-15: Capsules. 13: Calyptra. 19: Low cells of branch leaf. 20: Medial cells of branch leaf. 21: Upper cells of branch leaf. 22: Basal cells of branch leaf. 23: Low cells of stem leaf. 24: Medial transverse section of branch leaf. 25: Upper cells of stem leaf. 26: Medial cells of stem leaf. 27: Peristome. 28: Spores. 29: Upper cells of perichaetial leaf. 30: Medial cells of perichaetial leaf. 31: Basal cells of perichaetial leaf (all from holotype, US 00070266). Line scales: A = 2 mm (10-15); B = 1 mm (4-9, 16-18); C = 400 μm (1-3, 27); D = 100 μm (24, 28); E = 67 μm (19-23, 25-26, 29-31).

122. *Macromitrium longifolium* (Hook.) Brid., Bryol. Uni. 1: 309, 738. 1826. (Figure 143)

Basionym: *Orthotrichum longifolium* Hook., Hooker, W. J. 1818-1820. Musci Exot. 2 vol. Longman et al., London. Type protologue: Venezuela. Ad radices Bifariae glaucae altitudine 950 hexapod, regione temperate, in devexis montis Avilae prope Caraccas. *Humboldt et Bonpland.* Type citation: *Humboldt 38* (H630) (lectotype: BM 000873190); Venezuela. Caraccas, leg. *Humboldt et Bonpland* (isotype: E 00289633!).

Macromitrium denudatum A. Jaeger, Ber. Thätigk. St. Gallischen Naturwiss. Ges. 1872-73: 146. 1874, *fide* Grout, 1944.

**Macromitrium flexuosum* Schimp. ex Besch., Mém. Soc. Natl. Sci. Nat. Cherbourg 16: 189. 1872, *fide* Wijk et al., 1964.

Macromitrium haitense Thér., Rev. Bryol. Lichénol. 14: 15, 7. 1944, *fide* Allen, 2002. Type protologue and citation: Haiti: Massif de la Hotte western group, Les Roseaux, pineland, on stones, 2400. [Ekman] *no 10.645* (F-C 0001006F!).

Macromitrium homalacron Müll. Hal., Bull. Herb. Boissier 5: 197. 1897, *fide* Vitt, 1994. Type protologue: (Guatemala) Coban, 4300 ped. Alt., Jan. 1886: *H. v. Türckheim* in Hb. Brotheri (1894). Type citation: Guatemala, Aek Verapaz,....., Coban, 4300', legit *H. v. Türckheim* (holotype: H-BR 2629003!)

Macromitrium longifolium var. *viridissimum* Renauld & Cardot, Bull. Soc. Roy. Bot. Belgique 31(1): 156. 1893, *fide* Grout, 1944.

**Macromitrium muelleri* Schimp. ex Besch., Mém. Soc. Natl. Sci. Nat. Cherbourg 16: 189. 1872, *fide* Wijk et al., 1964.

***Macromitrium perundulatum* E.B. Bartram, Fieldiana, Bot. 25: 219. 1949. *fide* Allen, 2002.

Macromitrium scabrisetum Wilson, Trans. Linn. Soc. London 20: 163. 1847, *fide* Grout. 1944.

Macromitrium schimperi A. Jaeger, Ber. Thätigk. St. Gallischen Naturwiss. Ges. 1872-73: 129. 1874, *fide* Grout, 1944.

Macromitrium scleropelma Renauld & Cardot, Bull. Soc. Roy. Bot. Belgique 31(1): 157. 1893, *fide* Grout, 1944.

Schlotheimia longifolia (Hook.) Schwägr., Sp. Musc. Frond., Suppl. 2 2(2): 147. 1827, *fide* Valente *et al.*, 2020.

(1) Plants medium-sized to robust, yellow-green to rusty-brown; stems weakly creeping, branches up to 6 cm long. (2) Branch leaves strongly keeled, erect below, spreading, individually twisted and flexuose to contorted-flexuose above when dry, flexuose spreading above when moist, 3-5×0.7 mm, lanceolate; apices acuminate; margins undulate, serrate to serrulate, recurved below, plane above; costae percurrent; upper and medial cells 7-15(25) × 7-12 µm, rounded-quadrate, rounded-hexagonal to rhombic, occasionally porose, smooth, upper marginal cells not differentiated; basal cells 40-60 µm long, long-rectangular, thick-walled, porose, strongly tuberculate; enlarged marginall teeth-like cells at insertion not differentiated. (3) Setae papillose, 5-12 mm long, frequently with a whitish collar beneath the urn. (4) Capsule urns ovoid, ellipsoid to cylindric, furrowed; exostome teeth yellowish brown, papillose, united to form an erect membrane, endostome hyaline, basal membrane short, 50-75 µm high, segments stout, to 150 µm high; spores anisosporous, 17-25 µm, smooth and pellucid, or 25-40 µm and papillose. (5) Calyptrae mitrate, naked or sparsely to densely hairy, deeply lacerate.

Macromitrium longifolium (Hook.) Brid. is similar to *M. oblongum* (Taylor) Spruce in their leaf shape, furrowed capsules, but differs from the latter in having papillose setae. *Macromitrium crenulatum* Hampe is rather similar to *M. longifolium*, the former differs from the latter by its shorter branch leaves with crenulate margins above and smooth setae, spirally-twisted branch leaves when dry.

Müller (1851) described *M. longifolium* var. *brevifolium* Müll. Hal. based on the collection of Wagner from Columbia (Type: Galipan Columbia, ad arbores 6000' alt., m. Julii 1849; *Wagner s.n.*). The variety is different from the var. *longifolium* by its wider and shorter, dirty-green leaves, and shorter perichaetial leaves.

Distribution: Bolivia, Colombia, Ecuador, Guatemala, Haiti, Honduras, Mexico, Pamana (Allen, 2002), Peru, Venezuela, Vitenam (Zhang *et al.*, 2019).

Specimens examined: **BOLIVIA**. Larecaja, *G. Mandon* (BM 000720667). **COLOMBIA**. Arauca, *Antoine M. Cleel 10.120* (MO 3674226), Municipio Sonsón. *Steven Churchill et Inés Sastre-De Jesús 12956* (KRAM-B-072384). Musgos sobre roca, *Steven P. Churchill, Alba Luz Arbeláez y Wilson Rengifo 16284* (KRAM-B-090151), Antioquia, *J. Cuatrecasas et al. 24270* (MO 5382157). **ECUADOR**. Loja, Parque Nacional Podocarpus. *Benjamin Øllgaard & Jens Elgaard madsen 74724* (MO 6235311), Pichincha, *J. Jaramillo 8121* (MO 5281899). **HONDURAS**. Cortes, *B. Allen 14047* (MO 3973795). **PERU**. Cuzco, Machu Picchu: Inkapfad. Beim Tor., *P+E. Hegewald 8816* (MO 3670998), San Martin: Rioja, Buenos Aires. *Henk van der Werff et al. 15402* (MO 5243031). **VENEZUELA**. Merida, *D. Griffin, III & J. Dugarte PV-1126* (MO 5381264), Cerro da la Neblina, *William R. Buck 12656* (MO 3953995), Distritio Federal, *Gerrit Davidse & Ronald Liesner 127796* (MO 3061937), Colonia Tovar, Aragua, *H. Pittier 9973* (MO 5279787).

Taxonomy

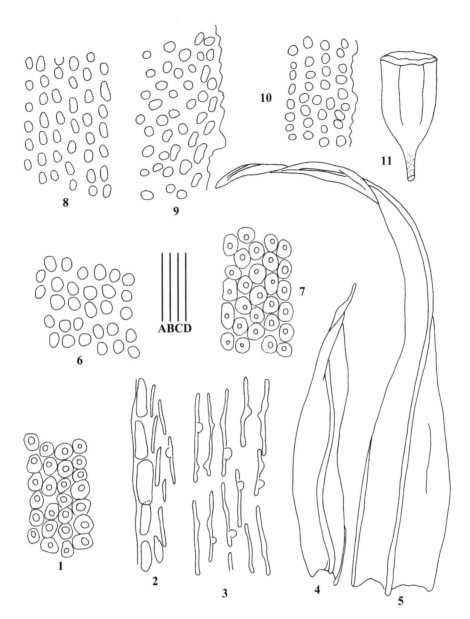

Fig. 143 *Macromitrium longifolium* (Hook.) Brid. 1, 6-7: Upper cells of branch leaf. 2: Basal marginal cells of branch leaf. 3: Basal cells of branch leaf. 4-5: Branch leaves. 8: Medial cells of branch leaf. 9: Apical cells of branch leaf. 10: Medial marginal cells of branch leaf. 11: Capsule (1, 4 from lectotype, BM 000873190; 2-3, 5-11 from isotype, E 00289633). Line scales: A = 4 mm (11); B = 1 mm (4); C = 400 μm (5); D = 67 μm (1-3, 6-10).

123. ***Macromitrium longipapillosum*** D.D. Li, J. Yu, T. Cao & S.L. Guo, Nordic J. Bot., 35(6): 711, 1-3. 2017. (Figure 144)

Type protologue and citation: JAPAN, Tochigi, Nikkō, *J. Bisset no.10,* Sept. 1886. (holotype: H-BR 2575009!). Paratypes: Honshu, Yamanashi-ken, Kita-koma-gun, Takane-cho, Kiyosato-kôgen plateau, about 1400 m alt., on tree trunks in deciduous forest. *Osada T & Suzuki N, no. 1530,* April 9, 1978. (KRAM-B-036764!); Awaji, Toshimura Tsuna, on rock, *Takata G, 456,* Dec. 18, 1917, H-BR 2577012!); Rikuzen, *Uematsu E. 609, 7/6 1908* (H-BR 2577019!); Hane, Tosa, *Shutai Okamura, 711*, Dec. 5, 1907 (H-BR 2578026); Awaji, Toshi, Leg. *G. Takata*, Com. *H. Sasaoka, no. 344,* H-BR 2577008!).

(1) Plants medium-sized, forming dense mats; stems long-creeping, densely with short, stout and erect branches, branches about 3.5-6.5 mm high, densely leaved, with a few branchlets. (2) Stem leaves different from branch leaves, irregularly flexuose-twisted when dry, spreading when moist, entire, narrow-triangular to ovate-lanceolate, with an acute apex; upper cells rounded to subquadrate, thin-walled, with a single large linear central papilla up to 18 μm; medial cells subquadrate to short rectangular, with a short tuberculate papilla, becoming clear, rectangular to elliptic and smooth farther down; basal cells thick-walled, smooth, distinctly porose. (3) Branch leaves keeled, curly and twisted, with the upper portion often adaxially inrolled when dry, flexuose-spreading when moist, 2.0-3.5 × 0.3-0.5 mm, ovate-oblong to oblong-lanceolate, somewhat incurved, slightly to moderately plicate below; apices obtuse,

acute or broadly acuminate; margins plane or recurved on one side; costae ending in or a few cells beneath the apex; upper and medial cells subquadrate, 8-11 μm wide, obscure, hexagonal, with rather thin walls, strongly bulging, with a single large, occasionally forked tuberculate papilla (up to 18 μm), becoming rectangular to elliptic and shorter farther down; low cells rectangular to sublinear, 12-22 × 5.5-6.5 μm, moderately thick-walled, with a single tuberculate papilla up to 12 μm long; basal cells along costa rectangular, thin-walled, 30-35 × 10-13 μm, smooth and pellucid, appearing as a "cancellina region", those near the margin elongate, rectangular to sublinear, thick-walled and slightly sinuous, often with a tuberculate papilla. (4) Perichaetial leaves ovate-lanceolate, often plicate below, up to 2.5 mm long, shorter than branch leaves, narrowly acuminate to long acuminate or aristate at apex; all cells oblong; upper cells with a single large tuberculate papilla, the papillae becoming weaker farther down; low and basal cells smooth and thick-walled. (5) Setae 1.2-3.0 mm, smooth, twisted to the left. (6) Capsule urns ellipsoid, contracted below the mouth when dry; peristome single, exostome of 16, well-developed, short-lanceolate, about 150 μm long, obtuse, papillose; endostome absent; operculum erect, conic-rostrate; spores anisosporous, 15-34 μm, finely papillose. (7) Calyptrae campanulate, about 2.0 mm long, with many long, yellowish hairs.

Distribution: Japan.

124. *Macromitrium longipes* (Hook.) Schwägr., Sp. Musc. Frond., Suppl. 2 2: 147. pl. 139. 1824. (Figure 145)
Basionym: *Orthotrichum longipes* Hook., Musci Exot. 1: 24. 1818. Type protologue and citation: In sinu Dusky Bay dicto, in Nova Zeelandia, *D. Menzies,* 1791 (isotype: H-BR!).
Macromitrium lonchomitrium Müll. Hal., Hedwigia 37: 148. 1898, *fide* Wijk *et al.*, 1964. Type protologue: Nova Seelandia, insula autralis, littore australi: *Richard Helms 1885* letit et misit e Greymouth.
Macromitrium pseudohemitrichodes Müll. Hal., Hedwigia 37: 150. 1898, *fide* Vitt, 1983. Type protologue: Nova Seelandia, insula spetentronalis, sine loco speciali: *F. M. Reader* 1882 legit, 1892 misit ex Dimboola Victoriae.
(1) Plants medium-sized to large, stems long and creeping, with erect branches, branches up to 5 cm high. (2) Stem leaves ovate-lanceolate, gradually tapered to a broadly subulate apex. (3) Branch leaves regularly and spirally twisted-curved in rows, upper portions curved-flexuose or curly, funiculate when dry, widely spreading when moist, 1.5-3.0 mm long, broadly lanceolate to ligulate-lanceolate, lamina occasionally bistratose at mid leaf; costae curving to one side above; all cells smooth and not porose; upper cells subquadrate to rounded, clear; low and basal cells long-rectangular to rectangular, lumens strongly curved to sigmoid. (4) Perichaetial leaves 3-4 mm long, longer than branch leaves, subsheathing at low portion, oblong-lanceolate, acute to acuminate-apiculate; all cells rounded to rectangular-elongate, sigmoid or curved. (5) Setae 7-25 mm long, smooth, twisted to the left. (6) Capsule urns broadly ovoid-ellipsoid, 8-plicate beneath the mouth, sometimes sharply constricted under the mouth; peristome single, exostome of 16, lanceolate, well-developed; spores highly variable in size. (7) Calyptrae long and very large, deeply lacerate, mitrate and smooth, naked.
Macromitrium longipes (Hook.) Schwägr. is similar to *M. microstomum* (Hook. & Grev.) Schwägr. in rope-like branches with spirally coiled leaves, all cells smooth and flat, but differing from the latter in having branch leaves with partially bistratose cells at midleaf, and sigmoid or curved basal cells.

Distribution: Australia, New Zealand.

Specimens examined: **NEW ZEALAND**. Nelson-Marlborough land district, *J. R. Shevock & W. M. Malcolm, 36511* (KRAM-B-199786); Gisborne land district, *J. R. Shevock & W. M. Malcolm, 39391* (KRAM-B-202670); South Island, *J. P. Frahm 12-7* (MO 5135008); North Island, *R. D. Svihla 5057* (H 3090510); Christchurch, Otago, *Naylor Beckett 676* (H-BR 2515016); Waiarapa lake Shore Scenic Reserve, *V. Stagsic 3003* (MEL 2160298); L. Wakatipu, Kinloch, *W. Bell* (H-BR 2509012).

125. *Macromitrium longipilum* A. Braun ex Müll. Hal., Syn. Musc. Frond. 2: 642. 1851.
Type protologue: Java: Blume.
(1) Plants large and robust, reddish brown, branches up to 8 cm long, simple or sparsely forked. (2) Branch leaves suberect and strongly flexuose when dry, recurved-patent and flexuose when moist, up to 4 mm long excluding arista and 1.5 mm wide, broadly lanceolate, oblong-lanceolate below, quickly narrowed to a broadly acute or obtuse apex; costae extending in a long, smooth and flexuose hair-point (up to 2 mm long); upper lamina soft-textured, plicate and slightly rugose; margins plane, entire or faintly and bluntly crenulate towards the apex; costa relatively narrow, reddish; upper cells oval-rhomboid with heavily thickened walls and very narrow, elongate and sinuate lumen, strongly conic-bulging or mamillate; low cells elongate, linear at base, tuberculate. (3) Inner perichaetial leaves lanceolate, acute, hair-points shorter than those of branch leaves. (4) Setae long, up to 25 mm, smooth, minutely papillose. (5) Capsules ovoid, smooth, plicated at the mouth; peristome double, very short and caducous. (6) Calyptrae plicated and naked.
Macromitrium longipilum A. Braun ex Müll. Hal. and *M. crinale* Broth. & Geh. are similar, only the setae of the former are minutely papillose, while those of the latter are smooth.

Distribution: Indonesia, Malaysia, Sumatra (Eddy, 1996).

Specimen examined: **MALAYSIA**. Sarawak, the Division, Gunong Mulu National park, *A. Touw 21038* (MO 6362356).

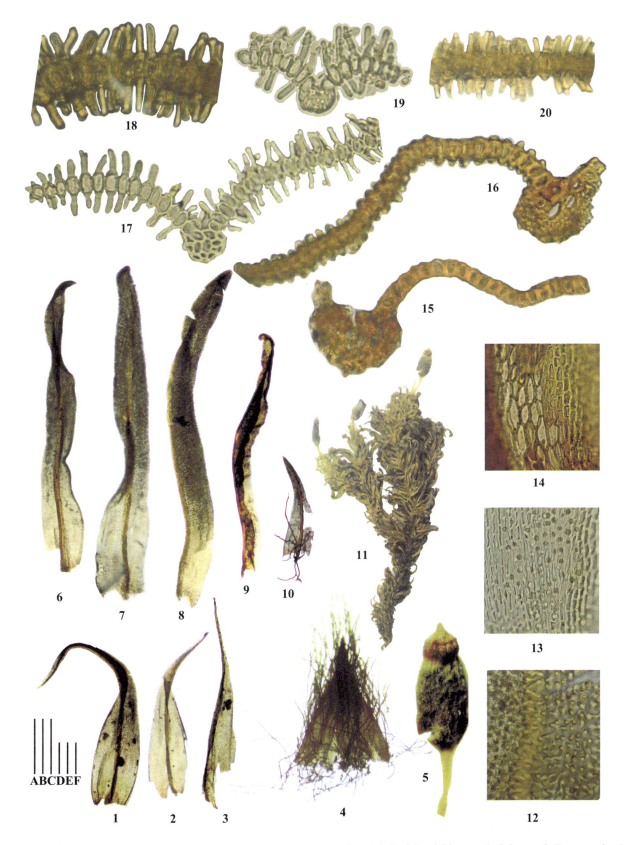

Fig. 144 *Macromitrium longipapillosum* D.D. Li, J. Yu, T. Cao & S.L. Guo 1-3: Perichaetial leaves. 4: Calyptra. 5: Dry capsule. 6-9: Branch leaves. 10: Stem leaf. 11: Plants. 12: Upper cells of branch leaf. 13: Low cells of branch leaf. 14: Basal cells of branch leaf. 15: Basal transverse section of branch leaf. 16: Low transverse section of branch leaf. 17-20: Upper and medial transverse sections of branch leaves (all from holotype, H-BR 2575009). Line scales: A = 2.0 mm (11); B = 0.5 mm (5); C = 0.75 mm (4); D = 0.2 mm (6-10); E = 40 µm (12-14); F = 20 µm (15-20). (Li *et al.*, 2017).

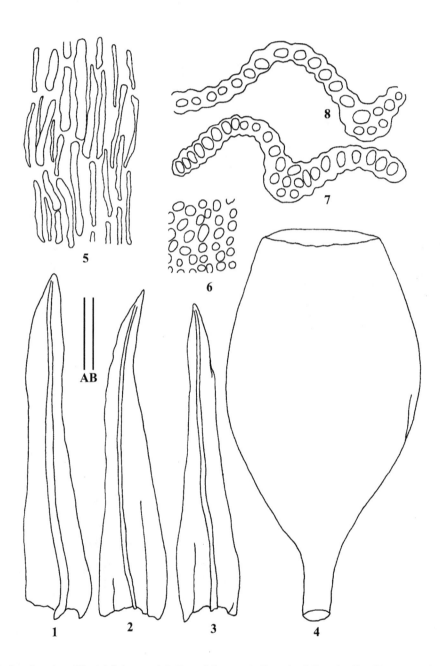

Fig. 145 *Macromitrium longipes* (Hook.) Schwägr. 1-3: Branch leaves. 4: Capsule. 5: Basal cells of branch leaf. 6: Upper cells of branch leaf. 7: Upper transverse section of branch leaf. 8: Basal transverse section of branch leaf (all from H-BR 2515016). Line scales: A = 0.44 mm (1-4); B = 44 μm (5-8).

126. *Macromitrium longirostre* (Hook.) Schwägr., Sp. Musc. Frond., Suppl. 2(1): 38. pl. 112. 1823. (Figure 146)

Basionym: *Orthotrichum longirostre* Hook., Musci Exot. 1: 25, 1818. Type protologue: In sinu Dusky Bay dicto, in Nova Zeelandia, *D. Menzies,* 1791 (isotype: H-BR!).

Macromitrium acutifolium (Hook. & Grev.) Brid., Bryol. Univ. 1: 735. 1826. - *Orthotrichum acutifolium* Hook. & Grev., Edinburgh J. Sci. 1: 118. 5. 1824. *fide* Vitt, 1983. Type protologue: Van Dieman's Land; *Dr. Spence* and *Mr. R. Meill.*

Macromitrium pertoruescense var. *torquatulum* Müll. Hal, Hedwigia 37: 148. 1898, *fide* Vitt, 1983. Type protologue: [Australia] Tasmania, Heuty River versus littus occidentale, Feb. 1891, *Weymouth*. Type citation: [Australia] on a log, Heuty River, West Coast, Leg. *W. A. Weymouth* (H-BR 2528002!).

Macromitrium torquatulum (Müll. Hal.) Müll. Hal. & Broth., Abh. Naturwiss. Vereins. Bremen 16(3): 501. 1900, *fide* Vitt, 1983.

Macromitrium rodwayi Dix. In Weym. & Rodw., Pap. & Proc. Roy. Soc. Tasmania 1921: 174. 1922, *fide* Vitt, 1983. Type protologue: Tasmania. On dripping rock at the entrance to Port Arhur.

(1) Plants robust, lustrous to dull, in loose spreading mats, olive-green to yellow-green above, chestnut-brown below; stems long creeping, with erect to ascending branches, branches up to 4 cm long. (2) Stem leaves flexuose-twisted when dry, erect-spreading when moist, 1.5-2.0 mm long, lanceolate, gradually narrowed to a long, slender acuminate apex. (3) Branch leaves loosely spirally twisted around the branch, with the apex flexuose-curved outwards when dry, erect and straight to twisted when moist, 2.3-4.0 mm long, narrowly lanceolate, keeled above; apices acuminate to long-cuspidate; margins plane and entire throughout; costa ending beneath the apex or forming a narrow cusp; upper cells rounded to elliptical, 5-12 × 4-10 μm, smaller at the margin, slightly bulging to smooth, partly bistratose in upper one-third of leaf near costa; medial cells frequently in longitudinal rows, quadrate to oblong, 10-15 × 8-10 μm, thick-walled, smooth or slightly bulging; basal inner cells elongate-rectangular, 20-40 × 8-10 μm, thick-walled, smooth, becoming longer towards the margin. (4) Perichaetial leaves not much differentiated, but elongate basal cells continuing higher than those of branch leaves, occasionally ovate-lanceolate and with a more stoutly cuspidate apex. (5) Setae 3.5-8 mm long, smooth, twisted to the right. (6) Capsules exserted, urns fusiform-ovoid to cylindric, 1.5-3.0 mm long, indistinctly broadly ribbed to smooth; peristome double; exostome of 16, irregular, blunt to coarsely papillose; endostome an irregular papillose membrane of 1-3 cells high; spores isosporous. (7) Calyptrae mitrate, strongly plicate, naked, deeply lacerate.

Distribution: Australia, New Zealand.

Specimens examined: **AUSTRALIA**. Macquarie Island, *R. D. Seppelt 12328* (KRAM-B-169757), *R. D. Seppelt 10389* (KRAM-B-169732); Tasmania, *D. H. Norris 30259* (H 3090698), *W. A. Weymouth* (H-BR 2528004), *D. H. Norris 30259* (H 3090698). **NEW ZEALAND**. Campbell Island, T. K. Crosby SA1180 (H 3090536); Lake Wilkie, *L. Visch 722* (MO 2567565); North Island, *J. Lewisky, 2001* (KRAM-B-050140); Otago Prov., Stewart Island, *W. Bell 683* (H-BR 2528014); West of Riverton, *D. H. Norris 80213* (KRAM-B-152759); Ulva Island, *J. Milne 316* (MEL 2371528).

127. *Macromitrium lorifolium* Paris & Broth., Rev. Bryol. 34: 30. 1907. (Figure 147)

Type protologue: [Vietnam] no detail type information, but the introduction of the paper included "Annam, Langbian, Pinus et Dipterocarpus, Par. M. le *Dr. Eberhardt*". Type citation: Annam, Langbian, *Pinus* et *Dipterocarpus*, leg. *Eberhardt, no. 433*, 1906 (with a mark: "sp n.!") (lectotype designated here: H-BR 2630007!; isolectotype: H-BR 2630006!).

(1) Plants medium-sized to large; stems creeping, covered with rusty-reddish rhizoids; erect branches to 20 mm long. (2) Branch leaves individually twisted-contorted and somewhat crisped when dry, spreading but undulate or rugose above when moist, long-lanceolate to lanceolate; apices acuminate or acute; margins irregularly and distinctly serrate above, entire below; costae percurrent; upper and medial cells rather small, subquadrate or irregularly quadrate, round-quadrate, moderately conic-bulging, smooth, thick-walled; low and basal cells elongate, thick-walled, pellucid, tuberculate. (3) Perichaetial leaves similar to branch leaves. (4) Setae erect, 3-7 mm, smooth, twisted to the left. (5) Capsule urns ovoid-ellipsoid, plicate towards the mouth when dry; peristome double, exostome densely papillose, yellowish, endostome a low membrane, smooth and hyaline; spores isosporous, small and smooth, about 20 μm. (6) Calyptrae mitrate, smooth and naked, deeply lacerate below.

Macromitrium lorifolium Paris & Broth., *M. sulcatum* (Hook.) Brid., *M. ellipticum* Hampe, and *M. turgidum* Dixon are similar in their robust plants, double peristome, naked and smooth calyptrae, and basal cells of branch leaves long-rectangular, thick-walled, strongly tuberculate. From *M. lorifolium*, *M. sulcatum* differs in having oblong-lanceolate to oblong-lingulate branch leaves with an acute apex, and its medial and upper cells smaller and arranged in diagonal rows, and its anisosporous spores; *M. ellipticum* in having oblong-lingulate branch leaves, smooth capsules, upper and medial cells of branch leaves unipapillose; *M. turgidum* in having long and narrowly lanceolate branch leaves, and upper and medial cells arranged in longitudinal rows.

Distribution: Philippines, Thailand, Vietnam.

Specimens examined: **PHILIPPINES**. Luzon, *M. R. Crosby 17601* (MO 6162743). **THAILAND**. *P. Sukkharak et al. 228-2* (MO). **VIETNAM**. Lam Dong, *S. He & K. Nguyen. 42900* (MO 6239367); Ha Giang, *K. Nguyen, 10-55*, Mar, 2010 (MO 6366640); Lam Dong, *Evrard 1383* (MO 3064950; PC 0083690); Lam Dong, *S. He & K. Nguyen. 42821* (MO 6239369); Ha Giang, *L. Averyanov, NTH B 024* (MO 5244297).

128. ***Macromitrium macrocomoides*** Müll. Hal., Hedwigia 37: 149. 1898. (Figure 148)

Type protologue: [Chile] Fuegia, Eden Harbour ad fretum Magellanicum occidentale. Hb. Horti Romani: Prof. *Pirotta* mis. 1885. Type citation: Fuürland: Eden Harbour *Pirotta s.n.* 1885. (isotypes: PC 0137796!, BM 000873305!).

(1) Plants small, brownish; stems weakly creeping, with erect and straight branches, branches up to 15 mm long and 1.5 mm thick, branches simple or forked. (2) Stem leaves inconpicuous and caducous. (3) Branch leaves funiculate, densely and spirally coiled around the branch when dry, widely spreading and flat when moist, oblong-lanceoalte to ovate-lanceolate, with an acute to acute-acuminate apex, narrowly recurved at one side, slightly plicate below; margins entire throughout; costae stout but vanishing several cells beneath the apex; lamina unistratose, all cells smooth and clear, thick-walled; upper cells small, subquadrate to round-quadrate, in diagonal rows; medial cell oblate and thick-walled, in longitudinal rows; short-elliptic cells confined to a small area near the base.

Distribution: Chile.

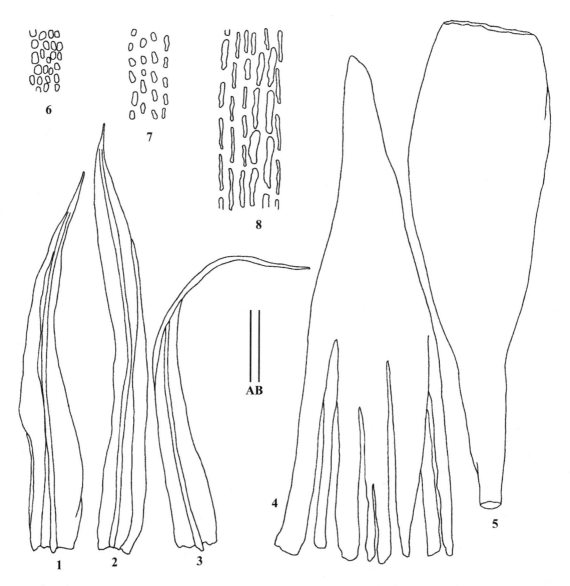

Fig. 146 *Macromitrium longirostre* (Hook.) Schwägr. 1-3: Branch leaves. 4: Calyptra. 5: Capsule. 6: Upper cells of branch leaf. 7: Medial cells of branch leaf. 8: Basal cells of branch leaf (all from H-BR 2528014). Line scales: A = 0.44 mm (1-5); B = 44 μm (6-8).

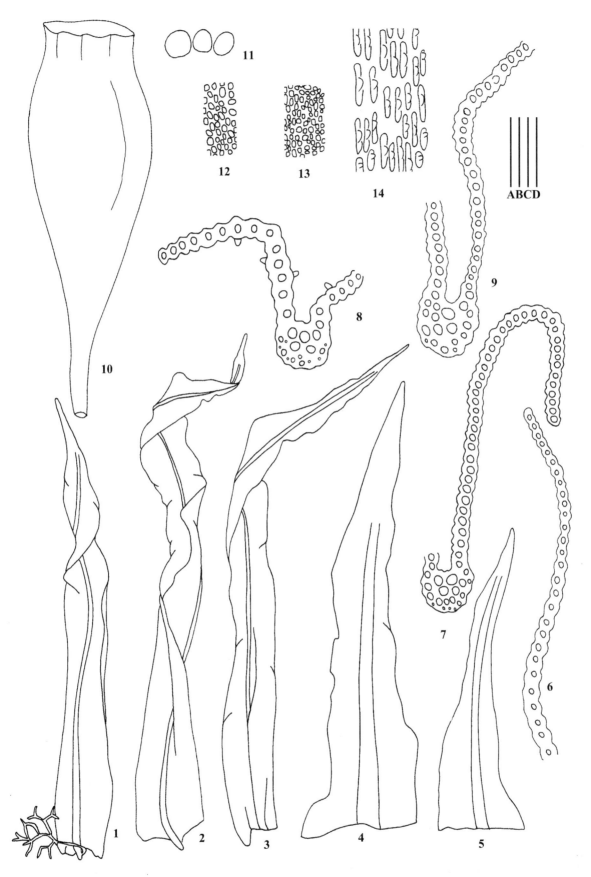

Fig. 147 *Macromitrium lorifolium* Paris & Broth. 1-3: Branch leaves. 4, 5: Apices of branch leaves. 6, 7, 9: Medial transverse sections of branch leaves. 8: Basal transverse section of branch leaf. 10: Capsule. 11: Spores. 12: Apical cells of branch leaf. 13: Medial cells of branch leaf. 14: Basal cells of branch leaf (all from lectotype, H-BR 2630007). Line scales: A = 0.44 mm (10); B = 176 μm (1-3); C = 70 μm (4, 5); D = 44 μm (6-9, 11-14).

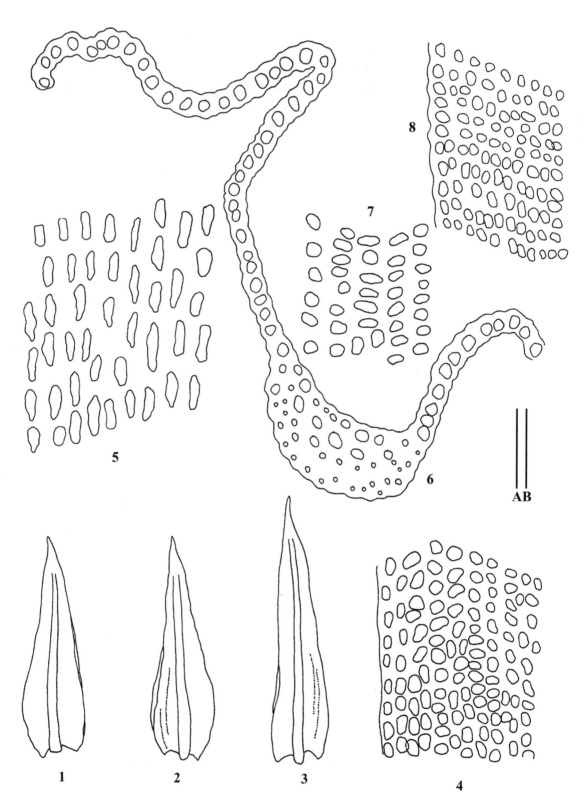

Fig. 148 *Macromitrium macrocomoides* Müll. Hal. 1-3: Branch leaves. 4: Medial cells of branch leaf. 5: Basal cells of branch leaf. 6: Basal transverse section of branch leaf. 7: Low cells of branch leaf. 8: Upper cells of branch leaf (all from isotype, BM 000873305). Line scales: A = 1 mm (1-3); B = 67 μm (4-8).

129. *Macromitrium macrosporum* Broth., Öfvers. Finska Vetensk.-Soc. Förh. 40: 168. 1898. (Figure 149)

Type protologue and citation: (Papua New Guinea) British New Guinea, 1896, Leg. *A. Giulianetti*. com. Sir W. Mac Gregor. (lectotype: H-BR 2532001!).

Macromitrium aspericuspis Dixon, Ann. Bryol. 12: 51. 1939, *fide* Vitt *et al.*, 1995. Type protologue: Indonesia. Sumatra, G, Losir, on bark of tree, 3250-3400 m, 4.II.1937 *van Steenis 10156* (type); 3250 m, among Hepaticae, *van Steenis 10140* (paratypes).

Macromitrium goniostomum Broth., Philipp. J. Sci. 5: 145. 1910, *fide* Vitt *et al.*, 1995. Type protologue and citation: Pauai (altitude about 2100 m), Province of Benguet, Luzon, coll. *R. C. McGregor*, June, 1909 (syntype: H-BR 2551015!); Flora of the Philippines, Luzon, *Elmer D. Merrill. No. 6401*, May, 1909 (syntype: H-BR 2551014!); *Bureau of Science no. 4551*, flora of the Philippines, herbarium of the bureau of science, Haight's in the oaks "1000 ft. alt.", Luzon, E. A. Mearns, Aug. 2, 1907 (syntype: H-BR 2551013!).

**Macromitrium hamatum* E.B. Bartram, Bryologist 48: 117. 1945, hom. illeg., *fide* Vitt *et al.*, 1995. Type protologue: (Papua New Guinea. Morobe) Rawlinson Range, 7000-12000 ft., VII. 1941 [*Clemens*] 12503.2.

Macromitrium morobense E.B. Bartram, Bryologist 48: 117. 1945, *fide* Vitt *et al.*, 1995. Type protologue: (Papua New Guinea. Morobe) Rawlinson Range, elev. 7000-12000 ft., [*Clemens*] 12445B.4, 12445B.13, *12448* type; *12503.3* (paratypes).

(1) Plant medium-sized to robust, in loose mats, rusty-reddish to chestnut-brown; stems longly and loosely prostrate, up to 12 cm; branches 3-4 cm long. (2) Branch leaves and stem leaves are similar, rusty-reddish or lustrous, individually flexuose-twisted to twisted-contorted from a loosely erect low portion when dry, spreading with an incurved apex when moist, 2.1-6 mm long, narrowly lanceolate to lanceolate from an oblong base, slender and acuminate above, sharply to bluntly and slenderly incurved at the apex; margins plane to reflexed on one side, reflexed to broadly recurved on one side below, entire to slightly crenulate above owing to protruding papillae; costae subpercurrent to percurrent; upper cells rounded, rounded-quadrate, shortly rhombic-oval, thick-walled, strongly conic-bulging and conic-papillose, clear; medial cells irregularly oval to rectangular rhombic, smooth, some slightly bulging to weakly conic-papillose, in distinct longitudinal rows; basal cells 8-10 × 30-70 µm, elongate rectangular, straight, thick-walled, lumens straight to curved-sigmoid, flat with scattered tuberculate or smooth. (3) Perichaetial leaves 2.5-4.6 mm long, erect sheathing, broadly lanceolate to ovate-lanceolate, gradually long acuminate to an acuminate arista; costae excurrent into the arista; upper cells smooth; setae 8-22 mm long, smooth. (4) Capsule urns ovoid-ellipsoid, lightly 8-ribbed above, wide-mouthed when old; peristome single, exostome of a low, plicate membrane, coarsely papillose; spores anisosporous. (5) Calyptrae mitrate, plicate, almost naked, sometimes deeply lacerate.

Macromitrium yuleanum Broth. & Geh. is similar to *M. macrosproum* Broth. in their rusty-reddish or brownish appearance, ellipsoid capsules with a peristome reduced to a low fused membrane on relatively long, slender, smooth setae, but *M. yuleanum* can be separated from *M. macrosporum* by its strongly abaxially curved leaves when moist, and its branch leaves with papillose-denticulate margins near the apex, and cells with one high papilla or 2-3 forked papillae in the medial and upper portions.

Distribution: China, New Caledonia, Papua New Guinea, Philippines.

Specimens examined: **CHINA**. Hainan, *Anonymous collector, 1730* (IBSC 09581). **NEW CALEDONIA**. *Lei Rell 1603* (H-BR 2551007), Mt Dzumac, *Le Rat 1909* (H-BR 2551003); *Lei Rell 1372* (H-BR 2551005); *Le Rat 1636* (H-BR 2551008); *Le Rat s.n.* (H-BR 2551010). **PAPUA NEW GUINEA**. Morobe Prov., *D.H. Norris 63347* (KRAM-B-114327), *T. Koponen 31877* (H 3195989), *31918* (H 3195982), *32046* (H 3195983), *32047* (H 3195986), *31868* (H 3195987), *32627* (H 3195988), *32724* (H 3195990), *32725* (H 3195991), *D. H. Horris 627310* (H 3195979); West Sepik, *A. Touw 16240* (H 3205271, H 3205168, MO 5375877), *A. Touw 16376* (MO 5375881, H 3205082), *16739* (H 3205236), *17205* (H 3205057), *17312* (MO 5371049, H 3205076), *17369* (H 3205256). **PHILIPPINES**. Luzon, *B. C. Tan 82-211* (MO 3071740).

130. *Macromitrium macrothele* Müll. Hal., Bot. Zeitung (Berlin) 6: 766. 1848. (Figure 150)

Type protologue: [Venezuela] Bei Galipan, 5000 Fuss hoch im Januar auf feuchten Steinen. No. 364. Dann noch; No. 933 von Lagunitta der merida, Merida, 6000'hoch, im September auf Baumen Gesammelt. Type citation: Venezuela, Galipan, *Funck Schlim 364* (isosyntypes: S-B 164068!, NY 01086494!).

(1) Plants small, brownish; stems creeping, with short branches, branches about 5-8 mm long, with branchlets. (2) Stem leaves caducous and inconspicuous. (3) Branch leaves loosely erect below, strongly twisted-flexuose to contorted-crisped, partially with a curly to circinate apex hidden in the inrolled cavity when dry, erect-incurved when moist, ligulate- to oblong-lanceolate; apices broadly acuminae to acute-acuminate; margins entire; upper and medial cells rounded to round-quadrate, bulging, strong collenchymatous; low cells elongate, thick-walled, not porose, distinctly tuberculate; enlarged marginal teeth-like cells not differentiated at insertion. (4) Perchiaetial leaves distinctly differentiated, broadly oblong to ovate-oblong, shorter and wider than branch leaves; all cells longer than

wide. (5) Setae smooth and slender, 3-6 mm long, straight or slightly twisted to the left. (6) Capsule urns ellipsoid or cupulate, distinctly furrowed when dry; peristome double, exostome united to form an erect membrane, yellowish, endostome pale, united and membraneous, higher than the exostome. (7) Calyptrae mitrate, naked, lobed at base.
Distribution: Venezuela.

131. *Macromitrium maolanense* Ze Y. Zhang, D.D. Li, J. Yu & S.L. Guo, J. Bryol. 41(3): 265, 2-4. 2019. (Figure 151)

Type protologue and citation: China, Guizhou Province, Libo County, National Natural Reserve, 25.25854920N, 108.05787620E, 477m, on trunk, *S. L. Guo & D. D. Li 171114076* (holotype: SHTU!); Hunan Province, Zhangjiajie, Tianzhishan, Dianjiangtai, Oct. 24, 1985, *D. K. Li, 18352* (paratype: SHM 12016!); Zhangjiajie, Tianzhishan, Xieqiao, Oct. 24, 1985, *D. K. Li, 18499* (paratype: SHM 12018!).

(1) Plants small to medium-sized, in dense, dark-green or reddish-brown mats; stems prostrate, creeping, with dense, short, erect branches, branches about 5.0-8.5 mm high, sparsely leaved, covered with reddish rhizoids. (2) Stem leaves 0.4-0.6 mm long, inconspicuous and caducous, much different from branch leaves, irregularly flexuose-twisted when dry, spreading when moist, ovate-triangular, gradually narrowed and acuminate, entire; upper and medial cells rounded to subquadrate, 3.5-4.5 µm wide, with moderately thick walls, becoming clear, rectangular to elliptic towards the base, all cells flat and smooth. (3) Branch leaves 2.5-3.0 × 0.25-0.30 mm, twisted-contorted and curly when dry, widely spreading when moist, lanceolate-ligulate, gradually narrowed to a variable subulate upper portion from an oblong low portion, keeled, slightly recurved at one side below, the subula and upper portion of most leaves easily broken off; costae prominent, ending obscurely in or just beneath the apex of the subula; upper and medial laminae partially and variably bistratose, cells rounded to quadrate, 4.5-5.5 µm, thin-walled, rather obscure, often coated with dark, plaque stains, densely pluripapillose, about 4-6 per cells, the papillae forming a continuous covering; basal cells linear, 8-25 µm long, with irregular thickened walls, smooth, those along costa thin-walled, smooth and pellucid, 30-45 × 10-15 µm, distinctly larger than their adjacent cells, appearing as a "cancellina region", those near the margin elongate, rectangular to sublinear, thick-walled and hyaline, somewhat differentiated from the inner cells. Sporophytes not seen.

Macromitrium maolanense Ze Y. Zhang, D.D. Li, J. Yu & S.L. Guo is similar to *M. gymnostomum* Sull. & Lesq. in their lanceolate, linear-lanceolate or oblong-lanceolate branch leaves, with rather obscure, densely pluripapillose upper and medial lamina cells, and clear, hyaline basal lamina cells. Under microscope, upper and medial portions of branch leaves were both seen to coat with dark and plaque stains. However, *M. maolanense* is distinctively different from *M. gymnostomum* by its easily broken off, subulate upper portion, and partially and variably bistratose laminae in the medial and upper portions of branch leaves.

Habitat: on trunks of broadleaved trees.
Distribution: China.

132. *Macromitrium masafuerae* Broth., Nat. Hist. Juan Fernandez 2: 423. pl. 26: f. 29, 32. 1924. (Figure 152)

Type protologue: [Chile] Masafuera: Quebrada del Mono; ad truncos Myrceugeniae; 475 m. s. m. (173). Type citation: Juan Fernandez: Masafuera, Quebrada del Mono, 20/2, 1917, *Carl o. Inga Skottsberg, no. 173*, Det. Brotherus (isotype: NY 01202047!).

(1) Plants rather small, in dense mats, light yellow-greenish; stems long creeping, densely with rusty rhizoids; branches single and short, only about 2-3 mm long, densely leaved. (2) Branch leaves appressed and spirally twisted and flexuose above when dry, spreading when moist, oblong, ligulate, oblong- to ligulate-lanceolate, distinctly plicate below; margins entire and plane; apices obtuse to obtuse-acute to acute; costae subpercurrent; upper cells strongly bulging and pluripapillose to varying degrees, rounded-quadrate, sometimes upper cells in radiating diagonal rows from the costa; medial cells rounded-quadrate, moderately bulging and frequently unipapillose; cells elongate towards the base, irregularly rectangular cells restricting to the low 1/5 to 1/4 of the leaf length, most smooth, some tuberculate and porose; cells becoming narrow and long towards the margin below. (3) Perichaetial leaves differentiated from branch leaves, broadly lanceolate, longer and wider than branch leaves, clasping or strongly incurved below or near the base, strongly plicate. (4) Setae short, about 3-4 mm long, smooth. (5) Capsules small, 0.75 × 0.35 mm, ellipsoid and smooth, slightly constricted beneath the brown-reddish mouth; peristome single, often rudimentary when old or absent. (6) Calyptrae mitrate, smooth and naked.

Distribution: Chile.

133. *Macromitrium mcphersonii* B.H. Allen, Monogr. Syst. Bot. Missouri Bot. Gard. 90: 599. 2002. (Figure 153)

Type protologue & citation: Panama. Chiriquí: Fortuna Dam region, along trail to Cerro Hornito (Pate de Macho) on southern ridge of watershed. 8°45'N, 82°15'W, 1800-1950 m, *McPherson 13595* (holotype: MO 3676011!).

(1) Plants robust, yellow-green and lustrous above, rusty-brown below; stems weakly creeping, branches to 4 cm long. (2) Branch leaves keeled, appressed-erect below, flexuose, spirally contorted and twisted above when dry, erect at base, flexuose spreading above when moist, long and narrowly lanceolate, 5.5-8 × 0.7 mm, slenderly acuminate to a subula; margins serrate to serrulate above, recurved below, plane above; costae percurrent; all cells longer than wide, thick-walled and strongly porose; upper interior cells rectangular, 8-24 × 4-6 μm, smooth; marginal cells linear, 30-50 × 2-3 μm, differentiated to form a distinct border; basal cells 30-60 μm long, linear-rectangular, smooth to slightly tuberculate. (3) Perichaetial leaves lanceolate from a broad oblong low part, narrowly acuminate, plicate below; all cells longer than wide, linear-rectangular, thick-walled, strongly porose, smooth. (4) Setae strongly papillose, 10-15 mm long, twisted to the right. (5) Capsule urns 1.5-2.0 mm long, subglobose to ellipsoid, smooth, occasionally with a conspicuous apophysis; peristome double, exostome teeth truncate, reddish, papillose, united to form an erect membrane, endostome hyaline, lightly papillose. (6) Calyptrae mitrate, 3 mm long, lacerate, smooth and naked (Allen, 2002).

Macromitrium mcphersonii B.H. Allen resembles *M. scoparium* Mitt. in their robust plants, long and narrowly lanceolate leaves and a distinctly differentiated border, but the setae of the latter are smooth.

Distribution: Colombia, Panama.

Specimens examined: **COLOMBIA**. Nariño, *S. P. Churchill et al. 18212*, *18249* (MO). **PANAMA**. Prov. Chiriqui, *Gordon Mcpherson 7250C* (MO 3651894).

134. *Macromitrium megalocladon* M. Fleisch., Hedwigia 50: 282. 1911.

Type protologue: (Indonesia, West Irian. Jayawijaya) Holländisch-Neu-Guinea: An den Abhangen des Hellwiggebirges, wahrscheinlich an Bäumen und jedenfalls über einer Höhe von 1 500 m durch Dajaksche Kulis gesammelt", X. 1909 *Römer 718*.

Macromitrium altipapillosum Bartr., Lloydia 5: 272. 33. 1942, *fide* Vitt *et al.*, 1995. Type protologue: (Indonesia. West Irian, Jayawijaya) Lake Habbema, on ground and epiphytic, 3 225 m, *Brass 9563* type: *Brass 9155a*.

Macromitrium submegalocladum Dix., Farlowia 1: 31. 1943, *fide* Vitt *et al.*, 1995. Type protologue: (Indonesia. West Irian, Jayawijaya) Nova Guinea, summit of Mt. Wichmann, 3 000 m", 8. II.1913 *A. Pulle 1032 p.p.*

(1) Plants large, dull, stiff, coarse, in loose mats; stems poorly differentiated; branches up to 3.5 cm long. (2) Stem leaves and branch leaves similar, erect below, irregularly squarrose-recurved to bent back with contorted-flexuose uppe portions when dry, stiffly wide-spreading to strongly abaxially curved from an erect base when moist, 3.5-5.0 mm long, narrowly lanceolate from an oblong base, ovate-lanceolate, narrowly acuminate; margins plane to reflexed below, entire except for projecting papillose in the upper portion; costa percurrent; all cell strong conic-bulging; upper cells rounded to rounded-short-elliptic, strongly unipapillose to high tuberculate or with a forked papilla, clear, with even firm walls, marginal cells somewhat smaller than their ambient cells; medial cells rounded-rhombic to short-elliptic, firm, thick-walled, bulging-conic, strongly unipapillose, distinctly porose, in longitudinal rows; low cells elliptic-rectangular, irregularly thick-walled, strongly tuberculate; basal cells near insertion elongate-rectangular, lumens straight, very irregularly thick-walled, distinctly porose. (3) Perichaetial leaves not much differentiated from but slightly longer than branch leaves, 4.5-7.0 mm long, loosely erect with erect-flexuose acumens, lanceolate, gradually short subulate; costa excurrent and filling the subula; margins irregularly denticulate near the apex. (4) Setae 11-13 mm long, papillose, twisted to the left. (5) Capsule urns 2.0-2.4 × 1.0 mm wide, ellipsoid to narrowly ovoid, smooth and somewhat 4-angled at the mouth to lightly 8-ribbed, sharply contracted to the seta, somewhat smaller at the mouth when old; peristome single; exostome a low, papillose membrane; spores 30-60 μm, anisosporous. (6) Calyptrae mitrate, sometimes plicate, densely hairy (Vitt *et al.*, 1995).

Distribution: Indonesia, Malaysia, Papua New Guinea.

Specimens examined: **MALAYSIA**. Sarawak, *A. Touw 19916* (MO 6366338). **PAPUA NEW GUINEA**. Southern Highlands, *H. Streimann 23813* (MO 5140806, KRAM-B-137713); West Sepik, *A. Touw 15926* (MO 5376035), *15768* (H 3205002, MO 5376041), *16555* (H 3205004, MO 5376000), *15589* (MO 5375861), *18008* (MO 5375806), *18143* (MO 5371015).

135. *Macromitrium melinii* Roiv., Ann. Bot. Soc. Zool.-Bot. Fenn. "Vanamo", 6(8): 16. f. 1-6. 1936. (Figure 154)

Type protologue and citation: Peru borealis, Roque (circiter 40 km ad SO versus e Moyobamba), ad ramum putridum. 15 Jul 1925, *Douglas Melin s.n.* (isotype: H 3090555!).

(1) Plants medium to large, rusty red, in dense mats; stems creeping, densely with rusty brown rhizoids and ascending branches; branches simple or with branchlets, densely leaved. (2) Branch leaves spirally twisted and flexuose when dry, spreading and slightly longitudinally undulate when moist, broadly lanceolate to ligulate-lanceolate; margins distinctly serrulate above, entire below; the apices acute to mucronate; costa strong and percurrent, often concealed in adaxial view by overlapping folds of laminae in upper portions; upper and medial cells subquadrate, round-quadrate, oblate, smooth and clear, thick-walled; low and basal cells elongate-rectangular, strongly porose and

thick-walled, strongly tuberculate; marginal enlarged teeth-like cells differentiated at insertion. (3) Perichaetial leaves broadly oblong-to ovate-lanceolate, costa keeled in abaxial surface and concealed in adaxial view, distinctly plicate and longitudinally undulate. (4) Setae erect and smooth, brown-red, about 15 mm long. (5) Capsule urns oblong to oblong-cylindric, smooth, with a constricted mouth; peristome double, short lanceolate. (6) Calyptrae conic-mitrate, smooth and naked, lacerate below, covering the whole capsule.

Distribution: Peru.

Specimen examined: **PERU**. Amazonas, Prov. Chachapoyas, let. *P+E, Hegewald 6672* (MO 5147613).

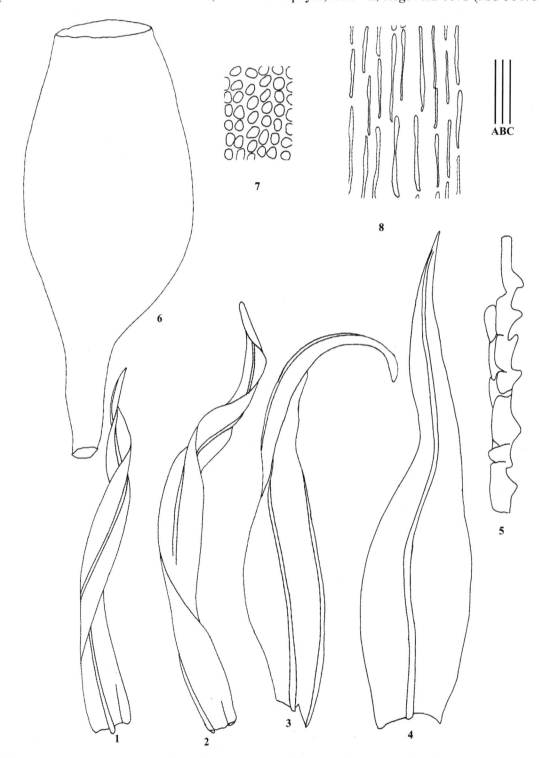

Fig. 149 *Macromitrium macrosporum* Broth. 1-4: Branch leaves. 5: Capsule. 6: Basal marginal cells of branch leaf. 7: Basal cells of branch leaf. 8: Medial cells of branch leaf (all from syntype of *M. goniostomum*, H-BR 2551014). Line scales: A = 0.88 mm (5); B = 0.44 mm (1-4, 6); C = 44 μm (7-8).

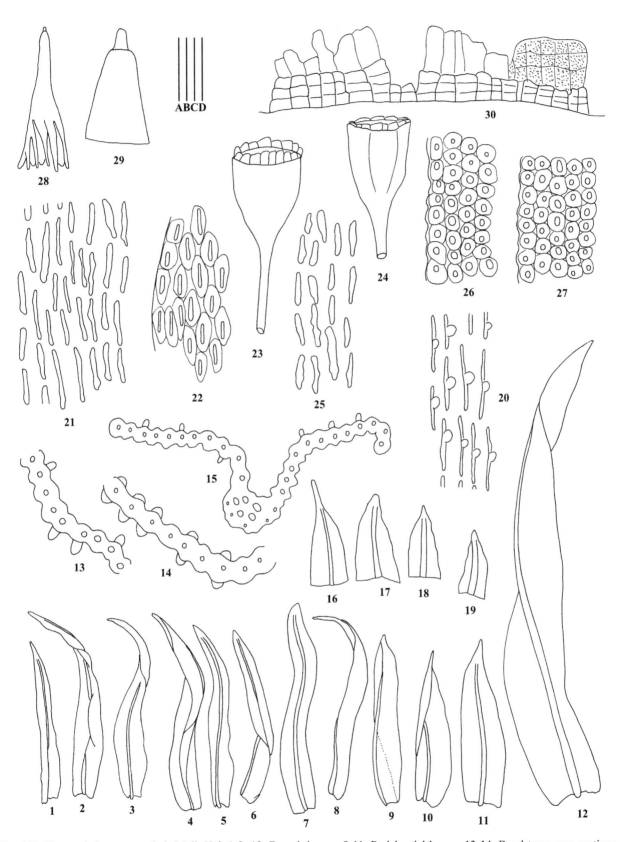

Fig. 150 *Macromitrium macrothele* Müll. Hal. 1-8, 12: Branch leaves. 9-11: Perichaetial leaves. 13-14: Basal transverse sections of branch leaves. 15: Low transverse section of branch leaf. 16-19: Apices of branch leaves. 20: Basal cells of branch leaf. 21: Basal cells of perichaetial leaf. 22: Upper cells of perichaetial leaf. 23-24: Capsules. 25: Medial cells of perichaetial leaf. 26: Upper cells of branch leaf. 27: Medial cells of branch leaf. 28: Calyptra. 29: Apex of calyptra. 30: Peristome (all from isosyntype, NY 01086495). Line scales: A = 2 mm (23-24, 28-29); B = 1 mm (1-11); C = 400 µm (12, 16-19, 30); D = 67 µm (13-15, 20-22, 25-27).

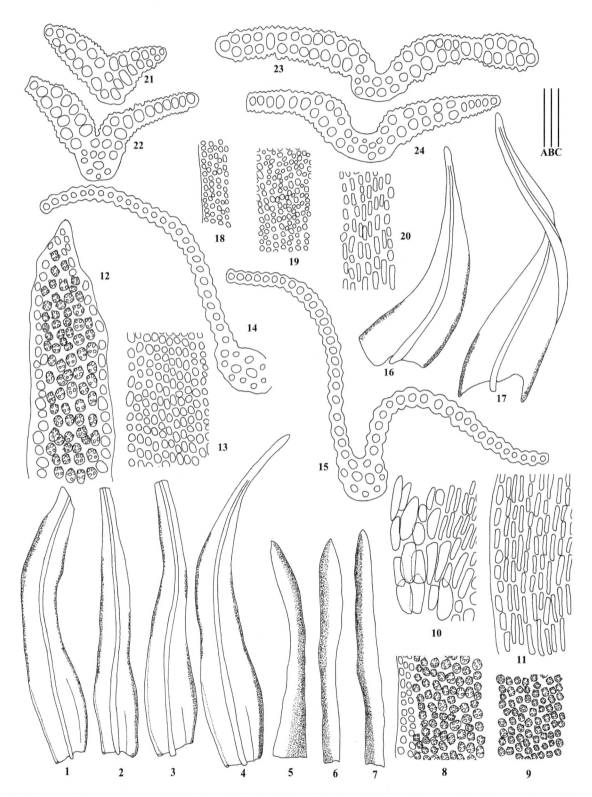

Fig. 151 *Macromitrium maolanense* Ze Y. Zhang, D.D. Li, J. Yu & S.L. Guo 1-4: Branch leaves. 5-7: Subulae of branch leaves. 8: Upper cells of branch leaf. 9: Medial cells of branch leaf. 10: Basal cells near costa of branch leaf. 11: Basal marginal cells of branch leaf. 12: Subula cells of branch leaf. 13: Low cells of branch leaf. 14: Basal transverse section of branch leaf. 15: Medial transverse section of branch leaf. 16: Stem leaves. 17: Upper cells of stem leaf. 18: Medial cells of stem leaf. 19: Basal cells of stem leaf. 21-22: Transverse sections of upper subulate part. 23-24: Upper transverse sections of branch leaves (all from holotype, SHTU 20171114046). Line scales: A = 500 μm (1-4); B = 200 μm (5-7, 16-17); C = 50 μm (8-15, 18-24). (Zhang et al., 2019).

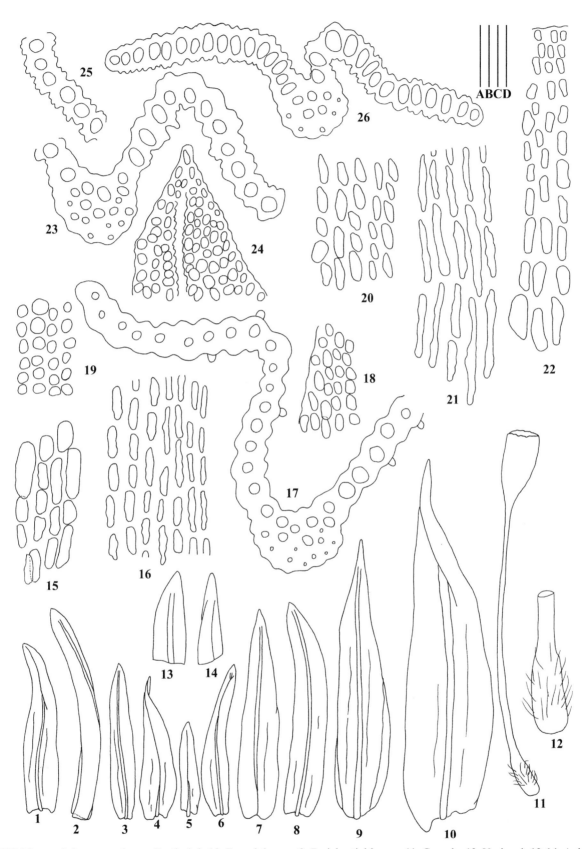

Fig. 152 *Macromitrium masafuerae* Broth. 1-8, 10: Branch leaves. 9: Perichaetial leaves. 11: Capsule. 12: Vaginual. 13-14: Apices of branch leaf. 15: Basal cells near costa of branch leaf. 16: Basal cells of branch leaf. 17: Medial transverse section of branch leaf. 18: Upper cells of perichaetial leaf. 19: Medial cells of branch leaf. 20: Medial cells of perichaetial leaf. 21: Basal cells of perichaetial leaf. 22: Cells of capsule mouth. 23: Medial transverse section of branch leaf. 24: Apical cells of branch leaf. 25-26: Upper transverse sections of branch leaves (all from isotype, NY 01202047). Line scales: A = 2 mm (11); B = 1 mm (1-9, 12); C = 400 μm (10, 13-14); D = 67 μm (15-26).

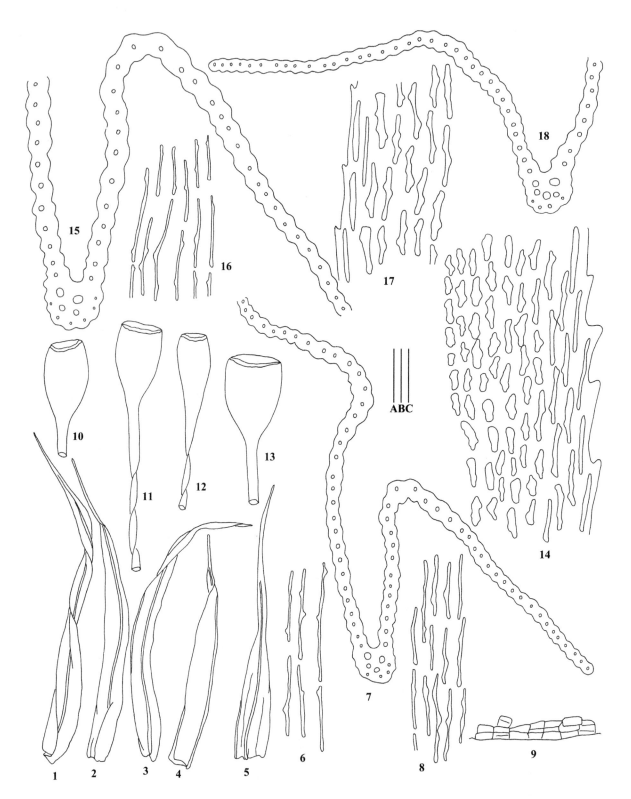

Fig. 153 *Macromitrium mcphersonii* B.H. Allen 1-4: Branch leaves. 5: Perichaetial leaves. 6: Basal cells of perichaetial leaf. 7, 15, 18: Basal transverse sections of branch leaves. 8: Upper cells of perichaetial leaf. 9: Peristome. 10-13: Capsules. 14: Upper cells of branch leaf. 16: Basal cells of branch leaf. 17: Medial cells of branch leaf (all from holotype, MO 3676011). Line scales: A = 2 mm (1-5, 10-13); B = 400 μm (9); C = 67 μm (6-8, 14-18).

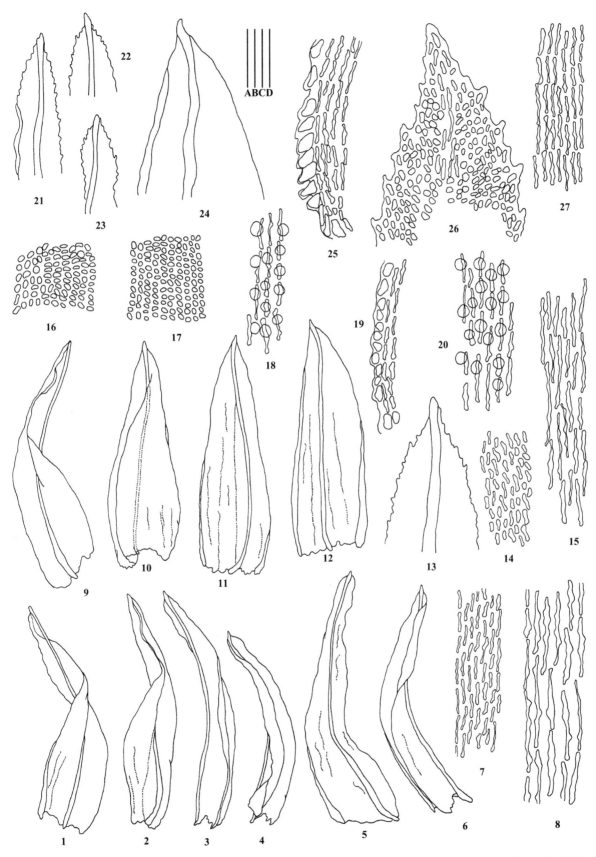

Fig. 154 *Macromitrium melinii* Roiv. 1-6, 9: Branch leaves. 7: Medial cells of perichaetial leaf. 8: Basal cells of perichaetial leaf. 10-12: Perichaetial leaves. 13, 21-23: Apices of branch leaves. 14: Upper cells of perichaetial leaf. 15: Low cells of perichaetial leaf. 16: Upper cells of branch leaf. 17: Medial cells of branch leaf. 18, 20: Basal cells of branch leaves. 19, 25: Basal marginal cells of branch leaves. 24: Apex of perichaetial leaf. 26: Apical cells of branch leaf. 27: Low cells of branch leaf (all from isotype, H 3090555). Line scales: A = 500 μm (1-6, 9-12); B = 200 μm (21-23); C = 100 μm (13, 24); D = 50 μm (7-8, 14-20, 25-27).

136. *Macromitrium menziesii* Müll. Hal., Bot. Zeitung (Berlin), 20: 361. 1862. (Figure 155)

Type protologue: [Tahiti, French Polynesia] Patria. Ex insula Otaheiti, ubi Menzies Legit, specimina mauca Swartziaua in Hb. Mohrii sub nom. "Ortothr. Crispi? Ex Otaheite: Menzies in Hb. Swartziano inveni". Type citation: Taiheiti, leg. *Menzies s. n* (isotype: H-BR 2558028!).

(1) Plants small, in dense mats, yellow-green above, dark brown below; stems long creeping, with short and erect branches, branches up to 5 mm tall. (2) Stem leaves densely leaved, short lanceolate. (3) Branch leaves appressed and twisted-flexuose when dry, spreading when moist, broadly lanceolate, oblong-lanceolate, unistratose throughout; apices mucronate; margins entire and plane throughout; costae percurrent, ending at the mucronate apex; upper cells rather small, pluripapillose and obscure; low and basal cells elongate-rectangular and thick-walled, often tuberculate. (4) Perichaetial leaves broadly lanceolate, acuminate. (5) Capsules not seen, calyptrae large, plicate and densely with yellow hairs.

Distribution: French Polynesia.
Specimen examined: **HAWAII**. Oahu, C. Lautarback 45^a (H-BR 2558012).

137. *Macromitrium microcarpum* Müll. Hal., Syn. Musc. Frond. 1: 727. 1849. (Figure 156)

Type protologue and citation: Chile, *Pöppig s.n.* (isotypes: JE 04008720!, NY 01202050!).

(1) Plants small in mats; stems creeping with inconspicuous and caducous leaves; branches 2-3 mm long, densely with rusty brown rhizoids. (2) Branch leaves tightly spirally appressed and the apices slightly flexuose when dry, distinctly funiculate, erecto-patent and imbricate when moist, oval-lanceolate, broadly oblong-lanceolate, lingulate-lanceolate to lingulate, somewhat hyaline; apices acute to apiculate; margins plane and entire throughout; all cells clear and smooth; most cells isodiametric, rounded-quadrate, quadrate to subquadrate; elongate, short rectangular to rhombic cells confine to a small area near the base, occupying 1/8 of the leaf length. (3) Perichaetial leaves not differentiated. (4) Setae smooth, 4-5 mm long, twisted to the left. (5) Capsule urns small, ovoid, ellipsoid to ellipsoid-cylindric, smooth or plicate beneath the mouth, forming 4-angled or 4-furrowed mouth; opercula conic-subulate, oblique; peristome single, short membranous, truncate above. (6) Calyptrae campanulate, smooth and naked.

Macromitrium microcarpum Müll. Hal. is similar to *M. microstomum* (Hook. & Grev.) Schwägr. in lingulate-lanceolate, broadly lanceolate leaves with acute or apiculate apices, funiculate branches, capsules with single peristome. However, *M. microcarpum* differs from *M. microstomum* in having short setae, isodiametric cells in medial and low portions of its branch leaves and smaller capsules.

Distribution: Chile.
Specimens examined: **CHILE**. Chepu, Chiloe, *E. J. Godley 311* (KRAM-B-120634), *E. J. Godley 144* (MO 3653345); Los Lagos (Region X) Valdivia Province, *M. Mahu & Harnell Mahu 12034* (MO 3987738), *M. Mahu & Harnell Mahu 12013* (MO 3987735); Los Lagos (Region X) Osorno Province, *J. Burgos 22144* (MO 3987737); Region VIII: Prov. Concepeion, Rocoto, *Ireland & Bellolio 30976* (MO 5639223); *Ireland & Bellolio 32376* (MO 5629607).

138. *Macromitrium microstomum* (Hook. & Grev.) Schwägr., Sp. Musc. Frond., Suppl. 2 2(2): 130. 1827. (Figure 157)

Basionym: *Orthotrichum microstomum* Hook. & Grev., Edinburgh J. Sci. 1: 114. 4. 1824. TYPE: [Australia] Van Dieman's Land [sic]; Dr. Spence. Type citation: Van Dieman's Land; Spence & Neill. (lectotype: E-Grev E 00011665!).

Leiotheca microstoma (Hook. & Grev.) Brid., Bryol. Univ. 1: 729. 1826, *fide* Valente *et al.*, 2020.

Dasymitrium borbonicum Besch., Ann. Sci. Nat., Bot., sér. 6, 9: 355, 1880, *fide* Wilbraham & Ellis, 2010. Type protologue: La Réunion: plaine des Chicots, Bory (herb. Cosson); Boivin, 1849 (in herb. Mus. Par.); plaine des Palmistes, G. DE L'ISLE, 1875, n°200 et 286; pas de Belcombe, P. LÉPERVANCHE, 1877 (hb. Mus. Par., sub *Macromitrio schizomitrioo* Nob. Prius), Madagascar (N.O.): PERVILLÉ, 1849, n°827.

Macromitrium acunae Thér., Mem. Soc. Cub. Hist. Nat. "Felipe Poey" 14: 353. 54 f. 2. 1940, *fide* Vitt & Ramsay, 1985. Type protologue: Cuba, Acuna, juin 1936, (407).

Macromitrium adstrictum Ångstr., Öfvers. Kongl. Vetensk.-Akad. Förh. 29(4): 19. 1872, *fide* Vitt & Ramsay, 1985. Type protologue: Honolulu (Sandwichs-öarnc), samlade I Juni 1852.

Macromitrium borbonicum (Besch.) Broth., Nat. Pflanzenfam. I(3): 481. 1903, *fide* O'Shea, 2006.

Macromitrium fasciculare Mitt. var. *fasciculare*, J. Proc. Linn. Soc., Bot., Suppl. 1: 51. 1859, *fide* Wilbraham & Ellis, 2010. Type protologue: Hab. In Ceylon ad Horton Plains et Newera Ellia, *Gardner (No. 225, 229, 230)*. Type specimen. Sri Lanka, Horton Plains, Feb 1846, *Gardner 230* (isosyntype: BM 000919522!), Ceylon, *Gardner 225* (isosyntype: BM 000919520!).

Macromitrium fasciculare var. *angustifolium* P. de la Varde, Rev. Bryol. Lichénol. 19: 152. 1950, *fide* Wilbraham & Ellis, 2010. Type protologue: Sommet oriental du massi du Marojejy a l'ouest de la haute Manantenina,

Gneiss et quartzite, Alt. 1850-2100 m. Type citation: Sommet oriental du massi du Marojejy (Nord-est), a l'ouest de la haute Manantenina, Affluent de la Lokoho, Gneiss et quartzite, Altitude: 1850-2137 m. Leg. H. Humbert & G. Cours (isotypes: PC 0105915!, PC 0105916!).

Macromitrium fasciculare var. *javense* M. Fleisch., Musci Buitenzorg 2: 432. 1904, *fide* Wilbraham & Ellis, 2010. Type protologue: Exsiccata: M. Fleischer, Musc. Archip. Ind., No. 281 (1902). Type citation: Mittel-Java, Diengplateau, April 1902, Leg. M. Fleischer 281 (isotypes: L 0060436!, GOET 012315!).

Macromitrium filicaule Müll. Hal., Syn. Musc. Frond. 1: 745. 1849, *fide* Yu *et al.*, 2018. Type protologue: '*M. aciculare* C. Müller in Gardneri Bras. No. 53. B. -*M. microstomum* Hsch. Fl. Bras. L. p. 21? Patria, Brasilia: Gardner'. Type citation: 'Brasilia, *Gardner 53B*' (lectotype designated here: BM 000873118!; isolectotypes: BM 000873119!, NY 01201996!).

Macromitrium flaccidisetum Müll. Hal., Hedwigia 37: 147. 1898, *fide* Vitt & Ramsay, 1985. Type protologue: Nova Seelandia, insula australis, littore australi prope Greymouth: Richard Helms 1885 legit et misit; insula septentriconalis loco non indicato: F. M. Reader leg. 1882, mis 1892.

Macromitrium hornschuchii Müll. Hal., Bot. Zeitung (Berlin) 3: 526. 1845, **syn. nov.** **Macromitrium microstomum* Hornsch, Fl. Bras. 1(2): 21. 1840. Type protologue: In truncis demortuis sylvarum prope Novo – Friburgum, prov. Sebastianopolitanae, Decembri. Beyrich. 4. Type citation: Brasil, H.Warming s.n. (lectotype designed here, BM 000873101!; isolectotype: BM 000873101!).

Macromitrium linearifolium Müll. Hal., Linnaea 37: 154. 1873[1872], *fide* Vitt & Ramsay, 1985. Type protologue: Australia, Nova Valesia australis, Mostland: *Vicary* legit, am, Hampe dedit 1869.

Macromitrium nitidum Hook. & Wilson, London J. Bot. 3: 156. 1844, *fide* Wilbraham, 2007. Type protologue: *Gardner 51*: On trees, Serra de Jaquari, Sept. 1840. Type citation: Brasil, *Gardner 51* (holotype: BM 000878023!), *Gardner 52* (paratype: BM 000878024!).

Macromitrium macropelma Müll. Hal., Bot. Zeitung (Berlin) 14: 420. 1856, *fide* Wilbraham, 2007. Type protologue: Prom. Bonae spei, ad Grootvaderbosch, Ecklon in Hb. Kunzeano. Type citation: Cape, Grootvaderbosch, *Zeyher s.n.* (isolectotype: BM 000982417!).

Macromitrium owahiense Müll. Hal., Bot. Zeitung (Berlin) 22: 359. 1864, *fide* Vitt & Ramsay, 1985. Type protologue: "Insula Oahu: Didrichsen in Exped. transatl. Danica; Jo. Lange amice donavit. Ex insulis Sandwicensibus et Societatis aliis Cl. Sullivant accepisse videtur." Type citation: Oahu, leg. *Didrichsen*, 1855 (lectotype designated by Vitt & Ramsay, 1985: H-SQL 1352025!).

Macromitrium pacificum Besch., Ann. Sci. Nat., Bot., sér. 5, 18: 209. 1873, *fide* Vitt & Ramsay, 1985. Type protologue: In summon monte Mi, ad arborum truncus (*Balansa, no. 917*); in monte Humboldt, 1200 m. alti. (*Balansa, no. 2536*); in monte Mou (*Balansa, no. 2974*). Type citation: New Caledonia, *Balansa, no. 2536* (isosyntype: H-BR 2519010!).

Macromitrium pacificum var. *brevisetum* Thér., Bull. Acad. Int. Géogr. Bot. 20: 99. 1910, *fide* Yu *et al.*, 2018. Type protologue: dumont Koghis. Type citation: New Caledonia. Mont Koghis, tronecs d'arbres, Jan. 11, 1909, *Franc s.n.* (lectotype designated by Yu *et al.*, 2018: PC0137851!, isolectotype: PC0137852!, PC0137853! JE 04006254!).

Macromitrium pinnulatum Herzog, Bibliot. Bot. 87: 65. 23. 1916, *fide* Churchill & Fuentes, 2005. Type protologue: Im Nebelwald uber Comarapa, about 2600 m, *No. 3945*; Quebrada de Pocona, about 2800 m, *No. 5146*. Type citation: Bolivia, *Herzog 5146* (syntype: JE 04008712!), *Herzog 3945* (syntype: JE 04008713!).

Macromitrium prolixum Bosw., J. Bot. 30: 97. 1892, *fide* Vitt & Ramsay, 1985. Type protologue: Wooded glen, Blue Mountains, New South Wales, Mr. Roper, Sent me by the Rev. C. H. Binstead.

Macromitrium pseudohemitrichodes Müll. Hal., Hedwigia 37: 150. 1898, *fide* Wijk *et al.*, 1964.

Macromitrium reinwardtii Schwägr., Sp. Musc. Frond., Suppl. 2 2(1): 69. pl. 173. 1826, *fide* Vitt & Ramsay, 1985. Type protologue: In Java leit et misit cl. *Prof. Reinwardt*. Type citation: Java, *Reinwardt* (lectotype: G00042114).

Macromitrium saxatile Mitt., J. Linn. Soc., Bot. 12: 200. 1869, *fide* Robinson, 1975. Type protolgoue: Hab. Ins. Juan Fernandez, ad saxa locis umbrosissimis collium, *Bertero s.n.*

Macromitrium scottiae Müll. Hal., Linnaea 35: 618. 1868, *fide* Vitt & Ramsay, 1985. Type protologue: Patria, Ash-Island (lectotype designated by Vitt & Ramsay, 1985: BM-Hampe).

Macromitrium seemannii Mitt., J. Proc. Linn. Soc., Bot., Suppl. 1: 51. 1859, *fide* Wilbraham, 2007. Type protologue: Hab. In India orient. Etiam in St Helean insula copiose legit cl. *Berthold Seemann*. Type citation: St. Helena, *Seemann* (isotypes: BM 000982412!, BM 000982410!, S-B 165028!).

Macromitrium stolonigerum Müll. Hal., Linnaea 42: 489. 1879, *fide* Allen, 2002. Type protologue and citation: Venezuela, [*Fendler*] *61* (isotype: FH 00213688!).

Macromitrium stratosum Mitt., J. Linn. Soc., Bot. 12: 199. 1869, *fide* Vitt & Ramsay, 1985. Type protologue: Jamaica, *Maxwell*.

Macromitrium subnitidum Müll.Hal., Linnaea 42: 488. 1879, *fide* Yu *et al.*, 2018. TYPE protologue: '(Fendler),

Nr. 56'. Type citation: '*Fendler 56, Macromitrium subnitidum* n. sp.' (lectotype designated by Yu *et al.*, 2018: BM 000873215!; isolectotypes: NY 01086589!; FH 00213694!).

Macromitrium tasmanicum Broth., Öfvers. Finska Vetensk.-Soc. Förh. 37: 162. 1895, *fide* Vitt & Ramsay, 1985. Type protologue: Tasmania Circular Head, South Road Forest, ad ligna (*W. A. Weymouth no. 846, 1040, 1041*, nec non in monte Wellington (*no. 121*, f. lutescens). Type citation: On wood South Road Forest, Circular Head, *W. A. Weymouth, 1040* (syntype: H-BR 2522003!), *1041* (syntype: H-BR 2522004!).

Macromitrium weymouthii Broth., Öfvers. Finska Vetensk.-Soc. Förh. 37: 161. 1895, *fide* Vitt & Ramsay, 1985. Type protologue: Tasmania, West Coast, Macquarie Harbhour, Queen River Road, Porteus Gully, ad ligna (*W. A. Weymouth no. 573, 574*), 575 nec non ad Henty River, ad ramulos, Myrtacearum (*no. 569*). Type citation: On wood, Porteur Gully, Queen River Road, Macquarie Hargbhour, West Coast, 1/2/1891, *W. A. Weymouth 575* (syntype: H-BR 2523016!).

(1) Plants small to medium-sized, olive-green, lustrous slender; stems creeping, prostrate, with numerous erect branches. (2) Stem leaves 1.0-1.5 mm long, irregularly flexuose-twisted when dry, recurved-twisted when moist, ovate-lanceolate, acuminate-subulate. (3) Branch leaves keeled, regularly and spirally twisted-curved in rows, the upper portions curved toward one side to curly, or spirally curved-twisted-coiled around the branch, forming a rope-like appearance (funiculate) to varying degrees when dry, spreading when moist, 1.2-2.5 mm long, lanceolate, gradually narrowed to a slender acute or short-acuminate apex; costae slender, ending 10-12 cells beneath the apex to short excurrent; all cells smooth and flat, not porose; upper cells rounded, subquadrate, shortly elliptic, or irregularly quadrate, firm-walled; medial cells elliptic to rectangular; basal and low cells elongate, with evenly thickened walls, not porose, occupying the low 1/2-1/3 of the leaf. (4) Perichaetial leaves oblong-lanceolate to ovate-oblong, quickly contracted to a slender acuminate-sharply cuspidate apex, with a subsheathing low portion. (5) Setae 15-25 mm long, straight or flexuose, smooth. (6) Capsule urns ovoid-ellipsoid to ellipsoid, 8-plicate, sharply puckered at the darker mouth; peristome single, exostome teeth 16, narrow, lanceolate, coarsely papillose; spore large, isosporous, 30-54 μm in diameter. (7) Calyptrae cucullate or plicate to different degrees, naked and lobed at the base.

Distribution: [Australia/Ocean] Australia, Hawaii, New Zealand, New Caledonia, Papua New Guinea, New Guinea; [Africa] Madagascar, South Africa, Réunion Island, Tanaznia, Malawi (Wilbraham, 2015); [Asia] China, Indonesia, Japan, Malaysia, Philippines, Sri Lanka, Vietnam; [America] Brazil, Costa Rica, Cuba, Guatemala, Honduras, Jamaica, Mexico, Venezuela, the Dominican Republic; [other region] St. Helena, Juan Fernández islands.

Specimens examined: **AUSTRALIA**. New South Wales, *W. W. Watts 1048* (H-BR 2521010); Queensland, *Stone19542* (ME2245721), *2050* (MEL 2249823), *22857* (MEL 2264767); Victoria, *H. Streimann 43671* (MO 3962348). **BOLIVIA**. Santa Cruz, *A. Fuentes et al. 1228* (MO 5647251). **BRAZIL**. Minas Gerais, *W. R. Buck 27003* (MO 6092991). **CHINA**. Yunan, *T. Cao & G. Y. Song 060194* (SHTU). Hainan, *W. D. Reese 17645, 17656* (MO), *Z. H. Li 85-47* (H3090673), *P. C. Wu, Wll85353* (H 3090674); Taiwan, *Guo & Cao 120913234, 120913274, 120913304, 120913316, 120913329*. **HAWAII**. Hualalai, *Carl Skottsberg 1386* (H-BR 2518016); E Molokai Mountains, *T. R. Herat 1138.1* (H 3090704); Kilauea, *Carl Skottsberg* 1297 (H-BR 2518014); Kauai, *Carl Skottsberg 1674* (H-BR 2518019), *1601* (H-BR 2518018); Oahu, *Carl Skottsberg* 1537 (H-BR 2518017); Sandwich, Zenselu, *J. Rock* (JE). **HONDURAS**. Atlantida, *Allen 17562* (MO 4461895). **INDONESIA**. Java, Pangerango, *Hochreutiner B. P. G. 931* (G00042116), *Max Fleischer* (H-BR 2520014). **JAMAICA**. Blue Mountain peak, *E. Jadertrolm 58* (H-BR 2512011), **JUAN FERNÁNDEZ**. *Gunther Kunkel 330* (H-BR). **MADAGASCAR**. Antananar, *Lowry II 4382* (MO 6491256). **NEW CALEDONIA**. *Rocx1003* (H-BR 2519001), *Le Rat 1913* (H-BR 2519007). **NEW ZEALAND**. Otago, *Bell (313)117* (H-BR 2516001), *Bell s.n.* (H-BR 2516013, 2516014). **PAPUA NEW GUINEA**. *W. E. trmit s.n.* (H-BR 2520034). **PHILIPPINES**. Prov. Bataan, *Elmer D. Merrill 3558* (H-BR 2513001); Mt. Halcon, Mindoro, *Elmer D. Merrill 5709* (H-BR 2513002). **SRI LANKA (Ceylon)**. Horton Plains, *Thwaites* (H-BR 2513006); *Thwaites 32* (MO); Central Provinces, *J. W. N. Beckett s.n.* (H-BR 2513003). **ST. HELENA**. *seeman 2633* (BM 000982411). **VENEZUELA**. Estado Lara, *Liesner et al. 8103* (MO 2861162). **VIETNAM**. Da Nang, *S. He & K. Nguyen 42599* (MO 6239379); Lam Dong, *S. He & K. Nguyen 42795* (MO 6239378).

139. *Macromitrium minutum* Mitt., J. Linn. Soc., Bot. 13: 303. 1873. (Figure 158)

Type protologue: [Indonesia] Java, in Mont. Megamendong, alt. 4000'-7000', *Motley*. Type citation: Java, 4000'-7000', *Motley* (isotype: E 00165190!).

(1) Plants small, in creeping mats, olive-green above, brownish below; branches slender, innovating, up to 3 cm long. (2) Branch leaves recurved-patent to abaxially curved when dry, spreading when moist, small, about 1.2 mm long, lingulate-lanceolate to lanceolate, subacute at the apex; margins plane; costae strong, pellucid to reddish, most excurrent to a short awn; all cells (except at insertion) quadrate-rotund and pluripapillose, but unipapillose to tuberculate below. (3) Perichaetial leaves not much differentiated from branch leaves, overtopping the vaginula. (4) Setae short, coarsely papillose, only 2 mm long. (5) Capsule urns very short, almost spherical, only 1.0 mm long and wide, wide-mouthed after dehiscence, smooth; peristome single but rather rudimentary, membranous and usually

Taxonomy

caducous in dried material. (6) Calyptrae hairy, especially below, not completely covering the capsule.

Macromitrium minutum Mitt. is similar to *M. orthostichum* Nees ex Schwägr. by its small plants, lingulate-lanceolate branch leaves with subacute apex, pluripapillose upper and medial quadrate lamina cells, the former differs from the latter by its rather small capsules.

Distribution: Indonesia, Malaysia.

Specimens examined: **INDONESIA**. Java, Barggartum, 1450 m, *M. Fleischer s.n.* (H-BR 2506002). **MALAYSIA**. Cameron Highlands, Schafer-Verwimp & Verwimp, Nr. 18908/A (MO).

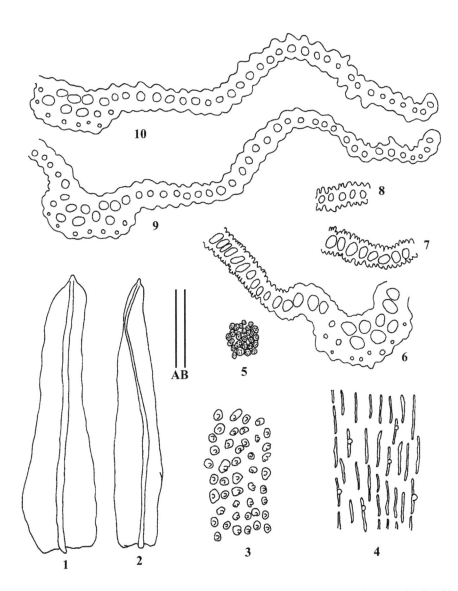

Fig. 155 *Macromitrium menziesii* Müll. Hal. 1-2: Branch leaves. 3: Medial cells of branch leaf. 4: Basal cells of branch leaf. 5: Uppee cells of branch leaf. 6-8: Upper transverse sections of branch leaves. 9-10: Basal transverse sections of branch leaves (all from isotype, H-BR 2558028). Line scales: A = 0.44 mm (1-2); B = 44 μm (3-10).

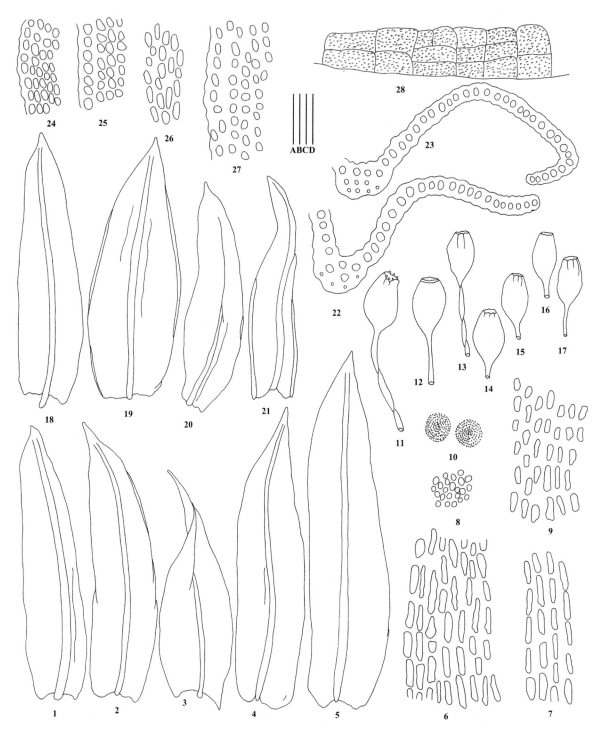

Fig. 156 *Macromitrium microcarpum* Müll. Hal. 1-5, 18-21: Branch leaves. 6-7: Basal cells of branch leaves. 8: Upper cells of branch leaf. 9: Low cells of branch leaf. 10: Spores. 11-17: Capsules. 22: Medial transverse section of branch leaf. 23: Basal transverse section of branch leaf. 24, 27: Upper marginal cells of branch leaves. 25: Medial marginal cells of branch leaf. 26: Medial cells of branch leaf. 28: Peristome (1-4, 6-7, 18-19, 22-27 from isotype, NY 01202050; 5, 15, 9-17, 20-21, 28 from isotype, JE 04008720). Line scales: A = 2 mm (11-17); B = 400 μm (1-5, 18-21); C = 100 μm (10, 28); D = 67 μm (6-9, 22-27).

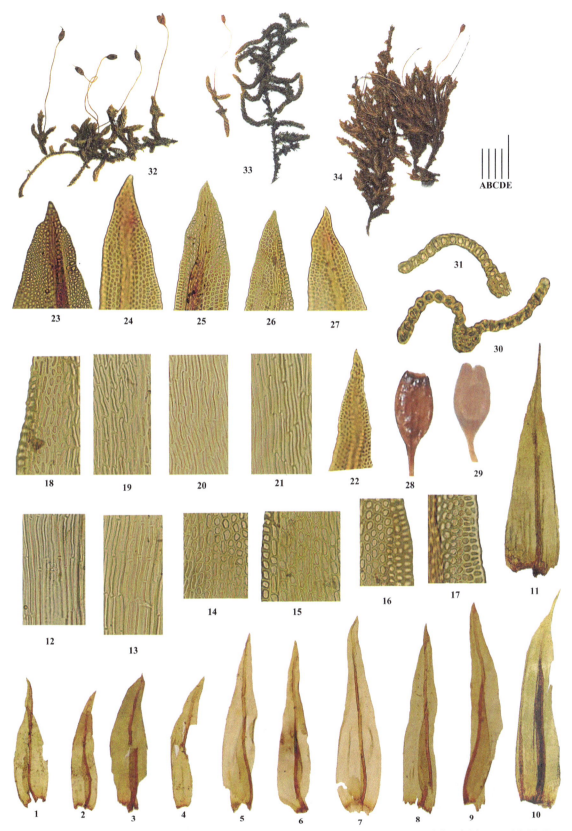

Fig. 157 *Macromitrium microstomum* (Hook. & Grev.) Schwägr. 1-9: Branch leaves. 10-11: Perichaetial leaves. 12-13: Low cells of branch leaves. 14-15: Medial cells of branch leaves. 16-17: Upper cells of branch leaves. 18: Upper cells of perichaetial leaf. 19: Medial cells of perichaetial leaf. 20: Low cells of perichaetial leaf. 21: Basal cells of perichaetial leaf. 22-27: Apices of branch leaves. 28-29: Capsules. 30-31: Medial cross-sections of branch leaves. 32-34: Dry habits (1-4, 23, 32 from isotype of *M. filicaule*, NY 01201996; 5-7, 10, 12, 15, 17, 22, 25-27, 29, 30, 33 from *M. filicaule*, NY 01201997; 8-9, 11, 13, 14, 16, 18-21, 24, 31, 28, 34 from isolectotype of *M. subnitidum*, NY 01086589). Line scales: A = 200 μm (1-11); B = 20 μm (12-21, 30-31), C = 500 μm (28-29), D = 50 μm (22-27), E = 5 mm (32-34). (Yu *et al.*, 2018).

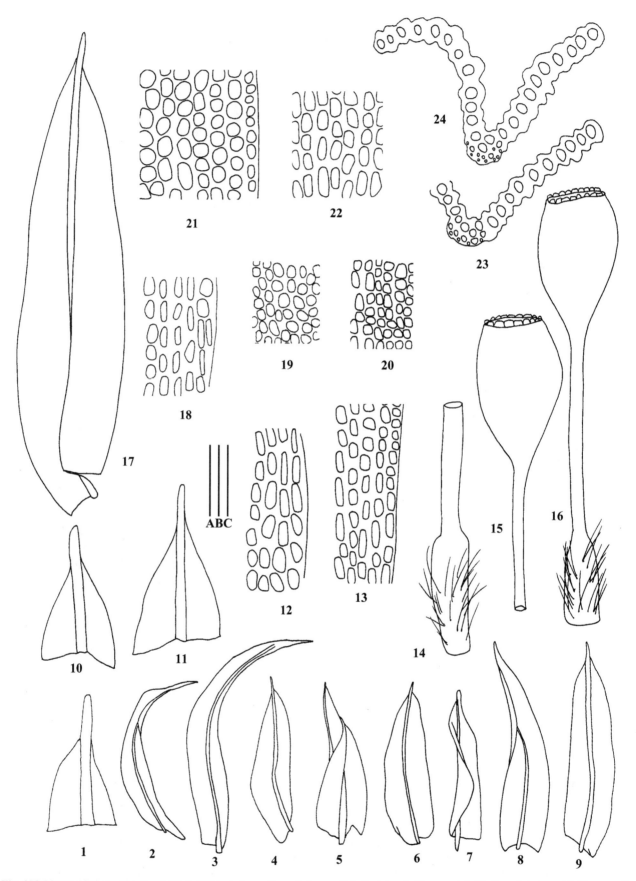

Fig. 158 *Macromitrium minutum* Mitt. 1,10,11: Apices of branch leaves. 2,3: Perichaetial leaves. 4-9,17: Branch leaves. 12: Upper cells near costa of branch leaf. 13: Upper marginal cells of branch leaf. 14: Vaginula. 15-16: Capsules. 18, 22: Basal cells of branch leaves. 19: Upper cells of branch leaf. 20, 21: Medial cells of branch leaves. 23-24: Medial transverse sections of branch leaves (all from isotype, E 00165190). Line scales: A = 0.44 mm (2-9, 14-16); B = 176 μm (10-11, 17); C = 44 μm (12, 13, 18-24).

140. *Macromitrium mittenianum* Steere, Bryologist, 51: 115. 1948. (Figure 159)

Type protologue and citation: [Ecuador] Andes Quitenses, in sylva Canelos (5000 ped.). *Spruce 138* (as *Zygodon crenulatus n. sp*) (isotype: BM 000720546!).

(1) Plants in loosely wefts, brown; primary stems prostrate, densely with rusty-brown rhizoids; secondary stems loosely ascending. (2) Branch and stem leaves are similiar, arranged in five rows around the stem; erecto-patent below, keeled, contorted-twisted and flexuose above, with an incurved apex when dry, widely spreading when moist, broad ligulate to lingulate; apices acute or somewhat mucronate; costae subpercurrent, ending several cells beneath the apex; margins crenulate throughout; all cells rounded and obscure, densely with pluripapillae (similar to those of *Zygodon*) except a few cells near the insertion.

The habits of *M. mittenianum* Steere is rather similar to those of *Zygodon fasciculatus* Mitt., but with creeping stems and dense rusty-brown rhizoids. Mitten (1869) first described this speceis as *Zygodon crenulatus* Mitt. Malta (1926) states definitely that *Zygodon crenulatus* was a member of *Macromitrium*, yet did not make taxonomic treatment. Steere (1948) moved *Z. crenulatus* into *Macromitrium*. Since there already existed a *Macromitrium crenulatum* Hampe (1863), he named the species *M. mittenianum* to dedicate to Mitten for his great contribution to the genus.

Distribution: Ecuador.

141. *Macromitrium moorcroftii* (Hook. & Grev.) Schwägr., Sp. Musc. Frond., Suppl. 2 2(1): 67. pl. 172. 1826. (Figure 160)

Basionym: *Orthotrichum moorcroftii* Hook. & Grev., Edinburg J. Sci. 1: 116. 4. 1824. Type protologue: Hab. Simirague in Nepaul, *Dr. Moorcroft*; communicated by Dr. Wallich. Type citation: Simirague in Nepaul, *Dr. Moorcroft s. n.* (isotype: E 004428995!).

(1) Plants small to medium-sized, forming dense, yellow-brownish mats; stems long creeping, with numerous erect branches, branches up to 10 mm long, with several short branchlets, densely leaved. (2) Stem leaves oblong-lanceolate to ovate-lanceolate, with an obtuse apex, 1.2-1.4 × 0.3-0.4 mm. (3) Branch leaves strongly keeled, individually twisted, contorted-crisped and undulate when dry, spreading or slightly twisted-spreading, adaxially incurved above when moist, yellow-brownish, ligulate to ligulate-lanceolate, 2.1-2.4 × 0.25-0.35 mm, with an acute apex, often slightly incurved at the apex, slightly plicated or plane, sometimes revolute in upper part; costae vanishing 2-3 cells beneath the apex; medial and upper cells inflate, quadrate to rounded-hexagonal, clear and thin-walled, 7-9 µm wide, distinctly pluripapillose; low cells sub-rectangular, 6-8 × 5-6 µm, rather clear, thick-walled, smooth; basal cells near costa larger, pellucid and thin-walled, 16-18 × 6-8 µm, smooth, the basal cells near the margin at one side often differentiated, rectangular, pellucid with thinner walls. (4) Perichaetial leaves up to 2.5 mm long, oblong-lanceolate with the widest at the base, acuminate, often plicate below, narrowly revoluted on both sides, most cells clear and smooth, longer than wide. (5) Setae 8-10 mm long, smooth, purple or brown, twisted to the right. (6) Capsules erect, urns 2.0-2.4 × 0.8-1.00 mm, ellipsoid, ellipsoid-cylindric to cylindric, moderate plicate or furrowed; opercula conic-rostrate, 0.7-0.8 mm long; peristome single, exostome teeth lanceolate and short acuminate, densely papillose throughout; spores minutely papillose to smooth, 20-30 µm in diameter. (7) Calyptrae campanulate, with numerous yellowish or brownish-yellow straight hairs.

Macromitrium moorcroftii (Hook. & Grev.) Schwägr. is similar to *M. japonicum* Dozy & Molk., the former differs from the latter by its long setae (up to 10 mm long), acute apices of branch leaves, and the long mitriform calyptrae with straight brownish hairs.

Macromitrium moorcroftii is rather similar to *M. giraldii* Müll. Hal., but the former could be separated from the latter by its larger calyptrae, longer setae, ligulate to ligulate-lanceolate branch leaves with the apices frequently hidden in a circinate or inrolled cavity.

Distribution: China, Burma, Nepal, Japan, India, Sri Lanka, Thailand.

Specimens examined: **CHINA**. Yunnan, *W. MICH olitz* (H-BR 2585002, H-BR 2582001, PC 0083693, S-B 115573, JE). **BURMA**. Shan State, *R. I. D. Svihla 3766a* (MO 5279127). **INDIA**. Mount. Khasian, *P. Decoly et Schaul, 2437* (JE); Sikkim-Himalana, *P. Decoly et Schaul 2438* (H-BR 2585009); Assam, *Griffith, s. n.* (BM 000852257); Khasia, *Griffith 165* (BM 000919492). **NEPAL**. Nepal, ex hb. Hooker (H-SQL 1353021); *W. A. Weber, B-99436* (MO 3955802); Centroid. **SRI LANKA**. *Th. Herbog* (JE). **THAILAND**. Payap, *A. Touw 8897* (MO 2163757); *A. Touw 8708* (MO 2163754); *A. Touw 9473* (MO 2140560).

142. *Macromitrium mosenii* Broth., Bih. Kongl. Svenska Vetensk.-Akad. Handl. 21 Afd. 3(3): 24. 1895. (Figure 161)

Type protologue: [Brazil] Prov. Minas Geraës, Caldas, ad arborem campi (*Mosén n. 231*) et Prope fonts dictos Pocos de Rio Verde, ad arbores (*n. 238*). Cum fructibus deoperculatis. Type citation: Brasilia, Prov. Minas Geraës, Caldas, Prope fonts dictos Pocos de Rio Varde, ad arbores, 1873, X. 15, leg. *Mosén 238* (syntype: H-BR 2628005!). Prov. Minas Geraes, Caldas, ad arborem campi, 1873, 5/10. Leg. *Mosén 231* (isosyntype: FH 00213651!).

(1) Plants large to medium-sized, forming loose mates; primary and secondary stems creeping, dark, up to 5 cm long, irregularly branched, branches prostrate and ascending, up to 15 mm long and 1.5 mm thick. (2) Stems inconspicuous, triangular lanceolate. (3) Branch leaves contorted-twisted and flexuose above when dry, spreading or slightly abaxially curved spreading when moist, ligulate- to oblong-lanceolate; apices mucronate, acute to broadly acuminate; margins entire below, distinctly and irregularly serrate near the apex; costae sometimes concealed in adaxial view by overlapping folds of laminae; upper and medial cells bulging, subquadrate, hexagonal, oblate, clear and smooth; low and basal cells elongate-rectangular, strongly porose, weakly to moderately tuberculate, marginal enlarged teeth-like cells distinctly differentiated at insertion; basal cells near costa at insertion large, thin-walled, forming a "cancellina region". (4) Perichaetial leaves oblong- to lingulate-lanceolate, shorter than branch leaves, with an acute apex. (5) Setae erect, straight below and twisted to the right above, about 12 mm long. (6) Capsule urns ellipsoid with a small mouth, smooth to plicate when dry; peristome double, the teeth often broke when old, with the low portion remaining and forming two fused low membranes. (7) Calyptrae mitrate, naked and smooth, lacerate below.

Distribution: Brazil.

143. *Macromitrium nanothecium* Müll. Hal. ex Cardot, Hist. Phys. Madagascar, Mousses 237: 1915. (Figure 162)

Type protologue: [Madagascar] Zone supérieure des forêts: Andrangoloakã (Hildebrandt; herb. Levier); Ankeramadinika (R.P. Camboue); plateau d'Ikongo (Dr Besson). Plateau central: pays Betsileo (R.P. Villaume; herb. Meylan). Type citation: Madagascar central, Imerina, Ankeramadinika, 1880, *leg. Hildebrandt s.n.* determ. C. Muller (lectotype designated by Wilbrahm, 2018: PC 0137518!); Madagascar central Imerina, leg. *J. M. Hildbrandt*, E. Levier (isolectotype: PC 0137519!); Madagascar, Betsileo, Imerina centrale, Villame, 1904 (isosyntype: PC 0137520!).

Macromitrium ancistrophyllum Cardot, Hist. Phys. Madagascar, Mousses, 247, 1915, *fide* Wilbraham, 2018. Type protologue: [Madagascar] Zone supérieuredes forêts: Ankeramadinikã (*Borgen, n. 202*; herb. Kiaer, comm. Geheeb.). Type citation: Madagascar, leg. *Borgen 202* (lectotype designated by Wilbraham, 2018: PC 0137485!).

(1) Plants small, stem long creeping, in dense mats, with short and erect branches, branches up to 4 mm tall. (2) Stem leaves abaxially curved, oblong-lanceolate, bluntly acuminate to acute. (3) Branch leaves individually twisted below, keeled, flexuose-crisped above, curved to circinate above when dry, spreading or incurved-spreading above when moist, oblong, ligulate to lanceolate-ligulate, recurved at one side; costae subpercurrent to percurrent, sometimes concealed in adaxial view by overlapping folds of laminae; upper cells relatively large, rounded quadrate, 8-12 μm wide, pluripapillose; medial cells rounded to rounded-quadrate, bulging, unipapillose, becoming elongate towards the base; basal cells long rectangular, strongly thick-walled, lumens linear rectangular, porose, smooth, becoming more elongate towards the margin. (4) Perichaetial leaves slightly differentiated from and shorter than branch leaves, oblong, oblong lingulate, broadly acuminate or acute; costae excurrent to form a short point; all cells longer than wide, smooth. (5) Setae smooth, up to 12 mm long. (6) Capsule urns ovoid, with a large mouth; peristome single, exostome of 16; spores large, up to 55 μm in diameter. (7) Calyptrae mitrate, plicate, lobed at base, sparsely hairy.

Distribution: Madagascar, Tanzania (Wilbraham, 2018).

Specimens examined: **MADAGASCAR**. Ikongo, Besson (PC 0137521, PC 0137522, PC 0137523).

144. *Macromitrium nematosum* E.B. Bartram, J. Washington Acad. Sci. 42(6): 181. 1952. (Figure 163)

Type protologue and citation: [Brazil] Rio Grande do SUl: Estacão São Salvador, ad arborem in silva, alt. 600 m, *A. Sehnem no. 2774* (holotype: FH 00213653!).

(1) Plants medium-sized, dark-brownish, stems weakly creeping. (2) Stem leaves caducous, branches to 30 mm. (3) Branch leaves strongly keeled, undulate and wrinkle, contorted-flexuose-crisped, apices often inrolled when dry, spreading and rugose above when moist, ligulate-lanceolate; margins serrulate near the apex; apices acute to acuminate-acute; costae percurrent; all cells clear, flat and smooth; upper and medial cells isodiametric, quadrate, subquadrate; low and basal cells rectangular, straight-walled, rarely porose; marginal enlarged teeth-like cells at the insertion distinctly differentiated; numerous rusty-reddish rhizoids present at base, and abundant reddish-brown multicellular, branched gemmae on the adaxial surface of branch leaves.

Macromitrium nematosum E.B. Bartram is somewhat similar to *M. argutum* Hampe in lanceolate leaves with quadrate, rounded-quadrate to subquadrate, smooth cells in upper and medial cells. Valente *et al.* (2020) considered *M. nematosum* as a synonym of *M. argutum*. However, *M. nematosum* distinctly differs from *M. argutum* in having leaves with all cells smooth, numerous rhizoids at base and abundant branched gemmae present on the adaxial surface.

Distribution: Brazil.

Taxonomy

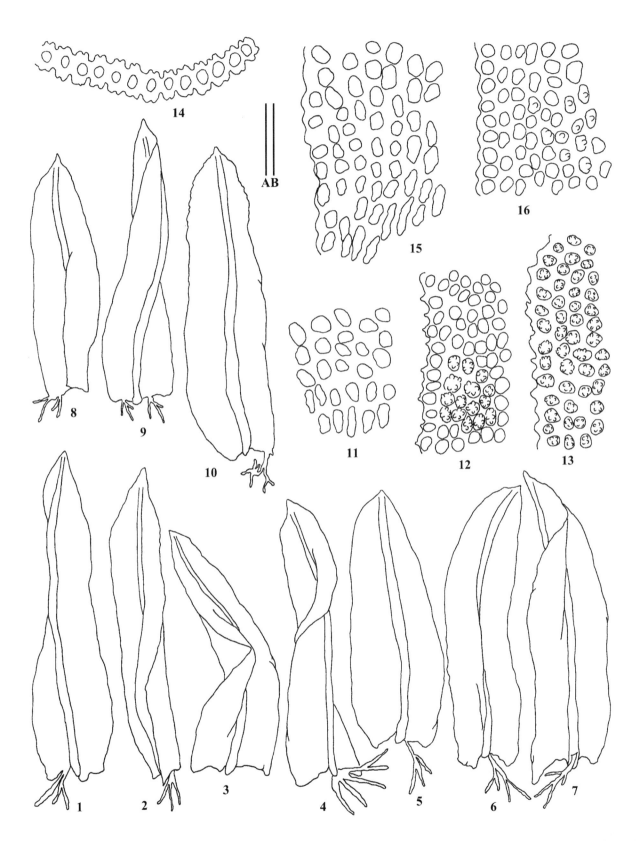

Fig. 159 *Macromitrium mittenianum* Steere 1-10: Branch leaves. 11: Basal cells near costa of branch leaf. 12: Medial cells of branch leaf. 13: Upper cells of branch leaf. 14: Medial transverse section of branch leaf. 15: Basal cells of branch leaf. 16: Low cells of branch leaf (all from isotype, BM 000720546). Line scales: A = 400 μm (1-10); B = 67 μm (11-16).

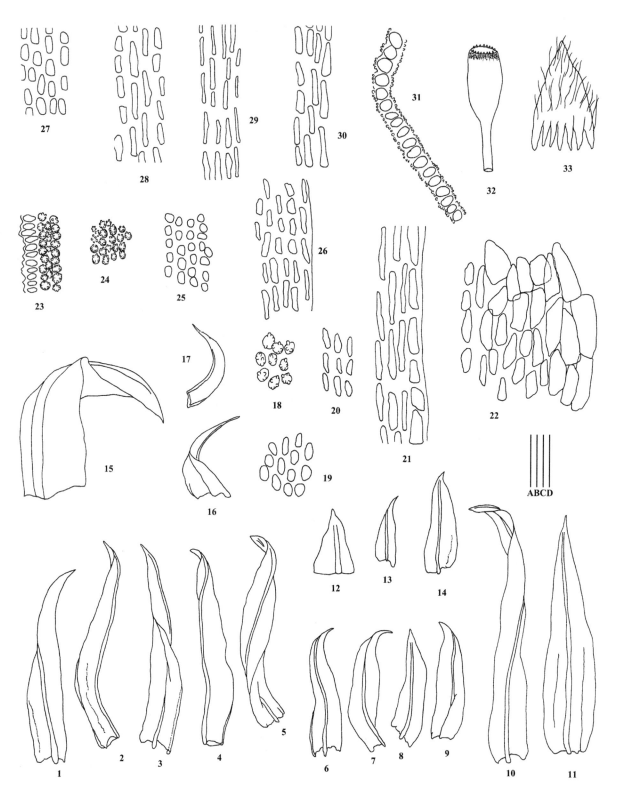

Fig. 160 *Macromitrium moorcroftii* (Hook. & Grev.) Schwägr. 1-10, 13-14: Branch leaves. 11: Perichaetial leaf. 12, 15: Apices of branch leaves. 16-17: Stem leaves. 18: Upper cells of stem leaf. 19: Medial cells of stem leaf. 20: Basal cells of stem leaf. 21, 26: Basal cells of branch leaves. 22: Basal cells near costa of branch leaf. 23: Upper marginal cells of branch leaf. 24: Upper cells of branch leaf. 25: Medial cells of branch leaf. 27: Upper cells of perichaetial leaf. 28: Medial cells of perichaetial leaf. 29: Basal cells of perichaetial leaf. 30: Basal marginal cells of perichaetial leaf. 31: Upper transverse section of branch leaf. 32: Capsule. 33: Calyptra (all from isotype, E 00428995). Line scales: A = 2 mm (32); B = 1 mm (1-11, 13-14, 16-17, 33); C = 400 μm (12, 15); D = 67 μm (18-31).

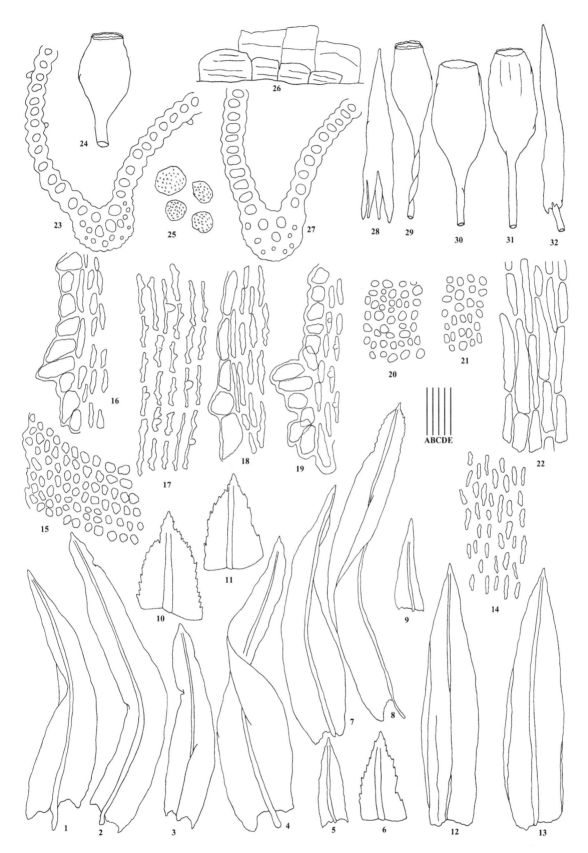

Fig. 161 *Macromitrium mosenii* Broth. 1-4, 7-8: Branch leaves. 5: Stem leaf. 6, 10-11: Apices of branch leaf. 9: Branch leaf in lower part. 12-13: Perichaetial leaves. 14: Medial cells of perichaetial leaf. 15: Upper cells of branch leaf. 16, 18-19: Basal marginal cells of branch leaf. 17: Basal cells of branch leaf. 20: Medial cells of branch leaf. 21: Upper cells of perichaetial leaf. 22: Basal cells of perichaetial leaf. 23: Basal transverse section of branch leaf. 24, 29-32: Capsules. 25: Spores. 26: Peristome. 27: Medial transverse section of branch leaf. 28: Calyptra (all from isosyntype, FH 00213651). Line scales: A = 2 mm (24, 28-32); B = 1 mm (1-5, 7-9, 12-13); C = 400 μm (6, 10-11); D = 100 μm (26); E = 67 μm (14-23, 25, 27).

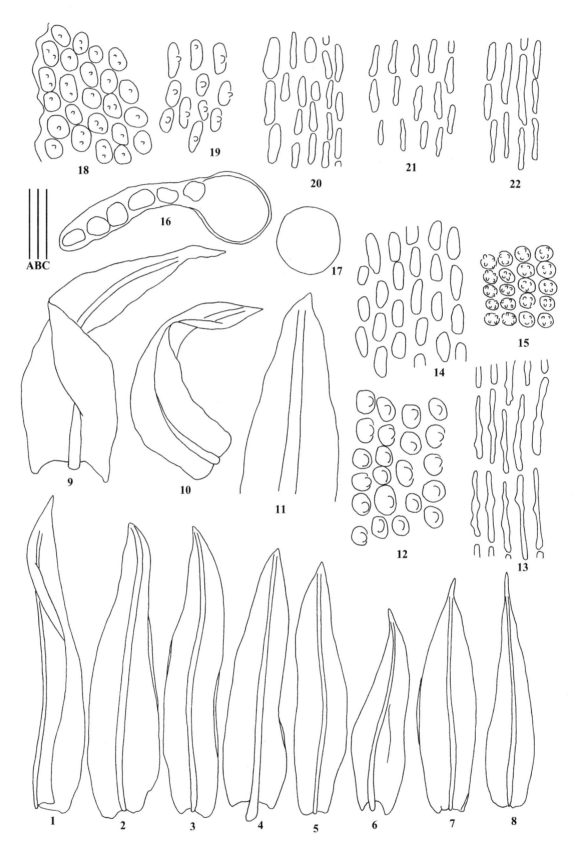

Fig. 162 *Macromitrium nanothecium* Müll. Hal. ex Cardot 1-6: Branch leaves. 7-8: Perichaetial leaves. 9-10: Stem leaves. 11: Apex of branch leaf. 12: Medial cells of branch leaf. 13: Basal cells of branch leaf. 14: Low cells of branch leaf. 15: Upper cells of branch leaf. 16: Protonema. 17: Spore. 18: Upper cells of stem leaf. 19: Medial cells of stem leaf. 20: Basal cells of stem leaf. 21: Upper cells of perichaetial leaf. 22: Low cells of perichaetial leaf (1-8, 11-17, 21-22 from lectotype, PC 0137518; 9, 10, 18-20 from PC 0137521). Line scales: A = 1 mm (1-10); B = 400 μm (11); C = 67 μm (12-22).

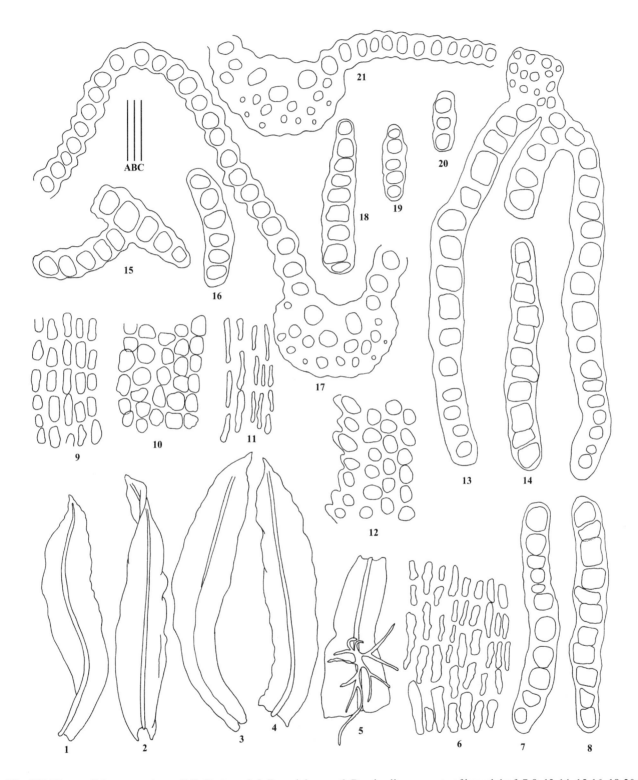

Fig. 163 *Macromitrium nematosum* E.B. Bartram 1-5: Branch leaves. 6: Basal cells near costa of branch leaf. 7-8, 13-14, 15-16, 18-20: Gemmae. 9: Low cells of branch leaf. 10: Medial cells of branch leaf. 11: Basal cells of branch leaf. 12: Upper cells of branch leaf. 17: Low transverse section of branch leaf. 21: Basal transverse section of branch leaf (all from holotype, FH 00213653). Line scales: A = 1 mm (1-5); B = 100 μm (7-8, 13-14, 15-16, 18-20); C = 67 μm (6, 9-12, 17, 21).

145. *Macromitrium nepalense* (Hook. & Grev.) Schwägr., Sp. Musc. Frond., Suppl. 2 2(2): 134. pl. 192. 1827. (Figure 164)

Basionym: *Orthotrichum nepalense* Hook. & Grev. Edinburgh Journal of Science 1: 117. pl. 4. 1824. Type protologue: Hab. Dhoopabasah and Beahico in Nepaul, *Dr. Wallich*. Type citation: Nepal, *Wallich s.n.* (lectotype designated by Guo & He, 2014: BM 000982533!).

Macromitrium incrustatifolium H. Rob., Bryologist 71: 90. f. 25-33. 1968, *fide* Guo & He, 2014. Type protologue and citation: Assam: Lushai Hills, Hmuntha, 5000 ft., on top limb of a forest tree, *Koelz 27473* (US!); Khasi Hills, Mongpoh, 2000 ft., on boulder in open, *Koelz 22682*.

Macromitrium longibrachteatum Dixon, Ann. Bryol. 9: 63. 1937, *fide* Guo & He, 2014. Type protologue: Hab. On fallen branch in evergrfeen forest, circa 400 m., Ta Wieng, Chiengkwang, 5 Apr. 1932 (*494*). Type citation: Ta Wieng, Chiengkwang, about 400 m., 5 Apr. 1932, *A.F.G. Kerr 494* (isotype: BM 000919514!).

(1) Plants small to medium-sized; primary stems long creeping, up to 5 cm long, abundantly covered with brown rhizoids; branches numerous, 5-10 mm long. (2) Stem leaves appressed below, squarrose above when dry, ovate-triangular to oblong-lanceolate, acuminate at the apex; upper leaf cells isodiametric, opaque, bulging, pluripapillose; medial leaf cells obliquely oval, smooth; low leaf cells ovate to irregularly rectangular, transparent and smooth; cells at the insertion elongate, thick-walled and smooth. (3) Branch leaves obliquely and spirally twisted to coiled around the branch, with a hook-like apex when dry, widely spreading when moist, lingulate-lanceolate to ovate-lanceolate, keeled above, slightly plicate below, 1.5-2.3 × 0.3-0.5 mm, acute to obtuse, occasionally mucronate at the apex; margins entire, narrowly recurved on one side below; costa percurrent; irregular 1-3-stratose proliferation with pluripapillose cells on both dorsal and ventral lamina surfaces; cells at the outermost marginal row small and oblate, smooth; low leaf cells oval, oblate, rounded-quadrate, shortly rectangular, strongly thick-walled, smooth or sometimes weakly unipapillose; cells near insertion row elongate, somewhat sinuous, smooth, thick-walled; cells at outermost row of basal leaf margin thin-walled, shortly rectangular. (4) Perichaetial leaves oblong-lanceolate to triangular-lanceolate, distinctly longer than branch leaves, 2.5-3.1 × 0.6-0.7 mm; costae percurrent; upper cells rhombic to shortly rectangular, more or less unipapillose; medial and low cells elongate-rectangular to linear, smooth, somewhat porose. (5) Setae 6-8 mm, smooth, twisted to the left when dry. (6) Capsules erect, urns ellipsoid-cylindric, smooth, 2.2-2.7 × 0.5-0.7 mm; peristome single, with 16 teeth, lanceolate, about 80 μm long, papillose, blunt at apex; spores anisosporous, 14-32 μm in diameter, finely papillose. (7) Calyptrae mitrate, large, up to 5 mm long, deeply lobed at base, cleft on one side, covering the whole capsule, with straight yellowish-brown hairs, moderately dense.

Macromitrium nepalense (Hook. & Grev.) Schwägr. is superficially similar to *M. tosae* Besch. and *M. aurescens* Hampe by having: 1) ligulate-lanceolate to ovate-lanceolate branch leaves with an acute to obtuse apex; 2) often obscure upper leaf cells; 3) oblong-lanceolate to triangular-lanceolate perichaetial leaves; 4) erect, ellipsoid-cylindric, smooth capsules; and 5) large hairy calyptrae, but differs by the presence of multi-stratose upper and medial lamina cells of the branch leaves. In contrast, the upper and medial leaf cells of *Macromitrium tosae* and *M. aurescens* are always uni-stratose.

In *Macromitrium*, *M. tongense* Sull. from Indonesia and *M. serpens* (Burch. ex Hook. & Grev.) Brid. from Africa also have multi-stratose, irregular proliferation of the upper leaf cells that are similar to *M. nepalense*. Both *M. tongense* and *M. serpens* can be distinguished from *M. nepalense* by elongate, curved to sigmoid low lamina cells of their branch leaves.

Distribution: Bhutan, China, India, Laos, Nepal, Sikkim.

Specimens examined: **BHUTAN**. Bhutan, *Griffith s.n* (BM 000876994). **CHINA**. Yunnan, Jinghong Co., *Crosby 14832, 15013, 15029* (all in MO), *Wu 21884* (MO); Menghai Co., *Crosby 15013A* (MO); Mengla Co., *Wang 864122* (MO), *Zhang s.n.* (NY); Mengyang Co., *Xu 6841* (MO). **INDIA**. Assam, Garo Hills, *Markh 5873* (H-BR). **LAOS**. *He 43782* (MO). **NEPAL**. *Wallich, s.n.* (BM 000982523). **SIKKIM**. *Kurz 2187* (H-BR).

146. *Macromitrium nigricans* Mitt., J. Proc. Linn. Soc., Bot. Suppl., 1: 53. 1859. (Figure 165)

Type protologue: [India] Coorg. In arborum ramulis (Herb. Van den Bosch.).

(1) Plants medium-sized to large, in loose mats; stems long creeping, with erect branches, branches simple or with short branchlets above, about 15 mm long and 1.1 mm thick. (2) Stem leaves caducous and inconspicuous. (3) Branch leaves individually contorted-twisted and circinate above when dry, slightly abaxially curved spreading at medial portion, often adaxially incurved and conduplicate above when moist, lanceolate from an oblong low portion, acuminate to the apex; margins entire throughout; costa percurrent and keeled; upper cells subquadrate to rounded-quadrate, pluripapillose; medial cells oblong-rectangular, clear and smooth, cells elongate towards the base, clear and smooth, moderately thick-walled. (4) Setae short, about 5 mm long. (5) Capsule urns ellipsoid and smooth when dry; peristome single with short lanceolate teeth. (6) Calyptrae conic-mitrate, smooth and naked, lobed at the base.

Macromitrium nigricans Mitt. is somewhat similar to *M. giraldii* Müll. Hal., the former could be separated from the latter by its smooth and naked calyptrae, and completely smooth low cells of branch leaves.

Distribution: Sri Lanka.

Specimen examined: **SRI LANKA**. (Ceylon), Suigaucally Malae, *J. W. M. Beckett c191* (H-BR 2630017).

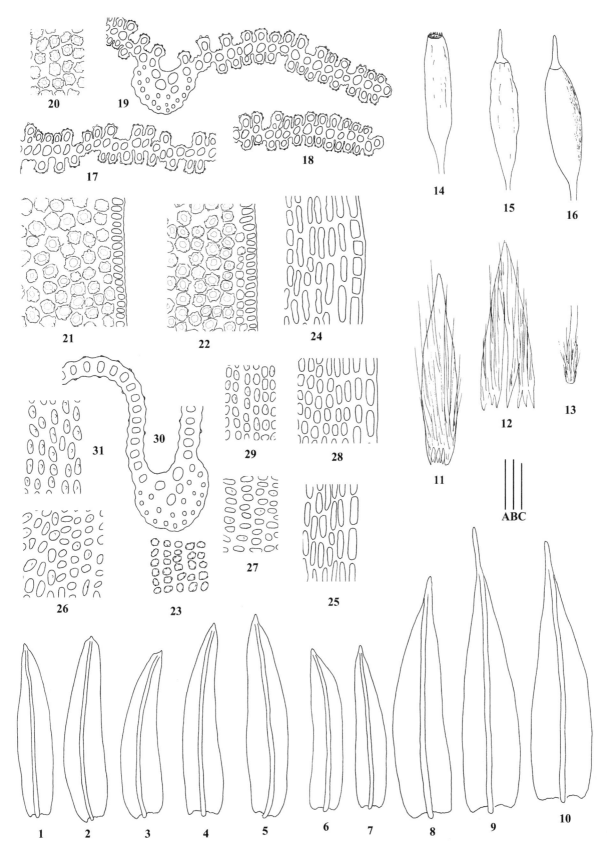

Fig. 164 *Macromitrium nepalense* (Hook. & Grev.) Schwägr. 1-7: Branch leaves. 8-10: Perichaetial leaves. 11-12: Calyptrae. 13: Vaginula. 14-15: Dry capsules. 16: Capsule when moist. 17-19: Upper transverse sections of branch leaves. 20-22: Upper leaf cells of branch leaves. 23, 26-27, 29: Low cells of branch leaves. 24-25, 28: Basal cells of branch leaves. 30: Low transverse section of branch leaf. 31: Upper leaf cells of perichaetial leaf (1-2, 8, 19, 21, 23, 25, 27-31 from the isotype of *M. nepalense*; 3-5, 9, 11-17, 22 from the holotype of *M. incrustatifolium*; 6-7, 10, 18, 20, 31 from the isotype of *M. longibrachteatum*). Line scales: A = 0.5 mm (1-10); B = 1 mm (11-16); C = 34 μm (17-31). (Guo & He, 2014).

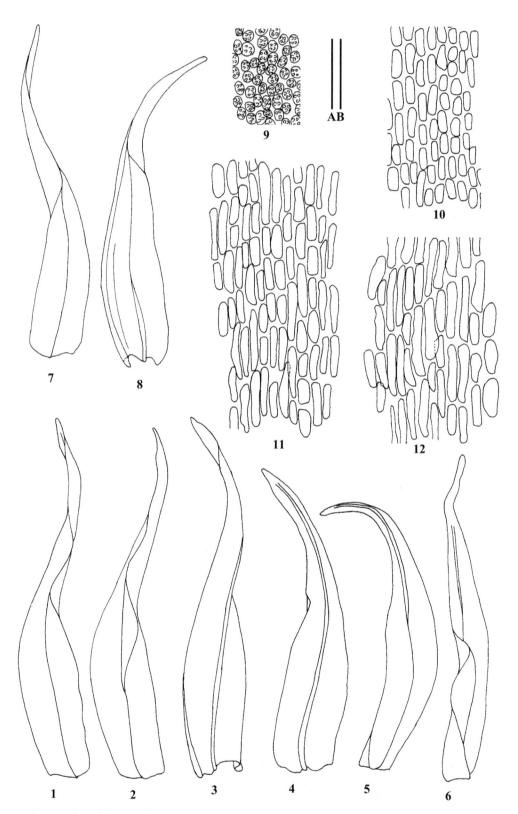

Fig. 165 *Macromitrium nigricans* Mitt. 1-8: Branch leaves. 9: Upper cells of branch leaf. 10: Medial cells of branch leaf. 11: Low cells of branch leaf. 12: Basal cells of branch leaf (all from H-BR 2630017). Line scales: A = 500 μm (1-8); B = 50 μm (9-12).

147. *Macromitrium noguchianum* W. Schultze-Motel, Taxon 11: 180. 1962.

Type: Papua New Guinea, Morobe, Mt. Sarawaket, 11000-12000 ft., 12.V. 1939 Clemens, (holotype: KUMA).
Macromitrium papuanum Nog., J. Hattori Bot. Lab. 10: 14. 1953, *fide* Vitt *et al*., 1995.

(1) Plants large, lustrous; stems creeping, covered with rusty rhizoids, with simple or forked branches, branches long, up to 6 cm. (2) Stem leaves small, inconspicuous, 2-4 mm long, lanceolate, erect, covered by rhizoids. (3) Branch leaves individually twisted and contorted, not funiculate when dry, erect-incurved to stiffly spreading when moist, 3-7 mm long, lanceolate to narrowly lanceolate, tapered to an acuminate apex, keeled; margins plane to broadly reflexed below, erect beneath the apex, entire in low 1/2, irregularly notched or serrulate in upper 1/3 to 1/2; costae subpercurrent to percurrent; all cells longer than wide, thick-walled and distinctly porose; upper cells large, 11-14 × 12-30 µm, rounded-quadrate, rhombic, elongate-oval to long-elliptic, smooth, becoming longer towards the apex and at the margin; medial cells 12-14 × 20-30 µm, rectangular, slightly bulging and smooth; basal cells long rectangular, 7-10 × 30-90 µm, most tuberculate, flat, straight, with irregularly thickened walls. (4) Perichaetial leaves not much differentiated from branch leaves, about 6 mm long, with an acuminate apex. (5) Setae shorter, 3-9 mm long, smooth or roughened above, twisted to the left; vaginulae naked or sparsely hairy. (6) Capsule urns 1.2-1.8 mm long, broadly ovoid to ellipsoid, wide-mouthed, smooth or slightly plicate at the mouth, sharply contracted to the seta through a papillose neck; peristome single, exostome of 16, papillose, erect, well-developed teeth, endostome absent; spores anisosporous, 20-44 µm wide. (7) Calyptrae mitrate, naked or sparsely hairy, slightly lobed at base.

Distribution: Papua New Guinea.

Specimens examined: **PAPUA NEW GUINEA**. Morobe Prov.: Mt. Sarawaket, *D.H. Norris 63293* (KRAM-B-114305). West Sepik Distr., Star Mts, *A. Touw 15255* (MO 5375859); *A. Touw 15922* (H 3205173).

148. *Macromitrium norrisianum* Vitt, Acta Bot. Fenn. 154: 53. f. 2c, 24a-k. 1995.

Type protologue and citation: Papua New Guinea. Southern Highlands Province, Lama Sawmill, 6 km SE of Ialibu, on *Elaeocarpus* trunk in grassland on gentle slope, 6°20'S, 144°01'E, 1 860 m elev., 11. XII.1982, *Streimann 26580* (holotype: H!; isotype: MO 4419069!).

(1) Plant large, dull, coarse, rigid, dark rusty, in loose mats; stems creeping; branches simple, short, about 1.5 cm long. (2) Branch leaves loosely erect to irregularly erect-spreading, with twisted, divergent apices when dry, abaxially curved when moist, 2.2-3.0 mm long, lanceolate, broadly acuminate, recurved; margins plane, distinctly serrate-denticulate in upper half, entire below; costae excurrent or ending just beneath the apex; upper and medial cells rounded to rounded-quadrate (9.0-12 µm), bulging, clear and smooth, with straight walls, marginal cells small and occasionally elliptic-rectangular; low and basal cells in inner portion elliptic-rounded to broadly rectangular, bulging, thick-walled and porose, mostly tuberculate, quickly grading to form a differentiated outer border with 5-10 rows of linear, smooth cells in the low third portion, cells at insertion elongate and smooth. (3) Perichaetial leaves longer than branch leaves, 3.0-3.8 mm long, ovate-lanceolate to narrowly lanceolate, gradually elongate and slenderly acuminate; margins denticulate above; cells longer than wide throughout, smooth. (4) Setae shorter, about 4.5 mm long, prorate, erect, twisted to the left; vaginulae sparsely hairy. (5) Capsule urns 1.5-1.9 mm long, narrowly ellipsoid, narrowed to a small, 4-angled blackish rim; peristome single; exostome a low membrane; spores 16-47 µm wide, anisosporous (Vitt *et al*., 1995).

Macromitrium norrisianum Vitt, *M. megalocladon* M. Fleisch., and *M. ochraceum* (Dozy & Molk.) Müll. Hal. are similar, but *M. norrisianum* can be separated from the latter two species by its branch leaves with a broader of 5-10 rows of linear, smooth cells in the low third of the branch leaf.

Distribution: Papua New Guinea.

149. *Macromitrium nubigenum* Herzog, Biblioth. Bot. 87: 67. 24. 1916. (Figure 166)

Type protologue: [Bolivia] Auf Baumasten an der Waldgrenze des Rio Saujana, about 3400 m, no. 3228; auf Baumästen im Nebelwald über Comarapa, about 2600 m, No. 4310. Type citation: Bolivia, Santa Cruz: Auf Baumästen im Nebelwald über Comarapa, about 2600 m, *Herzog, T. C. J., 4310* (syntypes: JE 04008710!, H-BR 2625011!); Bolivia, La Paz: An der Waldgrenze des Rio Saujana, about 3400 m, *Herzog, T. C. J., 3228* (syntype: JE 04008711!).

(1) Plants robust, green above, brown below; stems creeping, with inconspicuous and caduceus leaves; branches up to 30 mm long, thick. (2) Branch leaves keeled, erect and somewhat clasping at the base, individually twisted-flexuose above when dry, abaxially curved spreading and keeled above, plicate when moist, narrowly lanceolate, acuminate, decurrent at the insertion; margins almost entire except finely serrulate near the apex; costae stout, subpercurrent; upper cells in the apices short to long rectangular; upper cells rounded to rounded-quadrate, or oval-oblong, about 8 µm wide, conic-bulging to mammillose; medial and low cells rectangular, distinctly porose, papillose; basal cells long rectangular, strongly porose and tuberculate; marginal cells enlarged and hyaline at the insertion, but

not teeth-like. (3) Setae papillose, 5-7 mm long, twisted to the left, frequently with a whitish collar; vaginulae hairy. (4) Capsules ellipsoid to ellipsoid-cylindric, distinctly furrowed, slightly or not constricted beneath the mouth; opercula conic-subulate; peristome double, exostome of 16, lanceolate, united at the base, yellowish, endostome hyaline, with basal membrane and upper segments; spores globular, anisosporous, smooth or finely papillose. (5) Calyptrae mitrate, naked.

Macromitrium nubigenum Herzog is similar to *M. longifolium* (Hook.) Brid., but can be separated from the latter in having branch leaves with bulging to mammillose upper cells, and almost entire margins.

Distribution: Bolivia.

150. *Macromitrium oblongum* (Taylor) Spruce, Cat. Musc. 2. 1867. (Figure 167)

Basionym: *Schlotheimia oblonga* Taylor, London J. Bot. 5: 46. 1846. Type protologue and citation: (Ecuador) Andes of Quito, Prof. *William Jameson, n. 322*, 1845 (isotype: BM 000873146!).

(1) Plants robust, yellowish brown; stems weakly creeping, branches to 2-3 cm long, 2 mm thick, with rusty rhizoids. (2) Branch leaves strongly keeled, erect and appressed below, flexuose to spirally contorted and twisted above when dry, erecto-patent, some still twisted when moist, narrowly lanceolate, 4-6 × 0.5-1.0 mm; apices acuminate; margins undulate, crenulate to serrulate above, occasionally distantly and irregular serrate to dentate near the apex; upper and medial interior cells isodiametric to quadrate, short and irregularly rectangular, thick-walled, 6-10 μm wide, smooth to slightly bulging, marginal cells not differentiated; cells gradually elongate to the base; basal cells linear oblong to long rectangular, 30-50 μm, thick-walled and porose, strongly tuberculate; marginal enlarged teeth-like cells at insertion differentiated; costae percurrent to subpercurrent. (3) Perichaetial leaves not much different from branch leaves, but longer than branch leaves, long lanceolate, gradually narrowed to form a long arista; all cells longer than wide, almost not porose. (4) Setae 10-15 mm long, smooth, twisted to the left. (5) Capsule urns 2.0-2.5 mm long, ellipsoid-cylindric, strongly furrowed; peristome double, exostome teeth lanceolate, yellow-orange, striate-papillose, united to form an erect membrane, splitting into 8- 16 teeth when old, endostome hyaline, densely papillose, basal membrane 120 μm high, segments 120-150 μm high; spores anisosporous. (6) Calyptrae mitrate, deeply lacerate to the upper part, smooth and naked (Allen, 2002).

Macromitrium oblongum (Taylor) Spruce is similar to *M. sulcatum* (Hook.) Brid. and *M. fulvum* Mitt., but differs from the latter two species by its branch leaves with marginal enlarged teeth-like cells at the insertion.

Distribution: Bolivia, Colombia, Costa Rica, Ecuador, Panama, Peru (Allen, 2002).

Specimens examined: **BOLIVIA**. Santa Cruz, *S. Churchill et al. 21925* (MO 5926881). **COLOMBIA**. Dept. Sautander, *Pillip R. Smith 18463* (MO 3364994); Departamento de Antioquia, *SP Churchill et al. 15781* (MO 3374495). **PERU**. Prov. Azuay, *W.R. Buck 39317* (MO 6367777), *Eric Rodriguez R. et al. 144A* (MO 6091802).

151. *Macromitrium ochraceoides* Dixon, J. Linn. Soc., Bot., 50: 89. 2 f. 17. 1935. (Figure 168)

Type protologue: [Malaysia] Hab. Between Kamborangah and Pakka, Kinabalu, 2200-3100 m, 13 Nov. 1931; coll. Holttum (25481), type. Ibidem, below Pakka, c. 3100 m., 15 Nov. 1931 (25663).

(1) Plants large, in spreading and loose mats or weft, brownish; stems long creeping, with inconspicuous and caduceus leaves; branches simple to forked. (2) Branch leaves erect below, individually spreading-flexuose above when dry, spreading to abaxially curved when moist, lanceolate from an oblong base, acuminate; margins plane and entire below, occasionally rugose, weakly serrulate near the apex; costae often concealed by overlapping lamina from an adaxial view, long excurrent to form a long, flexuose, hyaline arista; upper cells regularly rounded-quadrate, arranged in longitudinal rows smooth; medial cells slightly elongate, short elliptic-rectangular; basal cells long rectangular to linear, tuberculate, with narrow lumens. (3) Perichaetial leaves slightly differentiated from branch leaves, broadly ligulate-lanceolate; costae long excurrent. (4) Setae 5-7 mm long, papillose, twisted to the left. (5) Capsule urns narrowly ellipsoid, 4-angled in upper half; peristome single. (6) Calyptrae mitrate, scarcely plicate, densely hairy, lobed at the base.

Macromitrium ochraceoides Dixon is similar to *M. ochraceum* (Dozy & Molk.) Müll. Hal., but differs from the latter by its denser habit, stronger and shorter branches; branch leaves much narrower (10 × 0.5 mm) than those of *M. ochraceum*, with very narrow smooth basal cells; upper cells usually isodiametric. The costae in *M. ochraceum* are very shortly excurrent in a reddish cuspidate point, while those of *M. ochraceoides* long-excurrent, forming a much longer, piliferous, flexuose and hyaline arista. The capsules of *M. ochraceoides* are similar to those of *M. ochraceum*, but a little longer and narrower, with a distinctly angled edge.

Distribution: Malaysia, Philippines (www.tropicos.org).

Specimen examined: **MALAYSIA**. Ranau, Kinabalu park, *M. Menzel, J. P. Frahm, W. Frey & H. Kurschner 3935* (MO 3955903).

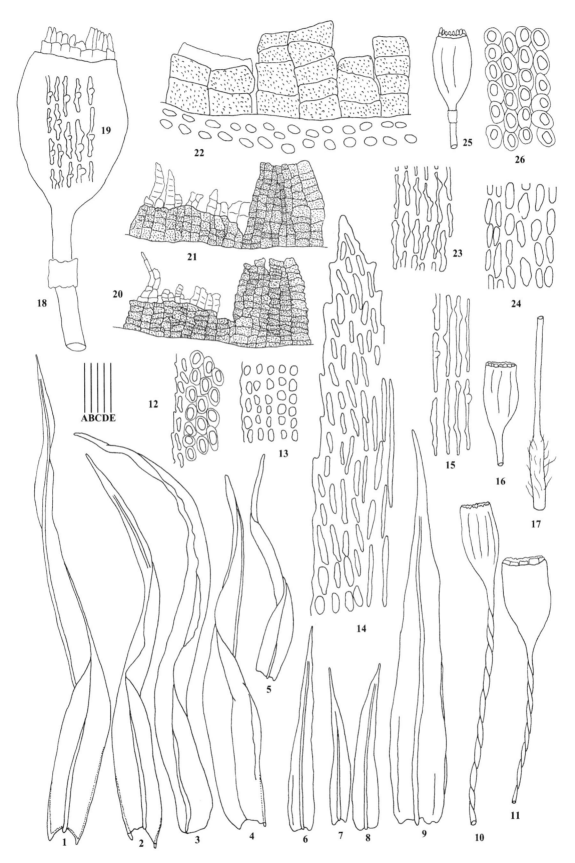

Fig. 166 *Macromitrium nubigenum* Herzog 1-5: Branch leaves. 6-9: Perichaetial leaves. 10-11, 16, 18, 25: Capsules. 12: Upper marginal cells of branch leaf. 13: Medial cells of branch leaf. 14: Apical cells of branch leaf. 15: Low cells of perichaetial leaf. 17: Vaginula. 19: Low cells of branch leaf. 20-22: Peristome. 23: Medial cells of perichaetial leaf. 24: Upper cells of perichaetial leaf. 26: Upper cells of branch leaf (1-17, 19, 22-26 from syntype, JE 04008711; 18, 20-21 from syntype, JE 04008710). Line scales: A = 2 mm (5-8, 10-11, 16-17, 25); B = 1 mm (1-4, 9); C = 400 μm (18, 20-21); D = 100 μm (22); E = 67 μm (12-15, 19, 23-24, 26).

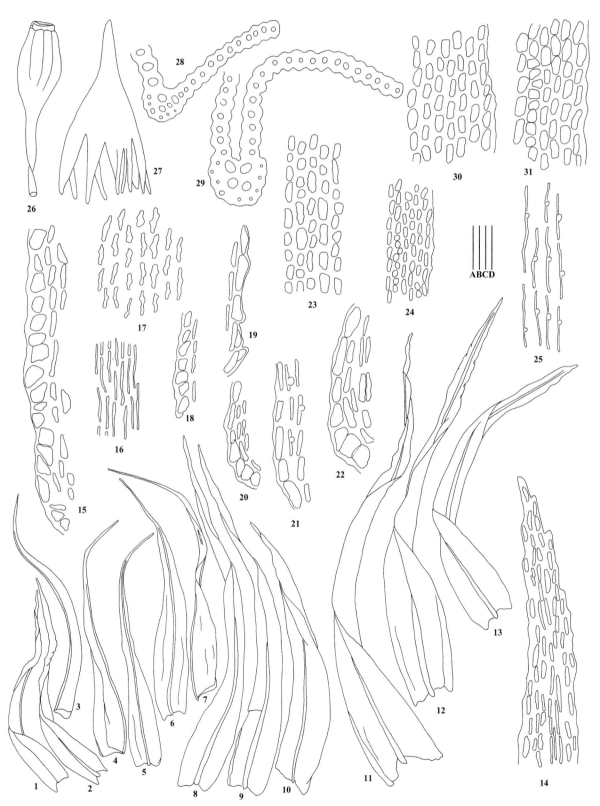

Fig. 167 *Macromitrium oblongum* (Taylor) Spruce 1-3, 8-13: Branch leaves. 4-7: Perichaetial leaves. 14: Apical cells of branch leaf. 15, 18-22: Basal marginal cells of branch leaves. 16: Upper cells of perichaetial leaf. 17: Low cells of branch leaf. 23: Medial cells of branch leaf. 24, 30-31: Upper cells of branch leaves. 25: Basal cells of branch leaf. 26: Capsule. 27: Calyptra. 28: Upper transverse section of branch leaf. 29: Medial transverse section of branch leaf (all from isotype, BM 000873146). Line scales: A = 2 mm (1-7, 26-27); B = 1 mm (8-13); C = 100 μm (14, 20, 24); D = 67 μm (15-19, 21-23, 25, 28-31).

Taxonomy

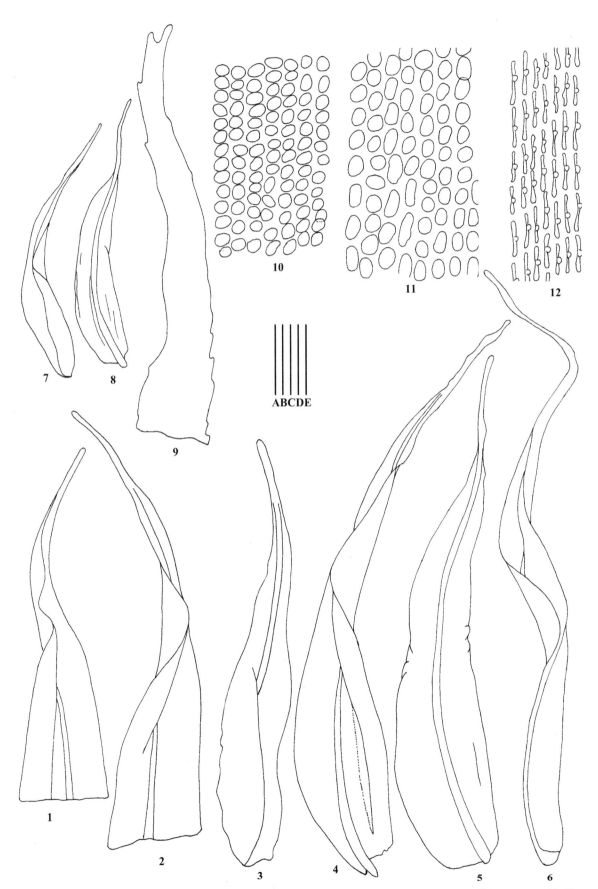

Fig. 168 *Macromitrium ochraceoides* Dixon 1-2: Apices of branch leaves. 3-8: Branch leaves. 9: Long piliferous apex. 10: Upper cells of branch leaf. 11: Medial cells of branch leaf. 12: Basal cells of branch leaf (all from MO 395503). Line scales: A = 1.0 mm (7, 8); B = 500 μm (3-6); C = 200 μm (1, 2); D = 100 μm (9); E = 50 μm (10-12).

152. *Macromitrium ochraceum* (Dozy & Molk.) Müll. Hal., Bot. Zeitung (Berlin) 3: 544. 1845. (Figure 169)
Basionym: *Schlotheimia ochracea* Dozy & Molk., Ann. Sci. Nat., Bot., sér. 3, 2: 314. 1844. TYPE: Indonesia. "Java" (lectotype designated by Vitt et al., 1995: L-Dozy & Molkenboer).
Macromitrium hallieri M. Fleisch. ex Broth., Nat. Pflanzenfam. (ed 2.) 11: 35. 1925, fide Vitt et al., 1995. Type protologue and citation: Indonesia. Borneo, Goenoeng Konopai, 1893/94 *Hallier* (holotype: H-BR 2551012!).
Macromitrium mindanaense Broth., Philipp. J. Sci. 3: 15. 1908, fide Wijk et al., 1964. Type protologue and citation: (Philipines) Mt. Malindang Prov. of Misamis, Mindanoao, coll.: *Maj. E. A. Mearus & W. G. Hutchinson, no. 4794*, May, 1906 (holotype: H-BR 2617022!; isotype: MO!).
Macromitrium rubricuspis Broth., Mitt. Inst. Allg. Bot. Hamburg 7(2): 123. 1928, fide Wijk et al., 1964. Type protologue: West-Borneo: Bukit Raja um 1100 m. (*Hans Winkler n 3146*).
Trichostomum neesii Sande Lac., Bryol. Jav. 2: 226. 1870, fide Wijk et al., 1969. Type protologue: Add. Celebes, herb. Lugd. Bat.

(1) Plants large, stiff, in spreading and loose mats or somewhat wefts; stems creeping, covered by rhizoidal tomentum; branches simple to forked, up to 4 cm long. (2) Branch leaves not very shriveled, densely set, erect below, individually spreading-flexuose above when dry, spreading to abaxially curved when moist, 3.6-4.6 mm long, oblong-lanceolate to ovate-lanceolate, acuminate; margins plane and evenly serrulate above, entire and sometimes broadly reflexed below; costae excurrent and filling the acumen; upper cells 8-11 × 8-20 μm, irregularly rounded to elliptic-rhombic, with irregularly thickened walls, bulging, smooth, intercellular spaces (appearing as 'dots' under a microscope) conspicuous, cells at marginal 2-4 rows often without intercellular spaces forming a differentiated border; medial cells 11-12 × 18-26 μm, irregularly elliptic to long-elliptic, bulging and tuberculate, lumens irregular, with irregular walls; basal cells 8-10 × 40-60 μm, elongate to elongate-rectangular, straight, strong incrassate, porose and distinctly tuberculate, thinner-walled, smooth at the insertion. (3) Perichaetial leaves differentiated, 2.5-4.5 mm long, broadly ovate-lanceolate to oblong, sharply acuminate; costae excurrent. (4) Setae 6-7 mm long, erect-curved, ridged, densely and coarsely prorate, twisted to the left. (5) Capsule urns 2.0-2.2 mm long, narrowly to broadly ellipsoid, 4-angled in upper half, smooth below, sharply contracted to the seta and papillose at the junction; peristome single, exostome fused to a low, pale, coarsely papillose membrane; spores 18-46 μm, distinctly anisosporous. (6) Calyptrae mitrate, scarcely plicate, densely hairy, lobed at the base.

Macromitrium ochraceum (Dozy & Molk.) Müll. Hal. is somewhat similar to *M. crinale* Broth. & Geh., but the latter can be separated from the former by its smooth setae, fusiform capsules, and naked calyptrae.

Distribution: Indonesia, Malaysia, Papua New Guinea, Philippines.

Specimens examined: **INDONESIA**. Java, *M. Fleischer 6* (H-BR 2642004), *A. Zippelius s. n.* (PC0137841), *A. Touw 19916* (MO 6366338); West Borneo, *H. Winkler 3146* (H-BR 2645001). **MALAYSIA**. Sarawak, *A. Touw 20989* (MO 6365235). **PAPUA NEW GUINEA**. Morobe, *H. Streimann 25154* (H 3196899), *H. Streimann 26143* (H 3196898). **PHILIPPINES**. Island of Negros, *A. D. E. Elmer 9592* (H-BR 2642001).

153. *Macromitrium onraedtii* Bizot, Rev. Bryol. Lichénol. 40(2): 123, pl. 14: A, C. 1974. (Figure 170)
Type protologue and citation: Madagascar. Entre Foulpointe et Andondabe, au N de Tamatave, *G. Cremers 2165*. (isotypes: PC 0137528!, PC 0137529!).

(1) Plants robust, in dense mats; stems long creeping, densely with short and erect branches, branches thick, up to 8 mm tall. (2) Stem leaves caducous, appressed below and flexuose above when dry, spreading when moist, lanceolate, keeled, acuminate, with serrulate margins above. (3) Branch leaves, brownish, appressed below, spirally twisted-curved in inconspicuous rows above, somewhat funiculate when dry, widely spreading when moist; oblong-lanceolate, with an acuminate, bluntly acuminate to acute apex; costae subpercurrent, occasionally plicate below; margin plane and entire; upper and medial lamina occasionally and irregularly bistratose to tristratose, uneven on both sides, cells rounded, rounded-quadrate, 7-10 μm wide, obscure and pluripapillose, chlorophyllose; cells becoming elongate towards the base; low and basal cells hyaline, rectangular, thick-walled, tuberculate, with sigmoid lumens. (4) Perichaetial leaves weakly differentiated, slightly shorter but wider than branch leaves, with acuminate-acute apices. (5) Setae smooth, 3-8 mm long, not twisted. (6) Capsule urns ellipsoid-cylindric, brownish, 1.6 × 0.6 mm, smooth to wealy plicate. (7) Calyptrae mitrate, yellow, smooth and naked.

Distribution: Madagascar.

154. *Macromitrium orthophyllum* Mitt., J. Proc. Linn. Soc., Bot 4: 79. 1859. 4: 79. 1859.
Type protologue: *Hab.* New Zealand, *Kerr, knight.*

(1) Plants robust, in spreading mats; stems distinctly prostrate and creeping, with dense erect branches, branches mostly simple, up to 3 cm high. (2) Stem leaves 1.2-1.5 mm long, ovate-lingulate to ovate, sharply narrowed to a long-acuminate, all cells smooth. (3) Branch leaves stiffy, erect-appressed below, nearly straight, non-twisted, but distinctly curved to one side above, forming a somewhat rope-like appearance when dry, widely spreading when

moist, 1.2-2.0 mm long, ligulate to ligulate-lanceolate, with an acute apex; unistratose, all cells flat and smooth; upper and medial cells irregularly small, rounded to isodiametric-quadrate or elliptic; basal cells elongate-rectangular to linear-rectangular, very thick-walled, longer and narrower near the margin, shorter near the costa, basal inner cells with curved to sigmoid lumens. (4) Perichaetial leaves longer than branch leaves, erect-sheathing, oblong to oblong-lanceolate, sharply acuminate to acuminate-long-cuspidate. (5) Setae 6-13 mm long, smooth, strongly twisted to the left. (6) Capsule urnsovoid, ellipsoid to ellipsoid -cylindric to narrowly fusiform, smooth below and 8-plicate and narrowed to a puckered, narrow mouth; peristome single, exostome of 16; spores anisosporous. (7) Calyptrae campanulate, mitrate, smooth and naked, strongly plicate, deeply 2-3 lacerate.

Macromitrium orthphyllum Mitt. is similar to *M. microstomum* (Hook. & Grev.) Schwägr., but the plants of the former are larger, its branch leaves loosely erect-appressed, straight or slightly curved to one side, and spores are anisosporous, while the plants of *M. microstomum* are small to medium-sized, branch leaves strongly twisted-curved, and spores are isosporus. Additionally, the branch leaves of *M. microstomum* regularly twisted forming a distinctly rope-like appearance.

Distribution: Australia, New Zealand (Vitt & Ramsay, 1985).

Specimens examined: **NEW ZEALAND**. North Island, *J.-P. Frahm 21.3.98, no: 35-6* (MO 5135005), *Pull Hill Ocli 80* (H-BR 2517005); Otago, *W. Bell 492* (H-BR 2517008), *316* (H-BR 2517007), *632* (H-BR 2517009); Bank's peninsula, *S. Berggren 1064* (H 3090594, H 3090595); Wairoa County, *Sainsbury, G.O.K s.n.* (MEL 0029450), Canterbury, *Beckett, T.W.N. s.n.* (MEL 1030346, MEL 1030347).

155. *Macromitrium orthostichum* Nees ex Schwägr., Sp. Musc. Frond., Suppl. 4 316. 1842. (Figure 171)

Type protologue: (Indonesia) In Java a Blumio lectum misit ill. Nees ab Esenbeck. Type citation: Java, *Nees s.n.* (isotypes: BM 000852247!, BM 000852248!); Java, marked as "*Macromitrium orthostichon*, n. sp" (isotype: BM 000852245!, BM 000852246!); Java, NY 00518229!, Java, Nees ab Esenbeck, 1830 (isotype: E 00165197!).

Macromitrium appressifolium Mitt., J. Linn. Soc., Bot. 13: 302. 1873, *fide* Vitt et al., 1995. *fide* Vitt et al., 1995. Type protologue: (Indonesia) Java (ex herb. Dozy et Molkenboer).

Macromitrium orthostichum subsp. *appressifolium* (Mitt.) M. Fleish., Musci Buitenzorg 2: 413. 1904.

Macromitrium fragilifolium Dixon, Ann. Bryol. 5: 28. 1932, *fide* Eddy, 1996.

Macromitrium seminudum Thwaites & Mitt., J. Linn. Soc., Bot. 13: 303. 1873. *fide* Vitt et al., 1995. Type protologue: (Sri. Lanka) Ins. Ceylon, *Dr. Thwaites*. Type citation: Ceylon, *Thwaites 43* (holotype: NY-Mitten!; isotype: BM 000919510!).

Macromitrium orthostichum subsp. *seminudum* (Thwaites & Mitt.) M. Fleish., Musci Buitennzorg 2: 412. 1904.

Macromitrium scleropodium Besch., Ann. Sci. Nat., Bot., sér. 6, 9: 357. 1880, *fide* Willbraham, 2008. Type protologue: La Réunion: Plaine des Palmistes, Sainte-Agathe, 9 Juillet 1875, *G. De L'Isle, 289*. Type citation: Réunio, 1875, *L'Isle 289* (holotype: BM 000982420!).

(1) Plants small, dull, in sparse to dense mats; stems creeping and prostrate, with numerous rhizoids and erect-ascending branches, branches up to 5 cm long. (2) Stem and branch leaves similar, stiffly appressed to erect below, loosely spirally arranged around the branch, individually flexuose-twisted with divergent to spreading-deflexed upper portion when dry, erect-spreading and spirally when moist, 0.9-1.7 mm long, broadly ligulate to ligulate-lanceolate; apices acuminate-apiculate, mucronate or broadly acute; margins plane, crenulate due to bulging of marginal cells; costae shortly excurrent to rarely percurrent; upper and medial cells irregularly rounded to rounded-quadrate, isodiametric, bulging, obscure, densely pluripapillose, conic or branched; basal cells 9-12 ×7-14 µm long, quadrate, rounded-quadrate to short-elliptic, clear, bulging, conic-tuberculate, small at the margin, those at the insertion irregularly thickened, rectangular, straight, basal marginal cells sometimes with adventitious, branched, papillose rhizoids, short-elliptic cells confined a small area near the base. (3) Perichaetial leaves similar to branch leaves. (4) Setae short, 2-6 mm long, distinctly papillose, twisted to the left. (5) Capsule urns 0.9-1.5 × 0.5-0.7 mm, obovoid to ellipsoid, smooth, 4-angled to rounded near rim, narrowed to mouth; peristome single; exostome a low papillose to nearly smooth membrane; spores 14-35 µm, distinctly anisosporous. (6) Calyptrae conic-mitrate, sometimes plicate, with dense and papillose hairs, lobed at base.

Macromitrium orthostichum Nees ex Schwägr., *M. erubescens* E.B. Bartram, and *M. angulatum* Mitt. are similar by their branched, papillose rhizoids attached to the basal 1-3 marginal leaf cells, most cells isodiametric except those near the base, prorate setae, and hairy calyptrae. *M. erubescens* can be separated from the other two species by its larger plants and longer branch leaves, and conic-bulging, unipapillose upper lamina cells. From *M. orthostichum*, *M. angulatum* can be distinguished by its calyptrae with hairs restricted to the basal area.

Macromitrium orthostichum includes three varieties:

Macromitrium orthostichum subsp. ***micropoma*** M. Fleisch., Musci Buitenzorg 2: 409, 414. 75. 1904.

Type protologue: An Rinde von Baumfarnen. West-Java: im Berggarten von Tjibodas an Alsophila 1450 m (F.) (isotype: JE 04008736!).

The variety mainly differed from var. *orthostichum* by its rather small capsules.

Macromitrium orthostichum var. ***burgeffii*** Herzog, Ann. Bryol. 5: 70. 1932.
    Type protologue and citation: Java: Tjibodas, untere Zone, n. *8126* (holotype: JE 04008739!).
    The variety differed from var. *orthostichum* by its leaves with recurved margins and elongated basal cells at auricles.

Macromitrium orthostichum var. ***siccosquarrosum*** Herzog, Ann. Bryol. 5: 70. 1932.
    Type protologue and citation: Philippinesn: haights places, leg. *Burgeff, n. 8082*. (holotype: JE 04008740!).
    The variety differed from var. *orthostichum* by its abaxially curved leaves when dry.

Distribution: (Africa) Cameroon, Gabon, Madagascar, Uganda, Zimbabwe, Democratic Republic of the Congo, Reunion, Uganda (Wilbraham, 2008); (Asia and Ocean) Indonesia, Malaysia, Papua New Guinea, Philippines, Solomon, Sri Lanka, Tahiti, Thailand (Vitt *et al.*, 1995).
    Specimens examined: **GUINEA**. Morobe Province, *H. Streimann 13382* (H 3196862), *17220* (H 3196863), *17471* (H 3196860), *13440* (KRAM-B-139868). **INDONESIA**. Celebes, O. *Warburg s.n.* (H-BR 2507015); Java, *Fleischer s. n.* (JE 04008736); West-Java, *Fleischer, s. n.* (JE 04008738); Java, *anonym s.n* (BM 000852242). **PAPUA NEW PHILIPPINES**. Haights places, *H. Burgeff. 8082* (JE 04008740). **RÉUNION**. *Georges B. De L'Isle, s.n.* (BM 000982419); *Georges B. De L'Isle, 289* (BM 000982420). **SOLOMON**. Manighai, *D. H. Norris & O. L. Roberts 49088* (H 3196858).

156. *Macromitrium osculatianum* De Not., Mem. Reale Accad. Sci. Torino, ser. 2, 18: 449, 9. 1859. (Figure 172)
    Type protologue: [Colombia] ad flumen Napo in Columbiae meridionalis, *Osculati s.n.* Type citation: [Colombia] Ad H. Napo Columbiae, leg. *Osculati s.n.* (lectotype designated by Florschütz-de & Florschütz, 1979: NY 01202207!; isolectotype: S-B 164387!).
    (1) Plants large and robust, brown-red; stems weakly creeping, densely with rusty rhizoids; branches erect, up to 15 mm long, often simple, occasionally with branchlets. (2) Branch leaves stiffy erect and appressed below, individually curved to twisted and flexuose above when dry, spreading or slightly abaxially curved, often conduplicate above; lanceolate from a broadly oblong base; margins distinctly serrulate above and entire below; the outmost marginal cells smaller than their ambient interior cells, forming a distinctly differentiated border; upper and medial cells small, rounded, subquadrate, smooth and clear; low cells elongate-rectangular, rather thick-walled and porose, tuberculate; basal cells narrowly rectangular, thick-walled, smooth; marginal enlarged teeth-like cells differentiated at the insertion; costa percurrent and keeled. (3) Perichaetial leaves broadly lanceolate, long acuminate and plicate below. (4) Setae smooth, up to 25 mm long. (5) Capsule urns elliposid to cylindric, strongly furrowed when dry; peristome double, exotome teeth short lanceolate with an obtuse apex; endostome lanceolate. (6) Calyptrae naked and smooth, mitrate, lacerate below.
    Distribution: Colombia.

157. *Macromitrium ousiense* Broth. & Paris, Rev. Bryol. 37: 2. 1910. (Figure 173)
    Type protologue and citation: [China] Ou Si, 16. 2. 1909. Leg. *R. P. Courtois* (isotype: PC 0083704!).
    Macromitrium heterodictyon Dixon, Hong Kong Naturalist, Suppl. 2: 12. 6. 1933, *fide* Yu *et al.*, 2013. Type protologue and citation: China, Hongkong, on Granite rock, 100-200 ft. alt., Amoy Island, coll.*G. A. C. Herklots (B.17 E)*, 11 July, 1931 (holotype: BM 000576133!); China, Hongkong, coll.: *Ah Nin (B.21 E)*, 12th July, 1931 (paratype: BM 000576132!); China, Hongkong, Granite rock, 1,000-1,500 ft. alt., Shaukiwan, Coll. *Ah Nin (H. 4I)*, 5 October, 1931 (paratype: BM 000576131!); China, Hongkong, Lan Yau Peak, Lan Tau I., Coll. *Herklots (359)* (paratype: BM 000576126!).
    (1) Plants small to medium-sized, forming yellowish brown mats; stems long creeping, up to 4 cm long, with erect and short branches, branches 5-6 mm long and 1.0-1.3 mm thick. (2) Branch leaves strongly twisted-contorted-crisped when dry, widely spreading when moist, entire, long ligulate or ligulate-lanceolate, 2.5-3.0 × 0.4-0.5 mm; apices acute or acuminate, somewhat incurved; margins entire, usually plane on one side and recurved on the other, slightly plicate below; costae single, keeled, prominent, ending beneath the apex or rarely in the apex; all cells unistratose, clear and smooth, moderately thick-walled; upper and medial cells small and clear, in diagonal rows, subquadrate-rounded to rounded-quadrate, 5-7 µm wide, gradually elongate from low part to base; basal cells along the costa rectangular, thin-walled, 7.5-12.5 × 20-30 µm, appearing as a "cancellina region", those near the margin elongate, rectangular to sublinear, 17.5-42 × 5-10 µm, thick-walled and slightly sinuous. (3) Perichaetial leaves similar to branch leaves, 2.5-3.0 × 0.5-0.6 mm, ovate-oblong, acuminate; the costa smooth, ending beneath the apex; cells quadrate or subquadrate, smooth and clear; paraphyses numerous. (4) Setae 2-3 mm, smooth, twisted to the

right. (5) Capsules ellipsoid-cylindric or ellipsoid, 1.6-2.0 × 0.6-0.9 mm, brown, constricted or slightly constricted below the mouth, forming a 4-angled or 4-furrowed shape, peristome absent. (6) Calyptrae large, campanulate, with many long, brown-yellowish hairs.

Macromitrium ousinese Broth. & Paris is somewhat similar to *M. schmidii* Müll. Hal., but differs from the latter by its smooth and clear cells from top to base, and capsules constricted below the mouth to form a 4-angled or 4-furrowed shape.

Distribution: China.

Specimens examined: **CHINA**. Anhui, *Z. L. Wan & J. X. Luo 9290* (MO 5276976), *9265* (MO 5276974), *9138* (MO 5363279); Fujian, *H. H. Chung. B 205* (FH), *H. H. Chung* (PC 0083705); Guangdong, *R. magill, P. Redfearn & M. Crobsy 8145* (MO 3983361). JAPAN, *Siebold P. F. von, s. n.* (PC 0083683).

158. *Macromitrium ovale* Mitt., J. Linn. Soc., Bot. 12: 209. 1869. (Figure 174)

Type protologue: [Ecuador] Hab. Andes Quitense, in sylva Canelos (3000 ped.), *Spruce, n. 98*. Type citation: Canelos, *Spruce s.n.* (isotypes: NY 01086487!, NY 01086488!); Ecuador, Andes Quitense, in wood Canelos, 3000 ft. *R. Spruce 98* (isotypes: E 00165156!, E 00165158!).

(1) Plants large, lustrous, yellowish-green to yellowish brown above, dark-brown below; stems creeping; branches to 2.5-3.0 cm long, thick. (2) Branch leaves loosely erect below, individually twisted, flexuose or contorted-crisped above when dry, abaxially curved spreading and longitudinally wrinkled when moist, lanceolate from an oblong low part, upper margins (1/3 to 2/5 of the leaf) distinctly and irregularly dentate; upper cells isodiametric, round-quadrate, 5-8 μm wide, flat, clear and smooth, some cells near the apices elongate; cells elongate from the medial part to the base; low and basal cells long or linear rectangular, 5-8 μm long, thick-walled, distinctly porose, tuberculate; marginal basal enlarged cells present but not teeth-like. (3) Perichaetial leaves are similar to branch leaves, lanceolate. (4) Setae smooth, 6.0-8.0 mm long. (5) Capsule urns ellipsoid, smooth or wrinkled; peristome double, exostome lanceolate, the apices truncate, united to a membrane, yellowish brown, striate-papillose; endstome hyaline, membranous. (6) Calyptrae mitrate, lacerate to the medial or upper part, smooth and naked.

Macromitrium ovale Mitt. is similar to *M. oblongum* (Taylor) Spruce, but differs from the latter in having smooth or wrinkly capsules, branch leaves without enlarged teeth-like cells at the insertion, and with distinctly and irregularly dentate margins in the upper portion.

Distribution: Colombia, Ecuador.

Specimens examined: **COLOMBIA**. Antioquia, *John James Pipoly, III 17301* (MO); Nariño, *Bernardo R. Ramírez Padilla, 4140* (MO).

159. *Macromitrium pallidum* (P. Beauv.) Wijk & Margad., Taxon 9: 190, 1960. (Figure 175)

Basionym: [Mauritius] *Orthotrichum pallidum* P. Beauv., Prodr. Aethéogam., 81: 1805. Type protologue: Ile de Bourbon; communique par M. Bory-St-Vincent. Communique par M. Dupetit-Thouars qui la recueillie aux Iles de France et de Bourbon. Type citation: Ile de Bourbon, Bory-St-Vincent (isosyntype: JE 00428860!).

Macromitrium voeltzkowii Broth., Reise Ostafr., Syst. Arbeit. 3: 55. 1908, *fide* Wilbraham & Ellis 2010. Type protologue: Mauritius [*Voeltzkow*]. Type citation. Mauritius, *Voeltzkow s.n.* (holotype: H-BR 2616005!).

*Lasia acicularis Spreng., Nomencl. Bot. 2: 231, 1824, *fide* Wijk *et al.*, 1964.

**Leiotheca acicularis* Brid., Bryol. Univ. 1: 730, 1826, *fide* Wijk *et al.*, 1964.

**Macromitrium aciculare* Brid., Muscol. Recent. Suppl. 4: 132. 1819 [1818], *fide* Crosby *et al.*, 1983.

**Orthotrichum aciculare* Hook. & Grev., Edinburgh J. Sci. 1: 114. 1824, *fide* O'Shea, 2006.

Orthotrichum breve P. Beauv., Prodr. Aethéogam. 80. 1805, *fide* Wijk *et al.*, 1964.

***Orthotrichum laeve* P. Beauv. ex Schrad., Neues J. Bot. 2(3): 325. 1807 [1808], *fide* O'Shea, 2006.

***Pterigynandrum aciculare* Schwägr., Nomencl. Bot. 2: 355. 1824, *fide* O'Shea, 2006.

**Schlotheimia acicularis* Brid., Muscol. Recent. Suppl. 2: 21. 1812, *fide* Wijk *et al.*, 1967.

Trichostomum arbustorum Brid., Muscol. Recent. Suppl., 1: 241. 1806, *fide* Wijk *et al.*, 1969.

(1) Plants small to medium-sized; stems long creeping, densely with short branches, branches shorter than 1.0 cm long. (2) Branch leaves keeled, erect and appressed below, curved to twisted-contorted above when dry, widely spreading when moist, ligulate-lanceolate, with an acute apex; margins entire throughout; costae subpercurrent; upper lamina occasionally bistratose, cells rounded, subquadrate, densely pluripapillose; basal cells rectangular, lumens curved to sigmoid, smooth to unipapillose. (3) Perichaetial leaves ovate-lanceolate, with bluntly acute the apices. (4) Setae highly varies in length, up to 15 mm long, smooth. (5) Capsule urns ovoid, smooth to weakly wrinkled; opercula long conic-rostrate; peristome absent. (6) Calyptrae mitrate, smooth and naked, lacerate below.

Distribution: Mauritius.

Specimens examined: **MAURITIUS**. Iles de France Dupetit-Thouars, 1806 (BM 000893859, BM 000982423, BM 000982424).

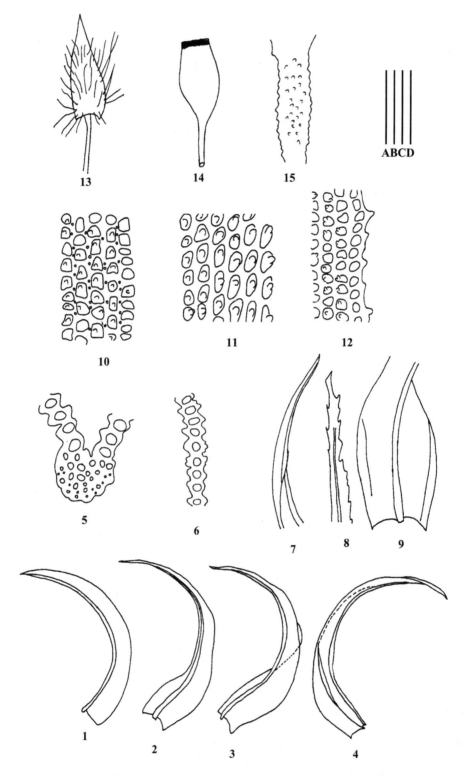

Fig. 169 *Macromitrium ochraceum* (Dozy & Molk.) Müll. Hal. 1-4: Branch leaves. 5, 6: Transverse sections of branch leaves. 7, 8: Apices of branch leaves. 9: Base of branch leaf. 10: Upper cells of branch leaf. 11: Medial cells of branch leaf. 12: Basal cells of branch leaf. 13: Calyptra. 14: Capsules. 15: Seta (all from holotype of *M. mindanaense*, H-BR 2617022). Line scales: A = 2 mm (13, 14); B = 1 mm (1-4); C = 400 μm (7-9, 15); D = 100 μm (5, 6, 10-12).

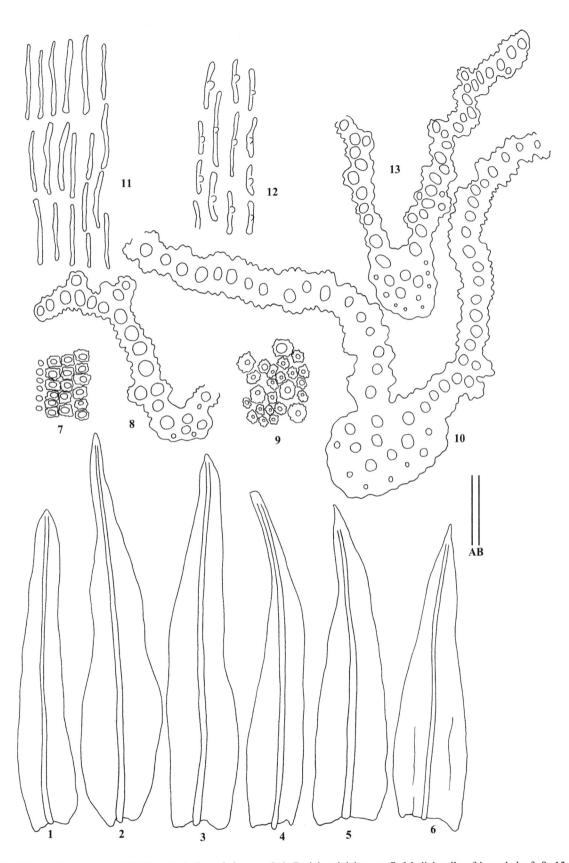

Fig. 170 *Macromitrium onraedtii* Bizot 1-4: Branch leaves. 5-6: Perichaetial leaves. 7: Medial cells of branch leaf. 8, 13: Upper transverse sections of branch leaves. 9: Upper cells of branch leaf. 10: Medial transverse section of branch leaf. 11: Basal cells of branch leaf. 12: Low cells of branch leaf (1-12 from isotype, PC 0137529; 13 from isotype, PC 0137528). Line scales: A = 1 mm (1-6); B = 67 μm (7-13).

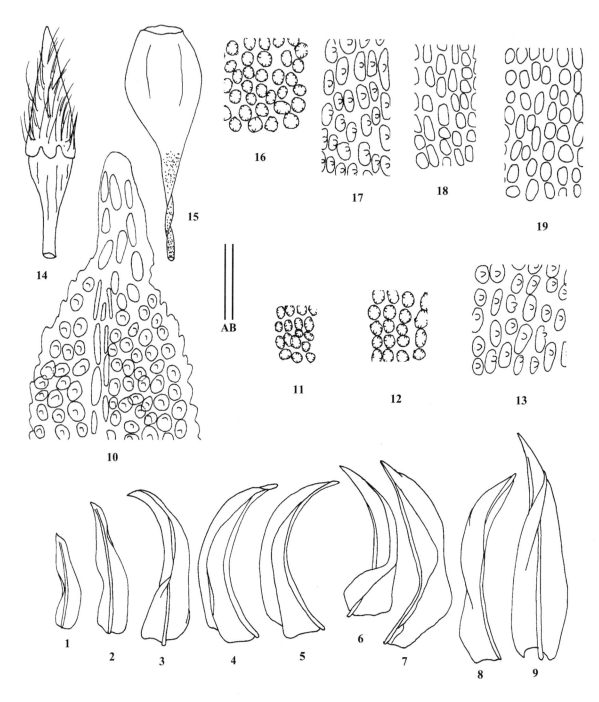

Fig. 171 *Macromitrium orthostichum* Nees ex Schwägr. 1-9: Branch leaves. 10: Apical cells of branch leaf. 11: Upper cells of branch leaf. 12, 16: Medial cells of branch leaves. 13, 17: Low cells of branch leaves. 14: Calyptra. 15: Capsule. 18: Basal marginal cells of branch leaf. 19: Basal cells of branch leaf (all from isotype, BM 000852248). Line scales: A = 0.44 mm (1-9, 14,15); B = 44 μm (10-13, 16-19).

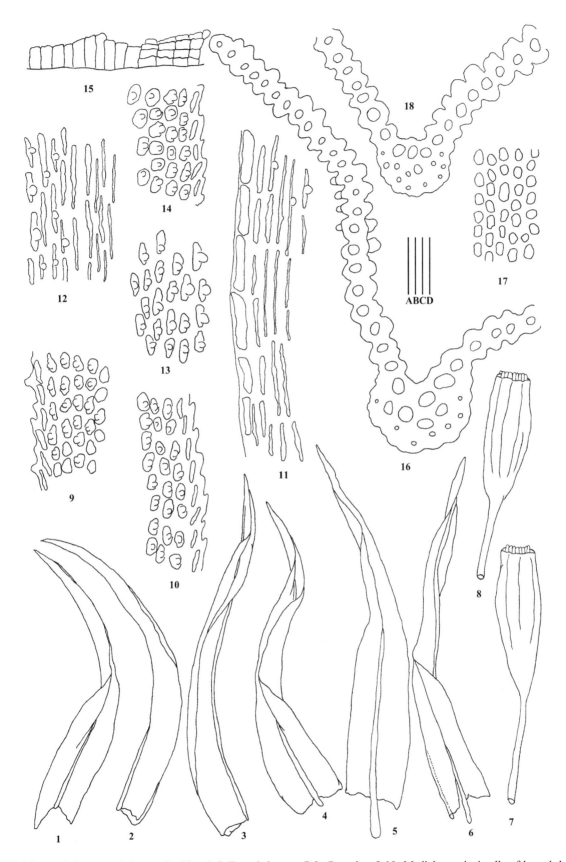

Fig. 172 *Macromitrium osculatianum* De Not. 1-6: Branch leaves. 7-8: Capsules. 9-10: Medial marginal cells of branch leaves. 11: Basal cells of branch leaf. 12: Basal cells near costa of branch leaf. 13: Low cells of branch leaf. 14: Upper cells of branch leaf. 15: Peristome. 16: Basal transverse section of branch leaf. 17: Medial cells of branch leaf. 18: Medial transverse section of branch leaf (1-4, 7-9, 11-18 from lectotype, NY 01202207; 5-6, 10 from isolectotype, S-B 164387). Line scales: A = 2 mm (7-8); B = 1 mm (1-6); C = 400 μm (15); D = 67 μm (9-14, 16-18).

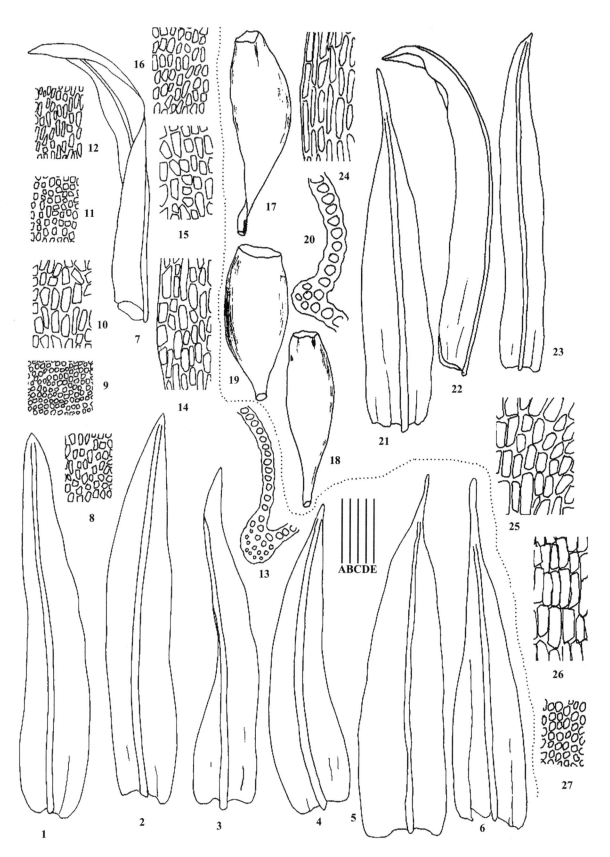

Fig. 173 *Macromitrium ousiense* Broth. & Paris 1-3, 7, 15-16: Branch leaves. 4-6, 14: Perichaetial leaves. 8: Upper cells of branch leaf. 9, 27: Upper cells of branch leaves. 10: Low cells near costa of branch leaf. 11: Medial cells of branch leaf. 12: Low cells of perichaetial leaf. 13, 20: Upper transverse sections of branch leaves. 17, 18: Dry capsules. 19: Wet capsule. 14, 24: Low marginal cells of branch leaves. 15, 25, 26: Low cells near costa of perichaetial leaves. 16: Medial cells of perichaetial leaf (1-16 from isotype of *M. ousiense*, PC 0083704; 17-27 from holotype of *M. heterodictyon*, BM 000576133). Line scales: A = 0.50 mm (1-7); B = 67 μm (8-13, 14-16); C = 0.50 mm (21-23); D = 50 μm (20, 24-27); E= 0.80 mm (17-19). (Yu *et al.*, 2013).

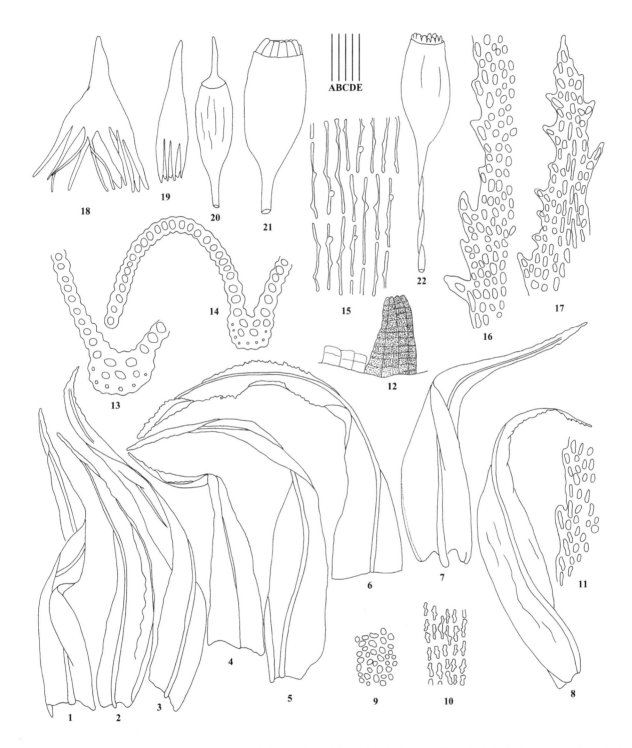

Fig. 174 *Macromitrium ovale* Mitt. 1, 3-5, 7-8: Branch leaves. 2: Perichaetial leaf. 6: Apex of branch leaf. 9: Upper cells of branch leaf. 10: Medial cells of branch leaf. 11: Low cells of branch leaf. 12: Peristome. 13: Basal transverse section of branch leaf. 14: Medial transverse section of branch leaf. 15: Basal cells of branch leaf. 16-17: Apical cells of branch leaves. 18-19: Calyptrae. 20-22: Capsules (all from isotype, NY 01086487). Line scales: A = 2 mm (18-22); B = 1 mm (1-5, 7-8); C = 400 μm (6, 12); D = 100 μm (17); E = 67 μm (9-11, 13-16).

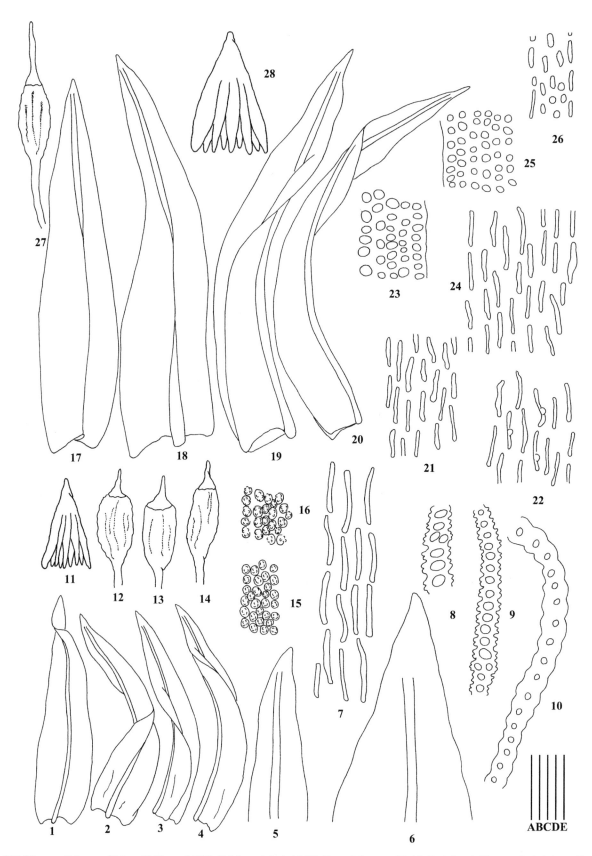

Fig. 175 *Macromitrium pallidum* (P. Beauv.) Wijk & Margad. 1-4, 19-20: Branch leaves. 17-18: Perichaetial leaves. 5-6: Apex of branch leaf. 7: Basal cells of branch leaf. 8-9: Upper transverse sections of branch leaves. 10: Low transverse section of branch leaf. 11, 28: Calyptrae. 12-14, 27: Capsules. 15: Medial cells of branch leaf. 16: Upper cells of branch leaf. 21: Basal cells of perichaetial leaf. 22: Basal cells of branch leaf. 23, 25: Medial cells of branch leaf. 24: Upper cells of perichaetial leaf. 26: Apical cells of perichaetial leaf (1-16 from isosyntype, JE 00428860; 17-28 from paratype, BM 000873859). Line scales: A = 4 mm (11-14, 27); B = 1 mm (1-4, 28); C = 400 μm (5); D = 100 μm (6); E = 67 μm (7-10, 15-16, 21-26).

160. *Macromitrium panduraefolium* Thouvenot, Crypt., Bryol. 39: 444, 1-24. 2018.

Type protologue: New Caledonia. South Province, Dumbéa, Montagne des Sources Nature Reserve, *Neocallitropsis* plateau, 745 m, 21, IX. 2016, *Thouvenot NC2329* (holotype: PC 0786119).

(1) Plants small, stems creeping with crowded erect branches, branches short and slightly fusiform, 3.0-4.0 mm long. (2) Branch leaves appressed and spirally coiled around the branch when dry, with a rope-like appearance, erect when moist, green and often rusty-reddish in upper portion, hyaline in basal portion 1/3 of the whole leaf length, 1.1-1.6×0.35-0.50 mm, oblong-ligulate, with an oval to oblong base, the upper parts elliptic to ligulate, constricted at medial part; rounded to obtuse or truncate at apex, more or less asymmetrically truncate or emarginate; costae excurrent in conspicuous hyaline aristae; all cells thick-walled; upper cells rounded or oblong, weakly pluripapillose; transitional parts short, hyaline, short rectangular, unipapillose; basal cells long rectangular to linear, smooth, porose. (3) Perichaetial leaves longer than branch leaves, the inner lanceolate, long acuminate. (4) Setae 7-8 mm long, smooth, twisted to the left; vaginulae naked or with sparse, short stiff paraphyses. (5) Capsule urns ellipsoid, narrowed and plicate in upper portion; peristome absent. (6) Caplytrade naked (Thouvenot, 2018).

Macromitrium panduraefolium Thouvenot, *M. pulchrum* Besch., *M. pulchrum* var. *neocaledonicaum* (Besch.) Thouvenot, *M. rufipilum* Cardot and *M. humboldtense* Thouvenot & Frank Müll. are similar in their large cells with external walls strongly thickened and protruding, bulging and bearing 1-3 small papillae, and excurrent costae and rusty-reddish coloration. However, *M. pulchrum* and *M. pulchrum* var. *neocaledonicum* differ from *M. panduraefolium* in their short setae and large plants; *M. humboldtense* can be separated from *M. panduraefolium* by its long, narrowly lanceolate to narrowly triangular branch leaves with a long acuminate apex; *M. rufipilum* differs from *M. panduraefolium* by its long setae (20-25 mm), long branch leaves (3-3.6 mm), long branches (5-13 mm), and its branches not rope-like branches when dry (Thouvenot, 2018).

Distribution: New Caledonia.

161. *Macromitrium paridis* Besch., Rev. Bryol. 25: 90. 1898. (Figure 176)

Basionym: **Macromitrium cacuminicola* Besch., Bull. Soc. Bot. France 45: 63. 1898, illegitimate, later homonym. Type protologue: [French Polynesia] Sur les arbres des erêtes, vers 900 mètres d'alt. RR. ([*Nadeaud*] n° *260*). Type citation: Tahiti, *Nadeaud 260*. (holotype: PC 0137866!; isotypes: PC 0137864!, PC 0137865!).

(1) Plants small in mats, light yellowish, densely with rusty rhizoids; stems long-creeping, with numerous short and erect branches, branches often shorter than 1.5 mm (when dry), simple or branched. (2) Branch leaves keeled, contorted and crisped, undulate when dry, erect-spreading below and still adaxially inrolled at the apex when moist, lanceolate from an oblong portion, or oblong-lanceolate, acuminate to acuminate-acute; costae rusty, percurrent; all cells clear and smooth, or weakly papillose; upper cells subrounded, rounded-quadrate, elongated towards the base; basal cells elongate, thick-walled, lumen somewhat crescent-shaped. (3) Perichaetial leaves not much differentiated from branch leaves, occasionally slightly longer than branch leaves, plicate below. (4) Setae 5 mm long, smooth. (5) Capsule urns small, ellipsoid, wrinkled beneath the mouth; opercula conic-rostrate to conic-subulate; peristome single, shorter, hyaline; spores large, globous, greenish, isosporous. (6) Calyptrae naked, mitrate, yellowish, scabrous at a dark red-brown beak, lacerate below.

Distribution: Tahiti in French Polynesia.

162. *Macromitrium parvifolium* Dixon, J. Bot. 80: 6. 1942.

Type protologue: Papua New Guinea. Central, Alola, forest tree, crica 1 800 m,11. II, 1936 *C.E. Carr 15049* (holotype: BM-Dixon; isotype: L).

Macromitrium brevirameum E.B. Bartram, Rev. Bryol. Lichénol. 30: 195. 1962. nom. nud., *fide* Vitt *et al.* 1995. Original collections: Papua New Guinea. Western Highlands, Wabag subdistrict, Yaki River valley, Poio Village, 7500 ft., epiphytic on *Cordline* in garden fence, *Hoogland & Schodde, no. 6825*, 29. Jun, 1960 (FH-Bartram, L), epiphytic on branches of tall tree in the open, *6959* (COLO, FH-Bartram).

**Macromitrium brevirameum* E.B. Bartram, Contr. U.S. Natl. Herb. 37: 54. 1965, *fide* Vitt *et al.*, 1995. Type protologue: Papua New Guinea. Eastern Highlands, Mt. Otto, south slopes. High epiphyte in *Castanopsis* forest, 2000 m, *Brass 31117* (isotype: L).

Macromitrium daymannianum E.B. Bartram, Brittonia 9: 43. 1957, *fide* Vitt *et al.* 1995. Type protologue: Papua New Guinea. Milne Bay, Mt. Dayman, north slopes, 2230 m, on branches of a tall *Araucaria*, *Brass 22306* (holotype: FH; Brass *22307, 22308b*, paratype: FH).

(1) Plants small, in spreading mats, rusty brown, with creeping stems and dense branches, branches up to 10 mm long. (2) Branch leaves funiculate to spirally-whorled around branch, frequently individually decurved-flexuose when dry, spreading and straight when moist, 1.3 -1.6 mm long, ligulate, ligulate-lanceolate, lanceolate, abruptly narrowed to an obtuse or asymmetric short-cuspidate apex; margins plane above, reflexed below, entire or erose at apex; costae ending in the cusp or excurrent and filling the cusp; upper and medial cells conic-bulging to strongly

conic-tuberculate, in longitudinal rows, rounded or rounded-quadrate to short-rectangular, marginal cells smaller; basal cells elongate to narrowly-rectangular, straight, thick-walled, most cells tuberculate above, smooth near the insertion. (3) Perichaetial leaves 1.0-1.7 µm long, loosely erect, broadly lanceolate, acuminate, upper cells smooth, longer than wide, basal cells mostly smooth to sparsely tuberculate. (4) Setae 9-12 mm long, smooth, twisted to the left, vaginulae not hairy. (5) Capsule urns 1.2-1.4 × 0.5-0.7 mm, broadly ellipsoid, smooth in low portion, 8-plicate in upper half, with a small, puckered mouth, sharply contracted to the seta; peristome single; exostome teeth fused to a low, coarsely papillose membrane; spores 14-42 µm, distinctly anisosporous. (6) Calyptrae plicate, mitrate, deeply lacerate, naked (Vitt et al., 1995).

Distribution: Papua New Guinea.

Specimens examined: **PAPUA NEW GUINEA**. Morobe, *D. H. Norris 62152* (H 3090615); Western Highlands, *R.D. Hoogland 6825* (MO 4463851).

163. *Macromitrium parvirete* E.B. Bartram, Bryologist 47: 16. 1944. (Figure 177)

Type protologue and citation: Costa Rica: Entre Rio Jesus y Calera de San Ramon, *A. M. Brenes 17075* (isotype: F-C 0001096F!).

(1) Plants small to medium-sized, greenish yellow; in prostrate wefts, stems weakly creeping, branches up to 10 mm long. (2) Stem leaves inconspicuous, much shorter than branch leaves, oblong-lanceolate, squarrose above when moist. (3) Branch leaves erect below, twisted-flexuose-curly, strongly undulate above, with an incurved apex when dry, erecto-patent to slightly recurved when moist, 2-3×0.8 mm, oblong-lanceolate to lingulate; apices obtuse, cuspidate; margins undulate, crenulate, recurved below, erect to plane above; costae excurrent to form a point (60-80 µm); upper interior cells 6-8 µm wide, round, rounded-quadrate, hexagonal, bulging mammillose, marginal cells not differentiated; basal cells long rectangular, 60-80 µm long, incrassate and porose, tuberculate; basal cells near costa enlarged, regularly rectangular to short rectangular, with straight walls, not porose; basal enlarged teeth-like cells not differentiated at the insertion. (4) Perichaetial leaves differentiated, shorter than branch leaves, lanceolate to triangularly lanceolate, plicate below. (5) Setae smooth, 5-12 mm long, twisted to the left. (6) Capsule urns cupulate, smooth above, wrinkled at neck; opercula conic-rostrate, with a short and stout beak; peristome double, exostome teeth short, truncate, papillose, united forming an erect membrane, endostome present as a rudimentary membrane or absent; spores anisosporus. (7) Calyptrae mitrate, 2-3 mm long, naked, deeply lacerated (Allen, 2002).

Macromitrium parvirete E.B. Bartram is similar to *M. sharpii* H.A. Crum ex Vitt, the latter can be distinguished from the former by its smaller plants, costae not excurrent, capsulee long ellipsoid, and with well-developed endostome segments.

Distribution: Costa Rica, Suriname.

Specimens examined: **COSTA RICA**. San Jose: Cantón de Turrubares. *Rodolfo Zúñiga 447* (MO 4428838). **PANAMA**. Herrera. *W. G. D'Arcy 14317* (MO 4424381). **SURINAME**. Sipaliwini, *Bruce Allen 25477* (MO).

164. *Macromitrium pellucidum* Mitt., J. Linn. Soc., Bot. 12: 203. 1869. (Figure 178)

Type protologue: [Brazil] Hab. Fl. Uaupes, Panuré ad arbores, *Spruce, n. 80.* Type citation: Panuré ad arbores, *Spruce* (isotypes: NY 01086500!, NY 01086501!); Panuré ad arbores, *R. Spruce 80* (isotype: BM 000720573!).

Macromitrium laevifolium Mitt., J. Linn. Soc., Bot. 12: 203. 1869, *fide* Grout, 1944. Type protologue: Hab. Ad fluv. Pacimoni, Venezuela, *Spruce, n. 103*; Guiana, Schomburgk in Herb. Hook. Type citation: Venezuela, by R. Pacimoni, Legit *Ric Spruce 103* (isosyntype: E 00165159!).

(1) Plants small to medium-sized, grey-yellow to yellowish-green, lustrous; stem long creeping, with inconspicuous and caduceus leaves; branches ascending and 3-5 mm long. (2) Branch leaves obliquely and tightly appressed, spirally appressed-curved, somewhat funiculate when dry, erecto-patent and rugose above when moist, lingulate to ligulate; the apices apiculate, mucronate; all cells clear, smooth, porose, and moderately bulging, longer than wide; cells elongate from the apex (18 µm) to the base (45 µm), weakly to strongly porose; marginal basal enlarged teeth-like cells at insertion not differentiated. (3) Setae smooth, 15-20 mm long, twisted to the left. (4) Capsules ellipsoid, ellipsoid-cylindric to cylindric, smooth to moderately furrowed, wide-mouthed, not constricted beneath the mouth; peristome absent. (5) Calyptrae mitrate, lacerate at base, plicate, with sparse and straight, brownish hairs.

Macromitrium pellucidum Mitt. is similar to *M. swaisonii* (Hook.) Brid. in leaf shape, but differs from the latter by its branch leaves with smooth leaf cells and without marginal enlarged teeth-like cells at insertion.

Valente *et al.* (2019) placed *M. pellucidum* in synonymy with *Schlotheimia trichomitria* Schwägr. However, the plants of *Schlotheimia* are reddish, while those of *M. pellucidum* are grey-yellow to yellowish-green; the basal cells in *Schlotheimia* are radiate from the costal region, while those in *M. pellucidum* arranged in longitudinal rows; the upper cells of branch leaves in *Schlotheimia* are often rather small, but those in *M. pellucidum* are rather large and distinctly bulging.

Grout (1944) placed *M. laevifolium* Mitt. in synonymy with *M. pellucidum*. The type of *M. laevifolium* showed that its calyptrae are long campanulate and almost not lobed at base, the peristome in its capsules are absent. However, the species of *Schlotheimia* have mitrate-campanulate calyptra with 4-5 broad lobs at base, the capsule with double peristome, and strongly reflex exostome. Therefore, more work is needed to clarify the classification status of *M. pellucidum*.

Distribution: Brazil, Suriname, Venezuela.

Specimens examined: **BRAZIL**. Rio Negro, *W. R. Buck 2621* (MO 2860887); Rio Marie, Manauna, *W. R. Buck 2382* (MO 2860017). **SURINAME**. Granhollo falls. *D.C. Geijskes 23* (MO 2122305), Sipaliwini, *Casado 534* (MO 5629990), *B Allen 23195* (MO 5644670); *Casado 432* (MO 5629988); Sipaliwini, *B. Allen 20509* (MO 5238387). **VENEZUELA**. Terr. Federal Amazonas: Depto. Atabapo, *F. Delascio Ch. & F. Guanchez 10769* (MO 2846372); Terr, Federal Amazonas. *Delascio Ch. & F. Guanchez 10769* (MO 3670921); Amazonas, Rio Negro. *Davidse Garrit 26811* (MO 4412277).

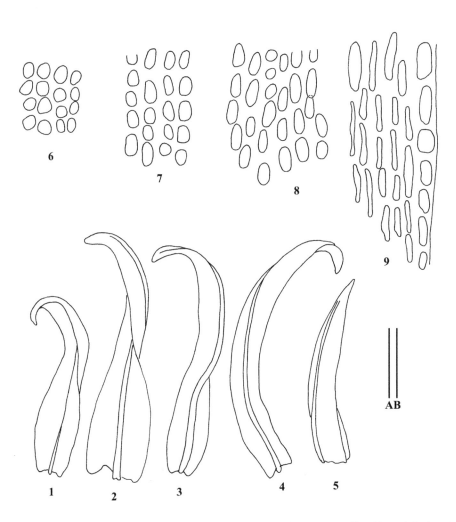

Fig. 176 *Macromitrium paridis* Besch. 1-4: Branch leaves. 5: Perichaetial leaf. 6: Upper cells of branch leaf. 7: Medial cells of branch leaf. 8: Low cells of branch leaf. 9: Basal cells of branch leaf (all from isotype, PC 0137864). Line scales: A = 1 mm (1-5); B = 67 μm (6-9).

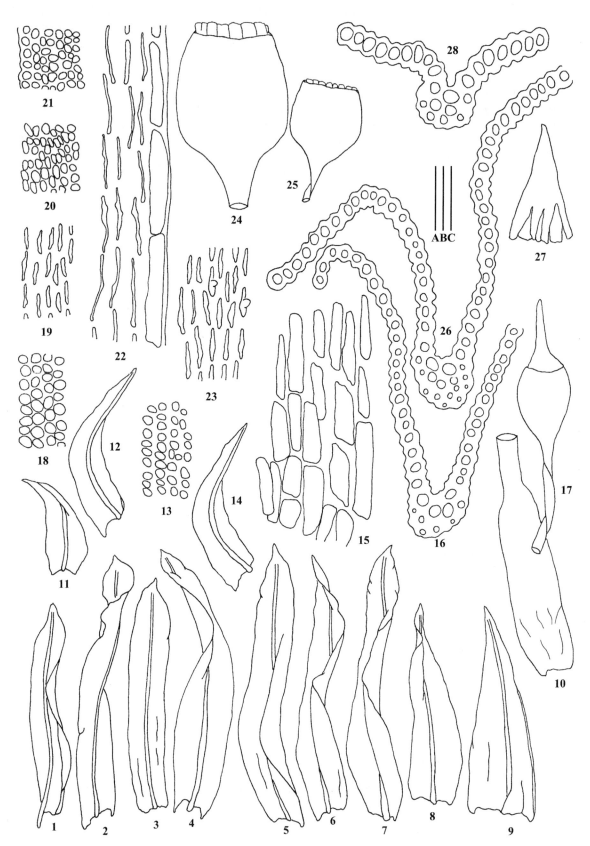

Fig. 177 *Macromitrium parvirete* E.B. Bartram 1-7: Branch leaves. 8-9: Perichaetial leaves. 10: Vaginula. 11-12, 14: Stem leaves. 13: Medial cells of branch leaf. 15: Basal cells near costa of branch leaf. 16, 26: Basal transverse sections of branch leaves. 17, 24-25: Capsules. 18: Upper cells of branch leaf. 19: Medial cells of perichaetial leaf. 20: Upper cells of perichaetial leaf. 21: Upper marginal cells of branch leaf. 22: Basal marginal cells of branch leaf. 23: Basal cells of branch leaf. 27: Calyptra. 28: Upper transverse section of branch leaf (all from isotype, F-C 0001096F). Line scales: A = 2 mm (17, 25, 27); B = 1 mm (1-12, 14, 24); C = 67 μm (13, 15-16, 18-23, 26, 28).

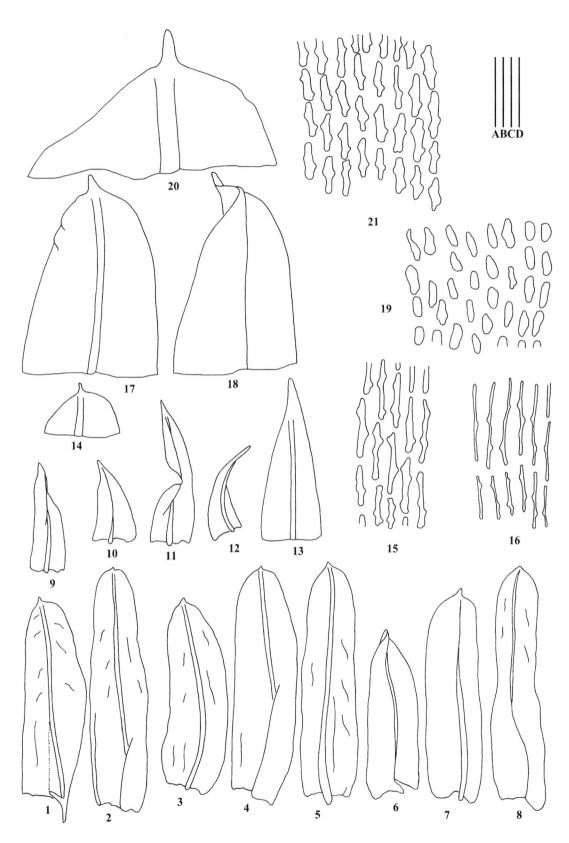

Fig. 178 *Macromitrium pellucidum* Mitt. 1-8: Branch leaves. 9-12: Stem leaves. 13: Apices of stem leaves. 14, 17-18, 20: Apices of branch leaves. 15: Medial cells of branch leaf. 16: Basal cells of branch leaf. 19: Apical cells of branch leaf. 21: Upper cells of branch leaf (all from isotype, NY 01086500). Line scales: A = 1 mm (1-12); B = 400 μm (13-14, 17-18); C = 100 μm (20); D = 67 μm (15-16, 19, 21).

165. *Macromitrium peraristatum* Broth., Öfvers. Finska Vetensk.-Soc. Förh. 35: 45. 1893.

Type protologue and citation: [Australia] Lord Howel Island, Mount Gower, ubi ad ramulos arborum m. Sept. 1887 Th. *Whitelegge (n.2)* (holotype: H-BR 2527001!).

(1) Plants robust, shiny to lustrous, in loose, spreading mats; stems creeping, with erect branches, branches up to 25 mm high, branching simple and regularly spaced; the branches (with leaves) gradually acuminate towards the upper terminal when dry. (2) Stem leaves flexuose-twisted when dry, 1.5-2.5 mm long, broadly ovate-ligulate to ovate-lanceolate, abruptly contracted to a reflexed acumen, costae excurrent. (3) Branch leaves curved-twisted around the branch, with an incurved apex, slightly funiculate when dry, straight and erect-spreading when moist, 2.6-3.0 mm long, oblong-lanceolate to broadly lanceolate, sharply acuminate to long-cuspidate, deeply keeled below; margins plane and entire; costae excurrent and filling the cusp or acumen; all cells smooth and clear; upper cells 9-10 μm wide, hexagonal with quadrate to rounded lumens, in distinct longitudinal rows, somewhat bulging, thick-walled, cells near costa elliptic; medial cells 7-13 × 7-8 μm, quadrate to short-rectangular, in longitudinal rows; transitional cells 15-35 μm long, rectangular, with strongly sigmoid to flexuose lumens; basal cells 40-60 × 7-10 μm, lumens about 2 μm wide, rectangular-elongate, irregularly thick-walled, lumens flexuose, curved to shallowly sigmoid. (4) Perichaetial leaves distinctly differentiated, much longer than branch leaves, 5-8 mm long, narrowly oblong to oblong-lanceolate, with a very long erect-flexuose arista (up to 1.3-3.0 mm long), loosely and distinctly curved around the branch; upper cells rounded-elliptic to long elliptic-rectangular, smooth, multi-stratose near the arista base, upper margins undulate-ruffled, bordered by bistratose, elongate cells in 2-4 rows; medial cells elongate with irregular thick walls, lumens straight to flexuose or slightly sigmoid; basal cells very long and thin-walled. (5) Setae only 2 mm long, stout, short, erect, smooth, straight or twisted to the right. (6) Capsule urns 1.5-2.5 mm long, narrowly ovoid-ellipsoid to ellipsoid-cylindric, smooth, with a darkened rim, wide-mouthed, usually smooth or rarely somewhat 8-plicate, gradually contracted to the seta through a wrinkled neck; peristome single; exostome of 16, linear-lanceolate, pale, well-developed teeth, coarsely papillose, erect-inflexed; endostome absent; opercula large, conic-rostrate; spores anisosporous, 17-35 μm, papillose. (7) Calyptrae about 4 mm long, large, long conic-mitrate, densely hairy, plications and uniformly lobed at the base.

Distribution: Australia.

Specimens examined: **AUSTRALIA**. New South Wales, *D. H. Vitt 28396* (H 3090621), *M.M. J.v.Balgooy 1129* (H 3090622), *Ring s.n.* (H-BR 2527002), *H. Streimann 56027* (KRAM-B-114566), *55835* (MO 5137842).

166. *Macromitrium perdensifolium* Dixon, J. Linn. Soc., Bot. 50: 90. 2 f. 18. 1935. (Figure 179)

Type protologue: [Indonesia] Hab. Kemoel. W. Koetai, c. 1700 m., Oct. 1925; Coll. *F. H. Endert (4535)*, Herb. Hort. Bot. Bog. (2522), type. On branch of small tree, white sand forest, c. 900 m., Ulu Koyan, Sarawak, 15 Sept. 1932; coll. Oxford Exped. (1875). Type citation: Borneo, Koetai occ., Kemoel, in silvis primig., in terra siliciosa. 1700 m, leg. *F. H. Endert*, X. 1925 det. H. H. Dixon. (isotypes: MO 406392!, MICH 525896!, JE 04008727!, GOET 011892!, F-C 0001131F!).

(1) Plants robust, reddish brown, primary stem far-creeping, thin; branches stout, tumid, up to 3 cm. (2) Branch leaves closely set, not very shriveled, erect below, individually to slightly spirally twisted and flexuose above when dry, widely spreading, undulate or rugose above when moist, 3-4 × 0.6-0.7 mm, oblong-lanceolate, ligulate-lanceolate, abruptly narrowed to a long-cuspidate to mucronate apex, deeply plicate below; margins plane, minutely crenulate above; costae excurrent, mostly concealed in adaxial view by overlapping folds of laminae; upper and medial cells isodiametric to short-ovoid, thick-walled, strongly conic-bulging, sometimes faintly mamillate or papillose, gradually becoming more elongate below, basal cells elongate except at insertion, tuberculate. (3) Perichaetial leaves not much differentiated, the outer about as long as branch leaves, the inner 2-4 leaves smaller, erect, lanceolate. (4) Setae smooth, up to 35 mm long. (5) Capsule urns ovoid, contracted above to a short-cylindric mouth, smooth or sulcate on drying. (6) Calyptrae cover the whole capsule, plicate, almost naked.

Macromitrium perdensifolium Dixon resembled *M. crinale* Broth. & Geh. and *M. ochraceum* (Dozy & Molk.) Müll. Hal. in habits. However, all leaf cells of *M. crinale* are longer than wide; the branch leaves of *M. ochraceum* ovate-lanceolate, acuminate, with intercellular spaces appearing as 'dots' under a microscope in upper portion.

Distribution: Indonesia, Malaysia.

167. *Macromitrium perfragile* E. B. Bartram, J. Washington Acad. Sci. 42(6): 181. 1952. (Figure 180)

Type protologue and citation: [Brasil] Rio Grande do Sul: Fazenda S. Borja, S. leopoldo, in arbore, alt. 50 m, *A. Sehnem no. 427* (syntype: FH 00290386!); Rio dos Sinos, S. Leopoldo, atl. 10 m, *A. Sehnem no. 432* (syntype: FH 00290387!); Aparados, Bom Jesús, in arbore, alt. 100m, *S. Sehnem no. 576* (syntype: FH 00290388!); Brasil, Campestre Montengegro, in arbore, at. 450m, *A. Sehnem no. 2175* (syntype: FH 00290389!), Vila Oliva, S. Franc. d. Paul, in arbore, alt. 750 m, A. Sehnem no. 2630, type 9 (syntype: FH 00290390!).

(1) Plants medium-sized to large, yellowish above, dark-brown below; stems long creeping, branches up to 20-25 mm long. (2) Branch leaves erect below, twisted-contorted spreading, flexuose and undulate above when dry,

spreading, somewhat abaxially curved when moist, long linear-lanceolate, 3-4 mm long, gradually narrowed to a fragile subulate upper portion from a long oblong low portion, easily broken in the upper portion when old; margins almost entire throughout, or crenulate from bulging cells; all cells smooth and clear; upper and medial cells bulging to conic-bulging, thin-walled, obscure and dark-coloration; low and basal cells slightly bulging or flat, thick-walled and porose; cells at the outermost marginal row enlarged, swollen, somewhat teeth-like and hyaline at the insertion, distinctly differentiated from their ambient cells; basal cells near the costa at insertion large, thin-walled, forming a "cancellina region". (3) Capsules hemispherical and smooth.

The species is characterized by its almost entire branch leaves with fragile subulate and easily broken upper portion, all smooth lamina cells, weakly differentiated teeth-like marginal cells at the insertion, smooth hemispherical and smooth capsules.

Macromitrium perfragile E.B. Bartram is similar to *M. sejunctum* B.H. Allen, but differs from the latter in having entire branch leaves. *Macromitrium frustratum* is similar to *M. perfragile*, but differs from the latter in having longer branch leaves, with differentiated border composing long and narrow cells in upper portion.

Macromitrium perfragile was synonymized with *M. longifolium* (Hook.) Brid. by Valente *et al.*, 2020. However, *M. perfragile* differs from *M. longifolium* in having long lanceolate leaves with fragile subulate upper portion, with cells at outermost marginal row near insertion differentiated, hyaline enlarged, and those near costa larger and hyaline, forming a "cancellina region".

Distribution: Brazil, Colombia.

Specimens examined: **BRAZIL**. Rs-Caxias do Sul-Ana Rech-Faxinal, *M. Rossato et al. 4264* (MO 3657299). **COLOMBIA**. Nariño, Ricaurte, *S. P. Churchill, Julio C. Betancur B. & Francisco J. Roldán 18233* (MO).

168. *Macromitrium perichaetiale* (Hook. & Grev.) Müll. Hal., Bot. Zeit. (Berlin) 3: 544. 1845. (Figure 181)

Basionym: *Orthotrichum perichaetiale* Hook. & Grev. Type protologue: [Saint Vincent and the Grenadines] Hab. In the island of St. Vincent, *D. Menzies*, Esq. and the Rev. L. Guilding. Type citation: S. Vincent, *D. Menzies s.n.* (isotypes: BM 000873186!, BM 000873187!); St. Vincent, Guilding (isotype: BM 000989778!).

Macromitrium truncatum Müll. Hal., Linnaea 17: 383. 1843, *fide* Grout, 1944. Type protologue: St. Vincent, Macromitgr. longifol. intermixtum. Celcb. Prof. Kunze Lips. benevole communicavit.

(1) Plants robust, lustrous brown; stems long creeping, branches to 20-25 mm long, 1.5 mm thick (with leaves). (2) Stem leaves caducous and inconspicuous. (3) Branch leaves erect and appressed below, slightly flexuose or spirally contorted above when dry, spreading when moist, lanceolate or long ligulate; the apices truncate with lamens extends on both sides of the costa, forming short asymmetric spiny protuberances; margins occasionally recurved on one side below; all cells longer than wide, moderately bulging, porose, thick-walled, smooth; upper and apical cells rhombic, gradually elongated towards the base; medial to basal cells linear elongate, thick-walled. (4) Perichaetial leaves distinctly larger and longer than branch leaves, and differ in their leaf shape; long- and linear-lanceolate, gradually narrowed to an acuminate apex, entire; all cells longer than wide, smooth, thick-walled, porose; costae excurrent to form an awn. (5) Setae smooth, 12-20 mm long, twisted to the left. (6) Capsules ellipsoid, weakly furrowed, slightly constricted below the mouth; peristome double, exostome of 16, lanceolate, united at base; endostome yellowish, with basal membrane and segments. (7) Calyptrae mitrate, with dense yellowish hairs.

Distribution: Caribbean, Guadeloupe, Saint Vincent and the Grenadines, Suriname, Venezuela.

Specimens examined: **CARIBBEAN**. Puerto Rico, *M. J. Price 856* (MO 5139249), *955* (MO 5139248). **GUADELOUPE**. *P. Allorge, no. 11* (H 3090630). **SURINAME**. Sipaliwini. *Christina Casado 547* (KRAM-B-165998); *Bruce Allen 23390* (MO 278180), *Christina Casado 558* (MO 5629992). **VENEZUELA**. Sierra San Luis, *R. Wingfield 13385* (MO 3062931); Zulia, *Dana Griffin III 342* (MO 2407886).

169. *Macromitrium perpusillum* Müll. Hal., Linnaea 38: 640. 1874. (Figure 182)

Type protologue and citation: Patria. Mexico, Mirador: Florentin Sartorius vere 1873 cum fructibus deoperculatis: Hb. C. Mohr 1873 (isotype: PC 0137875!).

(1) Plants medium-sized, stems weakly creeping, without rusty-red rhizoids; branches prostrately ascending, up to 10 mm long and 1.0 mm thick. (2) Branch leaves keeled, strongly contorted-crisped, undulate when dry, spreading and sometimes conduplicate or somewhat adaxially incurved above when moist, oblong-lanceolate or lanceolate from an ovate-oblong base; apices acute, obtuse-acute to slightly mucronate; margins lightly crenulate from protrudent conic-bulging cells above, entire below; costae occasionally concealed in adaxial view by overlapping folds of laminae below; upper and medial cells isodiametric, rounded-quadrate or subrounded, strongly conic-bulging to mammillose; low and basal cells elongate-rectangular, thick-walled and porose, tuberculate; marginal enlarged teeth-like cells not differentiated at insertion. (3) Perichaetial leaves slightly differentiated, oblong-lanceolate with the widest at the base, acuminate above. (4) Setae about 10 mm long, smooth and twisted to the left. (5) Capsule urns ellipsoid to ellipsoid-cylindric, furrowed above, peristome single and lanceolate. (6) Calyptrae mitrate, naked.

Distribution: Mexico.

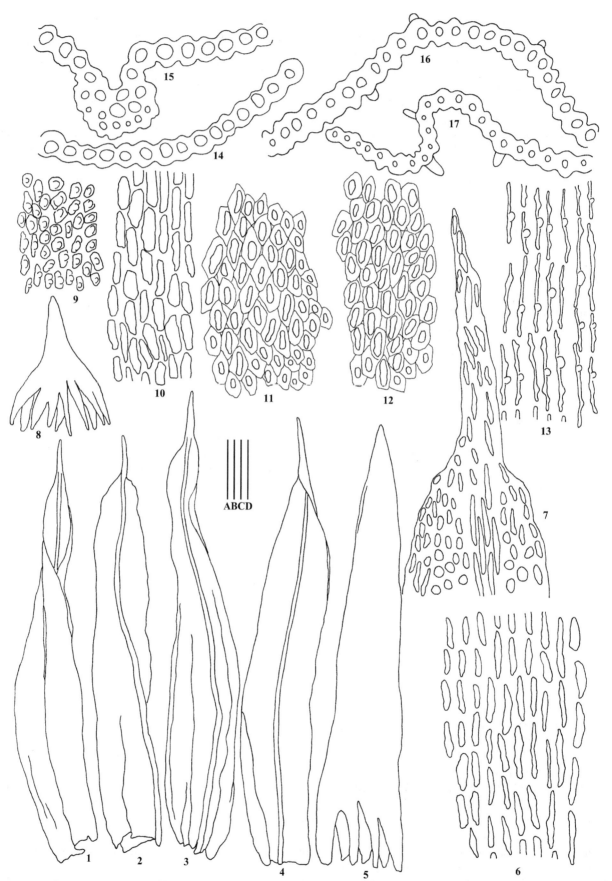

Fig. 179 *Macromitrium perdensifolium* Dixon 1-4: Branch leaves. 5, 8: Calyptrae. 6: Medial cells of branch leaf. 7: Apical cells of branch leaf. 9: Upper cells of branch leaf. 10: Low cells of branch leaf. 11-12: Medial cells of branch leaf. 13: Basal cells of branch leaf. 14-15: Medial transverse sections of branch leaves. 16-17: Basal transverse sections of branch leaves (all from isotype, MO 406392). Line scales: A = 2 mm (8); B = 1 mm (1-5); C = 100 μm (7); D = 67 μm (6, 9-17).

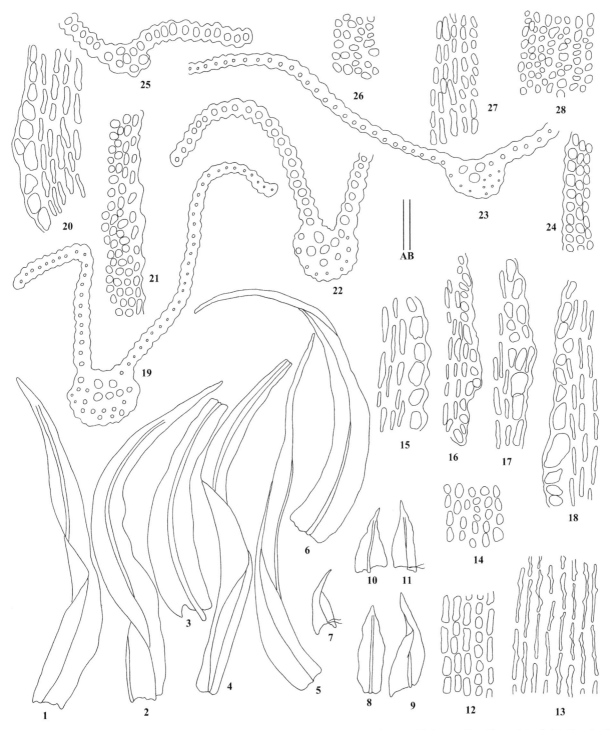

Fig. 180 *Macromitrium perfragile* E.B. Bartram 1-6: Branch leaves. 7-11: Stem leaves. 12: Low cells of branch leaf. 13: Basal cells of branch leaf. 14: Medial cells of stem leaf. 15-18: Basal marginal cells of branch leaves. 19: Low transverse section of branch leaf. 20: Basal marginal cells of stem leaf. 21, 24: Upper marginal cells of branch leaves. 22, 25: Upper transverse sections of branch leaves. 23: Basal transverse section of branch leaf. 26: Upper cells of branch leaf. 27: Basal near costa cells of stem leaf. 28: Medial cells of branch leaf (all from syntype, FH 00290387). Line scales: A = 1 mm (1-11); B = 67 μm (12-28).

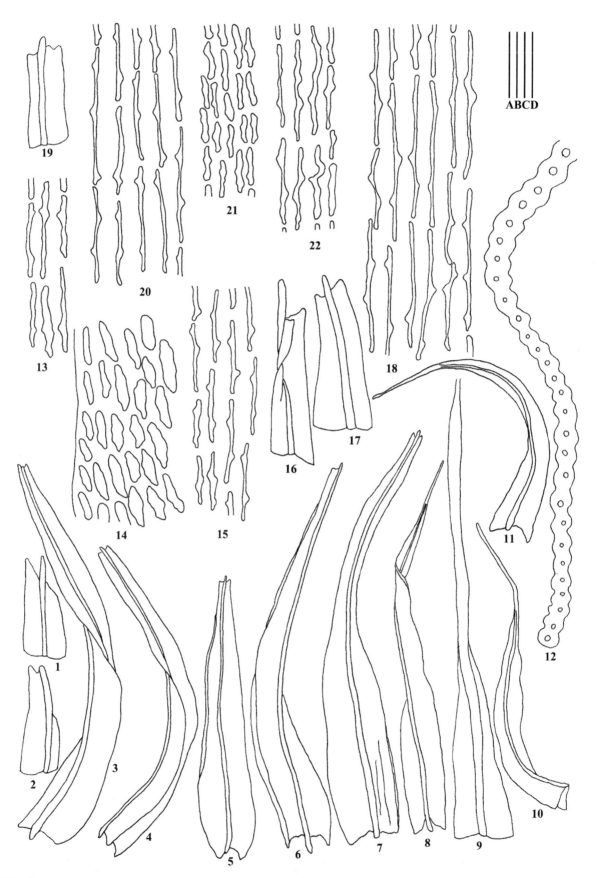

Fig. 181 *Macromitrium perichaetiale* (Hook. & Grev.) Müll. Hal. 1-2, 16-17, 19: Apices of branch leaves. 2-7: Branch leaves. 8: Perichaetial leaf. 9: Apex of perichaetial leaf. 10-11: Stem leaves. 12: Basal transverse section of branch leaf. 13: Upper cells of perichaetial leaf. 14: Upper cells of branch leaf. 15: Medial cells of branch leaf. 18: Basal cells of branch leaf. 20: Low cells of branch leaf. 21: Medial cells of stem leaf. 22: Basal cells of stem leaf (1-9, 12-20 from isotype, BM 000873186; 10-11, 21-22 from isotype, BM 000989778). Line scales: A = 2 mm (8, 10, 11); B = 1 mm (3-7); C = 400 μm (1-2, 9, 16-17, 19); D = 67 μm (12-15, 18, 20-22).

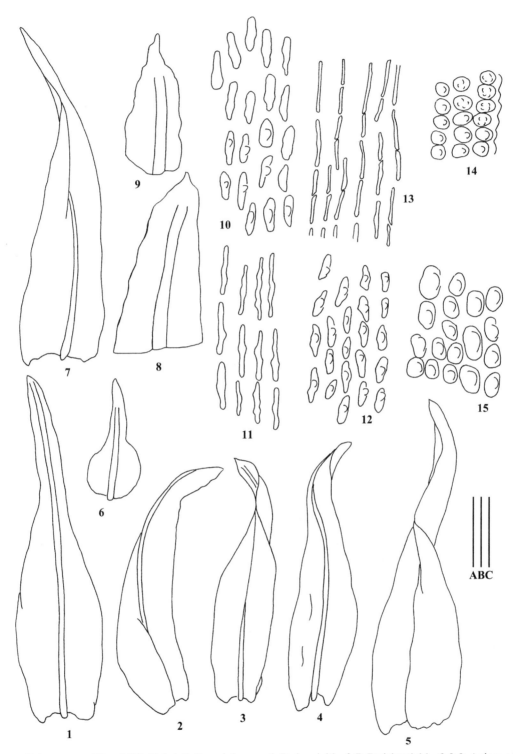

Fig. 182 *Macromitrium perpusillum* Müll. Hal. 1-5: Branch leaves. 6: Perigonial leaf. 7: Perichaetial leaf. 8-9: Apices of branch leaf. 10: Upper and medial cells of perichaetial leaf. 11: Low cells of branch leaf. 12: Medial cells of branch leaf. 13: Low cells of perichaetial leaf. 14: Upper marginal cells of branch leaf. 15: Upper cells of branch leaf (all from isotype, PC 0137875). Line scales: A = 1 mm (1-7); B = 400 μm (8-9); C = 67 μm (10-15).

170. *Macromitrium perreflexum* Steere, Brittonia 34: 437. f. 1-10. 1982. (Figure 183)

Type Protologue and citation: [Ecuador] Loja: Loma de Oro at Panamerican Hwy., 10 km S of Saraguro, in cloud forest, alt. 3300 m, epiphyte on Weinmannia sp., 24 Sep 1982, *Steere & Balslev 25555* (isotypes: MICH 525897!, US 00479341!).

(1) Plants medium-sized to large, rather stiff, reddish or yellowish green, darkening with age; the secondary stems up to 5 cm long, branching irregularly pinnate, branches and stems similar in size. (2) Leaves loosely and regularly arranged on the branch, clasping at the base, regularly and squarrosely spreading to bent back, weakly flexuose above when dry, strongly squarrosely spreading when moist, 1.5-2.0 mm long, lanceolate, acuminate above, broadest at base, strongly keeled and conduplicate above, plicate below; margins strongly and sharply serrulate from a long sharp papilla protruding in upper 1/5 of leaf, entire only at the extreme base; costae smooth, percurrent and ending in the apex or excurrent as a smooth, papillose or toothed spine; all cells longer than wide, in distinctly longitudinal rows; cells at the apex elongate, rectangular or elliptic-rectangular; upper cells rhomic, oblong to elliptic, bulging and strongly papillose; medial cells similar to upper cells but slightly elongate, rectangular, unipapillose; marginal cells narrower and longer than their ambient cells in upper and medial margin, forming a differentiated border; low and basal cells long rectangular, smooth, not porose. (3) Perichaetial leaves distinctly differentiated, straight and directed upwards, forming a conspicuous cluster, 5-6 mm long, much longer than branch leaves; oblong-lanceolate, sharply narrowed to a long arista, plicate below; margins sharply toothed; upper cells strongly papillose to tuberculate, costae excurrent into a long awn. (4) Setae 7-10 mm long, papillose throughout but more strongly upward, twisted to the left. (5) Capsule urns ovoid, smooth, dark brown to almost black; opercula conic-subulate; peristome double, fused into a cylinder; spores larger, 70-80 μm in diameter, irregularly papillose. (6) Calptrae up to 3 mm long, heavily clothed with erect yellowish hairs.

The species is characterized by 1) leaves strongly abaxially curved when dry, abaxially curved spreading when moist; 2) upper, medial and low cells all bulging and strongly papillose; basal cells smooth; 3) perichaetial leaves distinctly differentiated, much longer than branch leaves, with a long awn from excurrent costa; and 4) smooth capsules.

Distribution: Ecuador, Peru.

Specimens examined: **ECUADOR**. Zamora, *William C. Steere and Henrik Balslev 25791* (KEAM-B-058194, MO 3347529); Morona-Santiago, *William C. Steere 27783* (KRAM-B-072438, MO 3348025); Loja/Zamora-Chinchipe boundary, *L.J. Dorr & I. Valdespino 6594* (KRAM-B-090176); Loja, C. Luer, J. Luer, *A.Hirtz & W.Flcres 10782B* (MO 3077363); Prov. Zamora-Chinchipe, *Gunnar Harling 5965* (MO 3071504); Prov. Morona-Santago, *William C. Steere 27783* (H 3090634). **PERU**. Cajamarca, *J. Opisso et al. 806* (MO 5918532), *822a* (MO 5918533); San Ignacio, Jaen, Tabaconas, *Jasmin Opisso 806* (MO 5275205).

171. *Macromitrium pertriste* Müll. Hal., Hedwigia 37: 149. 1898. (Figure 184)

Type protologue: Habitatio, Chile, Valdivia: *Dr. H. Hahn* 1887, Hb. Jack. Type citation: Chile, Valdivia, 1887/1888, leg. *Dr. H. Hahn*, mist Jack (isotype: PC 0137885!).

(1) Plants small to medium-sized, stems weakly creeping, with prostrately ascending branches; branches varying in length, about 1.0 mm thick, simple or forked. (2) Stem leaves caducous and inconspicuous. (3) Branch leaves appressed below, spirally twisted arranged, keeled, with an adaxially incurve apex, somewhat funiculate when dry, spreading when moist; costae thin and subpercurrent, often partially concealed in adaxial view by overlapping folds of laminae under the apex when moist, broadly oblong-lanceolate; apices acute to broadly acuminate; margins entire and plane throughout; laminae most unistratose, occasionally scattered with bistratose cells; upper cells strongly bulging, sometimes weakly unipapillose, subquadrate, irregular short-rectangular; low and basal cells elongate-rectangular, distinctly porose and thick-walled. (4) Perichaetial leaves not much different from branch leaves, basal cells long-rectangular and not porose; (5) Setae smooth and about 5 mm long, twisted to the left. (6) Capsules ellipsoid to cylindric, smooth when dry; opercula conic-rostrate; peristome present and single, often caducous and rudimentary; isospores rather large, up to 70 μm in diameter, finely papillose. (7) Calyptrae small.

Distribution: Chile.

Specimens examined: **CHILE**. *P. Dusin A. 27* (BM 000873300, BM 000873301).

172. *Macromitrium petelotii* Tixier, Rev. Bryol. Lichénol. 34: 140. f. 9. 1966. (Figure 185)

Type protologue and citation: [Vietnam] Chapa, sur arbuste, 1500 m, février 1929 (*A. Petelot, no. 138*) (lectotype: PC 0083700!; isolectotype: PC 0083701!).

(1) Plants medium-sized, in dense mats, rusty-brown; stems creeping, densely with branches; branches erect, up to 20 mm long, often with several short branchlets, densely with reddish rhizoids. (2) Stem leaves loosely arranged, undulate to contorted when dry, often broken in upper portion, spreading when moist, ovate-lanceolate, long acuminate, 1.4 mm long; margins entire and plane; costa percurrent; all cells smooth and clear; upper and medial

cells subquadrate, rounded-quadrate or quadrate, gradually elongated towards the base, low and basal long rectangular, thick-walled. (3) Brancl leaves spirally contorted, twisted and flexuose above, with divergent apices when dry, spreading when moist, 5-6 × 0.4-0.5 mm, narrowly lanceoalte to linear-lanceolate, with a fragile upper portion, acuminate or sometimes with an awn, weakly plicate below; margins entire or weakly denticulate above, slightly recurved on one or both sides; costae slightly excurrent forming a short awn; upper and medial cells rounded-quadrate, conic-bulging or mammillate, transitional cells oblong to short rectangular; basal cell elongate, thick-walled, unipapillose or tuberculate. (4) Perichaetial leaves shorter than branch leaves, ovate lanceolate, long acuminate. (5) Setae short, 2.5-3.0 mm long, smooth. (6) Capsules erect, urns ovoid-urceolate, 2.0 × 1.5 mm, irregularly wrinkled when dry, gradually contracted towards the mouth; peristome single, exstome of 16, lanceolate, papillose; spores large, about 50 μm in diameter, papillose.

The species is similar to *M. taiheizanense* Nog., but the former is reddish rust, with longer setae, and the capsules with single peristome.

Distribution: Vietnam.

173. *Macromitrium picobonitum* B.H. Allen, Novon 8: 118. f. 4. 1998. (Figure 186)

Type protologue & citation: Honduras. Atlantida: El Porvenir. Pico Bonito National Park. Along trail at and above ridge camp. 15°38'N, 86°52'W, *Allen 17488* (holotype: MO 406397!); Honduras. Atlantida: El Porvenir, Pico Bonito National Park. Along trail from confluence to ridge camp. 1500 m. 15°38'-39'N, 86°51-52'W, 500-1500 m., 20 April 1996, *B. Allen 17366, 17370, 17373* (paratypes: MO!).

(1) Plants medium-sized, green to greenish-yellow above, brown below; stems long-creeping, branches up to 15 mm long. (2) Branch leaves keeled, erect below, flexuose to spirally contorted and twisted, undulate above when dry, erect-spreading when moist, 3-6 × 0.5 mm, linear-lanceolate; apices long-acuminate into a hyaline hair-tip; margins plane and serrate above, recurved and entire to serrulate below; costae excurrent into a hair tip; upper interior cells rounded to short-rectangular, 6-14 × 6-8 μm, thick-walled, bulging mammillose, in longitudinal rows; upper marginal cells longer and narrower than their interior cells, forming a weakly differentiated border; basal cells 25-50 μm, long-rectangular, thick-walled and porose, tuberculate; basal cells near costa enlarged, irregularly short rectangular, smooth; marginal enlarged teeth-like cells at insertion not differentiated. (3) Perichaetial leaves lanceolate from a triangular or oblong low part, recurved at one side below, abruptly narrowed in the upper to form a short or long cuspidate. (4) Setae smooth, 4-6 mm long, twisted to the left. (5) Capsule urns cupulate to hemispheric, smooth or weakly furrowed at the neck; peristome double, exostome teeth truncate, densely papillose, united to a membrane, discrete at tips, endostome yellowish to hyaline, weakly papillose; spores anisosporous. (6) Calyptrae mitrate, about 3 mm long, naked, deeply lacerated.

Macromitrium picobonitum B. H. Allen. is similar to *M. cirrosum* (Hedw.) Brid., differs from the latter in having branch leaves with a weakly differentiated upper border, and costa excurrent into a hyaline hair-tip.

Distribution: Colombia, Honduras.

Specimen examined: **COLOMBIA**. Nariño, *S. P. Churchill, Julio C. Betancur B. & Francisco J. Roldán 18233* (MO).

174. *Macromitrium pilicalyx* Dixon ex E. B. Bartram, Occas. Pap. Bernice Pauahi Bishop Mus. 19(11): 225. 1948. (Figure 187)

Type protologue and citation: [Fiji] Viti Levu: Nandarivatu, alt. about 3000 ft., *nos. 646* (syntype: BM 000982743!), 649 (syntype: BM 000982742!); Loma Lega [?] Mountain, *no. 648* (syntype: BM 000982741!).

(1) Plants slender, yellow-greenish in dense mats; stems long creeping, with ascending branches, branches 5-8 mm tall, densely leaved. (2) Branch leaves contorted-twisted and strongly crisped above, with an incurved to circinate apex often hidden in the inrolled cavity when dry, erect spreading and occasionally conduplicate above when moist, oblong-lanceolate, lingulate, acute; margins plane and finely crenulate above; costae excurrent; upper cells rounded-quadrate, pluripapillose; medial cells rounded-quadrate, slightly elongate, unipapillose; low cells elongate, papillose and porose; basal cells narrowly rectangular, clear and smooth, porose and thick-walled. (3) Perichaetial leaves not much differentiated, but slightly larger and longer than branch leaves. (4) Setae slender, erect and short, about 2.0 mm long, smooth; vaginulae with long hairs. (5) Capsule urns ovoid to ellipsoid, smooth, peristome single, exostome short and papillose; opercula conic-subulate. (6) Calyptrae mitrate, with numerous yellow straight hairs, lacerate at base.

Distribution: Fiji.

175. *Macromitrium piliferum* Schwägr., Sp. Musc. Frond., Suppl. 2 2(1): 66. pl. 172. 1826. (Figure 188)

Type protologue: [Hawaii] In insula Otaheite et Sandwicensibus *Gaudichaud*, D. Menzies 4. Type citation: Sandwia *Gaudichaud, 26,* Herbier Hedwig-Schwaegrichen (lectotypes: G 00046153!, G 00046154!).

(1) Plant small to medium-sized, in dense mats; primary stems long creeping, loosely with numerous, short secondary stems or branches, branches 3-15 mm long. (2) Stem leaves caducous, inconspicuous. (3) Branch leaves crowed, curled and spirally twisted when dry, erect-spreading when moist, oblong-lanceolate from an ovate base, strongly keeled, bluntly acute or obtuse, narrowed to a long, fragile, denticulate, twisted, yellowish or hyaline arista; the arista up to 2-3 mm or longer, easily broken in the upper portion; margins erect, entire below, minutely crenulate with papillae above; costae vanishing just below the base of the hair-point; lamina plicate or longitudinally undulate below; upper cells isodiametric, rounded, rather opaque or obscure upper, 5-7 μm, strongly pluripapillose; basal cells linear with narrow lumens and thick walls, unipapillose or nearly smooth, pellucid. (4) Perichaetial leaves erect, longer and broader than the branch leaves, ovate-lanceolate, gradually filiform acuminate; upper cells linear-rhombic, weakly papillose. (5) Setae smooth, highly vary, 5-50 mm long. (6) Capsules ovoid, reddish brown, small-mouthed, distinctly sulcate above when dry, peristome absent; opercula erect, rostrate, about 1.0 mm long; spores large, up to 65 μm in diameter. (7) Calyptrae mitrate, with numerous long, yellow hairs.

Distribution: Hawaii.

Specimens examined: **HAWAII**. Nuala Mounlaim, Oahw, Muwna Low, Hawaii, Sandw (FH); Honalcalu, Npsudcrron (H-SOL 1352018).

176. *Macromitrium pilosum* Thér., Bull. Acad. Int. Géogr. Bot. 17: 308. 1907. (Figure 189)

Type: New Caledonia, environs de Nouméa, 1906, *Franc s.n.* (lectotype designated by Thouvenot from Herbarium Thériot *s.n.* PC 0096514).

Macromitrium koghiense Thér., Diagn. Esp. Var. Nouv. Mouss. 8: Article 4. 1910, *fide* Thouvenot, 2019. Type protologue: New Caledonia, Mts Koghis, troncs d'arbre, alt. 500 m, *Franc s.n.* (lectotype designated by Thouvenot, 2019: PC 0096519).

Macromitrium koghiense var. *spiricaule* Broth. & Paris, Öfvers. Finska Vetensk.-Soc. Förh. 53A(11): 16. 1911, *fide* Thouvenot, 2019. Type protologue: New Caledonia, me Areimbo, *Le Rat s.n.* (lectotype designated by Thouvenot, 2019: PC 0096520).

Macromitrium pilosum var. *brevifolium* Thér., Bull. Acad. Int. Géogr. Bot. 20: 99. 1910, *fide* Thouvenot, 2019. Type protologue: New Caledonia, pied des Koghis, forêt, 300 m, *Franc s.n.* (lectotype designated by Thouvenot, 2019: PC 0083718).

Macromitrium subsessile Broth. & Paris, Öfvers. Finska Vetensk.-Soc. Förh. 51A (17): 17. 1909, *fide* Thouvenot, 2019. Type protologue: New Caledonia, Mt Dzumac, ad corticem arborum, *Le Rat s.n.* (lectotype designated by Thouvenot, 2019: REM 000099).

(1) Plants medium-sized, stems creeping, densely branched; branches thin. (2) Branch leaves hooked and appressed, funiculate, usually spirally arranged, keeled, with an adaxially incurved apex hidden between the neighboring leaves when dry, spreading recurved when moist, 0.8-1.6 × 0.3-0.4mm, lanceolate to ligulate from a wider basal part, the apices acute to obtuse and apiculate or mucronate, obscure in upper parts, gradually becoming translucent towards the basal parts; margins papillose crenulate; upper cells rounded to oval, small and thin-walled, bulging, strongly pluripapillose; transitional cells oval to short rectangular, unevenly thick-walled, unipapillose; low cells rectangular, thick-walled, basal cells unipapillose. (3) Perichaetial leaves similar to the branch leaves. (4) Setae shorter, 1-2 mm long, vaginulae with long hairs reaching the capsule. (5) Capsule urns ovoid to ellipsoid, smooth, rim plicate, small, peristome single. (6) Calyptrae mitrate, densely hairy.

Distribution: New Caledonia.

Specimens examined: **NEW CALEDONIA**. Mt. Koghis, leg. *Franc s.n.* 1. 11. 1909, marked as "*sp. n.*" (H-BR 2506008). Mt. Dzumac, Leg. *Le Rat*, 1906 (H-BR 2561004).

177. *Macromitrium proliferum* Mitt., J. Linn. Soc., Bot. 12: 217. 1869. (Figure 190)

Type protologue and citation: Andes Bogotenses. W*eir s.n.* Brasilia tropica, *Burchell n. 3959* (isosyntype: BM 000873087!); [Brazil] Parama, Prope Castro, alt. 2000 m, *J. Weir No. 92* (isosyntype: BM 000873088!).

Macromitrium runcinatella Müll. Hal., Linnaea 42: 488. 1879, *fide* Wijk *et al.*, 1964.

(1) Plants large, brown yellow; stems long creeping, branches up to 50 mm long, 1.5 mm thick. (2) Branch leaves strongly twisted-contorted, flexuose and rugose when dry, erect-spreading or slightly squarrose-curved below when moist, ligulate-lanceolate to narrowly lanceolate, bluntly acuminate the apices; margins strongly dentate above, entire below; upper, medial and low cells isodiametric, rather large, clear and smooth, 9-11 μm wide, strongly conic-bulging; basal cells short to long rectangular, thick-walled and porose, tuberculate, confined to 1/18-1/20 of the leaf; marginal enlarged teeth-like cells not differentiated at the insertion. (3) Setae smooth, 10-12 mm long. (4) Capsule urns ellipsoid, ellipsoid-cylindric, strongly furrowed; peristome double, both membranous. (5) Calyptrae mitrate, hairy.

Macromitrium proliferum Mitt. is similar to *M. punctatum* (Hook. & Grev.) Brid., but the former can be separated from the latter by its longer branch leaves with strongly dentate margin above, and rather large lamina cells in medial and upper portions. *Macromitrium proliferum* was synonymized with *M. argutum* Hampe by Valente *et al.*, 2020. However, *M. proliferum* differs from *M. argutum* by its branch leaves with strongly dentate margin above, and without marginal enlarged teeth-like cells at the insertion.

Distribution: Brazil, Colombia.

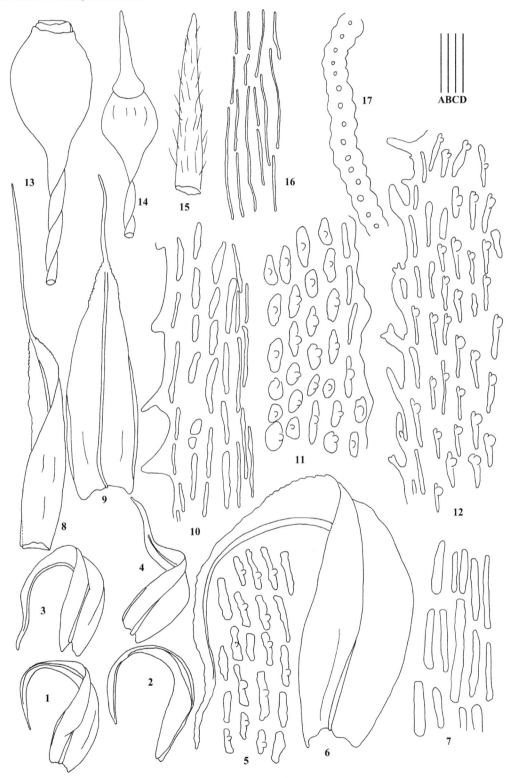

Fig. 183 *Macromitrium perreflexum* Steere 1-4, 6: Branch leaves. 5: Medial cells of branch leaf. 7: Basal cells of branch leaf. 8-9: Perichaetial leaves. 10: Apical cells of branch leaf. 11: Upper cells of branch leaf. 12: Medial marginal cells of branch leaf 13: Low cells of branch leaf. 14-15: Capsules. 16: Calyptra. 17: Basal transverse section of branch leaf (all from isotype, MICH 525897). Line scales: A = 2 mm (8-9, 13-15); B = 1 mm (1-4); C = 400 μm (6); D = 67 μm (5, 7, 10-13, 16).

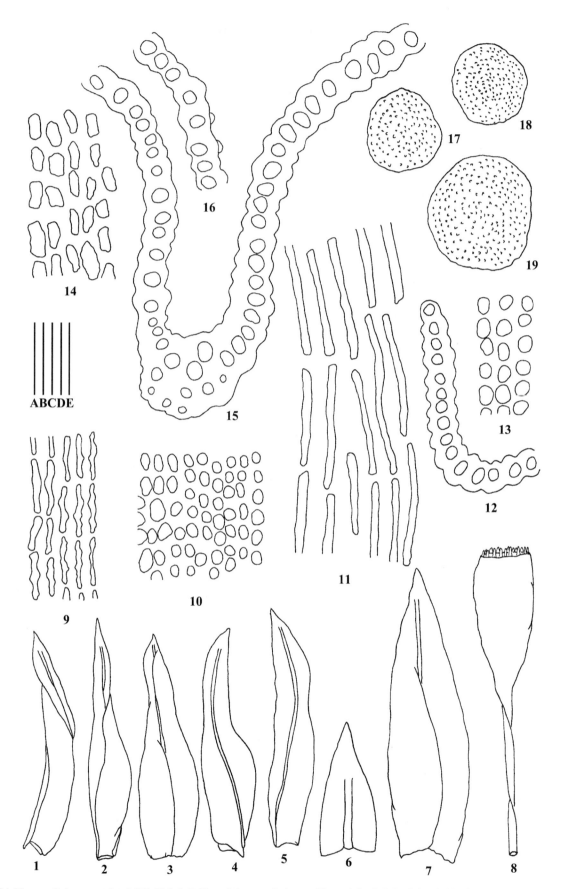

Fig. 184 *Macromitrium pertriste* Müll. Hal. 1-5: Branch leaves. 6: Apex of branch leaf. 7: Perichaetial leaf. 8: Capsule. 9: Low cells of branch leaf. 10: Upper cells of branch leaf. 11: Low cells of perichaetial leaf. 12, 16: Upper transverse sections of branch leaves. 13: Medial cells of branch leaf. 14: Upper cells of perichaetial leaf. 15: Basal transverse section of branch leaf. 17-19: Spores (all from isotype, PC 0137885). Line scales: A = 2 mm (8); B = 1 mm (1-5, 7); C = 400 μm (6); D = 100 μm (17-18); E = 67 μm (9-16, 19).

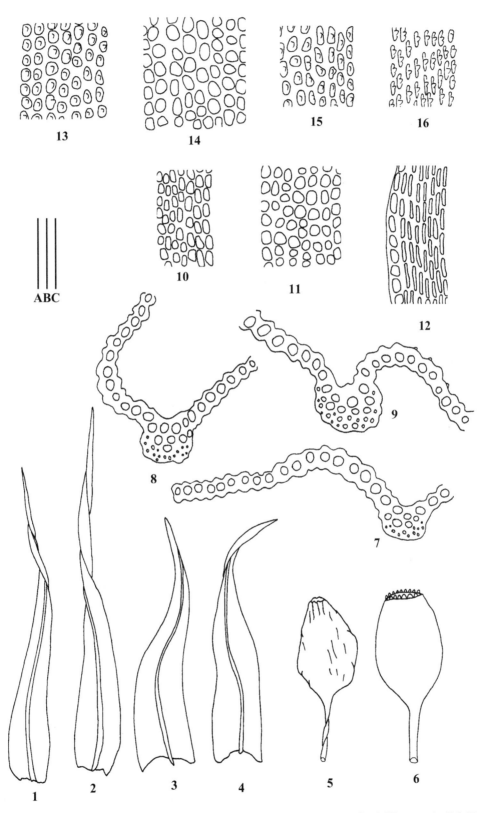

Fig. 185 *Macromitrium petelotii* Tixier 1, 2: Branch leaves. 3, 4: Stem leaves. 5: Dry capsule. 6: Wet capsule. 7-9: Transverse sections of branch leaves. 10: Upper cells of stem leaf. 11: Medial cells of stem leaf. 12: Basal marginal cells of stem leaf. 13: Upper cells of branch leaf. 14: Medial cells of branch leaf. 15: Low cells of branch leaf. 16: Basal cells of branch leaf (all from lectotype, PC 0083700). Line scales: A = 1 mm (1, 2, 5, 6); B = 400 μm (3, 4); C = 67 μm (7-16).

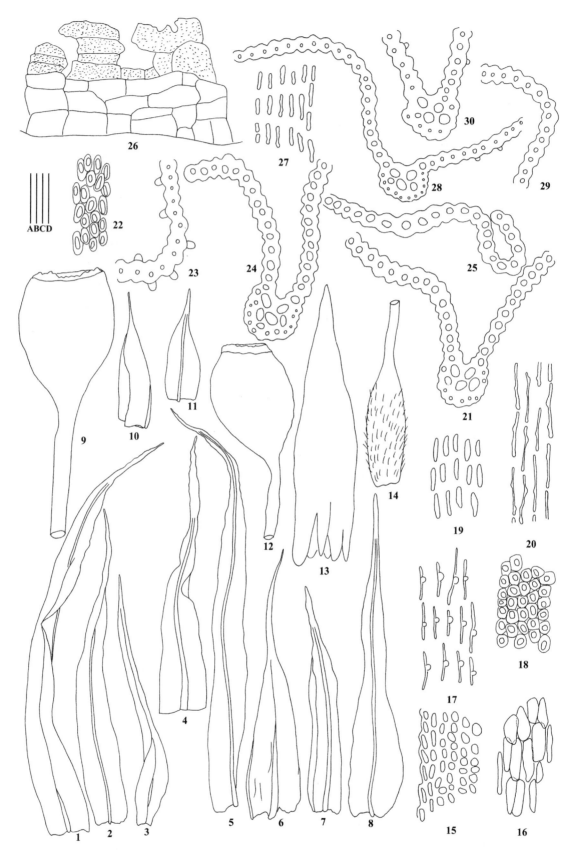

Fig. 186 *Macromitrium picobonitum* B. H. Allen. 1-3, 5: Branch leaves. 4: Apex of branch leaf. 6-8: Perichaetial leaves. 9, 12: Capsules. 10-11: Inner perichaetial leaves. 13: Calyptra. 14: Vaginula. 15: Low cells of branch leaf. 16: Basal cells near costa of branch leaf. 17: Basal cells of branch leaf. 18: Medial cells of branch leaf. 19: Medial cells of perichaetial leaf. 20: Basal cells of perichaetial leaf. 21, 24, 29, 30: Medial transverse sections of branch leaves. 22: Upper cells of perichaetial leaf. 23, 25: Basal transverse sections of branch leaves. 26: Peristome. 27: Low cells of branch leaf. 28: Low transverse section of branch leaf (all from holotype, MO 406397). Line scales: A = 1 mm (1-3, 5-14); B = 400 μm (4); C = 100 μm (26); D = 67 μm (15-25, 27-30).

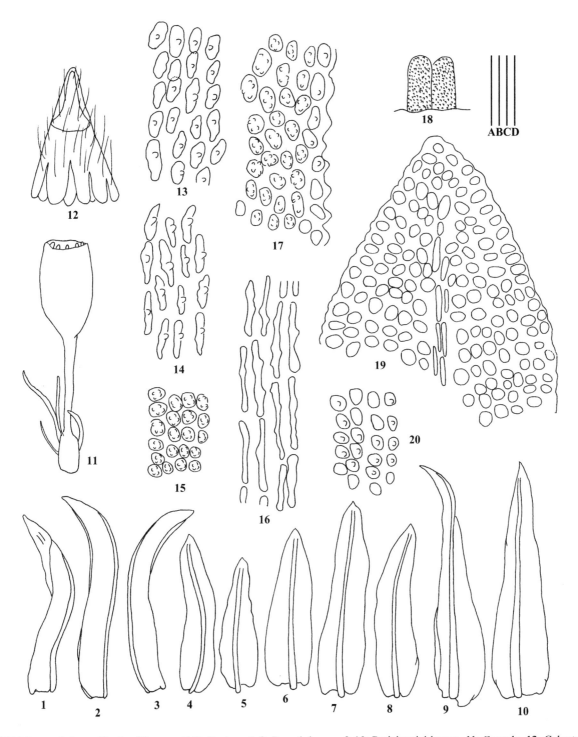

Fig. 187 *Macromitrium pilicalyx* Dixon ex E.B. Bartram 1-8: Branch leaves. 9-10: Perichaetial leaves. 11: Capsule. 12: Calyptra. 13: Medial cells of perichaetial leaf. 14: Low cells of perichaetial leaf. 15: Upper cells of branch leaf. 16: Basal cells of perichaetial leaf. 17: Upper cells of perichaetial leaf. 18: Peristome. 19: Apical cells of branch leaf. 20: Low cells of branch leaf (all from syntype, BM 000982741). Line scales: A = 2 mm (11-12); B = 1 mm (1-10); C = 100 μm (18); D = 67 μm (13-17, 19-20).

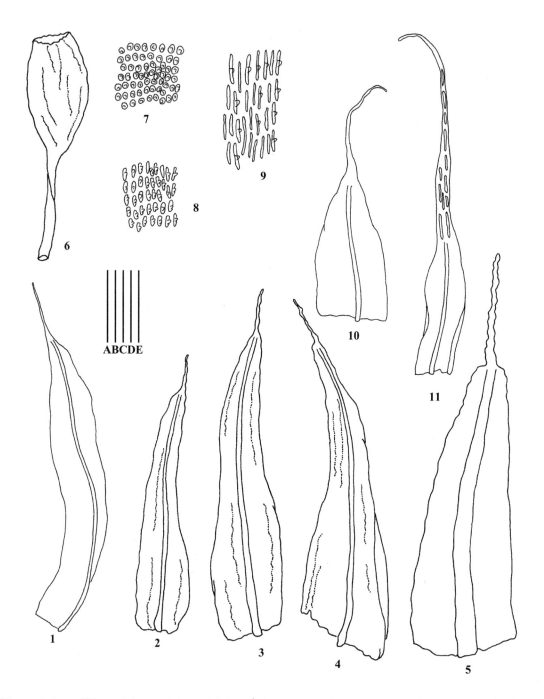

Fig. 188 *Macromitrium piliferum* Schwägr. 1-3: Branch leaves. 4: Perichaetial leaf. 5, 10-11: Apices of branch leaves. 6: Capsule. 7: Upper cells of branch leaf. 8: Low cells of branch leaf. 9: Basal cells of branch leaf (1, 7-11 from lectotype, G 00046153; 2-6 from FH). Line scales: A = 1 mm (6); B = 500 μm (2-4); C = 400 μm (1); D = 200 μm (5, 10-11); E = 67 μm (7-9).

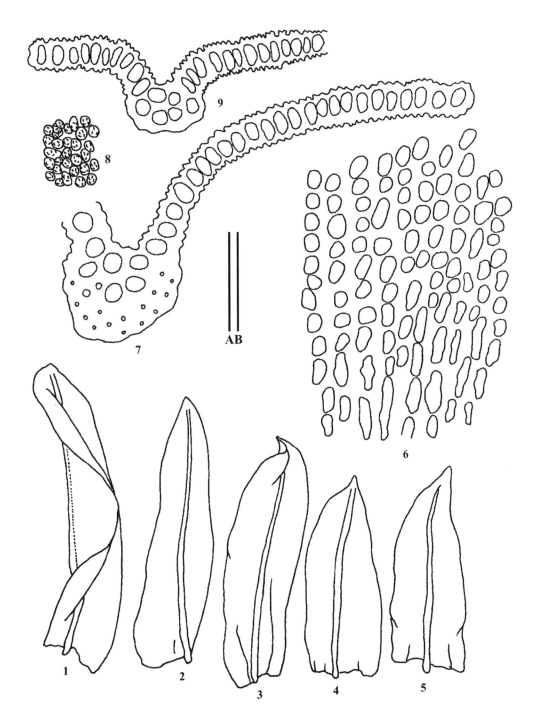

Fig. 189 *Macromitrium pilosum* Thér. 1-5: Branch leaves. 6: Basal cells of branch leaf. 7: Medial transverse section of branch leaf. 8: Upper cells of branch leaf. 9: Upper transverse section of branch leaf (all from H-BR 2561004). Line scales: A = 0.44 mm (1-5); B = 44 μm (6-9).

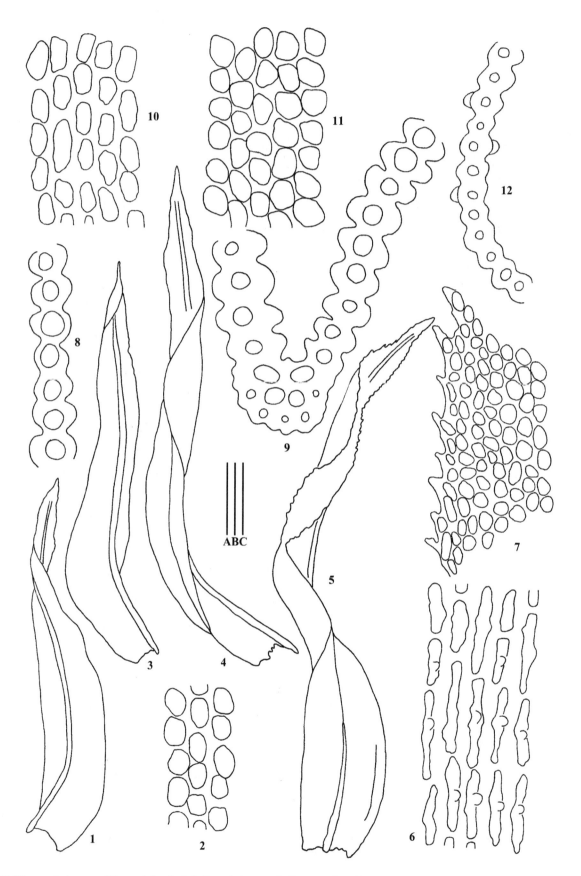

Fig. 190 *Macromitrium proliferum* Mitt. 1, 3-5: Branch leaves. 2: Upper cells of branch leaf. 6: Basal cells of branch leaf. 7: Upper marginal cells of branch leaf. 8-9: Medial transverse sections of branch leaves. 10: Medial cells of branch leaf. 11: Low cells of branch leaf. 12: Basal transverse section of branch leaf (all from isosyntype, BM 000873087). Line scales: A = 1 mm (1, 3-5); B = 100 μm (7); C = 67 μm (2, 6, 8-12).

178. *Macromitrium prolongatum* Mitt., Trans. Linn. Soc. London, Bot. 3: 162. 1891. (Figure 191)

Type protologue: [Japan] Kintoki, *Bisset s.n.* Type citation: Japan, Kintoki, 1876 (isotype: E 00165192!).

Macromitrium brachycladulum Broth. & Paris., Nat. Pflanzenfam. I(3): 1202. 1909, *fide* Noguchi & Iwatsuki, 1989. Type protologue: Japan. Type citation: (Japan) Im Shikoku, *Faurie, 1024*, 1900 (letotype designated here: H-BR 2573002!).

Macromitrium prolongatum var. *brevipes* Cardot, Bull. Soc. Bot. Genève 1: 122. 1909, *fide* Noguchi & Iwatsuki, 1989. Protologue: Japon: sommet du Miyogisan, 850 m. (*n. 2776*); Shikoku, Tsurugizan (*n. 1024*); Iyo, mont Ishizuchi (Gono). Corée: Ile Quelpaert (*n. 101, 519*).

(1) Plants small to medium-sized, yellowish green above, blackish green below, in dense mats; stems long creeping, prostrate, more or less slender above, or stoloniferous and ascending; branches up to 5 mm long, with many branchlets simple or branched. (2) Stem leaves appressed when dry, reflexed when moist, linear-lanceolate, gradually becoming narrowed towards the apex; leaves of slender stems oblong-ovate, slenderly acuminate, usually 2.0-2.5 mm long, keeled; margins entire, slightly recurved in the middle; costa stout, percurrent; medial cells hyaline, oblong, rectangular, papillose or smooth, thick-walled; low cells linear, strongly thick-walled. (3) Branch leaves keeled, contorted-twisted and curved when dry, spreading, often adaxially incurved, sometimes abaxially curved above when moist, 1.5-2.5 × 0.3-0.4 mm, long linear-lanceolate; apices acute or acuminate; margins entire throughout; costae stout, percurrent; upper and medial cells obscure, hexagonal to almost quadrate, small, 4.5-6.5 μm wide, densely well-developed pluripapillose, thin-walled, slightly inflated; low and basal cells larger, linear, 12-20 μm long, smooth, thick-walled. (4) Perichaetial leaves narrowly oblong, up to 2.5 mm long, gradually tapered to form a subulate apex; costa weak, ending beneath the apex; all cells hyaline, narrowly oblong, thick-walled and smooth. (5) Setae short, about 2 mm long. (6) Capsules erect, slightly exserted the perichaetial leaves, urns ellipsoid-ovoid, 1.2-1.5 × 0.8-1.0 mm, brown, slightly plicate when dry; opercula conic-rostrate, beak 0.4-0.6 mm long; peristome single, exostome teeth lanceolate, to 0.3 mm long, obtuse at the apex, hyaline; spores spherical, 20-35 μm in diameter, finely papillose. (7) Calyptrae mitrate, 1.8-2.5 mm long, densely hairy.

Macromitrium prolognatum Mitt. is similar to *M. giraldiii* Müll. Hal. The former could be separated from the latter by 1) upper and medial leaf cells obscure and with well-developed pluripapilliae; 2) basal leaf cells linear and smooth; and 3) setae rather short, and sporophytes slightly exserted the perichaetial leaves.

Distribution: Japan, South Korea.

Specimens examined: **JAPAN**. Pref. Kyoto, *M. Tagawa, 2241* (H 3090655); Pref. Yamanashi, *R. Watanabe 827* (H 3090652); Pref. Kagoshima, *Z. Iwatsuki, A.J. Sharp & Evelyn Sharp* (H 3090653); Kyushu, Kumamoto, *K. Mayebara* (KRAM-B-0777249), *K. Mayebara 72* (MO 2062870); Honshu, *H. Ochi 1561* (MO 2111615); Yakushima, *Z. Iwatsuki, A. J. Sharp & E. B. Sharp 1030* (MO 5269122), *K. Mayebara 22610* (MO), *K. Mayebara 283* (MO), *R. Watanabe 827* (H 3090652); Aki Miyajima, *J. Kasirira* (H-BR 2572009). **SOUTH KOREA**. Cheju-do, *J. R. Shevock 16381* (MO 4411809).

179. *Macromitrium prorepens* (Hook.) Schwägr., Sp. Musc. Suppl. 2 2(1): 62. 1826.

Basionym: *Orthotrichum prorepens* Hook., Musci Exotici 2: 120. 1819. Type protologue: In sinu Dusky Bay dicto apud Novam Zeelandiam. *D. Menzies.* 1791.

Macromitrium oocarpum Müll. Hal., Hedwigia 37: 157. 1898, *fide* Vitt, 1983. Type protologue: Nova Seelandia, Insula australis littore australsiaco prope Grey Mouth: Richard helms 1885 legit et misit, Type citation: New Zealand, leg. *R. helms* (lectotype: H-BR 2545013!).

Macromitrium prorepens var. *aristata* Allison, Trans. Roy. Soc. New Zealand 88: 10. 1960. *fide* Fife, 2017. Type protologue: New Zealand, Little Barrier Island, 300 feet 1958, *Dingley s.n.*

(1) Plants mostly slender and small, rusty brown, in dense and spreading mats; stems long creeping, with short and erect branches, branch up to 1.0 cm high. (2) Stem leaves ovate-lanceolate, stoutly acuminate. (3) Branch leaves irregularly twisted-flexuose, stiffly spreading when moist, 1.2-2.3 mm long, ligulate to broadly lanceolate-ligulate, abruptly contracted to a short, stout apiculus; margins broadly reflexed, subcrenulate above, entire below; costae percurrent or excurrent and filling the apiculus; upper cells quadrate-rounded to irregularly hexagonal-rounded, with firm walls, chlorophyllose, obscure, slightly to sometimes strongly bulging, densely pluripapillose, papillae mostly forked; medial cells elliptic-rounded, unipapillose and in longitudinal rows; basal cells long-rectangular, thick-walled, smooth or sometimes tuberculate, basal marginal cells thin-walled, wider than their ambient cells, form a short border. (4) Perichaetial leaves lanceolate-oblong to lanceolate-ovate, acute or cuspidate-apiculate. (5) Setae short, 2-4 mm, smooth, twisted to the left. (6) Capsule urns ovoid-obloid, smooth, lightly 4-plicate beneath mouth; peristome single, exostome of 16, well-developed. (7) Calyptrae conic-mitrate, distinctly plicate, smooth, sparsely to densely hairy, evenly lacerate.

Distribution: Australia, New Zealand (Vitt & Ramsay, 1985).

Specimens examined: **NEW ZEALAND**. South island, *H. Streimann 58181* (KRAM-B-131056, H 3090657), *Allan J. Fife 4999* (MO 3986989); Wellington prov. *Buchanan* (H-BR 2509013); Otago prov. Dunedin, *W. Bell 442a* (H-BR 2509025); *W. Bell 412* (H-BR 2509027).

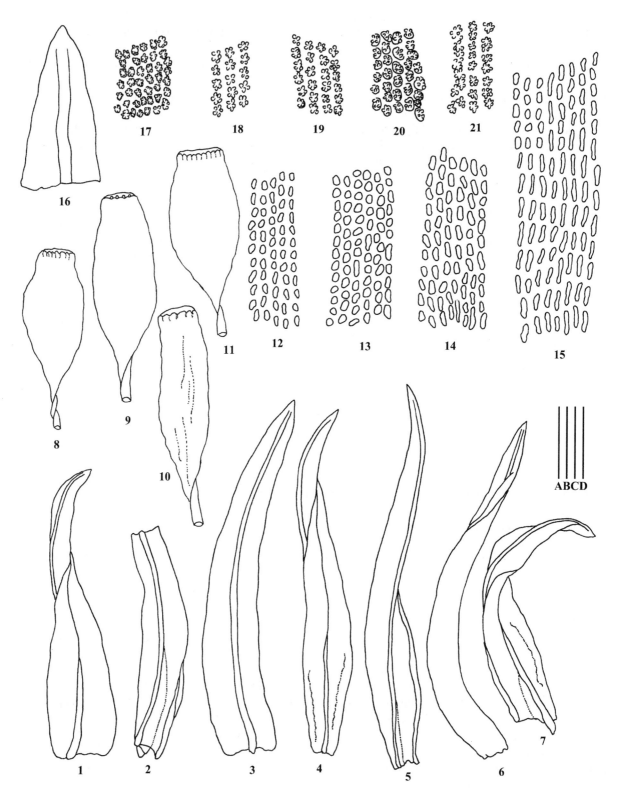

Fig. 191 *Macromitrium prolongatum* Mitt. 1-7: Branch leaves. 8-11: Capsules. 12-14: Low cells of branch leaves. 15: Basal cells of branch leaf. 16: Apex of branch leaf. 17-19: Upper cells of branch leaves. 20-21: Medial cells of branch leaves (all from isotype, E 00165192). Line scales: A = 1 mm (8-11); B = 500 μm (1-7); C = 200 μm (16); D = 50 μm (12-15, 17-21).

180. *Macromitrium proximum* Thér., Recueil Publ. Soc. Havraise Études Diverses 92: 140, f. 1-5. 1925. (Figure 192)

Type protologue and citation: Madgascar, Mont Tsaratanana, atl. 1200-2400 m., 04.1924, *Perrier de la Bathie* (lectotype designated here: PC 0073305!; isotype: H-BR 2626006!).

(1) Plants medium-sized, yellowish-green above, dark brown below; stems prostrate and creeping, with branches up to 20 mm long. (2) Stem leaves inconspicuous and caducous. (3) Branch leaves erect below, undulate, contorted-crisped and flexuose, with an curved to circinate apex hidden in the inrolled cavity when dry, spreading and weakly undulate and adaxially incurved above when moist, lanceolate from a broadly oblong base, conduplicate and curved in upper portion, acuminate; margins entire, plicate below; costae sometimes partially concealed in adaxial view by overlapping folds of laminae; upper cells rounded-quadrate to rounded, pluripapillose; medial cells short-rectangular; low and basal cells elongate-rectangular, thick-walled, not porose, smooth, narrow lumens curved. (4) Perichaetial leaves shorter than branch leaves, oblong-lanceolate to triangular lanceolate, abruptly narrowed to a weakly cuspidate apex. (5) Setae 5-10 mm long, smooth, weakly twisted to the left. (6) Capsules cylindric, furrowed, slightly constricted beneath the mouth; peristome present and single; opercula conic-subulate. (7) Calyptrae conic-mitrate, smooth and naked, lobed at the base.

Distribution: Madagascar.

181. *Macromitrium pseudofimbriatum* Hampe, Vidensk. Meddel. Dansk Naturhist. Foren. Kjøbenhavn, ser. 3, 6: 144. 1875. (Figure 193)

Type protologue and citation: [BRAZIL], Rio de Janeiro: *A. Glaziou 7053* (lectotype designated by Costa *et al.*, 2016: BM 000873081!).

Macromitrium doeringianum Hampe, Vidensk. Meddel. Dansk Naturhist. Foren. Kjøbenhavn ser. 4, 1–2: 96. 1879, *fide* Valente *et al.*, 2020. Type protologue: [BRAZIL], Rio de Janeiro: Petropolis, *Döring s.n.*

Macromitrium podocarpi Müll. Hal. Bull. Herb. Boissier 6: 96. 1898, *fide* Valente *et al.*, 2020. Type protologue: Brasilia, Minas Geraës, Serra Itabira de Campo, ad truncos arborum, Aprili 1892: E. Ule, Coll. No. 1066. Type citation: (Brazil) Minas Geraes, Serra Itabira de Campo, ad truncos arborum, Podocarpi, April 1892, *E. Ule* (marked with "*Macromitrium podocarpi* C. Müll, det: C. Müller, *n. sp.* distributed as Bryotheca brasiliansis 222) (isotypes: NY 01086627!, GOET 12309!).

Macromitrium podocarpi Müll. Hal. var. *falcifolium* Müll., Hal., Bull. Herb. Boissier 6: 96. 1898, *fide* Valente *et al.*, 2020.Type protologue: [Brazil] Minas Geraës: Caraça, in ramulis arborum, Martio 1892, *E. Ule, 1408*. Type citation: [Brazil] Minas Geraës: Caraça, *E. Ule 1408* (isotype: R 000014299).

Macromitrium portoricense R.S. Williams, Bryologist 32: 69. 5. 1929, *fide* Churchill & Linares, 1995 and Allen, 2002. Type protologue: On road between maricao and Monte Alegrillo, Porto Rico, at. 500-900 meters, on exposed rock. *No. 2638. Elizabeth G. Britton,* April 13, 1913.

(1) Plants small, yellow-green to dark-green; stems strongly creeping, densely covered with red-rusty rhizoids; branches 1-3 cm long. (2) Branch leaves erect below, individually to spirally twisted, flexuose-contorted to crisped above when dry, erect-spreading when moist, 1.5-2.2 × 0.5 mm, oblong to ligulate-lanceolate, with an acute to obtusely-apiculate apex; margins entire or crenulate, recurved or plane below, plane above; costae percurrent or excurrent into a short apiculus; upper and medial interior cells rounded to hexagonal-rounded, 6-12 µm, bulging mammillose, upper marginal cells not differentiated; basal cells long-rectangular, 40 µm long, thick-walled and porose, tuberculate, basal cells near costa enlarged and hyaline, short, irregularly rectangular, forming a "cancellina region", marginal enlarged teeth-like cells at the insertion slightly differentiated. (3) Perichaetial leaves larger and longer than vegetative, lanceolate from an oblong or broadly oblong low part, plicate below, widely acuminate; upper cells irregular rounded-quadrate, subquadrate; medial, low and basal cells elongate, long rectangular; setae 4-5 mm long, smooth, twisted to the left. (4) Capsule urns cylindric to ellipsoid-ovoid, weakly to strongly furrowed, or wrinkled; peristome double, exostome teeth lanceolate, united to form a membrane, endostome hyaline, with basal membrane and upper segments; spores anisosporous. (5) Calyptrae up to 2.5 mm long, mitrate, naked and deeply lacerate.

Distribution: Bolivia, Brazil, Colombia, French Guiana, Honduras, Peru, Suriname, Venezuela.

Specimens examined: **BOLIVIA**. La Paz, *A. Fuentes 4283* (MO 5647773), *5209* (MO 5647774), *9742* (MO 5910828), *A. Fuentes & C. Aldana 6457* (MO 5914036). **BRAZIL**. São Paulo, *D. M. Vital 7337* (MO 5368757); Serra de Itabira do Campo, *E. Ule s.n.* (H-BR 2603006); Minas Gerais, *Daniel M. Vital & William R. Buck 19633* (MO 6001425), *19655* (MO 6001427). **COLOMBIA**. Musgos sobre tronco, *S.P. Churchill et al., 14196* (MO 3374294), Aragua, Parque Nacional Macarao, *Morales Thalia 1489* (MO 6231629). **FRENCH GUIANA**. St-Laurent-du-Maroni, Canton de Maripasoula, *William R. Buck 18665* (MO 5136903). **HONDURAS**. Olancho Department, Sierra de Algalta, *B. Allen 12462* (KRAM-B 100803); Lempira Department, Montana de Celaque, *B. Allen 111231B* (MO 3965342). **PERU**. Cuzco, *P+E Hegewald 8768* (MO 4418592). **SURINAME**. Sipaliwini, Tafelberg, *B. Allen 23309* (MO 5644941), *23534* (MO 5644739), *23580* (MO 5282382), *23601* (MO 5282400), *23604* (MO 5644740). **VENEZUELA**. Aragua, *Morales Talia 1489* (MO 6231629); Monagas, *R.A. Pursell 8968* (MO 5134084).

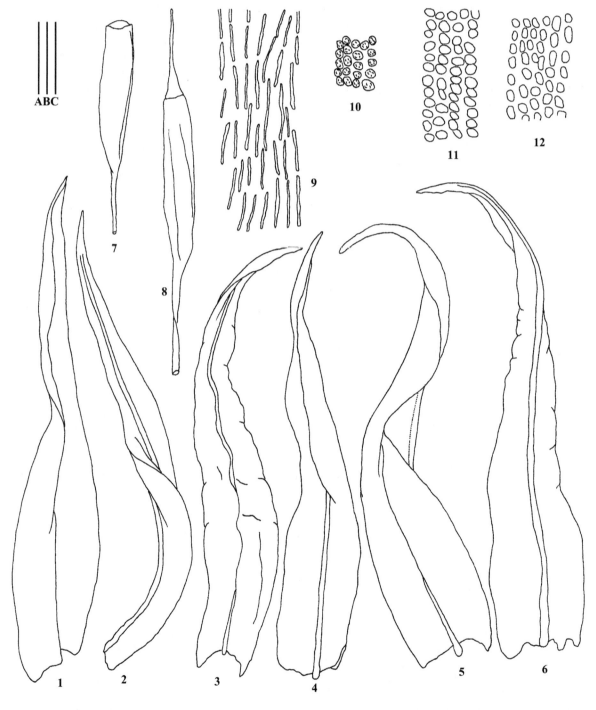

Fig. 192 *Macromitrium proximum* Thér. 1-6: Branch leaves. 7-8: Capsules. 9: Basal cells of branch leaf. 10-11: Upper cells of branch leaves. 12: Medial cells of branch leaf (all from isotype, H-BR 2626006). Line scales: A = 0.88 mm (7-8); B = 0.44 mm (1-6); C = 44 μm (9-12).

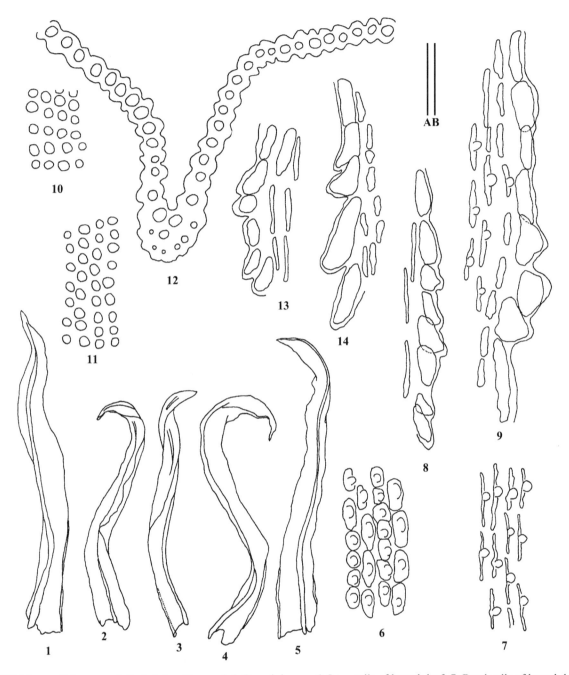

Fig. 193 *Macromitrium pseudofimbriatum* Hampe 1-5: Branch leaves. 6: Low cells of branch leaf. 7: Basal cells of branch leaf. 8-9, 13-14: Basal marginal cells of branch leaves. 10: Upper cells of branch leaf. 11: Medial cells of branch leaf. 12: Basal transverse section of branch leaf (all from lectotype, BM 000873081). Line scales: A = 1 mm (1-5); B = 67 µm (6-14).

182. *Macromitrium pseudoserrulatum* E.B. Bartram, Rev. Bryol. Lichénol. 33: 326. 1964 [1965]. (Figure 194)

Type Protologue: [Argentina] Tucuman: Estancia Los Pinos, La Mesada, 800 m, *Digilio & Grassi 8020*, holotype: herb. Bartram. Type citation: Argentina, Tucuman, Estancia "Los Pinos", "La Mesada", 1800 m, leg. *Digilio & Grassi, 8020*, 21. 1. 1947 (holotype: FH 00213665!; isotype: JE 04008743!).

(1) Plants large, dark-brownish; stems short and weak, with sparse reddish rhizoids; branches 20-25 mm long, 2 mm thick, ascending. (2) Branch leaves erect below, individually twisted, contorted-crisped and flexuose above when dry, spreading when moist, long lanceolate, short acuminate, keeled, conduplicate and curved above, undulate in medial and upper portion, and plicate below, decurrent at auricles; costae single and stout, percurrent; margins plane, weakly serrate to crenulate; cells near the apex rounded-quadrate, 8-10 µm wide, blackish triangular among

cells; medial and upper cells isodiametric, round-quadrate, pluripapillose; low cells rectangular, weakly porose; basal cells narrow rectangular, smooth; cells along the costa wider and larger, marginal cells wider, quadrate and rhombic, smooth; basal marginal teeth-like large cells not differentiated. (3) Perichaetial leaves lanceolate, acuminate to a long arista, plicate in medial and low portion, truncate and somewhat sheathing at the insertion; costae thin and weaker than those of branch leaves, vanished before the caudate apex; all cells smooth, longer than wide, those near the costa and margin not differentiated. (4) Setae smooth, 6-10 mm. (5) Capsule urns ellipsoid, ellipsoid-cylindric, or short cylindric, longitudinally wrinkled when dry, peristome double, both rudimentary and membraneous; opercula conic-subulate, 2 mm long. (6) Calyptrae large, mitrate, naked and smooth, deeply lobed below.

Distribution: Argentina.

183. *Macromitrium pulchrum* Besch., Ann. Sci. Nat., Bot., sér. 5, 18: 210. 1873. (Figure 195)

Type protologue: In moute Humboldt, 1200 m. alt. (*Balansa, no 2528*); in monte Mou (*Balansa, no 2980* partim). Type citation: New Caledonia, *Balansa 2528* (isosyntype: H-BR 2551011!).

Macromitrium pulchrum var. *aristatum* Thér., Rev. Bryol. 48: 16. 1921. *fide* Thouvenot, 2019. Type protologue: New Caledonia, *Franc s.n.* (lectotype designated by Thouvenot, 2019: PC 0083721!).

Macromitrium pulchrum var. *densirete* Thér., Rev. Bryol. 48: 16. 1921. *fide* Thouvenot, 2019. Type protologue: New Caledonia, Mt Koghis, XI 1909, *Franc s.n.* (lectotype designated by Thouvenot, 2019: PC 0083722!, JE!).

(1) Plants medium-sized to large, brown-reddish; stems long creeping, densely with branches, branches up to 10 mm tall and 1.5-2 mm thick. (2) Branch leaves erect below, individually twisted flexuose, keeled, with an incurved to circinate apex when dry, erect to patent when moist, 2.5-4.5 × 0.4-0.8 mm, lanceolate, ligulate from a wide ovate to oblong base; apices narrowly obtuse, mucronate to short aristate; upper parts obscure, basal translucent; costa excurrent, often partially concealed in the upper portion in adaxial view by overlapping folds of laminae; margins papillose crenulate, plane or recurved on one side near the base; upper cells quadrate, rounded to oblong elliptic, thick-walled, conic-bulging and weakly pluripapillose, the external walls strongly protruding; upper marginal cells smaller, pluripapillose; medial cells short rectangular, thick-walled, porose, unipapillose; low and basal cells elongate-rectangular to linear, very thick-walled, unipapillose, basal translucent part occupying 1/3-1/5 the leaf length. (3) Perichaetial leaves similar to branch leaves, lanceolate, mostly acuminate or acute, aristate translucent. (4) Setae long, up to 35 mm long, smooth, vaginulae naked. (5) Capsule urns ellipsoid to narrowly ellipsoid, sub-cylindric, smooth, plicate in the upper portion, peristome absent or reduced to a papillose membrane. (6) Calyptrae mitrate and naked.

Distribution: New Caledonia.

Specimens examined: **NEW CALEDONIA**. Nov Calédonie, *Franc 102*, 1907 (PC 0083723); Qic Du Suwus, *Le Rat s.n.* (H-BR 2551001); Mont Dzumac, *Le Rat s.n.* (H-BR 2551002), *1431* (H-BR 2551006); im Mou, *Le Rat 1228*, 1908 (H-BR 2551004).

Macromitrium pulchrum var. *neocaledonicum* (Besch.) Thouvenot, Cryptog. Bryol. 40(16): 203. 2019. (Figure 196)

Basionym: *Macromitrium neocaledonicum* Besch., Ann. Sci. Nat., Bot., sér. 5, 18: 211. 1873. Type protologue: New Caledonia, in monte Mou, 1200 m, *Balansa 2980* cum *M. pulchro socium* (lectotype designated by Thouvenot, 2019: BM 000982735; isolectotypes: PC 0137835, PC 0137836, PC 0096506, H-BR 2552010!).

Macromitrium pulchrum var. *neocaledonicum* (Besch.) Thouvenot differs from the type variety of *M. pulchrum* Besch. by the former having 1) leaf apices less tightly enrolled when dry; 2) leaves spreading recurved instead of erect-spreading when moist; 3) branches usually longer with fastigiated branchlets, up to 20 mm long, the perichaetia developing at the apices of branchlets or single main branches; 4) upper leaf shape narrowly ligulate, rounded to obtuse and mucronate at the apices instead of lanceolate acute; and 5) upper cells more evenly isodiametric, less thick-walled (Thouvenot, 2019).

Distribution: New Caledonia.

184. *Macromitrium pullenii* Vitt, Acta Bot. Fenn. 15: 63. f. 2e, 28. 1995.

Type protologue: Papua New Guinea. Milne Bay District, Nowata about 10 km west of Raboroba, forest on steep plateau side, on high-branches of forest trees, 340 m, 09°59′S, 149°43′E, *Pullen 7705* (holotype: L; isotype: COLO).

(1) Plants medium-sized, dull, stiff, in dense mats; stems creeping, prostrate; branches numerous, closely set, erect, up to 10 mm long; stems and branches with abundant rhizoids. (2) Stem leaves lanceolate, gradually acuminate. (3) Branch leaves tightly spirally twisted around the branch, flexuose-twisted when dry, erect to erect-spreading when moist, 1.7-2.0 mm long, lingulate, oblong-lingulate, upper portions often rugose; apices obtuse to broadly acute, some retuse, ending in a long hyaline arista; margins broadly recurved to reflexed above, plane below, entire throughout; costae excurrent and filling arista; upper and medial cells rounded to rounded-quadrate (5-11 µm), smooth, clear, thick-walled, bulging; inner basal cells 8-9 × 9-35 µm, irregularly thick-walled, straight, rounded to elliptic, short-

Taxonomy

rectangular, becoming shorter above, most cells strongly tuberculate, outer basal cells somewhat longer and narrower, flat at insertion, marginal row hyaline, thin-walled, 8-10 μm wide, forming a short border. (4) Perichaetial leaves 2.3-2.6 mm long, erect, sheathing, not much differentiated from branch leaves, inner basal cells longer, not tuberculate, upper cells slightly bulging. (5) Setae smooth, 8-11 mm long, twisted to the left; vaginulae not hairy. (6) Capsule urns broadly ovoid-ellipsoid to fusiform-cylindric, smooth, ridged when old, gradually contracted to the seta, with a firm mouth; peristome single, exostome of 16; spores 16-30 μm, indistinctly anisosporous. (7) Calyptrae plicate, conic-mitrate, naked (Vitt et al., 1995).

M. pullenii Vitt is similar to *M. tongense* Sull. and *M. similirete* E.B. Bartram in their branch leaves tightly spirally twisted around the branch, but the latter two species can be separated from *M. pullenii* by their densely pluripapillose upper and medial cells of branch leaves.

Distribution: Papua New Guinea.

185. *Macromitrium punctatum* (Hook. & Grev.) Brid., Bryol. Univ. 1: 739. 1826. (Figure 197)

Basionym: *Orthotrichum punctatum* Hook. & Grev. Edinburgh J. Sci. 1: 119. 5. 1824. Protologue: Brazil. Communicated from Brazil, together with *O. filiforme*, by Professor *Raddi*. Type citation: Brazil, *Raddi s.n.* (isotype: E 00011666!).

Macromitrium hirtellum E.B. Bartram, Contr. U.S. Natl. Herb. 26: 86. f. 25. 1928, *fide* Vitt, 1979. Type protologue: On tree, Quebrada Serena, southeast of Tilaran, province of Guanacaste, Costa Rica, altitude about 700 meters, *Paul C. Standley and Juvenal Valerio, no. 46257,* January 27, 1926.

Macromitrium liberum Mitt., J. Linn. Soc., Bot. 12: 214. 1869, *fide* Grout, 1944. Type protologue: Andes Novo-Granatenses, in arboribus ad viam inter pacho et Veragua (6500 ped.).

Macromitrium pentagonum Müll. Hal., Malpighia 10: 513. 1896, *fide* Vitt, 1979.

Macromitrium pentastichum Müll. Hal., Linnaea 21: 186. 1848, *fide* Vitt, 1979. Type protologue: Mai 1846 mit veralteren Früchten, *Hb. Kegel no 1405*.

Macromitrium reflexifolium Mitt., J. Linn. Soc., Bot. 12: 211. 1869. *fide* Bowers, 1974. Type protologue: Guatemala, Coban, *Godman & Salvin s.n.*

Macromitrium sartorii Müll. Hal., Linnaea 38: 641. 1874, *fide* Grout, 1944. Type protologue: Mexico, Mirador, in arboribus sylvestribus: *Florentin Sartorium* vere 1873 Hb. *C. Mohr.*

Macromitrium sumichrastii Duby, Mém. Soc. Phys. Hist. Nat. Genève 19: 297. 3 f. 1. 1868, *fide* Grout, 1944. Type protologue: Ad arbores in terries calidis mexicanis reperiit D. SuMICH rast.

Schlotheimia brachyrhyncha Schwägr., Sp. Musc. Frond., Suppl. 2 2(1): 53. pl. 168. 1826, *fide* Grout, 1944. Type protologue: In Brasilia prope Rio Ianeiro ad saxa, Ianuario mense legit dilig. Beyrich 4.

(1) Plants small to medium-sized, yellowish-green to rusty-brown; stems creeping, branches to 2-3 cm long. (2) Branch leaves erect below, flexuose and undulate, strongly twisted-crisped, with an circinate apex hidden in the inrolled cavity when dry, erect to flexuose-spreading, slightly abaxially curved when moist, 1.5-3 × 0.5-0.7 mm, ligulate-lanceolate to oblong-lanceolate; apices obtuse-acute, acute to shortly acuminate; margins serrulate to irregularly serrate and plane above, entire and recurved below; costae percurrent to shortly excurrent; upper interior cells rounded-quadrate to rounded-hexagonal, conic-bulging to mamimillose, thin-walled; elongated cells restricted to the low 1/5 of leaf, rectangular, 40 μm long, thick-walled and porose, rarely tuberculate; marginal enlarged teeth-like cells at the insertion not differentiated. (3) Perichaetial leaves linear-lanceolate, longer than branch leaves, up to 6 mm long. (4) Seate 3-12 mm long, smooth and straight below and ridge and slightly twisted to the right above. (5) Capsule urns cupulate to ellipsoid, smooth or furrowed; peristome double, exostome teeth lanceolate, densely papillose, united forming an erect membrane, endostome hyaline, membraneous. (6) Calyptrae mitrate, sparsely to densely hairy, lacerate.

Macromitrium punctatum (Hook. & Grev.) Brid. is similar to *M. microstomum* (Hook. & Grev.) Schwägr. and *M. richardii* Schwägr. in oblong-lanceolate to lingulate-lanceloate leaves with an acute or apiculate apex, but *M. microstomum* differs from *M. punctatum* by its elongate cells in the low 1/2 of the leaf, and its capsules with a puckered mouth and single peristome. *Macromitrium richardii* can be separated from *M. punctatum* by its pluripapillose upper cells of leaves.

Grout (1944) placed *M. sartorii* Müll. Hal. in synonymy with *M. punctatum*. When Müller (1987) described *M. sartorii*, he also described two varieties under the species, var. *gracilescens* and var. *robustius*. Müller thought that var. *gracilescens* was different from var. *sartorii* by its longer branches, leaves and setae; and var. *robustius* different from var. *sartorii* by its longer and thicker stems and shorter setae. Considering the variations of *M. punctatum* in lengthes of setae, leaves and branches, we guessed that these two varieties are conspecific with *M. punctatum*.

Distribution: Bolivia, Brazil, Colombia, Honduras, Suriname, Venezuela.

Specimens examined: **BOLIVIA**. La Paz: Franz Tamayo. *A.F. Fuentes & H. Huaylla 13108* (MO 6490222),

Mñecas, *A.F. Fuentes 8041* (MO 6363853); Cochabamba, *I. Jimenez & T. Kromer 2097* (MO 6094986); Chuquisaabout *R. Lozano, M. Serrano et al. 1721* (MO 5922879). **BRAZIL**. São Paulo, *Schafer-Verwimp & Verwimp 9577* (MO 5915877). **COLOMBIA**. Musgos sobre tronco de arbusto, *S.P. Churchill, E. Callejas, P. Acevedo y F. Saldarriaga 14867* (MO 3651965). **HONDURAS (as Belize)**. Toledo district, Don Owens-Lewis property, *A.T. Whittemore 5516B* (MO 5215661). **PERU**. Dpto. Pasco, *P+E Hegewald 8457* (MO 3670924). **SURINAME**. Sipaliwini, *B. Allen 19130* (MO 4410328). **VENEZUELA**. Monagas, *R. A. Pursell 8959* (MO 5134074), *R. A. Pursell 8986* (MO 5134075); Aragua, *Morales Thalia 1494* (MO 6230674).

186. *Macromitrium pyriforme* Müll. Hal., Syn. Musc. Frond. 2: 645. 1851. (Figure 198)

Type protologue: Venezuela, Galipan, 6000 pedes elevatum, ad arbores: *Wagner M*. Sept. 1849.". Type citation: Venezuela, Galipan, *Wagner, s.n.* (isotypes: NY 01086639!, NY 01086640!).

(1) Plants medium-sized, dark reddish brown; stems creeping, ascending branches with branchlets. (2) Branch leaves erect below, individually twisted, strongly spirally-contorted when dry, abaxially curved spreading, keeled and conduplicate when moist, about 5 mm long, lanceolate from an oval or oblong low part; the apices broadly acuminate; margins entire and recurved below; upper cells isodiametric, bulging, collenchymatous; medial cells elliptical to short rectangular, in longitudinal rows; cells elongate towards the base; basal cells linear rectangular, thick-walled, porose, tuberculate; costa excurrent to a short awn. (3) Perichaetial leaves shorter than branch leaves, lingulate-lanceolate; all cells longer than wide. (4) Setae smooth, up to 18 mm long, twisted to the left. (5) Capusle urns pyriform, obovate to cupulate, smooth or weakly to strongly furrowed; peristome teeth double, exostome lanceolate, discrete. (6) Calyptrae mitrate, naked.

Macromitrium pyriforme Müll. Hal. is similar to *M. cirrosum* (Hedw.) Brid., only the upper and medial cells strongly bulging, collenchymatous, and capsules pyriform.

Distribution: Colombia, Venezuela.

Specimen examined: **COLOMBIA**. Putumayo, Colón, *Bernardo R. Ramirez Padilla 10214* (MO).

187. *Macromitrium quercicola* Broth., Akad. Wiss. Wien Sitzungsber., Math.-Naturwiss. Kl. Abt. 1, 131: 212. 1923. (Figure 199)

Type protologue: [China] Prov. Yünnan: Prope vicos Hsinung 23°54' ([Handel-Mazzetti] *Nr. 508*), ca, 2000 m et Sanyingpan 26 °lat, about 2400 m ([Handel-Mazzetti] *Nr. 600*), ad septentr. urbis Yünnanfu, in regione calide temperata, ad truncos Quercuum.Type citation: [China, Yunnan] Y.: Hsinlung, 25°34', 10. III. 1914 (*Handel-Mazzetti 508*) (isosyntype: PC 0083724!); Sanyingpan, 26°, 14. III. 1914 (*Handel-Mazzetti 600*) n von Yünnanfu. (syntype: H-BR 2581011!; isosyntypes: S-B 115566!, PC 0083726!).

(1) Plants medium-sized, in loose mats; stems long creeping, densely with thick branches, branches 10-15 mm long and about 1.5 mm thick. (2) Stem leaves inconspicuous, appressed below and flexous above. (3) Branch leaves keeled, individually twisted, contorted-flexuose when dry, widely spreading when moist, about 2.5 × 0.5 mm, lanceolate-ligulate to lanceolate, with an acute or acuminate-acute apex; margins entire; costae subpercurrent, ending several cells beneath the apex; upper and medial cells isodiametric, rounded-quadrate, strongly bulging but clear, pluripapillose, gradually elongate from low part to the base; low cells short oblong, elliptic-rhombic, distinctly unipapillose; basal cells near costa thin-walled, smooth and pellucid, distinctly larger than their ambient cells, irregularly rectangular, various in size and shape, occasionally unipapillose, look like a "cancellina region". (4) Perichaetial leaves differentiated, lanceolate from a broadly oblong or ovate base, gradually narrowed to form a long aristate; cells near the apex rounded-quadrate, slightly elongate near the margin, clear and smooth; upper cells irregularly rounded-quadrate, ovate, elliptic, clear and smooth; medial and low cells oblong, elliptic, unipapillose. (5) Setae 5-12 mm long, smooth; vaginulae hairy. (6) Capsule urns ellipsoid-cylindric, smooth; peristome single, exstome of 16, short lanceolate. (7) Calyptrae mitrate, with long yellow-brown hairs.

Macromitrium quercicola Broth. was described as a new species based on the collection of Handel-Mazzetti from Yunnan province (Brotherus, 1923). The species was once treated as a synonym of *M. ferriei* Cardot & Thér. (Guo *et al*., 2007). However, we thought that *M. quercicola* had better be considered as an accepted species because it can be separated from *M. ferriei* in its 1) lanceolate-ligulate branch leaves with strongly conic-bulging, clear cells in upper and medial portions; 2) short elliptic-rhombic, thick-walled, porose cells in lower and basal portions; and 3) perichaetila leaves with a long arista.

Distribution: China.

Specimens examined: **CHINA**. Yunnan, *Handel-Mazzetti 3465* (H-BR 2581012, BM 000919469), *8784* (H-BR 2581015, PC 0083725, S-B 115562, BM 000919470, BM 000919471), *7201* (H-BR 2581013, PC 0083727; S-B 115582).

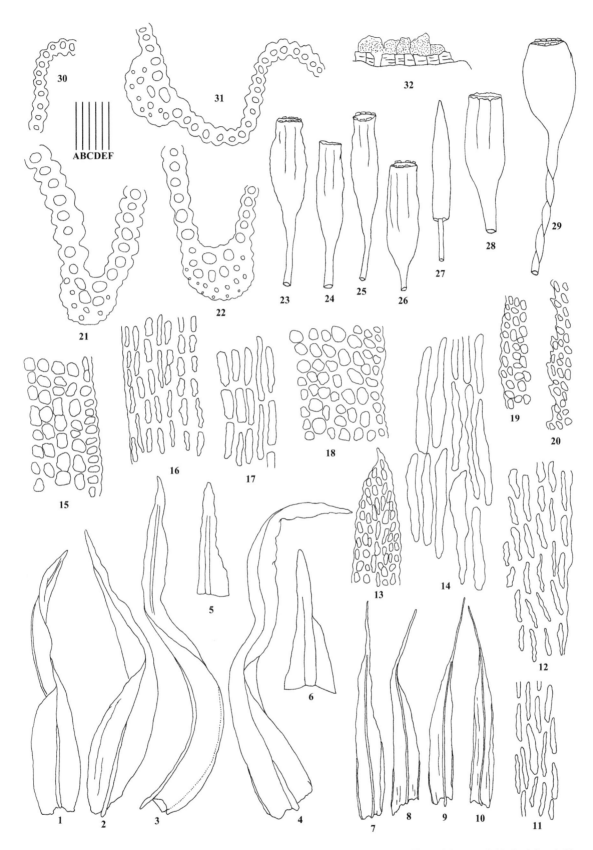

Fig. 194 *Macromitrium pseudoserrulatum* E.B. Bartram 1-4: Branch leaves. 5-6: Apices of branch leaves. 7-10: Perichaetial leaves. 11: Medial cells of perichaetial leaf. 12: Upper cells of perichaetial leaf. 13, 20: Apical cells of branch leaves. 14: Basal cells of perichaetial leaf. 15: Medial cells of branch leaf. 16: Low cells of branch leaf. 17: Basal cells of branch leaf. 18: Upper cells of branch leaf. 19: Upper marginal cells of branch leaf. 21-22: Medial transverse sections of branch leaves. 23-29: Capsules. 30-31: Basal transverse sections of basal leaves. 32: Peristome (all from holotype, FH 00213665). Line scales: A = 2 mm (7-10, 23-29); B = 1 mm (1-4); C = 400 μm (5-6); D = 200 μm (32); E = 100 μm (13, 19-20); F = 67 μm (11-12, 14-18, 21-22, 30-31).

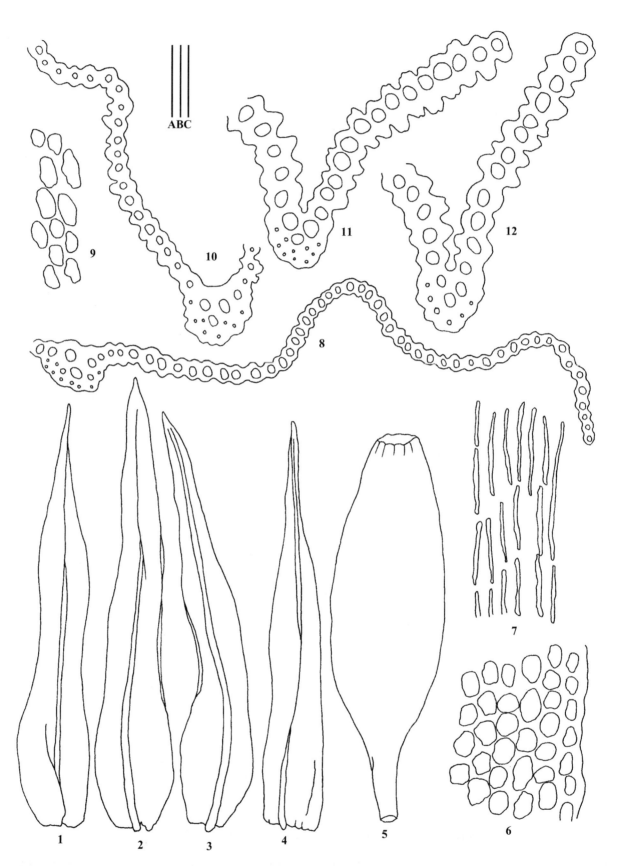

Fig. 195 *Macromitrium pulchrum* Besch. 1-4: Branch leaves. 5: Capsule. 6: Medial cells of branch leaf. 7: Basal cells of branch leaf. 8, 10: Basal transverse sections of branch leaves. 9: Upper cells of branch leaf. 11-12: Basal transverse sections of branch leaves. 9: Upper cells of branch leaf (all from isosyntype, H-BR 2551011). Line scales: A = 0.44 mm (1-5); B = 70 μm (8); C = 44 μm (6-7, 9-12).

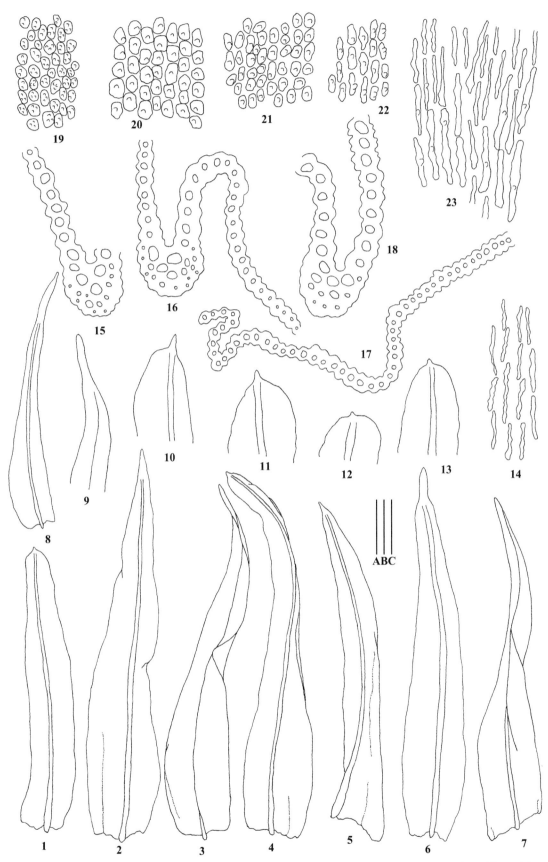

Fig. 196 *Macromitrium pulchrum* var. *neocaledonicum* (Besch.) Thouvenot 1-8: Branch leaves. 9-14: Apices of branch leaves. 14, 23: Basal cells of branch leaves. 15: Upper transverse section of branch leaf. 16: Basal transverse section of branch leaf. 17: Basal transverse section of branch leaf. 18: Medial transverse section of branch leaf. 19: Upper cells of branch leaf. 20-21: Medial cells of branch leaves. 22: Low cells of branch leaf (all from isolectotype, H-BR 2522010). Line scales: A = 0.44 mm (1-8); B = 176 μm (9-13); C = 70 μm (14-23).

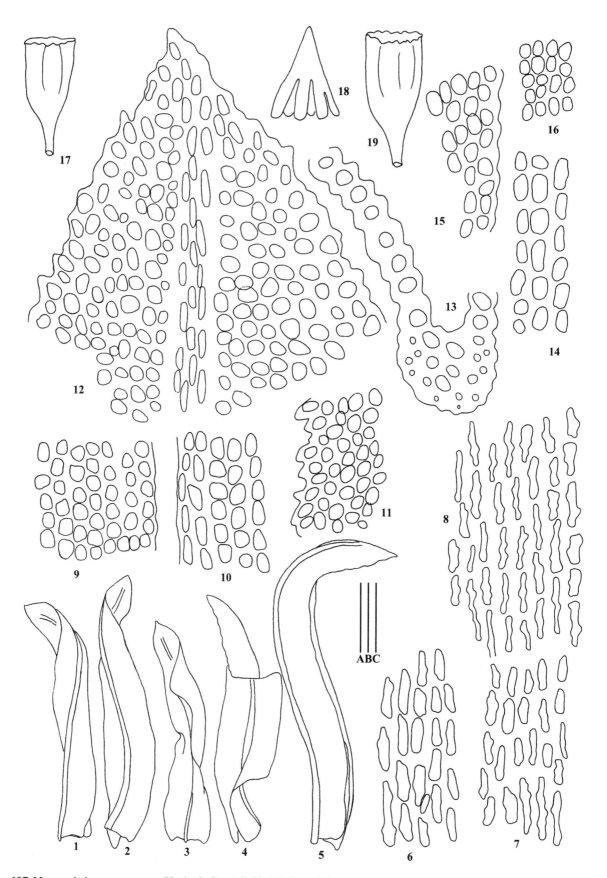

Fig. 197 *Macromitrium punctatum* (Hook. & Grev.) Brid. 1-5: Branch leaves. 6: Low cells of branch leaf. 7-8: Basal cells of branch leac. 9: Medial cells of branch leaf. 10, 14: Low cells of branch leaves. 11, 15: Upper marginal cells of branch leaves. 12: Apical cells of branch leaf. 13: Medial transverse section of branch leaf. 16: Upper cells of branch leaf. 17, 19: Capsules. 18: Calyptra (all from isotype, E 00002986). Line scales: A = 2 mm (17-19); B = 1 mm (1-5); C = 67 μm (6-16).

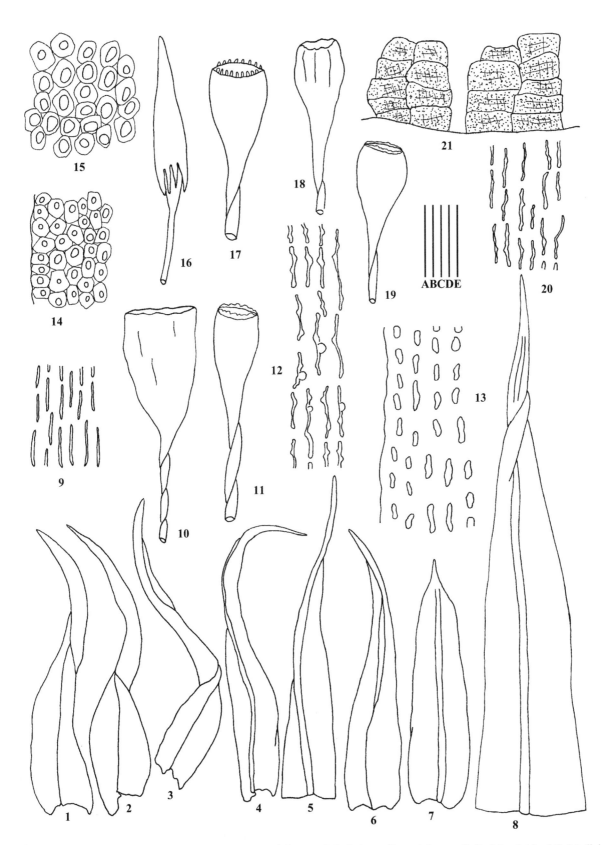

Fig. 198 *Macromitrium pyriforme* Müll. Hal. 1-4, 6: Branch leaves. 5, 8: Apices of branch leaves. 7: Perichaetial leaf. 9: Medial cells of perichaetial leaf. 10-11, 16-19: Capsules. 12: Basal cells of branch leaf. 13: Medial cells of branch leaf. 14: Upper cells of branch leaf. 15: Upper marginal cells of branch leaf. 20: Basal cells of perichaetial leaf. 21: Peristome (all from isotype, NY 01086639). Line scales: A = 2 mm (10-11, 16-19); B = 1 mm (1-4, 6-7); C = 400 μm (5, 8); D = 100 μm (21); E = 67 μm (9, 12-15, 20).

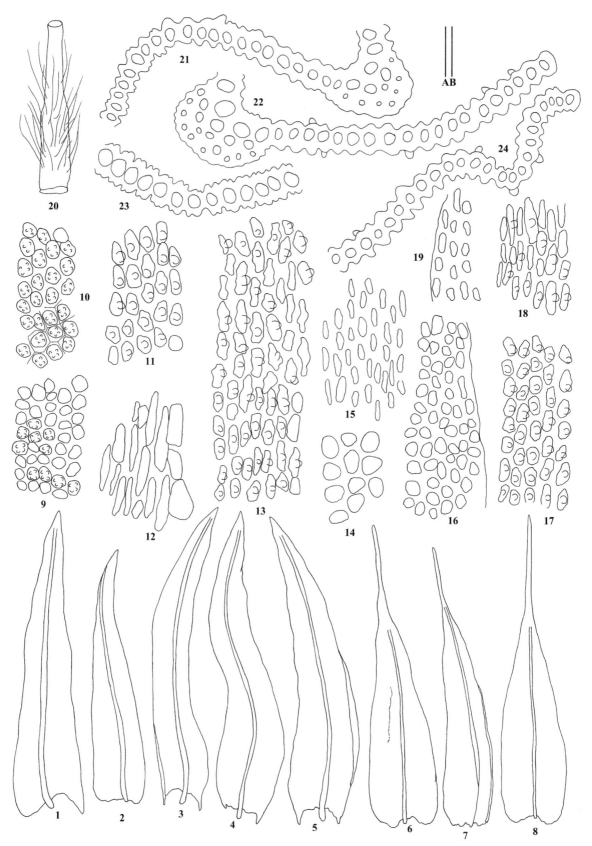

Fig. 199 *Macromitrium quercicola* Broth. 1-5: Branch leaves. 6-8: Perichaetial leaves. 9: Medial cells of branch leaf. 10: Upper cells of branch leaf. 11: Low cells of branch leaf. 12: Basal marginal cells of branch leaf. 13: Basal cells of branch leaf. 14: Exothecial cells of capsule. 15: Cells of capsule mouth. 16: Upper cells of perichaetial leaf. 17: Medial cells of perichaetial leaf. 18: Low cells of perichaetial leaf. 19: Apical cells of perichaetial leaf. 20: Vaginula. 21, 24: Basal transverse sections of branch leaves. 22: Low transverse section of branch leaf. 23: Upper transverse section of branch leaf (all from isosyntype, MO 2237138). Line sclaes: A = 1 mm (1-8, 20); B = 67 μm (9-19, 21-24).

188. *Macromitrium ramsayae* Vitt, J. Hattori Bot. Lab. 54: 14. f. 17-24, 26-29. 1983. (Figure 200)

Type protologue: [New Zealand] Chatham Island, Leg. *Gilpin*, June 1942 (holotype: WELT). Paratype: Chatham island, The Whanga lagoon at Cattle Point, *Horning NZ. 850* (H).

(1) Plants medium-sized, dull, in dense spreading mats; stems creeping, with erect branches up to 1.0 cm, branches simple or forked. (2) Stem leaves 1.0-1.5 mm long, spreading-curved to rect-flexuose when dry, widely spreading when moist, ovate-lanceolate, with acuminate the apices; costa percurrent to subpercurrent; upper cells 6-8 μm wide, rounded-quadrate or shortly rectangular to elongate, thin-walled, cholophyllose, slightly bulging, smooth. (3) Branch leaves irregularly and loosely spirally-twisted around branch, the upper portion flexuose and irregularly twisted outward when dry, erect-spreading when moist, 1.6-2.3 mm long, oblong to broadly lanceolate-oblong, strongly keeled below; apices acute, shorly cuspidate, to broadly acuminate-apiculate; margins broadly recurved to the apex, entire; costae prominent, ending in or a few cells beneath the apex; upper and medial cells 5-8 μm wide, rounded-quadrate, subquadrate or elliptic, plane to slightly bulging, smooth, chlorophyllose, unistratose; basal cells about 9-11×9-15 μm, a few cells up to 30 μm long, rounded-subquadrate, elliptic-shortly rectangular, not much different from the upper cells, bulging, weakly tuberculate or strongly papillose. (4) Perichaetial leaves shorter than branch leaves, ovate, shortly acuminate to cuspidate, elongate basal cells continuing to almost medial portion. (5) Setae shorter and stout, 2.5-4 mm long, straight or curved, smooth, twisted to the right. (6) Capsules shortly exserted, urns 1.2-2.3 mm long, narrowly ovoid to fusiform cylindric, weakly ribbed to almost smooth; peristome double, exostome of 16, irregular, blunt coarsely papillose teeth, the teeth often broken when old, with the low portion remaining and forming a fused low membrane; endostome an irregular, papillose membrane 1-3 cells high; opercula conic, with a long, erect rostrum; spores isosporous, 20-26 μm in diameter, papillose. (7) Calyptrae mitrate, naked, deeply lacerate (Vitt, 1983).

Fife (2017) treated *M. ramsayae* Vitt as a variety of *M. longirostre* (Hook.) Schwägr. However, *M. ramsayae* could be easily distinguished from *M. longirostre* by its branch leaves with fragile aristae, and always with unistratose cells. Therefore, *M. ramsayae* had better be kept at the species level.

Distribution: Australia, New Zealand.

Specimens examined: **AUSTRALIA**. Queensland, *D. H. Norris 42943* (KRAM-B-064983); New South Wales, *W. W. Watts* (H-BR 2539017); Tasmania, *H.P. Ramsay, 1379* (MEL 1035187).

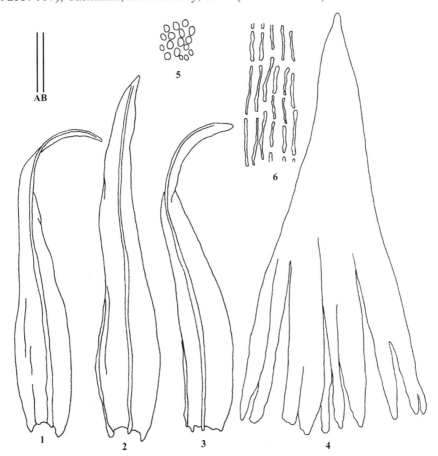

Fig. 200 *Macromitrium ramsayae* Vitt 1-3: Branch leaves. 4: Calyptra. 5: Medial cells of branch leaf. 6: Basal cells of branch leaf (all from H-BR 2539017). Line scales: A = 0.44 mm (1-4); B = 44 μm (5-6).

189. *Macromitrium raphidophyllum* Müll. Hal., Linnaea 42: 487. 1879. (Figure 201)

Type protologue: [Venezula, *A. Fendler*] *Nr. 53*. Type citation: Muscri Venezuelense, Hab. Prope Coloniam Tovar legit *A. Fendler, 53*, 1854-5 (isotypes: NY 01202318!, FH 00213669!).

(1) Plants large to medium-sized, brownish; stems long-creeping, with prostrately ascending branches; branches up to 20 mm, sometimes with branchlets. (2) Stems caducous and inconspicuous. (3) Branch leaves erect below, spreading, contorted-twisted, flexuosu or circinate above when dry, spreading or conduplicate above when moist, about 3-4 mm long, long ligulate-lanceolate, with an acuminate apex; margins entire or notched above, entire below, often recurved at one side below; costae percureent to excurrent, sometimes forming an subula; upper and medial cells isodiametric, subquadrate, rounded-quadrate, bulging and unipapillose; low and basal cells elongate-rectangular, porose and thick-walled, distinctly tuberculate; marginal enlarged teeth-like cells not differentiated at the insertion. (4) Setae 5-6 mm long, smooth, twisted to the left. (5) Capsules ovoid, ellipsoid to cupulate, furrowed to varying degrees when dry; opercula conic-rostrate; peristome double, both short membraneous. (6) Calyptrae mitrate, naked and smooth, lacerate below.

Distribution: Venezuela.

190. *Macromitrium refractifolium* Müll. Hal., Nuovo Giorn. Bot. Ital., n.s. 4: 123. 1897. (Figure 202)

Type protologue: Bolivia, Choquecamata provinciae Cochabamba, *Germain*, 1889, in the introducton of the original paper. Type citation: Bolivia, provincia Cochabamba prope Choquecamata, Jun. 1889 legit *Germain*, determ. O. C. Müller sub. *n° 1237* (lectotype: H-BR 2625020!).

(1) Plants medium-sized, golden-yellow to yellowish-brown above, rusty-brown below; stems long creeping, leaves inconspicuous and caducous; branches thin and slender, single or with short branchlets. (2) Branch leaves golden-yellow, clasping and erect below, individually twisted and flexuose-curly above when dry, strongly abaxially curved and often with conduplicate laminae when moist; narrowly lanceolate, gradually acuminate, vaginate at the base; margins erose-denticulate to serrate-serrulate above, entire below; costae gold-yellow, percurrent; apical cells irregular fusiform, somewhat porose; upper and medial cells subquadrate, rounded-quadrate, ovate-rhomboid, bulging and occasionally unipapillose, those in the outmost marginal row becoming elongate, narrow and irregular fusiform, different from their ambient cells to form a distinctive border; low cells rectangular, weakly porose; basal cells long-rectangular, thick-walled and porose, each with a linear lumen, strongly tuberculate. (3) Perichaetial leaves slightly longer than branch leaves. (4) Setae short and smooth. (5) Capsules ellipsoid, plicate; opercula conic-subulate; peristome double. (6) Calyptrae large, mitrate and plicate, lacinate below.

Macromitrium refractifolium is similiar to *M. rusbyanum*, and could be distinguished from the latter by its branch leaves with papillose cells throughout the leaf, and furrowed capsules.

Distribution: Bolivia.

191. *Macromitrium regnellii* Hampe, Syn. Musc. Frond. 1: 738. 1849. (Figure 203)

Type protologue and citation: Patria, Brasilia, Minas Geraës, ad Caldas: *Dr. Regnell* (isotypes: BM 000873240!, BM 000989741!, H-BR 2643007!, E 00011668!, E 00011669!).

Macromitrium contextum Hampe, Ann. Sci. Nat., Bot., sér. 5, 4: 331. 1865, *fide* Valente *et al.*, 2020. Type protologue and citation: Colombia, Rio negro, altit. 1200 metr., in sylvis ad arbores, cespites latos septb. *A. Lindig* legit (isotypes: PC 0137645!, PC 0137646!).

(1) Plants medium-sized, yellowish green, olive-green to brownish-yellow; stems creeping, 40 mm long, branches up to 20 mm long. (2) Stem leaves inconspicuous and caducous. (3) Branch leaves stiffy, erect below, spreading, twisted-contorted, or contorted-crisped above, with a circinate to enrolled apex when dry, unevenly spreading-recurved when moist, 2-3 mm long, oblong-lanceolate, rugous in upper portion; apices acuminate to acute, occasionally apiculate; margins irregularly and bluntly notched to serrate above, plane or reflexed below; costae excurrent into a blunt apiculus; all cells thick-walled, upper and medial cells rounded-quadrate, arranged in longitudinal rows, smooth or weakly papillose; basal cells narrowly rectangular, 20-26 μm long, porose, strongly tuberculate; marginal enlarged, swollen, sharply teeth-like cells distinctly differentiated at the insertion. (4) Perichaetial leaves broadly elliptical-lanceolate, entire; upper cells isodiametric, smooth; cells elongate towards the base, rectangular or long rectangular cells in low 2/3 of the perichaetial leaf. (5) Setae smooth, 7-18 mm long, not or slightly twisted to the left. (6) Capsule urns 1-1.5 mm long, ovoid or cupulate, smooth to weakly ribbed; opercula conic-rostrate; peristome double, exostome teeth united to a membrane, about 100 μm high, yellow and papillose; endostome segments hyaline, papillose; spores anisosporous, smooth or weakly papillose. (7) Calyptrae mitrate, deeply lacerate, naked.

Macromitrium atroviride R.S. Williams is rather similar to *M. regnellii* Hampe, only the latter has exostome teeth united to form a membrane.

Distribution: Brazil, Bolivia, Colombia, Mexico (Vitt, 1994) and Peru.

Specimens examined: **BRAZIL**. Minas Geraës, leg. Widgren (BM 000989742, BM 000989743, BM 000989744); *G. A. Lindberg* (H 3090691); Brasilia, Caldas, *G. A. Lindberg* (H 3090687); Caldas, Herachen (H-BR 2643010). **BOLIVIA**. La Paz: Franz Tamayo, *A. Fuentes 5215* (MO 5647769). **COLOMBIA**. Santander: San Gil Municipio, *S. P. Churchill 18776* (MO 4461976); Antioquia: Cocorna Municipio, *D. A. G. Canas 318* (MO 4461977). **PERU**. Dpto. Pasco, Prov. Oxapampa, *P+E Hegewald, 8445* (MO 3670912), *8455* (H 3090663, MO03670913)*, 8460 (MO 3670911)*, *8465* (MO 3670910); Dpto. Cutco, Prov. Urubamba, *P+E Hegewald 8779* (MO 3670915).

192. *Macromitrium renauldii* Thér., Bull. Acad. Int. Géog. Bot. 17: 307. 1907. (Figure 204)

Type protologue: "M. F. Renauld, …., Nouméa Nouvelle-Calédonie, en 1906, par *M. Franc*…."; Type citation: "Nouvelle-Calédonie, env. Noumea, leg. *Franc s.n.*" (lectotype designated by Thouvenot, 2019: PC 0083728!; isolectotype: PC 0083729!).

Macromitrium gracilipes Cardot, Bull. Herb. Boissier, sér. 2, 8: 168. f. 3. 1908, *fide* Thouvenot, 2019. Type protologue and citation: [New Caledonia], Balade (leg. *Vieillard*) (lectotype designated by Thouvenot, 2019: PC 0083666!; isolectotype: PC 0083665!).

Macromitrium leratioides Broth. & Paris, Öfvers. Finska Vetensk.-Soc. Förh. 51A(17): 15. 1909, *fide.* Thouvenot, 2019. Type protologue: [New Caledonia] Mont Dzumac, ad arbores (*A. Le Rat*). Type citation: Nouv Caledon, Mt Dzumac, leg. *A. Le Rat, 1011*, 1907 (lectotype designated by Thouvenot, 2019: H-BR 2618013!).

(1) Plants medium-sized, rusty-brown when old; stems long creeping, densely with ascending branches, branches up to 5 mm tall, and 1.3 mm thick. (2) Branch leaves tightly appressed below and slightly spirally coiled, curved or circinate above when dry, widely spreading when moist, 0.8-2.3 × 0.2-0.4 mm, lanceolate to long lanceolate, with an acuminate upper protion; apices acute, apiculate to a short arista; all cells thick-walled and in regularly longitudinal rows, with a longitudinally striated appearance; upper cells quadrate with rounded to oval lumens, smooth or weakly papillose; transitional part very short, with an inverted "V" shape, short rectangular, smooth or unipapillose; low cells rectangular elongate, unipapillose. (3) Perichaetial leaves larger than branch leaves, oblong to widely lanceolate or triangulate lanceolate, acuminate-cuspidate above, plicate below; costae occasionally concealed in adaxial view by overlapping folds of laminae; all cells longer than wide, unipapillose except those near the base; upper cells oval or rounded-elliptic; low and basal cells rectangular or irregularly rectangular. (4) Setae long, up to 25 mm long, smooth, twisted to the left; vaginulae with long hairs. (5) Capsules narrowly ellipsoid, ellipsoid-cylindric to sub-cylindric, smooth, rims small, plicate near the mouth; peristome single. (6) Calyptrae hairy.

Macromitrium renauldii Thér. is similar to *M. taoense* Thér., but differs from the latter in having hairy calyptrae and vaginulae.

Distribution: New Caledonia.

Specimens examined: **NEW CALEDONIA**. Hb. F. Renauld, leg. *Franc s.n.* 1906 (PC 0083729); Sur un petit arbre on forêt mesophile, pente droite de la vallée de la Pourina dans le tiers inférieur, leg. *H. Hürlimann 2636* (G 144177). Mt Dzumac, *Le Rat s.n.*, 1909 (H-BR 251005); Noumea, *Le Rat 1006*, 1907 (H 3090426).

193. *Macromitrium repandum* Müll. Hal., Bot. Jabrb. Syst. 5: 87. 1883. (Figure 205)

Type protologue: [Australia] Queensland (Naumann). Type citation: Morton Bay, Naumann (isotype: E 00165160!); Queensland, Naumann (isotype: H-BR 2524008!); Morton Bay, 1875, marked with "*Macromitrium repandum* C. Müll. *n.sp*, Herbarium Noumann" (isotype: JE 04008686!).

Macromitrium pallidovirens Müll. Hal., Hedwigia 37: 144. 1898, *fide* Vitt & Ramsay, 1985. Type protologue: Australia tropica Queensland, Sinc loco speciali: Bailey Kier, qui misit 1885.

Macromitrium pugionifolium Müll. Hal., Hedwigia 37: 145. 1898, *fide* Vitt & Ramsay, 1985. Type protologue: New South Wales, Eichmond River, ad arbores: Miss Hodgkinson in Hb. Melbourne 1880; Gosford: *Th. Whitelegge* in Hb. Brotheri 192. Type citation: Gosford: on trees, high side opposite station, Gosford (lectotype designated by Vitt & Ramsay, 1985: H-BR 2524002!).

Macromitrium whiteleggei Broth. & Geh., Öfvers. Finska Vetensk.-Soc. Förh. 37: 161. 1895, *fide* Vitt & Ramsay, 1985. Type protologue and citation: Queensland, Belenden ker Range et Mt Bartle Frere, 5000 p. (Stephen Johnson); New South Wales, Hurstville near Sydney (*Th. Whitelegge n. 301*) (lectotype designated by Vitt & Ramsay 1985: H-BR 2525001!).

(1) Plants small, rather dull to slightly lustrous, pale-green or olive-green above, dark olive-brown below, in compact and spreading mats; stems long creeping, with dense, short, stout and erect branches, branches up to 5-8 mm long, much smaller near margin of mats. (2) Stem leaves irregularly flexuose-twisted and curved upward at stem tips when dry, 1.0-1.4 mm long, narrowly ovate-lanceolate to lanceolate, gradually narrowly acuminate; costa excurrent, upper cells rounded, smooth and flat, basal cells elongate, smooth. (3) Branch leaves flexuose-curved to erect-curved, spirally and tightly curved around the branch, somewhat funiculate when dry, erect-spreading to spreading, straight, slightly wrinkled above when moist, 1.0-1.5 mm long, ligulate, broadly ligulate-lanceolate to

lanceolate-oblong or oblong, abruptly narrowed to a cuspidate to mucronate apex, sometimes retuse and asymmetric; margins plane to slightly reflexed, entire; costae strong, excurrent and forming a stout cusp or mucro; upper cells 5-7 μm wide, irregularly rounded, smooth, flat and clear; medial cells 5-7 μm wide, rectangular, flat and smooth, quickly grading into basal cells; basal cells 20-50 μm long, lumens 2-3 μm wide, elongate, straight, curved-sigmoid, mostly smooth, occasionally sparsely tuberculate; basal marginal cells differentiated. (4) Perichaetial leaves 1.5-2.0 mm long, erect and subsheathing, lanceolate-ovate, quickly contracted to a slenderly, long acuminate-sharply cuspidate apex. (5) Setae 4-12 mm long, smooth, twisted to the left; vaginulae not hairy. (6) Capsule urns broadly ovoid-ellipsoid, 8-plicate, gradually narrowed to a small, puckered mouth; peristome single, exostome of 16, well developed, white blunt, lanceolate teeth, finely papillose; opercula erect, conic-rostrate; spores distinctly anisosporous. (7) Calyptrae conic-mitrate, smooth, naked or occasionally with a few short hairs, evenly lobed.

Macromitrium repandum Müll. Hal. is similar to *M. brevicule* (Besch.) Broth. and *M. aurescens* Hampe, but differs from these two latter species by its smooth upper and medial leaf cells. *Macromitrium microstomum* (Hook. & Grev.) Schwägr. is similar to *M. repandum*, but differs from the latter in possessing lanceolate, more or less acute to acuminate leaves and completely smooth basal lamina cells.

Distribution: Australia.

Specimens examined: AUSTRALIA. New South Wales, *D. H. Vitt 27301* (H 3090685), *W. W. Watts 4464* (H-BR 2526019), *387* (H-BR 2525009), *W.W. Watts s.n.* (H-BR 2526019); Queensland, *H. Streimann 30756* (MO 4448376), *52322* (MO 4433965), *52235* (KRAM-B-147808), *52954* (KRAM-B-106991); *W.W. Watts 520* (H-BR 2524006), *495* (H-BR 2524004), *493* (H-BR 2524003), *I. G. Stone* (MEL 2369764, MEL 2256766, MEL 2246559).

194. *Macromitrium retusulum* Müll. Hal., Linnaea 42: 486. 1879. (Figure 206)

Type protologue: Venezuela. *s.n. 57*. Type citation: Musci Fendlerinni Venezuelenses, 57 (lectotype: FH 00213667!).

(1) Plants medium-sized to large, dark brown, stem weakly creeping, with erect branches, branches up to 20 mm tall and 1.5 mm thick. (2) Stem leaves inconspicuous and caudcuous. (3) Branch leaves obliquely appressed below, slighltly spirally twisted and flexuose above the apices when dry, widely spreading when moist; ovate-, oblong- to ligulate-lanceolate; margins finely serrulate near the apex, entire below, occasionally recurved at one side below; costae percurrent to form a short point; upper and medial cells isodiametric, rounded and bulging, unipapillose; low and basal cells elongate, slightly porose, with straightly lumens, tuberculate. (4) Perichaetial leaves differentiated from and wider than branch leaves. (5) Setae twisted to the left, smooth, up to 15 mm long. (6) Capsules ovoid, furrowed when dry, with an open and wide mouth; peristome possibly double (endostome caducous), exostome short and imperfect, arranged in pairs. (7) Calyptrae mitrate, smooth and naked, lacerate, covering the whole capsule.

Distribution: Venezuela.

195. *Macromitrium retusum* Hook. f. & Wilson, Fl. Nov.-Zel. 2: 79. 85 f. 6. 1854.

Type protologue: [New Zealand] Hab. Northern Island, *Colenso*.

Macromitrium aristatum Mitt., Handb. N. Zeal. Fl. 432. 1867. *fide* Vitt, 1983. Type protologue: [New Zealand] Northern Island: Auckland, *Knight*.

Macromitrium caducipilum Lindb, Öfvers. Kongl. Vetensk.-Akad. Förh. 21: 605. 1865. *fide* Vitt, 1983. *fide* Vitt, 1983. Type protologue: Inter *Leptostomum macrocarpum* (Hedw.) e Nova Zelandia (Collect. *Ralfs*) Paucissima specimina feminea decerpsi.

Macromitrium longirostre var. *caducipilum* (Lindb.) W. Martin & Sainsbury, Rev. Bryol. Lichénol. 21: 219. 1952.

(1) Plants small and slender; stems creeping, with stiff, erect to ascending branches, branches penicillate, up to 2 cm high. (2) Stem leaves lanceolate-acuminate. (3) Branch leaves curved-flexuose, spirally twisted around the branch when dry, erect-spreading when moist, 1.3-2.0 mm long, oblong or ligulate; apices acute, obtuse, often retuse, young leaves have very a long, linear, acute, green, smooth, stiffly flexuose arista, the arista fragile and often broken off before mature; all cells smooth and flat; upper and medial cells rounded-quadrate to rectangular-elliptic in longitudinal rows, thick-walled; basal cells elongate, occasionally rounded, elliptic or short-rectangular, thick-walled, smooth, not porose, lumens narrowly rectangular, straight. (4) Perichaetial leaves longer than branch leaves, ovate-lanceolate to oblong-lanceolate, with a fragile arista, cells longer than wide, smooth. (5) Setae 4-7 mm long, smooth, twisted to the right above. (6) Capsule urns narrowly ovoid to ovoid-obloid, ellipsoid-crylindric, almost smooth, peristome single, exostome of 16, lanceolate; spores isosporous. (7) Calyptrae mitrate, smooth and naked, deeply lacerate and strongly plicate.

In New Zealand, except *M. retusum* Hook. f. & Wilson, *M. gracile* (Hook.) Schwägr. and *M. helmsii* Paris also have fragile arista or subula. However, the upper cells of branch leaves for *M. gracile* and *M. helmsii* are obscure and pluripapillose.

Taxonomy

Sainsbury (1945) treated *M. retusum* Hook. f. & Wilson as a variety of *M. gracile* (*M. gracile* var. *retusum* (Hook. f. & Wilson) Sainsbury). However, the upper cells of branch leaves for *M. gracile* are obscure and pluripapillose. Therefore, *M. retusum* had better be kept at the species level (Vitt, 1983, Vitt & Ramsay, 1985).

Distribution: Australia, New Zealand.

Specimens examined: **NEW ZEALAND**. North Island: Cattle Riddge Track, *H. Streimann 58096* (H 3090695); Stewart island: Horeshoe point, deadman's beach, *Allan J. Fife 5981* (MO 3653698); Steward island: Paterson inlet, *Allan J. Fife 5958* (MO 3653697); Otago Prov., Otago Peninsula, Haori Kaik, *W. Bell 123/1* (H-BR 2529003).

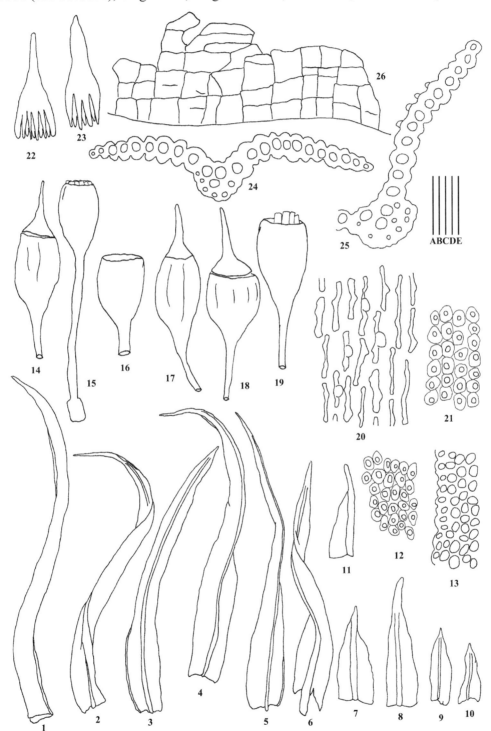

Fig. 201 *Macromitrium raphidophyllum* Müll. Hal. 1-6: Branch leaves. 7-8, 11: Apices of branch leaves. 9-10: Stem leaves. 12: Upper cells of branch leaf. 13: Upper marginal cells of branch leaf. 14-19: Capsules. 20: Basal cells of branch leaf. 21: Medial cells of branch leaf. 22-23: Calyptrae. 24: Upper transverse section of branch leaf. 25: Medial transverse section of branch leaf. 26: Peristome (1-16, 20-26 from isotype, FH 00213669; 17-19 from isotype, NY 01202318). Line scales: A = 2 mm (14-19, 22-23); B = 1 mm (1-6, 9-10); C = 400 μm (7-8, 11); D = 100 μm (26); E = 67 μm (12-13, 20-21, 24-25).

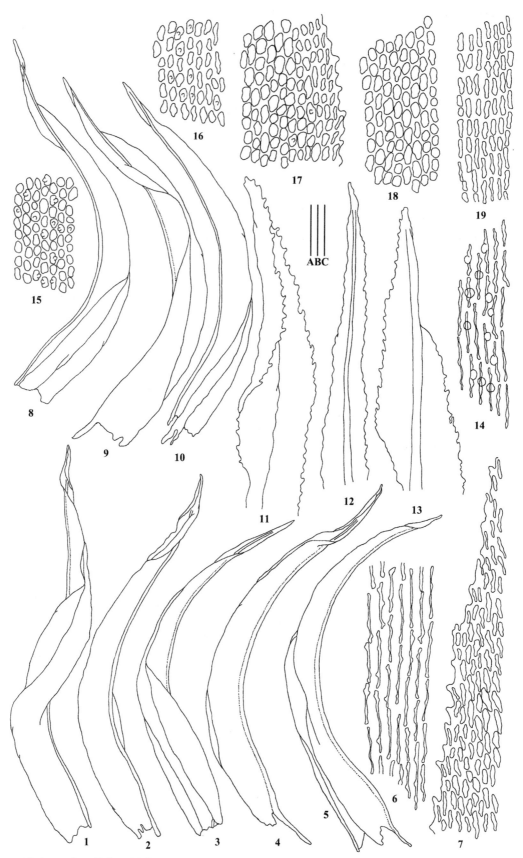

Fig. 202 *Macromitrium refractifolium* Müll. Hal. 1-5, 8-10: Branch leaves. 6, 14: Basal cells of branch leaves. 7: Apical cells of branch leaf. 11-13: Apices of branch leaves. 15, 18: Medial cells of branch leaves. 16: Upper cells of branch leaf. 17: Upper marginal cells of branch leaf. 19: Low cells of tranch leaf (all from lectotype, H-BR 2625020). Line Scales: A = 500 μm (1-5, 8-10); B = 100 μm (11-13); C = 50 μm (6, 7, 14-19).

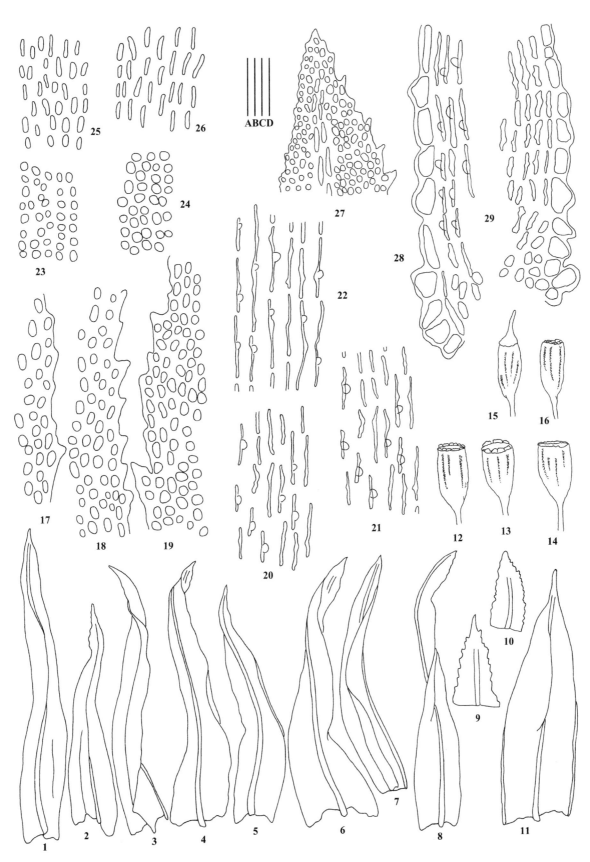

Fig. 203 *Macromitrium regnellii* Hampe. 1-8: Branch leaves. 9-10: Apices of branch leaves. 11: Perichaetial leaf. 12-16: Capsules. 17-19, 27: Apical cells of branch leaves. 20-21: Basal cells of branch leaves. 22: Basal cells of perichaetial leaf. 23: Medial cells of branch leaf. 24: Upper cells of branch leaf. 25: Upper cells of perichaetial leaf. 26: Medial cells of perichaetial leaf. 28-29: Basal marginal cells of branch leaves (all from isotype, E 00011668). Line scales: A = 2 mm (12-16); B = 1 mm (1-8, 11); C = 400 μm (9-10); D = 67 μm (17-29).

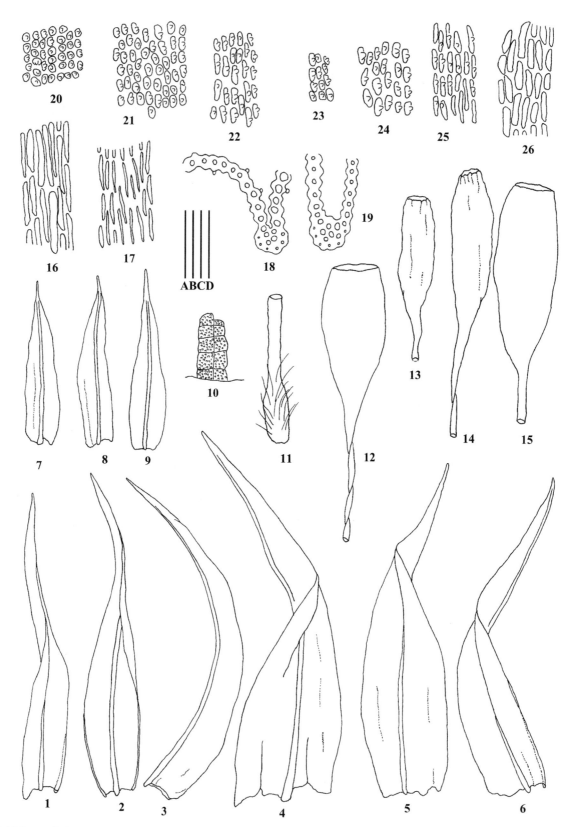

Fig. 204 *Macromitrium renauldii* Thér. 1-6: Branch leaves. 7-9: Perichaetial leaves. 10: Peristome. 11: Vaginual. 12, 15: Wet capsules. 13-14: Dry capsules. 16-17: Basal cells of branch leaves. 18: Upper transverse section of leaf cells. 19: Basal transverse section of leaf cells. 20: Upper cells of branch leaf. 21: Medial cells of branch leaf. 22: Low cells of branch leaf. 23: Upper cells of perichaetial leaf. 24: Medial cells of perichaetial leaf. 25: Low cells of perichaetial leaf. 26: Basal cells of perichaetial leaf (1-8, 10-22, 26 from lectotype, PC 0083728; 9, 23-25 from isolectotype, PC 0083729). Line scales: A = 1 mm (7-9, 11-15); B = 400 μm (1-6); C = 200 μm (10); D = 67 μm (16-26).

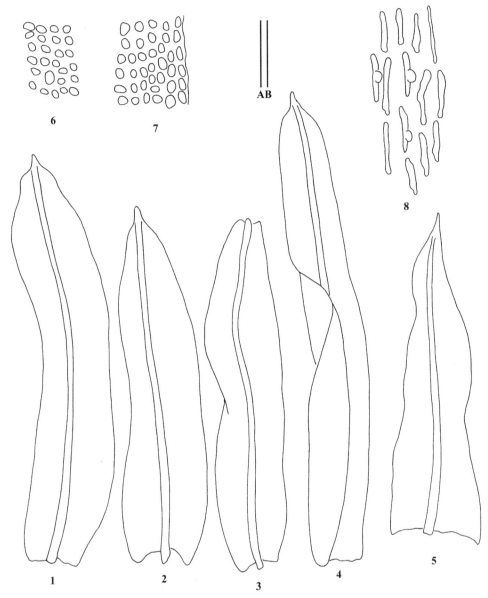

Fig. 205 *Macromitrium repandum* Müll. Hal. 1-4: Branch leaves. 5: Perichaetial leaf. 6: Upper cells of branch leaf. 7: Medial cells of branch leaf. 8: Basal cells of branch leaf (all from isotype, E 00165160). Line scales: A = 400 μm (1-5); B = 67 μm (6-8).

196. *Macromitrium rhacomitrioides* Nog., J. Sci. Hiroshima Univ., Ser. B, Div. 2, Bot. 3: 138. 2. 1938. (Figure 207)
Type protologue and citation: [China] Taiwan: Mt. Arisan, Prov. Tainan (*A. Noguchi, no. 1756* -type: in Herb. Hiros. Univ., July 1928) (holotype: HIRO!).

(1) Plants medium-sized, stems long creeping, 3-4 cm long, forming brownish mats; branches up to 25 mm long, with several short branchlets, densely leaved. (2) Stem leaves deflexed when moist, ovate-lanceolate to oblong-lanceolate, short acuminate; upper cells small, rounded-quadrate, obscure and densely pluripapillose; medial cells rounded quadrate, slightly larger than those in upper part, unipapillose, cells becoming elongate towards the base; low cells shorter and irregularly rectangular, thick-walled and smooth; basal cells elongate to linear rectangular, irregularly thick-walled, smooth. (3) Branch leaves keeled, twisted-contorted and flexuose when dry, spreading, squarrosely-recurved, or incurved and adaxially curved above when moist, 0.3-0.5 × 2.5-3.0 mm, long oblong-lanceolate, narrowly lanceolate, linear-lanceolate, narrowly acuminate, often plicated below; margins plane below and slightly denticulate owing to protruding papillae; upper cells smaller, 3-4 μm wide, rounded or round-quadrate, thin-walled, densely pluripapillose, rather obscure; medial cells elongate, short rectangular, unipapillose; low and basal cells narrowly elongate, thick-walled, distinctly unipapillose; basal cells near the costa thin-walled, smooth and pellucid, distinctly larger than their ambient cells, forming a "cancellina region". (4) Perichaetial leaves oblong-lanceolate, acuminate or narrowly acuminate, plicate below, up to 2.2 mm long; upper cells rounded-quadrate, basal and low cells elongate, thick-walled, unipapillose. (5) Setae smooth, 5-8 mm long, twisted to the left; vaginulae with

numerous paraphyses. (6) Capsules erect, obloid, ellipsoid-cylindric, smooth; peristome single, lanceolate. (7) Calyptrae mitrate, plicate, with long yellowish hairs.

The species is similar to *M. giraldii* Müll. Hal. However, the former could be separated from the latter by its linear-lanceolate branch leaves with rather small, obscure, densely pluripapillose upper lamina cells.

Distribution: China.

Specimens examined: **CHINA**. Taiwan, *C. C. Chuang 6074* (NY), *J. R. Shevock et al. 18149* (MO); Mt. Yuza, Sintiku, *Simada 3866* (BM 000919464).

197. *Macromitrium rimbachii* Herzog, Memoranda Soc. Fauna Fl. Fenn. 27: 109. 1952. (Figure 208)

Type protologue and citation: Ecuador: ostkordillere, ohne nähere Fundortsangabe, ca 3000 m, leg. *A. Rimbach* (holotype: JE 04008690!).

(1) Plants medium-sized to large; stems weakly creeping, loosely with erect forked branches, branches up to 30 mm long. (2) Branch leaves erect below, individually twisted, strongly contorted-flexuose, undulate and sometime circinate above when dry, abaxially curved and distinctly undulate and twisted above when moist, long linear lanceolate; apices acuminate to an arista; costae often partially concealed in adaxial view by overlapping folds of laminae; margins weakly serrate above and entire below; cells in and near the apex elliptical-oblong; upper cells subquadrate, rounded-quadrate to ovate-oblong, smooth and thick-walled; medial cells oblong-rectangular, longer than wide, smooth and thick-walled; low and basal cells elongate-rectangular, distinctly tuberculate and porose, thick-walled; marginal enlarged teeth-like cells not differentiated at the insertion. (3) Perichaetial leaves distinctly different from branch leaves, lanceolate from a broadly oblong base, acuminate to a long and fine arista; all cells longer than wide, thick-walled and smooth. (4) Setae long, up to 22 mm long, smooth. (5) Capsule urns ellipsoid to ellipsoid-cylindric, strongly furrowed, constricted beneath the mouth; peristome double, united to form low double membranes; opercula conic-rostrate, with a long beak; anisosporous, green, large spores 20 µm and small spores 12 µm in diameter. (6) Calyptrae cucullate, smooth and naked, lacerate below.

Distribution: Ecuador.

198. *Macromitrium rufipilum* Cardot, Bull. Herb. Boissier, sér. 2, 8: 169 f. 4. 1908.

Type porotlogue: [New Caledonia] Balade (leg. *Vieillard*). Type citation: Nounelle Caledonie, Balade, leg. *Vieillard no. 1735 p.p.* marked with *Macromitrium rufipilum* Card. *sp. nova.* (lectotype designated by Thouvenot, 2019: PC 0096531!).

(1) Plants medium-sized, red-brown, stems creeping. (2) Branch leaves loosely spiraled, erect to oblique, flexuose, keeled when dry, erect to patent when moist, 3-4.5 × 0.5-0.8 mm, ligulate; apices obtuse to truncate, shortly acute, with a very long, reddish and hyaline arista (0.4-1.5 mm); costae red and long excurrent; upper cells larger (10-20 × 10-20 µm), rounded, ovate to oblong, strongly bulging, the external walls strongly protruding, pluripapillose; marginal cells smaller in the outmost row; transitional part short, and cells rectangular, thick-walled, porose, unipapillose; low cells rectangular, elongate to linear, very thick-walled, smooth near base and numberous papillae near transitional parts. (3) Perichaetial leaves lanceolate, acuminate or acute, with a long arista. (4) Setae 20-25 mm long, vaginulae hairless but with a few short paraphyses. (5) Capsules narrowly ellipsoid, sub-cylindric, smooth, rims plicate; peristome absent or reduced to a white ridge. (6) Calyptrae naked (Thouvenot, 2019).

Distribution: New Caledonia.

199. *Macromitrium ruginosum* Besch., Bull. Soc. Bot. France 45: 63. 1898. (Figure 209)

Type protologue: [Tahiti in French Polynesia] Vallee de Puaa (1re herbor., *no 254*). Type citation: Tahiti, Legit *Dr. Nadeaud 254*, 1896 (isotypes: H-BR 2618024!, BM 000982762!).

(1) Plants small and dull, in loosely mats; stems long creeping, with short branches, branches stiff and thin, yellow-greenish or light rusty. (2) Stem and branch leaves spirally appressed around the branch, individually flexuose-twisted when dry, erecto-patent when moist, short oblong, oblong-lingulate, blunt acute; margins crenulate due to bulging of papillose cells; upper, medial cells large and isodiametric, rounded to rounded quadrate, strongly pluripapillose; low cells unipapillose, slightly larger than those of the upper and medial portion; basal cell elongate, thick-walled and smooth; costa percurrent. (3) Perichaetial leaves shorter than branch leaves. (4) Seta shorter, 1.5-3 mm long, papillose. (5) Capsules ovoid, smooth to weak wrinkled, constricted beneath the mouth; peristome single, exostome membranous. (6) Calyptrae mitrate, hairy, deeply lacerate.

Macromitrium ruginosum Besch. is rather similar to *M. orthostichum* Nees ex Schwägr., the latter could be separated from the former by the prostrate stems with dense and rusty-reddish rhizoids.

Distribution: Tahiti in French Polynesia.

Specimen examined: **FRENCH POLYNESIA**. Tahiti, *Nadeaud*, 1896, nr. II/2194 (S-B 164827).

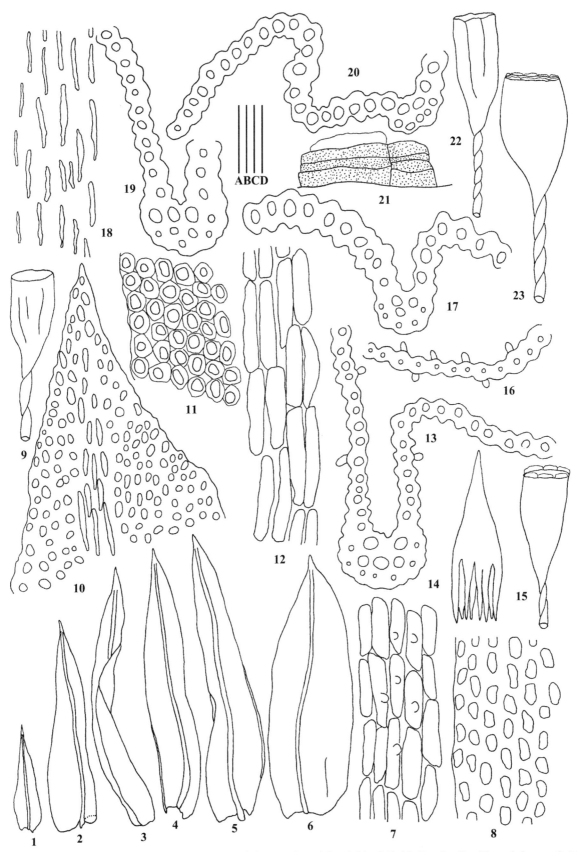

Fig. 206 *Macromitrium retusulum* Müll. Hal. 1-5: Branch leaves. 6: Perichaetial leaf. 7, 12: Basal cells of branch leaves. 8: Medial cells of branch leaf. 9, 15, 22-23: Capsules. 10: Apical cells of branch leaf. 11: Upper cells of branch leaf. 13, 16, 20: Basal transverse sections of branch leaves. 14: Calyptra. 17: Upper transverse section of branch leaf. 18: Upper cells of perichaetial leaf. 19: Medial transverse section of branch leaf. 21: Peristome (all from isotype, FH 00213668). Line scales: A = 2 mm (9, 14-15, 22-23); B = 1 mm (1-6); C = 400 μm (21); D = 67 μm (7-8, 10-13, 16-20).

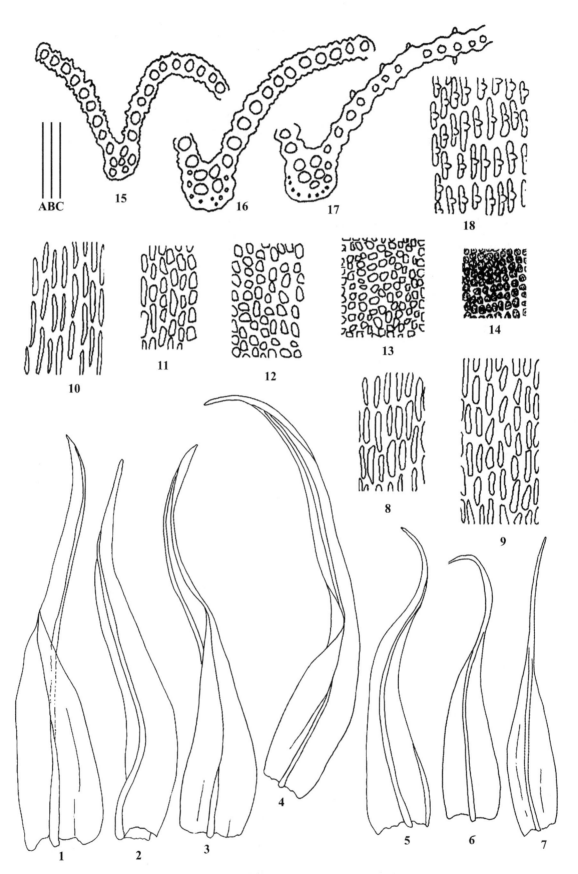

Fig. 207 *Macromitrium rhacomitrioides* Nog. 1-4: Branch leaves. 5: Stem leaves. 6, 7: Perichaetial leaves. 8: Medial cells of branch leaf. 9: Basal perichaetial leaf cells. 10: Basal cells of stem leaf. 11: Low cells of stem leaf. 12: Medial cells of stem leaf. 13: Upper cells of stem leaf. 14: Upper cells of branch leaf. 15: Sypper transverse section of branch leaf. 16: Medial transverse section of branch leaf. 17: Basal transverse section of branch leaf. 18: Basal cells of branch leaf (all from holotype, HIRO). Line scales: A = 1 mm (6, 7); B = 0.4 mm (1-5); C = 67 µm (8-18).

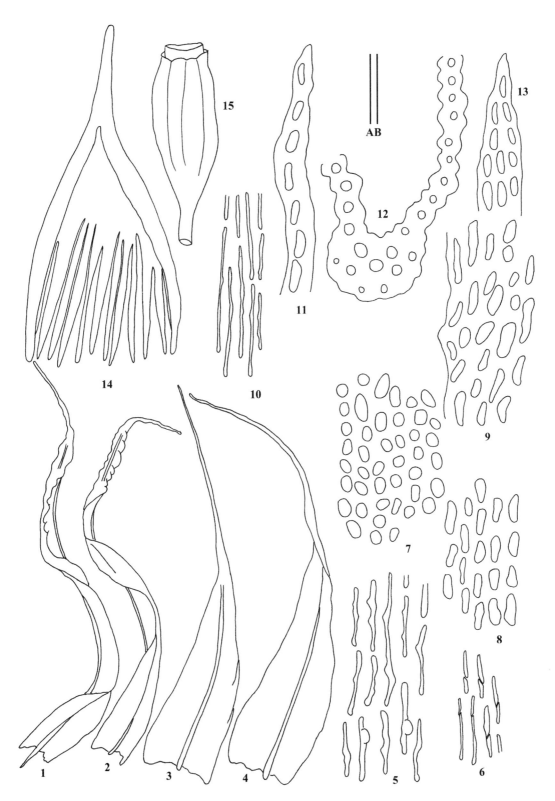

Fig. 208 *Macromitrium rimbachii* Herzog 1-2: Branch leaves. 3-4: Perichaetial leaves. 5: Basal cells of branch leaf. 6: Upper cells of perichaetial leaf. 7: Upper cells of branch leaf. 8: Medial cells of branch leaf. 9: Upper marginal cells of branch leaf. 10: Low cells of perichaetial leaf. 11, 13: Apical cells of branch leaves. 12: Medial transverse section of branch leaf. 14: Calyptra. 15: Capsule (all from holotype, JE 04008690). Line scales: A = 2 mm (1-4, 14-15); B = 67 μm (5-13).

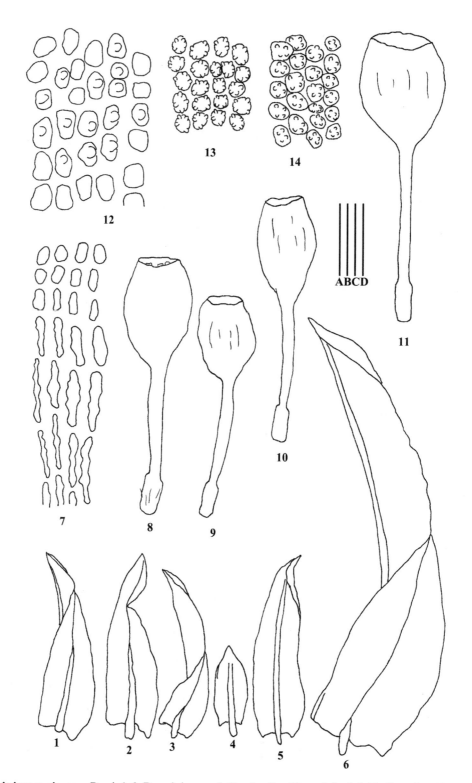

Fig. 209 *Macromitrium ruginosum* Besch. 1-6: Branch leaves. 7: Basal cells of branch leaf. 8-11: Capsules. 12: Low cells of branch leaf. 13: Upper cells of branch leaf. 14: Medial cells of branch leaf (all from S-B 164827). Line scales: A = 2 mm (8-11); B = 1 mm (1-5); C = 400 μm (6); D = 67 μm (7, 12-14).

200. *Macromitrium rugulosum* Ångstr., Öfvers. Kongl. Vetensk.-Akad. Förh. 33(4): 12. 1876. (Figure 210)

Type protologue: [Brazil] Widgren pauca specimina collegit.

(1) Plants medium-sized, in dense mats; stems weakly creeping, with inconspicuous and caducous leaves, and erect branches, branches about 15 mm long, branched or simple, densely with rusty-reddish rhizoids. (2) Branch leaves appressed below, twisted-crisped when dry, spreading but often conduplicate above when moist, broadly oblong-ligulate to lanceolate-ligulate; apices acute to acute-acuminate; margins distinctly serrate above and entire below; costae percurrent or short excurrent into a mucronate apex, often concealed in adaxial view by overlapping folds of laminae; upper cells subquadrate, oblate, rhomboid, thick-walled, in distinctly radiating diagonal rows from the costa; medial and low to basal cells thick-walled, strongly tuberculate, distinctly porose; medial cells rectangular, low and basal cells elongate-rectangualr; marginal enlarged teeth-like cells weakly differentiated at insertion; basal cells near costa large, rectangular to irregularly rectangular, smooth, form a "cancellina region". (3) Perichaetial leaves distinctly differentiated, broadly oblong-, triangular to lingulate-lanceolate, with the widest at base, plicate below, with a sheathing base; all cells smooth and clear, longer than wide. (4) Setae about 15 mm long, smooth, twisted to the left. (5) Capsules ellipsoid to cylindric, 4-furrowed and slightly twisted; opercula conic-rostrate with a long beak; peristome present, caducous. (6) Calyptrae naked.

Distribution: Brazil.

Specimens examined: **BRAZIL**. Brasilia, Prov. Minas Geraes Caldes, Leg. *A. F. Regnell* (H-BR 2628001, H-BR 2628002), Brasilia, Caldas, leg. *A. F. Regnell 497* (H-BR 2628003).

201. *Macromitrium rusbyanum* E. Britton, Bull. Torrey Bot. Club 23: 482. 1896. (Figure 211)

Type protologue and citation: [Bolivia] Unduavi 12000 ft. October, 1885, Leg. *H. H. Rusby 3188* (isotypes: NY 01086642!, NY 01086643!, NY 01086644!, NY 01086645!, PC 0138016!, PC 0138017!, FH 00213675!, FH 00213678!).

(1) Plants robust, light yellowish-green above, dark-brown below, forming a weft form; stems long creeping, with loose branches, branches up to 4-5 cm long. (2) Branch leaves erect below, widely spreading and flexuose above when dry, spreading and abaxially curved or twisted in medial upper portion when moist, 3-4 mm long, long lanceolate from a broadly oval or oblong base, somewhat clasping at base, plicate in the medial and low portion; acuminate to long setaceous-acuminate; upper and medial margins strongly serrate to serrulate, sometimes recurved at one side at medial and low portion; costae rusty reddish, vanishing beneath the apex; cells at the apex rectangular or elliptical-rectangular, smooth; upper cells rounded, rounded-rhombic, rounded-rectangular, bulging and unipapillose; medial cells slightly longer than upper cells, elliptical-rhombic, bulging and unipapillose; marginal cells in the outmost row of upper and medial portion narrowed and longer than their ambient cells, form a differentiated border; low cells elongate, rectangular, smooth and weakly porose; basal cells long rectangular, thick-walled and smooth, strongly porose. (3) Perichaetial leaves much longer than branch leaves, 7-8 mm, long oblong-lanceolate, setaceous-acuminate; distinctly plicate below, sheathing at base, upper cells papillose, and low and basal cells smooth. (4) Setae 4-6 mm long, smooth, twisted to the left. (5) Capsule urns globular, ovoid to ovoid-ellipsoid, or urceolate, smooth; peristome double, both membranous; spores rather large, 81-86 μm.

Distribution: Bolivia.

202. *Macromitrium saddleanum* Besch. ex Müll. Hal., Flora 68: 417. 1885. (Figure 212)

Type protologue: [Chile] Saddle island: *Hariot No. 160*. Type citation: Chile, Ile Saddle (Wollantas) T. de Feu, *Hariot P. A., 160* (isotypes: PC 0138018!, PC 0138019!, PC 0138022!).

(1) Plants large and robust, in loosely mats; stems weakly creeping, with inconspicuous and caduceus leaves; branches up to 55-60 mm long, thick, green above, light brown below, distinctly aggregating and somewhat dichotomously branching, in similar length. (2) Branch leaves individually to weakly spirally appressed-twisted when dry, widely spreading when moist; short oval-, oblong- to lingulate-lanceolate, plicate below; apices acuminate to acuminate-acute; margins entire and plane; laminae partially bistratose in upper portion; costae stout, vanishing beneath the apex; all cells flat, clear and smooth, not porose; upper and medial cells quadrate, subquadrate, rounded-quadrate, oblate or irregularly-triangular; ovate, elliptic or oblong-rectangular cells confined to a small area near the base. (3) Sporophytes not seen.

Macromitrium saddleanum Besch. ex Müll. Hal. is rather similar to *M. crassiusculum* Lorentz, only the branches of the former are longer.

Distribution: Chile.

Specimens examined: **CHILE**. Prov. Antarctica Chilena, *William R. Buck 46244* (KRAM-B-205420); Prov. Tierra del Fuego, *William R. Buck 47738* (KRAM-B-205423); Fuegie Ile Saddle, *Harrot 1883* (PC 0092549); berre-de-Feu: Ile Saddle, *Harrot, 1885* (PC 00138021); Fuegian, Dept. Ushuaia, Bahia, Buen Suceso, *Matteri-Schiavone, 3773* (MO 3684160).

Macromitrium saddleanum* var. *piliferum Herzog, Rev. Bryol. Lichénol. 23: 81. 1954.
Protologue: [Chile] Westpatagonien: Pto. Isla Magdalena, n° 20/d pp.
The variety differed from var. *saddleanum* by its piliform branch leaves.

203. *Macromitrium salakanum* Müll. Hal., Syn. Musc. Frond. 2: 646. 1851 ('*Salakanum*'). (Figure 213)
Type protologue and citation: [Indonesia] Java, ad arbores monist Salak: *Zollinger sub No. 1426. Z.*, cum *M. orthosticho* commixtum (isotype: JE 04008691!).
Macromitrium celebense Paris., Index Bryol. 773. 1897, *fide* Vitt *et al.*, 1995.—*Macromitrium salakanum* Müll. Hal. subsp. *celebense* (Paris) M. Fleisch., Musci Butenzorg 2: 446. 1904. Based on *M. reflexifolium* Sande Lac.
Macromitrium humile Bosch & Sande Lac., Bryol. Jav. 1: 128. 106. 1860, *fide* Vitt *et al.*, 1995. Type protologue: (Indonesia) Habitat insulam *Javae*; in montibus *Gedé* et Salak, sociis *Macromitrio goniorrhyncho* et *orthosticho*, legit *Teysmann*. Type citation: Java, *Anon s.n.* (isosyntype: H-BR 2617016!).
Macromitrium pungens Mitt. ex Bosch & Sande Lac., Bryol. Jav. 1: 122. 99. 1860, *fide* Vitt *et al.*, 1995. Type protologue: (Indonesia) Habitat insulam *Javae* Teysmann in herb. Dz. et Mb. Type citation: Java, Herb Dxy et Molk (isotype: BM 000852241!).
**Macromitrium reflexifolium* Sande Lac., Sp. Nov. Musc. Ind. 8, 5C. 1872, *fide* Vitt *et al.*, 1995. — *Macromitrium salakanum* var. *reflexifolium* M. Fleisch. ex E.B. Bartram, Philipp. J. Sci. 68: 179. 1939. Type protologue: (Indonesia) Celebes in prov. Menado, de Vriese (lectotype: L-Lacoste).
***Macromitrium nova-guinense* Broth., Musci Buitenzorg 2: 447. 1904, *fide* Vitt *et al.*, 1995.
(1) Plants small, in spreading mats; stems prostrate, creeping, with rusty rhizoids and numerous branches. (2) Stem leaves covered with rhizoids and inconspicuous. (3) Branch leaves individually curved-twisted to deflexed-twisted when dry, wide-spreading-recurved to abaxially curved when moist, 1.5-2.1 mm long, lanceolate from an ovate-oblong base, gradually narrowed to a short-cuspidate, mucronate-acuminate, or blunt-acuminate apex; margins plane, sometimes roughened above; costae excurrent, often filing the acumen; upper cells rounded to subquadrate (4-7 µm), evenly thick-walled, arranged in distinctly longitudinal rows, obscurely pluripapillose; medial cells elongate, 5-7 × 5-22 µm, rounded to rectangular, lumens straight-curved, hyaline, gradually elongate below; basal cells 8-10 × 20-50 µm, elongate with straight walls, lumens sigmoid-curved, hyaline and smooth. (4) Perichaetial leaves erect, sheathing, oblong to oblong-ligulate; apices obtuse, rounded to mucronate, rarely retuse; costae ending before the rounded apex or filling the mucro; all cells clear, longer than wide, with curved lumens. (5) Setae 5.5-9.0 mm long, erect-curved, smooth, twisted to the left; vaginulae without hairs. (6) Capsule urns 1.3-2.0 mm long, narrowly ellipsoid; peristome single; exostome of 16, pale, pale membrane; spores 12-35 µm, anisosporous. (7) Calyptrae plicate, mitrate, sparsely hairy, lacerate at base.
Macromitrium salakanum is somewhat similar to *M. angustifolium* Dozy & Molk., but the inner perichaetial leaves of *M. salakanum* are obtuse at the apex, while those of *M. angustifolium* narrowly acute or acuminate.
Distribution: Indonesia, New Caledonia (Thouvenot, 2019), Philippines, Papua New Guinea, Sri Lanka, the Solomon Islands.
Specimens examined: **INDONESIA**. West Java: Berggarten von Tjibodas am GTedeh (as *M. celebense*, H-BR 2605002, H-BR 2605003). **PAPUA NEW GUINEA**. Morbe Province, *H. Streimann & E. Tamba, 12580* (H 3196913); Southern Highlands Province, Nogoti, Tagari River, *H. Streimann 32565* (KRAM-B-107236). **PHILIPPINES**. Leyte, *B. C. Tan 16* (MO 3071733), Mindanao, *J. V. Paucho 2524* (FH); Mindanao, *M. S. Clemens 36936-2* (MO B757055); Leyte Island, Baybay, Mt. Pangasugan, *Tan BC, Navarez M & Raros L. 84-263* (KRAM-B-060633). **SRI LANKA**. Lebanon estate, *C. Ruinard 12/62* (MO 2859632); Samanala, *C. Ruinard 23/172* (MO 2859900). **THE SOLOMON ISLANDS**. Kolombangara, *D. H. Norris & G. L. Roberts 49380* (H 3196912).

204. *Macromitrium savatieri* Besch., Ann. Sci. Nat., Bot., sér. 7, 20: 25. 1894. (Figure 214)
Type protologue: [French Polynesia] Tahiti: Papeete, vallée de la Reine, septembre 1877, sur les lrones d'arbres (*Savatier, no 741*). Type citation: Tahiti, *Savatieri, no. 741*, Herb. Em. Bescherelle (isotypes: PC 0148121!, PC 0148122!).
(1) Plants large, in loose mats or somewhat wefts; stems creeping, with prostrate and ascending branches, branches up to 3-5 cm tall. (2) Branch leaves erect below, twisted-crisped above, partial apices hidden in the inrolled cavity when dry, spreading, conduplicate and adaxially incurved in the upper portion when moist, lanceolate from a long oblong base, acuminate to acuminate-acute; upper, medial and low cells rounded, round-quadrate, and oblate ovate, in longitudinal rows; upper cells obscure and pluripapillose; cells from the medial to the base clear and smooth; cells elongated towards the base, thick-walled; costa percurrent, often partially concealed in the upper portion in the adaxial view by the overlapping folds of laminae. (3) Perichaetial leaves longer than branch leaves, lanceolate from a broadly oblong base. (4) Setae 4-5 mm long, smooth; vaginulae hairy. (5) Capsule urns ovoid-ellipsoid, constricted beneath the mouth, smooth; peristome absent; spores anisosporus. (6) Calyptrae mitrate, with a few straight hairs,

Taxonomy

deeply lacerate.
  Distribution: Tahiti in French Polynesia.
  Specimens examined: **TAHITI**. *Savatieri s.n.* (PC 0148124); *anonym, 7008* (PC 0148123); Sur las Arbres, *Savatieri s.n*, 1877 (PC 0148125).

205. *Macromitrium schmidii* Müll. Hal., Bot. Zeitung (Berlin) 11: 61. 1853. (Figure 215)
  Type citation: India, Nilgherrsi, Leg. *G.S. Perrottet* (lectotype designated by Guo *et al*., 2007: H-BR 2595002!; isolectotype: H-BR 2595001!).
  Macromitrium benguetense R.S. Williams., Bull. New York Bot. Gard. 8(31): 343. 1914, *fide* Guo *et al.*, 2007. Type protologue and citation: (Philippines) Baguio, on tree, 1570 meters, Oct. 1904 (830) (isotype: H-BR 2572023!).
  (1) Plants medium-sized, forming yellowish brown mats, darkish below; stems long creeping, up to 10 cm long, with erect branches, branches 10-20 mm high and 1.5-2.0 mm thick. (2) Branch leaves individually twisted, moderately contorted-crisped and keeled when dry, widely spreading when moist, lanceolate, (2.5) 3.5-4.0 × 0.5-0.7 mm, plicate below; apices acuminate, somewhat incurved; margins entire throughout, plane on one side, reflexed-recurved on the other side, particularly in low portion; costae ending a few cells beneath the apex or in the apex; upper cells subquadrate-rounded to rounded-quadrate, 10-14 μm wide, obscure, densely pluripapillose; medial cells rounded-quadrate or slightly elongate, 10-15 ×15-20 μm, moderately bulging and mammillose or pluripapillose; basal cells brown-yellowish, rectangular to sublinear, 30-50 × 6-10 μm, strongly thick-walled and porose, bulging-unipapillose. (3) Perichaetial leaves ovate-oblong, long-acuminate in upper part, 2.8-3.2 mm long, shorter than branch leaves; costae ending far beneath the apex. (4) Setae 4-6 mm, smooth, twisted to the right. (5) Capsule urns ellipsoid-cylindric or ellipsoid, brown, plicate or ribbed under the mouth when dry, peristome absent. (6) Calyptrae medium-sized, 2.0-2.5 mm long, campanulate, with brownish hairs.
  Distribution: India, Philippines, Sri Lanka, Vietnam (Tan & Iwatsuki, 1993).
  Specimens examined: **INDIA**. Madras, *G. Foreau s. n.* (MO 5134067, MO 2559204); Nilghiri, *B. Suthi 7331* (H-BR 2595003). **SRI LANKA**. *Wichura, 2744* (H-BR 2595004); *C. Müeller 41* (H-BR 2595005); Malau, *T. W. N. Beckett c29/2* (H-BR 2595007); Central Province, *G. H. K. Thwaites, no. 41* (MO 5278951); Central Province, *G. H. K. Thwaites 38* (MO 5279545).

Macromitrium schmidii* var. *macroperichaetialium S. L. Guo & T. Cao, Gard. Bull. Singapore 58: 160. 2007. (Figure 216)
  Type protologue and citation: China, Kwangtung: Ngok Shing Shan, Sai-lin-shan Village, Sin-fung district; thicket on steep slope. 23-31 March 1938, *Y. M. Taam 402C* (holotype: NY!).
  Macromitrium schmidii var. *macroperichaetialium* differs from the type variety of *M. schmidii* in having 1) long lanceolate inner perichaetial leaves with a gradually long-acuminate upper portion, much longer than branch leaves; and 2) capsules distinctly constricted beneath the mouth to produce a 4-angled or 4-furrowed shape.
  Distribution: China.

206. *Macromitrium scoparium* Mitt., J. Linn. Soc., Bot. 12: 206. 1869. (Figure 217)
  Type protologue and citation: Hab. Ins. Jamaica, Wilds; Trinidad, *Crüger*. Type citation: [Jamaica] Trinidad, leg. *Crüger*. (isotype: S-B 165032!).
  Macromitrium palmense R.S. Williams, Torreya 14: 25. 1914, *fide* Allen, 2002. Type protologue: La Palma, Costa Rica, 480, Maxon, May 6, 1906.
  Macromitrium tonduzii Renauld & Cardot, Bull. Soc. Roy. Bot. Belgique 31(1): 155. 1893, *fide* Churchill & Linares, 1995 and Allen, 2002.
  Macromitrium williamsii E.B. Bartram, Contr. U.S. Natl. Herb. 26: 84. f. 22. 1928, *fide* Churchill & Linares, 1995 and Allen, 2002. Type protologue: On tree, La Hondura, Province of San Jose, Costa Rica, alt. 1300 to 1700 meters, *Paul C. Standley*, March 24, 1924, *no. 36405*.
  (1) Plants medium-sized to large, rusty-yellow to yellowish-green; stems creeping, branches up to 5 cm long. (2) Branch leaves keeled, erect below, contorted-twisted, flexuose and undulate above when dry, erect-spreading when moist, 4-6 × 0.5 mm, long lanceolate; apices acuminate to subulate; margins undulate, sharply serrate in upper half, recurved below, plane above; costa excurrent into the apex; all cells longer than wide, thick-walled, porose; upper and medial interior cells variable in size and shape, oval to rhomboidal, slightly bulging to papillose; upper marginal cells linear, forming a differentiated border; basal cells near margin narrowly rectangular, those near costa larger, straight-walled and tuberculate; marginal basal enlarged teeth-like cells at insertion not differentiated. (3) Perichaetial leaves broadly lanceolate to ligulate-lanceolate from a widely oblong low part, sharply narrowed to an acuminate or acuminate-acute apex, sheathing at base, plicate below; all cells long- to linear-rectangular, thick-walled and porose. (4) Setae up to 20 mm long, smooth. (5) Capsule urns subglobose, smooth, wrinkled at neck; peristome double, exostome teeth lanceolate, densely papillose, united to form an erect membrane, endostome strongly papillose,

with basal membrane and segments. (6) Calyptrae mitrate, lacerate, naked. *Macromitrium trinitense* R.S. Williams is similar to *M. scoparium* Mitt. in their long lanceolate leaves with cells all longer than wide, a differentiated border, and absence of basal enlarged teeth-like cells at insertion, but the former could be separated from the latter by the ellipsoid capsules and shorter setae.

Distribution: Colombia, Costa Rica, Dominica, Ecuador, French Guiana, Jamaica, Panama.

Specimens examined: **COLOMBIA**. Municipio Yarumal. *Ines Sastre-De Jesus y & Steven P. Churchill 1101* (MO 3374255). **COSTA RICA**. Puntarenas, M. J. Lyon 224 (MO 4455584). **DOMINICA**. Parish St. Paul. *E. P. Hegewald 9344* (H 3090709). **ECUADOR**. Prov. Loja: Loma de Oro. *W.C. Steere & Henrik Balslev 25902* (KRAM-B-058165, MO 5381270). **FRENCH GUIANA**. Canton de Approuague-Kaw. *W.R. Buck 38036* (MO 6001464). **JAMAICA**. Forest between Eastern Blue mountain Peak and Sugarloaf, *William R. Maxon 9953* (H-BR 2632016). **PANAMA**. Province Panama, *B. H. Allen 9037* (MO 3965356).

207. ***Macromitrium sejunctum*** B.H. Allen, Novon 8: 118. f. 5. 1998. (Figure 218)

Type protologue and citation: Honduras. Olancho: Sierra de Algalta: La Chorrerra below Montana Bibilonia, Rio lara, 15 km NNW of Catacamas, 14°59′N, 85°56′W, *Allen 12483* (holotype: MO!); Honduras. Atlántida: El Porvenir, Pico Bonito National Park, along trail from confluence to ridge camp. 15°38-39′N, 86°51-52′W, Allen 17367 (paratype: MO!).

(1) Plants medium-sized, rusty-brown, in somewhat wefts; stems creeping, with moderately dense rhizoids; branches 1.5-2.5 cm long. (2) Branch leaves erect below, individually twisted, contorted-flexuose and crisped above when dry, erect-spreading and undulate when moist, 2.5-3.0 × 0.8 mm, oblong, ligulate-lanceolate, broadly lanceolate; apices broadly acute or acuminate into a fragile leaf tip, easily broken in the upper portion when old; margins distinctly and strongly erose-dentate above, recurved below, plane above, with reddish rhizoids at base; costae subpercurrent to percurrent, dorsal costal surface in upper 1/3 covered by short, mammillose cells; upper and medial cells quadrate, rounded-quadrate or hexagonal, smooth, bulging to mammillose; basal cells short, 20-28 μm long, rectangular, incrassate, not or weakly porose, smooth or weakly tuberculate; marginal cells in the outmost rows enlarged, short-rectangular to quadrate, with straight lumen. (3) Sporophytes not seen.

Macromitrium sejunctum B.H. Allen, *M. fragilicuspis* Cardot and *M. frustratum* B.H. Allen are similar in their leaves with fragile leaf apices. *Macromitrium frustratum* differs from *M. sejunctum* by its larger plants with narrowly lanceolate leaves (to 7 mm long), and elongate cells on the dorsal surface of the costa. *Macromitrium fragilicuspis* can be distinguished from *M. sejunctum* by its entire leaf margin.

Distribution: Honduras.

208. ***Macromitrium semperi*** Müll. Hal., Linnaea 38: 559. 1874.

Type protologue: Patria, Insulae Philippinae, Luzon, Mariveals; Dr. *C. Semper.* (lectotype: BM-Hampe).

(1) Plants medium-sized to large, rusty-brown or reddish; stems long creeping, densely with branches, branches up to 2.5 cm long. (2) Stem leaves ovate-lanceolate, inconspicuous on old stems. (3) Branch leaves loosely, regularly and strongly contorted-twisted-flexuose, most the apices incurved to decurved-twisted, usually indistinctly funiculate when dry, wide-spreading to upper abaxially curved when moist, 2.0-3.5 mm long, narrowly lanceolate to ligulate-lanceolate from a broad oblong base; apices acuminate to bluntly acuminate; margins plane to reflexed below, entire; costae ending in the apex; upper cells rounded to subquadrate (6-8 μm), firm-walled, densely pluripapillose to nearly smooth; medial cells 8-11 × 7-21 μm, rounded-quadrate to rectangular, variable in length, with somewhat curved lumens, smooth; basal and low cells 9-10 × 18-38 μm, long-rectangular, strongly curved to sigmoid, smooth, flat, elongate near margin. (4) Perichaetial leaves 2.0-3.5 mm long, broadly lanceolate to oblong-lanceolate, with a cuspidate apex, plicate and sheath around the vaginula; costae excurrent to the cusp; all cells longer than wide, sigmoid-curved cells continuing into apex. (5) Setae 3.0-4.5 mm, smooth; vaginulae without hairs. (6) Capsule urns 1.1-1.5 mm long, broadly ovoid to ellipsoid-ovoid, lightly 8-plicate to smooth; peristome single; exostome of 16, pale to white teeth, well-developed; spores 12-55 μm, anisosporous. (7) Calyptrae mitrate and naked (Vitt *et al.*, 1995).

There are several species with sigmoid basal cells in the genus, but *M. semperi* Müll. Hal. is large and rusty-brown, having conspicuous, erect, sheathing, perichaetial leaves with a stoutly cuspidate apex. *Macromitrium semperi* is similar to *M. macrosporum* Broth., but the basal cells of the former are smooth without tubericulae. *Macromitrium semperi* is also morphologically similar to *M. leratii* Broth. & Paris, but the latter differs from the former by having larger plants, longer setae, more plicate capsules, and blunt perichaetial leaves.

Distribution: Papua New Guinea, Philippines.

Specimens examined: **PAPUA NEW GUINEA**. Morobe, *D. H. Norris, 64344* (H 3195961), *T. J. Koponen 33685* (MO 4435582); West Sepik, *A. Touw 15917* (H 3205124), *15470* (MO 5375860), *17788* (MO 5371048), *17790* (MO 5375804), *15917* (H 3205124); Eastern Highlands, *W. E. Wade B-33952* (MO 4411930), *H. Streimann 18094* (MO 4419005), *S. He 44910* (MO); Southern Highlands, *H. Streimann 26572* (KRAM-B-107265).

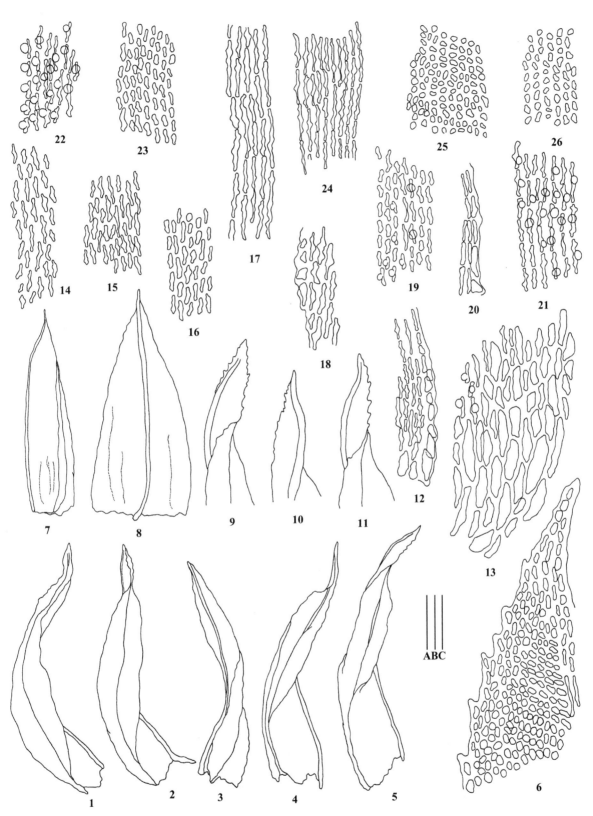

Fig. 210 *Macromitrium rugulosum* Ångström 1-5: Branch leaves. 6: Apical cells of branch leaf. 7-8: Perichaetial leaves. 9-11: Apices of branch leaves. 12, 20: Basal marginal cells of branch leaves. 13: Basal cells near costa of branch leaf. 14: Apical cells of perichaetial leaf. 15: Upper cells of perichaetial leaf. 16: Low cells of perichaetial leaf. 17: Basal cells of perichaetial leaf. 18: Basal marginal cells of perichaetial leaf. 19: Low cells of branch leaf. 21-22, 24: Basal cells of branch leaves. 23: Low cells of branch leaf. 25: Upper cells of branch leaf. 26: Medial cells of branch leaf (all from H-BR 2628001). Line scales: A = 500 μm (1-5, 7-8); B = 200 μm (9-11); C = 50 μm (6, 12-26).

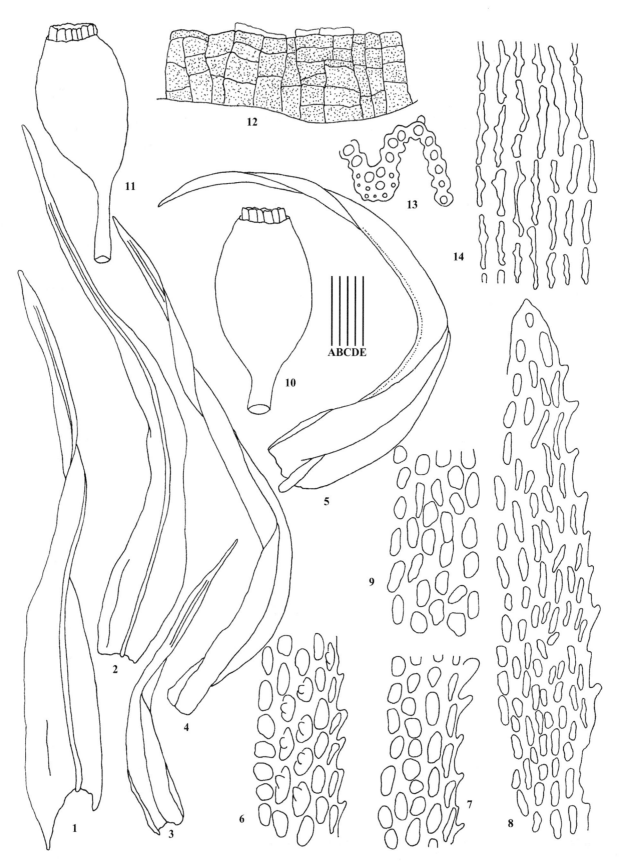

Fig. 211 *Macromitrium rusbyanum* E. Britton 1-5: Branch leaves. 6: Upper cells of branch leaf. 7: Medial marginal cells of branch leaf. 8: Apical cells of branch leaf. 9: Medial cells of branch leaf. 10-11: Capsules. 12: Peristome. 13: Upper transverse section of branch leaf. 14: Basal cells of branch leaf (all from isotype, FH 00213675). Line scales: A = 2 mm (3, 10-11); B = 1 mm (1-2, 4-5); C = 400 μm (12); D = 100 μm (13); E = 67 μm (6-9, 14).

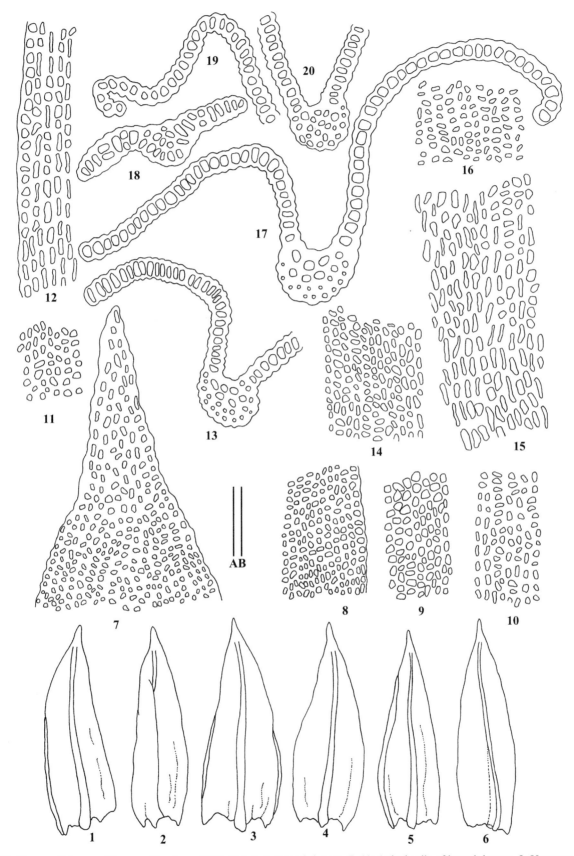

Fig. 212 *Macromitrium saddleanum* Besch. ex Müll. Hal. 1-6: Branch leaves. 7, 11: Apical cells of branch leaves. 8: Upper marginal cells of branch leaf. 9: Medial cells near costa of branch leaf. 10: Low cells of branch leaf. 12: Basal marginal cells of branch leaf. 13, 20: Medial transverse section of branch leaves. 14: Medial cells of branch leaf. 15: Basal cells of branch leaf. 16: Low cells near costa of branch leaf. 17, 19: Basal transverse sections of branch leaves. 18: Upper transverse section of branch leaf (all from isotype, PC 0138019). Line sclaes: A = 0.5 mm (1-6); B = 50 μm (7-20).

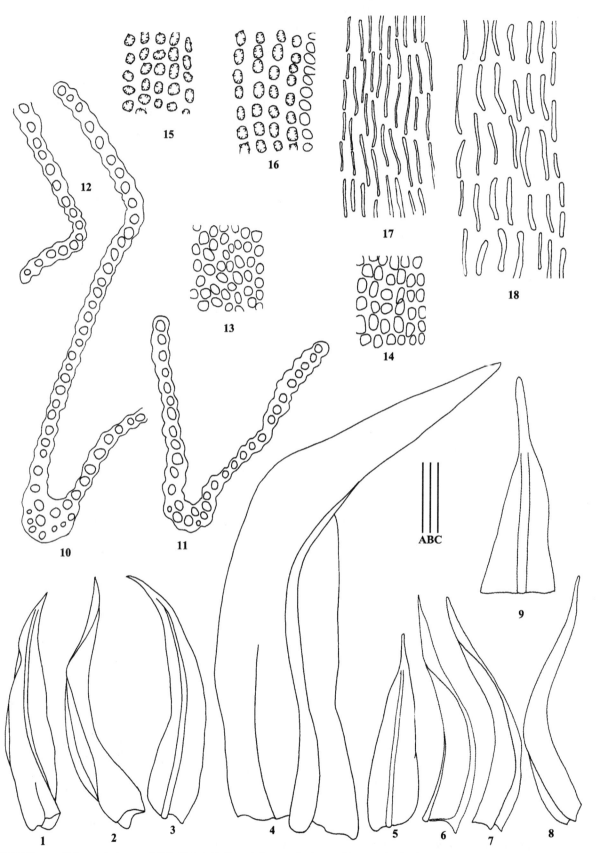

Fig. 213 *Macromitrium salakanum* Müll. Hal. 1-8: Branch leaves. 9: Apex of branch leaf. 10-12: Basal transverse sections of branch leaves. 13, 15: Upper cells of branch leaves. 14, 16: Medial cells of branch leaves. 17, 18: Basal cells of branch leaves (all from isotype, JE 04008691). Line scales: A = 0.44 mm (1-3, 5-8); B = 176 μm (4, 9); C = 44 μm (10-18).

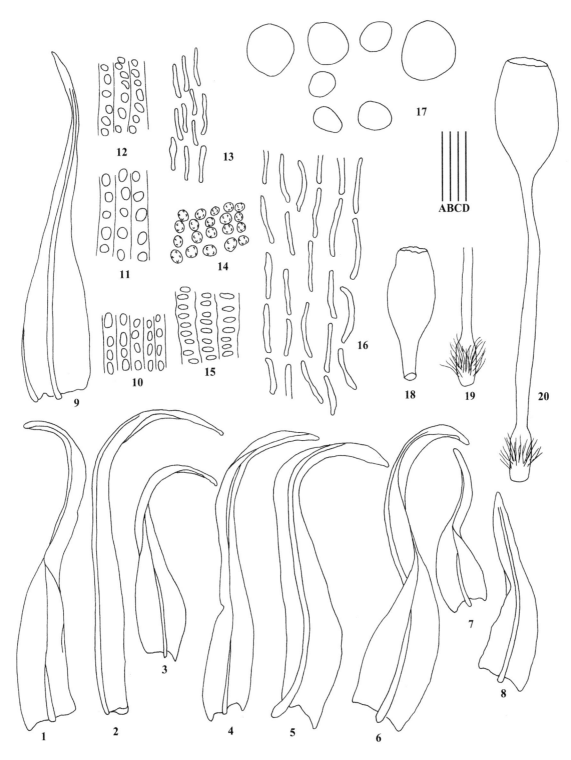

Fig. 214 *Macromitrium savatieri* Besch. 1-6: Branch leaves. 7-8: Stem leaves. 9: Perichaetial leaf. 10: Medial cells of perichaetial leaf. 11: Low cells of perichaetial leaf. 12: Upper cells of perichaetial leaf. 13: Basal cells of perichaetial leaf. 14: Upper cells of branch leaf. 15: Medial cells of branch leaf. 16: Basal cells of branch leaf. 17: Spores. 18, 20: Capsules. 19: Vaginula (all from isotype, PC 0148121). Line scales: A = 2 mm (18-20); B = 1 mm (1-9); C = 100 µm (17); D = 67 µm (10-16).

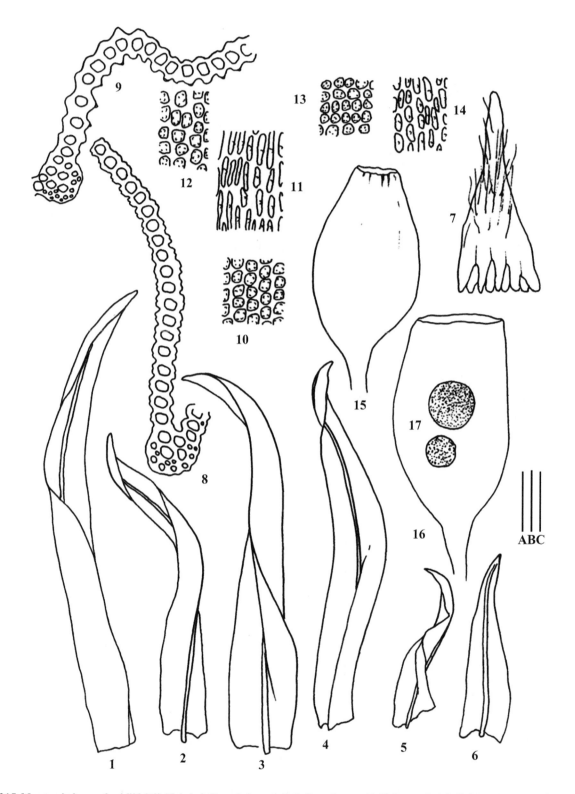

Fig. 215 *Macromitrium schmidii* Müll. Hal. 1-4: Branch leaves. 5-6: Stem leaves. 7: Calyptra. 8: Medial transverse section of branch leaf. 9: Basal transverse section of branch leaf. 10: Upper cells of branch leaf. 11: Medial cells of branch leaf. 12: Basal cells of branch leaf. 13: Upper cells of stem leaf. 14: Low cells of stem leaf. 15: Dry capsule. 16: Wet capsule. 17: Spores (all from isotype of *M. benguetense* R. S. William in H-BR). Line scales: A = 0.50 mm (1-6, 15-16); B = 50 μm (8-14, 17); C = 1 mm (7). (Guo & Cao, 2007).

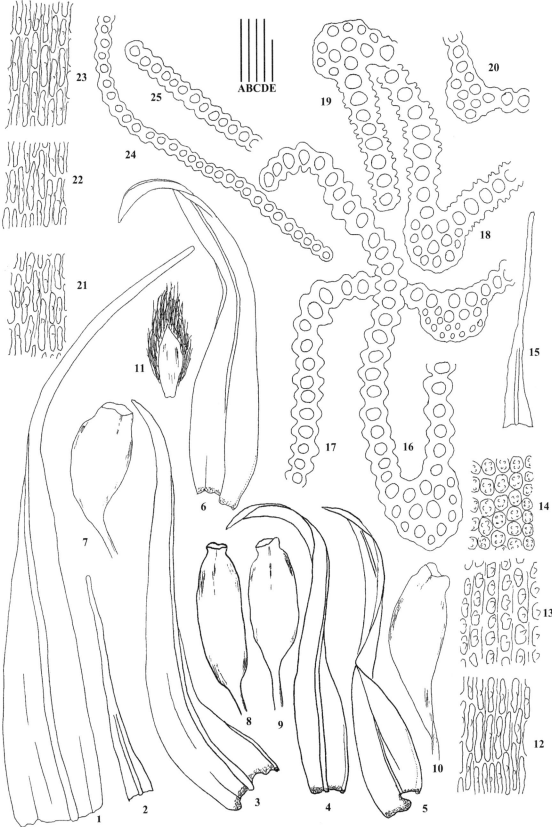

Fig. 216 *Macromitrium schmidii* var. *macroperichaetialium* 1: Perichaetial leaf. 2, 15: Apices of perichaetial leaves. 3-6: Branch leaves. 7-10: Dry capsules. 11: Calyptra. 12: Basal cells of branch leaf. 13: Medial cells of branch leaf. 14: Upper cells of branch leaf. 16: Medial transverse section of branch leaf. 17: Basal transverse section of branch leaf. 18, 19: Upper transverse sections of branch leaves. 20, 25: Upper transverse sections of perichaetial leaves. 21: Upper cells of perichaetial leaf. 21: Medial cells of perichaetial leaf. 23: Basal cells of perichaetial leaf. 24: Medial transverse section of perichaetial leaf (all from holotype, NY). Line scales: A = 0.50 mm (1, 3-6); B = 50 μm (12-14, 16-25); C = 2.0 mm (11); D= 1.0 mm (2, 15); E= 0.50 mm (7-10). (Guo & Cao, 2007).

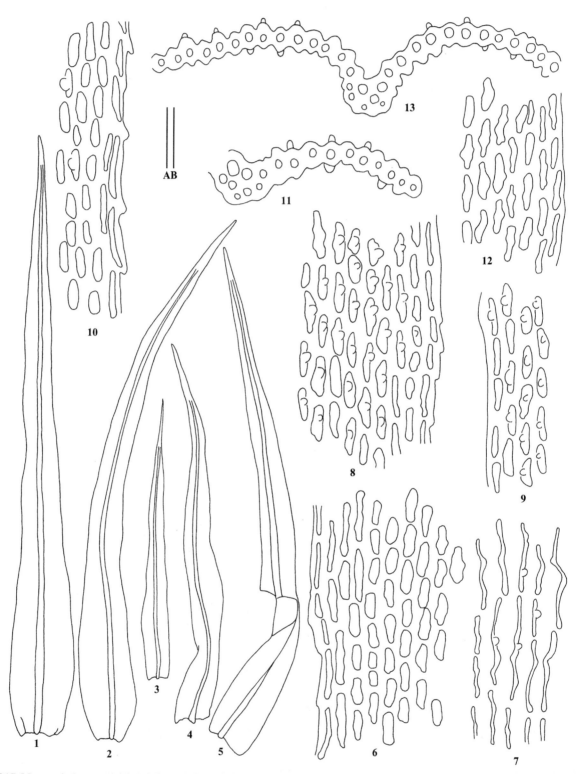

Fig. 217 *Macromitrium scoparium* Mitt. 1-5: Branch leaves. 6, 9: Medial cells of branch leaves. 7: Basal cells of branch leaf. 8, 10: Upper cells of branch leaves. 11: Upper transverse section of branch leaf. 12: Low cells of branch leaf. 13: Medial transverse section of branch leaf (all from isotype, S-B 165032). Line scales: A = 1 mm (1-5); B = 67 μm (6-13).

Taxonomy

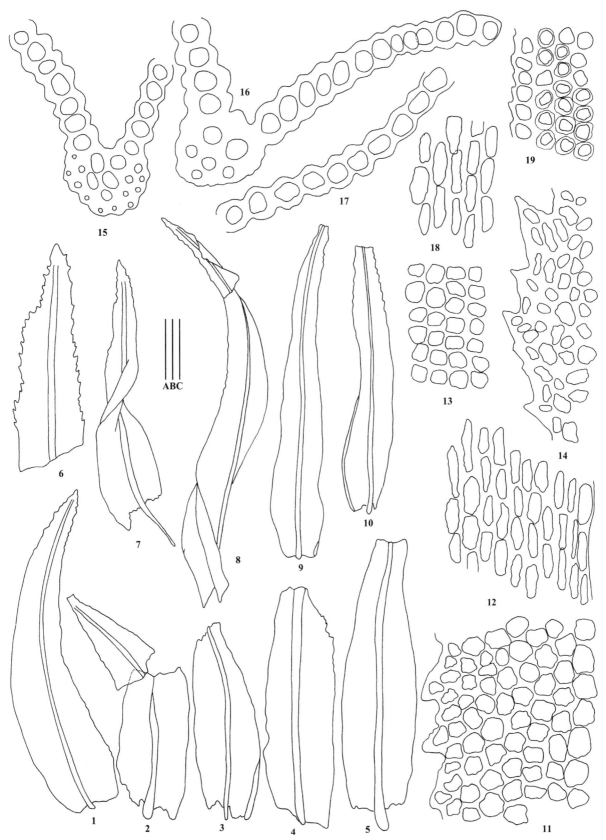

Fig. 218 *Macromitrium sejunctum* B.H. Allen 1-5, 7-10: Branch leaves. 6: Apex of branch leaf. 11: Upper cells of branch leaf. 12: Low cells of branch leaf. 13: Medial cells of branch leaf. 14: Apical cells of branch leaf. 15: Medial transverse section of branch leaf. 16: Upper transverse section of branch leaf. 17: Low transverse section of branch leaf. 18: Basal cells of branch leaf. 19: Medial marginal cells of branch leaf (1-7, 11-19 from holotype, MO 406401; 8-10 from paratype, MO 406402). Line scales: A = 1 mm (1-5, 7-10); B = 400 μm (6); C = 67 μm (11-19).

209. *Macromitrium serpens* (Burch. ex Hook. & Grev.) Brid., Bryol. Univ. 1: 736. 1826. (Figure 219)
Basionym: *Orthotrichum serpens* Hook. & Grev., Edinburgh J. Sci. 1: 119, pl. 5. 1824. Type protologue: Trunks and branches of trees in woods at Sylvas Station, near George's Plain, Anteniqua Land, Cape of Good Hope. W.J. Burchell, Esq. Type citation: [South Africa] Cape of Good Hope, *Burchell s. n.* (isotype: E 0007638!).
***Dasymitrium rehmannii* Müll. Hal., Rev. Bryol., 5: 70. 1878, *fide* O'Shea, 2006.
Macromitrium anomodictyon Cardot, Bull. Mus. Natl. Hist. Nat. 22: 345. 1916, *fide* Wilbraham, 2016. Type protologue and citation: (Madagascar) Province de Tananarive, district d'Andramasina; sur branches, dans un petit bois, vers 1,700 mètres d'altitude, entre Tsinjoarivo et Ambohimasina, 2 décembre 1912; n°. 1932. Type citation: Madagascar, Province de Tananarive, district d'Andramasina; entre Tsinjoarivo et Ambohimasina, vers 1700. d´ alt. sur branches, d´arbres dans un petit bois, 1912, *Viguier* et *Humbert 1932* (lectotype: PC 0106573!; isolectotype: PC 0106571!), Madagasccar, Prov. de Tananarive, distr, d'Andramasina, 2, 12, 1912, *Viguier & Humbert, n°1932* (isolectotypes: PC 0106570!, PC 0106572!).
Macromitrium astroideum Mitt., Philos. Trans.168: 390. 1879, *fide* Wilbraham & Ellis, 2010. Type protologue: Only at the top of Oyster River valley. Trailing over boulders. Type citation: Rodriguez, *Balfour s.n.,* isotype: BM 000873888!).
**Macromitrium elegans* Duby, Mém. Soc. Phys. Genève 19: 296. 1 f. 3. 1868, *fide* O'Shea, 2006.
***Macromitrium rehmannii* Paris, Index Bryol. 785. 1897, *fide* O'Shea, 2006.
Macromitrium serpentinum Mitt., Cape Monthly Mag. 17: 379. 1878, *fide* Wijk et al., 1964.
Macromitrium tristratosum Dixon, Kongel. Norske Vidensk. Selsk. Skr. (Trondheim) 1932(4): 10. 2. 1932, *fide* van Rooy & van Wyk, 1992. Type protologue and citation: Natal, Zululand, Eshowe, indigenous forest, *Höeg 121* (lectotype: BM!).

(1) Plants medium-sized, in dense mats; primary stems long creeping, up to 10 cm long, densely with short branches, branches up to 10 mm long, densely with reddish brown rhizoids. (2) Stem leaves strongly to moderately crowded, fragile on old stems; ovate-acuminate or ovate-subulate, 1.0-2.3 mm long. (3) Branch leaves keeled, erect below, twisted-flexuose to tightly contorted-crisped, apices curved to circinate, occasionally hidden in the inrolled cavity when dry, erect-spreading when moist, somewhat rugose, 1.7-3.0 mm long, narrowly ovate-lanceolate, narrowly lanceolate or narrowly lanceolate-subulate; apices acuminate and fragile; margins plane, entire to crenulate-papillose above, entire below; costae subpercurrent to percurrent; upper and medial portion irregular 1-3-stratose proliferation with pluripapillose cells on both dorsal and ventral surfaces; basal cells rectangular, longitudinal walls incrassate, straight to sigmoid, frequently unipapillose or tuberculate. (4) Perichaetia leaves oblong-lanceolate, 1.6-2.7 mm long, unistratose. (5) Setae 3.5-6 mm long, smooth and reddish brown, not twisted. (6) Capsule urns ovoid-ellipsoid to ovoid-cylindric, smooth to weakly ribbed above; peristome single, exostome of 16, narrowly oblong, blunt, frequently irregularly in outline, pale yellow or yellowish brown, erect to incurved dry, striate-papillose; spores 16-38 μm, minutely papillose, anisoporous. (7) Calyptrae 2.8-3.2 mm long, mitrate, essentially naked, plicate, lacerate below.

Distribution: Kenya, Madagascar, Malawi, Mauritius, Natal, Réunion, South Africa, Tanzania (O'Shea, 2006).
Specimens examined: **MAURITIUS**. Rodriguez, *John Hutton Balfour s.n.* (BM 000873888, BM 000982425). **NATAL**. Nkandla Forest Reserve, *R. E. Magill 5166* (H 3090718). **RÉUNION**. Above La Montagne, *T. Arts RÉU82/19* (BR-BRYO 318248-88). **SOUTH AFRICA**. *H. A. Wages 150* (H-BR 2552006); Eastern Cape Province, *Taylor 468* (BM 000873891), *Drege s.n.* (BM 000989758), *Burchell 5868* (BM 000868306-09); The Tsitsikamma, Cape, *J. Van Rooy 3323 DC* (MO 306006); Saasveld, Cape prov., *Theo Arts* (KRAM-B-123200); Transkei, *J. Van Rooy 3228 BD* (MO 3059983). **TANZANIA**. Usambara Mountains by Amany. *Kaj Rssmussen & Westy Esbensen, 30* (H 3090766).

210. *Macromitrium sharpii* H.A. Crum ex Vitt, Bryologist 82: 4. f. 10-20, 66. 1979. (Figure 220)
Type protologue & Type citation: Mexico. Durango: on Juniperus, 8850 ft. (2700 m.) elev., 45 km W of El Salto. 30 Dec. 1944. *Sharp 1859*" (isotype: MICH 525909!); Mexico, 40 km, west of El Salt, Dgo. 8850 ft., 30 Dec. 1944. *Sharp 1868* (paratype: MO 2561180!).

(1) Plants small, yellow-green, in dense mats; stems long creeping, branches up to 1-3 cm long, densely with rusty reddish rhizoids. (2) Branch leaves individually twisted below, twisted-flexuose-crisped above, partial apices hidden in the inrolled cavity when dry, erect-spreading when moist, 1.5-2.5 × 0.5 mm, lanceolate; apices apiculate, acute or acuminate; margins entire below and crenulate from bulging cells, recurved or plane below, plane above, often with reddish rhizoids at base; costae subpercurrent, percurrent or excurrent into an apiculus; upper cells rounded-hexagonal, bulging mammillose, marginal cells not differentiated; basal cells long-rectangular, to 40 μm long, thick-walled, porose, tuberculate, marginal basal enlarged teeth-like cells not differentiation at the insertion, those near costa large, straight-walled, rectangular to irregularly rectangular, smooth, forming a "cancellina region". (3) Perichaetial leaves triangular-lanceolate, shorter than branch leaves; upper cells rounded-hexagonal, bulging;

medial cells short-rectangular, papillose. (4) Setae about 10 mm long, smooth, not twisted. (5) Capsule urns ellipsoid-ovoid, ellipsoid-cylindric, smooth or wrinkled above, with a wide mouth, not constricted beneath the mouth; peristome double, exostome teeth lanceolate, reddish brown, densely papillose, united to form a membrane, endostome hyaline, with basal membrane and segments. (6) Calyptrae cucullate or mitrate, naked, slightly lacerate.

Macromitrium sharpii H.A. Crum ex Vitt is similar to *M. pseudofimbriatum* Hampe, but the capsules of the latter are furrowed, and its upper leaf cells are smooth.

Distribution: Colombia, Ecuador, Honduras, Mexico; Guatemala (Allen, 2002).

Specimens examined: **COLOMBIA**. Antioquia, *Margarita Escobar A. 851* (MO); Caquetá, *S. P. Churchill & Julio C. Betancur B. 16934* (MO); Cauca, *J.A. Uribe 3276* (MO), *S. P. Churchill 16496* (MO). **ECUADOR**. Tungurahua, *Robert M. King 6552A* (MO). **HONDURAS**. Lempira Department, *B. Allen 11882* (MO 3972919), *12202* (KRAM-B-101075), *12208* (MO 3965125, MO 3965330); Comayagua, *B. Allen 13985* (KRAM-B-104972). **MEXICO**. 40 km W of El Saito; Dgo, *A. J. Sharp 1868* (MO).

211. *Macromitrium similirete* E.B. Bartram, Svensk Bot. Tidskr. 47: 399. 1953.

Type protologue: Papua New Guinea. Western Highlands, Mt. Hagen district, Wahgi region, (S of Bismarck Mts.), Nondugl, 1600 m, *N. & G. Gyldenstolpe 37* (holotype: FH-Bartram).

(1) Plants small, dull, in compact, spreading mats; creeping stems with short branches, branches simple to forked, up to 1 cm long. (2) Stem leaves small, ovate-lanceolate, gradually and slenderly acuminate. (3) Branch leaves twisted-contorted, individually obscurely inrolled-twisted when dry, erect-spreading, incurved-flexuose when moist, 1.5-2.2 mm long, ligulate, ligulate-lanceolate to lanceolate; apices shortly cuspidate, mucronate-acuminate; margins plane to broadly reflexed, entire in upper portion, notched-crenulate because of protruding tuberculate in low portion; costae excurrent and filling the cusp; upper and medial cells rounded to rounded-subquadrate (6-11 μm), obscure, bulging, irregularly and weakly papillose to almost smooth, in longitudinal rows; basal cells short-rectangular-elliptic to rounded, strongly tuberculate, elongate towards the margin. (4) Perichaetial leaves 1.7-2.0 mm long, erect, short, ovate-lanceolate to ovate-oblong, gradually shortly acute to sharply acuminate-apiculate; costae shortly excurrent; all cells clear smooth and flat; upper and medial cells long-elliptic to round, basal cells elongate. (5) Setae 7.0-9.0 mm long, smooth, twisted to the left. (6) Capsule urns narrowly to widely ellipsoid, smooth; peristome single, with 16 well-developed exostome teeth, lanceolate, blunt; spores 14-30 μm, anisosporous. (7) Calyptrae mitrate, naked and plicate, slightly lobed at the base (Vitt *et al.*, 1995).

Distribution: Papua New Guinea.

Specimens examined: **PAPUA NEW GUINEA**. Morobe, *H. Streimann 18983* (H 3196884), *19277* (H 3196888).

212. *Macromitrium solitarium* Müll. Hal., Nuovo Giorn. Bot. Ital., n.s. 4: 125. 1897. (Figure 221)

Type protologue: [Bolivia] Cochabamba Bolivianae, *Germain*, 1889. Type citation: Bolivia, provincia Cochabamba prope Choquecamata, leg. *Germain s.n.*, Jun. 1889. [determ. D. C. Mueller sub n. 1261] (lectotype: NY 01086658!; isolectotype: H-BR 2625022!).

(1) Plants medium-sized, dark-brownish; branches up to 30 mm, with branchlets, densely with reddish rhizoids. (2) Branch leaves erect and appressed below, undulate, spirally twisted-contorted above when dry, spreading when moist, oblong-lanceolate to ligulate-lanceloate, ligulate, often conduplicate; apices acuminate, acute to apiculate; margins entire or irregularly serrulate near the apex; costae stout, light rusty, subpercurrent or excurrent; all cells smooth, collenchymatous, not porose; apical, upper and medial cells subquadrate, quadrate, round-rhombic, conic-bulging; low cells subquadrate to rectangular; basal cells rectangular, strongly porose; elongate cells confined to a small area near the base (in low 1/6 to 1/7 of the leaf length). (3) Perichaetial leaves lanceolate from an oval-oblong base, acuminate to acuminate-aristate, entire; costae stout, vanishing beneath the apex; upper and medial cells rounded-quadrate, rounded, elliptical or subquadrate, bulging, elongate towards the base; low and basal cells rectangular, long rectangular, not porose. (4) Setae smooth, 7-10 mm long, not twisted. (5) Capsule urns ellipsoid, reddish, smooth to weakly wrinkled or furrowed; peristome double, both membranous; opercula conic-rostrate, with an oblique beak. (6) Calyptrae mitrate, small, only covering half of the capsule, smooth and naked.

Distribution: Bolivia.

Macromitrium solitarium var. ***brevipes*** Broth., Biblioth. Bot. 87: 67. 1916.

Type protologue and citation: Im Bergwald von Tres Cruces (Cordillere von Santa Cruz) about 1400 m, [*Herzog*] no. *3497*. (isotype: JE04008701!).

The variety differs from the type variety by its shorter setae, which are shorter than 5 mm long.

Distribution: Bolivia.

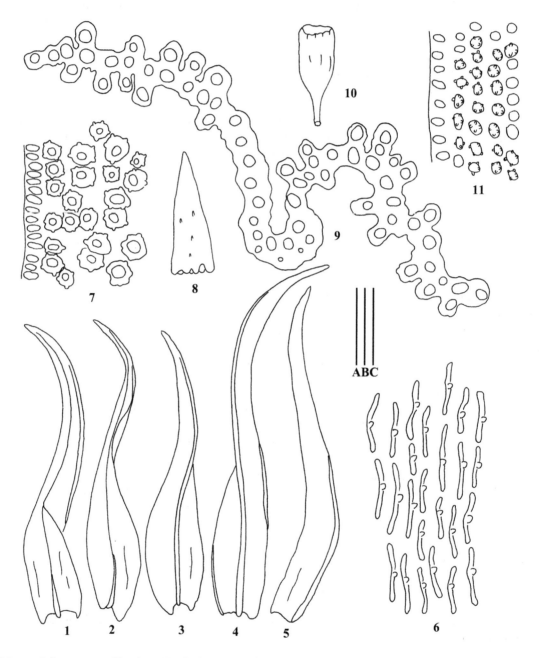

Fig. 219 *Macromitrium serpens* (Burch. ex Hook. & Grev.) Brid. 1-5: Branch leaves. 6: Low cells of branch leaf. 7: Upper cells of branch leaf. 8: Capsule. 9: Transverse section of branch leaf. 10: Calyptra. 11: Medial cells of branch leaf (all from BM 000868308). Line scales: A = 2 mm (8, 10); B = 1 mm (1-5); C = 67 μm (6, 7, 9, 11).

Taxonomy

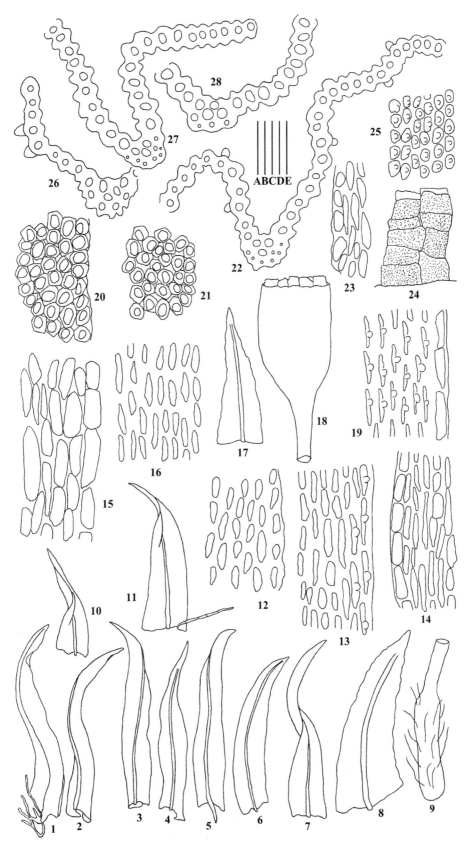

Fig. 220 *Macromitrium sharpii* H.A. Crum ex Vitt 1-7: Branch leaves. 8, 17: Branch leaf apex. 9: Vaginula. 10: Stem leaf. 11: Perichaetial leaf. 12: Upper cells of perichaetial leaf. 13: Medial cells of perichaetial leaf near margin. 14: Basal cells of perichaetial leaf. 15: Cancellate cells of branch leaf. 16: Medial cells of perichaetial leaf. 16: Capsules. 19: Basal cells of branch leaf. 20: Upper cells of branch leaf. 21, 25: Medial cells of branch leaves. 22, 28: Basal transverse sections of branch leaves. 23: Basal marginal cells of perichaetial leaf. 24: Peristome. 26, 27: Medial transverse sections of branch leaves (all from isotype, MICH 525909). Line scales: A = 1 mm (1-7, 9-11, 18); B = 400 µm (17); C = 200 µm (8); D = 100 µm (24); E = 67 µm (12-16, 19-23, 25-28).

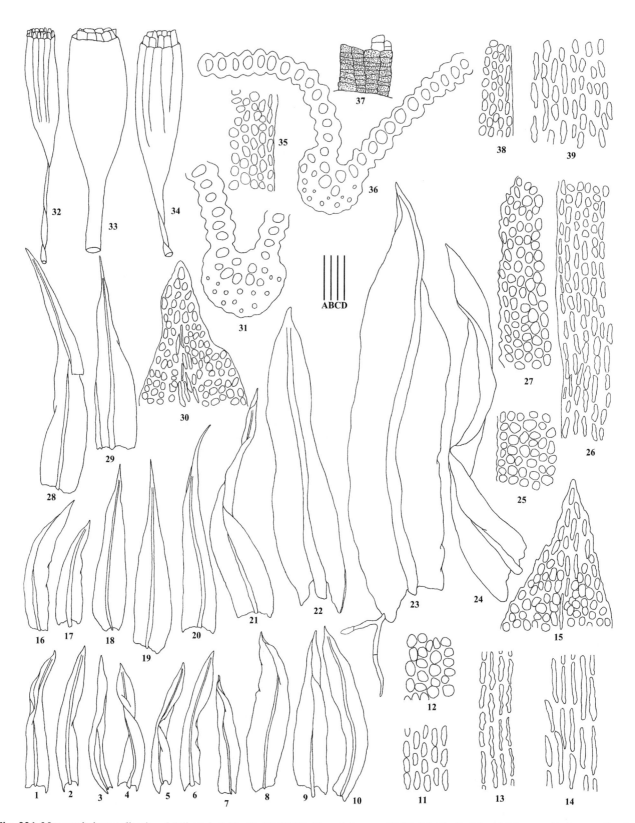

Fig. 221 *Macromitrium solitarium* Müll. Hal. 1-10, 16-17, 22-24: Branch leaves. 18-21, 28-29: Perichaetial leaves. 11: Low cells of branch leaf. 12: Medial cells of branch leaf. 13: Basal cells of branch leaf. 14: Low cells of perichaetial leaf. 15, 30: Apical cells of branch leaves. 25: Upper cells of branch leaf. 26: Low and basal cells of branch leaf. 27: Upper marginal cells of branch leaf. 31: Basal transverse section of branch leaf. 32-34: Capsules. 35: Medial marginal cells of branch leaf. 36: Medial transverse section of branch leaf. 37: Peristome. 38: Upper cells of perichaetial leaf. 39: Medial cells of perichaetial leaf (all from lectotype, NY 01086658). Line scales: A = 2 mm (32-34); B = 1 mm (1-10, 16-21, 28-29); C = 400 μm (22-24, 37); D = 67 μm (11-15, 25-27, 30-31, 35-36, 38-39).

213. *Macromitrium soulae* Renauld & Cardot, Rev. Bot. Bull. Mens. 9: 395. 1891. (Figure 222)

Type protologue: [Madagascar] Ambositra, Leg. *Rev. Soula*. Type citation: Madagascar, Ambositra (Betsileo), leg. *Rev. Soula*, 1890 (isotype: H-BR 2626003!).

**Macromitrium longisetum* Schimp. ex C.H. Wright, J. Bot. 26: 265. 1888.

(1) Plants robust, yellowish-brown above, dark-brown below; stems creeping, loosely with branches, branches up to 25 mm long, about 1.5-2.0 mm thick. (2) Stem leaves ovate-lanceolate to triangular-lanceolate, strongly abaxially curved above, often caducous. (3) Branch leaves appressed below, slightly twisted-contorted with a somewhat incurved apex, keeled above when dry; spreading and slightly undulate at the margin when moist, 2.0-2.5 × 0.5 mm, broad-lanceolate to oblong-lanceolalte, occasionally broken off near the apex, somewhat conduplicate above; apices acuminate-aucte; margins crenulate near the apex, entire below; costae often partially concealed in the adaxial view by the overlapping folds of the laminae; all cells flat and smooth; upper and medial cells rounded-quadrate, bulging or slightly bulging, clear and smooth; low and basal cells irregularly rectangular, thick-walled, almost smooth, those near the costa at the insertion enlarged, forming a distinct "cancellina region". (4) Perichaetial leaves broadly oblong, distinctly plicate, abruptly narrowed to form a thick subulate upportion. (5) Setae about 8 mm long, smooth and twisted to the left. (6) Capsules ovate-ellipsoid to ellipsoid-cylindric, strongly furrowed; peristome reduced and fragmentary. (7) Calyptrae mitrate, distinctly plicate, lacerate at the base, smooth and naked (Wilbraham, 2015).

Macromitrium soulae Renauld & Cardot is somewhat similar to *M. microstomum* (Hook. & Grev.) Schwägr. by oblong-lanceolate branch leaves, all cells smooth and clear, ovate-ellipsoid to ellipsoid-cylindrical capsules, mitrate calyptrae naked. However, *M. soulae* could be separated from *M. microstomum* by 1) robust plants; 2) brancl leaves not funiculate, and occasionally broken off near the apex, with a distinct "cancellina region" near the costa at the insertion; 3) broadly oblong perichaetial leaves with a thick subulate upportion; and 4) strongly plicate capsules.

Distribution: Madagascar.

Specimens examined: **MADAGASCAR**. Andringitra, leg. *Jessir* (H-BR 2626007, H-BR 2626008).

214. *Macromitrium speirostichum* Müll. Hal., J. Mus. Godeffroy 3(6): 68. 1874. (Figure 223)

Type protologue: Samoa-insulae, Upoln reg. montosa, ibidem quoqu in monte Tofua vulcanico. Type citation: Samoa, Upoln, *Dr. Ed. Graeffe 619* (lectotype designated here: JE 04008688!).

(1) Plants small to medium-sized, brownish; stems long creeping, with prostrate and ascending branches, branches with branchlets; branches about 25 mm long and 1.0 mm thick. (2) Stem leaves caducous and inconspicuous. (3) Branch leaves erect-twisted below, twisted-flexuose-crisped above, the apices curved to circinate, essentially not hidden in the inrolled cavity when dry, spreading and slightly abaxially curved, often partially conduplicate and adaxially incurved when moist, oblong-lanceolate to lanceolate, with an acuminate upper portion; margins entire below and weakly crenulate near the apex; costa percurrent, often partially concealed in the adaxial view by the overlapping folds of the laminae; upper and medial cells isodiametric, subquadrate, rounded, pluripapillose; low and basal cells elongate, thick-walled, not or weakly porose, smooth and hyaline, straight, curved or sigmoid; marginal enlarged teeth-like cells not differentiated at the insertion. (4) Perichaetial leaves longer than branch leaves, ligulate-lanceolate to broadly long triangular-lanceolate, all cells longer than wide. (5) Setae about 3-4 mm long, straight and smooth. (6) Capsule urns ellipsoid and smooth, with a small mouth; opercula conic-rostrate with a long fine beak; peristome simple with 16 long and pale lanceolate exostome teeth. (7) Calyptrae large, mitrate, covering the capsule completely, often with yellowish hairs.

Distribution: Samoa.

Specimens examined: **SAMOA**. Upolu, *Eduard O. Gräffe s.n.* (BM 000982767); *Eduard O. Gräffe 644* (JE 04008687); Saoaii: Lepaega (BM 000989763); Insel Sawai im Ivadisrikt an Palmen, *M. Fleischer* (BM 000989762).

215. *Macromitrium st-johnii* E.B. Bartram, Occas. Pap. Bernice Pauahi Bishop Mus. 15(27): 337, f. 7. 1940. (Figure 224)

Type protologue and citation: [French Polynesia] Rapa: Kopenena, on tree trunk, steep wooded gulch, alt. 125 m., *St. John and Maireau 15515* (type); Area, on tree trunk in woods, alt. 120 m., *St. John and Fosberg 15299* (isoparatype: BM 000982763!).

(1) Plants small to medium-sized, dull, yellowish green above, brown below, in dense mats; primary stems creeping, with numerous branches, branches about 5 mm long. (2) Stem leaves appressed below, squarrose above when dry, ovate-triangular to oblong-lanceolate, with an acuminate apex. (3) Branch leaves dense, spirally coiled with a hook-like apex, somewhat funiculate when dry, widely spreading when moist, 1.7-2.2 × 0.3-0.4 mm, ligulate; apices acute, obtuse-acute to minutely mucronate; margins entire below, finely crenulate above; costae percurrent; lamina uneven, 2-3-stratose near the apex, irregular 1-2-stratose at the upper poriton; upper cells rounded-quadrate, 7.0-8.0 μm wide, pluripapillose; medial and transitional cells oblate, oval to rounded-quadrate, smooth to weakly

unipapillose; low cells shortly rectangular to elongate-rectangular, smooth, thick-walled, 14-20 × 3-5 μm; cells at the insertion rectangular, 10-35 × 3-7 μm, smooth, thin-walled. (4) Perichaetial leaves oblong-lanceolate, longer than branch leaves. (5) Setae 4-5 mm, smooth, twisted to the left when dry. (6) Capsules erect, urns ovoid, smooth; peristome absent; spores about 30 μm in diameter, papillose. (7) Calyptrae mitrate, large, up to 4 mm long, lobed at base, covering the whole capsule, naked to very sparsely hairy.

Macromitrium st-johnii E.B. Bartram is similar to *M. nepalense* (Hook. & Grev.) Schwägr., especially in their leaf contour and surface in the microscopic view. However, the calyptrae of the type of *M. st-johnii* are smooth and naked or very sparsely hairy, and peristome is absent, while those of *M. nepalense* are hairy, and peristome is present, with 16 well-developed, lanceolate teeth. The transects of their branch leaves are also different.

Distribution: French Polynesia.

216. *Macromitrium standleyi* E.B. Bartram, Contr. U.S. Natl. Herb. 26: 85. f. 23. 1928. (Figure 225)

Type protologue and citation: On tree, Cerro de las Caricias, north of San Isidro, Province of Heredia, Costa Rica, altitude 2000 to 2400 meters, *Paul C. Standley and juvenal Valerio*, march 11, 1926, *no. 52147* (isotypes: JE 04008696!, NY 01243660!, H-BR 2625033!, US 00070278!); *Cerro de las Caricias, no. 52247* (isoparatypes: JE 04008695!, NY 01243659!, US 00070277!).

(1) Plants large and robust, rusty brownish to yellow green, in somewhat wefts; stems distinctly creeping, to 15 cm long; branches widely spaced along the primary stems,up to 4 cm long, with rusty-reddish rhizoids. (2) Stem leaves inconspicuous and caduceus. (3) Branch leaves loosely arranged, keeled, erecto-patent below, contorted-flexuose and undulate above when dry, flexuose-spreading when moist, rather large and long, 18 × 1.0 mm, long linear-lanceolate; apices long filiform acuminate; margins undulate, distinctly ciliate-dentate from the base, sharply spinose-denticulate near the apex, occasionally recurved at the extreme base below; costae excurrent; all cells longer than wide, thick-walled, and most cells porose; upper interior cells variable, 18-40 × 8 μm, somewhat rhomboidal, smooth or bulging to mammillose; upper marginal cells longer and narrower than their ambient cells, forming a distinct border; basal cells to 55 μm long, narrowly rectangular, tuberculate. (4) Setae smooth, to 25 mm long, twisted to the left. (5) Capsule urns globose-pyriform, globose or ovoid, smooth above, abruptly contracted to a short wrinkled neck; peristome double; spores anisosporous. (6) Calyptrae mitrate, naked, deeply lacerate.

Macromitrium trichophyllum Mitt. is similar to *M. standleyi* E.B. Bartram in their rather long linear-lanceolate leaves, but differs from the latter in having leaves with an oblong-ovate, tightly clasping leaf base, smooth basal cells, squarrose-recurved upper portion.

Distribution: Costa Rica, Honduras, Panama (Allen, 2002).

Specimens examined: **HONDURAS**. Olancho, Sierra de Algalta, *Bruce Allen 12707* (MO 3965655), *12743* (MO 3965337).

217. *Macromitrium stoneae* Vitt & H.P. Ramsay, J. Hattori Bot. Lab. 59: 400. f. 218-236. 1985.

Type protologue and citation: "Australia, New South Wales: northwest of Wauchope, Plateau Beech Preserve, Mount Boss State Forest, 31°11′S, 152°20′E, Eelevation 825 meters, *Vitt 27483*" (isotype: H 3090730!). Paratypes: *Vitt 27494, Vitt 27575, Vitt 27524, Vitt 27493, Vitt 27522, Vitt 28206*. Paratype citation: New South Wales: Queensland border area, Wiangaree Forest Drive, 2.0 km. Northeast of East boundary, Northeast of simes Road, *D. H. Vitt 28206* (isoparatype: KRAM-B-065696!).

(1) Plants medium-sized, rather dull, dark rusty-brown to chestnut-green, in dense spreading mats; stems creeping, with long and broad erect branches, branches up to 1.7 cm high. (2) Stem leaves broadly lanceolate to lanceolate-ligulate from a lanceolate-ovate base. (3) Branch leaves somewhat funiculate, individually strongly twisted-flexuose and crisped above, apices strongly twisted to one side or imperfectly inrolled when dry, wide-spreading and straight when moist, 2-2.5 mm long, narrowly lanceolate from a narrow-ovate base; apices acuminate-apiculate to acute; costae stout, percurrent, occasionally short excurrent, abaxial surface with elongate costal cells exposed throughout; upper cells clear and bulging, densely pluripapillose, papillae small and branched, with the rounded cells continuing proximally well past medial leaf; medial cells rounded but more strongly bulging; transitional cells subquadrate to short-rectangular, thick-walled and porose; basal cells elongate, flat and most smooth, thick-walled and porose, a few with a spinulose, central papilla. (4) Perichaetial leaves distinctly sheathing, stiff, ovate-lanceolate to lanceolate, shortly acuminate; costae excurrent; upper cells long-elliptic to rounded; basal cells elongate, smooth. (5) Setae 4-6 mm long, smooth, twisted to the left. (6) Capsule urns ovoid to narrowly ovoid-obloid to obloid-ellipsoid, smooth below, narrowed in upper 1/4 to a puckered, 4 or 8-plicated mouth; peristome single, exostome of 16, well developed but inconspicuous. (7) Calyptrae mitrate, conic plicate, smooth, hairy, evenly lacerate below (Vitt & Ramsay, 1985).

Macromitrium stoneae Vitt & H.P. Ramsay is superficially similar to *M. exsertum* Broth., *M. hemitrichodes* Schwägr., *M. incurififolium* (Hook. & Grev.) Schwägr., and *M. leratii* Broth. & Paris. *Macromitrium exertum* could

Taxonomy

be separated from *M. stoneae* by its completely smooth and flat upper leaf cells. *Macromitrium leratii* differs from *M. stoneae* by its larger plants, branch leaves with dense papillose and arranged in distinctly longitudinal rows, and by its sharp transition cell region of the branch leaves. *Macromitrium incurvifolium* is often olive-green and *M. hemitrichodes* golden-brown, its upper cells are densely papillose, while *M. stoneae* is rusty-brown, and its upper cells are lightly papillose.

Distribution: Australia.

Specimens examined: **AUSTRALIA**. Queensland, *D. H. Norris 34943* (H 3090732, KRAM-B-064914); *I. G. Stone* (MEL 2261926).

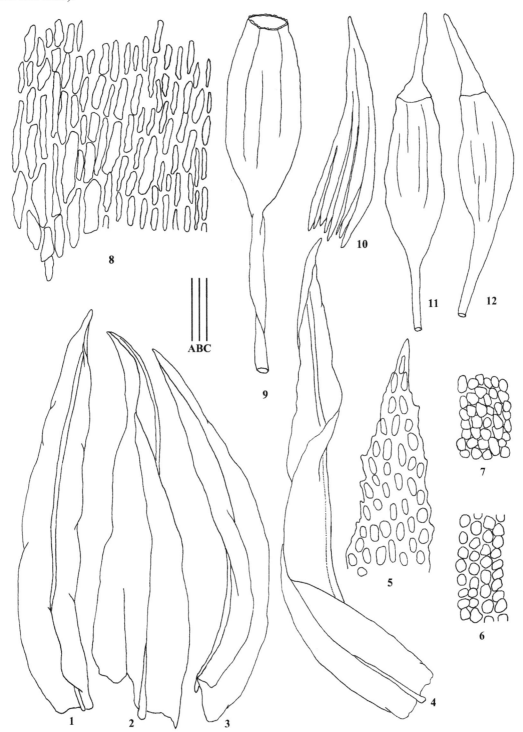

Fig. 222 *Macromitrium soulae* Renauld & Cardot 1-4: Branch leaves. 5: Apical cells of branch leaf. 6: Medial cells of branch leaf. 7: Upper cells of branch leaf. 8: Basal cells of branch leaf. 9, 11-12: Capsules. 10: Calyptra (1-10 from H-BR 2626007; 11-12 from H-BR 2626008). Line scales: A = 0.88 mm (9-12); B = 0.44 mm (1-4); C = 44 μm (5-8).

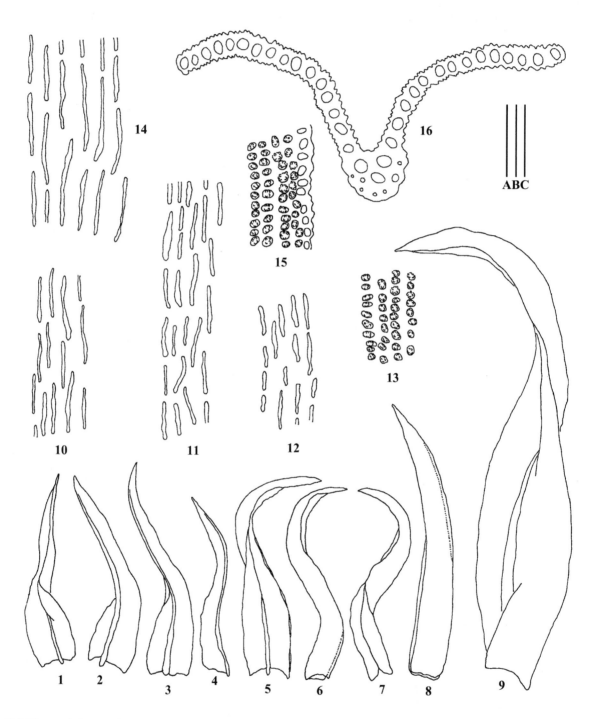

Fig. 223 *Macromitrium speirostichum* Müll. Hal. 1-7, 9: Branch leaves. 8: Perichaetial leaf. 10: Medial cells of perichaetial leaf. 11: Basal cells of branch leaf. 12: Upper cells of perichaetial leaf. 13: Medial cells of branch leaf. 14: Basal cells of perichaetial leaf. 15: Upper cells of branch leaf. 16: Medial transverse section of branch leaf (1-4 from JE 04008687; 5-16 from lectotype, JE 04008688). Line scales: A = 1 mm (1-8); B = 400 μm (9); C = 67 μm (10-16).

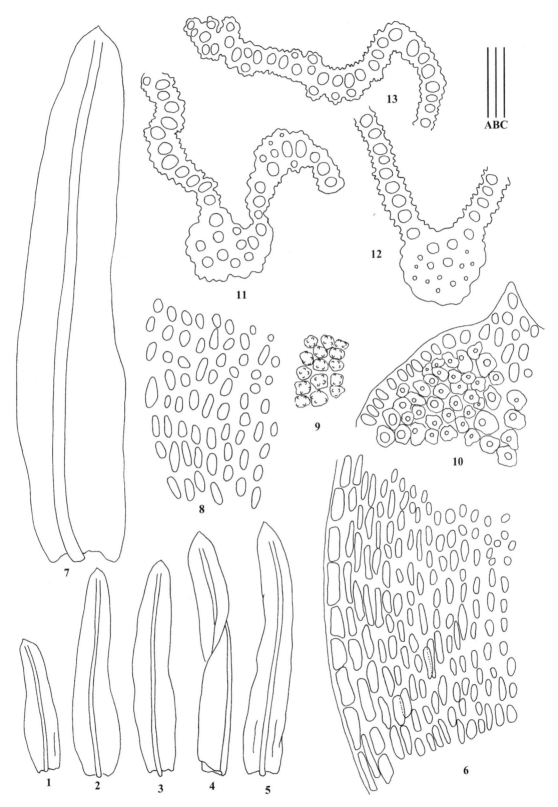

Fig. 224 *Macromitrium st-johnii* E.B. Bartram 1-5, 7: Branch leaves. 6: Basal marginal cells of branch leaf. 8: Basal cells of branch leaf. 9: Medial cells of branch leaf. 10: Apical cells of branch leaf. 11: Upper transverse section of branch leaf. 12-13: Medial transverse sections of branch leaves (all from isoparatype, BM 000982763). Line scales: A = 1 mm (1-5); B = 400 μm (7); C = 67 μm (6, 8-13).

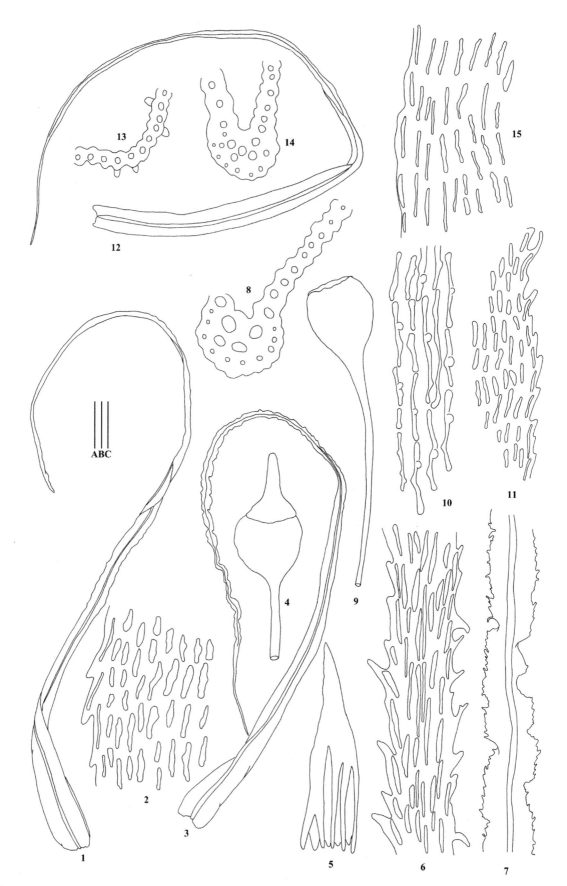

Fig. 225 *Macromitrium standleyi* E.B. Bartram 1, 3, 12: Branch leaves. 2: Upper cells of branch leaf. 4, 9: Capsules. 5: Calyptra. 6, 11: Apical cells of branch leaves. 7: Apex of branch leaf. 8, 14: Low transverse sections of branch leaves. 10: Basal cells of branch leaf. 13: Basal transverse section of branch leaf. 15: Medial cells of branch leaf (1-3, 6, 8-11, 13-15 from isoparatype, US 00070277; 4-5, 7 from isotype, US 00070278). Line scales: A = 2 mm (1, 3-5, 9, 12); B = 400 μm (7); C = 67 μm (2, 6, 8, 10-11, 13-15).

218. *Macromitrium streimannii* Vitt, Acta Bot. Fenn. 154: 72. f. 3b, 32. 1995.

Type protologue: Papua New Guinea. Morobe Province, Nauti Logging area, Upper Watut River, 11 km SW of Bulolo, *Araucaria* dominated low montane forest beside small gully on fallen tree branch, 7°17'S, 146°37'E,1 200 m, 6. III. 1982, *Streimann 17269* (holotype: ALTA; isotypes: CBG, H, LAE, NICH, NY).

(1) Plants medium-sized, dull, in stiff, loose mats; stems creeping, inconspicuous; branches erect, up to 1.0 cm long. (2) Branch leaves strongly twisted-contorted when dry, spreading with abn erect and inflexed upper portion when moist, 1.9-2.2 mm long, lanceolate, sharply acute to acuminate-mucronate; margins entire to roughened, plane to reflexed below; costae excurrent in the mucro to rarely percurrent; upper cells 7-8 × 5-7 µm, oblate to rounded, most wider than long, in bulging longitudinal rows, obscure, densely pluripapillose; regularly and gradually changing from upper to basal cells; medial cells 6-8 × 7-16 µm long, rounded, elliptic to elliptic-rectangular, clear, firm-walled, flat, smooth to tuberculate; basal cells 7-10 × 14-24 µm, long-rectangular, straight to curved, tuberculae, weakly porose. (3) Perichaetial leaves 2.8-3.0 mm long, oblong-lanceolate to ovate-lanceolate, acute to acuminate-shortly cuspidate, sheathing. (4) Setae 5.5-6.5 mm long, smooth, twisted to the left; vaginulae without hairs. (5) Capsule urns narrowly ellipsoid to ellipsoid-oblong, smooth, gradually contracted to setae; peristome single; exostome of 16, pale, densely papillose, erect, well-developed, blunt teeth; spores 12-37 µm, anisosporous. (6) Calyptrae mitrate, plicate, with a few thick, straight hairs, lacerate at the base. (Vitt *et al.*, 1995).

Distribution: Papua New Guinea.

Specimens examined: **PAPUA NEW GUINEA**. Morobe, *H. Streimann 19066* (H 3196826, MO 4433941); *F. Kos 22* (H 3196828); Western Highlands, *F. Kos 22* (KRAM-B-130955, MO 4462163); Eastern Highlands, *P. van Royen 4459* (MO 5375938).

219. *Macromitrium subbrevihamatum* Broth., Biblioth. Bot. 88: 12. 1920. (Figure 226)

Type protologue and citation: Bolivia, Cochabamba: An Bäumen im unteren Coranital, *Herzog T.C.J. 5065* (isotype: JE 04008732!).

(1) Plants medium-sized, green above, brown below; stems long creeping, moderately with reddish rhizoids, loosely with branches, branches up to 10 mm long. (2) Stem leaves triangular-lanceolate, strongly abaxially curved, all cells isodiametric except few cells confined to the base, buliging and mammillose. (3) Branch leaves erect below, twisted-contorted spreading and flexuose above when dry, erect-spreading or abaxially curved spreading, often conduplicate upper when moist, 2.5 × 0.4 mm, lanceolate, plicate below; apices acuminate to acuminate-acute; medial and upper margins serrulate from papilla protruding, crenulate near base, recurved below; costa stout, percurrent or vanishing beneath the apex; upper and medial cells rounded-quadrate, rounded-elliptical, conic-bulging to mammillose-papillose, marginal cells in the outmost row narrower than their ambient cells, forming a slightly differentiated border; low cell subquadrate or slightly elongate, short-rectangular to rectangular, thick-walled, smooth, occasionally porose; basal marginal enlarged teeth-like cells at the insertion slightly differentiated. (4) Perichaetial leaves oblong-lanceolate with the widest at base; acuminate-aristate, all cells longer than wide, rectangular. (5) Setae smooth, 6-10 mm, twisted to the left. (6) Capsule urns ellipsoid-cylindric, furrowed, slightly constricted beneath the mouth; opercula conic-subulate; peristome double, exostome teeth brown-reddish, united to form an erect membrane, endostome hyaline, segments lanceolate; spores anisosporous, globular, papillose.

Macromitrium subbrevihamatum Broth. is similar to *M. amboroicum* Herzog in leaf shape, capsules, but differs from the latter by having a differentiated border of branch leaves, and smooth basal and low cells.

Distribution: Bolivia.

220. *Macromitrium subcirrhosum* Müll. Hal., Bot. Zeitung (Berlin) 20: 373. 1862. (Figure 227)

Type protologue and citation: Patria, Costa Rica, in trachytaceis montis vulcanicis de Barba: Dr. Carl Hoffmann 26 Apr. 1855 cum fructibus matures pulcherrime legit. (isotypes: BM 000873221!, BM 000873222!).

(1) Plants large and robust, rusty-yellow to yellowish-green, conspicuously lustrous; stems slightly creeping, branches to 7 cm long, with sparse and rusty rhizoids. (2) Branch leaves large and long, strongly keeled, erect below, flexuose and twisted-contorted, undulate above when dry, abaxially curved to erect-spreading when moist, 6-8 × 1.0 mm, broadly lanceolate from a long oblong low part; apices subulate-acuminate with a stout arista; margins serrulate, undulate to plane above, entire and recurved below; all cells longer than wide; upper and medial interior cells (10-)20-48 µm, long-rhombic, smooth to slightly bulging, slightly thick-walled, porose, occasionally in distinctly radiating diagonal rows from the costa; upper marginal cells slightly narrower, sometimes not distinctly differentiated; cells gradually elongated from upper to low part; basal cells rectangular to long-rectangular, moderately thick-walled, porose, densely or sparsely tuberculate; marginal basal enlarged teeth-like cells not differentiated; costa excurrent into the arista (to 1 mm long). (3) Perichaetial leaves erect- spreading when moist, triangular-lanceolate, plicate below, with a long cuspidate apex. (4) Setae smooth and long, to 30 mm long. (5) Capsule urns 2-2.5 mm long, ovoid-ellipsoid to sub-cylindric, smooth; opercula erect to 2 mm long; peristome double, exostome lanceolate, reddish,

densely papillose, united an erect membrane, endostome light reddish, weakly papillose, with a basal membrane and segments; spores anisosporous, papillose. (6) Calyptrae 2-3 mm long, mitrate, smooth and naked, lacerate.

Macromitrium subcirrhosum Müll. Hal. is similar to *M. ulophyllum* Mitt. when some specimens of the former have upper leaf cells in radiating diagonal rows from the costa and basal cells weakly tuberculate. However, *M. ulophyllum* is more robust and its leaves are broader and strongly undulate and often without tuberculate basal cells.

Distribution: Colombia, Costa Rica, Ecuador, Panama, Venezuela; Nicaragua (Allen, 2002).

Specimens examined: COLOMBIA. Quindio, *S. Churchill, W. Rengifo M, C. Arbelaez, 17229* (MO 446169); Antioquia Guatapé, *S. P. Churchill 16378* (MO 3963617); Cauca, *S. P. Churchill & Wilson Rengifo M. 17306-C* (MO 4433002); Municipio de El Cerrito, *S. P. Churchill, con N. y H. Hollagnder 15312* (H 3090735, MO 3963689). **COSTA RICA.** Lara, *C. Hoffman s.n.*, 26, 5. 1885 (NY 01086559); Farit, *H. Cost n° 9943* (BM 000989722); Cartago, *Marshall R. & Carol A. Crosby 8590* (KRAM-B-162771). **ECUADOR.** Loja, *Jens E. madsen & L. Ellemann 85624* (MO 6235312). **PANAMA.** Province panama, *Bruce H. Allen 9042* (MO 3965352). **VENEZUELA.** Amazonas, *W.R. Buck 13016* (MO 3956254).

221. *Macromitrium subcrenulatum* Broth., Biblioth. Bot. 87: 66. 1916. (Figure 228)

Type protologue: [Bolivia] Rio Saujana, alt. 3400 m, [*Herzog, T.C.J*] *No. 3221*; Waldgrenze über Tablas, alt. 3400 m, *No. 2806a*. Type citatuon: Bolivia, An der Waldgrenze des Rio Saujana, *T. C. J. Herzog, 3221* (syntype: H-BR 2625005!; isosyntypes: JE 04008702!, NY 01086582!); Bolivia, Cochabamba: An der Waldgrenze über Tablas, about 3400 m, *T. C. J. Herzog, 2806/a* (isosyntype: JE 04008703!).

(1) Plants large and robust in clusters, brownish-green; stems weakly creeping, densely with erect branches, branches up to 40 mm long and 2 mm thick, with rusty rhizoids. (2) Branch leaves appressed-erect below, spreading, contorted-twisted and flexuose, with divergent apices when dry, spreading, often somewhat twisted above when moist, long and narrowly lanceolate from an oval base, plicate below, with an acuminate apex; margins often irregularly crenulate above, entire below, and rugose in the upper portion; costae subpercurrent, percurrent or excurrent; cells near the apex rhombic, oblong and oval, clear and smooth; upper and medial cells subhexagon, subquadrate, oval or oblong, bulging; low cells rectangular, elongate-rectangular, thick-walled and strong porose, unipapillose; basal cells linear-rectangular, thick-walled, strong porose, frequently unipapillose; basal cells near the costa and the margin distinctly differentiated, wider and shorter than their ambient cells, in the outmost row at the insertion not teeth-like. (3) Perichaetial leaves erect, lanceolate from an oblong low portion, with the widest at the base, longer than branch leaves, sheathing at the base, narrowly acuminate to form an arista in the upper portion, the upper portion fragile and easily broken; margins entire and involuted, strong plicate, truncated at base; costae single and stout, vanishing beneath the apex, all cell thick-walled and with a narrow lumen, cells near the margin and costa not differentiated. (4) Setae smooth, up to 10 mm, reddish and smooth. (5) Capsules red-brown, erect and ovoid, ellipsoid to ellipsoid-cylindric, weakly to moderately wrinkled or furrowed; opercula conic-subulate; peristome double. (6) Calyptrae mitrate, naked and lacerate.

Distribution: Bolivia, Ecuador.

Specimens examined: BOLIVIA. Depto. La Paz, Prov. Inquisivi: Cerro Lulini, *Marko lewis 89-966, d-3* (MO 3962554). **ECUADOR.** S. W. Azuay, Secus riasn inter Canas d Biblian, Prope hacienda, 3800m, Jan. 1909, *R. H. Allioui 8110* (H-BR 2625016), *8108* (H-BR 2625015).

222. *Macromitrium subdiscretum* R.S. Williams, Bull. New York Bot. Gard. 3(9): 130. 1903. (Figure 229)

Type protologue: [Bolivia] Santa Anna, Apolo region, 1950 meters, July 28, 1902 ([*Williams RS*] 1820). Type citation: Bolivia, Santa Anna, Apolo region, 6500ft, Jul 28, 1902, *Williams RS, 1820* (holotype: F-C 0001097F!; isotypes: JE 04008706!, MICH 525914, MICH 525915!, NY 01086583!).

(1) Plants medium-sized, rusty-reddish; stems weakly creeping, branches to 30 mm long, densely with rusty-reddish rhizoids. (2) Stem leaves inconspicuous and caducous. (3) Branch leaves not very shriveled, spirally twisted to coiled, flexuose above and somewhat funiculate when dry, erecto-patent when moist, 2.0-3.0 mm long, ligulate, oblong-lingulate to ligulate-lanceolate, occasionally rugose above, plicate below; apices acute; margins weakly serrulate in upper 1/3 of the leaf length; costae stout, vanishing beneath the apex; cells at the apex rhombic, bulging; upper cells rounded-hexagonal, elliptic, oblong or rhombic, bulging, weakly to moderately porose, occasionally papillose; marginal cells at the outmost row narrower than their ambient cells, forming an inconspicuous differentiated border; medial and low cells rectangular, strongly porose; basal cells elongate, long rectangular, strongly porose and tuberculate; basal marginal enlarged teeth-like cells at insertion not differentiated. (4) Perichaetial leaves broader than branch leaves, broadly oblong to lingulate-oblong or triangular-lanceolate, often with the widest at the base, plicate or slightly rugose; apices acute; upper cells elliptic to rhombic; medial and low cells long to linear rectangular, strongly porose; low and basal cells strongly tuberculate. (5) Setae smooth, 6-10 mm, twisted to the left. (6) Capsules ovoid to ellipsoid, distinctly furrowed, slightly constricted beneath mouth; opercula conic-rostrate;

peristome double, exostome teeth brown-yellow, lanceolate, united below, recurved when dry; endostome teeth hyaline. (7) Calyptrae mitrate, lacerate below, smooth and naked.

Macromitrium subdiscretum R.S. Williams is characterized by its erecto-patent branch leaves when moist, rhombic cells in upper and medial portion, strongly tuberculate basal lamina cells, and strongly furrowed capsules.The species is similar to *M. sharpii* H.A. Crum ex Vitt in their rusty-reddish plants, branch leaves with rhombic, bulging upper and medial cells, basal cells strongly tuberculate, but the former differs from the latter by its funiculate branches when dry, distinctly furrowed capsules. *Macromitrium subdiscretum* is also similar to *M. pseudofimbriatum* Hampe , but the latter has branch leaves with differentiated marginal enlarged teeth-like cells at insertion.

Distribtution: Bolivia.

223. *Macromitrium subhemitrichodes* Müll. Hal., Hedwigia 37: 144. 1898. (Figure 230)

Type protologue: Habitatio. New South Wales, Richmond River, *Miss Hodgkinson* in Hb. Melbourne 1880. Type citation: Richmond River, *Miss Hodgkinson* (isotype: JE 04008694!).

Macromitrium subhemitrichodes var. *hodgkinsoniae* Müll. Hal., Hedwigia 37: 144. 1898, **syn. nov.** Type protologue: [Australia] In iisdem locis: eadem 1880.

(1) Plants small, in dense, compact, spreading mats, light yellowish; stems creeping, densely with very short, erect branches, branches short, occasionally with multicellular clavate gemmaes. (2) Branch leaves obliquely appressed below, tightly curved to one side when dry, erect-spreading when moist, lingulate, oblong-ligulate; apices mucronate; margins entire or slightly crenulate above, occasionally narrowly recurved at one side and plicate below; costae very strong, excurrent into the mucro or ending just below the apex; upper and medial cells rounded-quadrate, obscure by low, dense pluripapillae; low and basal cells shorter, rounded to short elliptic-rectangular, thick-walled, clear, flat and smooth; the outmost marginal cells at the insertion not much different from their ambient cells. (3) Perichaetial leaves slightly different from branch leaves, with an acuminate apex. (4) Setae short, straight, smooth. (5) Capsule urns ovate-ellipsoid and small; opercula conic-rostrate, with an erect beck; peristome single, short, regularly lanceolate, pale. (6) Calyptrae small, conic-mitrate, smooth and naked, lightly plicate, deeply lacerate.

In the original publication, *M. subhemitrichodes* Müll. Hal. was considered similar to *M. hemitriodes* Schwägr. in their habits. The former differs from the latter by its small plants with short and globose branches. In fact, *M. subhemitrichodes* is rather similar to *M. brevicaule* (Besch.) Broth., but the latter differs from the former by its branch leaves with an inconspicuous border formed by several rows of longer marginal cells.

Macromitrium subhemitrichodes var. *hodgkinsoniae* differs from var. *subhemitrichodes* by the former having greener, wider leaves with a more rounded apex. We think that it was unnecessary to creat this variety.

Distribution: Australia.

224. *Macromitrium subincurvum* Cardot & Thér., Bull. Acad. Int. Géogr. Bot.16: 40. 1906. (Figure 231)

Type protologue and citation: China, Em. Bodinier, HongKong, 17 Janv. 1893 (isotype: S-B 115560!).

(1) Plants small to medium-sized, in dense mats; stems long creeping, with short and erect branches, branches up to 10 mm long, with short branchlets at upper part, densely leaved. (2) Branch leaves irregularly twisted-crisped and circinate and adaxially incurved above when dry, widely spreading when moist, ligulate, ligulate- to oblong-lanceolate; apices acute or mucronate; margins entire and plane; upper and medial cells flat, rounded-quadrate or rounded-hexagonal, obscure and pluripapillose; low cells oblong-rectangular, clear and thick-walled, often with a single large papilla; basal cells near margin at one side different from their ambient interior cells, rectangular, pellucid with thinner walls, often smooth; costa strong keeled, excurrent into a short mucro. (3) Perichaetial leaves different from branch leaves, lanceolate from a broadly oblong base, gradually acuminate to a subula; all cells longer than wide; medial and upper cells unipapillose, slightly porose; low cells rectangular to elongate-rectangular, thick-walled and smooth. (4) Setae about 5 mm long, smooth, twisted to the right. (5) Capsules erect, immature. (6) Calyptrae campanulate, large, covering the whole capsule, with brownish-yellow hairs.

Macromitrium subincurvum Cardot & Thér. is rather similar to *M. tosae* Besch., but the upper cells of the perichaetial leaves are *isodiametric* in *M. tosae*, but longer than wide in *M. subincurvum*.

Distribution: China.

225. *Macromitrium sublaeve* Mitt., J. Linn. Soc., Bot. 12: 208. 1869. (Figure 232)

Type protologue: [Ecuador] Andes Quitenses, fl. Pastasa superius ad pontem Agoyán, et An-tombos (5000 ped.), *Spruce, n. 105*. Type citation: *Spruce 105* (isotype: NY 01086584!); Andes Quitenses, fl Pastasa superitus Agoyan, *R. Spruce 105* (isotypes: MICH 525916!, E 00165164!); pontem Agoyán, *R. Spruce s. n.* (isotype: NY 01086585!).

(1) Plants medium-sized, yellow-brown above, dark brown below; stems weakly creeping, with caduceus and inconspicuous leaves; branches up to 30 mm long, with branchlets, densely with rusty-reddish rhizoids. (2) Branch

leaves undulate, strongly twisted-contorted and curly, often the apices twisted-flexuose or curved when dry, spreading with incurved and conduplicate above when moist, keeled, narrowly lanceolate to ligulate-lanceolate, occasionally slightly-rugose above, plicate below; apices acute, mucronate, apiculate; margins often crenulate from protruding bulging cells; costae yellowish, percurrent or excurrent to an apiculus; all cells not porose, marginal cells not differentiated; upper cells rounded, rounded-quadrate, quadrate to subquadrate, isodiametric, strongly bulging, obscure and unipapillose; medial and low cell elongate, rectangular, thick-walled, unipapillose; basal cells long-, linear- rectangular, strongly tuberculate, those near the costa at the insertion enlarged, becoming thin-walled, wider and shorter than their ambient cells, forming a distinct "cancellina region". (3) Perichaetial leaves shorter than vegetative, oval-lanceolate, acute-acuminate; upper cells subquadrate, rhombic or elliptical, bulging and unipapillose; medial and low cells rectangular to linear-elongate. (4) Setae smooth, 6-8 mm long, twisted to the left. (5) Capsule urns ovoid, ovoid-ellipsoid to ellipsoid-cylindric, smooth, occasionally weakly furrowed or wrinkled and slightly constricted beneath a wide mouth; opercula conic-subulate; peristome double, exostome teeth yellowish, lanceolate, papillose-striate, united to form an erect membrane, endostome hyaline, membranous. (6) Calyptrae mitrate, strongly lacerate, smooth and naked.

Macromitrium sublaeve Mitt. is similar to *M. sharpii* H.A. Crum ex Vitt in leaves with oblong-lingulate lanceolate outline, isodiametric bulging and papillose upper and medial cells and strongly tuberculate basal cells, smooth capsules with double peristome, as well as their branches densely with rusty-reddish rhizoids, but the former could be separated from the latter in having leaves inconspicuously porose.

Distribution: Bolivia.

Specimens examined: **BOLIVIA**. Antombed ad rupes, *Spruce s.n.* (NY 01086586, NY 01086588); Spurima, *Williams R. S., 1829* (H-BR 2632023).

226. ***Macromitrium sublongicaule*** E.B. Bartram, Bryologist 48: 116. 1945. (Figure 233)

Type protologue: [Papua New Guinea] Jan. 1940, no. [*Clemens*] *40860* type.-Bona, elev. 2500-4500 ft., *No. 12242.7*. Type citation: Clemens expedition to New Guinea, Jan 1940, *Mary Strong Clemens no. 40860* (isotypes: MICH 525917!, MICH 525918!).

(1) Plants large, robust and stiff, in loose mats or wefts; stems creeping, densely covered with rusty rhizoids; branches simple to forked, up to 4 cm long. (2) Stem leaves small, inconspicuous, lanceolate, erect, often with rhizoids at base. (3) Branch leaves sheathing at base, contorted-twisted-flexuose and divergent above when dry, , stiffly wide-spreading below and conduplicate above when moist, 4.0-6.0 mm long, long-lanceolate, gradually acuminate; margins plane to broadly reflexed below, entire to irregularly notched in upper portion; costae percurrent, often adaxially concealed by overlapping folds of laminae in the upper portion; upper cells uniform, regularly rounded, rounded-quadrate to short-elliptic (12-14 μm), clear, moderately bulging, smooth, those at upper margin slightly longer than their ambient inner cells; medial cells 11-14 × 12-20 μm, irregularly-elliptic, elliptic-rounded to elliptic-rhombic, bulging and clear, smooth to unipapillose; basal cells elongate, 9-11 × 40-80 μm, rectangular, thick-walled but not porose, lumens straight, flat, strongly tuberculate. (4) Perichaetial leaves 6.0-7.0 mm long, long-acuminate from a lanceolate-ovate low portion; low and basal cells unipapillose. (5) Setae smooth, 2.5-4.0 mm long, erect, twisted to the left; vaginulae sparsely hairy. (6) Capsule urns narrowly ellipsoid to ovoid-ellipsoid, smooth below, lightly 8-ribbed in upper 1/3, wide-mouthed, sharply contracted to the seta, sometimes papillose in low portion; peristome single; exostome of 16, well-developed; spores anisosporous. (7) Calyptrae mitrate, plicate, with a few, stiff, thick, erect hairs or naked, lacerate at the base (Vitt *et al.*, 1995).

Macromitrium sublongicaule E.B. Bartram is similar to *M. longicaule* Müll. Hal., but differs from the latter by its branch leaves with slightly differentiated marginal cells in upper and medial portions. *Macromitrium sublongicaule* also resembles *M. noguchianum* W. Schultze-Motel, but the branch leaves of the latter are slightly shorter, most upper cells at least two to three times as long as wide, with irregular thickened walls.

Distribution: Papua New Guinea.

Specimens examined: **PAPUA NEW GUINEA**. Simbu, *M. Toia 126* (H 3196713), *M. Toia 31* (H 3196716); West Sepik, *A. Touw 15009* (H 3205080); Morobe, *T. J. Koponen 34762* (MO 4428636), *D. H. Norris 62420* (MO 4435570); Eastern Highlands, *W. A. Weber B-32250* (MO 4411931).

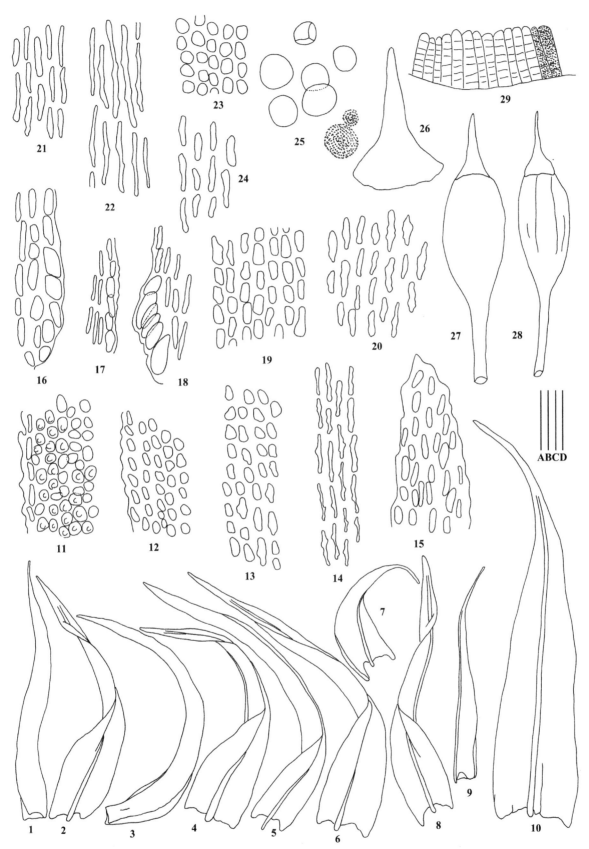

Fig. 226 *Macromitrium subbrevihamatum* Broth. 1-6, 8: Branch leaves. 7: Stem leaf. 9-10: Perichaetial leaf. 11: Upper cells of branch leaf. 12: Medial cells of branch leaf. 13: Low cells of branch leaf. 14: Basal cells of branch leaf. 15: Apical cells of branch leaf. 16-18: Basal marginal cells of branch leaves. 19: Basal cells of stem leaf. 20: Medial cells of perichaetial leaf. 21: Low cells of perichaetial leaf. 22: Basal cells of perichaetial leaf. 23: Upper cells of stem leaf. 24: Upper cells of perichaetial leaf. 25: Spores. 26: Operculum. 27-28: Capsules. 29: Peristome (all from isotype, JE 04008732). Line scales: A = 2 mm (9); B = 1 mm (1-8, 10, 26-28); C = 400 μm (29); D = 67 μm (11-25).

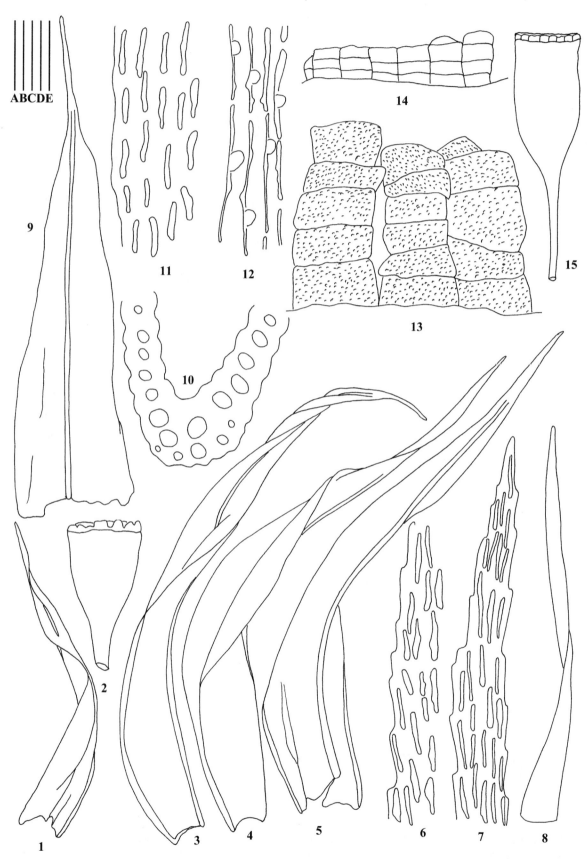

Fig. 227 *Macromitrium subcirrhosum* Müll. Hal. 1, 3-5: Branch leaves. 2, 15: Capsules. 6: Upper cells of branch leaf. 7: Apical cells of branch leaf. 8: Apex of branch leaf. 9: Perichaetial leaf. 10: Low transverse section of branch leaf. 11: Medial cells of branch leaf. 12: Basal cells of branch leaf. 13-14: Peristome (all from isotype, NY 01086558). Line scales: A = 2 mm (1-2, 15); B = 1 mm (3-5, 9); C = 400 μm (14); D = 100 μm (8, 13); E = 67 μm (6-7, 10-12).

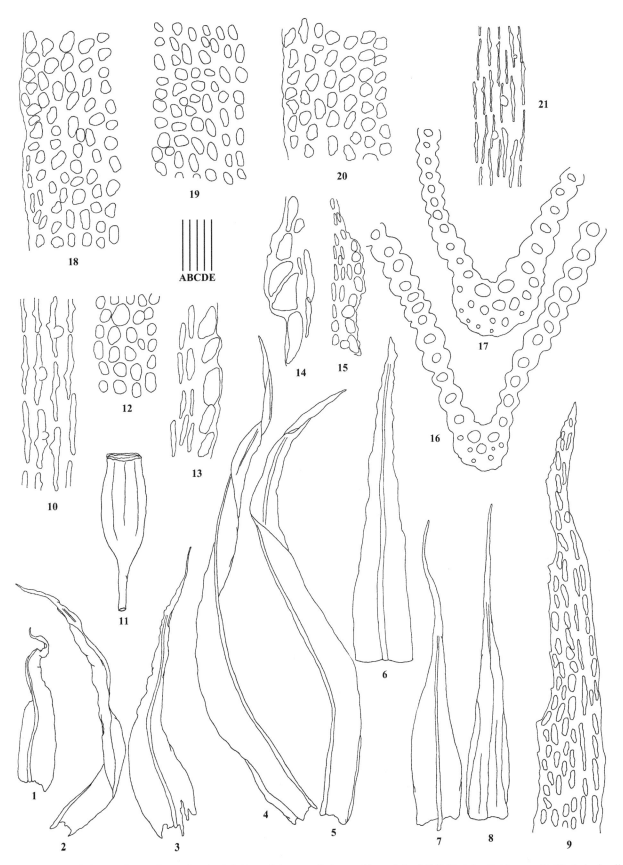

Fig. 228 *Macromitrium subcrenulatum* Broth. 1-5: Branch leaves. 6: Apex of branch leaf. 7-8: Perichaetial leaves. 9, 15: Apical cells of branch leaf. 10: Low cells of branch leaf. 11: Capsule. 12: Upper cells of branch leaf. 13-14: Basal marginal cells of branch leaves. 16: Medial transverse section of branch leaf. 17: Basal transverse section of branch leaf. 18-20: Medial cells of branch leaves. 21: Basal cells of branch leaf (1-3 from isosyntype, JE 04008702; 4-21 from isosyntype, JE 04008703). Line scales: A = 2 mm (1-3, 7-8, 11); B = 1 mm (4-5); C = 400 μm (6); D = 100 μm (9, 15, 21); E = 67 μm (10-14, 16-20).

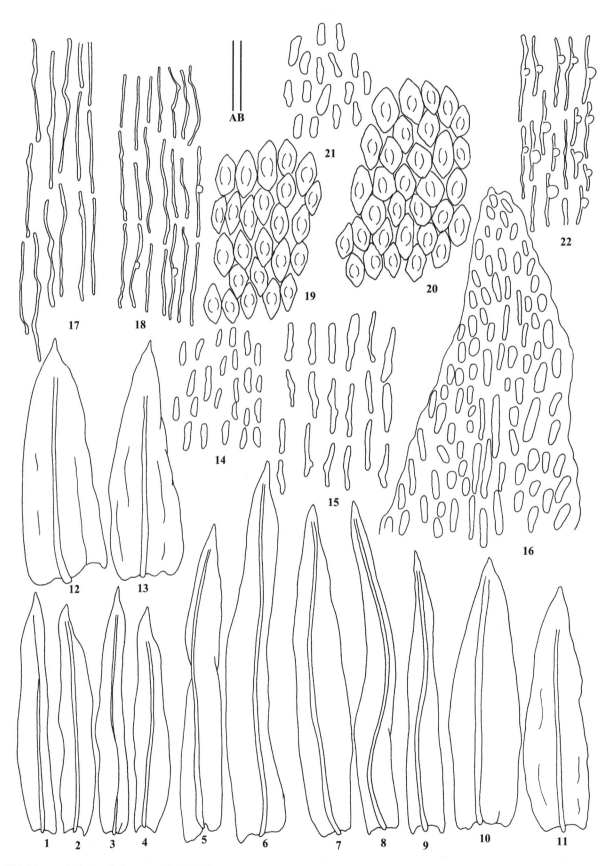

Fig. 229 *Macromitrium subdiscretum* R.S. Williams 1-9: Branch leaves. 10-13: Perichaetial leaves. 14: Upper cells of branch leaf. 15: Medial cells of branch leaf. 16, 19-20: Apical cells of branch leaves. 17: Medial cells of perichaetial leaf. 18: Basal cells of perichaetial leaf. 21: Upper cells of branch leaf. 22: Basal cells of branch leaf (1-7, 9-22 from holotype, F-C 0001097F; 8 from isotype, JE 04008706). Line scales: A = 1 mm (1-13); B = 67 μm (14-22).

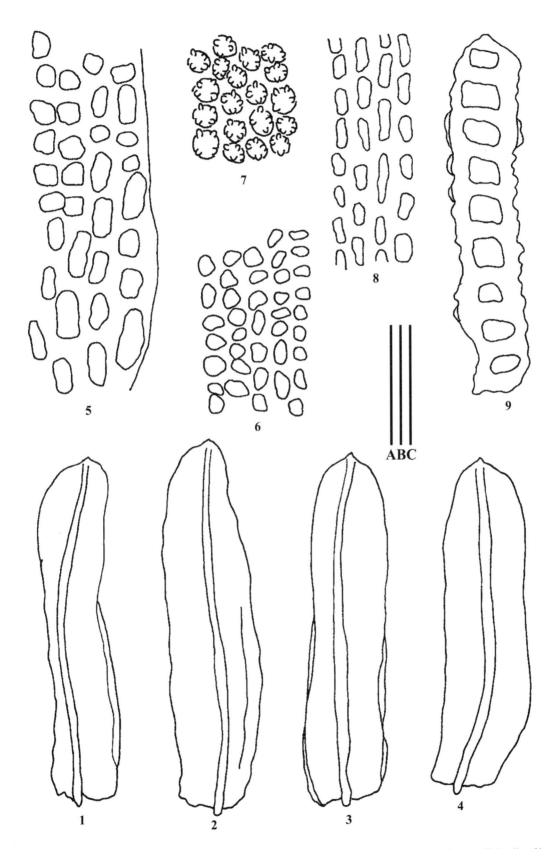

Fig. 230 *Macromitrium subhemitrichodes* Müll. Hal. 1-4: Branch leaves. 5: Basal cells of branch leaf. 6: Medial cells of branch leaf. 7: Upper cells of branch leaf. 8: Low cells of branch leaf. 9: Gemma (all from isotype, JE 04008694). Line scales: A = 1 mm (1-4); B = 100 μm (9); C = 67 μm (5-8).

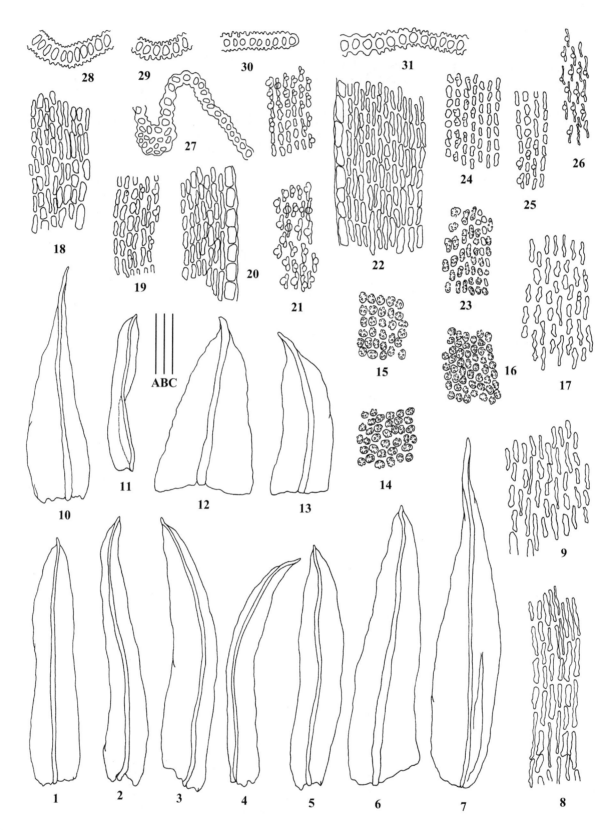

Fig. 231 *Macromitrium subincurvum* Cardot & Thér. 1-6, 11: Branch leaves. 7, 10: Perichaetial leaves. 8: Basal cells of perichaetial leaf. 9: Low cells of perichaetial leaf. 12-13: Apices of branch leaves. 14: Apical cells of branch leaf. 15-16: Upper cells of branch leaves. 17: Medial cells of perichaetial leaf. 18: Basal cells near costa of branch leaf. 19, 21, 25-26: Basal cells of branch leaves. 20, 22: Basal marginal cells of branch leaves. 23: Medial cells of branch leaf. 24: Low cells of branch leaf. 27: Low transverse section of branch leaf. 28-31: Medial transverse sections of branch leaves (all from isotype, S-B 115560). Line scales: A = 500 μm (1-7, 10-11); B = 200 μm (12-13); C = 50 μm (8-9, 14-31).

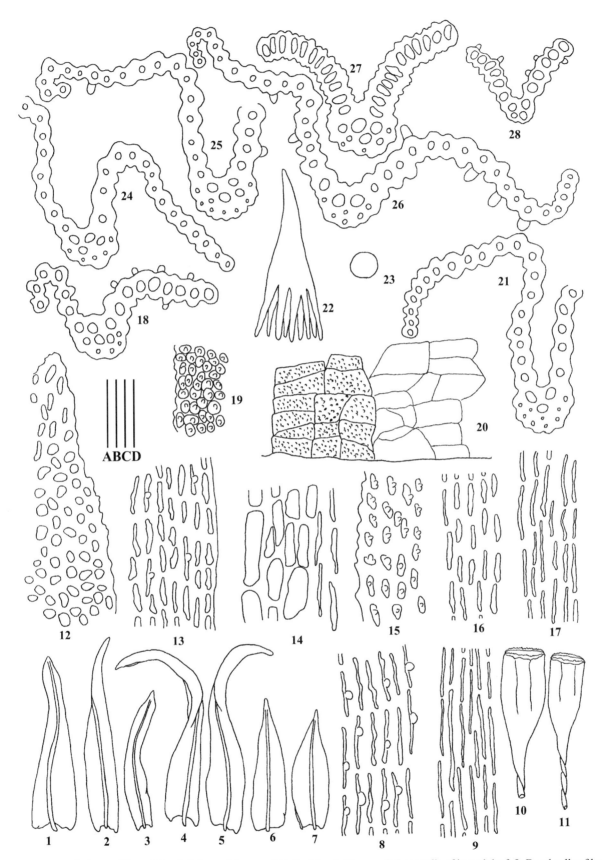

Fig. 232 *Macromitrium sublaeve* Mitt. 1-5: Branch leaves. 6-7: Perichaetial leaves. 8: Low cells of branch leaf. 9: Basal cells of branch leaf. 10-11: Capsules. 12: Apical cells of branch leaf. 13: Medial cells of branch leaf. 14: Cancellate cells of branch leaf. 15: Upper cells of perichaetial leaf. 16: Medial cells of perichaetial leaf. 17: Basal cells of perichaetial leaf. 18, 28: Upper transverse sections of branch leaves. 19: Upper cells of branch leaf. 20: Peristome. 21, 24: Basal transverse sections of branch leaves. 22: Calyptra. 23: Spore. 25-26: Low transverse sections of branch leaves. 27: Medial transverse section of branch leaf (all from isotype, MICH 525916). Line scales: A = 2 mm (10-11, 22); B = 1 mm (1-7); C = 100 μm (20); D = 67 μm (8-9, 12-19, 21, 23-28).

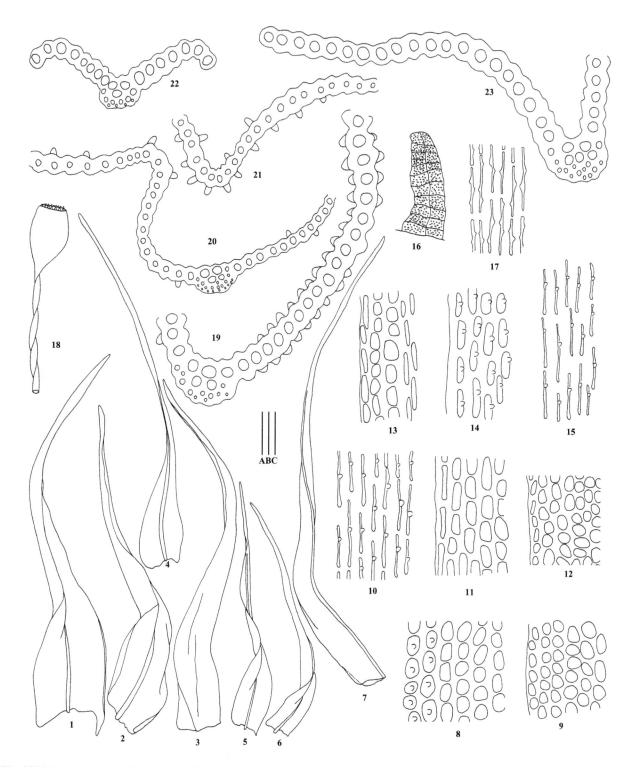

Fig. 233 *Macromitrium sublongicaule* E.B. Bartram 1-3: Branch leaves. 4-7: Perichaetial leaves. 8: Low cells of branch leaf. 9: Apical cells of branch leaf. 10, 17: Basal cells of perichaetial leaves. 11: Medial cells of perichaetial leaf. 12: Upper cells of branch leaf. 13: Apical cells of perichaetial leaf. 14: Low cells of perichaetial leaf. 15: Basal cells of branch leaf. 16: Peristome. 18: Capsule. 19: Low transverse section of perichaetial leaf. 20: Basal transverse section of perichaetial leaf. 21: Basal transverse section of branch leaf. 22: Upper transverse section of perichaetial leaf. 23: Medial transverse section of perichaetial leaf (all from isotype, MICH 525917). Line scales: A = 2 mm (5-6, 18); B = 1 mm (1-4, 7); C = 67 μm (8-17, 19-23).

Taxonomy

227. *Macromitrium submucronifolium* Müll. Hal. & Hampe, Linnaea 26: 499. 1853[1855].
Type protologue: Nova Seelandia.
Macromitrium erosulum Mitt., J. Proc. Linn. Soc., Bot. 4: 78. 1859, *fide* Vitt, 1983. Type protolgoue: New Zealand, *kerr*; near Wellington, Stephenson; middle Island, Bidwill; Waikeki, Milne.
Macromitrium papillifolium Müll. Hal., Hedwigia 37: 154. 1898, *fide* Wijk *et al.*, 1964. Type protologue: Nova Seelandia, sine loco speciali: ex Hb. Melbourneano in Hb. Goettingensi 1887.
Macromitrium coarctatulum Müll. Hal., Hedwigia 37: 153. 1898, *fide* Vitt, 1983. Type protolgoue: Nova Seelandia, insula australis, littore auralasiaco prope Greymouth: *Richard helms* 1885 legit et misit. Type citation: Nova Seelandia, insula, ralasiaco Greymouth, Helms, Hb. C. Müll. (lectotype: H-BR 2545012!).
Macromitrium petriei Dixon, Bull. Torrey Bot. Club 42: 101. 7. 1915, *fide* Vitt, 1983. Protologue: Clinton Valley, Te Anau, New Zealand, *D. Petrie*.

(1) Plants robust, dull, stems creeping, with erect branches up to 2 cm high. (2) Stem leves ovate-lanceolate, gradually acute-acuminate, upper cells rounded-elliptic, stoutly unipapillose, elongate-rectangular and most spinulose-papillose almost to base. (3) Branch leaves irregularly twisted-flexuose, decurved or twisted to one side above when dry, stiffly erect-spreading when moist, 1.5-3.0 mm long, lanceolate from an elliptic low portion to ligulate-lanceolate, acute to stoutly acuminate-apiculate; upper cells rounded to rounded-elliptic, with firm walls, rather clearer, very strongly bulging, each surface with 1 or 2 simple to forked, large conic papillae; medial cells elliptic-rectangular, each cell with one large, conical paplliae, transition to basal cells gradually; basal cells long-rectangular, with elongate lumens, thick-walled, smooth to tuberculate-papillose. (4) Perichaetial leaves ovate-lanceolate to oblong-lanceolate, acute, cells elliptic to irregularly elongate above, most strongly unipapillose. (5) Setae 3-9 mm long, slender and smooth. (6) Capsule urns ovoid-obloid, smooth, slightly 4-plicate beneath the mouth, peristome single, exostome of 16, well-developed. (7) Calyptrae mitrate, distinctly plicate, sparsely to densely hairy.

Macromitrium submucronifolium Müll. Hal. & Hampe is similar to *M. prorepens* (Hook.) Schwägr in their short, stout branches, non-funiculate leaves, excurrent costa, upper leaf cells bulging strongly, forked papillose or pluripapillose, setae relative shorter, capsules gradually narrow to a quadrate mouth, coarsely hairy calyptrae. However, *M. submucronifolium* differs from *M. prorepens* by its more robust plants, with leaves longer (1.5-3.0 mm) and setae longer (6-9 mm), the leaves mostly ligulate-lanceolate, and gradually narrowed to an apiculus, the upper leaf cells are clear, strongly bulging with one or two, mostly simple or forked papillae; the transitional cells are very thick-walled and long rectangular or long elliptic (Vitt, 1983).

Distribution: Australia, New Zealand (Vitt & Ramsay, 1985).
Specimens examined: **NEW ZEALAND**. North island, District Wellington, *Schäfer-Verwimp & Verwimp* (KRAM-B-101646); Nina Valley Track, *H. Streimann 51216* (KRAM-B-103625, H 3090656); South island, lake Ohau, *E. P. Hegewald 11306* (MO 5223207); South island, Lake Rotoitim, *H. Streimann 58208* (H 3090740).

228. *Macromitrium subperichaetiale* Thér., Mem. Soc. Cub. Hist. Nat. "Felipe Poey" 14: 350. 54 f. 1. 1940. (Figure 234)
Type protologue: [Cuba] Baracoa, leg. Jorge Nateson, com. J. Acuña. Type citation: Cuba: Baracoa, leg. Jorge Nateson, 1935, Herbier I. Thériot (isotypes: PC 0137724!, PC 0137725!).

(1) Plants large and rubost, lustrous and brown; stems weakly creeping, with erect branches, branches up to 50 mm long and 2.5 mm thick; branches with short branchlets above. (2) Stem leaves caducous and inconspicuous. (3) Branch leaves not very thriveled, erecto-appressed below, spreading and slightly flexuose above when dry, spreading when moist, lanceolate, long ligulate-lanceolate; apices acuminate, frequently truncate with laminae extending on both sides of the costa, forming short asymmetric spiny protuberances; margins finely serrulate above, entire below; all cells longer than wide, distinctly porose, thick-walled, smooth and clear; upper and apical cells rhombic, gradually elongated towards the base; medial to basal cells elongate-rectangular, thick-walled; the outmost marginal cells shorter and small than their ambient cells, forming a border. (4) Perichaetial leaves distinctly larger and longer than branch leaves. (5) Setae smooth.

Macromitrium subperichaetial Thér. differs from *M. perichaetiale* (Hook. & Grev.) Müll. Hal. by its robust and large plants, and longer branch leaves up to 4-5 mm long.
Distribution: Cuba.

229. *Macromitrium subscabrum* Mitt., J. Linn. Soc., Bot. 12: 215. 1869. (Figure 235)
Type protologue: [Ecuador] Andes Quitenses, in monte Tunguragua (8000 ped.), *Spruce, n. 84*. Type citation: Tunguragua, *Spruce*, Herbarium of William Mitten (lectotype: NY 01086591!).

(1) Plants large, brownish, stems long creeping, with erect to prostrately ascending branches, branches up to 20 mm long, and 1.5-2.0 mm thick, simple or occasionally with branchlets above. (2) Stems long creeping, with squarrose-recurved leaves when dry. (3) Branch leaves appressed below, widely spreading to abaxially curved,

twisted-flexuose above, spreading and abaxially curved when moist, narrowly lanceolate, recurved at one side below; margins finely notched or crenulate above near the apex; apices acuminate to long acuminate; upper and medial cells subquadrate, rounded-quadrate, oblate or oblong, clear and smooth; low and basal cells narrowly elongate-rectangular, thick-walled, strongly porose, sometimes tuberculate; the outmost marginal cells enlarged, forming a distinct border at insertion. (4) Perichaetial leaves different from branch leaves, longer than branch leaves, triangular-lanceolate with the widest part at base, long acuminate to an arista; plicate below, somewhat sheathing; all cells longer than wide. (5) Setae 7-10 mm long, smooth, twisted to the left. (6) Capsule urns ovoid-ellipsoid, strongly furrowed; opercula erect, conic-rostrate, with a long beak; peristome double, exostome teeth united to a short membrane, endostome teeth lanceolate, higher than exostome teeth. (7) Calyptrae mitrate, naked and smooth, lacerate below.

Macromitrium subscabrum Mitt. is similar to *M. sulcatum* (Hook.) Brid., the latter could be separated from the former by its perichaetial leaves shorter than branch leaves, and a "cancellina region" near the costa at the base.

Distribution: Bolivia, Ecuador.

Specimens examined: **BOLIVIA**. Paradiso, Sanjose-Apolo Trail, *R. S. Williams, no. 1824* (H-BR 2625025). **ECUADOR**. Prov. Bolivar, *L. J. Dorr & I. Valdespiao 6458A* (KRAM-B-090181, MO 3682412).

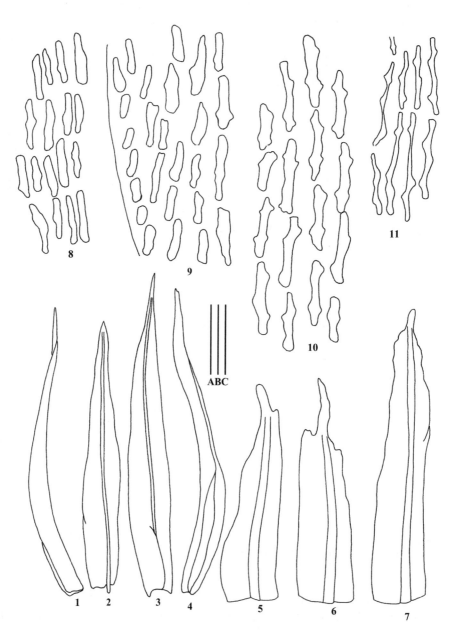

Fig. 234 *Macromitrium subperichaetiale* Thér. 1-4: Branch leaves. 5-7: Apices of branch leaves. 8: Upper cells of branch leaf. 9: Medial marginal cells of branch leaf. 10: Medial cells of branch leaf. 11: Basal cells of branch leaf (all from isotype, PC 0137724). Line scales: A = 1 mm (1-4); B = 400 μm (5-7); C = 67 μm (8-11).

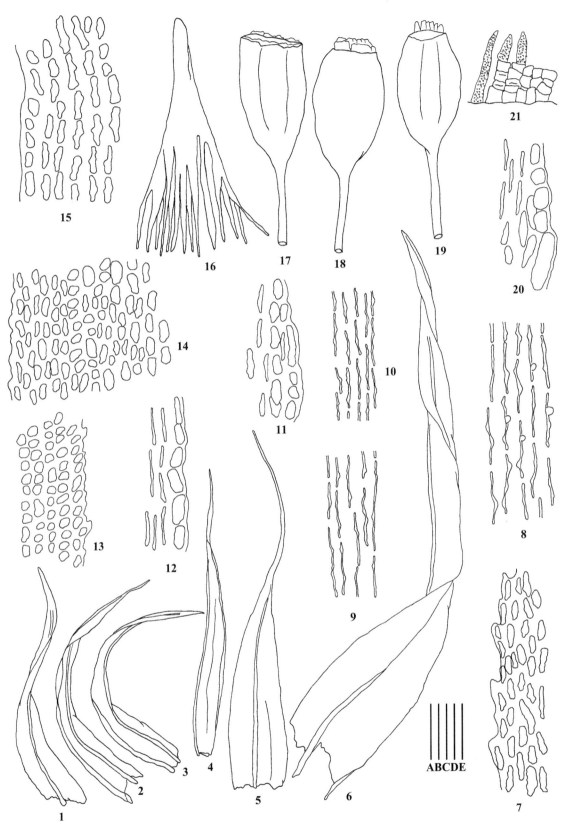

Fig. 235 *Macromitrium subscabrum* Mitt. 1-3, 6: Branch leaves. 4-5: Perichaetial leaves. 7: Apical cells of branch leaf. 8: Basal cells of branch leaf. 9: Basal cells of perichaetial leaf. 10: Medial cells of perichaetial leaf. 11-12, 20: Basal marginal cells of branch leaves. 13: Medial cells of branch leaf. 14: Upper cells of branch leaf. 15: Low cells of branch leaf. 16: Calyptra. 17-19: Capsules. 21: Peristome (all from lectotype, NY 01086591). Line scales: A = 2 mm (1-5, 16-19); B = 1 mm (6); C = 400 μm (21); D = 100 μm (9-12, 20); E = 67 μm (7-8, 13-15).

230. *Macromitrium subtortum* (Hook. & Grev.) Schwägr., Sp. Musc. Frond., Suppl. 2 2(2): 145. 1827. (Figure 236)

Basionym: *Orthotrichum subtortum* Hook. & Grev., Edinburgh J. Sci., 1: 117, 5. 1824. Type protologue: [India] Hab. Received from the East Indies by Arch. *Menzies*, Esq. Type citation: East Indies, leg. *Menzies 115* (letotype designated by Wilbraham, 2016: BM 000982489!; isolectotypes: BM 000982492!, BM 000982493!, BM 000982494!).

Macromitrium mauritianum Schwägr., Sp. Musc. Frond., Suppl. 2 2(2): 127. pl. 189. 1827, *fide* Wilbraham, 2016. Type protologue: In insula Mauritii lectum misit cl. *Sieber*. Type citation: Mauritius, leg. *Sieber s.n.* (lectotype designated by Wilbraham & Ellis, 2010: G 00046151; isolectotypes: BM 000873884!; BM 000873885!, BM 000873807!).

Macromitrium mauritianum var. *viride* Broth., Reise Ostafr., Syst. Arbeit. 3: 55. 1908, *fide* Wilbraham & Ellis, 2010. Type protologue: Mauritius. Type citation: Mauritius 1904, *Voeltzkow s.n.* (lectotype designated by Wilbraham & Ellis, 2010: H -BR 2608007!; isolectotype: S-B 164139!).

Macromitrium rhizomatosum Müll. Hal. ex Besch., Ann. Sci. Nat., Bot., sér. 6, 9: 360. 1880, *fide* Wilbraham & Ellis, 2010 and Schatz *et al.*, 2023. Type protologue and citation: (Madagascar) Nossi-bé: bord des ruisseaux, *Pervillé, n° 789*, April 1841. (isotypes: PC 0137536!, BM 000989746!, PC 0137538!, PC 0137539!, PC 0137537!, BM 000989745!, BM 000878259!).

Macromitrium sanctae-mariae Renauld & Cardot, Bull. Soc. Roy. Bot. Belgique 33(2): 120. 1895, *fide* Wilbraham, 2016. Type protologue: Hab. S^{te} marie de Madagascar (Charly Darbould).-(Renauld, Musci masc. mad. Exsicc., *n° 2170.*). Type citation: S^{te} marie de Madagascar (Charly Darbould, 1893) (isotypes: PC 0147385!, PC 0105700!, PC 0105701!, PC 0105698!, PC 0105699!).

Macromitrium subpungens Hampe ex Müll. Hal., Linnaea 40: 249. 1876, *fide* Schatz *et al.*, 2023. Type protologue: [Comoros] Comoro-insula Johanna, 1000 met. Supra mare, in lingo putrido sylvestri: *J. M. Hildebrandi* Junio-Aug. 1875. *Coll. No. 1814*, cum Macromitrio Hildebrandti associatum. Type citation: Type citation: Statio. Comoro-insula Johanna, 1000 met. ü. M. auf … im Walde Junio-Aug. 1875, leg. *J. M. Hildebrandi, 1814* (isotypes: BM 000868316!, BM 000868318!, S-B 165699!).

Macromitrium subpungens var. *madagassum* Cardot, Hist. Phys. Madagascar, Mousses 239. 1915, *fide* Wilbraham & Ellis, 2010. Type protologue: [Madagascar] Zone inférieure des forêts: Maroantsetra, dans la baie d'Antongil (Ch. Mathieu).

(1) Plants medium-sized to large, in dense mats; stems long creeping, with erect branches, branches to 25 mm long, 1.5 mm thick. (2) Stem leaves inconspicuously, appressed below and abaxially curved above. (3) Branch leaves erect below, individually curved, flexuose to twisted-contorted, incurved upper when dry, moderately clasping at base, abaxially curved spreading when moist, lanceolate from an ovate base and reflexes below, acuminate or somewhat cuspidate; costae percurrent to excurrent; the length to which the costa extends beyong the cuspidate-acuminate apex highly varies, ranging from the costa ending shortly above the acuminate apex to being stoutly excurrent; upper cells quadrate to rounded-quadrate, obscure and pluripapillose; medial cells rectangular, lumens straight, clear and smooth; low and basal cells elongate-rectangular, in the 1/3 low portion of the leaf, thick-walled, lumens curved to sigmoid (or arranged in sigmoid shape), smooth and clear, strongly porose. (4) Perichaetial leaves varied in size, from distinctly larger than, similar to, or smaller than branch leaves, oblong-lanceolate, widely spreading when moist, acuminate-acute. (5) Setae 4-12 mm long, smooth, twisted to the left. (6) Capsule urns ovoid, ellipsoid to ellipsoid-cylindric, smooth to weakly wrinkled, wide-mouth or becoming narrow beneath the mouth; peristome absent; spores anisosporous. (7) Calyptrae mitrate, smooth, naked to sparsely hairy.

Distribution: Comores, India, Madagascar, Mauritius and Réunion Island, Seychelles, Tanzania, the Agalega Island (Wilbraham & Ellis, 2010).

Specimens examined: COMOROS. Grande Comore, *R. E. Magill & T. Pócs 11779* (MO 4422631), *10918* (MO 4422621, KRAM-B-112250, H 3090740). **LA RÈUNION**. leg *Berssuir s.n.* (H-BR 2616015). **SEYCHELLES**. Mahé, *John Erikoton* (H 3090742).

231. *Macromitrium subulatum* Mitt., Trans. & Proc. Roy. Soc. Victoria 19: 64. 1882.

Type protologue: [Australia] Bass Straits, *Milne*.

(1) Plants rather robust, lustrous to shiny, in loose spreading mats; stems loosely creeping, with ascending to erect branches, branches up to 2.0 cm high and about 3.0 mm wide, regularly spaced. (2) Stem leaves flexuose-appressed to curved-flexuose when dry or moist; narrowly ovate-lanceolate to broadly lanceolate, gradually long-acuminate-subulate. (3) Branch leaves spirally contorted to contorted-flexuose when dry, flexuose-twisted, wide-spreading with a curved apex from an erect base when moist, 4.0-4.5 mm long, gradually narrowed to a slender acumen or subula from a broadly lanceolate to ovate-oblong low portion, strongly keeled; margins entire, plane to slightly reflexed below; costae excurrent to the subula, abaxial surface with elongate costal cells exposed along entire length; upper cells rounded to quadrate, thick-walled, in longitudinal rows, slightly bulging, densely pluripapillose,

the papillae forming a continuous covering (similar to those in *Zygodon*); medial cells short- to long-rectangular, irregularly thick-walled, lumens straight to curved, smooth; basal cells long near the costa, very irregularly thick-walled, lumens flexuose, straight or curved, smooth, becoming longer and straight near margin. (4) Perichaetial leaves 5.0-5.2 mm long, slightly longer than branch leaves, erect, sub-sheathing, long, lanceolate, subulate-aristate, all cells longer than wide, smooth and thick-walled, with straight lumens. (5) Setae 6.0-7.0 mm long, smooth, twisted to the left; vaginulae naked. (6) Capsule urns ellipsoid to narrowly ovoid-ellipsoid, smooth or wrinkled, not plicate at mouth; peristome single, exostome of 16, densely papillose; spores anisoporous (Vitt & Ramsay, 1985).

Distribution: Australia.

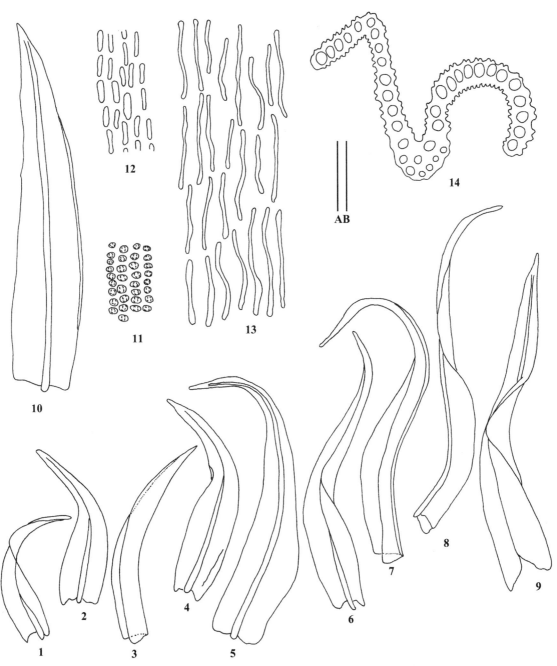

Fig. 236 *Macromitrium subtortum* (Hook. & Grev.) Schwägr. 1-9: Branch leaves. 10: Perichaetial leaf. 11: Upper cells of branch leaf. 12: Medial cells of branch leaf. 13: Basal cells of branch leaf. 14: Upper transverse section of branch leaf (all from isolectotype, BM 000982498). Line scales: A = 1 mm (1-10); B = 67 μm (11-14).

232. Macromitrium sulcatum (Hook.) Brid., Bryol. Univ. 1: 319. 1826. (Figure 237)
Schlotheimia sulcata Hook., Musci Exot. 2: 156. 1819. Type protologue and citation: Nepal, *Gardner* (isotype: FH!).

***Dasymitrium macleai* Rehmann ex Kindb, Enum. Bryin. Exot., Suppl. 1: 89. 1889, *fide* O'Shea, 2006.

Macromitrium belangeri Müll. Hal., Bot. Zeitung (Berlin) 20: 374. 1862, *fide* Wilbraham, 2016. Type protologue: Insula Bouboniae: *Belanger no. 4.* Type citation: Réunion, leg. *Belanger s.n.* (lectotype: PC: PC0137487; isolectotype: S-B 162961!).

Macromitrium bequaertii Thér. & Naveau, Bull. Soc. Roy. Bot. Belgique 60: 50. 27. 1927, *fide* Wilbraham, 2016. Type protologue: Kampianabulongo (Penghe→Irumu); forêt vierge de l'Ituri; sur tronc d'arbre. 25-2-14, cfr. (n. 2654). Kilo→Irumu; forêt vierge, sur branche dans la couronne d'un arbre. 30-6-14, cfr. (n. 4873). Tongo-Mukule; plaine de lave; sur un arbuste. 25-9-14, cfr. (n. 5872). Kilako (entre Masisi et Walikale); forêt vierge, sur un arbre, 2-1-15, crf. (n. 6463). Type citation: Democratic Republic of the Congo: Kampianabulongo (between Penghe and Irumu), 25 February 1914, leg. *Bequaert s.n.* (lectotype designated by Wilbraham, 2016: PC 0098353!). syntype: Democratic Republic of the Congo. Between Kilo and Irumu, 30 June 1914, leg. *Pobéguin s.n.* (PC 0098352!).

***Macromitrium boivini* Müll. Hal. in Besch. Nom. Nud., Rev. Bryol. 4: 15. 1877, *fide* Eills & Wilbraham, 2008.

Macromitrium ceylanicum Mitt. J. Proc. Linn. Soc., Bot., Suppl. 1: 52. 1859, *fide* Gangulee, 1976. Type protologue and citation: Hab. In Ceylon, *Walker et Gardner (no. 253)* (holotype: NY 00518287!).

Macromitrium sulcatum subsp. *ceylanicum* (Mitt.) M. Fleisch. Musci Buitenzorg, 2: 428. 1904.

***Macromitrium corrugatum* Wilson, Hooker's J. Bot. Kew Gard. Misc. 9: 328. 1857, *fide* Gangulee, 1976.

Macromitrium eckendorffii Thér. & P. de la Varde, Rev. Bryol. Lichénol. 11: 176. 11 f. 2. 1939, *fide* O'Shea, 2006. Type protologue: (Central African Republic) Chute Durzoro sur la Pendé, près village Woso. Type citation: [Central African Republic] Oubangui, Bocaranga, Chute Durzoro sur la Pendé, près village Woso. leg. *Eckendorff* 14, 3. 1936, *no. 6541.* (lectotype designated here: PC 0105919!, isolectotypes: PC 0105917, PC 0205918). ***Macromitrium insculptum* Mitt., Trans. Roy. Soc. South Africa 15: 281. 1926, *fide* O'Shea 2006.

***Macromitrium laevatum* Mitt. ex Broth., Nat. Pflanzenfam. I(3): 493. 1903, *fide* Tixier, 1989 and O'Shea, 2006.

Macromitrium levatum Mitt., J. Proc. Linn. Soc., Bot. 7: 152. 1863, *fide* O'Shea, 2006. Type protologue: (Cameroon) Cameroons Mountain, alt. 8000-10,000 feet, on trees and rocks, *Mann.*

***Macromitrium macleai* Paris, Index Bryol. 780. 1897, *fide* O'Shea, 2006.

Macromitrium mannii A. Jaeger, Ber. Thätigk. St. Gallischen Naturwiss. Ges. 1872-73: 147. 1874, *fide* O'Shea, 2006.

**Macromitrium menziesii* Mitt., J. Proc. Linn. Soc., Bot. 7: 152. 1863, *fide* O'Shea, 2006.

Macromitrium neelgheriense Müll. Hal., Syn. Musc. Frond. 1: 737. 1849, *fide* Gangulee, 1976. Type protologue: (India) Patria. Montes Neelgherienses: *Perrottet.* Type citation: India, Orientalis, Musci Neilgherriense, *Perrotte* (lectotype designated here: GOET 012318!). *Macromitrium muellerianum* Mitt., J. Proc. Linn. Soc., Bot., Suppl. 1: 51. 1859. **syn. nov.** Type protologue: [India] Hab. In mont. Nilghiri, Schmid; Gardner (no. 33). Nomen cl. Mülleri "uncinatum"a Bridelio jam abusum. Type citation: *Macromitrium uncinatum*, Nilghiries (lectotype designated here: E 00625564!).

***Macromitrium neilgherrense* Müll. Hal., Bot. Zeitung (Berlin) 11: 61. 1853, *fide* Gangulee, 1976.

***Macromitrium nilghiriense* Müll. Hal. ex Mitt., J. Proc. Linn. Soc., Bot., Suppl. 1: 52. 1859, *fide* Wijk *et al.*, 1964 and Gangulee, 1976.

Macromitrium perundulatum Broth., Wiss. Ergebn. Deut. Zentr.-Afr. Exped., Bot. 2: 149. 13 f. 20. 1910, *fide* O'Shea, 2006. Type protologue: (Rwanda) Rugege-Wald: An Ästen häufig, about 1900 m ü. M. (reichlich fruchtend im Aug. 1907-n. 857, 863); Steinblöcke im Rukarara, außer dem Wasser about 1900 m ü. M. (eine schwarzlichgrune, sterile Form im Aug. 1907-n. 830). Type citation: Afr. Rugege-Wald, Steinblocke in Rukarara, 1900m, 8 1907, *no. 830* (syntype: H-BR 2626001!); Afr. Rugege-Wald, An Ästen häufig 1900 m. *no. 857* (H-BR 2626002!).

Macromitrium pseudoramentosum Herzog, Hedwigia 66: 346. 1926. **syn. nov.** Type protologue: Ceylon: Auf Erde am Fuß Bäume im Urwald des Pidurutalagala, ca. 2100 m, Nr. 147, und auf Erde am Grund der Baumstämme in einer Schlucht bei Hakgala, ca. 1300 m, 151. Type citation: Ceylon: Auf Erde am Grund des Baumstämme in einer Schlucht bei Hakgala, ca. 1300 m, Febuar. 06. *Th. Herzog Nr. 151.* (syntype: JE 04008741!); Auf Erde am Urwald des Pidurutalagala, ca. 2100 m, Januar, 06, leg. *Th. Herzog Nr. 147* (syntype: JE 04008742!); Ceylon, Schlucht bei Hakgala, ca. 1300 m, leg *Th. Herzog* (isosyntype, BM 000982502!).

**Macromitrium rugifolium* Müll. Hal. Ex Broth., Bot. Jahrb. Syst. 24: 241. 1897, *fide* O'Shea, 2006.

Macromitrium rufescens Besch., Ann. Sci. Nat. Bot., sér. 6, 9: 362. 1880, *fide* Ellis & Wilbraham, 2008. Type protologue: La Réunion: Bory (herb. Cosson); Sur l' écorce des vieux arbres, lieux élevés, *Richard, n° 685;* Boivin (inb herb. Mus. Par.); plaine des L'lsle n° 254; Sainte-Agathe. P. Lepervanche, 1877. Grande Comore: mai 1850, Boivin (Macromitrium Boivinii C. Müll. In Rev. Bryol). Madagascar, N. O: Perville, 1841.

Macromitrium seriatum Paris & Broth., Rev. Bryol. 34: 95. 1907, *fide* Wilbraham, 2016. Type protologue: Ad ripas amnis Téné sup. [Pobéguin]. Guinea: on the bank of the river Téné, January 1907, leg. *Pobéguin s.n.* (lectotype designted by Wilbraham, 2016: PC 0137555!); isolectotypes: (PC: Hb. P. de la Varde-PC 0106736!; PC 0106738!; Hb. Theriot-PC 0106737!).

Macromitrium tortifolium Thér., Bull. Mus. Natl. Hist. Nat. 18(7): 477. 1912, *fide* O'Shea, 2006. Type protologue: Côte d'Ivoire.-Herbier Muséum Paris et herbier Thériot. Type citation: Côte d'Ivoire (holotype: PC 0106746!).

Macromitrium trollii Dixon, Repert. Spec. Nov. Regni Veg. 38: 104. 1935, *fide* Wilbraham, 2016. Type protologue: Kol. Kenia: Mt. Kenia, alte Bäume im Nebelwald, 2300 m (leg. *C. Troll, n. 5863*). Type citation: Mt. Kenia, alte Bäume im Nebelwald, 2300 m, 28 April 1934, Troll, africanum, *C. Troll 5863* (isotype: JE 04008697!).

Macromitrium undatifolium Müll. Hal., Flora 69: 278. 1886, *fide* O'Shea, 2006. Type protologue: S. Pedro, ad arbores montis Caffé, 1100 m. Alt., Junio; Bom Sucesso, 1050-1250 m. alt. ad arbores; [São Tomé and Príncipe] Pico de S. Thomé, 1500-2100 m. alt. ad arbores, Aug.; ubique cum fructibus supramaturis.

**Schlotheimia turbinata* Taylor, J. Bot. 75: 127. 1937, *fide* O'Shea, 2006.

(1) Plants medium to large, forming dense yellowish-green mats. Stems long creeping, with dense, erect branches, branches about 5-15 mm high, densely leaved, densely with reddish rhizoids. (2) Stem leaves different from branch leaves, irregularly flexuose-twisted when dry, spreading when moist, entire, 2.0-3.0 × 0.4-0.8 mm, narrowly triangular to ovate-lanceolate; apices acute; costae single, ending in or a few cells below the apex. (3) Branch leaves elect below, individually twisted, crisped, flexuose to contorted-crisped, occasionally undulate above when dry, flexuose-spreading when moist, 2.9-5.5 × 0.4-0.8 mm, linear lanceolate, lanceolate, oblong lanceolate, to ligulate; apices acuminate, broadly acuminate, acute, obtuse, occasionally cuspidate; margins entire to distinctly denticulate in upper portion, plane or occasionally narrowly recurved on one side below; costae single, ending a few cells below the apex; upper cells small, subquadrate, quadrate or quadrate-rotund, 2.5-4.0 μm wide, smooth, sometimes incrassate, flat to slightly bulging, often in regularly oblique rows in broad leaves and in longitudinal rows in narrow leaves; medial cells slightly elongate, subquadrate, short-rectangular, 4-10 × 3.0-4.0 μm, smooth, in longitudinal rows, gradually becoming elongate farther down; basal cells near costa rectangular to irregularly rectangular, thin-walled, smooth and pellucid, 15-25 × 8-10 μm, distinctly larger than their ambient cells, appearing as a "cancellina region", others elongate, rectangular to sublinear, 15-30 × 2.5-4 μm, thick-walled and porose, weakly to strongly tuberculate; outmost marginal cells at or near leaf insertion slightly differentiated, enlarged, short rectangular and hyaline, forming a more or less distinct border. (4) Perichaetial leaves differentiated, often shorter than branch leaves, oblong-ligulate to broadly oblong, often plicate in basal part, broadly acute at apex, with a percurrent to excurrent costa. (5) Setae 5.0 -17 mm long, smooth, twisted to the left. (6) Capsule urns ovoid to ellipsoid, occasionally ellipsoid cylindric to cylindric, narrowed at mouth, dark-purple, 1.5-2.0 mm long, furrowed when dry; peristome consists of a low double membrane, exostome yellowish, densely papillose, endostome hyaline; opercula erect, conic-rostrate; spores anisosporous, much different in size, 12-35μm, finely papillose. (7) Calyptrae mitrate, large and naked, deeply lacerate.

Though *Macromitrium sulcatum* (Hook.) Brid. is a highly variable species, it can be separated from all congeners by a combination of the following features: 1) upper cells of branch leaves smooth and somewhat bulging, subquadrate, quadrate or quadrate-rotund, often in regularly oblique rows; 2) basal cells near costa thin-walled, smooth and pellucid, distinctly larger than their ambient cells, appearing as a "cancellina region", others rectangular, longitudinal walls incrassate, straight to sinuous, weakly to strongly tuberculate; 3) capsules ovate-obloid, obloid-cylindric to cylindric, often strongly furrowed; 4) peristome consisting of a low double membrane; and 5) naked calyptrae campanulate, multiplicate and lacerate below (Yu *et al.*, 2018).

Distribution: (Africa) Cameroon, Central African Republic, Congo, Côte d'Ivoire, Guinea, Kenia Réunion, Rwanda, São Tomé and Príncipe; Malawi (Wilbraham, 2015), Uganda (Wilbraham, 2008); (Asia) India, Indonesia (Eddy, 1996), Iran, Nepal, Philippines, Sri Lanka, Tailand, Vietnam.

Macromitrium sulcatum includes the following three varieties:

Macromitrium sulcatum subsp. ramentosum (Thwaites & Mitt.) M. Fleisch., Musci Buitenzorg 2: 428. 1904.

Basionym: *Macromitrium ramentosum* Thwaites & Mitt., J. Linn. Soc., Bot., 13: 301. 1873. Type protologue: Hab. Ins. Ceylon, *Dr. Thwaites*. Type citation: Ceylon, Central Province, Dr. *Thwaites, no. 40* (lectotype designated here: MO 3653415!; isotype: H-BR 2630022!).

The varieity different from var. *sulcatum* by its hairy calyptrae.

Macromitrium sulcatum var. ***leptocarpum*** (Broth.) J. Yu, D.D. Li, Y. Li & S.L. Guo. Phytotaxa 361(3): 289. 2018. (Figure 238)

Basionym: *Macromitrium leptocarpum* Broth., Rec. Bot. Surv. India 1(12): 318. 1899. Type protologue and citation: Coorg: on trees in exposed situations near mercara (*n. 1, 7, 69*) (not seen); Mercara, exposed granite rocks (n. 68); dry jungle near Verajpet (*n. 132*) (isosyntype: MICH 525887!).

Macromitrium subleptocarpum Dixon & P. de la Varde, Ann. Cryptog. Exot. 3(4): 179, 4 f. 1. 1930, *fide* Yu et al., 2018. Protologue: Hab. Mahableshwar, Western Chats, Jan. 1928; coll. E. Blatter (376), type. Ibidem (379). Type citation: [India] Mahableshwar, Western Ghats, Jan. 1928; coll. E. Blatter 376, type, Herb. H. N. Dixon (isotype: BM 000982487!); Mahableshwar, W. Ghats, 1928, E. Blatter 379, co-type, Herb. H. N. Dixon (isoparatypes: BM 000982488!, US 00070280!).

The varieity different from var. *sulcatum* by its smooth, long or obloid-cylindric capsules, and undulate branch leaves when dry.

Macromitrium sulcatum var. ***torulosum*** (Mitt.) Tixier, Ceylon J. Sci., Biol. Sci. 11: 127. 1975. (Figure 239)

Basionym: *Macromitrium torulosum* Mitt. Type protologue: [Sri Lanka] Ceylon, *Thwaites*. Type citation: Ceylon, *Thwaites, 34.b* (lectotype: NY 00845354!; isolectotype: BM 000919512!, BM 000919513!, H-BR 2623001!); isotype: Ceylon *Thwaites s.n.* (isotype: NY 00845353!).

The varieity different from var. *sulcatum* by its short branches (about 4-6 mm tall) rather sparsely placed on the creeping stems.

Specimens examined:

Macromitrium sulcatum subsp. *ramentosum* — **SRI LANKA (CEYLON)**. *Dr. Thwaites, C. M. 40* (isotype: H-BR 2630022!, MO 3653415!); An Ärten in des Buschzone des Adamspcak, *Th. Herzog 151* (H-BR 2630028).

Macromitrium sulcatum var. *sulcatum* — **INDIA**. Madras, *G. Foreau s. n.* (MO 2559233), *W. Griffith 153* (MO1950193); Kerala, *M. E. Hale 47633* (MO 2489245); Tamil Nadu, *M. E. Hale. 47652* (MO 2489241), *M. Fleischer B 3162* (MO 3080548, MO 5222675, MO 3080510), *B 3161* (MO 3080549), *B 3160* (MO 5222664), *C. C. Townsend 73/572* (MO 5628440); Nilghiri, *Perrettet 16* (H-BR 2647025). **IRAN**. The Kundaks, *C. Fischer 23* (MO 3952053). **PHILIPPINES**. Luzon, *E. D. Merril 4936* (H-BR 2647001), *M. Ramos 5506* (H-BR 2647004), *M. L. Merritt & T. C. Zichokke 16424* (H-BR 2647007), *E. D. Merrill* (H-BR 2647036); Prov. Benguet, *E. B. Capeland 1347* (H-BR 2467033); Mindanao, *M. Ramon & G. Edano 37184* (H 3090746). **RÉUNION**. Bourbon, *L. H. Boivin s. n.* (BM 000872023). **SRI LANKA**. *H. Wright 3817* (H-BR 2647022); *W. Meijer 1799* (MO 5361179), *W. Meijer 1936* (MO 5361161); Central Province, Schäfer-Verwimp & Verwimp, *C372* (MO 5367894), *C269* (MO 5367817), *Gerrit Davidse 8292* (MO 2411379); Midlands, *C. Ruinard & A. H. M. Jayasuriya 10/144b* (MO 2859658); Hakgala Gardens and Forest Reserve, *C. Ruinard & A. H. M. Jayasuriya 19/146-2* (MO 2859563); Hinidoou Kanda hills, Ceylon S.W., *H. Wright 3817* (H-BR 2647022). **TAILAND**. Chiang Mai Province, *Schäfer-Verwimp Nr. 23767* (MO 5367893); Payap, *A. Touw. 8711* (MO 2163749); Northern, *A. Schäfer-Verwimp 24021* (MO 5367892); Udawn, *A. Touw 10544* (MO 2154138). **VIETNAM**. Lam Dông, Dalat, Corticola, *Evrard 1383* (H 3090748); Lam Dong lac Duong District, *S. He & Khang Nguyen 42842* (MO).

Macromitrium sulcatum var. *torulosum* — **SRI LANKA (CEYLON)**. *Dr. Thwaites* (Lectotype: NY 00845353); *Dr. Thwaites 34 G* (NY 00845354); Central Province, *C. M. 34-6* (BM 000919513), *Thwaites n. 34 B* (BM 000919512, H-BR 2623001); Hakgala Gardens and Forest Reserve, *C. Ruinard, A. H. M. Jayasuriya 21/16* (MO 2859739), *C. Ruinard, A. H. M. Jayasuriya 12/145* (MO 2859614); Im Hochland auf Horton plains im Urwald an Bäumen, *M. Fleischer 412* (BM 0009195, H-BR 2623002); Horton Plains, *R. Giasenergen, s.n.* (H-BR 2623004); *T. Herzog, 147.a* (H-BR 2623003).

233. *Macromitrium swainsonii* (Hook.) Brid., Bryol. Univ. 1: 318. 1826. (Figure 240)

Basionym: *Orthotrichum swainsonii* Hook., Musci Exot. 2: 127. 1819. Type protologue: Hab. Prope Rio janeiro Americae meridionalis. *D. Swainson*. Type citation: *Orthotrichum swainsonii*, Brazil, Bahia, 1818, *H718* (isotypes: BM 000873214!, BM 000989726! BM 000873226!).

Macromitrium altituberculosum E.B. Bartram, Bryologist 47: 17. 1944, *fide* Allen, 2002. Type protologue: Guatemala: Sierra de las Minas: Dept. Zacapa: oak-pine woods along the upper reaches of Rio Sitio Nuevo, between Santa Rosalia and first waterfall, alt. 1200-1500 m., on rock, Jan. 9, 1942, *Julian A. Steyermark, 42274*.

Macromitrium brotheri Müll Hal., Bull. Herb. Boissier 6: 97. 1898, *fide* Valente *et al.*, 2020. Type protologue: Habitatio. Brasilia. Goyaz, Serra Dourada, ad truncos arborum, Febr. 1893: E. Ule; Goyaz, Mossamedes, ad arbores sylvestres, Januario 1893: E. Ule, coll. No 1560, 1561; Goyaz, Passa Tempo, Septbr. 1892: idem, Coll. *No 1564*.

Macromitrium carionis Müll. Hal., Bull. Herb. Boissier 5: 199. 1897, *fide* Valente *et al.*, 2020. Type protologue: (Guatemala) Cuesta de Lovio, Aug. 1870. Coll. *No. 48*. Type citation: Cuesta de Lovio, Aug. 1870, Coll. *Bernoulli*

& *Cario no. 48* (isotype: GOET 011887!).

Macromitrium stellulatum (Hornsch.) Brid. Bryol. Univ. 1: 314, 1926. Basionym: *Schlotheimia stellulata* Hornsch. Horae Phys. Berol. 61, pl. 12. f. 1–6. 1820. *fide* Valente *et al.*, 2020. Type protologue: [Venezuela] sylvae ad *Orinoci* fluminis ripas. Type citation: [Venezuela] Banks of the River Orinoco, *Hornschuch C.F. s.n.* (E 00428901!)

Macromitrium vesiculatum (Herzog) Herzog, Hedwigia 57: 246. 1916. Basionym: *Schlotheimia vesiculata* Herzog. Beih. Bot. Centralbl., Abt. 2, 26(2): 68. f. 12. 1909, **syn. nov.** Type protologue and citation: An Bäumen im Urwald Des Rio Blanco (Prov. Velasco). Ca. 160 m. *Th. Herzog*, August 1907 (holotype: JE 04008705!).

(1) Plants small to medium-sized, dark-green to olive-green; stems weakly creeping, branches densely reddish tomentose below. (2) Branch leaves individually to spirally flexuose-twisted, weakly contorted-flexuose above when dry, spreading when moist, up to 2 mm long, lingulate, plicate below; apices rounded to obtuse, emarginated to short-mucronated; margins entire or crenulate above, plane or erect, basal marginal enlarged teeth-like large cells strongly differentiated at the insertion; costae shortly excurrent or percurrent; upper and medial cells 6-8 µm, rounded-hexagonal, strongly bulging to mammillose; basal cells strongly tuberculate, long-rectangular, incrassate and porose, up to 16 µm long. (3) Setae 3-9 mm long, smooth. (4) Capsule urns ovoid to cylindric, plicate; opercula conic-rostrate, 1-1.2 mm long; peristome double, exostome teeth lanceolate, not fused into a membrane, papillose-striate, 310 µm long, endostome basal membrane 117 µm high, segments 60 µm long high; spores anisosporous, small spores (about 20 µm) smooth, and large spores (40 µm) papillose. (5) Calyptrae larger, 3-3.5 mm long, mitrate, naked, deeply lacerate.

Valente *et al.* (2020) treated *Macromitrium stellulatum* (Hornsch.) Brid. as a synonym of *M. swainsonii*. However, according to the syntype E 00428901, the basal marginal cells of *M. stellulatum* are elongate-linear, the outmost cells not large teeth-like. *Macromitrium stellulatum* (Hornsch.) Brid. is possibly a member of the genus *Groutiella*. Churchill and Fuentes (2005) placed *M. vesiculatum* (Herzog) Herzog in synonymy with *M. stellulatum* (Hornsch.) Brid. Acutally, *M. vesiculatum* is conspecific with *M. swainsonii*. Therefore, the identity of M. stellulatum (Hornsch.) Brid. and M. stellulatum are still needed to study.

Distribution: Bolivia, Brazil, Guatemala, Hondruas, Mexico, Peru.

Specimens examined: **BOLIVIA**. Low Sorata River, *R. S. Williams, 1819* (H-BR 2571003). **BRAZIL**. *G. Gardner, 41* (BM 000989726), *G Gardner 60* (BM 000989723), *Buck WR, Araujo I, Steward WC, Ramos JR, & Ribamar J, 2052* (MO 2548626); Serra da Piedade *Anon s.n.* (H-BR 2571001). **GUATEMALA**. *Louis O. Williams, Antonio Molina R. & Terua P. Williams, no. 41418* (MO 2409730). **HONDRUAS**. Comayagua, Siguatepeque, *Yuncker TG, Dawson RF & Youse HR, 6521* (MO 1114881). **MEXICO**. Chiapas, *D. E. Breedlove & T. F. Daniel, 71051* (MO 3665104); Weimannia and Styrax, *Breedlove DE 24939* (MO 2408278). **PERU**. District of madre de Dios. *Piers majestyk 4307* (MO 5377514).

234. *Macromitrium taiheizanense* Nog., J. Sci. Hiroshima Univ., Ser. B, Div. 2, Bot. 3: 11. 1. 1936. (Figure 241)

Type protologue and citation: [China] Taiwan: Mt. Taiheiizan (ca 2000m), Prov. Taihoku (*A. Noguchi, no. 6548*- type: in Herb. Iros. Univ., Aug. 1932) (isotype: NICH 365243!).

(1) Plants robust, yellow-brownish above, dark brownish below, in loose mats; stems creeping, branches up to 25 mm high, densely covered with reddish rhizoids. (2) Stem leaves similar to branch leaves, up to 7 mm long, appressed when dry. (3) Branch leaves irregularly contorted-twisted, flexuose and divergent above when dry, spreading and often with an adaxially-curved upper portion when moist, 5.0-7.0 × 0.5-0.6 mm, long and narrowly lanceolate, gradually narrowed to a slender acuminate acumen or subula from a long oblong low portion, longitudinally plicate below, with the widest at the base; margins entire and plane throughout; costae ending in the apex; cells unistratose from the apex to the base; upper and medial cells smooth, flat and clear, subquadrate-rounded to rounded-quadrate, thick-walled, 8.0-10.0 µm wide; basal cells thick-walled and somewhat sinuous, gradually elongate from low part to base, 22-50 × 3.5-7.5 µm wide, bulging and unipapillose. (4) Perichaetial leaves 6-7 mm long, ovate-oblong, long-acuminate in upper part. (5) Setae rather short, 1-1.2 mm long, smooth and erect. (6) Capsules immersed, urns ovoid to ellipsoid, 1.6-1.9 ×1.4-1.5 mm, brown, slightly ribbed under the mouth; peristome absent. (7) Calyptrae campanulate, with many long, yellowish hairs.

Macromitrium taiheizanense Nog. is somewhat similar to *M. hainanense* S.L. Guo & S. He in short setae, immersed gymnostomous capsules, but the latter can be separated from the former by its exceeding short setae (0.2-0.4 mm), short oblong-lanceolate branch leaves (1.8-2.2 mm) with pluripapillose lamina cells in upper and medial portions.

Distribution: China.

Specimens examined: **CHINA**. Guizhou, Fanjingshan, *M. R. Crosby 15933* (MO 6166109); Sichuan, *P. L. Redfearn, et al. 34709* (MO 3965519), *B. Allen 6599* (MO 5133139), *6565* (MO 5133171).

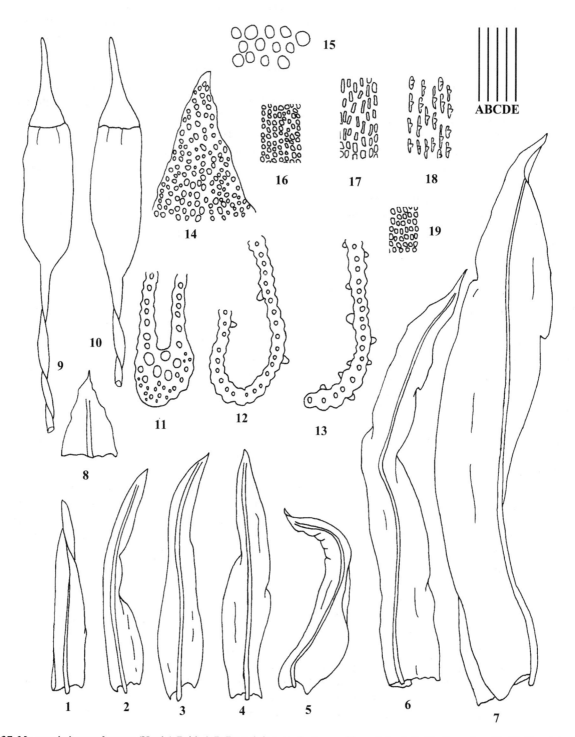

Fig. 237 *Macromitrium sulcatum* (Hook.) Brid. 1-7: Branch leaves. 8: Apex of branch leaf. 9, 10: Capsules. 11, 13: Basal transverse sections of branch leaves. 12: Medial transverse section of branch leaf. 14: Apical cells of branch leaf. 15: Spores. 16: Medial cells of branch leaf. 17: Low cells of branch leaf. 18: Basal cells of branch leaf. 19: Upper cells of branch leaf (all from isotype, FH). Line scales: A = 1 mm (1-5, 9, 10); B = 400 μm (6, 7); C = 200 μm (8); D = 100 μm (15); E = 67 μm (11-14, 16-19).

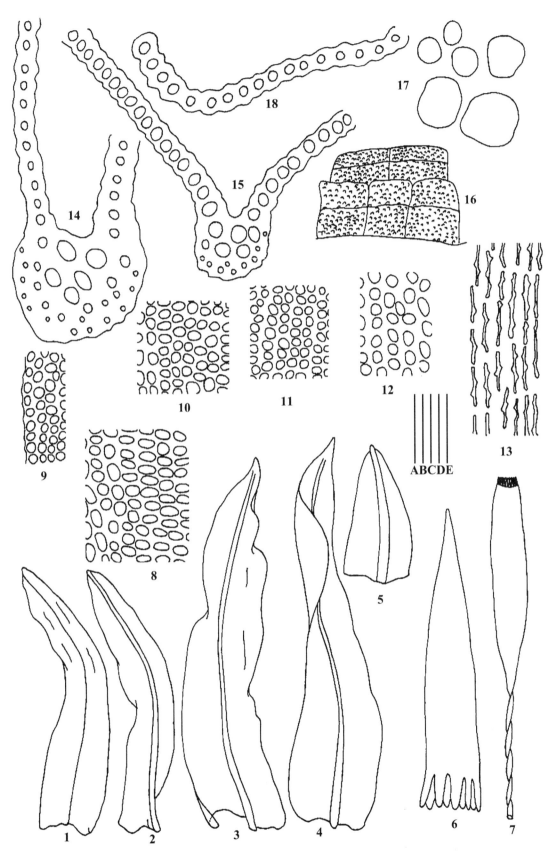

Fig. 238 *Macromitrium sulcatum* var. *leptocarpum* (Broth.) J. Yu, D.D. Li, Y. Li & S.L. Guo 1-4: Branch leaves. 5: Apex of branch leaf. 6: Calyptra. 7: Capsule. 8: Apical cells of branch leaf. 9: Upper marginal cells of branch leaf. 10, 11: Medial cells of branch leaves. 12: Low cells of branch leaf. 13: Basal cells of branch leaf. 14: Upper transverse section of branch leaf. 15: Medial transverse section of branch leaf. 16: Low transverse section of branch leaf. 17: Spores. 18: Peristome (all from isosyntype, MICH 525887). Line scales: A = 2 mm (6, 7); B = 1 mm (1-4); C = 400 μm (5); D = 100 μm (16, 17); E = 67 μm (8-15, 18).

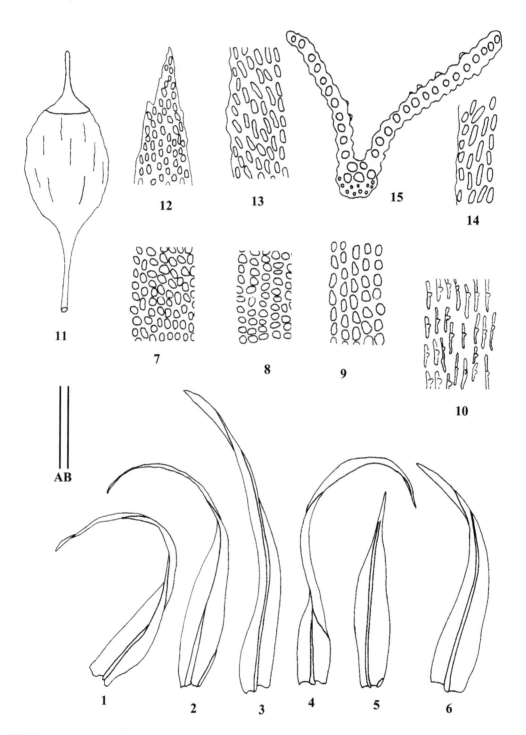

Fig. 239 *Macromitrium sulcatum* var. *torulosum* (Mitt.) Tix. 1-4: Branch leaves. 5, 6: Perichaetial leaves. 7: Upper cells of branch leaf. 8, 9: Medial cells of branch leaves. 10: Basal cells of branch leaf. 11: Capsule. 12: Apical cells of branch leaf. 13, 14: Apical marginal cells of branch leaves. 15: Basal transverse section of branch leaf (all from lectotype of *M. torulosum*, NY 00845354). Line scales: A = 0.44 mm (1-6, 11); B = 44 μm (7-10, 12-15).

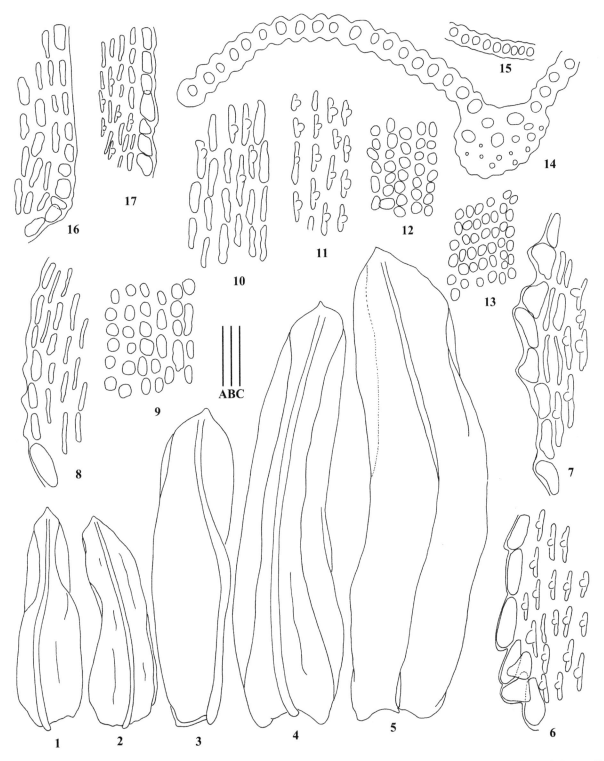

Fig. 240 *Macromitrium swainsonii* (Hook.) Brid. 1-5: Branch leaves. 6-8, 16-17: Basal marginal cells of branch leaves. 9: Low cells of branch leaf. 10-11: Basal cells of branch leaves. 12: Medial cells of branch leaf. 13: Upper cells of branch leaf. 14: Medial transverse section of branch leaf. 15: Upper transverse section of branch leaf (all from isotype, BM 000873214). Line scales: A = 1 mm (1-2); B = 400 μm (3-5); C = 67 μm (6-17).

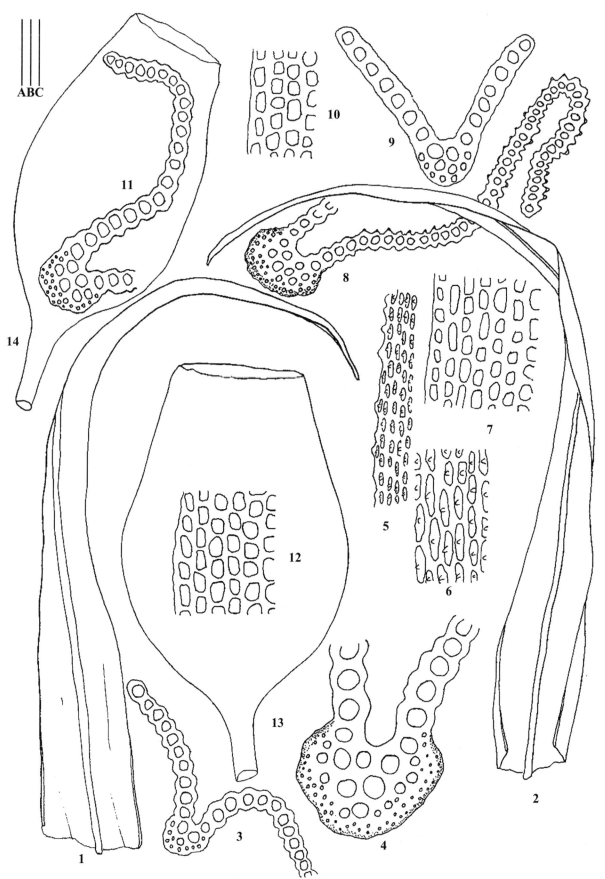

Fig. 241 *Macromitrium taiheizanense* Nog. 1-2: Branch leaves. 3-4, 8-9, 11: Transverse sections of leaves. 5, 7: Low marginal cells of branc leaves. 6: Low cells near costa of branch leaf. 10: Upper cells of branch leaf. 13-14: Capsules (all from isotype, NICH 365243). Line scales: A = 0.50 mm (1-2, 13-14); B = 80 μm (3, 8, 11); C = 50 μm (4, 6-7, 9-10, 12). (Guo & Cao, 2007).

235. *Macromitrium taiwanense* Nog., J. Sci. Hiroshima Univ. Ser. B, Div. 2, Bot. 3: 141. 1938. (Figure 242)

Type protologue and citation: [China] Taiwan: Mt. Taihei (MururoauTamonkei), prov. Taihoku (*A. NOGUCHI, no. 6621*-type: in Herb. Hiros. Univ., Aug. 1932). (holotype: HIRO; isotype: NICH 365244!).

(1) Plant small to medium-size, rusty-brown; stems long creeping, up to 40 mm, densely covered with brown-reddish rhizoids; branches numerous, 7.0-10 mm long, rather thick. (2) Stem leaves twisted-contorted when dry, spreading when moist, 2.0 mm long, ovate-lanceolate, with an acuminate apex; upper and medial cells round-quadrate, subquadrate (4.0-6.0 μm), strong conic-bulging, with a single large linear central papilla up to 18 μm long, somewhat obscure; basal cells elongate, linear, confined to a small area, thick-walled, clear, with a large tubericula per cell, the outmost marginal cell regularly rectangular, pellucid, forming a differentiated border. (3) Branch leaves individually twisted-contorted, sometimes spirally coiled when dry, spreading and adaxially incurved, keeled above when moist, plicate below, 0.3-0.4 × 2.0-2.6 mm, ligulate to oblong-lanceolate, sharply contracted to a long cuspidate or a long hyaline arista (up to 200 μm); costae excurrent to a long hyaline arista; upper, medial and low cells similar, isodiametric, round-quadrate, in longitudinal rows, strongly bulging to conic-bulging, with a single large, occasionally forked tuberculate papilla up to 20 μm high; basal cell somewhat elongate, but not much different from the above cells, with a large tubericula per cell. (4) Sporophytes not seen.

Macromitrium taiwanense Nog. is similar to *M. longipapillosum* D.D. Li, J. Yu, T. Cao & S.L. Guo in their bulging cells with a single large, occasionally forked tuberculate papilla up to 20 μm; the former can be separated from the latter by its branch leaves with a long hyaline arista.

Distribution: China.

Specimen examined: CHINA. Zhejiang, *P. C. Wu 305* (PE, MO 5922256).

236. *Macromitrium taoense* Thér., Diagn. Esp. Var. Nouv. Mouss. 8: 5. 1910. (Figure 243)

Type protologue: [New Caledonia] Tao, forêt, sur l'écorce des arbres, alt. 0 à 100 m. Type citation: Nouvelle-Calédonie, forêt, Tao, 100 m, Jaur 1910, *Franc s.n.* (lectotype designated by Thouvenot, 2019: PC 0096517; isotype: H-BR 2618016!).

(1) Plants small to medium-sized, stems creeping with dense branches, branches up to 10 mm tall. (2) Branch densely leaved, leaves regularly and spirally twisted-curved in rows, the upper portions curved to one side or curly, forming a rope-like appearance(funiculat), 1.1-1.5×0.25-0.4 mm, ligulate and hyaline, apiculate or abruptly acute; all cells thick-walled, hyaline and clear; upper and medial cells rounded to oblate, small (5-8 μm wide), smooth and flat; low and basal cells rectangular elongate, lumen narrow, distinctively tuberculate; abrupt transition between upper and basal parts. (3) Perichaetial leaves erect sheathing the seta, wider than branch leaves, oblong to wide lanceolate. (4) Setae 6-15 mm long, smooth. (5) Capsule urns ovoid-ellipsoid to ellipsoid, smooth, rims small, plicate when dry; peristome single, exostome of 16, lanceolate and papillose. (6) Calyptrae naked.

Macromitrium taoense Thér. is somewhat similar to *M. microstomum* (Hook. & Grev.) Schwägr. and *M. renauldii* Thér. However, *M. taoense* differs from *M. microstomum* by its tuberculate basal cells of branch leaves; from *M. renauldii*, *M. taoense* can be separated by its naked calyptrae and ligulate branch leaves.

Distribution: New Caledonia.

237. *Macromitrium tenax* Müll. Hal., Bot. Jahrb. Syst. 5: 83. 1883. (Figure 244)

Type protologue: no detail information.

Macromitrium coriaceum E.B. Bartram, Farlowia 2(3): 315. 1946, *fide* Seki, 1974. Type protologue: Chile: Prov. De Chiloe, Puerto Barroso de la Penins, Tres Montes, Nos 1563, 1570 type, 1588, 1628. Type citation: Prov. De Chiloe, Penins, Tres Montes, Puerto Barroso, leg. *H. Roviainen*, 1929. 2. IV. (lectotype: F-C 0001225F!; isolectotype: F-C 0001238F!).

Macromitrium tenax var. *theriotii* Cardot, Wiss. Ergebn. Schwed. Südpolar-Exped. 1901–1903, 4(8): 285. 1908, ***syn nov.*** Type protologue: Détroit de Magellan (Nadeaud; herb. I. Theriot).

(1) Plants small, stems weekly creeping, branches short. (2) Branch leaves obliquely appressed and slightly twisted-flexuose and incurved above, somewhat funiculate when dry, erecto-patent when moist, oblong- to oval-lanceolate, almost plane or slightly plicate; apices acute to acuminate-acute; margins entire; costae stout, vanished far beneath the apex; all cells clear, flat and smooth; upper and medial cells subquadrate, rounded-quadrate, oval, porose; cells gradually elongate from apex to base; low and basal cells elongate, thick-walled, strongly porose. (3) Perichaetial leaves broadly ligulate-lanceolate, larger and wider than branch leaves, entire and acuminate; all cells longer than wide; upper and medial cells long rectangular, long-oval and long oblong, smooth and clear. (4) Setae shorter, 3-5 mm long, smooth, not twisted, vaginulae hairy. (5) Capsules ellipsoid to ellipsoid-cylindric, smooth; opercula conic-subulate, oblique; peristome double, exostome of 16, lanceolate, papillose, endostome hyaline. (6) Calyptrae campanulate, lacerate at base, naked.

Macromitrium tenax var. *theriotii* Cardot differs from the typical form only by its more robust habit, longer branches, larger leaves, stronger costae and thickened cell walls. It is unnecessary to keep this variety.

Macromitrium tenax Müll. Hal. is similar to *M. campoanum* Thér., but the perichaetial leaves of *M. campoanum* shorter than branch leaves, broadly ovate-lanceolate, long cuspidate, while those of *M. tenax* slightly differentiated, larger than branch leaves, broadly oblong-lanceolate, with an acute apex. *Macromitrium tenax* is also somewhat similar to *M. microstomum* (Hook. & Grev.) Schwägr., but the latter could be distinguished from the former by its funiculate branch leaves and single peristome.

Distribution: Chile.

Specimens examined: CHILE. Magallanes, *B. Allen 26170* (MO 5626918); Prov. Antartica Chilena, *William R. Buck 45660* (MO 5626917), *57160* (MO 6367728, KRAM-B-205412), *57098* (KRAM-B-205680), *56857* (KRAM-B-205514), *47857* (KRAM-B-205470); Tueday Bay, Dr. *Naumann, s.n.* (BM 000982405).

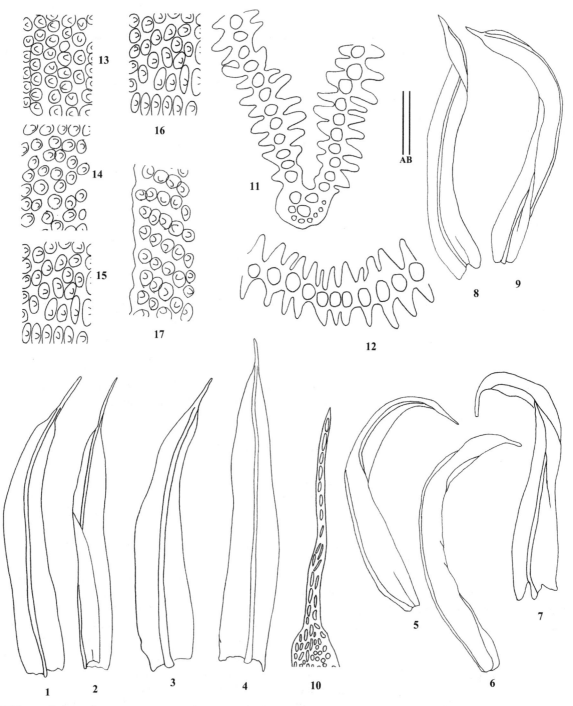

Fig. 242 *Macromitrium taiwanense* Nog. 1-9: Branch leaves. 10: Apex of branch leaf. 11-12: Upper transverse sections of branch leaves. 13-15: Upper and medial cells of branch leaves. 16-17: Low cells of branch leaves (1-4, 10 from MO 5922256; 5-9, 11-17 from isotype, NICH 365244). Line scales: A = 0.4 mm (1-9); B = 100 μm (10-17). (Li *et al.*, 2017).

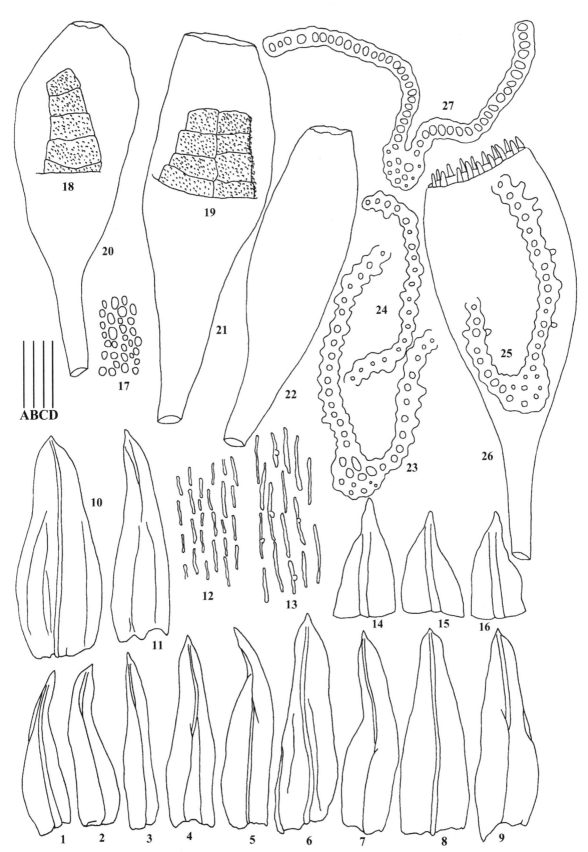

Fig. 243 *Macromitrium taoense* Thér. 1-9: Branch leaves. 10-11: Perichaetial leaves. 12: Medial cells of branch leaf. 13: Basal cells of branch leaf. 14-16: Apices of branch leaves. 17: Upper cells of branch leaf. 18-19: Peristomes. 20-22, 26: Capsules. 23-25: Basal transverse sections of branch leaves. 27: Medial transverse section of branch leaf (all from isotype, H-BR 2618016). Line scales: A = 0.44 mm (1-11, 20-22, 26); B = 176 μm (14-16); C = 70 μm (18-19); D = 44 μm (12-13, 17, 23-25, 27).

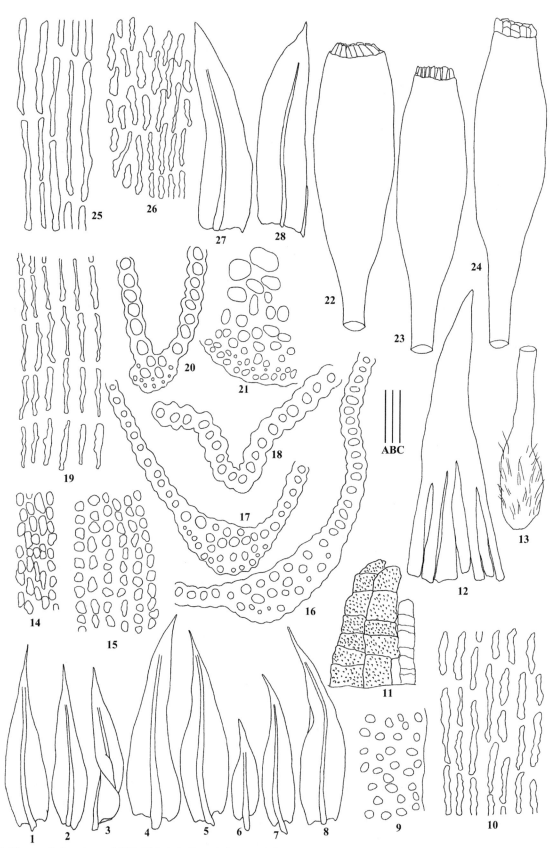

Fig. 244 *Macromitrium tenax* Müll. Hal. 1-4: Branch leaves. 5-8: Stem leaves. 9: Exothecial cells of capsule. 10: Medial cells of perichaetial leaf. 11: Peristome. 12: Calyptra. 13: Vaginula. 14: Upper cells of branch leaf. 15: Medial cells of branch leaf. 16-17: Low transverse sections of branch leaves. 18: Basal transverse section of branch leaf. 19: Basal cells of branch leaf. 20: Medial transverse section of branch leaf. 21: Transverse section of stem. 22-24: Capsules. 25: Basal cells of perichaetial leaf. 26: Upper cells of perichaetial leaf. 27-28: Perichaetial leaves (all from isolectotype, F-C 0001238F). Line scales: A = 1 mm (1-8, 12-13, 22-24, 27-28); B = 100 μm (9, 11, 21); C = 67 μm (10, 14-20, 25-26).

Taxonomy

238. *Macromitrium thwaitesii* Broth. ex M. Fleisch., Musci Buitenzorg 2: 420. 1904. (Figure 245)

Type protologue: no information. Type citation: Ceylon, Horton Plain, Leg. *Thwaites, H2, 2*/1846 (lectotype designated here: H-BR 2630024!).

(1) Plant medium-sized, in loose mats; stems long creeping, with branches up to 25-30 mm long and 1.5 mm thick. (2) Branch leaves moderately twisted-contorted, flexuose and divergent when dry, erect to patent, often abaxially curved below when moist, exceeding long, linear- lanocelate, gradually narrowed to a slender acuminate acumen or subula from oblong low portion, conspicuously revolute, all cells longer than wide, especially for those in medial and low portions, somewhat bulging, unipapillose or tuberculate. (3) Perichaetial leaves similar to branch leaves. (4) Setae smooth, 9-12 mm long, twisted to the left. (5) Capsule urns ovoid-ellipsoid, plicate when dry, with a contracted mouth; peristome double, exostome densely yellowish papillose, endostome pellucid; spores isosporous, smooth and finely papillose. (6) Calyptrae mitrate, smooth and naked, lacerate.

Macromitrium thwaitesii Broth. ex M. Fleisch. is similar to *M. sulcatum* (Hook.) Brid., *M. ceylanicum* Mitt., *M. ramentosum* Thwaites & Mitt., *M. pseudoramentosum* Herzog in elongate basal cells with tubericulae, ovate capsules with longititudinal plicates when dry, double peristome, naked calyptrae. However, *M. thwaitesii* could be separated from the other four species by its cells of branch leaves all longer than wide. *Macromitrium fulvum* Mitt. is rather similar to *M. thwaitesii*, but could be separated from the latter by its smooth cells of branch leaves.

Distribution: Sri Lanka.

239. *Macromitrium tocaremae* Hampe, Linnaea 31: 524. 1862. (Figure 246)

Type protologue and citation: [Colombia] Nova-Granada, Tacaremae ad arbores altit. 2300 metr., Aug. 1861, sub *no. 2156* (isotypes: BM 000873210!, BM 000873211!, BM 000873212!, GOET 011895!).

(1) Plants medium-sized; stems weakly creeping, branches up to 40 mm long. (2) Stem leaves inconspicuous and caducous. (3) Branch leaves erect below, spreading, contorted-crisped and circinate or flexuose when dry, keeled, abaxially curved spreading when moist, 2-3 mm long, oblong-lanceolate; apices acuminate-acute; margins irregularly and bluntly notched above, plane or recurved below; costae subpercurrent; upper and medial cells rounded-quadrate, smooth and clear; basal cells narrowly rectangular, 20-26 μm long, porose, strongly tuberculate; marginal enlarged, swollen, sharply teeth-like cells distinctly differentiated at insertion. (4) Perichaetial leaves broadly elliptical-lanceolate, entire; upper cells isodiametric, smooth; cells elongate towards to base, rectangular or long rectangular cells in low 2/3 of the perichaetial leaf. (5) Setae smooth, 7-18 mm long, not or slightly twisted to the left. (6) Capsule urns 1-1.5 mm long, ovoid or cupulate, smooth to weakly ribbed; opercula conic-rostrate; peristome double, exostome teeth united to a membrane, about 100 μm high, yellow and papillose; endostome segments hyaline, papillose; spores anisosporous, smooth or weakly papillose. (7) Calyptrae mitrate, deeply lacerate, naked.

Distribution: Colombia.

Specimens examined: **COLOMBIA**. Andes Bogotenses, Supra Pacho, *J. Welr 245* (BM 000989720); Bogota Pach, *A. Lindig s.n.* (BM 000989719).

240. *Macromitrium tongense* Sull., U.S. Expl. Exped., Musci 7, pl. 5B, f. 1-21. 1859[1860] (*"Tongense"*). (Figure 247)

Type protologue: Tonga. Tongatabu. Type citation: TONGA. Tongatabu: *Wilkes Expedition, s. n.* (isotypes: US 00070281!, PC 0695994!, PC 0695995!).

Macromitrium densifolium Thér., Bull. Acad. Int. Géogr. Bot., 18: 253, 1908, *fide* Thouvenot, 2019. Type protologue: Hab. Nouvelle-Caledonie. Type citation: Nou Caledonié, *Franc s.n.* (lectotype designated by Thouvenot, 2019: H-BR 2561002!).

Macromitrium elizabethae Dixon, Ann. Bryol. 7: 58. 1934, *fide* Wijk *et al.*, 1964.

Macromitrium ludovicae Broth. & Paris, Öfvers. Finska Vetensk.-Soc. Förh. 53A(11): 17. 1911, *fide* Thouvenot, 2019. Type protologue: (New Caledonia) Ile des Pins, Watchia, *L. Le Rat s.n.* Type citation: Ile des Pins, Watchia, Majo 1909, leg. *L. Le, Rat* (holotype: H-BR 2563009!).

Macromitrium subvillosum Broth. & Paris, Öfvers. Finska Vetensk.-Soc. Förh. 51A(17): 16. 1909, *fide* Thouvenot, 2019. Type protologue and citation: New Caledonia, Mt Dzumac, *A. Le Rat s.n.* (lectotype designated by Thouvenot, 2019: H-BR 2561008!, isotype: H-BR 2561005!).

Macromitrium villosum (Besch.) Broth., Nat. Pflanzenfam. I(3): 486. 1903. - *Dasymitrium villosum* (Besch.) A. Jaeger, Ber. Thätigk. St. Gallischen Naturwiss. Ges. 1877-78: 423. 1880. - *Drummondia villosa* Besch., Ann. Sci. Nat., Bot., sér. 5, 18: 207. 1873, *fide* Thouvenot, 2019. Type protologue: (New Caledonia) "Île des Pins", *Pancher 578*, Sud de la Nouvelle Calédonie, *Pancher 574*; Mt Mou, *Balansa 2979*.

Macromitrium villosum var. *elongatum* Thér., Bull. Acad. Int. Géogr. Bot. 17(217): 308. 1907, *fide* Thouvenot, 2019. Type protologue: no detail information. Type citation: New Caledonia, 1906, *Franc s.n.* com. *Renauld* (lectotype designated by Thouvenot, 2019: PC 0083740!).

Macromitrium villosum var. *intermedium* Thér., Bull. Acad. Int. Géogr. Bot. 17(217): 308. 1907, *fide* Thouvenot, 2019. Type protologue: no detail information. Type citation: New Caledonia, 1907, *Franc s.n.* (lectotype designated by Thouvenot, 2019: PC 0083741!).

Macromitrium villosum var. *longisetum* Thér., Bull. Acad. Int. Géogr. Bot. 17(217): 308. 1907, *fide* Thouvenot, 2019. Type protologue: no detail information. Type citation: New Caledonia, 1907, *Franc s.n.* (lectotype designated by Thouvenot, 2019: H-BR 2562007!).

(1) Plants small, dull, in dense compact, spreading mats; stems prostrate, creeping, giving rise to short, erect, stiff branches about 1 cm long. (2) Stem leaves ovate-lanceolate, acuminate. (3) Branch short and densely leaved, leaves stiffly curved-erect, appressed, some apices curved outward or deflexed, spirally curved around branch when dry, stiff, straight, wide-spreading when moist, 1.3-1.5 mm long, narrowly elliptic-oblong to ovate-oblong; apices retuse to obtuse, bluntly mucronate; margins reflexed to plane, minutely crenulate to subentire; costae excurrent and strong; upper and medial cells oblate to rounded (5-6 µm), obscure, densely pluripapillose, irregular proliferation up to 1-3-stratose thick, sharply grading to clear low cells; low and basal cells 7-9 × 18-35 µm, long-rectangular, hyaline and smooth, lumens sigmoid to curved, not porose. (4) Perichaetial leaves erect, sheathing, 1.5-1.9 mm long, oblong to oblong-ovate, mucronate to broadly acute, sometimes ± rugose, all cells longer than wide, curved-sigmoid. (5) Setae 6.0-8.0 mm long, smooth; vaginulae naked or with a few hairs. (6) Capsules narrowly ellipsoid, smooth and slightly ribbed at mouth, rim darker and purplish; peristome single; exostome of 16, pale, blunt, papillose, short teeth; spores anisosporous. (7) Calyptrae mitrate, plicate, oblong-conic, densely hairy.

Macromitrium tongense Sull. is similar to *M. nepalense* (Hook. & Grev.) Schwägr. and *M. serpens* (Burch. ex Hook. & Grev.) Brid. in the leaf upper parts in cross sections with irregularly, 1-3-stratose proliferation. From *M. nepalense*, *M. tongense* could be separated by its lumens sigmoid to curved in the low and basal portion of branch leaves. *Macromitrium tongense* differs from *M. serpens* in its shorter branch leaves spirally appressed-curved when dry, while the branch leaves of *M. serpens* longer, erect, curved to twisted, and the apices inrolled when dry.

Distribution: New Caledonia, Papua New Guinea, Tonga.

Specimens examined: **TONGA**. Tongatabu (BM 000989735, BM 000982773, BM 0000982774).

241. *Macromitrium tosae* Besch., J. Bot. (Morot) 12: 299. 1898. (Figure 248)

Type protologue: Japon: île de Shikoku, montagne de Tosa, 18. nov. 1893 (*Faurie, no. 11190*). Type citation: Japan, île de Shikoku, Tosa,18, nov. 1893. Legit *Faurie no. 11190* (isotype: H-BR 2581002!).

Macromitrium chungkingense P.C. Chen, Contr. Inst. Biol. Natl. Centr. Univ. 1: 7. 1943, *fide* Yu *et al.*, 2012. Type protologue: China Prov. Szechuan, Chungking, Tsuchi-Kou, ad rupes siccas, leg. *P. C. Chen* XII 1940. (isotypes: S-B 115591!, JE 04008728!).

Macromitrium courtoisii Broth. & Paris, Rev. Bryol. 36: 9. 1909["-i"], *fide* Yu *et al.*, 2012. Type protologue: "Chei long, ad rupes, 27.4.08"; Type citation: China, Chei long, ad rupes, 27.4.08, Leg. *R. P. Courtois, no. 1118* (holotype: H-BR 2581006!).

Macromitrium cylindrothecium Nog., J. Sci. Hiroshima Univ., Ser. B, Div. 2, Bot. 3: 137. 13 f. 10-20. 1936, *fide* Yu *et al.*, 2012. Type protologue and citation: [China] Taiwan: Urai, Prov. Taihoku (*S. Suzuki*, no. 9487-type: in Herb. Hiros. Univ., Jan. 1925) (holotype: NICH 365237!).

Macromitrium hamatum Dixon, J. Bombay Nat. Hist. Soc. 39: 777. 1937. *fide* Li *et al.*, 2020. Protologue: [India] On tree trunk, 1680 m., Pedi, Naga Hills; 1 August 1935 (284 bis). Type citation: On tree trunk, 5500', Pedi, Naga Hills; Assam, Coll. *N. L. Bor, no. 284 bis*, 1 August 1935 (lectotype: BM 000825426! from H.N. Dixon Herbarium).

Macromitrium lingulatum Cardot & P. de la Varde, Rev. Bryol. 50: f. 18. 1. 1923. *fide* Li *et al.*, 2020. Protologue: [India] Kodai Kanal. Leg. *André 1909 no 218*. Type citation: Muscinées récoltées par le R. F. Gilbert André, Mai 1909, à Kodikanel, en Maduré (Indes Orientales), dans les Gathes, à 7000 p. et plus d'altitude, no 218 a, marked with "*Macromitrium lingulatum* Card. *Sp. nova*)" (lectotype: PC 0137781!; isolectotypes: PC 0137782!, S-B 163898!, BM 000982486!).

Macromitrium melanostomum Paris & Broth., Rev. Bryol. 35: 126. 1908, *fide* Yu *et al.*, 2012. Type protologue: T'sang Zô, au N. de Sou Tcheou. Type citation: China, T'sang Zô, au N. de Sou Tcheou, 1908, leg. *R. P. Courtois, 1081, marked as Macromitrium melanostomum n. sp.* (holotype: H-BR 2581005!).

(1) Plants medium-sized, forming dense mats; stems long creeping, with numerous short and erect branches, branches 5-9 mm long, with short branchlets at upper part, densely leaved. (2) Stem leaves deflexed to suberect when moist, ovate-lanceolate to triangular-lanceolate, narrowly attenuate; basal cells rectangular, sinuous, thick-walled, brownish, confined to a small area; medial and upper cells quadrate or subquadrate. (3) Branch leaves irregularly twisted-curved, some regularly and stiffly spirally twisted to circinate around the branch when dry, most widely spreading, some incurved and keeled above when moist, brownish or yellowish, 1.6-2.3 × 0.3-0.4 mm, ligulate, oblong to broadly lanceolate oblong, slightly incurved at apex, plicated or slightly plicated to plane, often revolute at one side; apices acute or obtuse with a mucro, occasionally broadly acuminate; upper and medial cells flat, rounded-

quadrate or rounded-hexagonal, 7.0-9.0 μm wide, thin-walled, rather obscure, distinctly pluri-papillose; low cells rectangular, 12-14.0 × 4.0-6.0 μm, rather clear, thick-walled, often with a single large papilla; basal cells near margin at one side sometimes differentiated, rectangular, pellucid with thinner walls, those at leaf bottom brownish; costa yellowish-brown, percurrent. (4) Perichaetial leaves oblong-lanceolate, acuminate to short acuminate, up to 2.8 mm long, some plicate below; medial and upper cells rounded-hexagonal, unipapillose; low cells rectangular, clear and some unipapillose. (5) Setae 4-11 mm long, smooth, twisted to the right. (6) Capsules erect, urns ovoid-obloid, ellipsoid to ellipsoid-cylindric, 1.4-1.8 × 0.7-0.9 mm, often with a purplish tier of smaller exothecial cells surrounding the mouth; opercula conic-rostrate; peristome single, exostome teeth short-lanceolate, with obtuse the apices, densely papillose throughout; spores distinctly anisosporous, smooth or papillose, 19.0-40.0 μm in diameter. (7) Calyptrae campanulate, often cucullate, large, covering the whole capsule, with moderately or densely yellowish or brownish-yellow hairs.

Macromitrium tosae Besch. is similar to *M. comatum* Mitt., the latter could be separated from the former by its smooth basal and low cells.

Distribution: China, India, Japan, Thailand.

Specimens examined: **CHINA**. Fujian: *H. H. Chung B 311* (FH); *H. H. Chung B 159* (FH), *H. H. Chung B 6158* (FH); Guangdong, *R. Magill, P. Redfearn & M. Crosby 8148* (MO); *W. T. Tsang 26133* (FH, S-B115584); Guangxi: *S. He 40511* (MO), *C. Gao 1945* (MO); Guizhou, *Cavalerie 9691* (PC 0083626); Hainan: *P. C. Chen 869-b, 869-c* (MO, PE), *P. C. Chen et al. 517* (H 3090250); Sichuan: *B. Allen 6693, 6712* (MO); Taiwan: Schrabe-Beka 95 (JE); Xizang: *Su Yong ge 2186*; Yunnan, *W. X. Xu 6574, 6597* (MO); Zhejiang: Nissioreu 29 (S-B 115594); *P. C. Wu et al. 3332* (MO, PE). **INDIA**. Madras State, Palnt Hills, Kodaikanal, *G. Foreau* (MO 2559232). **THAILAND**. Phitsanulok, *Larsen, Smitinand & Warncke 760* (MO 3971158); Payap, granitic massive Doi (Mt.) Inthanon, *A. Touw 10277* (MO 2154144).

242. *Macromitrium trachypodium* Mitt., J. Linn. Soc., Bot. 12: 213. 1869. (Figure 249)

Type protologue: [Ecuador] Andes Quitenses, Cuenca (8000 ped.), Jameson; in monte Tunguragua (8000 ped.), *Spruce, n. 90*; Type citation: Mt. Cuenca, 8000ft, *H 1294* (syntype: NY 01086608!); Tunguragua, *Spruce, s.n.* (syntypes: NY 01086605!, NY 01086606!); Ecuador, Andes Quito; on Mt. Tunguragua, 8000ft. *Spruce no. 90* (isosyntypes: E 00165165!, BM 000873205!, BM 000873206!)); Cuenca (8000 ft.), *Jameson s.n.*, H1294 (isosyntype: BM 000873209!).

(1) Plants robust, lustrous, yellowish-green above, brown-yellow below; stems weakly creeping, branches to 55 mm long, with branchlets. (2) Branch leaves tightly appressed below, erecto-patent and weakly flexuose above when dry, erect-spreading when moist, keeled below, oblong-lanceolate, upper sharply narrowed to form a long, cuspidate, arista; margins finely serrulate in upper and medial portions, entire below, usually recurved at one side below; costae stout, excurrent into the apex to form a long arista; all cells longer than wide, moderately bulging, thick-walled and strongly porose; cells gradually elongated from upper to low parts; upper cells oblong-rectangular, 8-10 μm long, in distinctly radiating diagonal rows from the costa, cells longer and narrower near costa (to 15 μm long), slightly differentiated from their ambient cells; basal cells long rectangular, 25-32 μm, thick-walled, porose, cells in basal 1/6-1/7 of leaf tuberculate, the cells in the outmost row of margin enlarged and outward "convex", hyaline but not teeth-like. (3) Perichaetial leaves are similar to branch leaves. (4) Setae distinctly papillose, 10-12 mm long, frequently with a whitish collar. (5) Capsules cupulate, ovoid, obloid to ellipsoid, strongly and distinctly furrowed, with a widely open mouth, often constricted beneath the mouth; peristome double, exostome yellowish, united to a membrane, endostome hyaline, segments lanceolate. (6) Calyptrae mitrate, naked, deeply lacerated to the upper part.

Macromitrium trachypodium Mitt. is similar to *M. ulophyllum* Mitt. in medial and upper interior cells in distinctly radiating diagonal rows from the costa, but differs from the latter by its papillose setae, cupulate and hemispherical capsules with strong furrows, and upper marginal cells of branch leaves not differentiated.

Distribution: Colombia, Ecuador, Peru, Venezuela.

Specimens examined: **ECUADOR**. Prov. Zamora. *William C. Steere and Hernrik Balslev 25836* (KRAM-B-058242, H 3090760); Loja, *S. Laegaard, E. Terneus and A. Sanchez 19256A* (MO 5238688). **VENEZUELA**. Trujillo, *D. Griffin, III & M. Lopez F. Pv-1524* (MO 5381278), Estado Trujillo, *D. Griffin, III & M. Lopez F. Pv-929* (MO 3665868).

243. *Macromitrium trichophyllum* Mitt., J. Linn. Soc., Bot. 12: 207. 1869. (Figure 250)

Type protologue: [Ecuador] Andes Quitenses, in Cordillera occidentali, Chimborazo (3000 ped.), *Spruce 92*. Type citation: Chimborazo, *Spruce s.n.* (holotype: NY 01086621!); Andes Quitenses, *Ric. Spruce 92* (isotypes: BM 0008720593!, BM 0008720594!).

(1) Plants large, rather lustrous, brownish yellow to yellow-green; stems weakly creeping; branches to 3-4 cm long. (2) Branch leaves strongly keeled, erect-clasping below, squarrosely to widely-spreading, flexuose above when

dry, erect at base and strongly abaxially curved above when moist, rather long, up to 10- 15 mm long, long linear-lanceolate with a clasping base; apices linear to setaceous, long and finely acuminate; margins serrate above, broadly recurved and entire below, erect to plane above; costae long-excurrent; all cells longer than wide, slightly bulging, clear and smooth, thick-walled and porose, gradually elongated from the upper towards the base; upper interior cells long-rectangular to rhomboidal, 25-50 µm, marginal cells not differentiated; basal cells linear rectangular, 40-60 µm long. (3) Setae smooth, about 10 mm long, twisted to the right. (4) Capsule urns globose, subglobose, ovoid to ovoid-ellipsoid, smooth to weakly longitudinally wrinkled; peristome double, exostome teeth yellow brown, endostome hyaline, both membraneous.

Macromitrium trichophyllum Mitt. is similar to *M. standleyi* E.B. Bartram in their very long and setaceous leaves, but differs from the latter by its leaves with smooth laminal cells, clasping basal portion. The leaves of *M. standleyi* are undulate, not clasping at base, with tuberculate low and basal cells. *Macromitrium trichophyllum* is also similar to *M. fuscoaureum* E.B. Bartram in its linear lanceolate leaves with erect-clasping bases, and linear setaceous the apices. However, the leaves of *M. fuscoaureum* are much shorter (6-9 mm) than those of *M. trichophyllum* (10-15 mm long). Additionally, *M. scoparium* Mitt. is somewhat similar to *M. trichophyllum*, but the former can be separated from the latter by its leaves with a differentiated border.

Distribution: Colombia, Costa Rica (Allen, 2002), Ecuador, Panama.

Specimens examined: **COLOMBIA**. Dpto. De Nariño, *B.R. Ramirez P. & M.S. Gonazlez 9.518* (MO 4430634). **ECUADOR**. Pichincha, *Marshall R. Crosby 10543* (MO 2500447); Morona-Santiqgo, *Thomas B. Croat 86566* (MO 6231912). **PANAMA**. Prov. Chiriquí, *Gordon McPherson 8065* (MO 3657421), *Thomas B. Croat 66459* (MO 3652630).

244. *Macromitrium trinitense* R. S. Williams, Bryologist 24: 65, pl. 4. 1922. (Figure 251)

Type protologue and citation: [Trinidad and Tobago] Growing on tree trunks, El Valle to San Juan, Trinidad, Bri. West Indies, *J. R. Johnston*, July 6, 1903, number 163 (Type) (isotype: MICH 525920!).

(1) Plants medium-sized, rusty-yellow to yellowish-green; stems long creeping, branches up to 30 mm long. (2) Stem leaves inconspicuous and caducous. (3) Branch leaves loosely arranged, erecto-patent below, twisted-flexuose above when dry, spreading when moist, long and narrowly lanceolate; apices long acuminate, some into a long and fine arista; margins plane and entire; costae percurrent to subpercurrent; all cells longer than wide, thick-walled, distinctly porose, in longitudinal rows; upper and medial interior cells variable in size and shape, oval to rhomboidal, moderately bulging and unipapillose; upper marginal cells linear, forming a differentiated border; low and basal cells elong-rectangular, straight-walled and tuberculate; marginal basal enlarged teeth-like cells at insertion not differentiated. (4) Perichaetial leaves broadly lanceolate to ligulate-lanceolate from a widely oblong low part, sharply narrowed to a short arista, recurved below and sheathing at base, plicate below, marginal cells near the apex irregularly serrulate; all cells long- to linear-rectangular, thick-walled and porose; cells near upper margin in distinctly radiating diagonal rows. (5) Setae short and smooth, 6-9 mm long. (6) Capsule urns globose, smooth, tapering into a somewhat ribbed neck; peristome double; spores isosporous, 22-25 µm in diameter. (7) Calyptrae mitrate, naked and smooth, slightly scabrous above, lacerate below.

Distribution: British Guinana, Costa Rica, Suriname, Trinidad and Tobago.

Specimens examined: **BRITISH GUINANA**. Kaieteur. E.F. Noel 89 (BM 000873204). **COSTA RICA**. Limon, Gerrit Davidse et al. 28769 (MO 3090765). **SURINAME**. Sipaliwini. B. Allen 20544 (KRAM-B-135975, MO); Lelygebergte, Plateau V. Mountain savanna-forest, J. P. A. Florschutz 4851 (H 3090764).

245. *Macromitrium tuberculatum* Dixon, Hong Kong Naturalist, Suppl. 2: 13. 4. 1933. (Figure 252)

Type protologue and citation: [China] Tai Mo Shan, Hong Kong New Territories, 4th January, 1931; coll. *Youngsaye* (*Herklots 297 O*) (holotype: BM 000576127!); *Ibidem* (*297 B*) (syntype: BM 000576128!). White Cloud Mountains 800 ft. alt., Canton, 26th December, 1930, coll. *Youngsaye* (*herklots 302D*) (syntype: BM 000576129!).

(1) Plant medium-sized, yellowish brown upper, blackish below, in mats; stem long creeping, up to 10 cm long; branches erect, 10-15 mm high, 1.5-2.0 mm thick. (2) Branch leaves moderately crisped and contorted when dry, widely spreading when moist, 2.5-3.5 × 0.4-0.6 mm, lanceolate, slightly incurved and distinctively keeled, somewhat plicated below, ; apices acute or broadly acuminate; margins revolute at one side below; costae single, excurrent, percurrent or disappearing beneath the apex; lamina unistratose throughout; upper and medial cells isodiametric, rounded-quadrate (9.0-12.0 µm), moderately bulging to conic-bulging, slightly obscure, pluripapillose; low and basal cells long rectangular or linear, 22-30 × 6.0-8.0 µm, thick-walled, porose, strongly tuberculate. (3) Inner perichaetial leaves lanceolate, acuminate, shorter than branch leaves; outer perichaetial leaves longer than branch leaves, 5.5 mm long, all lamnial cells longer than wide, thick-walled and sinuous, conic-bulging, unipapillose. (4) Setae (4.0)7.0-8.0 mm long, smooth, twisted to the right; vaginulae densely hairy. (5) Capsule urns long ellipsoid or cylindric, 1.3-1.5

× 0.50-0.65 mm, brownish, constricted beneath the mouth in 4-angled or 4-furrowed shape; peristome absent; spores anisosporous. (6) Calyptrae large, cucullate, 2.0-2.5 mm long, with long and dense, brown-yellowish hairs.

Macromitrium tuberculatum Dixon is similar to *M. holomitrioides* Nog., *M. ousiense* Broth. & Paris, *M. gymnostomum* Sull. & Lesq., and *M. formosae* Cardot. However, the calyptrae of *M. gymnostomum* are naked; the medial cells of branch leaves for *M. holomitrioides* conic-bulging, larger, collenchymatous; the cells smooth from the apex to the base for *M. ousiense*; the basal cells of branch leaves are smooth for *M. formosae*. In fact, *M. tuberculatum* is similar to *M. schmidii* Müll. Hal., only slightly different in the capsules, and the degrees of papillosity in upper and medial portions. Thererfore, more work is needed about their relationship.

Distribution: China.

Specimens examined: **CHINA**. Anhui, *Guan ke-Jian, no. 11* (MO 5276970); Guangdong: Lin Fa Shan, *W. T. Tsang, no. 25592* (FH); Hei-yang District, *W. T. Tsang, no. 26042* (FH); Sam Kok Shan (Tsung-fa) District, *W. T. Tsang, no. 24994 a* (FH); Lin Fan Shan, *W. T. Tsang, no. 26032* (FH); Ngok Shing Shan, *Y. W. Taam 402* (NY, MO 5130513); Cong Hua County, *Peng Bao Xiong, no. 72* (MO 5918179); Hainan, Le Dong County, *W. D. Reese, no. 17682* (MO 5643689); Hunan, Yi Zhang County, *Chen Shao-qing, no. 2785* (MO 4440054).

246. ***Macromitrium turgidum*** Dixon, J. Siam Soc., Nat. Hist. Suppl. 9: 22. 1932. (Figure 253)

Type protologue and citation: [Thailand] *Chantaburi*. Krāt, Kao Kūap, on tree in evergreen forest, c. 700 m. alt., 27 Dec., 1929, Coll. *A. F. G. Kerr (438)*, type (holotype: BM 000825435!). Rāchaburī, Prachūap, Kao Lūang, on tree in open evergreen forest, 5 July, 1926, coll. *A. F. G. Kerr (151)* (paratype: BM 000825434!).

Macromitrium turgidum var. *laeve* Dixon, J. Bombay Nat. Hist. Soc. 39: 778. 1937. Type protologue: "Dafla Hills, 1,200m., March 1934 (90) Tako Senyak, 1,200 m.; October 1935 (303); Paora, Naga Hills, on tree trunk, 2,150 m., September 1935 (283 bis)." Type citation: Tako Senyak, Assam; 4000', *N. L. Bor 113*, 19 Mar. 1934, Herb H. N. Dixon (lectotype: BM 000825438!); Daffla Hills, Assam, 4000', *N. L. Bor 90*, 20 Mar. 1934, Herb. H.N. Dixon (syntype: BM 000825437!).

(1) Plants robust, brown-yellowish or yellowish-green above, dark brown below; stems long creeping, densely covered with brown-reddish rhizoids; branches up to 2.5 cm long. (2) Stem leaves inconspicuous and caducous, with rhizoids at base. (3) Branch leaves loosely erect, with spreading curved to somewhat deflexed and contorted apices when dry, not evenly spreading when moist, 4.5-6.5 mm long, from lanceolate or oblong base gradually tapered to a long narrowly acuminate apex, narrowly recurved at places, sometimes undulate at upper margin; costae percurrent; upper and medial cells rounded quadrate to elliptic, 4.0-7.0 × 4.0-7.0 μm, thick-walled, clear and smooth, in longitudinal rows; low cells long-rectangular, thick-walled, 15-30 × 5.0-7.0 μm, lumens straight, hyaline and distinctly tuberculate. (4) Perichaetial leaves ligulate-lanceolate or ovate-oblong, erect, acute or short-acuminate, with an excurrent costa; all cells longer than wide, smooth; upper and medial cells rectangular, 21.0-42.0 × 7.0-9.0 μm, with lumens 2.0-3.0 μm wide, thick-walled, those near costa at base rather large, 30.0-70.0 × 12.0-15.0 μm, straight and thin-walled, pellucid. (5) Setae 4.0-7.0 mm, smooth, twisted to the left. (6) Capsules erect, urns ovoid, obloid to ellipsoid, constricted under the mouth, moderately plicate; peristome double, often disintegrated and membraneous; exostome yellowish papillose; endostome pellucid; opercula conic-rostrate, 0.7-0.8 mm long; spores isosporous, 20-28 μm in diameter, smooth and somewhat pellucid. (7) Calyptrae large, mitrate, covering the whole capsule, naked and smooth, lacerate (Guo *et al.*, 2012).

Macromitrium turgidum Dixon is very similar to *M. sulcatum* (Hook.) Brid. These two species develop robust plants, sulcata capsules (ovate, constricted under the mouth, moderately plicate), double peristomes, naked and smooth calyptrae, and basal cells of branch leaves long-rectangular, thick-walled and tuberculate, onlythe branch leaves of *M. sulcatum* are often oblong-lanceolate with acute apices, and its medial and upper cells are smaller and arranged in diagonal rows in its type specimen, while those of *M. turgidum* are linear-lanceolate, lanceolate to ovate- or oblong- lanceolate, acuminate, or gradually tapered to long narrowly acuminate, and the medial and upper cells arranged in longitudinal rows. *Macromitrium lorifolium* Paris & Broth., which has been reported from Vietnam (Tan & Iwatsuki, 1993), is also similar to *M. turgidum*, only the branch leaves of the former are strongly curved even when moist, and with distinctly denticulate apices. More work is need to confirm the identity of *M. turgidum* and *M. lorifolium*.

Distribution: China, India (Gangulee, 1976), Thailand, Vietnam.

Specimens examined: **CHINA**. Xizang, *Y. G. Su 4833* (MO 3676218); *Y. G. Su 4641* (MO 3675104). **THAILAND**. Northern, Chiang Mai, *J. F. Maxwell B-148* (MO); Kao Bangto, Pang-nga, Siam, *A. F. G. Kerr 423* (BM 000825436). **VIETNAM**. Ha Giang, *L.V. Averyanov NT B024* (MO 5244297).

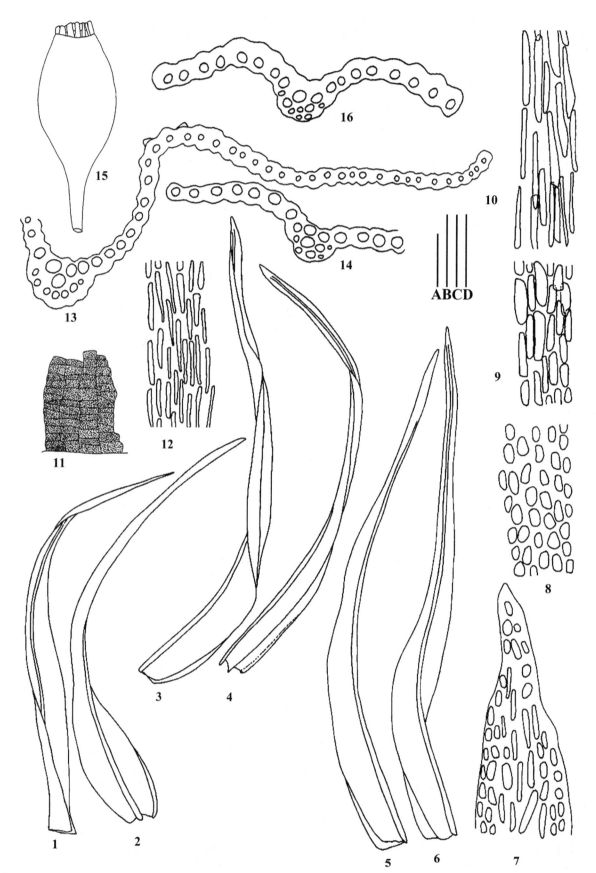

Fig. 245 *Macromitrium thwaitesii* Broth. ex M. Fleisch. 1-6: Branch leaves. 7: Apical cells of branch leaf. 8: Upper cells of branch leaf. 9: Medial cells of branch leaf. 10: Basal cells of branch leaf. 11: Peristome. 12: Low cells of branch leaf. 13: Basal transverse section of branch leaf. 14, 16: Upper transverse sections of branch leaves. 15: Capsules (all from lectotype, H-BR 2630024). Line scales: A = 0.44 mm (1-6); B = 0.88 mm (15); C = 140 μm (11); D = 44 μm (7-10, 12-14, 16).

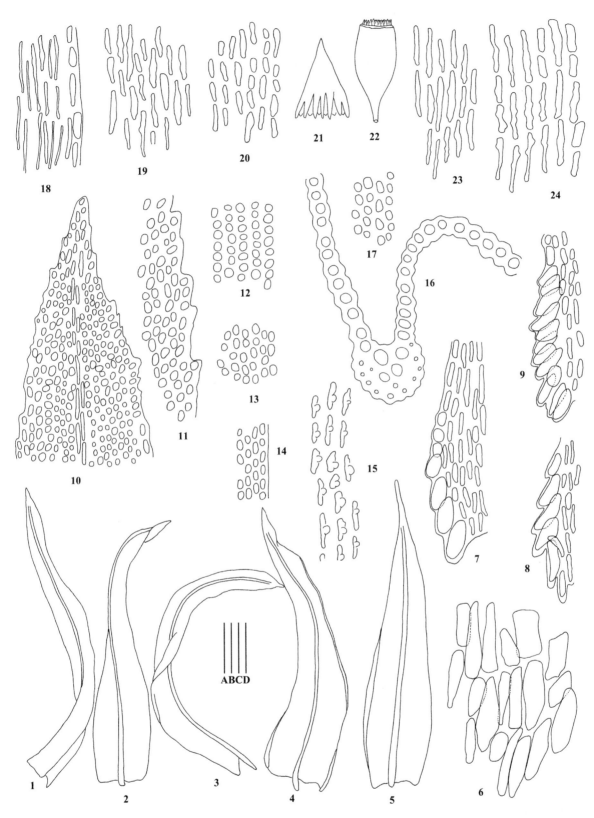

Fig. 246 *Macromitrium tocaremae* Hampe 1-4: Branch leaves. 5: Perichaetial leaf. 6: Basal cells near costa of branch leaf. 7: Basal marginal cells of perichaetial leaf. 8-9: Basal marginal cells of branch leaves. 10: Apical cells of branch leaf. 11: Apical cells near maegin of branch leaf. 12: Medial cells of branch leaf. 13: Upper cells of branch leaf. 14: Upper marginal cells of branch leaf. 15: Basal cells of branch leaf. 16: Basal transverse section of perichaetial leaf. 17: Upper cells of perichaetial leaf. 18: Basal marginal cells of perichaetial leaf. 19: Low cells of perichaetial leaf. 20: Medial cells of perichaetial leaf. 21: Calyptra. 22: Capsule. 23: Basal cells of perichaetial leaf. 24: Basal cells of branch leaf (all from isotype, BM 000873213). Line scales: A = 4 mm (21-22); B = 1 mm (1-5); C = 100 µm (6, 10); D = 67 µm (7-9, 11-20, 23-24).

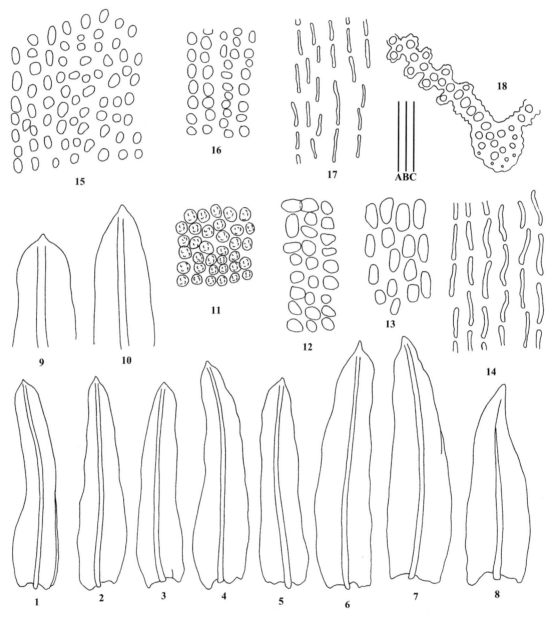

Fig. 247 *Macromitrium tongense* Sull. 1-5: Branch leaves. 7-8: Perichaetial leaves. 9-10: Apices of branch leaves. 11: Upper cells of branch leaf. 12: Medial marginal cells of branch leaf. 13: Medial cells near costa of branch leaf. 14: Basal cells of branch leaf. 15: Upper cells of perichaetial leaf. 16: Medial cells of perichaetial leaf. 17: Basal cells of perichaetial leaf. 18: Medial transverse section of branch leaf (all from isotype, US 00070281). Line scales: A = 1 mm (1-8); B = 400 μm (9-10); C = 67 μm (11-18).

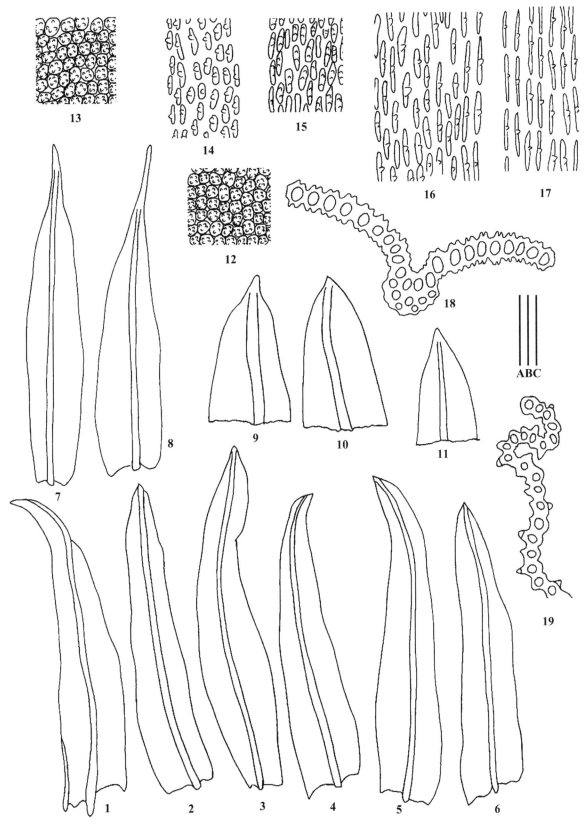

Fig. 248 *Macromitrium tosae* Besch. 1-6: Branch leaves. 7-8: Perichaetial leaves. 9-11: Apices of branch leaves. 12: Upper cells of branch leaf. 13: Medial cells of branch leaf. 14-15: Low cells of branch leaves. 16-17: Basal cells of branch leaves. 18: Upper transverse section of branch leaf. 19: Low transverse section of branch leaf (all from isotype, H-BR 2581002). Line scales: A = 0.44 mm (1-8); B = 44 μm (12-19); C= 0.176 mm (9-11).

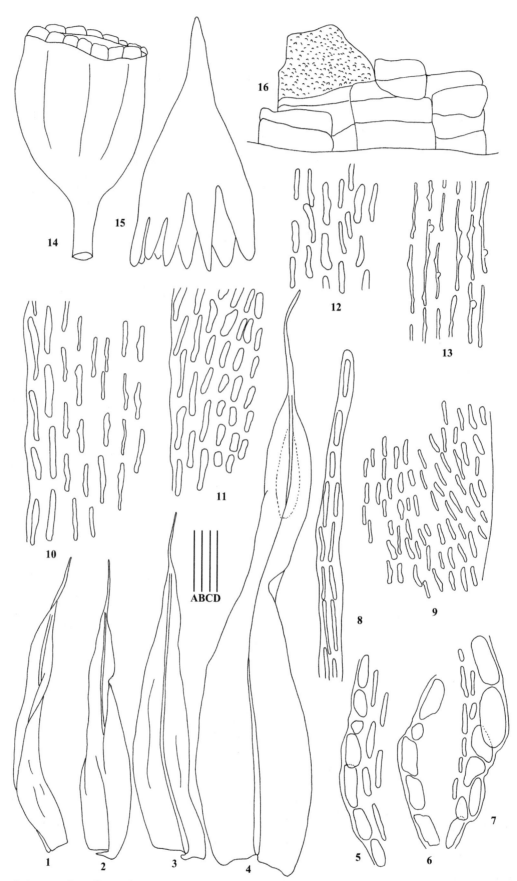

Fig. 249 *Macromitrium trachypodium* Mitt. 1-4: Branch leaves. 5-7: Basal marginal cells of branch leaves. 8: Apical cells of branch leaf. 9: Apical marginal cells of branch leaf. 10: Low cells of branch leaf. 11: Upper cells of branch leaf. 12: Medial cells of branch leaf. 13: Basal cells of branch leaf. 14: Capsule. 15: Calyptra. 16: Peristome (1-13 from syntype, NY 01086605; 14-16 from NY 01086607). Line scales: A = 2 mm (1-3, 14-15); B = 1 mm (4); C = 100 μm (8, 16); D = 67 μm (5-7, 9-13).

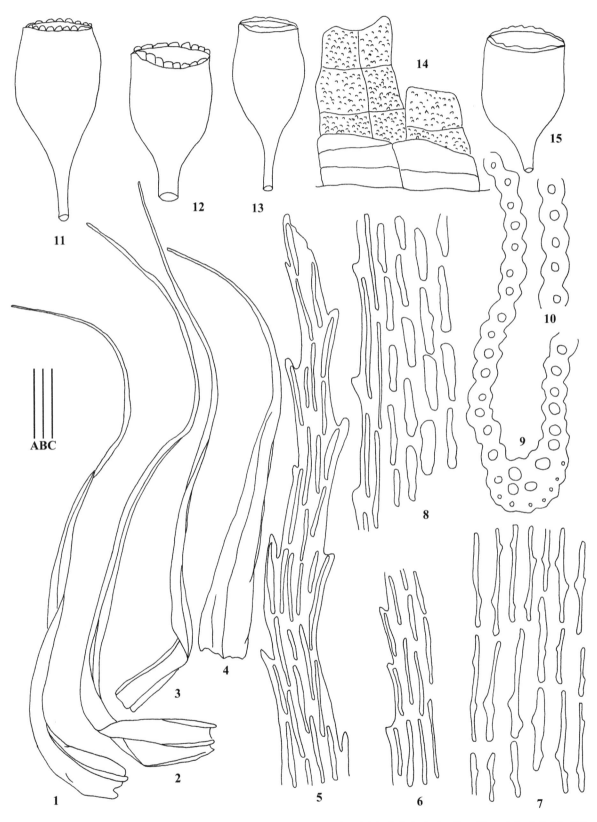

Fig. 250 *Macromitrium trichophyllum* Mitt. 1-3: Branch leaves. 4: Perichaetial leaf. 5: Apical cells of branch leaf. 6: Medial cells of branch leaf. 7: Basal cells of branch leaf. 8: Upper cells of branch leaf. 9, 10: Basal transverse sections of branch leaves. 11-13, 15: Capsules. 14: Peristome (all from isotype, NY 01086622). Line scales: A = 2 mm (1-4, 11-13, 15); B = 100 μm (14); C = 67 μm (5-10).

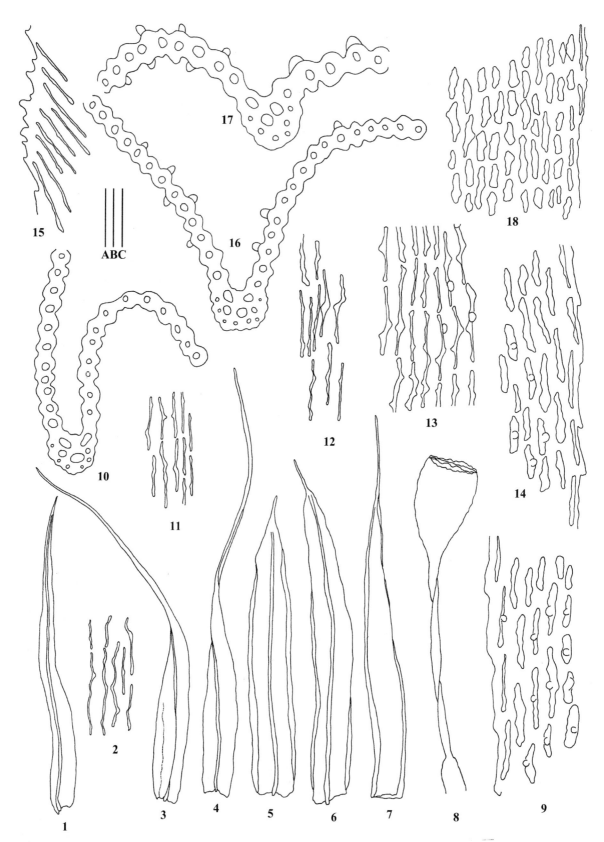

Fig. 251 *Macromitrium trinitense* R. S. Williams 1, 3-4: Branch leaves. 2: Medial cells of perichaetial leaf. 5-7: Perichaetial leaves. 8: Capsule. 9: Apical cell of branch leaf. 10: Medial transverse section of branch leaf. 11: Basal cells of perichaetial leaf. 12: Upper cells of perichaetial leaf. 13: Basal cells of branch leaf. 14: Upper cells of branch leaf. 15: Apical cells of perichaetial leaf. 16-17: Upper transverse sections of branch leaves. 18: Medial cells of branch leaf (all from isotype, MICH 525920). Line scales: A = 2 mm (1, 3-4, 8); B = 1 mm (5-7); C = 67 μm (2, 9-18).

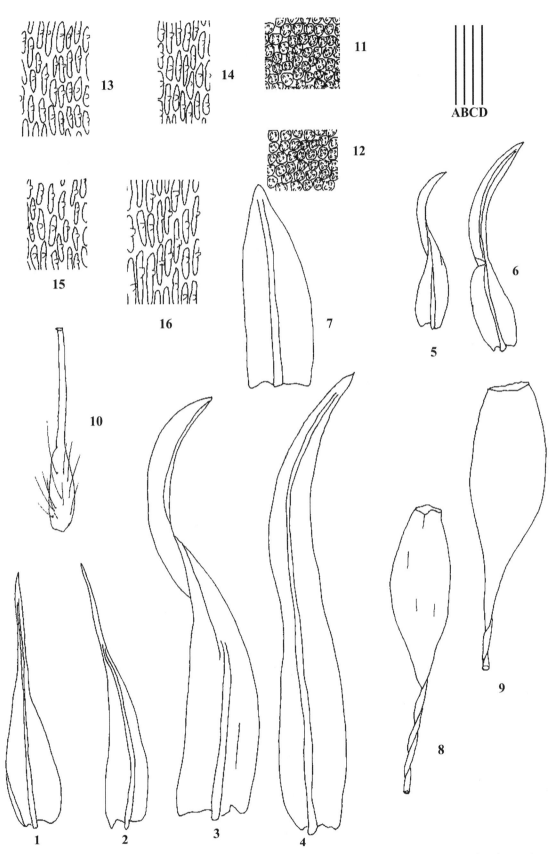

Fig. 252 *Macromitrium tuberculatum* Dixon 1-6: Branch leaves. 7: Branch leaf apex. 8-9: Capsules. 10: Vaginual. 11: Upper cells of branch leaf. 12: Medial cells of branch leaf. 13: Low cells of branch leaf. 14: Upper cells of perichaetial leaf. 15: Medial cells of perichaetial leaf. 16: Low cells of perichaetial leaf (all from holotype, BM 000576127). Line scales: A = 1 mm (5-6, 8-10); B = 0.4 mm (1-4); C = 0.2 mm (7); D = 67 μm (11-16).

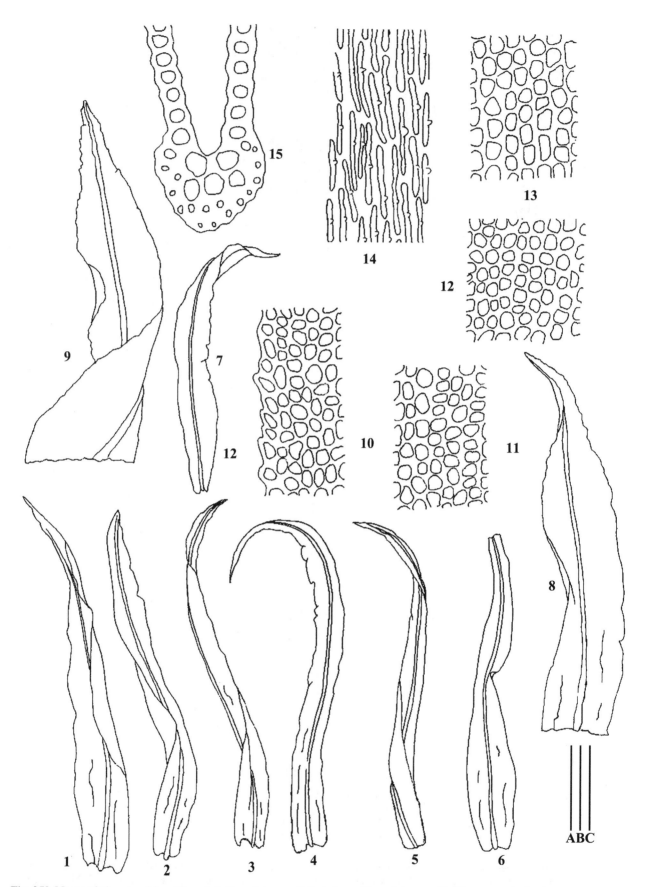

Fig. 253 *Macromitrium turgidum* Dixon 1-7: Branch leaves. 8-9: Apices of branch leaves. 10: Upper marginal cells of branch leaf. 11-12: Upper cells of branch leaf. 13: Medial cells of branch leaf. 14: Low cells of branch leaf. 15: Medial transverse section of branch leaf (all from *P. Z. Zhen, 29713*. IBSC). Line scales: A = 1 mm (1-7); B = 200 μm (8-9); C = 40 μm (10-15). (Guo *et al.*, 2012).

Taxonomy

247. *Macromitrium tylostomum* Mitt. ex Broth & Sande Lac., Bryol. Jav. 1: 131. 109. 1860.

Type protologue: [Indonesia] Habitat insulam *Javae* Zollinger itin. Jav. secundi *coll. no. 3717.*

Macromitrium hooglandii E.B. Bartram, Brittonia 11: 92. 1954, *fide* Vitt *et al.*, 1995. Type protologue: Western Highlands; Nondugl, on low branch of isolated tree, about 1 600 m, *R. D. Hoogland 3195.*

(1) Plants medium-sized; stems prostrate, becoming indistinct when old, frequently broken to form separate simple to rarely forked branches, branches up to 20 mm long. (2) Branch leaves loosely and irregularly twisted-contorted, with an inrolled to involute upper portion, sometimes strongly spirally-twisted around the branch when dry, loosely erect with incurved, keeled upper portion from a spreading low portion when moist, 4.1-5.0 mm long, ligulate to lanceolate-ligulate; apices bluntly acuminate to acute, stoutly mucronate; margins plane to broadly reflexed, entire; costae shortly excurrent, percurrent or ending several cells beneath the apex; all cells rounded except for a small area at insertion; upper cells 9-12 µm long and wide, rounded-quadrate, bulging, clear, smooth; medial cells 11-13 × 9-16 µm, irregularly rounded-elliptic to rounded-quadrate, bulging, clear, smooth; basal cells 12-15 × 12-24 µm, short-rhombic to elliptic-rhombic, unevenly thick-walled and porose, distinctly tuberculate. (3) Perichaetial leaves 4.0-4.5 mm long, narrowly lanceolate, long-acuminate, cells similar to those of branch leaves. (4) Setae shorter, 1.8-4.0 mm long, straight, not twisted. (5) Capsule urns narrowly ellipsoid- cylindric to ovoid-ellipsoid, smooth; peristome single, exostome of 16, rudimentary; spores anisosporous. (6) Calyptrae oblong-conic, mitrate, plicate, densely hairy (Vitt *et al.*, 1995).

Distribution: Indonesia, Papua New Guinea, Philippines (Tan & Shevock, 2015).

Specimens examined: **INDONESIA**. Java, *R. S. Williams 24456* (H-BR 2596005). **PAPUA NEW GUINEA**. Morobe, *H. Streimann 19318* (H 3196871); Eastern Highlands, *H. Streimann 18034* (H 3196872).

248. *Macromitrium ulophyllum* Mitt., J. Linn. Soc., Bot. 12: 206. 1869. (Figure 254)

Type protologue and citation: [Ecuador] Andes Quitenses, Abitagua (6000 ped.), *Spruce 94* (holotype: NY 00322438!).

(1) Plants large and robust, lustrous, brown yellow, in somewhat wefts; stems creeping, branches prostrate, up to 2 cm long and 1.8 mm thick, sparsely with reddish rhizoids. (2) Branch leaves not very shriveled, slightly keeled, erecto-appressed below, spreading and flexuose, undulate above when dry, erecto-patent when moist, sometimes rugose above, 6-7 × 1-1.5 mm, broadly oblong-lanceolate to broadly lanceolate, narrowed towards the apex to form an arista (varying in length); costae excurrent into a long arista or cuspidate; margins undulate, finely and evenly serrulate above; all cells longer than wide, rectangular to long rectangular, thick-walled, distinctly porose; upper interior cells 40-60 × 12 µm, oblong-rectangular, in distinctly radiating diagonal rows from the costa (appearing as V- shape), elongate-rhombic, thick-walled and porose; upper marginal cells narrower than their ambient cells, forming a distinct border; basal cells 30-70 µm long, long-rectangular, thick-walled and porose, smooth or weakly tuberculate. (3) Setae 15-20 mm long, smooth. (4) Capsules 1.5-2 mm long, obovoid to ellipsoid, smooth; peristome double, exostome teeth lanceolate, densely papillose, brown, united to form a membrane, endostome hyaline, weakly papillose, basal membrane low or absent, segments linear-lanceolate; spores isosporous. (5) Calyptrae large and mitrate, about 4 mm long, naked, deeply lacerate.

Macromitrium ulophyllum Mitt. is rather similar to *M. trachypodium* Mitt. in their robust plants, broadly oblong-lanceolate leaves with a long cuspidate or arista, upper interior cells in distinctly radiating diagonal rows from the costa, all cells long than wide, thick-walled and porose, However, *M. trachypodium* differs from *M. ulophylum* by its papillose setae, furrowed capsules.

Distribution: Brazil, Colombia, Costa Rica, Ecuador, Guyana, Panama (Allen, 2002), Peru, Venezuela.

Specimens examined: **BRAZIL**. *Bassett Maguire, Celia K. Maguire 60499M* (MO 2554213). **COLOMBIA**. Antioquia, *Ricardo Callejas & et al. 10088, 10616* (MO), *S. P. Churchill & Julio C. Betancur B. 18661* (MO); Antioquia, *R. Callejas 10616* (MO 4424126). **COSTA RICA**. Guanacaste, *Barry hammel 17649* (MO 3686996); Prov. Alajuela, *Gerardo Herrera Ch. 519* (MO 3653442). **ECUADOR**. Prov. Azuay, *L. Holm-Nielsen et al. 29513* (MO 3655546); Prov. Tungurahua, *L. Holm-Nielsen & J. Jaramillo 28409* (MO 3655940). **GUYANA**. Mazaruni-Potaro, *Ronald Liesner 23299A* (MO 4411472); Upper Mazaruni District, *S. R. Gradstein 5406* (MO 5215142). **PERU**. Cajamarca, San Ignacio, *E. Rodriguez R. 1790A* (MO 6095408); *Eric Rodriguez R. et al. 2842* (MO 6091956). **VENEZUELA**. Cerro de la Neblina, *Bassett Maguire, John J. Wurdack & George S. Buating 37068* (MO 2554218); Territorio Federal Amazonas, *Julian Steyermark 129619* (MO 3082417); Amazonas, *Ronald Liesner, German Carnevali 22954* (KRAM-B-136219); Ilu-Tepui, *Bassett Maguire 33555* (MO 2551218).

249. *Macromitrium uraiense* Nog., J. Sci. Hiroshima Univ., Ser. B, Div. 2, Bot. 3: 140. 3. 1938. (Figure 255)

Type protologue and citation: [China] Taiwan: Urai, Prov. Taihoku (*A. Noguchi, no. 573-type*: in Herb. Hiros. Univ., July 1928) (holotype: HIRO; isotype: NICH 365246!).

(1) Plant medium-size to robust; stems creeping, with many branches, densely covered with reddish rhizoids; branches with leaves 2.5 mm thick, terminal obtuse. (2) Stem leaves oblong-lanceolate, acuminate, costae excurrent; basal marginal cells usually with reddish, branched rhizoids. (3) Branch leaves twisted-contorted-crisped when dry, widely spreading when moist, 2.7-4.0 mm long, linear-lanceolate from an oblong low portion, acuminate in upper portion, often fragile and broken; basal marginal cells usually with reddish, branched rhizoids; upper cells isodiametric, rounded-elliptic to rounded quadrate, clear, conic-bulging, mammillose; medial cells rounded-elliptic, short rectangular, bulging, unipapillose; low cells rectangular, thick-walled, porose, strongly tuberculate. (4) Inner perichaetial leaves shorter, triangular-lanceolate or oblong-lanceolate, acuminate, with a wide low base, 2.5 mm long. (5) Setae smooth and slender, 4-5 mm long. (6) Capsules ellipsoid, smooth. (7) Calyptrae mitrate, naked, scabrous at beak, lacerate at the base.

Macromitrium uraiense Nog. is similar to *M. taiwanense* Nog., but the branch leaves of the former long lanceolate, acuminate in upper portion with mammillose medial and upper cells, and without an awn.

Distribution: China (Taiwan).

250. *Macromitrium urceolatum* (Hook) Brid., Bryol. Univ. 1: 312. 1826. (Figure 256)

Basionym: *Orthotrichum urceloatum* Hook., Musc. Exot. 2: t. 124, f. 1-8. 1819. Type protologue: [St. Helena] In insula Sanctae helenae, *D. Menzies, 1795*, Type citation: St. Helena, leg. *D. Menzies 1795* (isotype: E 00011670!).

(1) Plants medium-sized, primarily stems inconspicuous, branches stout and thick, with obtuse terminal. (2) Branch leaves erect to obliquely appressed below, spirally curved to twisted and curly above when dry, widely spreading when moist, narrowly lanceolate, bluntly acuminate to acute, occasionally broken in upper portion when old; upper and medial cells rounded to rounded-quadrate, pluripapillose, becoming elongate towards the base; low and basal cells narrowly rectangular, thick-walled, occasionally curve-sigmoid, thick-walled, smooth; costa percurrent, partially concealed in adaxial view by overlapping folds of laminae in the upper portion when moist. (3) Perichaetial leaves shorter than branch leaves, erect and acute. (4) Setae about 15 mm long, smooth, twisted to the left. (5) Capsule urns smooth, ovoid to urceolate, constricted and slightly plicate beneath the mouth, forming a dark-brown, tubuliform upper portion when dry; peristome single, hyaline, lanceolate, about 200 μm. (6) Calyptrae mitrate, plicate and naked, lacerate at the base.

Distribution: St. Helena.

251. *Macromitrium validum* Herzog, Biblioth. Bot. 87: 66. 1916. (Figure 257)

Type protologue: [Bolivia] Im Bergwald zwischen San mateo und Sunchal, about 2500 m, *No. 4465*. Type citation: Bolivia, Santa Cruz: Im Bergwald zwischen San Mateo und Sunchal, about 2500 m, *Herzog, T. C. J.*, *4465* (holotype: JE 04008704!).

(1) Plants robust, brownish; stems weakly creeping, branches to 20 mm long, about 2 nn thick. (2) Branch leaves erect below, squarrosely to abaxially-curved spreading, twisted-flexuose above when dry, spreading and slightly abaxially curved and occasionally conduplicate above when moist, long linear-lanceolate, 5-6 mm long, often plicate below; apices acuminate to setaceous-acuminate; margins serrulate above, entire below; costae slender, rusty-reddish, subpercurrent or vanishing beneath the apex; all cells longer than wide, thick-walled, strongly porose from the apex to the base; upper cells elliptical, rhombic, rectangular, strongly bulging, unipapillose; upper marginal cells narrower and longer than their ambient cells, slightly differentiated to form a border; medial cells rectangular to long rectangular, bulging and unipapillose; low and basal cells elongate, strongly porose and tuberculate. (3) Perichaetial leaves shorter but broader than branch leaves, lingulate-lanceolate; apices cuspidate to acuminate; all cells strongly thick-walled, linear elongate, porose, strongly tuberculate at low and basal portion.

Macromitrium validum Herzog is similar to *M. echinatum* B.H. Allen and *M. fuscoaureum* E.B. Bartram in linear-lanceolate branch leaves with a setaceous-acuminate acumen, all cells longer than wide. However, *M. validum* differs from *M. echinatum* in having leaves with strongly bulging and papillose upper cells, and strongly tuberculate basal and low cells; all cells strongly porose from the apex to the base. *Macromitrium standleyi* E.B. Bartram and *M. trichophyllum* Mitt. also have linear-lanceolate leaves with all cells longer than wide, but their leaves are much longer (10-15 mm for *M. trichophyllum*, 18 mm for *M. standleyi*) than those of *M. validum* (5-6 mm long).

Distribution: Bolivia.

Specimen examined: **BOLIVIA**. La Paz: Franz Tamayo, *A. Fuentes 5124* (MO 5647776).

252. *Macromitrium vesiculosum* Tixier, Rev. Bryol. Lichénol. 34: 143. f. 10. 1966. (Figure 258)

Type protologue: [Vietnam] Chapa, sur branches d'arbres, 1 500 m, janvier 1928 (*A. Petelot 86*) (holotype: PC 0083737!; isotypes: PC 0083738!, PC 0083739!).

(1) Plant robust, rust-brownish or brown-reddish; stem long creeping, up to 6.5 cm long, in dense mats or wefts, densely with numerous reddish rhizoids; branches erect, irregularly branched, 15-20 mm long. (2) Stem leaves slightly crisped when dry, spreading or contorted-spreading when moist, oblong-lanceolate, acuminate or acute, 2.6-

3.2 × 0.3-0.5 mm, with many reddish rhizoids at the base; margins plane and entire, sometimes narrowly recurved below; all cells isodiametric (8.0-10.0 μm), conic-bulging or mammillate; costae percurrent or excurrent. (3) Branch leaves spirally coiled, crisped and contorted when dry, spreading or weakly contorted spreading, 2.6-3.8 × 0.3-0.5 mm, long lanceolate, plicate below; margins entire and plane throughout; upper and medial cells isodiametric, round, rounded-quadrate, 12.0-14.0 μm, thin-walled and clear, conic-bulging, mammillate; transitional cells slightly elongate, bulging and strong tuberculate; basal cells rounded-elliptic to sub-rectangular, short rectangular, strongly tuberculate. (4) Perchiaetial leaves ovate, gradually acuminate to a long arista up to 2.4 mm long; all cells rounded-quadrate, subquadrate, clear, unipapillose. (5) Setae erect, smooth, 6-8 mm long, twisted to the left. (6) Capsule urns obloid or ovoid-obloid, contracted at the mouth, plicate above when dry; peristome single, exostome of 16, short lanceolate, papillose; spores about 30.0 μm, papillose. (7) Calyptrae mitrate, plicate, with numerous thick hairs.

Macromitrium vesiculosum Tixier is similar to *M. petelotii* Tixier in rust-brownish habits, creeping stems densely with reddish rhizoids, conic-bulging or mammillate upper and medial lamina cells, tuberculate basal cells, but the latter can be separated from the former by its longer branch leaves (5.0-6.0mm), shorter setae (2.5-3.0 mm).

Distribution: Vietnam.

Specimen examined: **VIETNAM**. Chapa, *Petelot*, Janv. 1926 (PC 0083739).

253. *Macromitrium viticulosum* (Raddi) Brid., Bryol. Univ. 1: 738. 1826.

Basionym: *Schlotheimia viticulosa* Raddi, Critt. Bras. 4. 1822. Type protologue: (Brazil: Rio de janeiro) Corcovado, *Raddi s.n.*

Macromitrium didymodon Schwagr., Sp. Musc. Frond., Suppl. 2 2(2): 138. pl. 190. 1827, *fide* Grout, 1944. Type protologue: Brazil. Legit ad vicum Novo Friburgo Brasiliae capsulis matures mense Decembri dilig. Beyrich. Type citation: Brasilia, Novo Friburgo, leg. *Beyrich*. Dec. 1822 (isotype: JE 04008689!).

***Macromitrium dentatum* Schimp., Mém. Soc. Natl. Sci. Nat. Cherbourg 16: 190. 1872 (invalid name), *fide* Vitt, 1994.

Macromitrium domingense A. Jaeger, Ber. Thätigk. St. Gallischen Naturwiss. Ges. 1872-73: 149. 1874, *fide* Grout, 1944.

***Macromitrium glaucescens* Schimp. Mém. Soc. Natl. Sci. Nat. Cherbourg 16: 190. 1872, *fide* Vitt, 1994.

Macromitrium goniopodium Mitt., J. Linn. Soc., Bot. 12: 198. 1869, *fide* Grout, 1944. Type protologue: Brasilia tropica, ad cortices. *Burchell 1003*.

Macromitrium insularum Mitt., J. Linn. Soc., Bot. 12: 200. 1869, *fide* Grout, 1944. Type protologue: Ins. Guadelupe, *Parker s.n.*, herb. Hook.

Macromitrium intortifolium Hampe, Bot. Zeitung (Berlin) 20: 362. 1862, *fide* Valente *et al.*, 2020. Type protologue: Sta. Catharina Brasilia, *Dr. Blumenau*. Type citation: Brasilia, Sa Catharina (isotypes: BM 000873097!, BM 000873098!); Hab Sta Catharina, Brasilia, *Blumenau* (isotype: NY 01202032!).

Macromitrium leptophyllum Besch., Mém. Soc. Natl. Sci. Nat. Cherbourg, 16: 190. 1872, *fide* Grout, 1944.

Macromitrium rhabdocarpum Mitt., J. Linn. Soc., Bot. 12: 199. 1869, *fide* Grout, 1944. Type protologue: Andes Quitenses, Attobos, *Spruce s.n.*

Macromitrium richardii Schwägr., Sp. Musc. Frond., Suppl. 2 (2,1): 70-71, pl. 173 [bottom]. 1826, *fide* Valente *et al.*, 2020. Type protologue: In arboribus Guianae lectum ded. beatus Claud. Richardus, 4.

Macromitrium tenellum Cardot, Rev. Bryol. 36: 109. 1909, *fide* Grout, 1944. Type protologue: Etat de Vera Cruz: pres de Jalapa, en mélange avee l'espèce suivante, 1908 (n. 15146 in parte).

(1) Plants small to medium-sized, yellowish green above, brown below; stems long creeping with inconspicuous and caduceus leaves, branches short, 5-8 mm long. (2) Branch leaves erect below, twisted-flexuose-curved above, the apices curved to circinate above when dry, erecto-patent when moist, 1.5-2.0 × 01.3-0.5 mm, oblong, ligulate, oblong-lanceolate; apices acute to obtuse or apiculate; margins crenulate, plane or reflexed below, plane above, enlarged basal teeth-like cells not differentiated at insertion; costae subpercurrent to percurrent; upper and medial cells rounded hexagonal to rounded quadrate, 7-12 μm, pluripapillose in upper 1/2 to 2/3 of the leaf, bulging mammillose below, upper marginal cells not differentiated; basal cells 15-30 μm long, rectangular, thick-walled, smooth. (3) Capsules urns ellipsoid to ovoid, furrowed when dry, puckered at mouth; peristome single, exostome rudimentary, teeth lanceolate truncate, hyaline papillose; spores isosporous, papillose. (4) Calyptrae mitrate, lacerate, naked to sparsely hairy.

Macromitrium glaziovii Hampe was synonymized with *M. viticulosum* (Raddi) Brid. by Valente *et al.* (2020). However, the basal cells of branch leaves for the former unipapillose and those of the latter smooth.

Distribution: Brazil, Colombia, Ecuador, French Guiana, Mexico, South Africa (van Rooy, 1990), United States, Venezuela.

Specimens examined: **BRAZIL**. RS-Mun. de Bom jesus-Fazenda do Cilho, *R. Wasum 2324 a* (MO 5367040); RS-Mun. Sao Francisco de Paula-Floresta Nacional, *R. Wasum et alii, 9948* (MO 5360359). **COLOMBIA**. Cauca

Valley, thicket, *F.W. Pennell 6327* (MO 5279732); Cauca: Popayan Municipio, *S. Churchill & W. Rengifo M. 17351* (H 3194588); Santander: Oiba Municipio, musgos sobre troco, en la sombra, *S.P. Churchill 18764* (MO 4461970), *S. P. Churchill, P. Franco & J. D. Parra 18764* (MO 4434733). **ECUADOR**. Sobre rocas en lugares humedos, Zamora, *Alberto Ortega, 542* (MO 2555404), effito, cutucu, *Alberto Ortega, 470* (MO 2555659). **MEXICO**. San Luis Potosi, *D. H. Vitt 1484* (H 3090697). **SOUTH AFRICA**. Cape Prov., *R. E. Magill 6074* (MO 3986640). **UNITED STATES**. Louisiana, Washington County, W.D. & D. C. Reese 16874 (MO 5269127); Florida, Wakulla County, *B. Allen 9949* (MO 3688270). **VENEZUELA**. Monagas, *R.A Pursell 8987* (MO 5134078), *R.A Pursell 9210a* (MO 5134066), *R.A Pursell 8963* (MO 5134079); Distrito Federal, *Julian A. Steyermark, 94798* (MO 2562028).

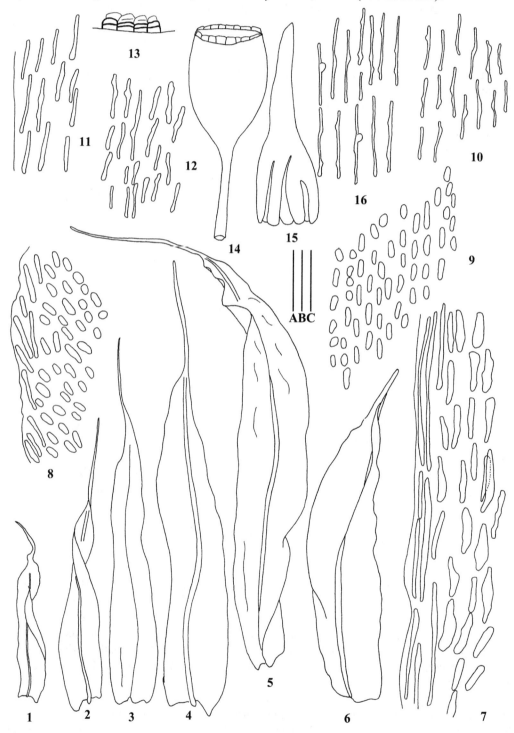

Fig. 254 *Macromitrium ulophyllum* Mitt. 1-5: Branch leaves. 6: Perichaetial leaf. 7: Basal cells of branch leaf. 9: Upper cells of perichaetial leaf. 10: Basal cells of perichaetial leaf. 11: Apical cells of branch leaf. 12: Upper cells of branch leaf. 13: Peristome. 14: Capsule. 15: Calyptra. 16: Basal cells of branch leaf (all from holotype, NY 00322438). Line scales: A = 2 mm (1-3, 14, 15); B = 1 mm (4-6); C = 400 μm (13); D = 67 μm (7-12, 16).

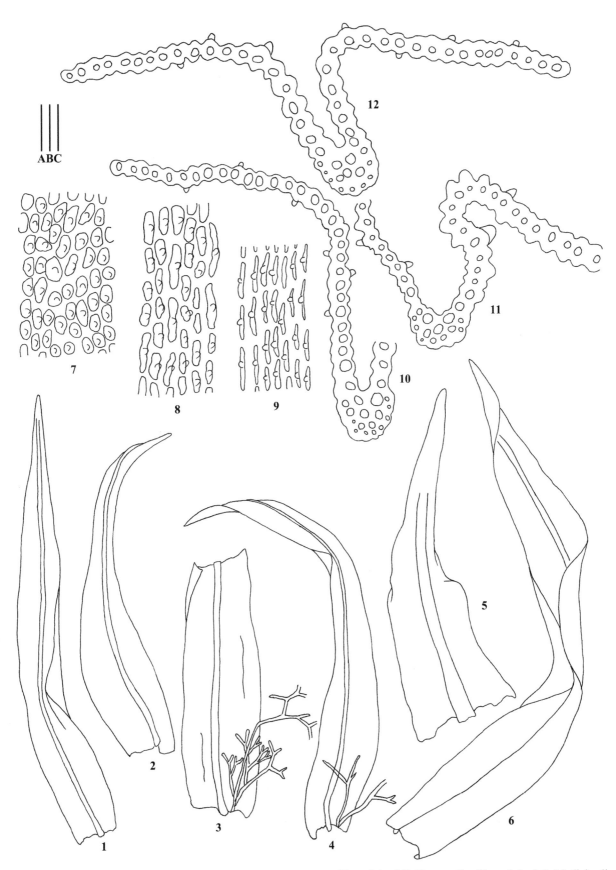

Fig. 255 *Macromitrium uraiense* Nog. 1-4, 6: Branch leaves. 5: Apex of branch leaf. 7: Upper cells of branch leaf. 8: Medial cells of branch leaf. 9: Basal cells of branch leaf. 10: Basal transverse section of branch leaf. 11: Upper transverse section of branch leaf. 12: Medial transverse section of branch leaf (all from isotype, NICH 365246). Line scales: A = 0.4 mm (1-4, 6); B = 160 μm (5); C = 40 μm (7-12).

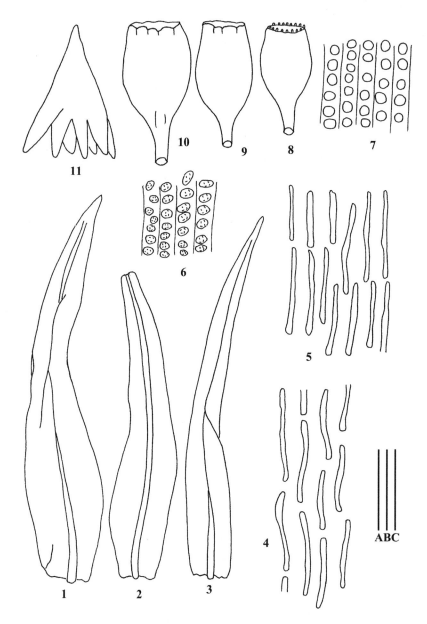

Fig. 256 *Macromitrium urceolatum* (Hook.) Brid. 1-3: Branch leaves. 4: Basal cells of branch leaf. 5: Low cells of branch leaf. 6: Upper cells of branch leaf. 7: Medial cells of branch leaf. 8-10: Capsules. 11: Calyptra (all from isotype, E 00011670). Line scales: A = 4 mm (8-11); B = 1 mm (1-3); C = 67 μm (4-7).

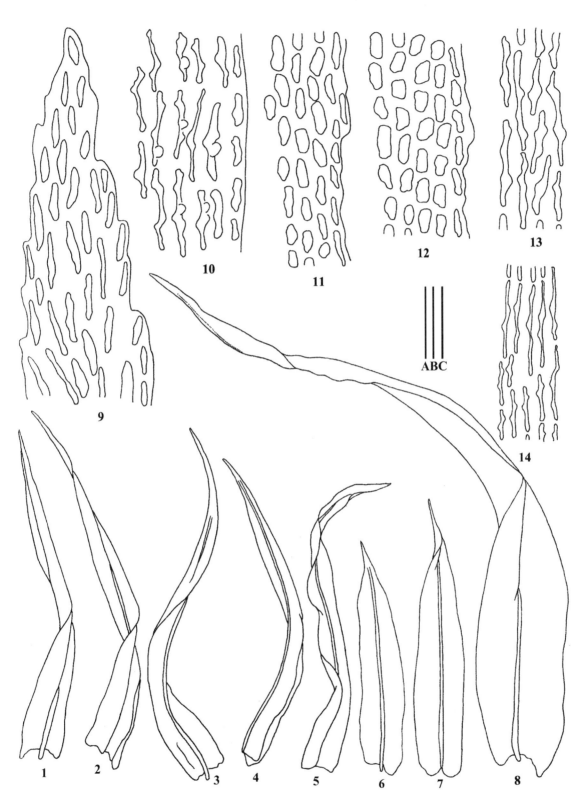

Fig. 257 *Macromitrium validum* Herzog 1-5, 8: Branch leaves. 6-7: Perichaetial leaves. 9: Apical cells of branch leaf. 10: Low cells of branch leaf. 11: Upper cells of branch leaf. 12: Medial cells of branch leaf. 13: Upper cells of perichaetial leaf. 14: Low cells of perichaetial leaf (all from holotype, JE 04008704). Line scales: A = 2 mm (1-7); B = 1 mm (8); C = 67 μm (9-14).

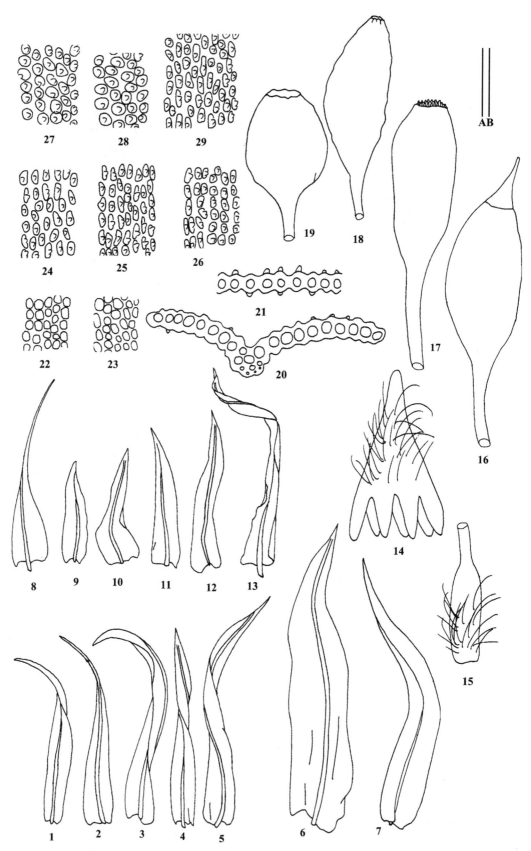

Fig. 258 *Macromitrium vesiculosum* Tixier 1-6: Branch leaves. 7, 9-13: Stem leaves. 8: Perichaetial leaf. 14: Calyptra. 15: Vaginula. 16-19: Capsules. 20: Upper transverse section of branch leaf. 21: Basal transverse section of branch leaf. 22: Apical cells of stem leaf. 23: Basal cells of stem leaf. 24: Upper cells of perichaetial leaf. 25: Low cells of perichaetial leaf. 26: Basal cells of perichaetial leaf. 27: Upper cells of branch leaf. 28: Medial cells of branch leaf. 29: Basal cells of branch leaf (all from holotype, PC 0083737). Line scales: A = 1 mm (1-19); B = 67 μm (20-29).

254. *Macromitrium xenizon* B.H. Allen & W.R. Buck, Mem. New York Bot. Gard. 76(3): 82. f. 78. 2003. (Figure 259)

Type protologue: [French Guiana] Canton de Maripasoula: commune de Saül, along trail to Pic Matécho, near km 2, between the Route de Bélizon near Crique Sant-Éloi and camp at km 6.5 on Crique Saint-Éloi, 434"N, 513'23"W, about 200 m. Sep 23, 2000, *William R. Buck 38090* (holotype: NY).

(1) Plants small to medium-sized, stems creeping, to 5 cm long, densely reddish rhizoids, branches dark green, short, 2-4 mm long. (2) Branch leaves sprially twisted, contorted or inrolled when dry, spreading, undulate and conduplicate when moist; 2-2.8 × 0.5-0.75 mm, oval- to oblong-lanceolate, broadly lanceolate, acute to broadly acute, fragile and often broken off at the medial part; margins entire above, enlarged teeth-like cells distinctly differentiated at the insertion; costae vanishing beneath the apex, occasionally concealed in adaxial view by overlapping folds of laminae; upper and medial cells rounded hexagonal, rounded-quadrate, 8-10 μm wide, mammillose, slightly elongate towards the base; basal interior cells short rectangular, thick-walled and porose, strongly tuberculate. (3) Perichaetial leaves are similar to vegetative leaves; setae smooth, 9-11 mm long, reddish. (4) Capsule ovoid, smooth or wrinkled at neck, peristome membranous; spores anisosporous, small ones 18-22 μm, large ones 24-28 μm, papillose.

Macromitrium xenizon B.H. Allen & W.R. Buck is similar *to M. punctatum* (Hook. & Grev.) Brid. in leaf shape, but the former can be distinguished from the latter by its branch leaves with fragile upper portion, marginal basal differentiated enlarged teeth-like cells at the insertion, and basal cells strongly tuberculate.

Distribution: French Guiana, Suriname.

Specimens examined: SURINAME. Sipaliwini, Eilerts De Haan Gebergte, *B. Allen 25473* (MO 5628135); Sipaliwini, Kayserberg Airstrip Area, *B. H. Allen 25396* (MO 5628132), *B. H. Allen 25307* (MO 5628137).

255. *Macromitrium yuleanum* Broth. & Geh., Öfvers. Finska Vetensk.-Soc. Förh. 37: 160. 1895.

Type protologue: *Patria*. Nova Guinea, Mt Yule, ad ramulos, ubi legit Sir W. mac Gregor.

**Macromitrium clemensiae* Nog., J. Hattori Bot. Lab. 10: 16. 8 f. 1-7. 1953, *fide* Vitt *et al.*, 1995. Type protologue: Papua New Guinea. Morobe, on big *Pygeum* in mossy bush, Mt. Sarawaket, about 11 000 ft., 29.V.1939 *M. S. Clemens 10192c* (lectotype: KUMA; isotypes, L); 27.V.1939 *Clemens* (syntypes: KUMA, L).

Macromitrium deflexum E.B. Bartram, Lloydia 5: 270. 29. 1942, *fide* Vitt *et al.*, 1995. Type protologue: Indonesia. West Irain, Jayawijaya. Mossy forest, common in the tree tops, 18 km SW of Bernhard Camp, Idenburg River, 2 150 m, *Brass 12610* (holotype: FH-Bartram; isotype: F), *Brass 12609* (paratypes, FH-Bartram, L).

Macromitrium habbemense E.B. Bartram, Lloydia 5: 271. 31. 1942, *fide* Vitt *et al.*, 1995. Type protologue: Indonesia. West Irain, Jayawijaya. Enveloping upper branches of tall trees, 9 km NE of Lake Habbema, 2 800 m, *Brass 10687* (isotypes, FH-Bartram, BRI, F).

Macromitrium hattorii Nog., J. Hattori Bot. Lab. 10: 14. 7 f. 1-10. 1953, *fide* Vitt *et al.*, 1995. Type protologue: Papua New Guinea. Morobe, growing on the trunk of *Cyathea*, Mt. Sarawaket, 12 000-13 000 ft., 12.VI.1939 *M. S. Clemens* (lectotype: KUMA; isotype: L); 8 000-9 000 ft., 28.II.1939 *Clemens 9935B* (syntypes, KUMA, L).

Macromitrium novoguinense E.B. Bartram, Lloydia 5: 271. 32. 1942, *fide* Vitt *et al.*, 1995. Type protologue: Indonesia. West Irain, Jayawijaya. Mt. Wilhelmina; epiphyte, subalpine forest, 3 400 m, *Brass 9692* (holotype: FH-Bartram; isotypes, BRI, L), *9701, 9702a* (paratypes, F, FH-Bartram).

Macromitrium sarawaketense Nog., J. Hattori Bot. Lab. 17: 31. 1956, *fide* Vitt *et al.*, 1995.

Macromitrium speirophyllum E.B. Bartram, Lloydia 5: 270. 30. 1942, *fide* Vitt *et al.*, 1995. Type protologue: Indonesia. West Irain, Jayawijaya. Lake Habbema, abundant in *Libocedrus* trees of open forest communities, 3 225 m, *Brass 9336* (isotypes: FH-Bartram, BRI, L).

(1) Plants moderate to large, rust-brownish, in loose, spreading mats; creeping stems covered with numerous rusty reddish rhizoids. (2) Stem leaves wide-spreading when dry, abaxially curved when moist, ovate-lanceolate, gradually acuminate. (3) Branch leaves not very shriveled, erecto-patent below, individually twisted and contorted-flexuose when dry, wide-spreading-recurved to abaxially curved when moist, 2.0-3.2 mm long, ovate-lanceolate, oblong-lanceolate, or ligulate lanceolate, sharply acute to mucronate-acuminate, occasionally rugose or undulate near upper margin, keeled; margins plane and papillose-denticulate above, broadly reflexed and entire below; costae shortly excurrent to percurrent; upper and medial cells rounded-quadrate, or short-elliptic, to rectangular, strongly bulging, unipapillose, papillae occasionally 2-3 forked, becoming smaller and oblate towards the margin in upper portion; basal cells 6-10 × 20-70 μm, elongate-rectangular, tuberculate, frequently porose. (4) Perichaetial leaves larger and longer than branch leaves, 3.0-4.5 mm long, ovate-lanceolate, sharply narrowed to a long, flexuose subentire or denticulate arista, upper cells tuberculate. (5) Setae long, up to 20 mm, smooth, twisted to the left; vaginulae frequently hairy. (6) Capsule urns broadly ellipsoid to obloid-ellipsoid, 8- plicate in upper portion, with a narrow mouth, gradually contracted to the seta; peristome single; exostome teeth fused to a coarsely papillose membrane; spores 18-58 μm, distinctly anisosporous. (7) Calyptrae mitrate, plicate, naked or sparsely hairy at upper portion.

Distribution: Indonesia, Papua New Guinea, Philippines (Tan & Shevock, 2015).

Specimens examined: **PAPUA NEW GUINEA**. Southern Highlands, *H. Streimann 24233* (MO 5140804, KRAM-B-137731); *H. Streimann 24224* (H 3196819), *A. Touw 16155* (H 3205152, MO 5375872), *A. Touw 15821* (H 3205289, MO 5371067).

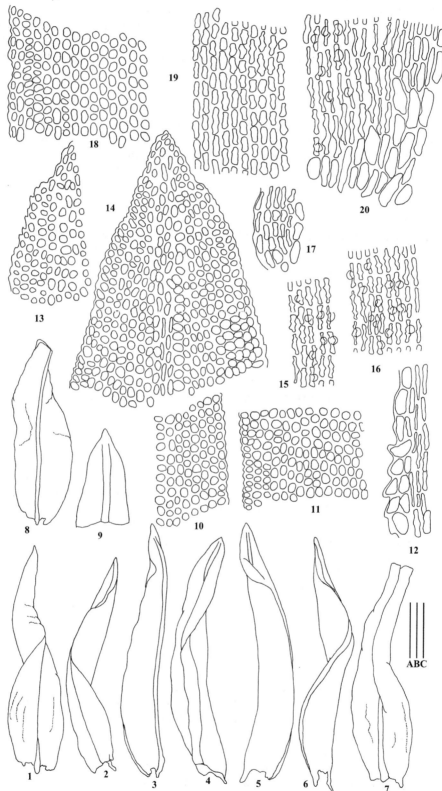

Fig. 259 *Macromitrium xenizon* B.H. Allen & W.R. Buck 1-8: Branch leaves. 9: Apex of branch leaf. 10-11: Upper cells of branch leaves. 12, 17: Basal marginal cells of branch leaves. 13-14: Apical cells of branch leaves. 15-16: Basal cells of branch leaves. 18: Medial cells of branch leaf. 19: Low cells of branch leaf. 20: Basal cells near costa of branch leaf (all from *B. Allen 19265A*, MO 4431784). Line scales: A = 0.5 mm (1-8); B = 200 μm (9); C = 50 μm (10-20).

256. *Macromitrium zimmermannii* M. Fleisch., Musci Buitenzorg 2: 432. 80. 1904. (Figure 260)

Type protologue: [Indonesia] An den Aesten der Rasamalabäume. West-Java: im Urwald bei Tjibodas, einmal sparlich aufgefunden, 1450 m. (isotype: JE 04008698!).

(1) Plants small, yellowish, in dense mats; stem long creeping, densely covered with reddish rhizoids; branches simple, rarely with branchlets, irregularly arranged in stems, 0.5-1.5 cm long. (2) Branch leaves contorted-twisted and crisped, with the apex hidden in the inrolled cavity when dry, spreading when moist, keeled, about 1.5 mm long, oblong-lanceolate to ligulate-lanceolate, with obtuse or acute apices, distinctly plicate below, recurved on one or both sides; margins plane throughout; costae vanishing below the apex, percurrent or excurrent; upper cells clear, regularly round-quadrate, moderately thick-walled, densely pluripapillose; medial cells subquadrate to elliptic rectangular, smooth or unipapillose, arranged in longitudinal rows; low cells elongate, clear and unipapillose; basal cells pellucid, elongate, thick-walled, with a large tubercula per cell. (3) Perichaetial leaves smaller than branch leaves, ovate lanceolate, gradually acuminate; sheathing at base, costa slender. (4) Setae longer, smooth, 14-18 mm, erect, twisted to the right. (5) Capsules ovoid, 8-plicated; peristome single, exostome of 16, lanceolate, densely papillose on both sides; spores 27-33 μm, greenish. (6) Calyptrae smooth and naked.

Distribution: Indonesia.

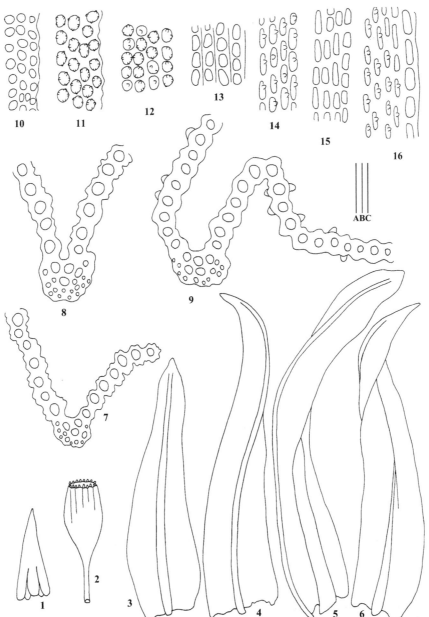

Fig. 260 *Macromitrium zimmermannii* M. Fleisch. 1: Calyptra. 2: Capsule. 3-6: Branch leaves. 7: Upper transverse section of branch leaf. 8: Low transverse section of branch leaf. 9: Medial transverse section of branch leaf. 10: Apical cells of branch leaf. 11, 12: Upper cells of branch leaf. 13: Medial cells of branch leaf. 14: Low cells of branch leaf. 15: Basal cells of branch leaf. 16: Basal marginal cells of branch leaf (all from isotype, JE 04008698). Line scales: A = 1 mm (1, 2); B = 400 μm (3-6); C = 67 μm (7-16).

4.4 Doubtful species

The following species are listed as doubtful taxa because we could not obtain their type specimens. According to the original literature about these species, we preliminarily and briefly described them as follows.

1. *Macromitrium acutissimum* Müll. Hal., Flora 83: 333. 1897.

Type protologue: Habitatio, Venezeula, Tovar, 1800 m altum, Octobri 1890.

Plants in widely dense mats, green; stems slender; leaves curly, curly-crispate when dry, still curled when moist, long and narrowly lanceolate, with a long and curly hair; carinate and deeply channeled; margins plate above, recurved below, entire throughout; costae goldenish, percurrent; cells small and rounded, greenish from the base to the apex, scarcely larger, rectangular, thickened and more transparent. Perichaetial leaves similar to branch leaves. Setae short and glabrous, yellow and erect; Capsules narrowly ellipsoid, with a constricted angled mouth; opercula conic-rostrate, with a long straight beak; peristome double, caducous and rudimentary, interrupted; Calyptra covering the whole capsule, sulcate, furrowed, shiny goldenish.

2. *Macromitrium altum* Müll. Hal., Flora 82: 453. 1896.

Type protologue: Habitatio. Insulae Hawaiieae, Oahu, ubi Menzies legit. Swartz in Hb. Mohr. Sub Orthotricho. Type citation: Hawaiieae, Oahu, leg. *Menzies s.n.* (isotype, H-BR 2530001!).

(1) Plants small, stems slender, the secondary stems with dichotomous branchlets. (2) Branch leaves erect below and contorted-crisped above when dry, spreading when moist; lanceolate from a long oblong base, easily broken off in the upper portion; apices acuminate; margins entire throughout; costae percurrent; low and basal lamina cells elongate, lumens curved, thick-walled, porose. (3) Perichaetial leaves longer than branch leaves, long acuminate above. (4) Setae rather short. (5) Capsules small, ellipsoid to cylindric.

3. *Macromitrium brachycarpum* Mitt., J. Linn. Soc., Bot. 12: 204. 1869.

Type protologue: Andes Peruvianae, Tarapoto, *Spruce s. n.*

Plants small, greenish-red, branches 0.25-1.5 inch. Branches densely leaved, leaves carinate, appressed-contorted when dry; margins entire throughout; upper cells small and rounded, low and basal cells slightly elongate, strongly tuberculate, those near costa becoming wider and smooth. Perichaetial leaves broader, acute. Setae long, 0.25 inch. Capsules with a wide and open mouth; opercula conic-subula; peristome short and pale. Calyptrae naked.

4. *Macromitrium braunii* Müll. Hal., Syn. Musc. Frond. 2: 643. 1851.

Type protologue: Java: Blume.

Eddy (1996) once remarked that the species resembles *M. angustifolium*, with leaves up to 4 mm long and inner perichaetial bracts finely acuminate. Apart from its robust plants, it differs from *M. angustifolium* mainly in possessing an inner peristome with delicate, elongate processes in addition to the 'angustifolium'-like exostome.

5. *Macromitrium bujongolanum* O'Shea, Trop. Bryol. 29: 5. 2008.

Replaced synonym: *Macromitrium fragile* G. Negri 1908. Type protologue: in M. Ruwenzori (Africa centr.) prope Bujongolo m. 3800.

Plants robust, in dense mats, olive-green, densely with tomenta; branches erect and thick, stiffy, up to 25 mm tall, divided into bundles at the apex; Branch leaves curled-crisped when dry, erecto-patent when moist; lanceolate, 2-3 mm long, with a very long loriform hair; costae percurrent; basal cells irregularly elongate, pellucid and smooth, the others rounded-angled or irregularly quadrate, subopaque, greenish, often slightly papillose.

6. *Macromitrium clavatum* Schimp. ex Grout, Bryologist 47: 16. 1944.

Type protologue: Guadeloupe (*L'Herminier*). Type at the New York Botanical Garden.

Branch leaves lingulate-lanceolate, 1.5-2 mm long; margins recurved at the base, entire; apices obtuse to acute, keeled; costae subpercurrent or vanishing beneath the apex. Cells thick-walled, upper cells round, 6-8 μm wide, medial and low cells elongate, linear. Perichaetial leaves with an acute apex, cells elongate. Setae 8-9 mm long. Capsules ovoid; opercula 2 mm long. Peristome obtuse, papillose.

Macromitrium clavatum Schimp. ex Grout is similar to *M. richardii*, but differs from the latter in having linear cells from the medial part to the base, and upper cells clear and small.

Taxonomy 407

7. *Macromitrium clemensiae* Bartr. Philipp. J. Sci. 61: 242. 1936.

Type protologue: Borneo, Mount Kinabalu, Penibukan, elevation 4000 feet, jungle, east ridge, on twigs of tree, *Clemens 40531* type; Penibukan, elevation 4500 feet, jungle ridge above camp, on tree, Clemens *50305ᵃ*, Masilan River, elevation 7000 feet, on Vernonia, magnolia, ect., *Clemens 51487b*.

(1) Plants moderately robust, branch up to 20 mm tall. (2) Branch leaves densely leaved. Leaves keeled, erect below, crisped above when dry, arcuate-patent to somewhat squarrose when moist, narrowly lanceolate, up to 2.8 mm long and 0.5 mm; margins mainly flat, strongly and irregularly dentate; apices narrowly acuminate; costae percurrent to slightly excurrent; upper lamina cells quadrate, about 10 μm wide, densely pluripapillose and obscure, internal walls not strongly thickened; lower lamina tissue dimorphic, the interior cells rather uniformly elongate, about 20 μm long, the majority conspicuously tuberculate, several marginal rows longer, narrow and smooth, forming a strongly differentiated pellucid border (which is similar to that of *Groutiella*). (3) Perichaetial leaves strongly differentiated, narrowly lanceolate and finely acuminate, longer than normal leaves, up to 3.4 mm long. (4) Setae ca. 8 mm long, scabrid. (5) Capsule urns ovoid, peristome is a papillose membrane about 50 μm tall (Eddy, 1996).

8. *Macromitrium complicatulum* Müll. Hal., Monsunia, 1: 175. 1903.

Type protologue: [Indonesia] Batjan: Mt. Sibella.

Plants in dense mats, brown below, yellowish above, dull. Stems long creeping, densely with branches, branches up to 15-30 mm, erect to curved, densely leaved, simple or branched at the apex. Branch leaves crisped but not twisted when dry, spreading when moist; keeled, lanceolate, shortly acuminate; margins plane and entire throughout; costae reddish, subpercurrent; upper cells subquadrate, pellucid, weakly papillose; basal cells clear and pellucid, long elongate, with thicken walls and narrow lumens, curved and smooth. Perichaetial leaves similar to branch leaves. Setae about 5 mm, erect, reddish and smooth. Capsules erect, ovoid to ellipsoid, smooth, plicate near the mouth. Calyptrae small, with erect hairs.

Eddy (1996) thought that this species is similar to *M. angustifolium*, but the branch leaves of *M. complicatum* shorter, the capsules smooth, plicate near the neck and mouth.

9. *Macromitrium cucullatulum* Müll. Hal., Malpighia 7: 296. 1893.

Type protologue: In monte Dongollo prope Ghinda.

According to the original description, *M. cucullatulum* was similar to *Macrocoma abyssinica* (Müll. Hal.) Vitt. (as *Macromitrium abyssinicum* Müll. Hal.), but different by its ligulate-lanceolate leaves and hooded apices.

10. *Macromitrium eriomitrium* Müll. Hal., Bull. Herb. Boissier 6: 98. 1898.

Type protologue: Habitatio, Brasilia, Serra Italiaia, 1800 m. alt., ad rupes: *E. Ule*, martio 1894. Coll. No *1834*. Type citation: Brasilia, Serra Itatiaia, *E. Ule 1834* (type: R 000081757).

(1) Plants large and loosely, somewhat in wefts, golden-yellowish, brown; stems long creeping, with inconspicuous and caduceus leaves; branches simple or branched, up to 20 mm tall. (2) Branch leaves strongly keeled, with branched rhizoids, erecto-apressed below, individually twisted and flexuose, undulate above when dry, spreading and occasionally twisted (around the costa) above when moist; lanceolate, acuminate, plicate below; margins entire, recurved; costae golden-yellow, percurrent; almost of upper and medial cells rounded, pluripapillose, elongate toward the base, basal cells rectangular, unipapillose. (3) Perichaetial leaves longer than branch leaves. (4) Setae short and reddish, erect. (5) Capsules small, narrowly ellipsoid, furrowed, constricted beneath the slightly puckered mouth; opercula conic-rostrate; peristome double, exostome shorter and coarse, papillose, pale; endostome lanceolate, united in pairs. (6) Calyptrae large, covering the whole capsule, mitrate, deeply lacerate below, densely hairy.

Comments: Valente *et al*. (2020) recognized this species, but didn't give detailed information about its features.

Specimens examined: BRAZIL. Rio de Janeiro, Minas Geraes, *V. Schiffner, 753* (H-BR 2638003); Serra de Slatiaia, *P. Dusen, 601* (H-BR 2638010).

11. *Macromitrium fragile* Hampe, Linnaea 20: 71. 1847.

No type information and description in detail.

12. *Macromitrium globirameum* Müll. Hal., Bot. Zeitung (Berlin) 15: 779. 1857.

Type protologue: Nova Irlandia Oceani pacifici: collect. Cuming. Type citation: New Zealand, leg. Cuming (lectotype designed by Vitt: H-BR 2560002).

Plants in spreading mats, stems long creeping, densely with globular branches, golden-yellow. Branch leaves erecto-appressed when dry, spreading when moist; leaves short, oblong with a wide base and an acuminate upper portion, plicate, with a somewhat asymmetrical apex; margins entire throughout, often recurved below; costae rusty-red, excurrent to form a short point; basal cells hyaline, thicken-walled, in crescent shape, tuberculate, porose.

13. *Macromitrium kinabaluense* J. Froehl., Rev. Bryol. Lichénol. 31: 92. 1962.

Type protologue: North-Borne, Kinabalu, summit region, 10000-30000 feet, 9.3. 1961.

Plants slightly robust, stems long creeping, densely with thick branches, branch erect, up to 20 mm tall, simple or with short branchlets, densely leaved, rusty-reddish above, red to orange-yellow, weakly lustrous. Branch leaves keeled, twisted-curled when dry, widely spreading when moist, lanceolate from an oblong low portion, tapering to a long acuminate, hyaline arista with a row of cells; margins somewhat plate, crenulate below, serrulate above; costae light reddish, subpercurrent; apical cells oblong, thick-walled and smooth; upper cells rounded-quadrate, short rectangular, rhomboid to hexagonal, thicken and porose; medial cells unipapillose; basal cells linear long elongate, thick-walled and strongly porose, unipapillose; Vaginulae densely hairy. Setae 8-10 mm long, red and smooth. Capsule urns ellipsoid, plicate at the neck and beneath the constricted mouth. Calyptrae mitrate, brownish, hairy.

14. *Macromitrium liliputanum* Müll. Hal., Flora 73: 483. 1890.

Type protolegue: Patria. Africa or. Trop., Leikipia in occidente montis Kenia, ad pedem der Aberdare-Kette, Novbr. 1887: L. Höhnel in Exped. Telekiana.

Stems long creeping, densely with branches, branches multiple branching, thin and light yellow. Branch leaves densely, appressed when dry, slightly spreading when moist; lanceolate or narrowly lanceolate, plicate below; margins entire throughout, recurved below; costae rusty-red, subpercurrent to percurrent; cells relatively small, rounded but obscure; capsules slightly large, erect, ellipsoid, erect, opercula conic-rostrate.

15. *Macromitrium lonchomitrioides* Müll. Hal., Hedwigia 37: 155. 1898.

Type prtologue: Habitatio. Australia subtropica, insula Norfolk: Robinson in Hb. Melbourne, 1884, 1885 misit. Type citation: Norfolk, leg. *Robinson* (isotype: H-BR 259001).

Plants relatively robust, in loose mats; stems creeping, with ascending branches, branches thick and short. Branch leaves keeled, loosely arranged, twisted-crisped when dry, spreading when moist; oblong-lanceolate, gradually acuminate, with an acute-acuminate apex; margins entire throughout. Costae narrow, percurrent; cells rounded, in longitudinal rows, almost smooth to slightly unipapillose; basal cells elongate, thick-walled, with curved lumens. Perichaetial differentiated, larger than branch leaves. Setae short. Capsules small, erect, narrowly cylindric and smooth, opercula conic-subula, peristome absent; calyptrae covering the whole capsule, yellow, lobed at base, with long and appressed hairs.

16. *Macromitrium marginatum* Dixon, Ann. Bryol. 7: 25. 1934.

Type protologue: [Indonesia] Hab. Lombesang, alt. 600 m., 30 May, 1921; coll. *Bunnemeijer (11798)*, Herb. Hort. Bot. Bog. (2163b).

Plants small, in mats, brown, branches short. Branch leaves keeled, twisted-crisped, inrolled or curled when dry; oblong-lanceolate, broadly acute; upper cells rather small, rounded, smooth; medial, low and basal cells similar, ovate, mostly thick-walled, often horizontally arranged, papillose, those near the costa at the insertion large and thin-walled, forming a 'cancellina region'; the marginal cells differentiated in one or several rows, narrower than their ambient cells. Setae short, shorter than 2 mm long. Capsule urns turgid, ellipsoid and smooth, constricted beneath the angled-mouth, opercula conic-rostrat, peristome absent. Calyptrae reddish brown, hairy.

17. *Macromitrium nietneri* Müll. Hal., Linnaea 30: 39. 1869.

Type protologue: no information.

Plants in mats, golden-yellow or light brown, stiffy and robust; stems long creeping, densely with short and thick branches; branch leaves spirally coiled but not inrolled when dry, spreading when moist; lanceolate from an oblong base, gradually tapering to an acuminate apex; incurved near the apex, plicate; costae strong, percurrent; margins entire throughout, recurved below; cells green, regularly arranged, strongly papillose; basal cells elongate.

18. *Macromitrium novorecurvulum* B.C. Tan & B.C. Ho, J. Hattori Bot. Lab. 98: 228. 2005.

Replaced synonym: * *Macromitrium recurvulum* Cardot, Rev. Bryol. 28: 113. 1901.

According to the original description of *M. recurvulum* Cardot, *Macromitrium novorecurvulum* is similar to *M. salakanum* Müll. Hal. (as *M. reflexifolium* Sande Lac), but *M. novorecurvulum* has short leaves with mucronate apices, excurrent costae, denticulate margins near the apex, shorter basal cells and all with large papillae on the dorsal surface.

19. *Macromitrium okabei* Sakurai, Bot. Mag. (Tokyo) 57: 250. 4. 1943.

Type protologue: [Micronesia] Ponape: Leg. *M. Okabe* Typus in Herb. *K. Sakurai Nr. 14475* Jan. 1941.

(1) Plants medium-sized, in dense mats, dull yellowish-brown above; stems creeping, ca. 50 mm long, densely with reddish tomenta below; branches simple or forked, densely leaved. (2) Branch leaves strongly twisted-contorted

and crisped, incurved above when dry, erect-spreading when moist; linear or long lanceolate, acuminate, 2.5 × 0.2 mm at base, margins entire; costae excurrent to form a hyaline hair point; median and upper cells subquadrate, obscure, densely papillose; cells elongate toward the base, oblong to rectangle, smooth and clear. (3) Perichaetial leaves broad lanceolate, with an acute apex, translucent; costa percurrent; vaginulae cylindric, hairless. (4) Setae up to 5 mm long, reddish. (5) Capsule urns ovoid, 0.8×0.5 mm, plicate or wrinkled when dry, smooth when moist; peristome single, exostome short and obtuse, hyaline, densely papillose; opercula erect; spores papillose. (6) Calyptrae deeply lacerate, slightly hairy.

Distribution: Pohnpei (formerly Ponape) island of Micronesia.

20. *Macromitrium papillisetum* Dixon, Linn. Soc., Bot. 50: 89. 1935.

Protologue: Bettotan, Sandakan, 23 July 1927; coll. *C. Boden Kloss 18742*.

Plants similar to *M. blumei* in habit but branches are shorter and more compact, in dense mats. Branch leaves spirally contorted when dry, much more obtuse, shortly apiculate; margins plane above and minutely crenulate; costae strong, reddish, excurrent into a conspicuous murcro; upper lamina cells turgid, densely pluripapillose and obscure; lower lamina cells becoming rectangular-oval, most with a tall tubercula; at extreme base elongate, smooth and pellucid. Perichatial leaves with an acute apex. Setae very short, ca. 3 mm long, coarsely papillose; capsule short, widely urceolate, peristome absent. Caplytra densely and shortly pilose.

The branch leaves appear to be unranked, but in other respects *M. papillisetum* appears to be very close to *M. orthostichum* and its allies. The species is characterized by its short, stout, highly papillose seta, and the short, wide, urceolate capsule.

Distribution: Borneo.

21. *Macromitrium patens* Wilson in Seemann, Bot. Voy. Herald 244. 1854.

Type protologue: no information.

Branches erect, simple and long; leaves arranged in five rows around the stem; twisted-flexuose when dry, lanceolate-liguate, acute, keeled; margins slightly plane, minutely denticulate. Perichaetial leaves erect, long acuminate. Capsules ellipsoid and smooth; peristome unknown.

22. *Macromitrium recurvifolium* (Hook. & Grev.) Brid., Bryol. Univ. 1: 740. 1826.

Basionym: *Orthotrichum recurvifolium* Hook. & Grev., Edinburgh J. Sci. 1: 120. 5. 1824. Type protologue: Communicated from Java to Mr. Dr. Dickson.

Branch leaves keeled, spirally twisted when dry, ovate and long acuminate, recurved; perichaetial leaves convolutions; capsules ovoid-ellipsoid, slightly furrowed.

23. *Macromitrium sclerodictyon* Cardot in Grandidier, Hist. Phys. Madagascar, Mousses 246. 1915.

Type protologue: Zone du Plateau central: Ankadivavală (*R.P. Camboué*).

Stems creeping, densely with branches and brown tomenta; branch leaves keeled, arranged in five rows, squarrose-flexuose when dry, spreading and recurved when dry, decurrent at base; linear lanceolate, costa light reddish, excurrent into a short point; 1.5-3.0×0.35-0.4 mm; margins broadly recurved at one side below, undulate and entire, occasionally crenulate below the apex; medial and upper cells small and rounded-quadrate or slightly elongate, clear, densely with weak papilla; basal cells linear elongate, smooth, thick-walled.

24. *Macromitrium sinuatum* (Hornsch.) Müll. Hal., Bot. Zeitung (Berlin) 3: 543. 1845.

Basionym: *Schlotheimia sinuata* Hornsch., Fl. Bras. 1(2): 31. 1840. Type protologue: no information.

Leaves overlapping, appressed and twisted when dry, ligulate, slightly sinuate and corrugated in the medial portion, very short; perichaetial leaves oblong; capsules cylindric, calyptrae long.

25. *Macromitrium stephanodictyon* J. Fröhlich, Rev. Bryol. Lichénol. 31: 93. 1962.

Type protologue: Borneo. Mt. Kinabalu, summit area, 10000-13000 feet; 9.3.1961.

Plants medium-sized to large, in dense mats, dark brown to organ-brown; stems long spreading, densely with branches; branches erect, up to 2 cm tall, densely leaved, simple or with short branchlets. Stem leaves ovate, with a short acuminate upper portion; costa percurrent; cells elongate-rectangular, thick-walled. Branch leaves appressed below and flexuose above when dry, spreading when moist, long lanceolate, gradually acuminate, 3.5×0.5 mm; margins entire throughout, narrowly recurved at one side below; costae somewhat strong, light reddish, subpercurrent; upper cells rounded or quadrate, in longitudinal rows, thick-walled, densely papillose; basal cells smooth, short rectangular or elongate rectangular, thick-walled. Perichaetial leaves longer than branch leaves, rectangular or oblong, strongly thick-walled; vaginulae densely papillose. Setae 8-10 mm long, smooth and reddish. Capsules ellipsoid, constricted beneath the mouth and narrowed towards the neck, plicate; peristome single, lanceolate, densely papillose; opercula conic-subulate. Calyptrae mitrate-campanulate, brown-yellow, plicate, naked, lacerate below, occasionally with sparse, short and straight brown hairs near the beak.

26. *Macromitrium striatum* Mitt. ex Bosch & Sande Lac., Bryol. Jav. 1: 134. 1860.

Type protologue: Habitat insulam Borneo; in monte Kina Balloo legit H. Low herb. Hooker.

Plants dark brown; stems long creeping, densely with erect and short branches; branches densely leaved; branch leaves erect-spreading below, spirally twisted above when dry, oblong lanceolate, obtuse-acute, plicate; costae excurrent to a short point; upper cells small, rounded, thick-walled, almost smooth; basal cells narrowly elongate; Perichaetial leaves smaller, ovate and sharply narrowed to the apex; costae percurrent. Setae rather long. Capsule urns ellipsoid, plicate when dry, peristome truncate, operculum conic-rostrate and erect. Calyptrae mitrate, plicate, smooth and naked, lacerated below.

27. *Macromitrium strictfolium* Müll. Hal., Bull. Herb. Boissier 6: 99. 1898.

Type porotologue: Habitatio. Brasilia, Serra dos Orgaos, ad rupes vigens: Dr. H. Schenck, Martio 1887 lg. Hb. Brotheri mis.

Plants robust, in dense mats, light golden-yellow; stems creeping, branches up to one inch tall, simple and thick, branch leaves twisted-flexuose when dry, erect spreading when moist; ligulate-lanceolate, with an acuminate apex; margins minutely serrulate, slightly recurved below; costae narrow, golden or rusty-red, excurrent; plicate or somewhat folded; upper cells small, short rectangular, thick-walled, papillose (?); basal cells elongate, smooth or sparsely unipapillose, with curved lumens, porose. Perichaetial leaves similar to branch leaves. Setae short. Capsules small and ovoid, smooth. Calyptrae golden-yellow, smooth and naked.

28. *Macromitrium stricticuspis* Müll. Hal., Flora 83: 334. 1897.

Type protologue: Habitatio. Venezuela, Toyur, 1800 m alt., Octobri 1890.

Plants small, stems creeping, with short branches. Branch leaves appressed erect and recurved, linear lanceolate, long and linear acuminate, plicate; margins serrulate; costae golden-yellow, percurrent, keeled; cells small, narrowly rectangular, becoming longer towards the base, with a curved lumen, thick-walled, smooth. Perichaetial leaves similar to or slightly longer than branch leaves. Setae short. Capsules small ellipsoid, smooth; peristome double.

29. *Macromitrium subdiaphanum* Renauld & Cardot, Rev. Bot. Bull. Mens. 10: 707. 1892.

Type protologue: Hab. Madagascar: plateau d'Ikongo, Lg. Dr Besson.

The species is similar to *Macromitrium soulae* Ren., but different from the latter by its shorter, oblong, non-wavy, very entire leaves, by the leaves with distinctly large, pale, transparent cells. The cells of leaves elliptical, oblique along the costa, becoming smaller towards the margins.

30. *Macromitrium subpaucidens* Müll. Hal., Flora 83: 334. 1897.

Type protologue: Habitatio. Venezuela, Tovar, 1800 m atl., Octobri 1890

Plants small and greenish, in dense mats, with brown rhizoidal tomenta; branches short, leaves strongly crisped when dry, squarrose spreading when moist, oblong-lanceolate, acuminate, flexuose, plicate; margins sparsely dentate, narrowly recurved below; costae golden-yellow, percurrent; upper cells small and rounded, basal cell rectangular, strongly thick-walled, porose, coarsely papillose. Perichaetial leaves similar to branch leaves. Seta short and reddish, smooth, slightly curved. Capsules turgid ellipsiod, constricted at the mouth, plicate; peristome double, rudimentary; Calyptrae lacerate, golden-yellow, plicate, naked and smooth.

31. *Macromitrium tahitisecundum* Margad., Lindbergia 1: 128. 1972.

Macromitrium tahitisecundum Margad. Nom. Nov. Typon. *Dasymitrium nadeaudii* Besch., Bull. Soc. Bot. France 45: 65. 1898: syn. *Macromitrium tahitense* Broth. 103. hom. illeg. Type protologue: Sur les arbres des crêtes seches, de 800 à 900 mètres d'altitude (1re herbor, n° 266; 3° herbor. n° 267). Type citation: Mouseese de Thaiti, coll N° 1, Dr. Nadeaud, 1896, n° 266 (syntype: H-BR 2561007!).

Plants small, in dense mats, stems long creeping, densely with erect, short and forked branches. Branch leaves greenish above, dark brown to rusty-reddish below, spirally twisted when dry, spreading when moist; narrowly ligulate with a little wider base; apices acute-acuminate; margins entire; costae disappearing beneath the apex, canceling due to overlapping by the laminae from the adaxial view; upper cells quadrate, chlorophyllous, papillose on the dorsal surface; low cells subquadrate; Perichaetial leaves longer than branch leaves, ovate-lanceolate, long acuminate. Seta smooth, twisted to the left, vaginula hairy. Capsules ovoid and smooth; peristome single, with broad teeth, papillose. Calyptrae cucullate, densely hairy.

Taxonomy

32. *Macromitrium undatum* Müll. Hal., Bulletin de l'Herbier Boissier 6: 97. 1898.

Type protologue: Brasilia, Serra Itatiaia, in Capao ad truncus arborum, 1100 m. alt. Martio 1894. *E. Ule n°. 1832* (isotype: R 000081758).

Plants in widely spreading mats, green above, gloden below, loosely with branches, curved, simple or branched, with rhizoidal tomenta; leaves strongly keeled, flexuose-crisped when dry, widely spreading when moist; oblong-lanceolate; apices acuminate to acute; margins entire, broadly recurved, irregularly undulate; costae narrow and rusty-red, percurrent; upper cells small and rounded, greenish and, with thickened walls; cells became larger and elliptic towards the base, papillose.

The species is characterized by its broad and undulate branch leaves. Valente et al. (2020) recognized this species but didn't give detailed information.

Specimens examined: BRAZIL. Minas Gerais, *William R. Buck 26995* (KRAM-B- 178775, MO6092984).

33. *Macromitrium urceolatulum* Müll. Hal., Abh. Naturwiss. Vereins Bremen 7: 208. 1881.

Protologue: Wald von Ambatondrazaka, 6. Decbr. 1877. There was almost no information about the species in its original publication.

34. *Macromitrium venezuelense* Paris, Index Bryol. 791. 1897.

Replaced * *Macromitrium serrulatum* Müll. Hal., Linnaea, 42: 490. 1879.

The species is similar to *Macromitrium longifolium*, but its plants much more slender, branch leaf margins regularly serrate above (from the medial portion to the apex), cells hexagonal-rounded, darker greener; setae densely covered with conical or truncated papillae, peristome simple, teeth fused together, membranaceous.

35. *Macromitrium vitianum* E.B. Bartram, Occas. Pap. Bernice Pauahi Bishop Mus. 19 (11): 225. 1948.

Type protologue: Viti Levu: Nandarivatu, alt. 2700-3000 ft., nos. 590, 754, 756 (type); Serua hills, dripping clay bank, alt. about 700 ft., no. 980; Xandarivatu, Ba Road, wet bank, alt. about 3000 ft., no. 1182; Navai near Xandarivatu, on bank, alt. about 2500 ft., no. 1211; wet bank near Nandarivatu, alt. about 3000 ft., *nos. 1221, 1236*.

Plants large and robust, in dense mats, light brown below, greenish above; branches up to 15 mm tall, densely leaved. Branch leaves erect below, twisted above when dry, erect-spreading when moist, 3.5-4.5 mm long, oblong-ligulate, obtuse-acute with a short point; margins minutely crenulate, narrowly recurved at one side below, plane above; costae percurrent; upper cells rounded, 10-12 μm, coarsely papillose, low and basal cells elongate, with a narrow lumen, tuberculate. Setae up to 20-25 mm long, slender and smooth, reddish. Capsule urns ellipsoid-cylindric, plicate when dry, peristome double; spores smooth, 15-38 μm in diameter; Calyptrae campanulate, naked and smooth, brown at the beak, plicate.

36. *Macromitrium weissioides* Müll. Hal., Linnaea 37: 153. 1873[1872].

Type protologue: Patria. Brisbane River ubi inter Hypnum (Tamariscella) suberectum Hpe. Amalie Dietrich 1864 legit.

Stems creeping, branches rarely divided. Leaves green and curly, erecto-patent, narrowly lanceolate to linear from a narrowly oblong base, strongly carinated-concave; costae yellowish and excurrent; basal cells elongate and narrow, upper cells small and rounded. Perichaetial leaves oblong, acuminate. Setae long and slender, yellowish, slightly curved and erect, not twisted. Capsule cylindric, peristome simple, lanceolate. Calyptrae mitrate-campanulate, yellow, slightly folded, multi-lobed at base.

Chapter 5

Results and Discussion

5.1 Taxonomy

By the end of the 20th century, a total of *Macromitrium* 959 species names (including subspecies and variety) were reported, of which 365 taxa were recognized. Having examined the ordinary and type specimens as could as possible, and referred to the regional taxonomic revisions of the genus in recent years in the Central America by Allen (2002), New Caledonia (Thouvenot, 2019), Brazil (Valente, 2020), and partial African countries (Wilbraham, 2007, 2008, 2015, 2016, 2018; Wilbraham & Ellis, 2010), as well as our taxonomic works since 2003, we confirmed 271 species (including 12 varieties and 3 subspecies) in this book.

Considering the huge number of *Macromitrium* names, the type specimens of partial species that are still unavailable to us, and about 30 taxonomically doubtful species, the present taxonomical revision of the genus is not final, especially for its systematics and the division within the genus.

5.2 Habitats

Almost all species of the genus *Macromitrium* have been reported from tropical and subtropical regions except a few species from the southern-temperate region such as *M. campoanum* Thér., *M. krausei* Lorentz, *M. microcarpum* Müll. Hal., *M. tenax* Müll. Hal. and *M. saddleanum* Besch. ex Müll. Hal. These species were mainly reported from the temperate region of Chile. Additionally, some species were mainly reported from subtropical region, but their distribution regions extend to the transitional region between the subtropical and temperate regions in the northern hemisphere. For example, *M. viticulosum* (Raddi) Brid. (as *M. richardii* Schwägr.) reported from Sampson, North Carolina (34.78°N, 78.42°W), which is the only species that was reported from the North America (Am 1). In East Asia, *M. giraldii* Müll. Hal. was once reported from the Taibai Mountain of Qinling, Shaanxi province, China (33.98°N, 107.76°E), but the species has a wide distribution region in the subtropical region (Guo & He, 2008b; Luo et al., 2014). According to prediction, *M. giraldii* may extend to temperate regions at higher latitudes than what were recorded.

Macromitrium species are typically epiphytic in forests and open habitats around forests. In East Asia, they mainly occur in warm-temperate evergreen hardwood forests, or temperate deciduous hardwood forests, mixed evergreen and deciduous hardwood forests, as well as deciduous hardwood forests (Guo et al., 2007a). *Macromitrium* species are also rather common in the man-made habitats such as trees along roads and young plantations of various trees.

Vitt & Ramsay (1985) described the habitats of Australian species of *Macromitrium* in detail. In Australia, *Macromitrium* species are recorded from temperate, subtropical, and tropical rain forests, but mostly confined to rain forest habitats, especially at higher elevations in moist tropical rain forests. In the littoral rain forest along the sea coast; the species of the genus are common and rich on both rocks and tree trunks (Vitt & Ramsay, 1985). According to Vitt and Ramsay (1985), the habitats of Australian species of *Macromitrium* were divided into six groups along elevation and drought gradients. For example, *M. brevicaule* and *M. longirostre* are confined to situations with high amounts of salt spray. These two species are most common on rocks but also recorded from tree trunks in littoral rain forests close to the beach. The greatest species diversity is found in the lower elevation, ravine rain forests in Australia (Vitt & Ramsay, 1985). Host specificity is not very obvious for *Macromitrium* species although *Macromitrium* species occur more commonly as epiphytes on certain tree species (Vitt & Ramsay, 1985).

Macromitrium on the Huon Peninsula were reported from 400 m up to 3350 m, from closed primary and secondary forests, mossy and montane rain forests, disturbed rainforests, also from garden areas or badly distributed vegetation or even in tree fern savannahs (Vitt *et al.*, 1995).

Macromitrium species mainly grow on tree trunks, branches, twigs and basal trunks, on fallen or rotten trunks and barks, logs, branches and twigs. A few species of the genus occasionally grow on stones and cliffs such as *M. giraldii*, *M. gymnostomum* and *M. japonicum*, even on soil (*M. angulatum*, Vitt *et al.*, 1995).

5.3 Geographic distribution

According to "*Index Muscorum*" (Wijk *et al.*, 1959), the world was divided into 21 moss geographical units (MGUs). *Macromitrium* has a typical pan-tropical distribution pattern, with the richest species around the Pacific Ocean. The 271 confirmed species (including subspecies and varieties) have been reported from fifteen MGUs (Figure 261).

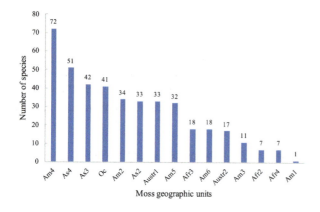

Fig. 261 Species number of *Macromitrium* in fifteen moss geographical units in the world

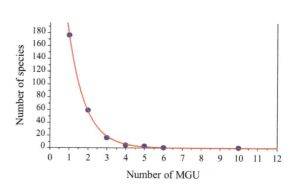

Fig. 262 Relationship between the number of *Macromitrium* species and MGUs with their distribution

Tropical America (Am 4) and tropical Asia (As 4) are the first two most species-rich geographic units, with 72 and 51 species, respectively. The next species-rich regions are South Asia (As 3) and Tropical Ocean (Oc), with 42 and 41 species, respectively. Central America (Am 2), East Asia (As 2), Australia (Austr 1) and the tropical America dominated by the Amazon basin (Am 5) are similar in species richness of the genus, with 34, 33, 33 and 32 species, respectively. Madagascar and its adjacent Indian Ocean islands (Afr 3), Southern South America (Am 6, dominated by temperate and plateau mountain climates) and New Zealand (Austr 2) are not very rich in Macromitrium species, with 18, 18 and 17 species, respectively. There are only seven *Macromitrium* species in both Central Africa (Afr 2) and South Africa (Afr 4). *Macromitrium viticulosum* is the only species that was reported from North America (Am 1). No species has been reported from the remaining moss geographic units (Ant, Eur, As1, Af1, As5). The southern Hemisphere is much richer in *Macromitrium* species than the Northern hemisphere (Figure 261). Figure 262 visually showed that most species are restricted to one to three MGU.

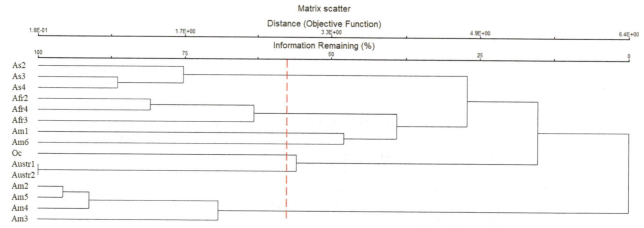

Fig. 263 Cluster dendrogram of 15 moss geographic units based on the distribution of 271 species (with Jaccard coefficient as similarity measurement and ward's linkage method)

Results and Discussion

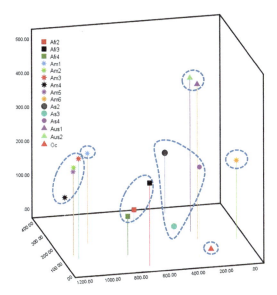

Fig. 264 Three-dimensional ordination of 15 moss geographic units based on distribution of 271 species by using DCA

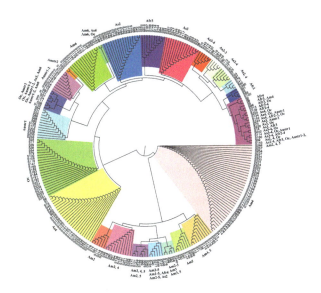

Fig. 265 Cluster dendrogram of 271 species based on their distribution in 15 MGUs

According to the presence/absence data of the species, by using cluster analysis and principal component analysis, we divided the fifteen MGUs into six regions (Figures 263 & 264):

Region 1 includes three Asian MGUs (As 2, As 3 & As 4). There are 92 species of the genus in these three MGUs. Among them 77 species are endemic to this region, accounting for 83.7% of the total. Representative species are *M. angustifolium*, *M. blumei*, *M. blumei* var. *zollingeri*, *M. cuspidatum* var. *gracile*, *M. fortunatii*, *M. nepalense* and *M. microstomum*, which have been recorded in all these three MGUs. Among these three Asian MGUs, East Asia (As 2) is obviously different from As 3 and As 4 in *Macromitrium* flora. A total of sixteen species are endemic to this region, accounting for 48.48% of the total. Representative species are *M. comatum*, *M. ferriei*, *M. hainanense*, *M. holomitrioides*, *M. longipapillosum*, *M. maolanense*, *M. ousiense*, *M. prolongatum*, *M. quercicola*, *M. rhacomitrioides*, *M. schmidii* var. *macroperichaetialium*, *M. subincurvum*, *M. taiheizanense*, *M. taiwanense*, *M. tuberculatum*, and *M. uraiense*. These species are characterized by their small to medium-sized plants, often contorted-flexuose-crisped leaves when dry and pluripapillose upper cells of branch leaves, and ovoid-ellipsoid capsules with single or absent peristome.

A total of 76 *Macromitrium* species have been recorded in As 3 and As 4, 46 species are endemic to this region (accounting for 60.5% of the total), most of them have a typical tropical distribution pattern, particularly for those recorded in As 4 such as *M. acuminatum*, *M. archboldii*, *M. austrocirrosum*, *M. concinnum*, *M. crinale*, *M. erubescens*, *M. longicaule*, *M. longipilum*, *M. megalocladon*, *M. norrisianum*, *M. ochraceum*, *M. parvifolium*, *M. perdensifolium*, *M. pullenii*, *M. semperi*, *M. similirete*, *M. streimannii*, *M. sublongicaule*, *M. tylostomum*, *M. yuleanum*, and *M. zimmermannii*. Some of them have their own morphological characteristics, obviously different from those in As 2. For example, the branch leaves of *M. nepalense* and *M. tongense* have irregular 1-3-stratose proliferation with pluripapillose cells on both dorsal and ventral lamina surfaces, the capsules of *M. sulcatum* have double peristomes, the setae of *M. orthostichum* are papillose.

Region 2 includes three African MGUs (Afr 2, Afr 3 & Afr 4), with 32 species of the genus. Among them, 17 species are endemic to this region, accounting for 53.13% of the total. *Macromitrium serpens* and *M. sulcatum* are the only two species which was recorded in all these three MGUs. Most of the species are characterized by their branch leaves with pluripapillose cells, and some with irregular 1-3-stratose proliferation with pluripapillose cells.

Region 3 only includes Am 1 (North America) with *M. viticulosum*. The species has been recorded from the United States (North Carolina, Florida, Louisiana, Mississippi and Texas, www.tropicos.org), which is the only species that was reported from the North America (Am 1).

Region 4 includes Am 6 (Chile-Patagonia), with eighteen *Macromitrium* species. Among them, fourteen species are endemic to this region, accounting for 77.78% of the total. The fourteen species are characterized by their relatively small plants, appressed and twisted branch leaves (*M. bifasciculare*, *M. bifasciculatum*, *M. campoanum*, *M. pertriste*, *M. saddleanum* and *M. tenax*), or partially and sporadically bistratose laminae in branch leaves. These morphological characteristics are related to the adaptation of these species to the relatively cold and dry climate of the region.

Region 5 includes Austr 1 (Australia), Austr 2 (New Zealand) and Oc (Islands on the Pacific Ocean), with 66 species of the genus. Among them 52 are endemic to the region, accounting for 78.79% of the total species. *Macromitrium brevicaule*, *M. hemitrichodes* and *M. ligulaefolium* were reported in all these three geographic units, and also endemic to this region. Partial species of the genus in this region, such as *M. diaphanum*, *M. peraristatum* and *M. retusum*, are rather special in their morphological features. For example, *M. diaphanum* is characterized by its stiffy and appressed branch leaves, which are broadly oblong-ovate to oblong-lanceolate, irregularly and abruptly narrowed to an irregular and notched tip, and with a hyaline, flexuose, occasionally broken off awn, and with rather obscure and bulging upper cells with irregularly 1- to 4-stratose laminae with simple or irregularly forked papillae. In *M. peraristatum*, its setae are rather short, only 2 mm long. However, its perichaetial leaves are distinctly differentiated, much longer than branch leaves, up to 8 mm long, and with a very long arista up to 3.0 mm long. In *M. retusum*, its young leaves have a very long, linear, green, smooth, stiffly flexuose arista, which is fragile and easily broken off before mature.

Region 6 includes Am 2 (includes Mexico, and some countries in Central America (Guatemala, Nicaragua, El Salvador, Honduras, Panama, Costa Rica, Belize), Am 3 (Caribbean island countries in Central America), Am 4 (Venezuela, Colombia, Ecuador with Galapagos Islands, Peru and Bolivia), Am 5 (Brazil, Paraguay, Guiana, Trinidad and Tobago). Among the six regions, this is the species-richest region for *Macromitrium*, with up to 105 species of the genus. Among them up to 98 species are endemic to this region, accounting for 93.33% of the total in this region. *Macromitrium cirrosum*, *M. punctatum*, *M. scoparium* and *M. swainsonii* are not only endemic to this region, but were also recorded in all these four MGUs. Partial species of the genus in this region are rather special in their morphological features. For example, *M. perreflexum* is characterized by its branch and perichaetial leaves. The branch leaves of the species are loosely and regularly arranged, regularly and strongly squarrose-recurved and often back bent, and its perichaetial leaves with a long awn are distinctly different from and much longer than its branch leaves, being straight and directed upwards to form a conspicuous cluster. *Macromitrium echinatum*, *M. trinitense* and *M. standleyi* are large and robust, with rather long, threadlike linear lanceolate branch leaves, some up to 18 mm long. Their dry habits are much different from those of the species in the other five regions. Most species of the genus in this region have capsules with double peristomes, many have branch leaves with conic-bulging to mammillose upper cells, and some have branch leaves with enlarged teeth-like basal cells at the outmost basal margins. Therefore, *Macromitrium* in Region 6 is not only rather rich in species, but also highly varies in their morphological features, its floristic characteristics are particularly obvious.

Among the 271 recognized species, *M. microstomum* has the widest geographic distribution range, covering all the six regions and ten MGUs. The next is *M. viticulosum*, which has been recorded from six MGUs in North, Central and South America. However, as many as 177 species of the genus are confined to only one MGU, and 59 species are confined to two MGUs (Figure 265). Obviously geographic differentiation of *Macromitrium* species existed among different MGUs (Figure 265). According to Figure 268, at the MGU level, Am 4 is the richest in species endemic to this MGU, next are As 4 and Oc. Am 6, As 2, As 3 and Austr 1 are also rich in regional endemic species. The number of species (S) decreased significantly with the increase of the geographical units of their distribution (N), following: $S = 548.9 * e^{-1.13*N}$ ($R^2 = 0.999$, $N = 7$, $S = 1.584$) (Figure 263). This reflects a high degree of geographic differentiation and fast evolution of the genus, and may be also due to inadequate collection and taxonomic revision in some regions such as Africa and ocean as there are still many species that are known only from their type locality.

Macromitrium microstomum has been recorded from As 2-4, Afr 3, Oc, Austr 1-2, Am 2, 4-5. Therefore, the species is important in revealing floristic relationship of the genus among these MGUs. The Indian subcontinent in As 3 had been geographically connected with Africa until Eocene. Three species were recorded in both these two regions (*M. microstomum*, *M. sulcatum* and *M. subtortum*), which reveals the floristic relationship of the genus between As 3 and Africa in the genus.

5.4 Morphological variations at the intraspecific level

The morphological features are highly variable in nearly all features, not only at the interspecific level, but also at the intraspecific level for partial species, particularly for some widely distributed species. *Macromitrium cirrosum* is the common species of the genus in Central America. Typically it has medium-sized, lanceolate leaves, with an acute to acuminate-acute apex, excurrent costa, more or less isodiametric upper cells that are arranged in distinct longitudinal rows, and tuberculate basal leaf cells, and differentiated, one-cell-thick limbidium above, and enlarged, swollen and hyaline basal marginal cells. However, the size, outline, apex, lamina cells and papillosity highly vary among populations. Moreover, obvious variations of the morphological features even exist among leaves from the same branch or the same mat.

Using macroclimatic models, the potential geographic distribution of *M. giraldii* (as *M. cavaleriei*) was shown to range throughout central, southern and southwestern China to northern Vietnam, Myanmar, Bhutan, Nepal and India, as well as northwards to Japan and the Korean Peninsula (Lou *et al*., 2014). Owing to this wide range, *M. giraldii* has a highly variable branch leaf morphology among different populations, in addition to the considerable variation found within a single population (Guo & He, 2008b; Lou *et al*., 2014). The branch leaf shape varies from linear-lanceolate, lanceolate, ovate-lanceolate to oblong-lanceolate, and can be plane or plicate below, with apices mostly acuminate or narrowly acuminate, but sometimes broadly acuminate or obtuse-acute and incurved (Figure 269). The branch leaf margins are revolute on one side near the leaf base, and are entire, or sometimes indistinctly serrulate or crenulate above due to the protruding papillae on the cells. The branch leaf medial and upper cells are quadrate to subquadrate, slightly to moderately bulging, thin-walled, hyaline or slightly obscure, and often moderately pluripapillose. The branch leaf lower cells are rhombic, rectangular to sublinear, slightly thick-walled and porose, and smooth to weakly or moderately unipapillose, while the basal leaf cells near the margin are often differentiated and rectangular, pellucid, and with thin walls. Compared with *M. giraldii*, *M. japonicum* has a relatively stable morphological features (Figures 266-269).

Macromitrium microstomum is the most widely distributed species of the genus, with ten MGUs (Afr 3, Am 2, 4, 5, As 2-4, Oc, Austr 1-2) (Vitt, 1983; Vitt & Ramsay, 1985; Vitt *et al*., 1995; Cao *et al*., 2008; Guo *et al*., 2017). *Macromitrium microstomum* is well-known and characterized by a combination of the following features: 1) plants small, olive-green, lustrous; 2) branch leaves spiral, funiculate when dry; 3) leaf cells all smooth, flat and clear; 4) dry capsules with a puckered mouth; 5) peristome single, with 16 well-developed teeth and 6) calyptrae mitrate or cucullate, and naked.

In *Macromitrium*, many species exhibit intraspecific variation in laminal papillosity (Guo & He, 2008b, 2014; Yu *et al*., 2012). In contrast, the laminal cells of *M. microstomum* are relatively stable, being smooth and flat throughout the leaves. For this reason, Vitt & Ramsay (1985) considered *M. microstomum* to be a morphologically stable species. As a result of many synonyms and extension of its distribution range, *M. microstomum* has been shown to vary in setae length, the shape of branch and perichaetial leaves as well as in the shape of the capsules. The setae continuously vary from 4.3 to 22.6 mm in length. The branch leaves vary from indistinctly to very distinctly funiculate when dry, and from widely spreading, erecto-patent or erect-inflexed to occasionally curved flexed. The shapes of branch leaves are sometimes shortly ovate-lanceolate with acute to acuminate apices (Figure 270) although they are often ligulate to oblong/ligulate-lanceolate with cuspidate apices. The length of branch leaves short, from 0.5 to 1.5 mm, mostly shorter than 1.2 mm and variation even occurs within the same specimen. The size and shape of the perichaetial leaves differ greatly among specimens from very short (0.6 mm) to rather long (2.0 mm) and from oblong-ovate to oblong-lanceolate (Figure 271). The capsules vary not only in shape but also in size (Figure 272). The presence of all this variation in *M. microstomum* might be responsible for the description of numerous species in the past from different regions of the world as evident by the current inclusion of 24 synonyms under the name (Yu *et al*., 2018).

Quantitative methods including cluster analysis and ordination were used to reveal the relationship between morphologically similar species. *Macromitrium zollingeri* and *M. blumei* are two confusing taxa, and *M. annamense* is a poorly known species. Cluster Analysis and Principle Component Analysis were applied to evaluate the taxonomic distinctiveness of these three species based on eleven character indices of the morphology of leaves and setae sampled from three type specimens and 38 other specimens preserved at H-BR and H (Figures 273-274) (Guo *et al*., 2006). The results showed that *Macromitrium zollingeri* is better recombined as *M. blumei* var. *zollingeri*, which differs from *M. blumei* var. *blumei* in having longer leaf with longer awn, and a ligulate leave shape with higher ratio of leaf length/leaf width, and *M. annamense* was a synonym of *M. blumei* var. *zollingeri* (Figures 275-276; Guo *et al*., 2006).

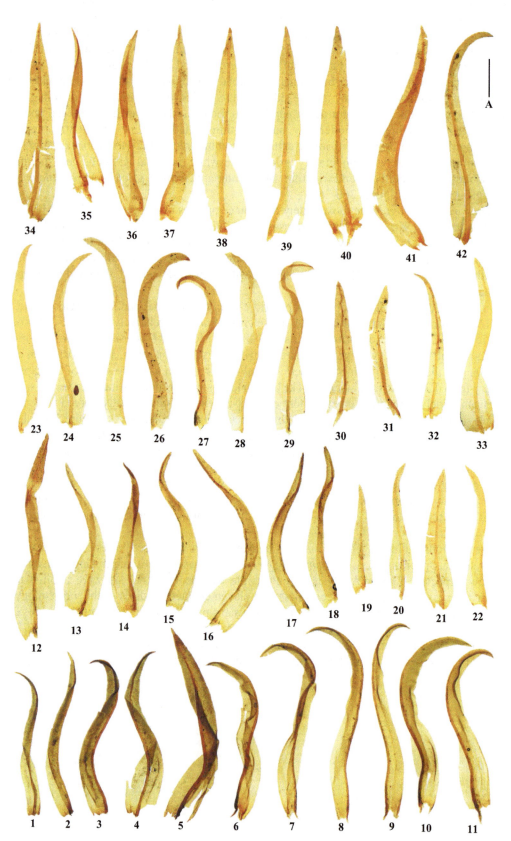

Fig. 266 Branch leaves of *M. giraldii* from types of seven synonyms (1-5 from *M. gebaueri* H-BR 2572001; 6-11 from *M. rigbyanum* BM 000825431; 12-18 from *M. giraldii* var. *acrophylloides* H-BR 2576003; 19-22, 32-33 from *M. syntrichophyllum* PC 0083731; 23-24 from *M. handelii* PC 0083672; 25-31 from *M. handelii* syntype JE; 34-41 from *M. sinense* PC 0083730; 42 from *M. syntrichophyllum* var. *longisetum* PC 0083734). Line scales: A = 500 μm (1-42). (Li *et al.*, 2024).

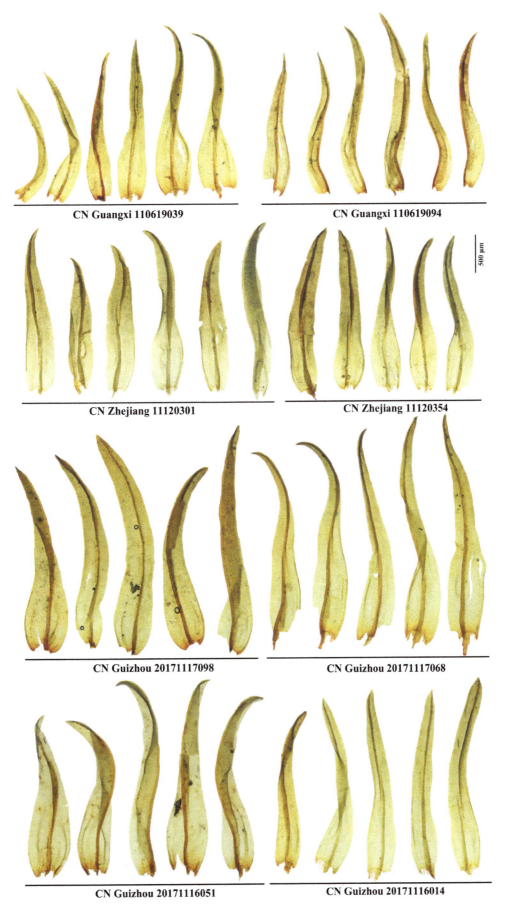

Fig. 267 Branch leaves of *M. giraldii* from eight populations. (Li *et al*., 2024).

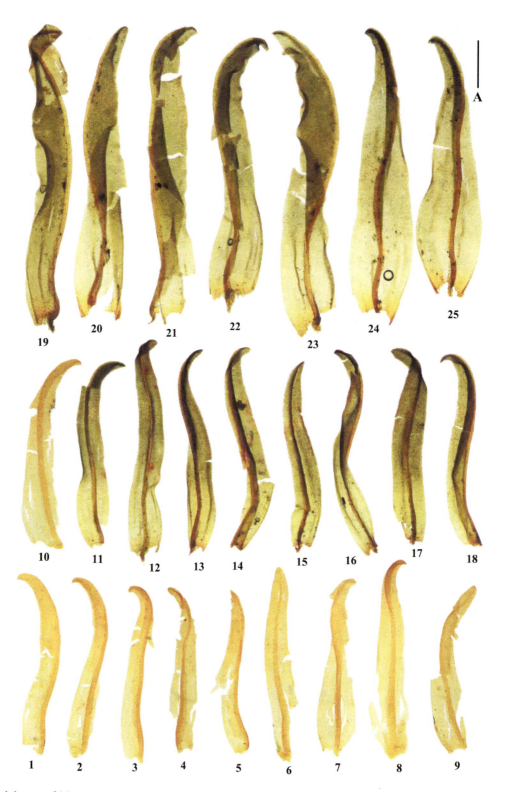

Fig. 268 Branch leaves of *M. japonicum* from the types of three synonyms (1-10 from *Dasymitrium incurvum* PC 0083675; 11-18 from *M. bathyodontum* H-BR 2572017; 19-25 from *M. japonicum* var. *makinoi* H-BR 2572002). Line scales: A = 500 μm (1-25). (Li *et al.*, 2024).

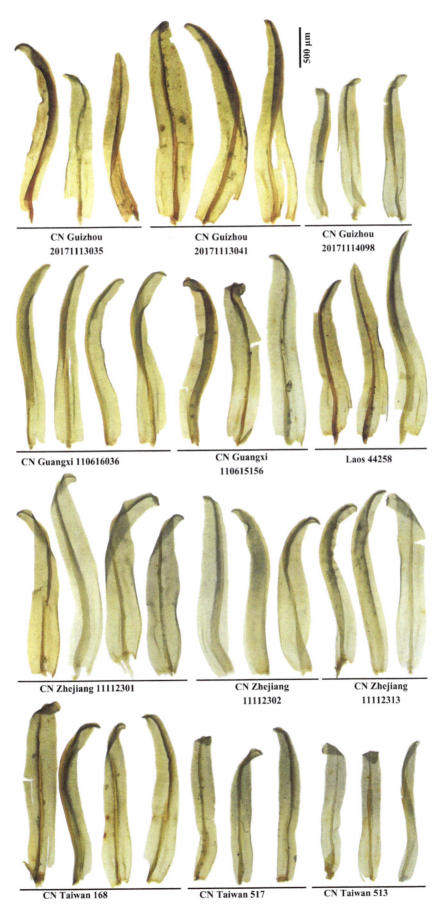

Fig. 269 Branch leaves of *M. japonicum* from twelve populations. (Li *et al.*, 2024).

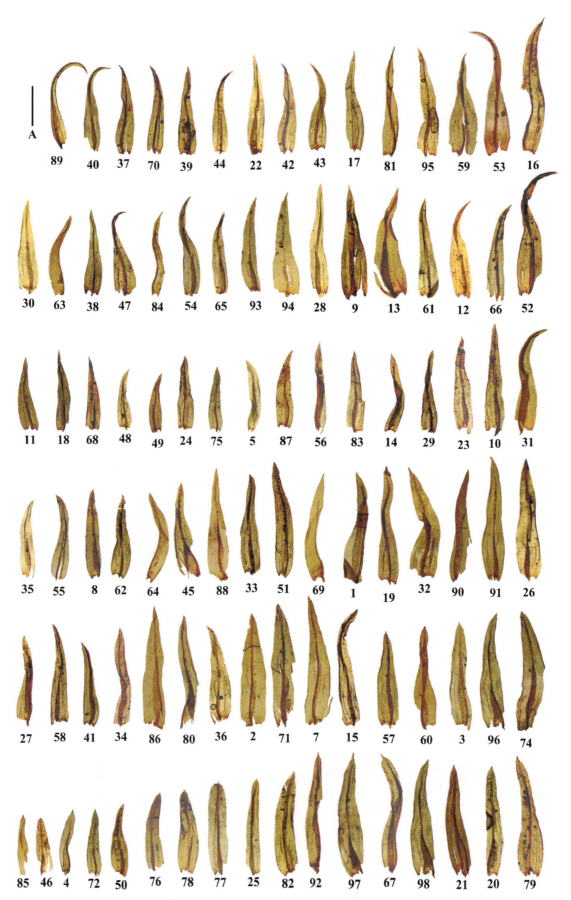

Fig. 270 Branch leaves of *M. microstomum* (Line scale: A = 0.4 mm)
Note: the numbers in the Figure correspond to the specimens listed in Appendix (Yu *et al*., 2018, J. Bryol. 40: 324-332).

Fig. 271 Perichaetial leaves of *M. microstomum* (Line scale: A = 0.4 mm)
Note: the numbers in the Figure correspond to the specimens listed in Appendix (Yu *et al*., 2018, J. Bryol. 40: 324-332).

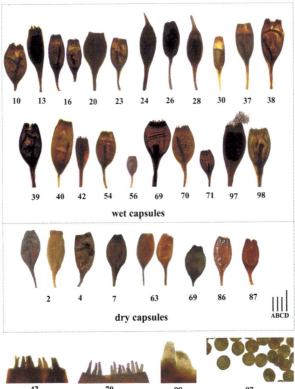

Fig. 272 Capsules of *M. microstomum*. Wet capsules (10, 13, 16, 20, 23, 24, 26, 28, 30, 37, 38, 39, 40, 42, 54, 56, 69, 70, 71, 97, 98); dry capsules (2, 4, 7, 63, 69, 86, 87); peristome (42, 70, 98); spores (97). Line scales: A = 20 μm (spores); B = 500 μm (capsules); C = 100 μm (peristome, 42, 70); D = 50 μm (peristome, 98). The numbers in the Figure correspond to the specimens listed in Appendix (Yu *et al*., 2018, J. Bryol. 40: 324-332).

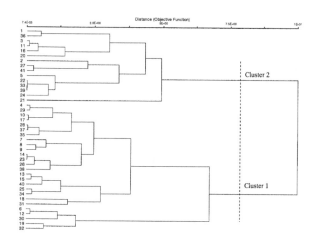

Fig. 273 Cluster dendrogram of 41 specimens of *M. zollingeri*, *M. blumei* and *M. annamense* based on 11 character indices (Guo *et al.*, 2006)

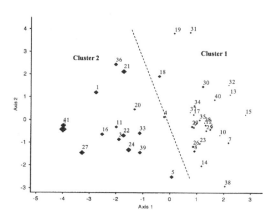

Fig. 274 Ordination plot of 41 specimens of *M. zollingeri*, *M. blumei* and *M. annamense* based on 11 character indices (Guo *et al.*, 2006)

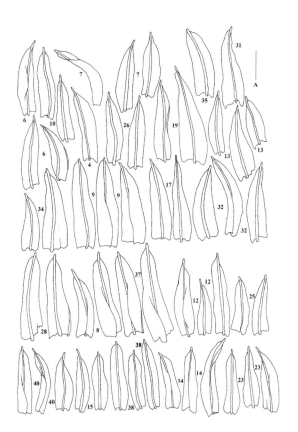

Fig. 275 The variation of leaf shape of *M. blumei* var. *blumei*. Line scale: A = 0.4 mm. (Guo *et al.*, 2006)

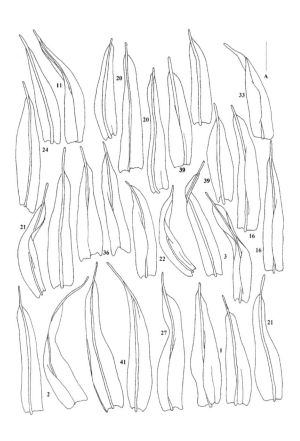

Fig. 276 The variation of leaf shape in *M. blumei* var. *zollingeri*. Line scale: A = 0.4 mm. (Guo *et al.*, 2006)

Results and Discussion

5.5 Systematics and classification of *Macromitrium* – a preliminary result

5.5.1 Molecular phylogeny of Macromitrium

The tree topology of maximum likelihood phylogenetic inference based on the combined dataset with three molecular markers was shown in Figure 277, together with bootstrap (BS) values. All fourty species of *Macromitrium*, *Groutiella and macrocoma* formed a robust clade (BS=100), which belongs to Macromitrieae, with Schlotheimieae (including *Schlotheimia grevilleana* and *S. macgregorii*) as a sister clade. All 122 samples and 40 species, which were assigned to *Macromitrium*, were located in the same clade, indicating that the presently recognized *Macromitrium* was resolved as a monophyletic supported by maximum likelihood bootstrap value (BS=75).

All fourty species of the genus *Macromitrium* in the phylogenetic tree were further divided into two big clades (Clades I and II). Clade I, which supported by maximum likelihood bootstrap value (BS=39), included 17 species being characterized by (1) their capsules with double peristomes; (2) smooth or conic-bulging to mammillose upper cells, smooth to tuberculate low and basal cells; and (3) central and south American distribution patterns (except *M. lorifolium*, *M. turgidum*, *M. sulcatum* in South and Southeast Asia and Africa). Clade II, which supported by maximum likelihood bootstrap value (BS=51), essentially corresponded to the 23 species being characterized by (1) their capsules with single peristome or without peristome; (2) leaf upper and medium cells smooth, conic-bulging to mammillose, unipapillose to pluripapillose or with forked papillae, smooth or unipapillose cells; and (3) East, South, and Southeast Asian, Australia and Ocean distribution patterns.

Sixteen clades, which were supported by Bayesian posterior probabilities to varying degrees, were identified as follows:

Clade 1 (BS=83) including *M. longifolium*, *M. subscabrum* and *M. greenmanii*. The clade is characterized by 1) medium-sized to large plants; 2) branch leaf cells isodiametric in upper and medial portions, elongate towards the base; 3) setae papillose to varying degrees; 4) capsules with double teeth. These nine samples were collected from Bolivia, Columbia and Costa Rica (Am 2 and Am 4).

Clade 2 (BS=75) including *M. trachypodium* and *M. subcirrhosum*. These two species are characterized by 1) large and robust plants; 2) all leaf cells longer than wide, distinctly porose, smooth and arranged in radiating diagonal rows from the costa in the upper portion, weakly to strongly tuberculate in low and basal portions; 3) costa of branch leaves excurrent into the apex to form an arista; 4) capsules with double peristome; and 5) calyptrae mitrate, smooth and naked, lacerate below. These eight samples were from Columbia, Costa Rica, Panama and Peru (Am 2 and Am 4).

Clade 3 (BS=43) including *M. regnellii*, *M. fragilicuspis*, *M. xenizon*, *M. pseudofimbriatum* and *M. swainsonii*. The clade is characterized by 1) small to medium-sized plants; 2) branch leaf cells almost isodiametric in upper portion, elongate and tuberculate towards the base, those at the outmost margin near the insertion enlarged, swollen and teeth-like; and 3) capsules with double peristomes. These fifteen samples were collected from Bolivia, Brazil, Columbia, Honduras, Mexico and Venezuela (Am 2, Am 4 and Am 5).

Clade 4 (BS=95) including *M. lorifolium*, *M. turgidum* and *M. sulcatum*. These species are characterized by 1) medium-sized plants; 2) branch leaf cells rounded-quadrate, smooth or bulging in upper and medial portions, elongate, porose, weakly to strongly tuberculate in low and basal portions; 3) capsules ovoid to ellipsoid, constricted beneath the small mouth, strongly furrowed when dry, with double peristomes; and 4) calyptrae large, smooth and naked, lacerate below. These eleven samples were collected from Guinea, Kenya, Laos, Philippines, and Vietnam (As 3, As 4 and Afr 2).

Clade 5 (BS=86) including *M. scoparium* and *M. standleyi*. These two species are characterized by 1) rather large and robust plants; 2) branch leaves very long, linear-lanceolate to threadlike; 3) branch leaf cells all longer than wide, strong porose, longer and narrower than their ambient cells at upper margins to form a distinct border; 4) smooth and long setae up to 20 mm in length; 5) capsules with double teeth; and 5) calyptrae mitrate, naked and smooth, lacerate. These five samples are collected from Suriname and Honduras (Am 2 and Am 4).

Clade 6 (BS=100) included one species, *M. perichaetiale*. The species is characterized by 1) robust and lustrous plants; 2) branch leaves lanceolate, with a truncate apex having lamin extends on both sides of the costa, forming short asymmetric spiny protuberances; 3) all leaf cells longer than wide, porose, thick-walled, and smooth; 4) distinctly differentiated perichaetial leaves, which are larger and longer than branch leaves, and differ in their leaf shape; 5) smooth setae; 6) ellipsoid capsules with weak furrow and double peristomes; and 7) mitrate and hairy calyptrae. The species was recorded in Saint Vincent and the Grenadines, Guadeloupe, Suriname, and Venezuela (Am 3 and Am 4).

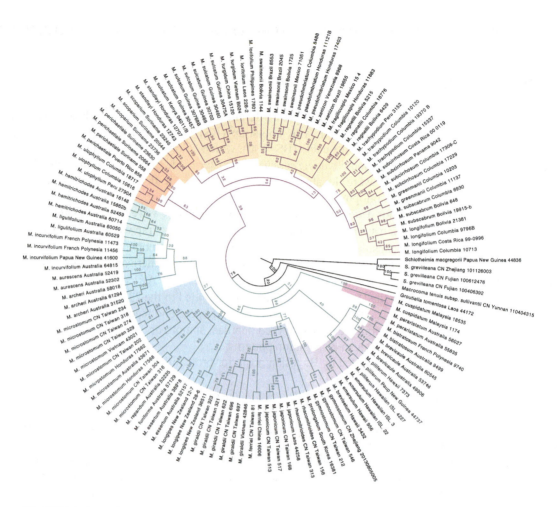

Fig. 277 Maximum likelihood phylogenetic tree based on the combined datasets of *trn*L-F, *trn*G and ITS2
(The maximum likelihood bootstrap values are labelled above the branches.)

Clade 7 (BS=100) including *M. ulophyllum*. The species is characterized by 1) large and robust, lustrous plants; 2) branch leaves large, broadly oblong-lanceolate to broadly lanceolate, with a costa excurrent to form a long arista or cuspidate, and undulate, finely and evenly serrulate upper margins; 3) branch leaf cells all longer than wide, thick-walled, distinctly porose, those in upper interior region elongate-rhombic, thick-walled and porose, arranged in distinctly radiating diagonal rows from the costa, appearing as V- shape, while those near upper margin narrower than their ambient cells, forming a distinct border; 4) setae long and smooth; 5) capsules with double peristomes; and 6) calyptrae large, mitrate, naked and lacerate. The three samples were collected from Peru and Columbia (Am 2 and Am 4).

Clade 8 (BS=100) including *M. ligulaefolium* and *M. hemitrichodes*. They are characterized by 1) small to medium-sized plants; 2) branch leaves with the apices hidden in the inrolled cavity when dry, ovate-lanceolate, ligulate-lanceolate to ligulate, upper and medial cells almost isodiametric, strongly conic-bulging to pluripapillose or unipapillose, low and basal cells elongate, smooth to tuberculate; 3) setae smooth; and 4) capsules with single peristome. These xix samples were all collected from Australia (Austr 1).

Clade 9 (BS=64) including *M. incurvifolium*. The species is characterized by 1) medium-sized plants; 2) branch leaves ligulate-lanceolate, lanceolate to narrowly lanceolate, acuminate; 3) branch leaf cells rounded-quadrate to subquadrate, densely pluripapillose in upper portion, short- to long-rectangular, curved to shallow-sigmoid, smooth and clear in low and basal portions; 4) setae smooth; 5) capsules oblong-ovoid, smooth to lightly 8-plicate, with single peristome; and 6) calyptrae conic mitrate, sparsely to densely hairy. These three samples were collected from Australia, French Polynesia and Papua New Guinea (Austr 1 and Oc).

Clade 10 (BS=76) including *M. archeri* and *M. aurescens*. These two species are characterized by 1) small to medium-sized plants; 2) branch leaves ligulate to ligulate-lanceolate; 3) branch leaf cells rounded to round-quadrate, bulging and pluripapillose in upper and medial portions; 4) setae smooth; 5) capsules 4- or 8-plicate beneath the mouth, with single peristome; and 6) calyptrae conic-mitrate, naked, sparsely to densely hairy. These five samples were all collected from Australia (Austra 1).

Clade 11 (BS=100) including *M. microstomum*, *M. longipes*, *M. repandum*, *M. funiforme*, and *M. exsertum*. These five species are characterized by 1) small to medium-sized plants; 2) branch leaves twisted and funiculate when dry; 3) perichaetial leaves strongly differentiated; 4) basically flat and smooth upper and medial cells; 5) long and smooth setae supporting ovoid capsules with narrowed 8-plicate mouth and well-developed exostome. These nineteen samples were from Australia, China (Taiwan), Honduras, New Zealand and Vietnam (As 2, As 3, Austr 1, Austr 2).

Clade 12 (BS=77) including *M. giraldii*, *M. ferriei*, *M. japonicum*, *M. rhacomitrioides* and *M. prolongatum*. These five species are characterized by 1) small to medium-sized plants; 2) leaf upper pluripapillose cells and unipapillose to smooth low and basal cells; 3) smooth dry capsules with single peristome and exostome of 16 teeth; and 4) hairy calyptrae. These fourteen samples were from China, Laos, Korea and Vietnam (As 2, As 3).

Clade 13 (BS=100) including *M. gymnostomum*. This species is characterized by 1) small to medium-sized plants; 2) ligulate-lanceolate, oblong-lanceolate, lanceolate, or linear-lanceolate branch leaves; 3) branch leaf cells isodiametric, obscure, densely pluripapillose in upper and medial portions, and hyaline and smooth, elongate in low and basal portions; 4) dry capsules deeply plicate and constricted at the mouth, without peristome. These three samples were all collected from China (As 2).

Clade 14 (BS=100) including *M. emersulum*, *M. piliferum* and *M. brevisetum*. They are characterized by 1) small to medium-sized plants with long creeping stems; 2) branch leaves crisped, twisted-contorted when dry, cell isodiametric, rounded to rounded-hexagonal, unipapillose to pluripapillose in the upper portion, and smooth or almost smooth in low and basal portions; 3) perichaetial leaves gradually filiform acuminate to an arista; 4) capsules longitudinally wrinkled beneath the mouth, without peristome, spores large; and 5) calyptrae mitrate, with dense hairs. These seven samples were from Hawaii and Papua New Guinea (Oc).

Clade 15 (BS=100) including *M. brevicaule*. The species is characterized by 1) slender plants in dense and compact mats, with rather short branches; 2) stem leaves with brown, multicellular clavate gemmae; 3) stiff ligulate to broadly oblong branch leaves with dense pluripapillose upper cells, and several rows of longer marginal cells below; 4) capsule urns ovate-ellipsoid smooth to slightly wrinkled, not constricted below mouth; and 5) small, naked and smooth calyptrae, only covering half of the capsule. The species is somewhat similar to *Groutiella* but its place in the phylogenetic tree indicated that the species is a member of the genus *Macromitrium*. These three samples were all from Australia (Austr 1).

Clade 16 (BS=100) including *M. bistratosum*, which is characterized by 1) small plants; 2) the branch leaves often with bistratose laminae in the upper portion, upper leaf cells smooth to weakly pluripapillose; basal cells elongate, clear and smooth, with curved or slightly sigmoid lumens; 3) setae smooth; 4) capsules smooth, with single peristome. The two samples were from Australia and French Polynesia (Austr 1 and Oc).

Clade 17 (BS=100) including *M. peraristatum*. The species is characterized by 1) robust, shiny to lustrous plants; 2) perichaetial leaves distinctly differentiated from, much longer and larger than branch leaves, with a bistratose border and a very long arista; and 3) rather short setae only 2 mm long. The two samples were from Australia (Austr 1).

Clade 18 (BS=100) including *M. cuspidatum*. The species is characterized by 1) robust plants; 2) branch leaves with an asymmetric apical area, sharply contracted to a long cuspidate or a long aristate point, costa long excurrent; and 3) all leaf cells longer than wide, flat and smooth. The two samples were from Malaysia (As 4).

5.5.2 Toward a new classification of Macromitrium

According to the above phylogenetic analysis, we suggest to divide the genus *Macromitrium* in the world into two subgenera: Subgen. *Diplohymenium* and Subgen. *Haplodontiella*.

Subgen. *Diplohymenium* corresponds to Clade I in the above phylogenetic tree (including Clades 1-7). This subgenus is characterized by their capsules with double peristomes and a distribution pattern in South and Central American regions (except clade 4). The representative species of the subgenus is *M. longifolium*.

Subgen. *Haplodontiella* corresponds to the species in Clade II in the above tree (including Clades 8-18). Species in this subgenus have capsules with single peristome or without peristome, distributed in tropical and subtropical Asian, Australian and Ocean regions. The representative species of the subgenus is *M. japonicum*.

Six groups under Subgen. *Diplohymenium* (considering morphological similarity, Clades 2 and 7 were merged into a new group) and 11 groups under Subgen. *Haplodontiella*, each sharing some morphological features and with molecular phylogenetic support, were preliminarily identified. These groups correspond to the clades in the above phylogenetic tree except Group 2 (including Clades 2 and 7), respectively.

The above 17 groups included 40 *Macromtirium* species. Our taxonomical revisions basically confirmed 271 species. Among these confirmed species, 231 were not included in the phylogenetic analysis due to their samples unavailable for us to extract DNA materials. Moreover, the sporophytes in the specimens of many species were immature or unavailable to us. Therefore, not all species could be assigned to the above groups. According to morphological similarity, partial taxonomically-confirmed species were assigned to corresponding groups.

Additionally, According to morphological similarity, we also proposed other five morphological groups, three under Subgen. *Diplohymenium* and two under Subgen. *Haplodontiella*. There is no phylogenetic information yet for these five morphological groups. Finally, a total of 22 morphological groups, nine under the subgen. *Diplohymenium* and thirteen under the subgen. *Haplodontiella*, were proposed here. Among them, seventeen were phylogenetic support to varying degrees.

Group 1 (corresponding to Clade 1) including *M. greenmanii, M. huigrense, M. longifolium, M. mcphersonii, M. nubigenum, M. perrefexum* and *M. subscabrum*. The diagnostic features: setae papillose to varying degrees; capsules with double teeth.

Group 2 (corresponding to Clade 2 & 7) including *M. subcirrhosum, M. trachypodium* and *M. ulophyllum*. The diagnostic features: plants large and robust; all leaf cells longer than wide, distinctly porose, upper and medial cells in radiating diagonal rows from the costa, smooth.

Group 3 (corresponding to Clade 3) including *M. amboroicum, M. atratum, M. atroviride, M. attenuatum, M. brevihamatum, M. cataractarum, M. crassirameum, M. crispatulum, M. fendleri, M. fragilicuspis, M. guatemalense, M. herzogii, M. hildebrandtii, M. laevisetum, M. melinii, M. mosenii, M. nematosum, M. oblongum, M. osculatianum, M. pseudofimbriatum, M. regnellii, M. ruglosum, M. rugulosum, M. subbrevihamatum, M. swainsonii, M. tocaremae* and *M. xenizon*. The diagnostic features: enlarged teeth-like cells differentiated at basal margins of branch leaves; setae smooth, peristome double.

Group 4 (corresponding to Clade 4) including *M. ellipticum, M. fulvum, M. lauterbachii, M. lorifolium, M. sulcatum* (*M. levatum, M. seriatum*), *M. thwaitesii* and *M. turgidum*. The diagnostic features: branch leaves with isodiametric (rarely elongate), smooth to bulging upper and medial cells, low (basal) cells elongate, tuberculate, porose; setae smooth; capsules ovoid to ellipsoid, constricted towards the small mouth, strongly furrowed when dry, with double peristomes; setae smooth; calyptrae large, smooth and naked, lacerate below. This group somewhat corresponded to Sect. *Epilimitrium* M. Fleisch. (Fleischer, 1904).

Group 5 (corresponding to Clade 5) including *M. catharinense, M. cylindricum, M. dubium, M. echinatum, M. fulgescens, M. fuscoaureum, M. leprieurii, M. rimbachii, M. scoparium, M. standleyi, M. trichophyllum* and *M. trinitense*. The diagnostic features: rather large and robust plants with rather long and linear lanceolate branch leaves, all or most leaf cells longer than wide, strong porose; peristome double, setae smooth. This group partially corresponded to Sect. *Reverberatum* W.R. Buck (Buck, 1990).

Group 6 (corresponding to Clade 6) including *M. perichaetiale* and *M. subperichaetiale*, The diagnostic features: lanceolate branch leaves having a truncate apex with lamin extends on both sides of the costa, forming short asymmetric spiny protuberances, or long linear-lanceolate with a gradually acuminate upper portion; all leaf cells longer than wide, porose, thick-walled, and smooth; setae smooth; capsules with double peristome.

Group 7 including *M. aureum, M. caldense, M. cirrosum, M. constrictum, M. crosbyorum, M. frustratum, M. harrisii, M. ovale, M. picobonitum, M. pseudoserrulatum, M. pyriforme, M. raphidophyllum, M. rimbachii, M. sharpii, M. subcrenulatum, M. subdiscretum* and *M. sublaeve*. The diagnostic features: upper cells isodiametric, conic-bulging to mammillose, low (basal) elongate, porose, tuberculate, enlarged teeth-like cells not differentiated at basal margins of branch leaves; capsules with double peristomes; setae smooth.

Group 8 including *M. divaricatum, M. flavopilosum, M. frondosum, M. gigasporum* and *M. rusbyanum*. The diagnostic features: branch leaves lanceolate to linear lanceolate, with all cells clear and smooth, longer than wide or isodiametric in the upper and medial portion; setae smooth; peristome double. This group partially corresponded to Subg. *Teichodontium* (Müll. Hal.) Chen. (Müller, 1845; Chen, 1978)

Group 9 including *M. lomasense, M. macrothele, M. parvirete, M. proliferum, M. punctatum, M. ramsayae* and *M. solitarium*. The diagnostic features: branch leaves ligulate, oblong-ligulate to lanceolate-ligulate, with all cells clear and smooth, isodiametric in the upper and medial portion and slightly elongate near the base; setae smooth; peristome double.

Group 10 (corresponding to Clade 8) including *M. hemitrichodes* and *M. ligulaefolium*. The diagnostic features: branch leaves with the apices hidden in the inrolled cavity when dry, ovate-lanceolate to ligulate-lanceolate, upper and medial cells almost isodiametric, strongly conic-bulging to pluripapillose or unipapillose, low and basal cells elongate, smooth to tuberculate; setae smooth; peristome single.

Group 11 (corresponding to Clade 9) including *M. acuminatum, M. angustifolium, M. incurvifolium, M. laevigatum, M. proximum, M. salakanum, M. semperi, M. speirostichum, M. streimannii, M. subtortum, M. subulatum* and *M. urceolatum*. The diagnostic features: leaves with unistratose laminae, upper pluripapillose cells and low (basal) curved-sigmoid, smooth cells; setae smooth; peristome single (a few absent). This group well corresponded to Sect. *Campylodictyon* Mitt. (Mitten, 1873b).

Results and Discussion

Group 12 (corresponding to Clade 10) including *M. archeri, M. aurescens, M. brachypodium, M. francii, M. fuscescens, M. glaziovii, M. masafuerae, M. nanothecium, M. pilicalyx, M. pilosum, M. prorepens, M. rufipilum, M. subincurvum, M. viticulosum* and *M. zimmermannii*. The diagnostic features: branch leaves ligulate-lanceolate, ligulate to lingulate leaves, with upper and medial almost isodiametric, pluripapillose cells and low (basal) elongate, smooth to unipapillose cells; setae smooth; peristome single.

Group 13 (corresponding to Clade 11) including *M. antarcticum, M. caloblastoides, M. campoanum, M. dielsii, M. exsertum, M. funiforme, M. krausei, M. longipes, M. macrocomoides, M. microcarpum, M. microstomum, M. orthophyllum, M. paridis, M. repandum, M. retusum, M. soulae* and *M. tenax*. The diagnostic features: all leaf cells basically smooth and clear, differentiated perichaetial leaves, often funiculate leaf arrangement; capsules with single peristome; setae smooth. This group partially corresponded to Sect. *Goniostoma* Mitt. (Mitten, 1869).

Group 14 (corresponding to Clade 12) including *M. calocalyx, M. calomicron, M. calymperoideum, M. cardotii, M. chloromitrium, M. clastophyllum, M. comatum, M. falcatulum, M. ferriei, M. giraldii, M. involutifolium, M. japonicum, M. laosianum, M. longicaule, M. longipapillosum, M. moorcroftii, M. nigricans, M. prolongatum, M. quercicola, M. rhacomitrioides, M. sublongicaule* and *M. tosae*. The diagnostic features: branch leaves oblong-lanceolate, lanceolate, to narrowly linear lanceolate, with unistratose laminae, almost isodiametric, pluripapillose upper and medial cells and elongate, straight unipapillose to smooth low (basal) cells; setae smooth; capsules smooth with single peristome; hairy or naked calyptrae. This group partially corresponded to Sect. *Leiostoma* Mitt. (Mitten, 1869).

Group 15 (corresponding to Clade 13) including *M. fimbriatum, M. formosae, M. gymnostomum, M. hainanense, M. holomitrioides, M. ousiense, M. panduraefolium, M. pulchrum, M. savatieri, M. schmidii, M. soulae, M. taiheizanense* and *M. tuberculatum*. The diagnostic features: capsules constricted and puckered beneath the mouth; upper and medial cells of branch leaves often pluripapillose and low (basal) cells smooth to unipapillose; setae smooth; peristome absent.

Group 16 (corresponding to Clade 14) including *M. brevisetum, M. emersulum* and *M. piliferum*. The diagnostic features: branch leaves with isodiametric, unipapillose to pluripapillose upper cells, and smooth or almost smooth low and basal cells; perichaetial leaves gradually filiform acuminate to an arista; capsules longitudinally wrinkled beneath the mouth; setae smooth; peristome absent, spores large.

Group 17 (corresponding to Clade 15) including *M. brevicaule*. The diagnostic features: stem leaves with brown, multicellular clavate gemmaes, several rows of longer marginal cells of branch leaves; small, naked and smooth calyptrae; setae smooth; peristome single.

Group 18 (corresponding to Clade 16) including *M. bifasciculare, M. bifasciculatum, M. bistratosum, M. crassiusculum, M. diaphanum, M. funicaule, M. gracile, M. lebomboense, M. leratii, M. longirostre, M. maolanense, M. nepalense, M. onraedtii, M. pallidum, M. pertriste, M. saddleanum, M. serpens, M. st-johnii* and *M. tongense*. The diagnostic features: laminae of branch leaves partially, sporadically or frequently bistratose, or irregularly 1-(3)4-stratose; setae smooth; peristome single. This group partially corresponded to Sect. *Argyrothrix* Müll. Hal. (Müller, 1872).

Group 19 (corresponding to Clade 17) including *M. peraristatum*. The diagnostic features: distinctly differentiated perichaetial leaves, which are much larger and longer than branch leaves, and with a very long arista (up to 1.3-3.0 mm long); setae smooth; peristome single.

Group 20 (corresponding to Clade 18) including *M. archboldii, M. crinale, M. cuspidatum* and *M. larrainii*. The diagnostic features: leaves with cells all longer than wide, flat and smooth or unipapillose, and sharply contracted to a long cuspidate or a long arista, or with a narrowly acuminate upper portion; setae smooth; peristome single.

Group 21 including *M. angulatum, M. erubescens, M. longipilum, M. megaloclodon, M. minutum, M. noguchianum, M. norrisianum, M. ochraceum, M. orthostichum* and *M. ruginosum*. The diagnostic features: setae papillose, capsules with single peristome. This group partially corresponded to Sect. *Cometium* Mitt. (1873a).

Group 22. including *M. binsteadii, M. concinnum, M. densum, M. dusenii, M. evrardii, M. fortunatii, M. hortoniae, M. macrosproum, M. parvifolium, M. perpusillum, M. petelotii, M. renauldii, M. tylostomum, M. uraiense, M. vesiculosum* and *M. yuleanum*. The diagnostic features: plants often reddish to chestnut-brown; branch leaves mostly oblong-ligulate, ligulate to lingulate, a few lanceolate to narrowly lanceolate, with unistratose laminae, upper conic-bulging, mammillose to unipapillose cells and low (basal) straight, smooth to papillose cells; setae smooth; peristome single.

Among the above 22 morphological groups, Groups 1 to 9 are placed under Subgen. *Diplohymenium*, and Groups 10 to 22 under Subgen. *Haplodontiella*.

The subgeneric classification of the genus *Macromitrium* is based on phylogenetic relationships among the 40 species in the phylogenetic tree. Fleischer (1904) also proposed Subgen. *Haplodontiella* with single peristome, and Subgen. *Diplohymenium* with double peristome. However, the features and species of these two subgenera proposed here are different from those proposed by M. Fleischer. According to the original publication of Fleischer (1904), his

Diplohymenium mainly included the species with double peristome, papillose setae and round, papillose upper cells and tuberculate low (basal) cells. The subgenus *Diplohymenium* proposed here included all species with double peristome, and most of them with smooth setae, and many with clear and smooth cells from apex to base. The Subgen. *Haplodontiella* proposed by Fleischer (1904) included the species with single peristome, smooth basal leaf cells, and smooth setae. While in our subgenus *Haplodontiella*, all the species are included in this subgenus as long as they have capsules with single peristome.

The Subgen. *Macromitrium* suggested by Brotherus (1902) is too broad, including nearly all species of the genus, while the Sugben. *Orthophyllina* (Müll. Hal.) Broth. included species either with single peristome or with double peristome, not supported by the present work.

5.6 Preliminary thought about phylogeny and biogeography of *Macromitrium*

5.6.1 About the morphology of *Macromitrium*

Based on *rbc*L sequence of three species of *Macromitrium* and other twenty species of the family Orthotrichaceae, by using neighbor-joining and maximum parsimony, Goffinet and Vitt (1998) found that *Macromitrium* was not monophyletic. The genus appeared to be an artificial assemblage. However, when forty species of the genus *Macromitrium*, which represented different morphological groups and geographical taxa, were included as in-group, and *Schlotheimia macgregorii* Broth. & Geh., *S. grevilleana* Mitt., *Macrocoma tenuis* subsp. *sullivantii* (Müll. Hal.) Vitt, *Groutiella tomentosa* (Hornsch.) Wijk & Margad. as out-group, based on nuclear marker (ITS1-5.8S-ITS2) and two plastid markers (*trn*G and *trn*L-F), by using Maximum likelihood (ML), *Macromitrium* appeared to be a monophyletic genus, with relatively high phylogenetic support (Fig. 278).

Mitten (1869) divided the genus *Macromitrium* into four sections. Two of these have been raised to the genus level, namely *Micromitrium* (*Groutiella* Steere) and *Macrocoma* Grout. Morphologically, *Macromitrium* is utmost similar to *Groutiella* and *Macrocoma*. *Groutiella* differs from *Macromitrium* by the marginal limbidium of hyaline elongate cells at base and a small calyptra covering only the upper portion of the capsule. Although *Macromitrium brevicaule* is somewhat similar to *Groutiella* in its gametophytes by its small calyptrae and several rows of slightly elongate and narrow marginal cells below, the species was well located between the other species of the genus *Macromitrium* in the phylogenetic tree (Fig. 278). Allen (2002) once remarked that some species of the genus *Groutiella* have enlarged, teeth-like marginal cells at the insertion, which links *Groutiella* to *Macromitrium* via *Macromitrium guatemalense*-group (with enlarged, teeth-like marginal cells at the insertion). Unfortunately, the species of *Groutiella* with such features were not included in our phylogenetic analyses because their samples are unavailable to us.

The genus *Macrocoma* could be distinguished from *Macromitrium* by its slender and filiform plants in tangled mats. The primary stems are long creeping, irregularly to subpinnately branched, with slender and widely spaced, erect-ascending branches often with secondary branchlets. The leaves are small, straight, closely and stiffly erect-appressed to imbricate when dry. The laminal cells are small, rounded to oval throughout the leaf or only elliptic and slightly elongate near the base, smooth to mammillose and plane to bulging but never papillose. Some species of *Macromitrium* are similar to those of *Macrocoma* by the appressed and stiffy branch leaves, such as *M. orthophyllum* and *M. saddleanum*, but they could be separated from *Macrocoma* either by their elongate to linear basal cells of branch leaves, or by distinctly aggregating branches.

5.6.2 About general evolutionary directions and geographic region of origin

The function of the peristome is to regulate the dispersal of spores. The peristome of most mosses is positioned so that spores can be released during episodes of drying and contained within the urn during extended periods of dampness. Xerophytic mosses often reduced peristomes (Vitt, 1971, 1981, 1983). The possible reason is that since the habitat is dry for a large percentage of time, spore dispersal need not to be regulated as precisely as would be necessary under mesic forest environments where wet-dry cycles would be much more frequent (Vitt, 1983). Within the family Orthotrichaceae, peristome has overall reduced in evolution, which has been reported in other genus such as *Ulota* (Wang *et al.*, 2020), *Orthotrichum* (Lewinsky, 1993) and *Zygodon* (Calabrese, 2006). The overall trend in evolution of the peristome in *Macromitrium* seems to be from double to single, finally to absent (Vitt, 1983).Moreover, such a trend is unidirectional and irreversible. Therefore, the feature of peristome is rather important for reconstructing the phylogeny of the genus *Macromitrium*. Overall, in *Macromitirum*, the species with double peristome are primeval and ancestral, while those with signal or without peristome are descendent.

According to Fig. 278, nearly all of the species in Subgen. *Diplohymenium* have been recorded from tropical America (Am 2-5), while most of the species in Subgen. *Haplodontiella* have been reported from other geographical regions including Australasia (Austr 1-2), Africa (Afr 2-4), Oc, Asia (As 2-4), the Southern South America (Am 6) and a narrow geographical range of North America (Am 1). Therefore, tropical America is very likely the original region of the genus *Macromitrium*. We speculated that all the characteristics adapted to mesophilic and humid, as well as warm environments are plesiotypic, and those adapted to arid and cold environments are apotypic.

Generally speaking, moss gametophytes in mesic and tropical rain forests should be larger than those in dry forest habitats. Therefore, large and robust plants with long branches and leaves are plesiotypic, while small plants with short branches and leaves are apotypic. The species in tropical America, such as *M. frondosum*, *M. trachypodium* and *M. ulophyllum*, are rather large and robust. Their branch leaves are often 6-7 mm long. However, many species in the MGUs excluding tropical America are rather small with short branch leaves, such as *M. angulatum* in As 3-4, Austr 1. And *M. giraldii*, *M. gymnostomum*, *M. japonicum* and *M. ousiense* in As 2. For example, the branch leaves of *M. angulatum* are only 0.7-1.5 mm long.

There are various habits of gametophytes for the genus when dry. We thought that fluffy and loose, not or less flexuose, crisped or contorted branch leaves are plesiotypic, while the leaves, which are tightly appressed, tightly spirally twisted, strongly and tightly contorted-twisted-crisped, are apotypic because they are more able to hold water in arid environments. Therefore, leaf sets with Types 1-12, 14 are typically apotypic, and species with these habits were often recorded from the MGUs excluding tropical America. The representative species with such features are *M. bifasciculatum*, *M. crispatulum*, *M. giraldii*, *M. gymnostomum*, *M. japonicum*, *M. moorcroftii*, *M. nepalense*, *M. saddleanum* and *M. sulcatum*. The former seven species are often recorded from dry forests and latter two species from relatively cold environments in the southern South America (Am 6). Contrary to the above situation, leaf set with Types 16-18 are plesiotypic, species with such habits were mainly recorded from tropical America. The representative species are *M. catharinense*, *M. cylindricum*, *M. echinatum*, *M. flexuosum*, *M. frondosum*, *M. gigasporum*, *M. herzogii*, *M. rusbyanum*, *M. scoparium*, *M. standleyi*, *M. trichophyllum* and *M. trinitense*.

The leaves with unistratose laminae are not well adapted to arid environment compared with those having bistratose or irregularly 1-3-stratose laminae of branch leaves. Nearly all species in tropical America have leaves with unistratose laminae, while species with bistratose or irregularly 1-3-stratose laminae are mainly from other MGUs, such as *M. bifasciculare*, *M. bifasciculatum*, *M. bistratosum*, *M. lebomboense*, *M. leratii*, *M. longirostre*, *M. maolanense*, *M. nepalense*, *M. onraedtii*, *M. pertriste*, *M. saddleanum*, *M. serpens* and *M. tongense*. Therefore, unistratose laminae are plesiotypic, while bistratose or irregularly 1-3-stratose laminae are apotypic.

Hyaline leaf tips are related to water holding capacity. Species with hyaline leaf tips are better adapted to dry conditions. Thus branch leaves with hyaline tips are apotypic. Representative species are *M. diaphanum*, *M. ochraceoides*, *M. panduraefolium*, *M. piliferum*, *M. pullenii* and *M. taiwanense*, these species have been reported from As 2, As 4 and Am 6.

According to out-group comparisons, oblong, oblong-lanceolate and lanceolate are plesiotypic, representative species are *M. punctatum*, *M. microstomum*, *M. cirrosum*, *M. longifolium*, *M. guatemalense*, *M. sharpie*, *M. pesudofimbriatum* and *M. viticulosum*, while rather long and threadlike such as *M. standleyi*, *M. trichophyllum*, or very short and lingulate leaves are apotypic such as *M. densum*, *M. swainsonii*.

Some species, such as *M. clastophyllum*, *M. fernandezianum*, *M. fragilicuspis*, *M. helmsii*, *M. perfragile*, *M. petelotii*, *M. retusum*, *M. uraiense* and *M. xenizon*, with branch leaves which are fragile and easily broken off in upper portions. This is an ecological adaptation strategy to reproduce via gametophyte fragments in dry habitats and apotypic.

Branch leaves with an arista or long and narrowly acuminate upper portion are apotypic, because plants with such leaves, which are contorted, crisped and flexuose when dry, are good for water retention in arid conditions.

Branch leaves with well-developed papillae, unipapillae, tuberculate or pluripapillae, or conic-bulging are apotypic, because the leaves with such features help them to hold water in dry environments, while those with smooth and flat cells are pleisotypic because they are related to more mesic habitats.

Cells of branch leaves with pores and sigmoid curved lumens are apotypic, while those without pores and with straight lumens are plesiotypic according to out-group comparison. Moreover, the former are good for lateral (horizontal) transport of water between cells in dry habitats.

Branch leaves with brood-bodies are apotypic, otherwise plesiotypic. Gametophytes of some mosses are able to reproduce via asexual reproduction by brood-bodies, which is good for them to survive in dry habitats. A few species of the genus, such as *M. aurescens*, *M. brevicaule*, *M. gymnostomum*, *M. maolanense*, *M. nematosum* and *M. subhemitrichodes* are able to produce multicellular clavate gemmae.

Perichaetial leaves differentiated from branch leaves, particularly for those which are much larger and longer than branch leaves, are apotypic, because long and large perichaetial leaves are good for protect young sexual reproductive organ in dry habitats. According to out-group comparison, distinctly differentiated perichaetial leaves are also apotypic.

Short setae, particularly for extremely short setae, are plesiotypic, because short setae are characteristic of taxa able to tolerate xerophytic habitats. Representative species with rather short setae are *M. brachypodium*, *M. brevisetum*, *M. chloromitrium*, *M. francii*, *M. hainanense*, *M. prolongatum* and *M. taiheizanense*. They all have been reported from the MGUs excluding tropical America.

Among the 271 species recognized here, only twenty-one species have papillose setae. A smooth seta is common in the Bryidae and thus is judged as plesiotypic. According to out-group here, setae with papillae are apotypic, while smooth setae are plesiotypic.

As stated above, capsules with peristome, particularly with double peristome are plesiotypic, while those with single peristome or without peristome are apotypic. Most of *Macromitirum* species reported from tropical America are characterized by their capsules with double peristome, but almost all *Macromitrium* species in the MGUs excluding tropical America have capsules with single peristome or their peristome are absent.

Species of *Macromitrium* highly vary in capsule shape and plication. In tropical America, many species have ellipsoid-cylindric and furrowed capsules with double peristome, while in other MGUs, species often have smooth capsules with single or caduceus peristome. Therefore, we speculated that irregularly plicate, or 8-plicate capsule with puckered near the mouth are plesiotypic, while smooth capsules are apotypic.

Among the 33 species of the genus recorded in As 2, almost all species have capsules with hairy calyptrae, only *M. gymnostomum* and *M. microstomum* have naked calyptrae. In Am 4, on the contrary, among 77 species most with double peristome, only *M. catharinense*, *M. gigasporum*, *M. longifolium*, *M. perreflexum*, *M. punctatum* and *M. viticulosum*, have hairy capsules. Therefore, hairy calyptrae are apotypic, while naked calyptrae are pleisotypic.

5.6.3 About the endemism, disjunction and speciation time

A total of 111 *Macromitrium* species have been recorded so far from Central and South America (Am 2-6), of which 103 are endemic to this region, indicating a high percentage of endemism in South and Central America of the genus (up to 92.79%). In Africa (Afr 2-4), twenty-six *Macromitrium* species have been recorded, among them 19 are endemic to this region, with 73.08% of endemism in Africa of *Macromitrium* species. In Australasia (Austr 1 and Austr 2), 35 species have been reported so far, among them 23 are endemic to this region, indicating 65.71% of endemism at the species level in this region. In East Asia (As 2), a total of 33 species have been reported so far, and 17 are endemic to this region, accounting for 51.5% of endemism of *Macromitrium* species. The high percentage of endemism in the above regions suggests a potential for relatively rapid evolution and diversification of the genus.

However, only three species (including *M. chloromitrium*, *M. microstomum* and *M. viticulosum*) have been recorded in both Central, South America and Africa, indicating a very low percentage of disjunction element of the genus (2.24%) between these two regions. Likewise, also only three species (*M. gracile*, *M. involutifolium* and *M. microstomum*) have been recorded in both Central, South America and Australasia, a very low percentage of disjunction element at the species level for the genus (2.09%). Similar situations also existed between Asia and Africa, between Asia and South America.

Based on the flora of *Macromitrium* in New Zealand, Vitt (1983) thought that *Macromitrium* originated in Gondwanaland after the split of Pangaea, but before the breakup of the Gondwanaland continental mass. Among the 17 species recorded in New Zealand, 15 were also reported from Australia. Therefore, the species of the genus *Macromitrium* in Australasia originated after the breakup of Gondwanaland but before the division of the Australian plate. Vitt (1983) thought that the modern *Macromitrium* species in Australasia have evolved from these mostly extinct ancestral taxa.

East Asia (As 2) and South, Southeast Asia (As 3) are two adjacent MGUs on land. A total of 63 species of the genus reported from these two regions. However, their flora are obviously different, only twelve species were shared by them, indicating different origin of their flora. Considering a high percentage of endemism of their species, a low percentage of disjunction distribution element between As 2 and As 3, as well as with other MGUs, we thought most of *Macromitrium* species in As 2 and As 3 mainly independently originated. Their common species, such as *M. giraldii*, *M. japonicum*, *M. microstomum*, *M. nepalense* and *M. toase* resulted from dispersal between each other.

According to the flora of the genus in the world, and the percentage of endemism and disjunction distribution elements of the species among 15 MGUs, as well as their morphological characters and molecular phylogeny, we thought the genus originated in tropical America of Gondwanaland after the split of Pangaea, and most species of the genus independently originated after the breakup of the continental plates.

Acknowledgements

The first author is deeply grateful to Dr. He Si, Bruce Allen, John Atwood at Missouri Botanical Garden (MO), to Jaakko Hyvönen, Timo Koponen, He Xiaolan, Johannes Enroth, Viivi Virtanen, Sirkka Sälkinen and Juhani Heino at the Herbarium of Natural History Museum of the University of Helsinki, for their hospitality and great help during his visit to these herbaria. We wish to thank the curators of ALTA, B, BM, E, FH, FI, FIELD, G, GACP, GOET, H, H-BR, HIRO, Hattori, HSNU, IBSC, IFP, JE, KRAM-B, KUN, L, MEL, MICH, MO, NICH, NY, PC, PE, S, SHM, SHTU and US for their sending us specimens on loan during our study. We also received some specimens as a gift from Zhang Li at Fairy Lake Botanical Garden and Vítězslav Plášek at University of Ostrava (Czech). Without their help on specimens, we could not finish the present revision of this problematic genus with such a large number of species.

We are especially grateful to Prof. Cao Tong at Shanghai Normal University, He Si at Missouri Botanical Garden for their valuable suggestions and inspired ideas about the taxonomic treatments of many problematic species. Thanks are also due to Wang Youfang and Zhu Ruiliang (East China Normal University), Jia Yu (Institute of Botany, Chinese Academy of Sciences), Xiong Yuanxin (Guizhou University), Wu Yuhuan (Hangzhou Normal University), Zhao Zuntian (Shandong Normal University), Fang Yanming (Nanjing Forestry University), Ji Mengcheng (Zhejiang Agriculture and Forestry University), Sha Wei (Qiqihar University), Yi Yanjun (Qingdao Agricultural University), Zhao Jiancheng (Hebei Normal University), Sulayman Mamtimin (Xinjiang University), Ma Wenzhang (Kunming Institute of Botany, Chinese Academy of Sciences), He Lin (Zunyi Normal University), Ryszard Ochyra (Polish Academy of Science), Bernard Goffinet (University of Connecticut, USA) and Joanna Wilbraham (Natural History Museum, UK), for their help in various forms during our work on the project. We also thank Li Sha, Liu Xiaohui, Zhan Ling, Wu Qianqian, Li Yan, Li Yuehan, Ren Yan, Liu Shutong and Yang Jun (Shanghai Normal University) for their participation in the project to varying degrees. Here the first author also deeply remembers the late professor, Prof. Benito C. Tan (National University of Singapore), who also helped us a lot in the present project.

We gratefully acknowledge the support by the National Fund for Academic Publication in Science and Technology (2019), National Natural Science Foundation of China (Grants: 32100171, 32071643, 31370233, 30970184, 30570121), Shanghai Sailing Program (No. 20YF1435500). The first author also thanks the financial support by the State Scholarship Fund of China for his visit to the University of Helsinki in 2003-2004, and the Finnish-Chinese Botanical Foundation during that time.

References

Allen, B. H. 1998. Five new species of *Macromitrium* (Musci: Orthotrichaceae), with a key to the species of *Macromitrium* in Central America. Novon, 8(2): 113-123.

Allen, B. H. 2002. Moss flora of Central America, Part 2. Encalyptaceae-Orthotrichaceae. Monographs in Systematic Botany from the Missouri Botanical Garden, 90: viii + 699 pp.

Ångström, J. 1876. Prima lineae muscorum cognoscendorum, qui ad Caldas Brasilia sunt collecti. Öfversigt af Kongl. Vetenskaps-Akademiens Förhandlingar, 33(4): 3-55.

Bartram, E. B. 1928. Costa Rican mosses collected by Paul C. Standley in 1924-1926. Contributions from the United States National Herbarium, 26: i-x, 51-114.

Bartram, E. B. 1940. Mosses of southeastern Polynesia. Occasional Papers of the Bernice Pauahi Bishop Museum of Polynesian Ethology and Natural History, 15(27): 323-349.

Bartram, E. B. 1942. Third Archbold Expedition mosses from the Snow Mountains, Netherlands New Guinea. Lloydia, 5: 245-292. 4 pl.

Bartram, E. B. 1945. Mosses of the Morobe District, northeast New Guinea. The Bryologist, 48(3): 110-126.

Bartram, E. B. 1948. Additional Fijian mosses. Occasional Papers of the Bernice Pauahi Bishop Museum of Polynesian Ethology and Natural History, 19(11): 219-231.

Bartram, E. B. 1952. New mosses from southern Brazil. Journal of the Washington Academy of Sciences, 42(6): 178-182.

Bartram, E. B. 1953. Additional mosses from northeast New Guinea. Svensk Botanisk Tidskrift, 47: 397-401.

Bescherelle, É. 1872. Prodromus bryologiae Mexicanae ou énumération des mousses de Mexique. Mémoires de la Société Nationale des Sciences Naturelles de Cherbourg, 16: 144-256.

Bescherelle, É. 1873. Florule bryologique de la Nouvelle-Calédonie. Annales des Sciences Naturelles; Botanique, sér. 5, 18: 184-245.

Bescherelle, É. 1879. Florule bryologique de la Réunion, de Maurice et des autres îles austro-africaines de l'océan Indien [first part]. Annales des Sciences Naturelles; Botanique, sér. 6, 9: 291-380.

Bescherelle, É. 1895. Florule bryologique de Tahiti et des iles de Nukahiva et Mangareva. Annales des Sciences Naturelles; Botanique, sér. 7, 20: 1-62.

Bescherelle, É. 1898a. Bryologiae Japonicae supplementum I. Journal de Botanique (Morot), 12: 280-300.

Bescherelle, É. 1898b. Florule bryologique de Tahiti (supplément). Bulletin de la Société Botanique de France, 45: 52-67, 116-128.

Bowers, F. D. 1974. The mosses reported from Costa Rica. The Bryologist, 77(2): 150-171.

Bridel, S. É. v. 1819[1818]. Muscologiae Recentiorum Supplementum Pars IV seu Mantissa Generum Specierumque Muscorum Frondosorum Universa. Gotha: C. G. Ettinger, 220 pp., pls. 1 and 2.

Bridel, S. É. v. 1826. Bryologia Universa. Leipzig: J. A. Barth, 1: [i*-ii*], [1]-746.

Britton, E. G. 1896. An enumeration of the plants collected by H. H. Rusby, in Bolivia, 1885-1886.–II. Bulletin of the Torrey Botanical Club, 23: 471-499.

Brotherus, V. F. 1893. Some new species of Australian mosses. II. Öfversigt af Finska Vetenskaps-Societetens Förhandlingar, 35: 34-56.

Brotherus, V. F. 1895a. Some new species of Australian mosses. III. Öfversigt af Finska Vetenskaps-Societetens Förhandlingar, 37: 149-172.

Brotherus, V. F. 1895b. Beitrage zur Kenntniss der Brasilianischen Moosflora. Hedwigia, 34: 117-131.

Brotherus, V. F. 1895c. Nouvelles contributions à la flore bryologique du Brésil. Bihang till Kongliga Svenska Vetenskaps-Akademiens Handlingar, 21 Afd. 3(3): 1-76.

Brotherus, V. F. 1897. Musci Africani II. Botanische Jahrbücher für Systematik, Pflanzengeschichte und Pflanzengeographie, 24: 232-284.

Brotherus, V. F. 1898. Some new species of Australian mosses IV. Öfversigt af Finska Vetenskaps-Societetens Förhandlingar, 40: 159-193.

Brotherus, V. F. 1899. Contributions to the bryological flora of southern India. Records of the Botanical Survey of India, 1(12): 311-329.

Brotherus, V. F. 1902. Musci. Orthotrichaceae. In H. G. A. Engler & K. Prantl (eds.), Natuerlichen Pflanzenfamilien. Leipzig: Engelmann, I(3): 456-498.

Brotherus, V. F. 1908. Musci Voelzkowiani. In A. Voeltzkow, eise in Ostafrika, Systematische Arbeiten. Stuttgart: E. Nägele, 3: pp. 49-64, pl. 7-9.

Brotherus, V. F. 1924. The Musci of the Juan Fernandez Islands. In C. J. F. Skottsberg, The Natural History of Juan Fernandez and Easter Island. Uppsala: Almquist & Wiksells, 2: pp. 409-448, pl. 26-27.

Brown, R. R., Wright, C. H. & Darbishire, O. V. 1905. The Botany of Gough Island.—II. Cryptogams (excluding Ferns and Unicellular Algæ). Botanical Journal of the Linnean Society, 37(259): 263-267.

Buck, W. R. 1990. Contributions to the moss flora of Guyana. Memoirs of the New York Botanical Garden, 64: 184-196.

Buck, W. R. 2003. Guide to the plants of central French Guiana. Part 3. Mosses. Memoirs of the New York Botanical Garden, 76(3): vi + 167 pp.

Calabrese, G. M. 2006. A taxonomic revision of *Zygodon* (Orthotrichaceae) in southern South America. The Bryologist, 109: 453-509.

Cao, T., Shi, C. L., Guo, S. L., *et al.* 2008. On *Macromitrium microstomum*, a notable moss species and its geographic distribution. Guihaia, 28(3): 336-339.

Cardot, J. 1897. Contribution á la flore bryologique de Java, mousses récoltées par M. J. Massart. Annales du Jardin Botanique de Buitenzorg, Supplément 1: 1-31.

Cardot, J. 1904. Premiere contribution a la flore bryologique de la Coree. Beihefte zum Botanischen Centralblatt, 17: 1-44.

Cardot, J. 1905. Mousses de l'île Formose. Beihefte zum Botanischen Centralblatt, 19: 85-148.

Cardot, J. 1908. Notes bryologiques III. Sur une petite collection de mousses de la Nouvelle-Calédonie. Bulletin de l'Herbier Boissier, sér. 2, 8: 166-172.

Cardot, J. 1909. Diagnoses préliminaires de mousses mexicaines. Revue Bryologique, 36: 67-77, 81-88, 105-115.

Chen, P. C. 1978. Genera Muscorum Sinicorum. Pars Secunda. Beijing: Science Press, viii + 331 pp. Fig. 192-400.

Churchill, S. P. & Fuentes, A. F. 2005. Additions, combinations, and synonyms for the Bolivian moss flora. Tropical Bryology, 26: 119-131.

Churchill, S. P. & Linares Castillo, E. L. 1995. Prodromus bryologiae Novo-Granatensis: introducción a la flora de musgos de Colombia. Parte 2: Grimmiaceae a Trachypodaceae. Biblioteca José Jerónimo Triana, 12: 455-924.

Churchill, S. P. 2016. Bryophyta (Mosses). In R. Bernal González, S. R. Gradstein & M. Celis (eds.) Catálogo de Plantas y Líquenes de Colombia. Universidad Nacional de Colombia (Sede Bogotá), Bogotá: Instituto de Ciencias Naturales: 353-441.

Costa, D. P. da, Peralta, D. F., Carvalho-Silva, *et al.* 2016. Types of the moss names based on Glaziou's collections from Brazil. Taxon, 65(4): 839-861.

Costa, D. P. da, Pôrto, K. C., Luizi-Ponzo, A. P., *et al.* 2011. Synopsis of the Brazilian moss flora: checklist, distribution and conservation. Nova Hedwigia, 93(3-4): 277-334.

Crosby, M. R. & Magill, R. E. 1981. A Dictionary of Mosses. Third Printing. St. Louis: Missouri Botanical Garden, 43 pp.

Crosby, M. R., Schultze-Motel, U. & Schultze-Motel, W. 1983. Katalog der Laubmoose von Madagaskar und den umliegenden Inseln. Willdenowia, 13: 187-255.

Crum, H. A. & Anderson, L. E. 1981. Mosses of Eastern North America. New York: Columbia University Press, 2: pp. 665-1328.

Crum, H. A. & Steere, W. C. 1950. Additions to the moss flora of Panama. The Bryologist, 53(2): 139-152.

De Notaris, G. 1859. Musci Napoani sive muscorum ad Flumen Napo in Columbia. Memorie della Reale Accademia delle Scienze di Torino, ser. 2 18: 437-455.

Dixon, H. N. & Potier de la Varde, R. A. L. 1930. Nouvelle contribution à la flore bryologique de l'Inde. Annales de Cryptogamie Exotique, 3(4): 168-193.

Dixon, H. N. 1932. On the moss flora of Siam. Journal of the Siam Society, Natural History Supplement 9: 1-51.

Dixon, H. N. 1933. Mosses of Hong Kong. With other Chinese mosses. The Hong Kong Naturalist, Supplement 2: 1-31.

Dixon, H. N. 1934. Mosses of Celebes. Annales Bryologici, 7: 19-36.

Dixon, H. N. 1935. A contribution to the moss flora of Borneo. Botanical Journal of the Linnean Society, 50(333): 57-140.

Dixon, H. N. 1942. Papuan mosses. Journal of Botany, British and Foreign, 80: 1-11, 25-35.

Dozy, F. & Molkenboer, J. H. 1844. Musci frondosi ex archipelago Indico et Japonia. Annales des Sciences Naturelles; Botanique, sér. 3, 2(5): 297-316.

Dozy, F. & Molkenboer, J. H. 1861. Bryologia Javanica. Leiden: A. W. Sythoff, 1: 161 pp.

Draper, I. & Hedenäs, L. 2009. Circumscription of European taxa within the *Sciuro-hypnum reflexum* complex (Brachytheciaceae, Bryophyta), based on molecular and morphological data. Taxon, 58(2): 572-584.

Dusén, P. 1903. Patagonian and Fuegian mosses. In G. Macloskie, Reports of the Princeton University Expeditions to Patagonia. Princeton: The University of Princeton, 8(3): 63-104.

Eddy, A. 1996. Splachnobryaceae to Leptostomataceae. In A Handbook of Malesian Mosses. London: Natural History Museum Publications, 3: [iv] + 277 pp.

Ellis, L. T. & Wilbraham, J. 2008. New synonymy in *Macromitrium* (Musci, Orthotrichaceae) and *Syrrhopodon* (Musci, Calymperaceae) in the bryoflora of Réunion Island. Cryptogamie, Bryologie, 29: 23-31.

Fedosov, V. E., Shkurko, A. V., Ignatova, E. A., *et al.* 2022. A review of the genus *Glyphomitrium* Brid. (Rhabdoweisiaceae, Bryophyta) in the Russian Far East. Journal of Bryology, 44(3): 226-246.

Fife, A. J. 2017. Orthotrichaceae. Flora of New Zealand - Mosses, 31: [i–iii] 1-115.

Fleischer, M. 1904. Die Musci der Flora von Buitenzorg. 2: I-XVIII + 381-643. Brill, Leiden.

References

Fleischer, M. 1911. Neue Laubmoose aus Hollandisch neu Guinea. Hedwigia, 50: 279-286.

Flora of North America Editorial Committee. 2014. Bryophyta, part 2. In Flora of North America North of Mexico. New York: Oxford University Press, 28: i-xxii, 1-702.

Florschütz, P. A. 1964. Musci. In A. A. Pulle (ed.). Flora of Suriname. Amsterdam: Koninlijke Vereeninging Indisch Instituut, 6(1): 1-271.

Florschütz-de Waard, J. & Florschütz, P. A. 1979. Estudios sobre criptógamas Colombianas III. Lista comentada de los musgos de Colombia. The Bryologist, 82(2): 215-259.

Forzza, R. C. 2010. Lista de espécies Flora do Brasil http://floradobrasil.jbrj.gov.br/2010. Jardim Botânico do Rio de Janeiro, Rio de Janeiro.

Froehlich, J. 1962 [1963]. Musci novi malesiani collecti a Dre. Guil. Meijer. Revue Bryologique et Lichénologique, 31: 91-94.

Gangulee, H. C. 1976. Mosses of Eastern India and Adjacent Regions. Calcutta: Privately published, 5: pp. (xxvii-xxxv) + xvii + 1135-1462.

Geheeb, A. 1898. Weitere Beiträge zur Moosflora von Neu-Guinea. Bibliotheca Botanica, 44: 1-25.

Goffinet, B. & Vitt, D. H. 1998. Revised generic classification of the Orthotrichaceae based on a molecular phylogeny and comparative morphology. In Bryology for the Twenty-first Century. Leeds: U.K. Maney Publishing and the British Bryological Society, 143-159.

Goffinet, B. 1993. Taxonomic and floristic notes on neotropical Macromitrioideae (Orthotrichaceae). Tropical Bryology, 7: 149-154.

Grout, A. J. 1944. Preliminary synopsis North American Macromitriae. The Bryologist, 47(1): 1-22.

Guo, S. L. & He, S. 2008a. A new species of *Macromitrium* (Orthotrichaceae) from Hainan, China. The Bryologist, 111(3): 505-509.

Guo, S. L. & He, S. 2008b. *Macromitrium cavaleriei*, a little known moss from China with four new synonyms (Musci: Orthotrichaceae). Journal of Bryology, 30(4): 264-270.

Guo, S. L. & He, S. 2014. Toward a new understanding of *Macromitrium nepalense* (Orthotrichaceae), with two new synonyms. The Bryologist, 117(1): 15-21.

Guo, S. L., Tan, B. C. & Virtanen, V. 2006. Taxonomic and morphometric comments on *Macromitrium blumei*, *M. zollingeri* and *M. annamense* (Orthotrichaceae, Bryophyta). Nova Hedwigia, 82: 467-482.

Guo, S. L., Cao, T., Tan, B. C., et al. 2007a. Taxonomic notes on Asian species of Orthotrichaceae (Bryopsida): *Macromitrium* with gymnostomous capsules. Gardens' Bulletin, Singapore, 58: 155-177.

Guo, S. L., Enroth, J. & Koponen, T. 2007b. Bryophyte flora of Hunan province, china. Orthotrichaceae, Annual Botanici Fennici, 44: 1-34.

Guo, S. L., Yu, J. & Ma, Y. H. 2013. Typification of an endangered moss *Macromitrium fortunatii* (Orthotrichaceae, Musci). Chenia, 11: 95-101.

Guo, S. L., Ma, Y. H., Cao, T., et al. 2012. A synopsis of *Macromitrium* (Orthotrichaceae) in China. Cryptogamie, Bryologie, 33(4): 341-355.

Guo, S. L., Wu, Q. Q., Yu, J., et al. 2017. Geographical distribution pattern of *Macromitrium* in the world and its biogeographical significance. Bulletin of Botanical Research, 37(2): 164-173.

Hampe, E. 1844. Icones Muscorum Novorum vel Minus Cognitorum (Hampe). Bonnae: Sumptibus Henry & Cohen, pl. 1-30.

Hampe, E. 1847. Ein Referat über die Columbischen Moose, welch von Herrn Moritz gesammelt wurden. Linnaea, 20: 65-98.

Hampe, E. 1849. Musci frondosi in F. A. W. Miquel, Plantae Regnellianae. Linnaea, 22: 581-583.

Hampe, E. 1860. Muscorum frondosorum florae Australasiae. Linnaea, 30: 623-646.

Hampe, E. 1862. Species novas muscorum ab Dr. Alexandro Lindigio in Nova-Granada mensibus Julio et Augusto a. 1861 collectas. Linnaea, 31: 518-532.

Hampe, E. 1863. Species novas muscorum ab Alexandro Lindigio in Nova-Granada collectas, amplius proposuit. Linnaea, 32: 127-164.

Hampe, E. 1865. Prodromus florae Novo-granatensis ou énumétion des plantes de la Nouvelle-grenade. Annales des Sciences Naturelles; Botanique, sér. 5, 4: 324-378.

Hampe, E. 1872. Musci frondosi in insularis Ceylon et Borneo a Dr. O. Beccari lecti. Nuovo Giornale Botanico Italiano, 4: 273-291.

Hampe, E. 1875. Musci frondosi. In E. Warming, Symbolae ad floram Brasiliae centralis cognoscendam. Videnskabelige Meddelelser fra Dansk Naturhistorisk Forening i Kjøbenhavn, ser. 3, 6: 129-178.

Hartmann, F. A., Wilson, R., Gradstein, S. R., et al. 2006. Testing hypotheses on species delimitations and disjunctions in the liverwort Bryopteris (Jungermanniopsida: Lejeuneaceae). International Journal of Plant Sciences, 167(6): 1205-1214.

Hedenäs, L. 2012. Molecular differentiation within European *Cratoneuron filicinum*, and differences from Asiatic and American populations. Plant Systematics and Evolution, 298(5): 937-945.

Hedwig, J. 1801. Species Muscorum Frondosorum. Lipsiae: Joannis Ambrosii Barthii, vi + 352 pp.

Herzog, T. & Hosseus, C. C. 1938. Contribucion al conocimiento de la Flora Briofita del sur de Chile. Archivos de la Escuela de Fármacia de la Facultad de Ciencias Médicas de Córdoba, 7: 3-95.

Herzog, T. 1909. Beiträge zur Laubmoosflora von Bolivia. Beihefte zum Botanischen Centralblatt. Zweite Abteilung, Systematik, Pflanzengeographie, angewandte Botanik, 26(2): 45-102.

Herzog, T. 1916. Die Bryophyten meiner zweiten Reise durch Bolivia. Bibliotheca Botanica, 87: 347 pp.

Herzog, T. 1921. Die Bryophyten meiner zweiten Reise durch Bolivia. Nachtrag. Bibliotheca Botanica, 88: 1-31.

Herzog, T. 1925. Neue Bryophyten aus Brasilien. Repertorium Specierum Novarum Regni Vegetabilis, 21: 22-33.

Herzog, T. 1932. Neue und bemerkenswerte Bryophyten von H. Burgeff 1927/28 auf Java und Philippinen gesammelt. Annales Bryologici, 5: 69-82.

Herzog, T. 1935. Beiträge zur Kenntnis der Bryophytenflora von Ostafrika. Repertorium Specierum Novarum Regni Vegetabilis, 38: 100-105.

Herzog, T. 1952. Miscellanea bryologica III. Memoranda Societatis pro Fauna et Flora Fennica, 27: 92-110.

Hill, M. O. 1979. DECORANA-A FORTRAN program for detrended correspondence analysis and reciprocal averaging. New York: Cornell University, Ithaca.

Hill, M. O. & Gauch Jr, H. G. 1980. Detrended correspondence analysis: an improved ordination technique. Vegetatio, 42(1): 47-58.

Holl, F. 1830. Verzeichniss der auf der Insel Madeira beobachteten Pflanzen, nebst Beschreibung einiger neuen Arten. Flora, 13(24): 369-392.

Hooker, J. D. 1855. Flora Novae-Zelandiae. London: Lovell Reeve, 2: 378 pp.

Hooker, J. D. 1860 [1859]. Flora Tasmaniae. London: Reeve, 2: iii + 422 pp, pl. 101-200.

Hooker, W. J. & Greville, R. K. 1824. Sketch of the characters of the species of mosses, belonging to the genera *Orthotrichum*, (including *Schlotheimia*, *Micromitrion* [sic] and *Ulota*), *Glyphomitrion*, and *Zygodon*. Edinburgh Journal of Science. 1: 110-133.

Hooker, W. J. 1818. Musci Exotici. London: Longman et al., 1: viii + pl. 1-96.

Hooker, W. J. 1820. Musci Exotici. London: Longman et al., 2: pl. 97-176 + 31 pp.

Ignatov, M. S. & Afonina, O. M. 1992. Checklist of mosses of the former USSR. Arctoa, 1: 1-85.

Lewinsky, J. 1993. A synopsis of the genus *Orthotrichum* Hedw. (Musci, Orthotrichaceae). Bryobrothera, 2: 1-59.

Li, D. D., Guo, S. L. & Fang, Y. M. 2019. Taxonomic notes on *Macromitrium catharinense* (Bryopsida, Orthotrichaceae) with its new synonyms. Nordic Journal of Botany, 37(6): 1-8.

Li, D. D., Li, Y. H., Ren, Y., et al. 2024. Three new synonyms of *Macromitrium japonicum* Dozy & Molk. (Bryophyta, Orthotrichaceae) based on morphological and molecular evidence. Cryptogamie, Bryologie, 45(3): 23-36.

Li, D. D., Yu, J., Cao T., et al. 2017. *Macromtrium longipapillosum* sp. nov. (Bryophyta, Orthotrichaceae) from Japan, with comments on *M. comatum*. Nordic Journal of Botany, 35(6): 711-718.

Li, D. D., Yu, J., Shen, L., et al. 2018. Predictive modelling of the distribution and evaluation of the conservation status with a taxonomic clarification of *Macromitrium fortunatii* Thér. (Orthotrichaceae, Bryophyta) in China and adjacent regions. Cryptogamie, Bryologie, 39(4): 499-513.

Li, D. D., Zhang, Z. Y. & Guo, S. L. 2020. Two new synonyms of *Macromitrium tosae* (Orthotrichaceae) with a key to Indian species of *Macromitrium*. Phytotaxa, 474(3), 250-260.

Lindberg, S. O. 1864. *Dasymitrium*, Novum genus Orthotrichearum. Journal of Botany, British and Foreign, 2: 385-387.

Lorentz, P. G. 1866. Musci frondosi in Chile prope Valdiviam et prope Corral lecti per Dr. Krause. Botanische Zeitung (Berlin), 24(24): 185-189.

Lorentz, P. G. 1868. Musci frondosi a clarissimo H. Krause in Ecuador, prov. Loja collecti. Botanische Zeitung (Berlin), 26: 793-800.

Lou, Y. X., He, S. & Guo, S. L. 2014. Using macroclimatic models to estimate the distribution ranges of taxonomically challenging taxa, an example with *Macromitrium cavaleriei* Cardot & Thér. (Orthotrichaceae). Journal of Bryology, 36(4): 271-278.

Malta, N. 1926. Die Gattung *Zygodon*. Riga: Armijas spiestuve, 184 pp.

Margadant, W. D. 1972. Notes on the nomenclature of Musci. Lindbergia, 1: 121-129.

Miller, M. A., Pfeiffer, W., & Schwartz, T. 2010. Creating the CIPRES Science Gateway for inference of large phylogenetic trees. In Proceedings of the Gateway Computing Environments Workshop (GCE). –Louisiana: New Orleans, pp. 1-8.

Mitten, W. 1856. A list of some mosses and Hepaticae, collected by the Rev. Charles Parish, at Moulmein, and communicated to Sir W. J. Hooker. Hooker's Journal of Botany and Kew Garden Miscellany, 8: 353-357.

Mitten, W. 1859a. Musci Indiae Orientalis, an enumeration of the mosses of the East Indies. Journal of the Proceedings of the Linnean Society, Botany, Supplement 1: 1-171.

Mitten, W. 1859b. Descriptions of some new species of Musci from New Zealand and other parts of the Southern Hemisphere, together with an enumeration of the species collected in Tasmania by William Archer, arranged upon the plan proposed in the Musci Indiae Orientalis (1859). Journal of the Proceedings of the Linnean Society, Botany, 4: 64-100.

Mitten, W. 1868. A list of the Musci collected by the Rev. Thomas Powell in the Samoa or Navigator's Islands. Journal of the Linnean Society, Botany, 10: 166-195.

Mitten, W. 1869. Musci Austro-Americani. Journal of the Linnean Society, Botany, 12: 1-659.

Mitten, W. 1873a. New species of Musci collected in Ceylon by Dr. Thwaites. Journal of the Linnean Society, Botany, 13: 293-326.

Mitten, W. 1873b. Musci. In B. C. Seemann, Flora Vitiensis. London: L. Reeve and Co., pp. 378-404.

Mitten, W. 1876. The Musci and Hepaticae collected by H. N. Moseley, naturalist to H. M. S. Challenger. Journal of the Linnean Society, Botany, 15: 59-73.

Mitten, W. 1883. Australian mosses. Transactions and Proceedings of the Royal Society of Victoria, 19: 49-96.

Mitten, W. 1891. On the species of Musci and Hepaticae recorded from Japan. Transactions of the Linnean Society of London, 2nd series: Botany, 3: 153-206.

References

Montagne, C. 1840. Seconde centurie de plantes cellulaires exotiques nouvelles, Décades VI, VII et VIII. Annales des Sciences Naturelles; Botanique, sér. 2, 14: 321-350.

Müller, C. & Geheeb, A. 1881. Laubmoose. In F. Buchenau, Reliquiae Rutenbergianae. III. Abhandlungen herausgegeben vom Naturwissenschaftlichen Vereins zu Bremen, 7(2): 203-214.

Müller, C. & Hampe, E. 1853. Musci frondosi Australasiae ab Dr. Ferd. Muller lecti. Linnaea, 26: 489-505.

Müller, C. 1845. Nachtragliche Bemerkungen über die von Gardner in Brasilien gesammelten Laubmoose. Botanische Zeitung (Berlin), 3: 89-94, 105-111.

Müller, C. 1848. Ueber die Laubmoose der von Herren Funck und Schlim in Columbien veranstalteten, kauflichen Sammlung des Herrn J. Linden in Luxemburg. Botanische Zeitung (Berlin), 6: 761-768.

Müller, C. 1849. Synopsis Muscorum Frondosorum omnium hucusque Cognitorum. Berlin: Alb. Foerstner, Vol. 1, fasc. 5, pages 641-812.

Müller, C. 1851. Synopsis Muscorum Frondosorum omnium hucusque Cognitorum. Berlin: Alb. Foerstner, 2(9-10): 511-772.

Müller, C. 1854. Musci Neilgherrenses. Botanische Zeitung (Berlin), 11: 17-24, 33-40, 57-62.

Müller, C. 1857a. Decas muscorum Oceani Pacifici. Botanische Zeitung (Berlin), 15: 777-782.

Müller, C. 1857b. Manipulus muscorum e Flora Novae Granadae. Botanische Zeitung (Berlin), 15: 577-583.

Müller, C. 1862. Addimenta ad Synopsin Muscorum nova. Botanische Zeitung (Berlin), 20: 327-329, 337-339, 348-350, 361-362, 373-374, 381-382, 392-393.

Müller, C. 1870 [1868]. De muscorum Ceylonsium collectione. Linnaea, 36: 1-40.

Müller, C. 1872. Musci Australici praesertim Brisbanici novi. Linnaea, 37: 143-162.

Müller, C. 1874a. Novitates bryothecae Müllerianae. 1. Musci Philippinenses praesertim Wallisiani adjectis nonnullis musciis aliis Indicis. Linnaea, 38: 545-572.

Müller, C. 1874b. Novitates bryothecae Müllerianae. 3. Musci Mexicani praesertim a Cl. C. Mohr et Sartorius collecti. Linnaea, 38: 620-660.

Müller, C. 1874c. Musci polynesiaci praesertim Vitiani et Samoani Graeffeani. Journal des Museums Godeffroy, 3(6): 51-90.

Müller, C. 1876. Musci Hildebrandtiani in Archipelago Comorense et in Somalia littoris Africani anno 1875 ab I. M. Hildebrandt lecti. Linnaea, 40: 225-300.

Müller, C. 1879. Musci Fendleriani Venezuelenses. Linnaea, 42: 461-502.

Müller, C. 1883. Die auf der Expedition S. M. S. "Gazelle" von Dr. Naumann gesammelten Laubmoose. Botanische Jahrbücher für Systematik, Pflanzengeschichte und Pflanzengeographie, 5: 76-88.

Müller, C. 1885. Bryologia Fuegiana. Flora, 68: 391-429.

Müller, C. 1890. Die moose von vier kilimandscharo-Expeditionen. Flora, 73: 465-499.

Müller, C. 1896. Bryologia Hawaiica adjectis nonnullis muscis oceanicis. Flora, 82: 434-479.

Müller, C. 1897a. Musci Venezuelenses novi a Prof. C. Goebel collecti. Flora, 83: 327-341.

Müller, C. 1897b. Prodromus bryologiae Bolivianae. Nuovo Giornale Botanico Italiano, n.s. 4(1): 5-50.

Müller, C. 1897c. Prodromus bryologiae Argentinicae atque regionum vicinarum III. Hedwigia, 36: 84-144.

Müller, C. 1898a. Symbolae ad bryologiam Australiae II. Hedwigia, 37: 76-171.

Müller, C. 1898b. Bryologia serrae Itatiaiae (Minas Geraës Brasiliae) adjectis nonnullis speciebus affinibus regionum vicinarum. Bulletin de l'Herbier Boissier, 6: 18-48, 89-126.

Müller, K., Müller, J., Neinhuis, C., et al. 2010. PhyDE: Phylogenetic data editor, version 0.9971. Program distributed by the authors. http://www. phyde. de.

Naveau, R. 1927. Musci Bequaerti. Bulletin de la Société Botanique de Belgique, 60: 11-56.

Noguchi, A. & Z. Iwatsuki. 1989. Illustrated Moss Flora of Japan. Nichinan: Hattori Botanical Laboratory, Part 3: pp. 492-742.

Noguchi, A. 1936. Studies of the Japanese mosses of the orders Isobryales and Hookeriales I. Journal of Science of the Hiroshima University, Series B, Division 2 (Botany), 3: 11-26.

Noguchi, A. 1938. Studies of the Japanese mosses of the orders Isobryales and Hookeriales III. Journal of Science of the Hiroshima University, Series B, Division 2 (Botany), 3: 135-152.

Noguchi, A. 1967. Musci japonici. VII. The genus *Macromitrium*. Journal of Hattori Botanical Laboratory, 30: 205-230.

Onraedt, M. 1976. Bryophytes des îles mascaréno-malgaches et Seychelles I et II. Bulletin du Jardin Botanique National de Belgique, 46: 351-378.

O'Shea, B. J. 2006. Checklist of the mosses of sub-Saharan Africa (version 5, 12/06). Tropical Bryology Research Reports, 6: 1-252.

Pacak, A. & Szweykowska-Kulińska, Z. 2000. Molecular data concerning alloploid character and the origin of chloroplast and mitochondrial genomes in the liverwort species *Pellia borealis*. Journal of Plant Biotechnology, 2(2): 101-108.

Palisot de Beauvois, A. M. F. J. 1805. Prodrome des Cinquième et Sixième Familles de l'Aethéogamie. Paris: Pournier Fils, ii + 114 pp.

Paris, J. É. G. N. 1894-1898. Index Bryologicus. 1380 pp. Paris.

Paris, J. É. G. N. & Brotherus, V. F. 1907a. Muscinées de l'Asie Orientale (5e article). Revue Bryologique, 34: 29-33.

Paris, J. É. G. N. & Brotherus, V. F. 1907b. Muscinées de l'Asie Orientale (6e article). Revue Bryologique, 34: 41-56.

Paris, J. É. G. N. & Brotherus, V. F. 1908. Muscinées de l'Asie Orientale (7e article). Revue Bryologique, 35: 40-55.

Paris, J. É. G. N. 1900. Index Bryologicus. Supplementum Primum iv + 334 pp.

Paris, J. É. G. N. 1910. Muscinées de l'Asie orientale (11e article). Revue Bryologique, 37: 1-4.

Potier de la Varde, R. A. L. 1930. Musci novi africani. Annales de Cryptogamie Exotique, 3: 43-49.

Potier de la Varde, R. A. L. 1955. Mousses récoltées par M. le Dr. Olov Hedberg, en Afrique orientale, au cours de la mission suédoise de 1948. Arkiv för Botanik, n.s. 3: 125-204.

Raddi, G. 1822. Crittogame Brasiliane. Modena: Dalla Società Tipografica, 33 pp.

Rambaut, A. 2018. FigTree v1.4.4. https://github.com/rambaut/figtree

Ramsay, H. P., Cairns, A. & Meagher, D. A. 2017. *Macromitrium erythrocomum* (Bryophyta: Orthotrichaceae), a new species from tropical Queensland, Australia. Telopea, 20: 261-268.

Ramsay, H. P., Vitt, D. H.& Lewinsky-Hapasaari, J. 2006. Orthotrichaceae. In (eds.), Flora of Australia. Melbourne: ABRS/CSIRO, 51: xvii–xix.

Redfearn Jr., P. L., Tan, B. C. & He, S. 1996. A newly updated and annotated checklist of Chinese mosses. Journal of the Hattori Botanical Laboratory, 79: 163-357.

Reese, W. D. 1993. Calymperaceae. Flora Neotropica, Monograph, 58: 102 pp.

Reese, W. D. 1997. Identity and lectotypification of *Orthotrichum undulatum* (Orthotrichaceae, Musci). Taxon, 46: 249-252.

Reinwardt, C. G. C. & Hornschuch, C. F. 1829. Musci frondosi Iavanici, reddidi coniunctis studiis et opera. Nova Acta Physico-medica Academiae Caesareae Leopoldino-Carolinae Naturae Curiosorum Exhibentia Ephemerides sive Observationes Historias et Experimenta, 14(2): 697-732.

Renauld, F. & Cardot, J. 1915. Histoire Physique, Naturelle et Politique de Madagascar, Mousses. Paris: Librairie Hachette, viii + 563 pp.

Robinson, H. E. 1971. Four new species of mosses from Peru. Phytologia, 21: 389-393.

Robinson, H. E. 1975. The mosses of Juan Fernandez Islands. Smithsonian Contributions to Botany, 27: iv + 88 pp.

Roivainen, H. 1936. *Macromitrium melini* n. sp. ex Peru meridionali. Annales Botanici Societatis Zoologicae-Botanicae Fennicae "Vanamo", 6(8): 16-17.

Sainsbury, G. O. K. 1945. New and critical species of New Zealand mosses. Transactions of the Royal Society of New Zealand, 75: 169-186.

Sakurai, K. 1943. Bryoflora von Micronesia. (II). Botanical Magazine, Tokyo, 57: 249-257.

Schatz, G. E., Andriambololonera, S., Lowry II, P. P., *et al.* 2023. Catalogue of the Plants of Madagascar.

Schultze-Motel, W. 1962. Zur Nomenclatur der Laubmoose *Macromitrium noguchianum* und *Pohlia rubripila* aus Neuguinea. Taxon, 11: 179-180.

Schultze-Motel, W. 1975. Katalog der Laubmoose von West-Afrika. Willdenowia, 7: 473-535.

Schwägrichen, C. F. 1823. Species Muscorum Frondosorum, Supplementum Secundum. Leipzig: Barth, sect. 1, pages i-vi + 1-86, plates 101-125.

Schwägrichen, C. F. 1824. Species Muscorum Frondosorum, Supplementum Secundum. Leipzig: Barth, sect. 2, pages [87]-186 + pl. 126-150.

Schwägrichen, C. F. 1826. Species Muscorum Frondosorum, Supplementum Secundum. Leipzig: Barth, vol. 2, sect. 1, pages 1-79 + plates 151-175.

Schwägrichen, C. F. 1827. Species Muscorum Frondosorum, Supplementum Secundum. Leipzig: Barth, vol. 2, sect. 2, pages 81-210 + plates 176-200.

Schwägrichen, C. F. 1842. Species Muscorum Frondosorum, Supplementum Quartum. Leipzig: Barth, sect. 1, plates 301-325.

Sehnem, A. 1978. Musgos Sud-Brasileiros V. Pesquisas, Botânica, 32: 170 pp.

Seki, T. 1974 [1976]. A moss flora of Provincia de Aisén, Chile. Results of the Second Scientific Expedition to Patagonia by Hokkaido and Hiroshima Universities, 1967. Journal of Science of the Hiroshima University, Series B, Division 2 (Botany), 15(1): 9-101 + 6 pl.

Stamatakis, A. 2006. RAxML-VI-HPC: maximum likelihood-based phylogenetic analyses with thousands of taxa and mixed models. Bioinformatics, 22(21): 2688-2690.

Stamatakis, A. 2014. RAxML version 8: a tool for phylogenetic analysis and post-analysis of large phylogenies. Bioinformatics, 30(9): 1312-1313.

Staples, G. W., Imada, C. T., Hoe, W. J., *et al.* 2004. A revised checklist of Hawaiian mosses. Tropical Bryology, 25: 35-70.

Steere, W. C. 1948. Contribution of the bryogeography of Ecuador. I. A review of the species of Musci previously reported. The Bryologist, 51: 65-167.

Steere, W. C. 1982. Four new species of Musci from the Andes of Ecuador and Colombia. Brittonia, 34: 435-441.

Stöver, B. C. & Müller, K. F. 2010. TreeGraph 2: combining and visualizing evidence from different phylogenetic analyses. BMC bioinformatics, 11: 1-9.

Streimann, H. & Curnow, J. 1989. Catalogue of mosses of Australia and its external territories. Australian Flora and Fauna Series, 10: viii + 479 pages.

Sullivant, W. S. & Lesquereux, L. 1859. Characters of some new Musci collected by Charles Wright in the North Pacific Exploring Expedition, under the command of Captain John Rodgers. Proceedings of the American Academy of Arts and Sciences, 4: 275-282.

References

Sullivant, W. S. 1859. United States Exploring Expedition, Musci. Philadelphia: C. Sherman & Son, 1-112.

Swartz, O. 1788. Nova genera & species plantarum seu Prodromus. In O. Swartz Prodr. Holmiae, Upsaliae, & Aboae: M. Swederi, 97.

Taberlet, P., Gielly, L., Pautou, G., et al. 1991. Universal primers for amplification of three non-coding regions of chloroplast DNA. Plant Molecular Biology, 17: 1105-1109.

Tan, B. C. & Shevock, J. R. 2015. Species of *Macromitrium* (Orthotrichaceae) new to the Mindanao region and the Philippines with one species new to science. Proceedings of the California Academy of Sciences, Series 4, 62(3): 541–549.

Tan, B. C. & Iwatsuki, Z. 1991. A new annotated Philippine moss checklist. Harvard Papers in Botany, 3: 1-64.

Tan, B. C. & Iwatsuki, Z. 1993. A checklist of Indochinese mosses. Journal of the Hattori Botanical Laboratory, 74: 325-405.

Taylor, T. 1846. The distinctive characters of some new species of Musci, collected by Professor William Jameson, in the vicinity of Quito, and by Mr. James Drummond at Swan River. London Journal of Botany, 5: 41-67.

Thériot, I. 1906. Diagnoses du quelques mousses nouvelles. Bulletin de l'Académie Internationale de Géographie Botanique, ser. 3, 16: 40.

Thériot, I. 1907. Diagnoses d'especes et de variétés nouvelles de muscinées. Bulletin de l'Académie Internationale de Géographie Botanique, 17: 306-308.

Thériot, I. 1908. Diagnoses d'espèces et de variétés nouvelles de mousses (5e article). Bulletin de l'Académie Internationale de Géographie Botanique, 18: 250-254.

Thériot, I. 1909. Diagnoses d'espèces et de variétés nouvelles de mousses (6e article). Bulletin de l'Académie Internationale de Géographie Botanique, 19: 17-24.

Thériot, I. 1910. Diagnoses d'espèces et de variétés nouvelles de mousses (7e article). Bulletin de l'Académie Internationale de Géographie Botanique, 20: 96-104.

Thériot, I. 1925. Cinquième contribution à la flore bryologique de Madagascar. Recueil des Publications de la Société Havraise d'Études Diverses, 92: 122-151.

Thériot, I. 1931. Mousses de l'Annam, 3e contribution (1). Revue Bryologique, nouvelle série, 3: 181-185.

Thériot, I. 1940. Complement au catalogue des mousses de Cuba. III. Memorias de la Sociedad Cubana de Historia Natural "Felipe Poey", 14: 349-372.

Thouvenot, L. & Müller, F. 2016. *Macromitrium humboldtense* (Orthotrichaceae, Bryophyta), a new species from New Caledonia. Cryptogamie, Bryologie, 37(3): 295-303.

Thouvenot, L. & Yong, K. T. 2015. *Macromitrium larrainii*, a new species of *Macromitrium* (Orthotrichaceae, Bryophyta) from New Caledonia. Cryptogamie, Bryologie, 36(4): 343-348.

Thouvenot, L. 2018. *Macromitrium panduraefolium* (Orthotrichaceae, Bryophyta), a new species from New Caledonia, with a key to the aristate *Macromitrium* species in the Pacific, Malesia and Australasia regions. Cryptogamie, Bryologie, 39(4): 443-450.

Thouvenot, L. 2019. A review of the genus *Macromitrium* Brid. (Orthotrichaceae, Bryophyta) in New Caledonia. Cryptogamie, Bryologie, 40(16): 167-217.

Tixier, P. & Guého, J. 1997. Introduction to Mauritian Bryology: A Check-List of Mosses and Liverworts. Réduit: Mauritius Sugar Industry Research Institute, 233 pp.

Tixier, P. 1966. Bryophytes du Vietnam. Récoltes de A. Petelot et V. Demange au Nord Vietnam (Relictae Henryanae). Revue Bryologique et Lichénologique, 34: 127-181.

Tixier, P. 1989. Bryophyta exotica --- 8. Récoltes de J.-F. Brunel au Togo (1983--1985). Candollea, 44: 493-511.

Touw, A. 1992. A survey of the mosses of the Lesser Sunda Islands (Nusa Tenggara), Indonesia. Journal of the Hattori Botanical Laboratory, 71: 289-366.

Valente, D. V., Peralta, D. F., Prudêncio, R. X. A., et al. 2020. Taxonomic notes and new synonyms on Brazilian *Macromitrium* Bridel (Bryophyta, Orthotrichaceae). Phytotaxa, 454(3): 213-225.

van Rooy, J. & Wyk, A. E. v. 1992. A conspectus of the subfamily Macromitrioideae (Bryopsida: Orthotrichaceae) in southern Africa. The Bryologist, 95: 205-215.

van Rooy, J. 1990. A new species and a new record of *Macromitrium* (Orthotrichaceae) from southern Africa: *M. lebomboense* sp. nov. and *M. richardii* Schwaegr. Journal of Bryology, 16: 209-214.

Vitt, D. H. & Crum, H. A. 1970. *Groutiella tomentosa* new to the United States. The Bryologist, 73(1): 145-149.

Vitt, D. H. & Ramsay, H. P. 1985. The *Macromitrium* complex in Australasia (Orthotrichaceae: Bryopsida). Part I. Taxonomy and phylogenetic relationships. Journal of Hattori Botanical Laboratory, 59: 325-451.

Vitt, D. H. 1973. A revisionary study of the genus *Macrocoma*. Revue Bryologique et Lichénologique, 39(2): 205-220.

Vitt, D. H. 1979. New taxa and new combinations in the Orthotrichaceae of Mexico. The Bryologist, 82(1): 1-19.

Vitt, D. H. 1980 [1981]. The genus *Macrocoma* I. Typification of names and taxonomy of the species. The Bryologist, 83(4): 405-436.

Vitt, D. H. 1981. The genera *Leiomitrium* and *Cardotiella* gen. nova (Orthotrichaceae). Journal of the Hattori Botanical Laboratory, 49: 92-113.

Vitt, D. H. 1983. The New Zealand species of the pantropical genus *Macromitrium* (Orthotrichaceae: Musci): taxonomy, phylogeny and phytogeography. Journal of Hattori Botanical Laboratory, 54: 1-94.

Vitt, D. H. 1989. The genus *Schlotheimia* (Orthotrichaceae: Bryopsida) in Australia and New Zealand. The Bryologist, 92(3): 282-298.

Vitt, D. H. 1990. *Desmotheca* (Orthotrichaceae): Gondwanan fragmentation and the origin of a Southeast Asian genus. Tropical Bryology, 3: 78-88.

Vitt, D. H. 1994. Orthotrichaceae. – In A. J. Sharp et al. (editors), Moss Flora of Mexico. Memoirs of the New York Botanical Garden, 69: 590-656.

Vitt, D. H., Koponen, T. & Norris, D. H.1995. Bryophyte flora of the Huon Peninsula, Papua New Guinea. LV. *Desmotheca, Groutiella, Macrocoma* and *Macromitrium* (Orthotrichaceae, Musci). Acta Botanica Fennica, 154: 1-94.

Werner, O., Patiño, J., González–Mancebo, J. M., *et al.* 2009. The taxonomic status and the geographical relationships of the Macaronesian endemic moss *Fissidens luisieri* (Fissidentaceae) based on DNA sequence data. The Bryologist, 112(2): 315-324.

Whittier, H. O. 1976. *Mosses of the Society Islands*. x + 410 pp. Gainesville: The University Presses of Florida.

Wijk, R. van der & Margadant, W. D. 1959. New combinations in mosses II. Taxon, 8: 70-75.

Wijk, R. van der & Margadant, W. D. 1960. New combinations in mosses V. Taxon, 9: 189-191.

Wijk, R. van der, Margadant, W. D. & Florschütz, P. A. 1959. Index Muscorum. 1 (A-C). Regnum Vegetabile, 17: 548 pp.

Wijk, R. van der, Margadant, W. D. & Florschütz, P. A. 1962. Index Muscorum. 2 (D–Hypno). Regnum Vegetabile, 26: 535 pp.

Wijk, R. van der, Margadant, W. D. & Florschütz, P. A. 1964. Index Muscorum. 3 (Hypnum–O). Regnum Vegetabile, 33: 529 pp.

Wijk, R. van der, Margadant, W. D. & Florschütz, P. A. 1967. Index Muscorum. 4 (P–S). Regnum Vegetabile, 48: 604 pp.

Wijk, R. van der, Margadant, W. D. & Florschütz, P. A. 1969. Index Muscorum. 5 (T–Z, Appendix). Regnum Vegetabile, 65: xii + 922 pp.

Wilbraham, J. & Ellis, L. T. 2010. Further taxonomic studies on the families Calymperaceae (Musci) and Orthotrichaceae (Musci) in the bryoflora of Réunion Island, with notes on taxa from other islands in the western Indian Ocean. Cryptogamie, Bryologie, 31(1): 31-66.

Wilbraham, J. & Price, M. J. 2013. A lectotype for *Macromitrium cirrosum* (Hedw.) Brid. (Orthotrichaceae). Journal of Bryology, 35(2): 119-122.

Wilbraham, J. 2007. Taxonomic notes on the pantropical genera *Macromitrium* and *Macrocoma* (Bryopsida: Orthotrichaceae). Journal of Bryology, 29: 54-59.

Wilbraham, J. 2008. Bryophyte flora of Uganda. 8. Orthotrichaceae Part 1 --- Macromitrioideae. Journal of Bryology, 30: 201-207.

Wilbraham, J. 2015. Annotated checklist and keys to the Orthotrichaceae of Malawi, together with new country records for East Africa. Journal of Bryology, 37(2): 87-95.

Wilbraham, J. 2016. Taxonomic notes on African Orthotrichaceae I. New synonymy in *Macromitrium*. Journal of Bryology, 38(2): 87-93.

Wilbraham, J. 2017. Taxonomic notes on African Orthotrichaceae 2: *Macrocoma* – new synonymy and first report of multicellular spores. Journal of Bryology, 39(4): 324-330.

Wilbraham, J. 2018. Taxonomic notes on African Orthotrichaceae 3. New synonymy in Madagascan *Macromitrium*. Journal of Bryology, 40(4): 393-398.

Williams, R. S. 1903. Bolivian mosses, Part 1. Bulletin of the New York Botanical Garden, 3(9): 104-134.

Williams, R. S. 1911. Panama mosses. Bulletin of the Torrey Botanical Club, 38: 33-36.

Williams, R. S. 1922. Mosses from British Guiana and Dominica, Lesser Antilles collected by Miss E. F. Noel in 1914. The Bryologist, 24: 65-67.

Williams, R. S. 1927. Mosses from Ecuador, collected in 1918 by Dr. J. N. Rose. Journal of the Washington Academy of Sciences, 17: 491-497.

Wilson, W. 1857 [1854]. Flora of the Isthmus of Panama, Musci. In B. C. Seemann(ed), The Botany of the Voyage of H.M.S. Herald. London: Lovell Reeve, 244-245.

Yano, O. 1981. A checklist of Brazilian mosses. Journal of the Hattori Botanical Laboratory, 50: 270-456.

Yu, J., Guo, S. L., Ma, Y. H., *et al.* 2012. Taxonomic and morphometric comments on *Macromitrium tosae* Besch. (Orthotrichaceae), with its four new synonyms. The Bryologist, 115(3): 388-401.

Yu, J., Guo, S. L., Ma, Y. H., *et al.* 2013. *Macromitrium ousiense*, a neglected Chinese moss species (Orthotrichaceae, Bryopsida) with new synonym and records. Nordric Journal of Botany, 31(3): 339-343.

Yu, J., Li, D. D., Li, Y., *et al.* 2018. On taxonomic status of *Macromitrium leptocarpum* and *M. subleptocarpum*, with comments on *M. sulcatum* (Bryophytea, Orthotrichaceae). Phytotaxa, 361(3): 287-293.

Yu, J., Yong, K. T. & Guo, S. L. 2014. Taxonomic notes on *Macromitrium densum* (Bryopsida) with a new synonym, *M. brevissimum*. Nordic Journal of Botany, 32(4), 437-440.

Yu, N. N. & Jia, Y. 2023. A world revisionary study of the genus *Groutiella* Steere (Orthotrichaceae, Bryopsida). Cryptogamie, Bryologie, 44(7): 161-182.

Zander, R. H. 1972. Revision of the genus Leptodontium (Musci) in the New World. The Bryologist, 75: 213-280.

Zhang, Z. Y., Li, D. D., Yu, J., *et al.* 2019. *Macromitrium maolanense* Zeyou Zhang, D.D. Li, Jing Yu & S.L. Guo, a new species from China based on molecular and morphological evidence. Journal of Bryology, 41(3): 263-273.

Index to Scientific Names

M. acuminatum 5, 18, 29, 46, 47, 61, 65, 66, 415, 428
    Schlotheimia acuminata
    **M. elongatum*
M. acutirameum 8, 17, 21, 40, 60, 65, 67
M. amaniense 8, 18, 27, 49, 65, 66, 68
M. amboroicum 7, 18, 24, 40, 46, 49, 69, 71, 351, 428
M. angulatum 9, 11, 18, 31, 41, 54, 69, 72, 267, 414, 429, 431
M. angulosum 5, 46, 59, 69, 70, 73
M. angustifolium 5, 18, 20, 29, 61, 65, 70, 74, 204, 244, 328, 406, 407, 415, 428
M. antarcticum 8, 18, 25, 53, 70, 75, 429
M. archboldii 5, 18, 30, 52, 76, 77, 85, 415, 429
    M. ruberrimum
M. archeri 6, 46, 60, 76, 78, 426, 429
    M. asperulum
    M. muelleri
    M. pusillum
M. argutum 8, 19, 45, 47, 64, 78, 79, 252, 291
M. atratum 8, 18, 28, 40, 49, 80, 81, 428
M. atroviride 7, 18, 30, 40, 49, 80, 82, 124, 314, 428
M. attenuatum 7, 19, 37, 50, 80, 83, 428
M. aurantiacum 5, 40, 55, 84, 86
M. aurescens 6, 17, 23, 47, 56, 84, 85, 87, 258, 316, 426, 429, 431
    M. cylindromitrium
    M. sordidevirens
M. aurescens var. *caledonicum* 5, 84, 88
M. aureum 7, 17, 18, 32, 41, 58, 85, 89, 124, 428
M. austrocirrosum 5, 18, 27, 54, 85, 90, 415
M. bifasciculare 8, 18, 34, 40, 48, 88, 91, 415, 429, 431
M. bifasciculatum 8, 18, 34, 40, 48, 88, 92, 415, 429, 431
M. binsteadii 5, 59, 93, 95, 429
M. bistratosum 9, 18, 32, 40, 44, 46, 47, 49, 93, 96, 427, 429, 431
M. blumei 5, 11, 17, 20, 40, 54, 93, 94, 97, 119, 409, 415, 417, 424
    M. assimile
    M. copelandii
    M. horridum
    M. teres
M. blumei var. *zollingeri* 9, 46, 47, 54, 94, 97, 415, 417, 424
    M. zollingeri
    M. contortum
    M. magnirete
    M. annamense
M. brachypodium 5, 17, 22, 40, 41, 62, 84, 94, 429, 432
    M. brevisetaceum
M. brevicaule 5, 17, 21, 40, 41, 60, 84, 98, 99, 353, 413, 416, 427, 429, 430, 431
    Micromitrium brevicaule
    M. mucronatulum
    M. subbrevicaule
    M. subfragile
    M. wattsii
M. brevihamatum 7, 18, 31, 40, 50, 98, 100, 210, 428
M. brevisetum 9, 17, 41, 62, 101, 103, 427, 429, 432
    M. aristocalyx
M. caldense 8, 40, 63, 101, 104, 428
M. caloblastoides 6, 18, 25, 55, 101, 102, 105, 429
    M. dimorphum
M. calocalyx 9, 18, 28, 53, 102, 106, 429
    M. semipapillosum
M. calomicron 9, 59, 102, 107, 429
M. calymperoideum 5, 45, 47, 56, 108, 109, 429
M. campoanum 8, 17, 18, 34, 46, 52, 108, 110, 378, 413, 415, 429
M. cardotii 5, 18, 32, 55, 108, 111, 143, 429
M. cataractarum 7, 19, 38, 50, 112, 115, 428
M. catharinense 8, 19, 36, 37, 44, 53, 112, 116, 428, 431, 432
    M. prolongatum
    M. catharinense var. *gracilius*
    M. prolongatum var. *gracilius*
    M. profusum
    Teichodontium catharinense var. *gracilior*
    M. drewii
    M. schiffneri
M. chloromitrium 9, 18, 62, 113, 429, 432
    M. fimbriatum var. *chloromitrium*
M. cirrosum 3, 7, 18, 19, 33, 37, 59, 85, 113, 114, 117, 119, 191, 210, 289, 306, 416, 428, 431
    Anictangium cirrosum
    Anoectangium cirrosum
    Hypnum cirrhatum
    M. barbense
    M. costaricense
    M. cirrosum var. *stenophyllum*
    M. cubensicirrhosum
    M. erectopatulum
    M. hoehnei
    M. mammillosum
    M. microtheca
    M. praelongum
    M. pseudocirrosum
    M. schwaneckeanum
    M. stenophyllum
    M. substrictifolium
    M. werckleanum
M. clastophyllum 4, 41, 60, 114, 118, 429, 431
M. comatum 4, 46, 47, 62, 119, 120, 383, 415, 429
    M. nipponicum
M. concinnum 5, 11, 45, 58, 119, 121, 131, 415, 429
M. constrictum 7, 18, 19, 64, 119, 122, 428

M. crassirameum   7, 18, 30, 50, 123, 125, 428
M. crassiusculum   8, 17, 21, 40, 41, 46, 48, 123, 126, 327, 429
M. crinale   5, 63, 123, 124, 127, 214, 228, 266, 282, 415, 429
   M. mindorense
M. crispatulum   7, 18, 28, 50, 124, 128, 428, 431
M. crosbyorum   18, 32, 40, 57, 124, 129, 176, 428
M. cuspidatum   5, 18, 33, 40, 41, 51, 76, 130, 132, 214, 415, 427, 429
   M. elongatum
M. cuspidatum var. gracile   18, 33, 130, 415
M. cylindricum   7, 19, 36, 40, 52, 130, 133, 428, 431
M. densum   5, 11, 17, 21, 40, 46, 47, 58, 130, 131, 134, 429, 431
   M. brevissimum
M. diaphanum   6, 11, 17, 21, 40, 49, 131, 135, 410, 416, 429, 431
   M. circinicladum
M. dielsii   6, 18, 27, 52, 136, 147, 214, 429
M. divaricatum   7, 19, 36, 45, 53, 136, 137, 428
M. diversifolium   8, 59, 136, 138
   M. divortiarum
M. dubium   7, 11, 18, 40, 52, 139, 140, 215, 428
   M. vernicosum
M. echinatum   6, 17, 18, 35, 40, 41, 46, 51, 130, 139, 142, 172, 396, 416, 428, 431
M. ecrispatum   8, 17, 19, 41, 61, 143, 144
M. eddyi   55, 143
M. ellipticum   5, 61, 143, 145, 231, 428
M. emersulum   9, 46, 47, 62, 146, 148, 427, 429
M. erubescens   5, 41, 54, 69, 146, 267, 415, 429
M. erythrocomum   57, 146, 147
M. evrardii   5, 60, 147, 149, 429
M. exsertum   6, 18, 26, 41, 64, 147, 346, 427, 429
M. falcatulum   6, 46, 56, 150, 151, 429
   M. merrillii
   M. winkleri
M. fendleri   7, 18, 24, 47, 50, 150, 152, 316, 428
M. fernandezianum   8, 17, 18, 23, 63, 150, 153, 431
M. ferriei   4, 29, 47, 56, 114, 154, 156, 178, 306, 415, 427, 429
   M. comatulum
   *M. comatulum
   M. inflexifolium
   M. nipponicum
M. fimbriatum   4, 8, 18, 31, 61, 113, 154, 157, 429
   Orthotrichum fimbriatum
   M. nanothecium var. sublaeve
   M. uncinatum
   Orthotrichum uncinatum
   Weissia uncinata
M. flavopilosum   6, 17, 19, 38, 40, 51, 155, 158, 428
M. flexuosum   19, 41, 63, 155, 159, 226, 431
M. formosae   4, 62, 160, 162, 385, 429
M. fortunatii   4, 17, 23, 58, 147, 160, 163, 415, 429
   M. fortunatii var. nigrescens
M. fragilicuspis   6, 18, 26, 40, 48, 160, 161, 164, 330, 425, 428, 431
M. francii   5, 41, 56, 161, 165, 429, 432
   M. contractum
M. frondosum   7, 17, 19, 37, 38, 40, 51, 161, 166, 428, 431
M. frustratum   6, 18, 35, 37, 40, 41, 63, 161, 167, 168, 283, 330, 428

M. fulgescens   6, 18, 30, 40, 52, 167, 169, 428
   M. standleyi var. subundulatum
   *M. fuscescens
M. fulvum   5, 52, 167, 170, 262, 381, 428
M. funicaule   9, 17, 20, 40, 46, 49, 171, 173, 429
M. funiforme   17, 20, 64, 171, 214, 427, 429
M. fuscescens   9, 18, 27, 41, 61, 65, 167, 171, 174, 204, 429
   M. calvescens
   M. semipellucidum
   M. calvescens
   M. miquelii
   M. glaucum
   M. eurymitrium
   M. subsemipellucidum
M. fuscoaureum   6, 17, 19, 36, 40, 41, 45, 46, 47, 52, 130, 139, 155, 172, 175, 384, 396, 428
M. galipense   18, 32, 58, 176, 179
M. gigasporum   7, 10, 11, 17, 19, 37, 41, 44, 51, 161, 176, 180, 428, 431, 432
M. giraldii   4, 18, 29, 41, 46, 56, 154, 176, 177, 178, 181, 251, 258, 299, 322, 413, 414, 417, 418, 419, 427, 429, 431, 432
   M. cavaleriei
   M. sinense
   M. syntrichophyllum
   M. syntrichophyllum var. longisetum
   M. gebaueri
   M. giraldii var. acrophylloides
   M. rigbyanum
   M. handelii
   M. cancellatum
M. glabratum   7, 17, 18, 30, 47, 64, 136, 178, 182
M. glaziovii   8, 18, 28, 40, 55, 178, 183, 397, 429
M. gracile   6, 17, 18, 20, 33, 48, 130, 184, 187, 204, 305, 316, 317, 415, 429, 432
   Orthotrichum gracile
   *M. appendiculatum
   M. mossmanianum
   M. gracile var. proboscideum
M. greenmanii   6, 18, 33, 41, 45, 47, 54, 184, 188, 425, 428
M. grossirete   58, 114, 184, 185
   M. papillifolium
   M. rigescens
M. guatemalense   6, 18, 19, 28, 37, 40, 50, 98, 185, 186, 189, 428, 430, 431
   M. liberum
   M. negrense
   M. paucidens
   M. penicillatum
   M. rhystophylllum
   M. serrulatum
   M. subreflexum
   M. tortuosum
   M. trianae
   M. verrucosum
M. gymnostomum   4, 17, 18, 25, 46, 47, 61, 186, 190, 236, 385, 414, 427, 429, 431, 432
   M. brevituberculatum
   M. robinsonii
   M. rupestre
   M. gymnostomum var. brevisetum
   M. gymnostomum var. robustum
M. hainanense   4, 9, 18, 29, 41, 61, 191, 192, 371, 415, 429,

Index to Scientific Names

432
M. harrisii   7, 18, 30, 64, 191, 193, 428
   M. peraristatum
M. helmsii   6, 40, 55, 191, 192, 194, 316, 431
   M. appendiculatum
M. hemitrichodes   5, 6, 17, 40, 41, 46, 57, 194, 195, 196, 346, 347, 416, 426, 428
   M. amoenum
   M. baileyi
   M. intermedium
M. hemitrichodes var. **sarasinii**   5, 17, 195
   M. sarasinii
M. herzogii   7, 19, 37, 44, 50, 195, 197, 428, 431
M. hildebrandtii   8, 19, 36, 50, 195, 198, 428
M. holomitrioides   4, 25, 41, 58, 199, 200, 385, 415, 429
M. hortoniae   6, 41, 62, 199, 201, 429
M. huigrense   7, 19, 37, 41, 54, 201, 202, 428
M. humboldtense   5, 40, 55, 203, 277
M. incurvifolium   5, 6, 11, 41, 44, 47, 61, 203, 204, 347, 426, 428
   Orthotrichum incurvifolium
   M. beecheyanum
   M. cumingii
   M. javanicum
   M. kaernbachii
   M. leucoblastum
   M. planocespitosum
   M. powellii
   M. subtile
   M. subtile subsp. subuligerum
   M. subuligerum
   M. zippelii
   Orthotrichum undulatum
M. involutifolium   6, 18, 24, 25, 56, 102, 147, 161, 204, 205, 207, 429, 432
   Orthotrichum involutifolium
   M. daemelii
   M. incurvulum
   M. malacoblastum
   M. noumeanum
M. involutifolium subsp. **ptychomitrioides**   147, 205
   M. ptychomitrioides
   M. carinatum
   M. platyphyllaceum
   M. plicatum
   M. plicatum var. aristatum
   M. plicatum var. obtusifolium
   M. suberosulum
   M. viridissimum
M. japonicum   4, 10, 18, 24, 27, 40, 62, 70, 108, 119, 177, 205, 208, 214, 251, 414, 417, 420, 421, 427, 429, 431, 432
   Dasymitrium japonicum
   M. japonicum var. makinoi
   Dasymitrium makinoi
   Dasymitrium incurvum
   M. incurvum
   M. bathyodontum
   M. dickasonii
   M. polygonostomum
   M. insulanum
M. krausei   8, 18, 33, 40, 52, 206, 209, 413, 429
M. laevigatum   5, 61, 210, 211, 428
M. laevisetum   7, 41, 47, 50, 210, 212, 428
M. lanceolatum   9, 18, 31, 52, 210, 213
M. laosianum   5, 62, 213, 214, 429
M. larrainii   5, 52, 215, 429
M. lauterbachii   5, 11, 64, 215, 217, 428
M. lebomboense   9, 17, 46, 47, 48, 215, 218, 429, 431
M. leprieurii   8, 18, 33, 40, 52, 139, 215, 219, 428
   M. crumianum
   M. dussii
M. leratii   49, 216, 220, 330, 346, 347, 429, 431
   M. leratii var. erectifolium
   M. salakanum var. majus
M. ligulare   6, 18, 26, 55, 221, 223
   M. luehmannianum
M. ligulaefolium   6, 18, 25, 40, 55, 76, 102, 161, 221, 222, 416, 426, 428
   M. brevipilosum
   M. cucullatum
   M. ligulatulum
   M. perminutum
   M. rapaense
   M. woollsianum
   M. woollsianum var. chlorophyllosum
M. lomasense   7, 17, 23, 40, 46, 52, 224, 225, 428
M. longicaule   5, 17, 56, 224, 354, 415, 429
   M. brachystele
M. longifolium   7, 8, 168, 37, 41, 54, 85, 184, 201, 226, 227, 262, 283, 411, 425, 427, 428, 431, 432
   Orthotrichum longifolium
   M. denudatum
   *M. flexuosum
   M. haitense
   M. homalacron
   M. longifolium var. viridissimum
   *M. muelleri
   **M. perundulatum
   M. scabrisetum
   M. schimperi
   M. scleropelma
   Schlotheimia longifolia
M. longipapillosum   4, 9, 41, 58, 185, 227, 229, 377, 415, 429
M. longipes   6, 11, 17, 20, 40, 43, 48, 228, 230, 427, 429
   Orthotrichum longipes
   M. lonchomitrium
   M. pseudohemitrichodes
M. longipilum   5, 11, 41, 53, 57, 214, 228, 415, 429
M. longirostre   6, 11, 40, 43, 46, 48, 231, 232, 313, 316, 413, 429, 431
   Orthotrichum longirostre
   M. acutifolium
   M. pertoruescense var. torquatulum
   M. torquatulum
   M. rodwayi
M. lorifolium   5, 44, 47, 63, 143, 231, 233, 385, 425, 428
M. macrocomoides   8, 18, 34, 40, 41, 51, 232, 234, 429
M. macrosporum   5, 17, 18, 59, 235, 238, 330
   M. aspericuspis
   M. goniostomum
   *M. hamatum
   M. morobense
M. macrothele   7, 18, 24, 41, 46, 58, 235, 239, 428
M. maolanense   4, 9, 17, 40, 48, 236, 240, 415, 429, 431
M. masafuerae   8, 40, 55, 236, 241, 429

M. mcphersonii   6, 41, 53, 155, 236, 237, 242, 428
M. megalocladon   5, 19, 39, 54, 237, 261, 415, 429
   M. altipapillosum
   M. submegalocladum
M. melinii   7, 40, 50, 237, 243, 428
M. menziesii   9, 53, 244, 247
M. microcarpum   8, 11, 17, 22, 40, 41, 46, 52, 244, 248, 413, 429
M. microstomum   6, 11, 17, 18, 20, 22, 24, 40, 41, 45, 46, 47, 52, 108, 136, 147, 171, 206, 228, 244, 249, 267, 305, 316, 345, 377, 378, 415, 416, 417, 422, 423, 427, 429, 431, 432
   Orthotrichum microstomum
   Leiotheca microstoma
   Dasymitrium borbonicum
   M. acunae
   M. adstrictum
   M. borbonicum
   M. fasciculare var. *fasciculare*
   M. fasciculare var. *angustifolium*
   M. fasciculare var. *javense*
   M. filicaule
   M. flaccidisetum
   M. hornschuchii
   *M. microstomum
   M. linearifolium
   M. nitidum
   M. macropelma
   M. owahiense
   M. pacificum
   M. pacificum var. *brevisetum*
   M. pinnulatum
   M. prolixum
   M. pseudohemitrichodes
   M. reinwardtii
   M. saxatile
   M. scottiae
   M. seemannii
   M. stolonigerum
   M. stratosum
   M. subnitidum
   M. tasmanicum
   M. weymouthii
M. minutum   5, 11, 54, 221, 246, 247, 250, 429
M. mittenianum   7, 54, 251, 253
M. moorcroftii   5, 18, 28, 62, 251, 254, 429, 431
   Orthotrichum moorcroftii
M. mosenii   8, 17, 19, 38, 41, 44, 45, 46, 50, 251, 255, 428
M. nanothecium   9, 18, 25, 41, 62, 154, 252, 256, 429
M. nematosum   8, 17, 18, 28, 40, 49, 252, 257, 428, 431
M. nepalense   5, 17, 22, 23, 40, 42, 49, 108, 258, 259, 346, 382, 415, 429, 431, 432
   Orthotrichum nepalense
   M. incrustatifolium
   M. longibrachteatum
M. nigricans   5, 62, 258, 260, 429
M. noguchianum   5, 41, 53, 261, 354, 429
   *M. papuanum
M. norrisianum   5, 40, 41, 53, 261, 415, 429
M. nubigenum   7, 19, 38, 41, 46, 53, 261, 262, 263, 428
M. oblongum   40, 50, 80, 226, 262, 264, 269, 428
   Schlotheimia oblonga
M. ochraceoides   6, 54, 214, 262, 265, 431
M. ochraceum   5, 11, 18, 33, 41, 54, 124, 261, 262, 266, 270, 282, 415, 429
   Schlotheimia ochracea
   M. hallieri
   M. mindanaense
   M. rubricuspis
   Trichostomum neesii
M. onraedtii   9, 17, 20, 40, 49, 266, 271, 429, 431
M. orthophyllum   6, 11, 18, 35, 40, 51, 266, 429, 430
M. orthostichum   5, 11, 17, 18, 31, 41, 54, 69, 247, 267, 268, 272, 322, 409, 415, 429
   M. appressifolium
   M. fragilifolium
   M. orthostichum subsp. *appressifolium*
   M. orthostichum subsp. *seminudum*
   M. scleropodium
   M. seminudum
M. orthostichum subsp. *micropoma*   5, 267
M. orthostichum var. *burgeffii*   5, 268
M. orthostichum var. *siccosquarrosum*   6, 268
M. osculatianum   7, 19, 37, 40, 49, 268, 273, 428
M. ousiense   4, 18, 25, 41, 53, 268, 274, 385, 415, 429, 431
   M. heterodictyon
M. ovale   7, 11, 18, 19, 30, 38, 40, 63, 269, 275, 428
M. pallidum   3, 4, 8, 18, 29, 48, 269, 276, 429
   Orthotrichum pallidum
   M. voeltzkowii
   *Lasia acicularis
   *Leiotheca acicularis
   *M. aciculare
   *Orthotrichum aciculare
   Orthotrichum breve
   **Orthotrichum laeve
   **Pterigynandrum aciculare
   *Schlotheimia acicularis
   Trichostomum arbustorum
M. panduraefolium   5, 60, 277, 429, 431
M. paridis   9, 18, 28, 53, 277, 279, 429
   *M. cacuminicola
M. parvifolium   5, 60, 277, 415, 429
   M. brevirameum
   *M. brevirameum
   M. daymannianum
M. parvirete   6, 18, 28, 44, 59, 186, 278, 280, 428
M. pellucidum   8, 17, 21, 40, 51, 171, 172, 278, 279, 281
   M. laevifolium
M. peraristatum   6, 21, 40, 41, 51, 191, 214, 282, 416, 427, 429
M. perdensifolium   5, 17, 18, 32, 47, 60, 282, 284, 415
M. perfragile   8, 19, 38, 40, 53, 282, 283, 285, 431
M. perichaetiale   40, 41, 51, 283, 286, 363, 425, 428
   Orthotrichum perichaetiale
   M. truncatum
M. perpusillum   6, 18, 28, 60, 283, 287, 429
M. perreflexum   7, 19, 39, 40, 41, 53, 288, 291, 416, 432
M. pertriste   8, 18, 34, 40, 41, 48, 288, 292, 415, 429, 431
M. petelotii   5, 40, 58, 288, 293, 397, 429, 431
M. picobonitum   6, 47, 57, 289, 294, 428
M. pilicalyx   9, 56, 289, 295, 429
M. piliferum   9, 58, 214, 289, 296, 328, 427, 429, 431
M. pilosum   5, 56, 290, 297, 429
   M. koghiense
   M. koghiense var. *spiricaule*
   M. pilosum var. *brevifolium*

Index to Scientific Names

 M. subsessile
M. proliferum 8, 17, 64, 290, 291, 298, 428
M. prolongatum 4, 46, 62, 112, 299, 300, 415, 427, 429, 432
 M. brachycladulum
 M. prolongatum var. *brevipes*
M. prorepens 6, 57, 299, 363, 429
 Orthotrichum prorepens
 M. oocarpum
 M. prorepens var. *aristata*
M. proximum 9, 18, 27, 62, 301, 302, 428
M. pseudofimbriatum 8, 18, 49, 69, 301, 303, 341, 353, 425, 428
 M. doeringianum
 M. podocarpi
 M. podocarpi var. *falcifolium*
 M. portoricense
M. pseudoserrulatum 8, 18, 30, 62, 136, 178, 303, 307, 428
M. pulchrum 5, 41, 55, 277, 304, 308, 429
 M. pulchrum var. *aristatum*
 M. pulchrum var. *densirete*
M. pulchrum var. **neocaledonicum** 5, 277, 304, 309
 M. neocaledonicum
 M. pulchrum var. *neocaledonicum*
M. pullenii 5, 63, 304, 305, 415, 431
M. punctatum 8, 18, 28, 62, 69, 224, 291, 305, 310, 403, 416, 428, 431, 432
 Orthotrichum punctatum
 M. hirtellum
 M. liberum
 M. pentagonum
 M. pentastichum
 M. reflexifolium
 M. sartorii
 M. sumichrastii
 Schlotheimia brachyrhyncha
M. pyriforme 7, 18, 30, 41, 64, 306, 311, 428
M. quercicola 18, 29, 45, 56, 178, 306, 312, 415, 429
M. ramsayae 6, 64, 313, 428
M. raphidophyllum 7, 19, 59, 314, 317, 428
M. refractifolium 7, 57, 314, 318
M. regnellii 8, 11, 18, 19, 30, 38, 40, 50, 101, 124, 314, 319, 425, 428
 M. contextum
M. renauldii 5, 41, 60, 315, 320, 377, 429
 M. gracilipes
 M. leratioides
M. repandum 6, 11, 17, 22, 63, 315, 316, 321, 427, 429
 M. pallidovirens
 M. pugionifolium
 M. whiteleggei
M. retusulum 18, 32, 58, 316, 323
M. retusum 6, 40, 52, 192, 316, 317, 416, 429, 431
 M. aristatum
 M. caducipilum
 M. longirostre var. *caducipilum*
M. rhacomitrioides 4, 56, 178, 321, 324, 415, 427, 429
M. rimbachii 7, 18, 30, 63, 322, 325, 428
M. rufipilum 5, 55, 277, 322, 429
M. ruginosum 9, 18, 31, 40, 54, 322, 326, 429
M. rugulosum 8, 41, 49, 327, 331, 428
M. rusbyanum 7, 10, 11, 19, 41, 60, 314, 327, 332, 428, 431
M. saddleanum 8, 17, 18, 34, 40, 48, 327, 328, 333, 413, 415, 429, 430, 431

M. saddleanum var. **piliferum** 328
M. salakanum 5, 18, 31, 61, 65, 70, 204, 215, 328, 334, 408, 428
 M. celebense
 M. humile
 M. pungens
 **M. reflexifolium*
 ***M. nova-guinense*
M. savatieri 9, 18, 29, 62, 143, 328, 329, 335, 429
M. schmidii 5, 18, 29, 46, 47, 55, 269, 329, 336, 385, 429
 M. benguetense
M. schmidii var. **macroperichaetialium** 4, 9, 329, 337, 415
M. scoparium 7, 11, 19, 38, 57, 167, 237, 329, 330, 338, 384, 416, 425, 428, 431
 M. palmense
 M. tonduzii
 M. williamsii
M. sejunctum 6, 18, 30, 40, 58, 161, 283, 330, 339
M. semperi 6, 19, 39, 61, 65, 330, 415, 428
M. serpens 9, 17, 18, 23, 24, 40, 49, 258, 340, 342, 382, 415, 429, 431
 Orthotrichum serpens
 ***Dasymitrium rehmannii*
 M. anomodictyon
 M. astroideum
 **M. elegans*
 ***M. rehmannii*
 M. serpentinum
 M. tristratosum
M. sharpii 6, 18, 24, 40, 59, 278, 340, 341, 343, 353, 354, 428
M. similirete 5, 60, 305, 341, 415
M. solitarium 7, 19, 41, 45, 53, 341, 344, 428
M. solitarium var. **brevipes** 7, 19, 341
M. soulae 45, 52, 345, 347, 410, 429
 ***M. longisetum*
M. speirostichum 9, 18, 26, 61, 345, 348, 428
M. st-johnii 9, 40, 48, 66, 345, 346, 349, 429
M. standleyi 6, 18, 35, 40, 57, 167, 346, 350, 384, 396, 416, 425, 428, 431
M. stoneae 6, 18, 25, 41, 56, 346, 347
M. streimannii 5, 56, 351, 415, 428
M. subbrevihamatum 19, 38, 40, 49, 69, 351, 355, 428
M. subcirrhosum 6, 19, 41, 63, 351, 352, 356, 425, 428
M. subcrenulatum 7, 19, 38, 45, 59, 352, 357, 428
M. subdiscretum 7, 18, 32, 47, 57, 352, 353, 358, 428
M. subhemitrichodes 6, 17, 40, 60, 353, 359, 431
 M. subhemitrichodes var. *hodgkinsoniae*
M. subincurvum 57, 353, 360, 415, 429
M. sublaeve 7, 11, 18, 26, 40, 41, 59, 154, 353, 354, 361, 428
M. sublongicaule 5, 19, 39, 64, 224, 354, 362, 415, 429
M. submucronifolium 6, 57, 363
 M. erosulum
 M. papillifolium
 M. coarctatulum
 M. petriei
M. subperichaetiale 7, 18, 33, 40, 41, 51, 363, 364, 428
M. subscabrum 7, 19, 38, 40, 44, 45, 64, 120, 363, 364, 365, 425, 428
M. subtortum 5, 18, 19, 27, 29, 39, 61, 143, 366, 367, 416, 428
 Orthotrichum subtortum

M. mauritianum
M. mauritianum var. viride
M. rhizomatosum
M. sanctae-mariae
M. subpungens
M. subpungens var. madagassum
M. subulatum 6, 60, 366, 428
M. sulcatum 5, 18, 19, 24, 28, 29, 41, 45, 64, 167, 195, 231, 262, 364, 368, 369, 370, 372, 381, 385, 415, 416, 425, 428, 431
    Schlotheimia sulcata
    **Dasymitrium macleai
    M. belangeri
    M. bequaertii
    **M. boivini
    M. sulcatum subsp. ceylanicum
    **M. corrugatum
    M. eckendorffii
    **M. insculptum
    **M. laevatum
    M. levatum
    **M. macleai
    M. mannii
    *M. menziesii
    M. neelgheriense
    M. muellerianum
    **M. neilgherrense
    **M. nilghiriense
    M. perundulatum
    M. pseudoramentosum
    M. ramentosum
    *M. rugifolium
    M. rufescens
    M. seriatum
    M. tortifolium
    M. trollii
    M. undatifolium
    **Schlotheimia turbinata
M. sulcatum var. **leptocarpum** 9, 18, 26, 47, 370, 373
M. sulcatum subsp. **ramentosum** 19, 39, 369, 370
M. sulcatum var. **torulosum** 64, 370, 374
M. swainsonii 8, 11, 17, 23, 40, 50, 131, 370, 371, 375, 416, 425, 428, 431
    Orthotrichum swainsonii
    M. altituberculosum
    M. brotheri
    M. carionis
    M. stellulatum
    M. vesiculatum
M. taiheizanense 4, 41, 63, 289, 371, 376, 415, 429, 432
M. taiwanense 4, 41, 57, 377, 378, 396, 415, 431
M. taoense 5, 17, 41, 44, 63, 315, 377, 379
M. tenax 8, 18, 34, 52, 108, 377, 378, 380, 413, 415, 429
    M. coriaceum
    M. tenax var. theriotii
M. thwaitesii 5, 41, 63, 381, 386, 428
M. tocaremae 7, 19, 37, 47, 50, 381, 387, 428
M. tongense 9, 17, 21, 40, 49, 258, 305, 381, 382, 388, 415, 429, 431

M. densifolium
M. ludovicae
M. subvillosum
M. villosum
Dasymitrium villosum
Drummondia villosa
M. villosum var. elongatum
M. villosum var. intermedium
M. villosum var. longisetum
M. tosae 4, 17, 23, 57, 93, 108, 119, 258, 353, 382, 383, 389, 429
    M. chungkingense
    M. courtoisii
    M. cylindrothecium
    M. hamatum
    M. lingulatum
    M. melanostomum
M. trachypodium 7, 18, 34, 40, 41, 44, 47, 51, 383, 390, 395, 425, 428, 431
M. trichophyllum 4, 7, 11, 19, 36, 40, 41, 46, 51, 172, 346, 383, 384, 391, 396, 418, 428, 431
M. trinitense 8, 18, 35, 40, 57, 330, 384, 392, 416, 428, 431
M. tuberculatum 4, 55, 186, 190, 384, 385, 393, 415, 429
M. turgidum 5, 64, 231, 385, 394, 425, 428
    M. turgidum var. laeve
M. tylostomum 5, 64, 395, 415, 429
    M. hooglandii
M. ulophyllum 6, 11, 18, 33, 40, 41, 51, 155, 352, 383, 395, 398, 426, 428, 431
M. uraiense 4, 18, 25, 40, 58, 395, 396, 399, 415, 429, 431
M. urceolatum 8, 17, 22, 60, 396, 400, 428
    Orthotrichum urceloatum
M. validum 7, 19, 36, 41, 57, 396, 401
M. vesiculosum 5, 17, 59, 396, 397, 402, 429
M. viticulosum 11, 18, 27, 62, 397, 413, 414, 415, 416, 429, 431, 432
    Schlotheimia viticulosa
    M. didymodon
    **M. dentatum
    M. domingense
    **M. glaucescens
    M. goniopodium
    M. insularum
    M. intortifolium
    M. leptophyllum
    M. rhabdocarpum
    M. richardii
    M. tenellum
M. xenizon 8, 40, 49, 403, 404, 425, 428, 431
M. yuleanum 5, 17, 18, 32, 59, 235, 403, 415, 429
    *M. clemensiae
    M. deflexum
    M. habbemense
    M. hattorii
    M. novoguinense
    M. sarawaketense
    M. speirophyllum
M. zimmermannii 5, 18, 26, 40, 55, 405, 415, 429

List of all names in the genus *Macromitrium*

271 confirmed species, including subspecies, and varieties

1. *Macromitrium acuminatum* (Reinw. & Hornsch.) Müll. Hal.
2. *Macromitrium acutirameum* Mitt.
3. *Macromitrium amaniense* P. de la Varde
4. *Macromitrium amboroicum* Herzog
5. *Macromitrium angulatum* Mitt.
6. *Macromitrium angulosum* Thwaites & Mitt.
7. *Macromitrium angustifolium* Dozy & Molk.
8. *Macromitrium antarcticum* C.H. Wright
9. *Macromitrium archboldii* E.B. Bartram
10. *Macromitrium archeri* Mitt.
11. *Macromitrium argutum* Hampe
12. *Macromitrium atratum* Herzog
13. *Macromitrium atroviride* R.S. Williams
14. *Macromitrium attenuatum* Hampe
15. *Macromitrium aurantiacum* Paris & Broth.
16. *Macromitrium aurescens* Hampe
    Macromitrium aurescens var. *caledonicum* (Thér.) Thouvenot
17. *Macromitrium aureum* Müll. Hal.
18. *Macromitrium austrocirrosum* E.B. Bartram
19. *Macromitrium bifasciculare* Müll. Hal. ex Dusén
20. *Macromitrium bifasciculatum* Müll. Hal.
21. *Macromitrium binsteadii* Dixon
22. *Macromitrium bistratosum* E.B. Bartram
23. *Macromitrium blumei* Nees ex Schwägr.
    Macromitrium blumei var. *zollingeri* (Mitt. ex Bosch & Sande Lac.) S.L. Guo, B.C. Tan & Virtanen
24. *Macromitrium brachypodium* Müll. Hal.
25. *Macromitrium brevicaule* (Besch.) Broth.
26. *Macromitrium brevihamatum* Herzog
27. *Macromitrium brevisetum* Mitt.
28. *Macromitrium caldense* Ångstr.
29. *Macromitrium caloblastoides* Müll. Hal.
30. *Macromitrium calocalyx* Müll. Hal.
31. *Macromitrium calomicron* Broth.
32. *Macromitrium calymperoideum* Mitt.
33. *Macromitrium campoanum* Thér.
34. *Macromitrium cardotii* Thér.
35. *Macromitrium cataractarum* Müll. Hal.
36. *Macromitrium catharinense* Paris
37. *Macromitrium chloromitrium* (Besch.) Wilbraham
38. *Macromitrium cirrosum* (Hedw.) Brid.
39. *Macromitrium clastophyllum* Cardot
40. *Macromitrium comatum* Mitt.
41. *Macromitrium concinnum* Mitt. ex Bosch & Sande Lac.
42. *Macromitrium constrictum* Hampe & Lorentz
43. *Macromitrium crassirameum* Müll. Hal.
44. *Macromitrium crassiusculum* Lorentz
45. *Macromitrium crinale* Broth. & Geh.
46. *Macromitrium crispatulum* Mitt.
47. *Macromitrium crosbyorum* B.H. Allen & Vitt
48. *Macromitrium cuspidatum* Hampe
    Macromitrium cuspidatum var. *gracile* Dixon
49. *Macromitrium cylindricum* Mitt.
50. *Macromitrium densum* Mitt.
51. *Macromitrium diaphanum* Müll. Hal.
52. *Macromitrium dielsii* Broth. ex Vitt & H.P. Ramsay
53. *Macromitrium divaricatum* Mitt.
54. *Macromitrium diversifolium* Broth.
55. *Macromitrium dubium* Schimp. ex Müll. Hal.
56. *Macromitrium dusenii* Müll. Hal. ex Broth.
57. *Macromitrium echinatum* B.H. Allen
58. *Macromitrium ecrispatum* Dixon
59. *Macromitrium eddyi* B.C. Tan & Shevock
60. *Macromitrium ellipticum* Hampe
61. *Macromitrium emersulum* Müll. Hal.
62. *Macromitrium erubescens* E.B. Bartram
63. *Macromitrium erythrocomum* H.P. Ramsay, A. Cairns & Meagher
64. *Macromitrium evrardii* Thér.
65. *Macromitrium exsertum* Broth.
66. *Macromitrium falcatulum* Müll. Hal.
67. *Macromitrium fendleri* Müll. Hal.
68. *Macromitrium fernandezianum* Broth.
69. *Macromitrium ferriei* Cardot & Thér.
70. *Macromitrium fimbriatum* (P. Beauv.) Schwägr.
71. *Macromitrium flavopilosum* R.S. Williams
72. *Macromitrium flexuosum* Mitt.
73. *Macromitrium formosae* Cardot
74. *Macromitrium fortunatii* Cardot & Thér.
75. *Macromitrium fragilicuspis* Cardot
76. *Macromitrium francii* Thér.
77. *Macromitrium frondosum* Mitt.
78. *Macromitrium frustratum* B.H. Allen
79. *Macromitrium fulgescens* E.B. Bartram
80. *Macromitrium fulvum* Mitt.
81. *Macromitrium funicaule* Schimp. ex Besch.
82. *Macromitrium funiforme* Dixon
83. *Macromitrium fuscescens* Schwägr.
84. *Macromitrium fuscoaureum* E.B. Bartram
85. *Macromitrium galipense* Müll. Hal.
86. *Macromitrium gigasporum* Herzog
87. *Macromitrium giraldii* Müll. Hal.
88. *Macromitrium glabratum* Broth.
89. *Macromitrium glaziovii* Hampe
90. *Macromitrium gracile* (Hook.) Schwägr.
91. *Macromitrium greenmanii* Grout
92. *Macromitrium grossirete* Müll. Hal.
93. *Macromitrium guatemalense* Müll. Hal.
94. *Macromitrium gymnostomum* Sull. & Lesq.
95. *Macromitrium hainanense* S.L. Guo & S. He
96. *Macromitrium harrisii* Paris
97. *Macromitrium helmsii* Paris
98. *Macromitrium hemitrichodes* Schwägr.
    Macromitrium hemitrichodes var. *sarasinii* (Thér.) Thouvenot
99. *Macromitrium herzogii* Broth.
100. *Macromitrium hildebrandtii* Müll. Hal.
101. *Macromitrium holomitrioides* Nog.
102. *Macromitrium hortoniae* Vitt & H.P. Ramsay

103. *Macromitrium huigrense* R.S. Williams
104. *Macromitrium humboldtense* Thouvenot & Frank Müll.
105. *Macromitrium incurvifolium* (Hook. & Grev.) Schwägr.
106. *Macromitrium involutifolium* (Hook. & Grev.) Schwägr.
    Macromitrium involutifolium subsp. *ptychomitrioides* (Besch.) Vitt & H.P. Ramsay
107. *Macromitrium japonicum* Dozy & Molk.
108. *Macromitrium krausei* Lorentz
109. *Macromitrium laevigatum* Thér.
110. *Macromitrium laevisetum* Mitt.
111. *Macromitrium lanceolatum* Broth.
112. *Macromitrium laosianum* Paris & Broth.
113. *Macromitrium larrainii* Thouvenot & K.T. Yong
114. *Macromitrium lauterbachii* Broth. ex M. Fleisch.
115. *Macromitrium lebomboense* van Rooy
116. *Macromitrium leprieurii* Mont.
117. *Macromitrium leratii* Broth. & Paris
118. *Macromitrium ligulaefolium* Broth.
119. *Macromitrium ligulare* Mitt.
120. *Macromitrium lomasense* H. Rob.
121. *Macromitrium longicaule* Müll. Hal.
122. *Macromitrium longifolium* (Hook.) Brid.
123. *Macromitrium longipapillosum* D.D. Li, J. Yu, T. Cao & S.L. Guo
124. *Macromitrium longipes* (Hook.) Schwägr.
125. *Macromitrium longipilum* A. Braun ex Müll. Hal.
126. *Macromitrium longirostre* (Hook.) Schwägr.
127. *Macromitrium lorifolium* Paris & Broth.
128. *Macromitrium macrocomoides* Müll. Hal.
129. *Macromitrium macrosporum* Broth.
130. *Macromitrium macrothele* Müll. Hal.
131. *Macromitrium maolanense* Ze Y. Zhang, D.D. Li, J. Yu & S.L. Guo
132. *Macromitrium masafuerae* Broth.
133. *Macromitrium mcphersonii* B.H. Allen
134. *Macromitrium megalocladon* M. Fleisch.
135. *Macromitrium melinii* Roiv.
136. *Macromitrium menziesii* Müll. Hal.
137. *Macromitrium microcarpum* Müll. Hal.
138. *Macromitrium microstomum* (Hook. & Grev.) Schwägr.
139. *Macromitrium minutum* Mitt.
140. *Macromitrium mittenianum* Steere
141. *Macromitrium moorcroftii* (Hook. & Grev.) Schwägr.
142. *Macromitrium mosenii* Broth.
143. *Macromitrium nanothecium* Müll. Hal. ex Cardot
144. *Macromitrium nematosum* E.B. Bartram
145. *Macromitrium nepalense* (Hook. & Grev.) Schwägr.
146. *Macromitrium nigricans* Mitt.
147. *Macromitrium noguchianum* W. Schultze-Motel
148. *Macromitrium norrisianum* Vitt
149. *Macromitrium nubigenum* Herzog
150. *Macromitrium oblongum* (Taylor) Spruce
151. *Macromitrium ochraceoides* Dixon
152. *Macromitrium ochraceum* (Dozy & Molk.) Müll. Hal.
153. *Macromitrium onraedtii* Bizot
154. *Macromitrium orthophyllum* Mitt.
155. *Macromitrium orthostichum* Nees ex Schwägr.
    Macromitrium orthostichum subsp. *micropoma* M. Fleisch.
    Macromitrium orthostichum var. *burgeffii* Herzog
    Macromitrium orthostichum var. *siccosquarrosum* Herzog
156. *Macromitrium osculatianum* De Not.
157. *Macromitrium ousiense* Broth. & Paris
158. *Macromitrium ovale* Mitt.
159. *Macromitrium pallidum* (P. Beauv.) Wijk & Margad.
160. *Macromitrium panduraefolium* Thouvenot
161. *Macromitrium paridis* Besch.
162. *Macromitrium parvifolium* Dixon
163. *Macromitrium parvirete* E.B. Bartram
164. *Macromitrium pellucidum* Mitt.
165. *Macromitrium peraristatum* Broth.
166. *Macromitrium perdensifolium* Dixon
167. *Macromitrium perfragile* E.B. Bartram
168. *Macromitrium perichaetiale* (Hook. & Grev.) Müll. Hal.
169. *Macromitrium perpusillum* Müll. Hal.
170. *Macromitrium perreflexum* Steere
171. *Macromitrium pertriste* Müll. Hal.
172. *Macromitrium petelotii* Tixier
173. *Macromitrium picobonitum* B.H. Allen
174. *Macromitrium pilicalyx* Dixon ex E.B. Bartram
175. *Macromitrium piliferum* Schwägr.
176. *Macromitrium pilosum* Thér.
177. *Macromitrium proliferum* Mitt.
178. *Macromitrium prolongatum* Mitt.
179. *Macromitrium prorepens* (Hook.) Schwägr.
180. *Macromitrium proximum* Thér.
181. *Macromitrium pseudofimbriatum* Hampe
182. *Macromitrium pseudoserrulatum* E.B. Bartram
183. *Macromitrium pulchrum* Besch.
    Macromitrium pulchrum var. *neocaledonicum* (Besch.) Thouvenot
184. *Macromitrium pullenii* Vitt
185. *Macromitrium punctatum* (Hook. & Grev.) Brid.
186. *Macromitrium pyriforme* Müll. Hal.
187. *Macromitrium quercicola* Broth.
188. *Macromitrium ramsayae* Vitt
189. *Macromitrium raphidophyllum* Müll. Hal.
190. *Macromitrium refractifolium* Müll. Hal.
191. *Macromitrium regnellii* Hampe
192. *Macromitrium renauldii* Thér.
193. *Macromitrium repandum* Müll. Hal.
194. *Macromitrium retusulum* Müll. Hal.
195. *Macromitrium retusum* Hook. f. & Wilson
196. *Macromitrium rhacomitrioides* Nog.
197. *Macromitrium rimbachii* Herzog
198. *Macromitrium rufipilum* Cardot
199. *Macromitrium ruginosum* Besch.
200. *Macromitrium rugulosum* Ångstr.
201. *Macromitrium rusbyanum* E. Britton
202. *Macromitrium saddleanum* Besch. ex Müll. Hal.
    Macromitrium saddleanum var. *piliferum* Herzog
203. *Macromitrium salakanum* Müll. Hal.
204. *Macromitrium savatieri* Besch.
205. *Macromitrium schmidii* Müll. Hal.
    Macromitrium schmidii var. *macroperichaetialium* S.L. Guo & T. Cao
206. *Macromitrium scoparium* Mitt.
207. *Macromitrium sejunctum* B.H. Allen
208. *Macromitrium semperi* Müll. Hal.
209. *Macromitrium serpens* (Burch. ex Hook. & Grev.) Brid.
210. *Macromitrium sharpii* H.A. Crum ex Vitt
211. *Macromitrium similirete* E.B. Bartram
212. *Macromitrium solitarium* Müll. Hal.
    Macromitrium solitarium var. *brevipes* Broth.
213. *Macromitrium soulae* Renauld & Cardot
214. *Macromitrium speirostichum* Müll. Hal.
215. *Macromitrium st-johnii* E.B. Bartram
216. *Macromitrium standleyi* E.B. Bartram
217. *Macromitrium stoneae* Vitt & H.P. Ramsay
218. *Macromitrium streimannii* Vitt
219. *Macromitrium subbrevihamatum* Broth.
220. *Macromitrium subcirrhosum* Müll. Hal.
221. *Macromitrium subcrenulatum* Broth.

Appendix. List of all names in genus Macromitrium

222. **Macromitrium subdiscretum** R.S. Williams
223. **Macromitrium subhemitrichodes** Müll. Hal.
224. **Macromitrium subincurvum** Cardot & Thér.
225. **Macromitrium sublaeve** Mitt.
226. **Macromitrium sublongicaule** E.B. Bartram
227. **Macromitrium submucronifolium** Müll. Hal. & Hampe
228. **Macromitrium subperichaetiale** Thér.
229. **Macromitrium subscabrum** Mitt.
230. **Macromitrium subtortum** (Hook. & Grev.) Schwägr.
231. **Macromitrium subulatum** Mitt.
232. **Macromitrium sulcatum** (Hook.) Brid.
   Macromitrium sulcatum subsp. **ramentosum** (Thwaites & Mitt.) M. Fleisch.
   Macromitrium sulcatum var. **leptocarpum** (Broth.) J. Yu, D.D. Li, Yan Li & S.L. Guo
   Macromitrium sulcatum var. **torulosum** (Mitt.) Tixier
233. **Macromitrium swainsonii** (Hook.) Brid.
234. **Macromitrium taiheizanense** Nog.
235. **Macromitrium taiwanense** Nog.
236. **Macromitrium taoense** Thér.
237. **Macromitrium tenax** Müll. Hal.
238. **Macromitrium thwaitesii** Broth. ex M. Fleisch.
239. **Macromitrium tocaremae** Hampe
240. **Macromitrium tongense** Sull.
241. **Macromitrium tosae** Besch.
242. **Macromitrium trachypodium** Mitt.
243. **Macromitrium trichophyllum** Mitt.
244. **Macromitrium trinitense** R.S. Williams
245. **Macromitrium tuberculatum** Dixon
246. **Macromitrium turgidum** Dixon
247. **Macromitrium tylostomum** Mitt. ex Bosch & Sande Lac.
248. **Macromitrium ulophyllum** Mitt.
249. **Macromitrium uraiense** Nog.
250. **Macromitrium urceolatum** (Hook.) Brid.
251. **Macromitrium validum** Herzog
252. **Macromitrium vesiculosum** Tixier
253. **Macromitrium viticulosum** (Raddi) Brid.
254. **Macromitrium xenizon** B.H. Allen & W.R. Buck
255. **Macromitrium yuleanum** Broth. & Geh.
256. **Macromitrium zimmermannii** M. Fleisch

20 new synonmys and 1 new combination

Macromitrium himalayanum Dixon = *Macrocoma orthotrichoides* (Raddi) Wijk & Margad., **syn. nov.**
Macromitrium cancellatum Y.X. Xiong = *Macromitrium giraldii* Müll. Hal., **syn. nov.**
Macromitrium cavaleriei Cardot & Thér. = *Macromitrium giraldii* Müll. Hal., **syn. nov.**
Macromitrium handelii Broth. = *Macromitrium giraldii* Müll. Hal., **syn. nov.**
Macromitrium rigbyanum Dixon = *Macromitrium giraldii* Müll. Hal., **syn. nov.**
Macromitrium gebaueri Broth. = *Macromitrium giraldii* Müll. Hal., **syn. nov.**
Macromitrium gigasporum fo. *brevipes* Herzog = *Macromitrium gigasporum* Herzog, **syn. nov.**
Macromitrium giraldii var. *acrophylloides* Müll. Hal. = *Macromitrium giraldii* Müll. Hal., **syn. nov.**
Macromitrium gymnostomum var. *brevisetum* Thér. = *Macromitrium gymnostomum* Sull. & Lesq., **syn. nov.**
Macromitrium gymnostomum var. *robustum* Broth. = *Macromitrium gymnostomum* Sull. & Lesq., **syn. nov.**
Macromitrium sinense E.B. Bartram = *Macromitrium giraldii* Müll. Hal., **syn. nov.**
Macromitrium syntrichophyllum Thér. & P. de la Varde = *Macromitrium giraldii* Müll. Hal., **syn. nov.**
Macromitrium syntrichophyllum var. *longisetum* Thér. & Reimers = *Macromitrium giraldii* Müll. Hal., **syn. nov.**
Macromitrium crenulatum Hampe = *Macromitrium longifolium* (Hook.) Brid., **syn. nov.**
Macromitrium hornschuchii Müll. Hal. = *Macromitrium microstomum* (Hook. & Grev.) Schwägr., **syn. nov.**
Macromitrium muellerianum Mitt. = *Macromitrium sulcatum* (Hook.) Brid., **syn. nov.**
Macromitrium pseudoramentosum Herzog = *Macromitrium sulcatum* (Hook.) Brid., **syn. nov.**
Macromitrium subhemitrichodes var. *hodgkinsoniae* Müll. Hal. = *Macromitrium subhemitrichodes* Müll. Hal., **syn. nov.**
Macromitrium vesiculatum (Herzog) Herzog = *Macromitrium swainsonii* (Hook.) Brid., **syn. nov.**
Macromitrium woollsianum var. *chlorophyllosum* Müll. Hal. = *Macromitrium ligulaefolium* Broth., **syn. nov.**
Macromitrium homaloblastum Herzog = *Groutiella homalbolastum* (Herzog) Guo & Li, **comb. nov.**

508 species to be synonmys

*Macromitrium aciculare Brid. = *Macromitrium pallidum* (P. Beauv.) Wijk & Margad., *fide* Crosby et al., 1983
Macromitrium acunae Thér. = *Macromitrium microstomum* (Hook. & Grev.) Schwägr., *fide* Vitt & Ramsay, 1985
Macromitrium acutifolium (Hook. & Grev.) Brid. = *Macromitrium longirostre* (Hook.) Schwägr., *fide* Fife, 2017
Macromitrium adelphinum Cardot = *Macrocoma tenuis* (Hook. & Grev.) Vitt subsp. *tenuis*, *fide* Wilbraham, 2017
Macromitrium adnatum Müll. Hal. = *Groutiella chimborazensis* (Spruce ex Mitt.) Florsch., *fide* Valente et al., 2020
Macromitrium adstrictum Ångstr. = *Macromitrium microstomum* (Hook. & Grev.) Schwägr., *fide* Vitt & Ramsay, 1985
Macromitrium altipapillosum E.B. Bartram = *Macromitrium megalocladon* M. Fleisch., *fide* Vitt et al., 1995
Macromitrium altipes Müll. Hal. = *Macromitrium cirrosum* var. *jamaicense* (Mitt.) Grout., *fide* Grout, 1944
Macromitrium altituberculosum E.B. Bartram = *Macromitrium carionis* Müll. Hal., *fide* Goffinet, 1993
Macromitrium amoenum Hornsch. ex Müll. Hal. = *Macromitrium hemitrichodes* Schwägr., *fide* Vitt & Ramsay, 1985
Macromitrium anacamptophyllum Müll. Hal. = *Macrocoma sullivantii* (Müll. Hal.) Grout., *fide* Grout, 1944
Macromitrium ancistrophyllum Cardot = *Macromitrium nanothecium* Müll. Hal. ex Cardot., *fide* Wilbraham, 2018
**Macromitrium andamaniae Müll. Hal. = *Groutiella tomentosa* (Hornsch.) Wijk & Margad., *fide* Vitt & Crum, 1970
**Macromitrium andamanum Müll. Hal. = *Groutiella tomentosa* (Hornsch.) Wijk & Margad., *fide* Vitt & Crum, 1970
**Macromitrium aneurodictyon Cardot ex P. de la Varde = *Macromitrium anomodictyon* Cardot., *fide* Crosby et al., 1983
Macromitrium angulicaule Müll. Hal. = *Macrocoma orthotrichoides* (Raddi) Wijk & Marg., *fide* Vitt, 1980 [1981]
*Macromitrium angulicaule var. *phyllorhizans* (Müll. Hal.) Paris = *Macrocoma orthotrichoides* (Radd.) Wijk & Marg., *fide* Vitt, 1980 [1981]
Macromitrium annamense Broth. & Paris = *Macromitrium blumei* var. *zollingeri* (Mitt. ex Bosch & Sande Lac.) S.L. Guo, B.C. Tan & Virtanen, *fide* Guo et al., 2006
Macromitrium anomodictyon Cardot = *Macromitrium serpens* (Burch. ex Hook. & Grev.) Brid., *fide* Wilbraham, 2016
Macromitrium apiculatum (Hook.) Schwägr. = *Groutiella apiculata* (Hook.) H.A. Crum & Steere., *fide* Crum & Steere, 1950
Macromitrium appendiculatum (Renauld & Cardot) Paris = *Cardotiella appendiculata* (Renauld & Cardot) Vitt., *fide* Vitt, 1981
*Macromitrium appendiculatum Müll. Hal. = *Macromitrium helmsii* Paris., *fide* Fife, 2017
Macromitrium appressifolium Mitt. = *Macromitrium orthostichum* Nees ex Schwägr., *fide* Vitt et al., 1995
Macromitrium aristatum Mitt. = *Macromitrium retusum* Hook. f. & Wilson., *fide* Fife, 2017
Macromitrium aristocalyx Müll. Hal. = *Macromitrium brevisetum* Mitt., *fide* Wijk et al., 1964
Macromitrium aspericuspis Dixon = *Macromitrium macrosporum* Broth., *fide* Vitt et al., 1995

Macromitrium asperulum Mitt. = *Macromitrium archeri* Mitt., *fide* Vitt & Ramsay, 1985
Macromitrium assamicum (Griff.) Mitt. = *Macromitrium nepalense* (Hook. & Grev.) Schwägr., *fide* Gangulee, 1976
Macromitrium assimile Broth. = *Macromitrium blumei* Nees ex Schwägr., *fide* Wijk *et al.*, 1964
**Macromitrium assimile* Broth. & Dixon = *Macromitrium binsteadii* Dixon., *fide* Wijk *et al.*, 1964
Macromitrium astroideum Mitt. = *Macromitrium serpens* (Burch. ex Hook. & Grev.) Brid., *fide* Wilbraham & Ellis, 2010
Macromitrium baileyi Mitt. = *Macromitrium hemitrichodes* Schwägr., *fide* Vitt & Ramsay, 1985
***Macromitrium barbatum* Mitt. = *Macrocoma tenuis* (Hook. & Grev.) Vitt., *fide* Fife, 2017
Macromitrium barbense Renauld & Cardot = *Macromitrium cirrosum* (Hedw.) Brid., *fide* Grout, 1944
Macromitrium bathyodontum Cardot = *Macromitrium japonicum* Dozy & Molk., *fide* Noguchi & Iwatsuki, 1989
Macromitrium beecheyanum Mitt. = *Macromitrium incurvifolium* (Hook. & Grev.) Schwägr., *fide* Whittier, 1976
Macromitrium belangeri Müll. Hal. = *Macromitrium sulcatum* (Hook.) Brid., *fide* Wilbraham, 2016
Macromitrium benguetense R.S. Williams = *Macromitrium schmidii* Müll. Hal., *fide* Guo *et al.*, 2007
Macromitrium bequaertii Thér. & Naveau = *Macromitrium sulcatum* (Hook.) Brid., *fide* Wilbraham, 2016
**Macromitrium bescherellei* H. Whittier & B. Whittier = *Macromitrium tahitisecundum* Marg., *fide* Margadant 1972
***Macromitrium boivinii* Müll. Hal. = *Macromitrium sulcatum* (Hook.) Brid. subsp. *sulcatum*, *fide* Ellis & Wilbraham, 2008
Macromitrium bolivianum Müll. Hal. = *Macrocoma sullivantii* (Müll. Hal.) Grout, *fide* Gangulee, 1976
Macromitrium borbonicum (Besch.) Broth. = *Macromitrium serpens* (Burch. ex Hook. & Grev.) Brid., *fide* O'Shea, 2006
Macromitrium brachiatum Hook. & Wilson = *Desmotheca brachiata* (Hook. & Wilson) Vitt, *fide* Vitt, 1990
Macromitrium brachycladulum Broth. & Paris = *Macromitrium prolongatum* Mitt., *fide* Noguchi & Iwatsuki, 1989
Macromitrium brachyrhynchum (Schwägr.) Schimp. = *Macromitrium punctatum* (Hook. & Grev.) Brid., *fide* Grout, 1944
Macromitrium brachystele Dixon = *Macromitrium longicaule* Müll. Hal., *fide* Vitt *et al.*, 1995
Macromitrium braunioides Cardot = *Macrocoma tenuis* (Hook. & Grev.) Vitt., *fide* Allen, 2002
Macromitrium brasiliense Mitt. = *Macrocoma brasiliensis* (Mitt.) Vitt, *fide* Vitt, 1980 [1981]
Macromitrium brevipes Müll. Hal. = *Groutiella apiculata* (Hook.) H.A. Crum & Steere., *fide* Goffinet, 1993
Macromitrium brevipilosum Thér. = *Macromitrium ligulaefolium* Broth., *fide* Vitt & Ramsay, 1985
**Macromitrium brevirameum* E.B. Bartram = *Macromitrium parvifolium* Dixon., *fide* Vitt *et al.*, 1995
Macromitrium brevirameum E.B. Bartram = *Macromitrium parvifolium* Dixon., *fide* Vitt *et al.*, 1995
Macromitrium brevisetaceum Hampe = *Macromitrium brachypodium* Müll. Hal., *fide* Vitt & Ramsay, 1985
Macromitrium brevissimum Dixon = *Macromitrium densum* Mitt., *fide* Yu *et al.*, 2014
Macromitrium brevituberculatum Dixon = *Macromitrium gymnostomum* Sull. & Lesq., *fide* Guo *et al.*, 2007
Macromitrium brotheri Müll. Hal. = *Macromitrium swainsonii* (Hook.) Brid., *fide* Valente *et al.*, 2020
Macromitrium brownii (Schwägr.) Müll. Hal. = *Schlotheimia brownii* Schwägr., *fide* Vitt, 1989
**Macromitrium cacuminicola* Besch. = *Macromitrium paridis* Besch., *fide* Wijk *et al.*, 1964
Macromitrium cacuminicola Müll. Hal. = *Macromitrium stratosum* Mitt., *fide* Grout, 1944
Macromitrium caducipilum Lindb. = *Macromitrium retusum* Hook. f. & Wilson., *fide* Fife, 2017
***Macromitrium caespitans* Müll. Hal. = *Macrocoma lycopodioides* (Schwägr.) Vitt., *fide* O'Shea, 2006
Macromitrium calvescens Bosch & Sande Lac. = *Macromitrium fuscescens* Schwägr., *fide* Vitt *et al.*, 1995
Macromitrium calycinum Mitt. = *Glyphomitrium calycinum* (Mitt.) Cardot, *fide* Redfearn *et al.*, 1996
Macromitrium canum Müll. Hal. = *Macromitrium owahiense* Müll. Hal., *fide* Wijk *et al.*, 1964
Macromitrium capillicaule Müll. Hal. ex Broth. = *Macrocoma brasiliensis* (Mitt.) Vitt., *fide* Vitt, 1980 [1981]
Macromitrium carinatum Mitt. = *Macromitrium involutifolium* subsp. *ptychomitrioides* (Besch.) Vitt & H.P. Ramsay., *fide* Vitt & Ramsay, 1985
Macromitrium carionis Müll. Hal. = *Macromitrium swainsonii* (Hook.) Brid., *fide* Valente *et al.*, 2020
Macromitrium catharinense var. *gracilius* (Müll. Hal.) Paris = *Macromitrium catharinense* Paris., *fide* Li *et al.*, 2019
Macromitrium celebense Paris = *Macromitrium salakanum* Müll. Hal., *fide* Vitt *et al.*, 1995
Macromitrium ceylanicum Mitt. = *Macromitrium sulcatum* (Hook.) Brid., *fide* Gangulee, 1976
Macromitrium chamissonis (Hornsch.) Müll. Hal. = *Schlotheimia chamissonis* Hornsch., *fide* Sehnem, 178
Macromitrium chimborazense Spruce ex Mitt. = *Groutiella chimborazensis* (Spruce ex Mitt.) Florsch., *fide* Valente *et al.*, 2020
Macromitrium chrysomitrium Müll. Hal. = *Macrocoma orthotrichoides* (Raddi) Wijk & Margad., *fide* Valente *et al.*, 2020
Macromitrium chungkingense P.C. Chen = *Macromitrium tosae* Besch., *fide* Yu *et al.*, 2012
Macromitrium circinicladum Müll. Hal. = *Macromitrium diaphanum* Müll. Hal., *fide* Streimann & Curnow, 1989
Macromitrium cirrosum var. *stenophyllum* (Mitt.) Grout = *Macromitrium cirrosum* (Hedw.) Brid., *fide* Churchill, 2016
Macromitrium clavellatum (Hook. & Grev.) Schwägr. = *Drummondia prorepens* (Hedw.) Trevis., *fide* Flora of North America Editorial Committee, 2014
**Macromitrium clemensiae* Nog. = *Macromitrium yuleanum* Broth. & Geh., *fide* Vitt *et al.*, 1995
Macromitrium coarctatulum Müll. Hal. = *Macromitrium prorepens* (Hook.) Schwägr., *fide* Wijk *et al.*, 1964
***Macromitrium coarctatum* Schimp. ex C.H. Wright = *Macromitrium fasciculare* Mitt., *fide* Crosby *et al.*, 1983
Macromitrium comatulum Broth. = *Macromitrium ferriei* Cardot & Thér., *fide* Noguchi & Iwatsuki, 1989
Macromitrium confusum Mitt. = *Macrocoma tenuis* (Hook. & Grev.) Vitt subsp. *tenuis*, *fide* Vitt, 1980 [1981]
Macromitrium consanguineum Cardot = *Macrocoma sullivantii* (Müll. Hal.) Grout., *fide* Vitt, 1973
Macromitrium contextum Hampe = *Macromitrium regnellii* Hampe., *fide* Valente *et al.*, 2020
Macromitrium contortum Thwaites & Mitt. = *Macromitrium zollingeri* Mitt. ex Bosch & Sande Lac., *fide* Wijk *et al.*, 1964
Macromitrium contractum Thér. = *Macromitrium francii* Thér., *fide* Thouvenot, 2019
Macromitrium copelandii Broth. = *Macromitrium blumei* Nees ex Schwägr., *fide* Wijk *et al.*, 1964
Macromitrium coriaceum E.B. Bartram = *Macromitrium tenax* Müll. Hal., *fide* Seki, 1974 [1976]
***Macromitrium corrugatum* Wilson = *Macromitrium sulcatum* (Hook.) Brid., *fide* Gangulee, 1976
Macromitrium costaricense E.B. Bartram = *Macromitrium cirrosum* (Hedw.) Brid., *fide* Florschütz, 1964
Macromitrium courtoisii Broth. & Paris = *Macromitrium tosae* Besch., *fide* Yu *et al.*, 2012
Macromitrium crumianum Steere & W.R. Buck = *Macromitrium leprieurii* Mont., *fide* Goffinet, 1993
Macromitrium cubensicirrhosum Müll. Hal. = *Macromitrium cirrosum* (Hedw.) Brid., *fide* Wijk *et al.*, 1964
Macromitrium cucullatum Thér. = *Macromitrium ligulaefolium* Broth., *fide* Vitt & Ramsay, 1985
Macromitrium cumingii Müll. Hal. = *Macromitrium incurvifolium* (Hook. & Grev.) Schwägr., *fide* Staples *et al.*, 2004
***Macromitrium cylindricum* Schimp. = *Groutiella tomentosa* (Hornsch.) Wijk & Margad., *fide* Vitt & Crum, 1970
***Macromitrium cylindricum* Schimp. ex C.H. Wright = *Ulota fulva* Brid., *fide* O'Shea, 2006
Macromitrium cylindromitrium Müll. Hal. = *Macromitrium aurescens* Hampe., *fide* Streimann & Curnow, 1989
Macromitrium cylindromitrium var. *caledonicum* Thér. = *Macromitrium aurescens* var. *caledonicum* (Thér.) Thouvenot, *fide* Thouvenot, 2019

Macromitrium cylindrothecium Nog. = *Macromitrium tosae* Besch., *fide* Yu *et al*., 2012
Macromitrium daemelii Müll. Hal. = *Macromitrium involutifolium* (Hook. & Grev.) Schwägr., *fide* Vitt & Ramsay, 1985
***Macromitrium dausonifolium* Müll. Hal. ex Paris = *Macrocoma tenuis* (Hook. & Grev.) Vitt., *fide* O'Shea, 2006
***Macromitrium dawsanomitrium* Müll. Hal. = *Macrocoma tenuis* (Hook. & Grev.) Vitt., *fide* O'Shea, 2006
Macromitrium dawsoniomitrium Müll. Hal. = *Macrocoma tenuis* (Hook. & Grev.) Vitt., *fide* O'Shea, 2006
Macromitrium daymannianum E.B. Bartram = *Macromitrium parvifolium* Dixon., *fide* Vitt *et al*., 1995
Macromitrium deflexum E.B. Bartram = *Macromitrium yuleanum* Broth. & Geh., *fide* Vitt *et al*., 1995
**Macromitrium densifolium* Cardot = *Groutiella chimborazensis* (Spruce ex Mitt.) Florsch. subsp. chimborazensis., *fide* Vitt, 1994
Macromitrium densifolium Thér. = *Macromitrium tongense* Sull., *fide* Thouvenot, 2019
Macromitrium dentatulum Müll. Hal. = *Macromitrium pentastichum* Müll. Hal., *fide* Grout, 1944
***Macromitrium dentatum* Schimp. = *Macromitrium richardii* Schwägr., *fide* Vitt, 1994
Macromitrium denudatum A. Jaeger = *Macromitrium longifolium* (Hook.) Brid., *fide* Grout, 1944
Macromitrium dickasonii E.B. Bartram = *Macromitrium japonicum* Dozy & Molk., *fide* Li *et al*., 2024
Macromitrium didymodon Schwägr. = *Macromitrium richardii* Schwägr., *fide* Grout, 1944
Macromitrium diffractum Cardot = *Groutiella tomentosa* (Hornsch.) Wijk & Margad., *fide* Vitt, 1970
Macromitrium dimorphum Müll. Hal. = *Macromitrium caloblastoides* Müll. Hal., *fide* Vitt & Ramsay, 1985
Macromitrium divortiarum Sehnem = *Macromitrium diversifolium* Broth., *fide* Valente *et al*., 2020
Macromitrium doeringianum Hampe = *Macromitrium pseudofimbriatum* Hampe., *fide* Valente *et al*., 2020
Macromitrium domingense A. Jaeger = *Macromitrium richardii* Schwägr., *fide* Grout, 1944
Macromitrium dregei Hornsch. = *Macrocoma sullivantii* (Müll. Hal.) Grout., *fide* Grout, 1944
Macromitrium drewii H. Rob. = *Macromitrium catharinense* Paris., *fide* Li *et al*., 2019
Macromitrium durandii Renauld & Cardot = *Groutiella chimborazensis* (Spruce ex Mitt.) Florsch., *fide* Allen, 2002
Macromitrium drummondii (Hook. & Grev.) Hampe = *Ulota drummondii* (Hook. & Grev.) Brid., *fide* Crum & Anderson, 1981
Macromitrium dussii Broth. = *Macromitrium leprieurii* Mont., *fide* Florschütz, 1964
Macromitrium eckendorffii Thér. & P. de la Varde = *Macromitrium sulcatum* (Hook.) Brid., *fide* Tixier, 1989
**Macromitrium elegans* Duby = *Macromitrium serpens* (Burch. ex Hook. & Grev.) Brid., *fide* O'Shea, 2006
Macromitrium elegans Hornsch. = *Groutiella apiculata* (Hook.) H.A. Crum & Steere., *fide* Wijk *et al*., 1964
Macromitrium elizabethae Dixon = *Macromitrium tongense* Sull., *fide* Wijk *et al*., 1964
Macromitrium elongatum Dozy & Molk. = *Macromitrium cuspidatum* Hampe., *fide* Wijk *et al*., 1964
**Macromitrium elongatum* Dozy & Molk. ex Bosch & Sande Lac. = *Macromitrium acuminatum* (Reinw. & Hornsch.) Müll. Hal., *fide* Vitt *et al*., 1995
Macromitrium emarginatum Broth. = Schlotheimia *merkelii* Hornsch., *fide* Valente *et al*., 2020
Macromitrium erectopatulum Müll. Hal. = *Macromitrium cirrosum* (Hedw.) Brid., *fide* Grout, 1944
Macromitrium erectopatulum var. *grossirete* Müll. Hal. = *Macromitrium cirrosum* (Hedw.) Brid., *fide* Grout, 1944
Macromitrium erosulum Mitt. = *Macromitrium prorepens* (Hook.) Schwägr., *fide* Fife, 2017
Macromitrium eucalyptorum Müll. Hal. & Hampe = *Macrocoma tenuis* (Hook. & Grev.) Vitt., *fide* Vitt, 1973
Macromitrium eucalyptorum var. *recurvulum* (Müll. Hal.) Sainsbury = *Macrocoma tenuis* (Hook. & Grev.) Vitt subsp. *tenuis*, *fide* Fife, 2017
Macromitrium eurymitrium Besch. = *Macromitrium* fuscescens Schwägr., *fide* Vitt *et al*., 1995
Macromitrium fasciculare Mitt. = *Macromitrium microstomum* (Hook. & Grev.) Schwägr., *fide* Wilbraham & Ellis, 2010
Macromitrium fasciculare var. *angustifolium* P. de la Varde = *Macromitrium microstomum* (Hook. & Grev.) Schwägr., *fide* Wilbraham & Ellis, 2010
Macromitrium fasciculare var. *javense* M. Fleisch. = *Macromitrium microstomum* (Hook. & Grev.) Schwägr., *fide* Wilbraham & Ellis, 2010
Macromitrium ferrugineum (Burch. ex Hook. & Grev.) Müll. Hal. = *Schlotheimia ferruginea* (Burch. ex Hook. & Grev.) Brid. , *fide* Bridel, 1826
Macromitrium filicaule Müll. Hal. = *Macromitrium microstomum* (Hook. & Grev.) Schwägr., *fide* Yu *et al*., 2018
**Macromitrium filiforme* Schwägr. = *Macrocoma orthotrichoides* (Raddi) Wijk & Margad., *fide* Vitt, 1973
Macromitrium filiforme var. *squarrulosum* Hampe = *Macrocoma orthotrichoides* (Raddi) Wijk & Margad., *fide* Yano, 2011
Macromitrium fimbriatum var. *chloromitrium* Besch. = *Macromitrium chloromitrium* (Besch.) Wilbraham., *fide* Wilbraham & Ellis, 2010
Macromitrium fitzgeraldii Lesq. & James = *Macrocoma sullivantii* (Müll. Hal.) Grout., *fide* Grout, 1944
Macromitrium flaccidisetum Müll. Hal. = *Macromitrium microstomum* (Hook. & Grev.) Schwägr., *fide* Vitt & Ramsay, 1985
**Macromitrium flexuosum* Schimp. ex Besch. = *Macromitrium longifolium* (Hook.) Brid., *fide* Wijk *et al*., 1964
Macromitrium fortunatii var. nigrescens Tixier = *Macromitrium fortunatii* Cardot & Thér., *fide* Li *et al*., 2018
Macromitrium foxworthyi Broth. = *Macromitrium subtile* Schwägr., *fide* Eddy, 1996
**Macromitrium fragile* G. Negri = *Macromitrium bujongolanum* O'Shea, *fide* O'Shea, 1908
**Macromitrium fragile* Mitt. = *Groutiella tomentosa* (Hornsch.) Wijk & Margad., *fide* Vitt & Crum, 1970
Macromitrium fragilifolium Dixon = *Macromitrium orthostichum* Nees ex Schwägr., *fide* Eddy, 1996
Macromitrium frigidum Müll. Hal. = *Macrocoma frigida* (Müll. Hal.) Vitt., *fide* Vitt, 1973
Macromitrium fruhstorferi Cardot = *Macromitrium angustifolium* Dozy & Molk., *fide* Wijk *et al*., 1964
**Macromitrium fuscescens* E.B. Bartram = *Macromitrium fulgescens* E.B. Bartram, *fide* Grout, 1944
Macromitrium fuscoviride (Hornsch.) Müll. Hal. = *Schlotheimia fuscoviridis* Hornsch., *fide* Forzza, 2010
Macromitrium geheebii Müll. Hal. = *Macrocoma tenuis* (Hook. & Grev.) Vitt subsp. *tenuis*, *fide* Vitt, 1980 [1981]
Macromitrium ghiesbreghtii Besch. = *Macrocoma tenuis* (Hook. & Grev.) Vitt., *fide* Allen, 2002
Macromitrium ghiesbreghtii var. brevifolium Besch. = *Macrocoma sullivantii* (Müll. Hal.) Grout., *fide* Vitt, 1973
Macromitrium gimalacii Bizot & Onr. = *Macrocoma lycopodioides* (Schwägr.) Vitt., *fide* Arts, 2005
Macromitrium glaucum Mitt. = *Macromitrium fuscescens* Schwägr., *fide* Vitt *et al*., 1995
Macromitrium goniopodium Mitt. = *Macromitrium richardii* Schwägr., *fide* Grout, 1944
Macromitrium goniorrhynchum (Dozy & Molk.) Mitt. = *Groutiella tomentosa* (Hornsch.) Wijk & Margad., *fide* Touw, 1992
Macromitrium goniorrhynchum var. *denticulatum* Dixon = *Groutiella tomentosa* (Hornsch.) Wijk & Margad, *fide* Touw, 1992
Macromitrium goniorrhynchum var. *glaucescens* Müll. Hal. = *Groutiella tomentosa* (Hornsch.) Wijk & Margad, *fide* Touw, 1992
Macromitrium goniostomum Broth. = *Macromitrium macrosporum* Broth., *fide* Vitt *et al*., 1995
Macromitrium gracile var. *proboscideum* Dixon = *Macromitrium gracile* (Hook.) Schwägr., *fide* Fife, 2017
Macromitrium gracilipes Cardot = *Macromitrium renauldii* Thér., *fide* Thouvenot, 2019
Macromitrium gracillimum (Besch.) Broth. = *Matteria gracillima* (Besch.) Goffinet , *fide* Goffinet & Vitt, 1998
Macromitrium habbemense E.B. Bartram = *Macromitrium yuleanum* Broth. & Geh., *fide* Vitt *et al*., 1995
Macromitrium haitense Thér. = *Macromitrium longifolium* (Hook.) Brid., *fide* Allen, 2002

Macromitrium hallieri M. Fleisch. ex Broth. = *Macromitrium ochraceum* (Dozy & Molk.) Müll. Hal., *fide* Vitt *et al*., 1995
Macromitrium hamatum Dixon = *Macromitrium tosae* Besch., *fide* Li *et al*., 2020
**Macromitrium hamatum* E.B. Bartram = *Macromitrium macrosporum* Broth., *fide* Vitt *et al*., 1995
Macromitrium hariotii Besch. ex Müll. Hal. = *Macromitrium hymenostomum* Mont., *fide* Gangulee, 1976
Macromitrium hattorii Nog. = *Macromitrium yuleanum* Broth. & Geh., *fide* Vitt *et al*., 1995
Macromitrium hectorii Mitt. = *Schlotheimia campbelliana* Müll. Hal., *fide* Fife, 2017
Macromitrium heterodictyon Dixon = *Macromitrium ousiense* Broth. & Paris, *fide* Yu *et al*., 2013
Macromitrium hirtellum E.B. Bartram = *Macromitrium punctatum* (Hook. & Grev.) Brid., *fide* Vitt, 1979
Macromitrium hispidulum Thwaites & Mitt. = *Macromitrium minutum* Mitt., *fide* Wijk *et al*., 1964
Macromitrium hoehnei Herzog = *Macromitrium cirrosum* (Hedw.) Brid., *fide* Valente *et al*., 2020
Macromitrium homalacron Müll. Hal. = *Macromitrium longifolium* (Hook.) Brid., *fide* Vitt, 1994
Macromitrium hooglandii E.B. Bartram = *Macromitrium tylostomum* Mitt. ex Bosch & Sande Lac., *fide* Vitt *et al*., 1995
Macromitrium horridum Dixon = *Macromitrium blumei* Nees ex Schwägr., *fide* Eddy, 1996
Macromitrium humile Bosch & Sande Lac. = *Macromitrium salakanum* Müll. Hal., *fide* Vitt *et al*., 1995
Macromitrium humillimum (Mitt.) Paris = *Glyphomitrium humillimum* (Mitt.) Cardot, *fide* Fedosov *et al*., 2022
Macromitrium husnotii Schimp. ex Besch. = *Groutiella husnotii* (Schimp. ex Besch.) H.A. Crum & Steere., *fide* Crum & Steere, 1950
Macromitrium hyalinum Broth. = *Macrocoma tenuis* (Hook. & Grev.) Vitt subsp. *tenuis*, *fide* Vitt, 1980 [1981]
Macromitrium hymenostomum Mont. = *Macrocoma sullivantii* (Müll. Hal.) Grout., *fide* Grout, 1944
Macromitrium incrustatifolium H. Rob. = *Macromitrium nepalense* (Hook. & Grev.) Schwägr., *fide* Guo & He, 2014
Macromitrium incurvulum Müll. Hal. = *Macromitrium involutifolium* (Hook. & Grev.) Schwägr., *fide* Vitt & Ramsay, 1985
Macromitrium incurvum (Lindb.) Mitt. = *Macromitrium japonicum* Dozy & Molk., *fide* Ignatov & Afonina, 1992
Macromitrium inflexifolium Dixon = *Macromitrium ferriei* Cardot & Thér., *fide* Noguchi & Iwatsuki, 1989
Macromitrium insulanum Podp. = *Macromitrium japonicum* Dozy & Molk., *fide* Noguchi, 1967
**Macromitrium insularum* Mitt. = *Macromitrium richardii* Schwägr., *fide* Grout, 1944
Macromitrium insularum Sull. & Lesq. = *Macromitrium japonicum* Dozy & Molk., *fide* Noguchi & Iwatsuki, 1989
Macromitrium intermedium Mitt. = *Macromitrium hemitrichodes* Schwägr., *fide* Vitt & Ramsay, 1985
Macromitrium intortifolium Hampe = *Macromitrium viticulosum* (Raddi) Brid., *fide* Valente *et al*., 2020
Macromitrium intricatum Müll. Hal. = *Macrocoma tenuis* subsp. *sullivantii* (Müll. Hal.) Vitt., *fide* Vitt, 1980 [1981]
Macromitrium jamaicense Mitt. = *Macromitrium cirrosum* var. *jamaicense* (Mitt.) Grout., *fide* Grout, 1944
Macromitrium jamesonii (Arn.) Müll. Hal. = *Schlotheimia jamesonii* (Arn.) Brid., *fide* Allen, 2002
Macromitrium japonicum var. *makinoi* (Broth.) Nog. = *Macromitrium japonicum* Dozy & Molk., *fide* Li *et al*., 2024
Macromitrium javanicum Bosch & Sande Lac. = *Macromitrium incurvifolium* (Hook. & Grev.) Schwägr., *fide* Vitt *et al*., 1995
Macromitrium julaceum (Hornsch.) Müll. Hal. = *Schlotheimia rugifolia* (Hook.) Schwägr., *fide* Wijk *et al*., 1964
Macromitrium kaernbachii Broth. = *Macromitrium incurvifolium* (Hook. & Grev.) Schwägr., *fide* Vitt *et al*., 1995
Macromitrium kegelianum Müll. Hal. = *Schlotheimia kegeliana* (Müll. Hal.) Müll. Hal., *fide* Wijk *et al*., 1964
Macromitrium koghiense Thér. = *Macromitrium pilosum* Thér., *fide* Thouvenot, 2019
Macromitrium koghiense var. *spiricaule* Broth. & Paris = *Macromitrium pilosum* Thér., *fide* Thouvenot, 2019
***Macromitrium laevatum* Mitt. ex Broth. = *Macromitrium sulcatum* (Hook.) Brid., *fide* O'Shea, 2006
Macromitrium laevifolium Mitt. = *Macromitrium pellucidum* Mitt., *fide* Grout, 1944
Macromitrium lamprocarpum Müll. Hal. ex Renauld & Cardot = *Groutiella wagneriana* (Müll. Hal.) H.A. Crum & Steere., *fide* Bowers, 1974
Macromitrium lampromitrium Müll. Hal. = *Macrocoma tenuis* subsp. *sullivantii* (Müll. Hal.) Vitt., *fide* Vitt, 1980 [1981]
Macromitrium laxotorquatum Müll. Hal. ex Besch. = *Groutiella tomentosa* (Hornsch.) Wijk & Margad., *fide* Tixier & Guého, 1997
Macromitrium laxum (Hornsch.) Müll. Hal. = *Schlotheimia rugifolia* (Hook.) Schwägr., *fide* Wijk *et al*., 1964
Macromitrium leiboldtii Hampe = *Macrocoma tenuis* (Hook. & Grev.) Vitt subsp. Sullivantii (C. Müll.) Vitt, *fide* Allen, 2002
Macromitrium leptocarpum Broth. = *Macromitrium sulcatum* var. *leptocarpum* (Broth.) J. Yu, D.D. Li, Y. Li & S.L. Guo., *fide* Yu *et al*., 2018
Macromitrium leptophyllum Besch. = *Macromitrium richardii* Schwägr., *fide* Grout, 1944
Macromitrium leratii var. *erectifolium* Thér. = *Macromitrium leratii* Broth. & Paris., *fide* Thouvenot, 2019
Macromitrium leratioides Broth. & Paris = *Macromitrium renauldii* Thér., *fide* Thouvenot, 2019
Macromitrium leucoblastum Müll. Hal. ex Broth. = *Macromitrium incurvifolium* (Hook. & Grev.) Schwägr., *fide* Vitt *et al*., 1995
Macromitrium levatum Mitt. = *Macromitrium sulcatum* (Hook.) Brid., *fide* Tixier, 1989
Macromitrium liberum Mitt. = *Macromitrium guatemalense* Müll. Hal., *fide* Vitt, 1994
Macromitrium ligulatulum Müll. Hal. = *Macromitrium ligulaefolium* Broth., *fide* Vitt & Ramsay, 1985
Macromitrium limbatulum Broth. & Paris = *Groutiella laxotorquata* (Müll. Hal. ex Besch.) Wijk & Margad., *fide* Schultze-Motel, 1975
Macromitrium linearifolium Müll. Hal. = *Schlotheimia linearifolia* (Müll. Hal.) Wijk & Margad., fide Wijk & Margadant, 1960
**Macromitrium linearifolium* Müll. Hal. = *Macromitrium microstomum* (Hook. & Grev.) Schwägr., *fide* Vitt & Ramsay, 1985
Macromitrium lingulatum Cardot & P. de la Varde = *Macromitrium tosae* Besch., *fide* Li *et al*., 2020
Macromitrium lonchomitrium Müll. Hal. = *Macromitrium longipes* (Hook.) Schwägr., *fide* Wijk *et al*., 1964
Macromitrium longibrachteatum Dixon = *Macromitrium nepalense* (Hook. & Grev.) Schwägr., *fide* Guo & He, 2014
Macromitrium longifolium var. *viridissimum* Renauld & Cardot = *Macromitrium longifolium* (Hook.) Brid., *fide* Grout, 1944
Macromitrium longirostre var. *acutifolium* (Hook. & Grev.) Wilson = *Macromitrium longirostre* (Hook.) Schwägr., *fide* Fife, 2017
Macromitrium longirostre var. *caducipilum* (Lindb.) W. Martin & Sainsbury = *Macromitrium retusum* Hook. f. & Wilson., *fide* Fife, 2017
Macromitrium longirostre var. *ramsayae* (Vitt) Fife = *Macromitrium ramsayae* Vitt, *fide* Fife, 2017
***Macromitrium longisetum* Schimp. ex C.H. Wright = *Macromitrium soulae* Renauld & Cardot., *fide* Crosby *et al*., 1983
Macromitrium ludovicae Broth. & Paris = *Macromitrium tongense* Sull., *fide* Thouvenot, 2019
Macromitrium luehmannianum Müll. Hal. = *Macromitrium ligulare* Mitt., *fide* Vitt & Ramsay, 1985
Macromitrium lycopodioides Schwägr. = *Macrocoma lycopodioides* (Schwägr.) Vitt., *fide* Vitt, 1973
***Macromitrium macleai* Paris = *Macromitrium sulcatum* (Hook.) Brid., *fide* O'Shea, 2006
Macromitrium macropelma Müll. Hal. = *Macromitrium microstomum* (Hook. & Grev.) Schwägr., *fide* Wilbraham, 2007
Macromitrium macropyxis Broth. = *Macrocoma orthotrichoides* (Raddi) Wijk & Marg., *fide* Vitt, 1980 [1981]
Macromitrium macrorrhynchum Mitt. ex Bosch & Sande Lac. = *Groutiella macrorrhyncha* (Mitt. ex Bosch & Sande Lac.) Wijk & Margad., *fide* Wijk & Margadant, 1960
**Macromitrium macrosporum* Herzog = *Macromitrium gigasporum* Herzog., *fide* Wijk *et al*., 1964
Macromitrium macrostomum Schwägr. = *Groutiella apiculata* (Hook.) H.A. Crum & Steere., *fide* Vitt, 1979
Macromitrium magnirete Dixon = *Macromitrium zollingeri* Mitt. ex Bosch & Sande Lac., *fide* Eddy, 1996

Macromitrium makinoi (Broth.) Paris = *Macromitrium japonicum* Dozy & Molk., *fide* Li *et al.*, 2024
Macromitrium malacoblastum Müll. Hal. = *Macromitrium involutifolium* (Hook. & Grev.) Schwägr., *fide* Vitt & Ramsay, 1985
Macromitrium mammillosum E.B. Bartram = *Macromitrium cirrosum* (Hedw.) Brid., *fide* Churchill & Linares, 1995
Macromitrium mannii A. Jaeger = *Macromitrium sulcatum* (Hook.) Brid., *fide* O'Shea, 2006
Macromitrium martianum (Hornsch.) Müll. Hal. = *Schlotheimia rugifolia* (Hook.) Schwägr., *fide* Wijk *et al.*, 1964
Macromitrium mauritianum Schwägr. = *Macromitrium subtortum* (Hook. & Grev.) Schwägr., *fide* Wilbraham, 2016
Macromitrium mauritianum var. *viride* Broth. = *Macromitrium mauritianum* Schwägr., *fide* Wilbraham & Ellis, 2010
Macromitrium megalosporum Thér. & Naveau = *Macrocoma abyssinica* (Müll. Hal.) Vitt var. *abyssinica*, *fide* Wilbraham, 2017
Macromitrium merkelii (Hornsch.) Müll. Hal. = *Schlotheimia merkelii* Hornsch., *fide* Valente *et al.*, 2020
Macromitrium melanostomum Paris & Broth. = *Macromitrium tosae* Besch., *fide* Yu *et al.*, 2012
**Macromitrium menziesii* Mitt. = *Macromitrium sulcatum* (Hook.) Brid., *fide* O'Shea, 2006
Macromitrium merrillii Broth. = *Macromitrium falcatulum* Müll. Hal., *fide* Vitt *et al.*, 1995
Macromitrium mexicanum Mitt. = *Macrocoma tenuis* subsp. *sullivantii* (Müll. Hal.) Vitt, *fide* Allen, 2002
Macromitrium microphyllum (Hook. & Grev.) Brid. = *Macrocoma tenuis* (Hook. & Grev.) Vitt, *fide* Vitt, 1973
**Macromitrium microstomum* Hornsch. = *Macromitrium hornschuchii* Müll. Hal., *fide* Wijk *et al.*, 1964
Macromitrium microtheca Mitt. = *Macromitrium cirrosum* (Hedw.) Brid., *fide* Grout, 1944
Macromitrium mindanaense Broth. = *Macromitrium ochraceum* (Dozy & Molk.) Müll. Hal., *fide* Wijk *et al.*, 1964
Macromitrium mindorense Broth. = *Macromitrium crinale* Broth. & Geh., *fide* Eddy, 1996
Macromitrium miquelii Mitt. ex Bosch & Sande Lac. = *Macromitrium fuscescens* Schwägr., *fide* Vitt *et al.*, 1995
Macromitrium morobense E.B. Bartram = *Macromitrium macrosporum* Broth., *fide* Vitt *et al.*, 1995
Macromitrium mossmanianum Müll. Hal. = *Macromitrium gracile* (Hook.) Schwägr., *fide* Wijk *et al.*, 1964
Macromitrium mucronatulum Müll. Hal. = *Macromitrium brevicaule* (Besch.) Broth., *fide* Vitt & Ramsay, 1985
Macromitrium mucronifolium (Hook. & Grev.) Schwägr. = *Groutiella mucronifolia* (Hook. & Grev.) H.A. Crum & Steere., *fide* Crum & Steere, 1950
***Macromitrium mucronifolium* var. *squarrosum* Thér. = *Groutiella mucronifolia* (Hook. & Grev.) H.A. Crum & Steere., *fide* Allen, 2002
Macromitrium mucronulatum Müll. Hal. = *Macromitrium wattsii* Broth., *fide* Wijk *et al.*, 1964
Macromitrium muelleri Hampe = *Macromitrium archeri* Mitt., *fide* Vitt & Ramsay, 1985
**Macromitrium muelleri* Schimp. ex Besch. = *Macromitrium longifolium* (Hook.) Brid., *fide* Wijk *et al.*, 1964
Macromitrium nadeaudii Besch. = *Macromitrium subtile* subsp. *subuligerum* (Bosch & Sande Lac.) M. Fleisch., *fide* Wijk *et al.*, 1964
Macromitrium nakanishikii Broth. = *Macromitrium japonicum* var. *makinoi* (Broth.) Nog., *fide* Noguchi & Iwatsuki, 1989
Macromitrium nanothecium var. *sublaeve* Thér. = *Macromitrium fimbriatum* (P. Beauv.) Schwägr., *fide* Wilbraham, 2018
Macromitrium neelgheriense Müll. Hal. = *Macromitrium sulcatum* (Hook.) Brid., *fide* Gangulee, 1976
Macromitrium negrense Mitt. = *Macromitrium guatemalense* Müll. Hal., *fide* Valente *et al.*, 2020
***Macromitrium neilgherrense* Müll. Hal. = *Macromitrium sulcatum* (Hook.) Brid., *fide* Gangulee, 1976
Macromitrium neocaledonicum Besch. = *Macromitrium pulchrum* var. *neocaledonicum* (Besch.) Thouvenot., *fide* Thouvenot, 2019
Macromitrium nigrescens Kunze = *Ptychomitrium nigrescens* (Kunze) Wijk & Margad., *fide* Wijk & Margadant, 1959
Macromitrium nipponicum Nog. = *Macromitrium comatulum* Broth., *fide* Noguchi & Iwatsuki, 1989
Macromitrium nitidum Hook. & Wilson = *Macromitrium microstomum* (Hook. & Grev.) Schwägr., *fide* Valente *et al.*, 2020
Macromitrium noumeanum Besch. = *Macromitrium involutifolium* (Hook. & Grev.) Schwägr. subsp. *involutifolium*., *fide* Wijk *et al.*, 1964
Macromitrium novae-valesiae Müll. Hal. = *Macrocoma tenuis* (Hook. & Grev.) Vitt subsp. *tenuis*., *fide* Vitt, 1980 [1981]
Macromitrium novoguinense E.B. Bartram = *Macromitrium yuleanum* Broth. & Geh., *fide* Vitt *et al.*, 1995
Macromitrium obtusum Mitt. = *Micromitrium mucronifolium* (Hook. & Grev.) Grout., *fide* Grout, 1944
Macromitrium okamurae Broth. = *Macromitrium hymenostomum* Mont., *fide* Gangulee, 1976
Macromitrium oocarpum Müll. Hal. = *Macromitrium prorepens* (Hook.) Schwägr., *fide* Wijk *et al.*, 1964
Macromitrium orthotrichaceum Müll. Hal. = *Micromitrium wagnerianum* (Müll. Hal.) Paris., *fide* Bartram, 1949
Macromitrium orthostichum subsp. *appressifolium* (Mitt.) M. Fleisch. = *Macromitrium orthostichum* Nees ex Schwägr., *fide* Vitt *et al.*, 1995
Macromitrium orthostichum subsp. *seminudum* (Thwaites & Mitt.) M. Fleisch. = *Macromitrium orthostichum* Nees ex Schwägr., *fide* Vitt *et al.*, 1995
Macromitrium ottonis (Schwägr.) Müll. Hal. = *Schlotheimia torquata* (Sw. ex Hedw.) Brid. , *fide* Grout, 1944
Macromitrium owahiense Müll. Hal. = *Macromitrium microstomum* (Hook. & Grev.) Schwägr., *fide* Vitt & Ramsay, 1985
Macromitrium pacificum Besch. = *Macromitrium microstomum* (Hook. & Grev.) Schwägr., *fide* Vit & Ramsay, 1985
Macromitrium pacificum var. *brevisetum* Thér. = *Macromitrium microstomum* (Hook. & Grev.) Schwägr., *fide* Yu *et al.*, 2018
Macromitrium pacificum var. *longisetum* Thér. = *Macromitrium microstomum* (Hook. & Grev.) Schwägr., *fide* Thouvenot, 2019
Macromitrium pallidovirens Müll. Hal. = *Macromitrium repandum* Müll. Hal., *fide* Vitt & Ramsay, 1985
Macromitrium palmense R.S. Williams = *Macromitrium scoparium* Mitt., *fide* Allen, 2002
Macromitrium papillifolium Müll. Hal. = *Macromitrium prorepens* (Hook.) Schwägr., *fide* Fife, 2017
Macromitrium papillosulum Thér. = *Matteria papillosula* (Thér.) Goffinet, *fide* Goffinet & Vitt, 1998
Macromitrium papuanum Dixon = *Macromitrium angulatum* Mitt., *fide* Vitt *et al.*, 1995
**Macromitrium papuanum* Nog. = *Macromitrium noguchianum* W. Schultze-Motel, *fide* Vitt *et al.*, 1995
Macromitrium paraphysatum Mitt. = *Macrocoma tenuis* subsp. *sullivantii* (Müll. Hal.) Vitt., *fide* Gangulee, 1976
Macromitrium paraphysatum var. *chilense* Thér. = *Macrocoma tenuis* subsp. *sullivantii* (Müll. Hal.) Vitt., *fide* Vitt, 1973
Macromitrium paucidens Müll. Hal. = *Macromitrium guatemalense* Müll. Hal., *fide* Grout, 1944
Macromitrium penicillatum Mitt. = *Macromitrium guatemalense* Müll. Hal., *fide* Grout, 1944
Macromitrium pentagonum Müll. Hal. = *Macromitrium punctatum* (Hook. & Grev.) Brid., *fide* Vitt, 1979
Macromitrium pentastichum Müll. Hal. = *Macromitrium punctatum* (Hook. & Grev.) Brid., *fide* Vitt, 1979
**Macromitrium peraristatum* Müll. Hal. = *Macromitrium harrisii* Paris., *fide* Grout, 1944
Macromitrium perminutum Broth. & Paris = *Macromitrium ligulaefolium* Broth., *fide* Vitt & Ramsay, 1985
Macromitrium perrottetii Müll. Hal. = *Macrocoma tenuis* subsp. *sullivantii* (Müll. Hal.) Vit., *fide* Vitt, 1980 [1981]
Macromitrium pertorquescens Müll. Hal. = *Macromitrium longirostre* (Hook.) Schwägr., *fide* Wijk *et al.*, 1964
Macromitrium pertorquescens var. *torquatulum* Müll. Hal. = *Macromitrium longirostre* (Hook.) Schwägr., *fide* Vitt & Ramsay, 1985
Macromitrium perundulatum Broth. = *Macromitrium sulcatum* (Hook.) Brid., *fide* O'Shea, 2006
***Macromitrium perundulatum* E.B. Bartram = *Macromitrium longifolium* (Hook.) Brid., *fide* Allen, 2002
***Macromitrium pervillei* Schimp. ex C.H. Wright = *Ulota fulva* Brid., *fide* Crosby *et al.*, 1983
Macromitrium petriei Dixon = *Macromitrium prorepens* (Hook.) Schwägr., *fide* Fife, 2017

Macromitrium phyllorhizans Müll. Hal. = *Macrocoma orthotrichoides* (Radd.) Wijk & Marg., *fide* Vitt, 1980 [1981]
***Macromitrium pileatum* Wilson = *Macromitrium moorcroftii* (Hook. & Grev.) Schwägr., *fide* Gangulee, 1976
Macromitrium pilosum var. *brevifolium* Thér. = *Macromitrium pilosum* Thér., *fide* Thouvenot, 2019
Macromitrium pinnulatum Herzog = *Macromitrium microstomum* (Hook. & Grev.) Schwägr., *fide* Churchill & Fuentes, 2005
Macromitrium planocespitosum Müll. Hal. = *Macromitrium incurvifolium* (Hook. & Grev.) Schwägr., *fide* Vitt et al., 1995
Macromitrium platyphyllaceum Müll. Hal. = *Macromitrium involutifolium* subsp. *ptychomitrioides* (Besch.) Vitt & H.P. Ramsay., *fide* Vitt & Ramsay, 1985
Macromitrium plebejum Müll. Hal. = *Macromitrium piliferum* Schwägr., *fide* Wijk et al., 1964
Macromitrium pleurosigmoideum Paris & Broth. = *Groutiella laxotorquata* (Müll. Hal. ex Besch.) Wijk & Margad., *fide* Schultze-Motel, 1975
Macromitrium plicatum Thér. = *Macromitrium involutifolium* subsp. *ptychomitrioides* (Besch.) Vitt & H.P. Ramsay., *fide* Thouvenot, 2019
Macromitrium plicatum var. *aristatum* Thér. = *Macromitrium involutifolium* subsp. *ptychomitrioides* (Besch.) Vitt & H.P. Ramsay., *fide* Thouvenot, 2019
Macromitrium plicatum var. *obtusifolium* Thér. = *Macromitrium involutifolium* subsp. *ptychomitrioides* (Besch.) Vitt & H.P. Ramsay., *fide* Thouvenot, 2019
Macromitrium pobeguinii Paris & Broth. = *Groutiella laxotorquata* (Müll. Hal. ex Besch.) Wijk & Margad., *fide* Schultze-Motel, 1975
Macromitrium podocarpi Müll. Hal. = *Macromitrium pseudofimbriatum* Hampe, *fide* Valente et al., 2020
Macromitrium podocarpi var. *falcifolium* Müll. Hal. = *Macromitrium pseudofimbriatum* Hampe., *fide* Valente et al., 2020
Macromitrium poeppigii Duby = *Zygodon pentastichus* (Mont.) Müll. Hal., *fide* Wijk et al., 1964
Macromitrium polygonostomum Dixon & P. de la Varde = *Macromitrium japonicum* Dozy & Molk., *fide* Li et al., 2024
Macromitrium portoricense R.S. Williams = *Macromitrium podocarpi* Müll. Hal., *fide* Churchill & Linares, 1995
Macromitrium powellii Mitt. = *Macromitrium incurvifolium* (Hook. & Grev.) Schwägr., *fide* Schultze-Motel, 1974
Macromitrium praelongum Mitt. = *Macromitrium cirrosum* (Hedw.) Brid., *fide* Wijk et al., 1964
Macromitrium pringlei Cardot = *Macrocoma frigida* (Müll. Hal.) Vitt., *fide* Vitt, 1979
Macromitrium progressum Hampe = *Macrocoma tenuis* subsp. *sullivantii* (Müll. Hal.) Vitt., *fide* Vitt, 1980 [1981]
Macromitrium prolixum Bosw. = *Macromitrium microstomum* (Hook. & Grev.) Schwägr., *fide* Vitt & Ramsay, 1985
**Macromitrium prolongatum* Müll. Hal. = *Macromitrium catharinense* Paris, *fide* Wijk et al., 1964
Macromitrium prolongatum var. *brevipes* Cardot = *Macromitrium prolongatum* Mitt., *fide* Noguchi & Iwatsuki, 1989
Macromitrium prolongatum var. *gracilius* Müll. Hal. = *Macromitrium catharinense* Paris, *fide* Li et al., 2019
Macromitrium prorepens var. *aristata* Allison = *Macromitrium prorepens* (Hook.) Schwägr., *fide* Fife, 2017
Macromitrium protractum Broth. = *Macrocoma abyssinica* (Müll. Hal.) Vitt., *fide* Vitt, 1980 [1981]
Macromitrium pseudocirrosum Müll. Hal. = *Macromitrium cirrosum* (Hedw.) Brid., *fide* Wijk et al., 1964
Macromitrium pseudohemitrichodes Müll. Hal. = *Macromitrium longipes* (Hook.) Schwägr., *fide* Fife, 2017
Macromitrium ptychomitrioides Besch. = *Macromitrium involutifolium* subsp. *ptychomitrioides* (Besch.) Vitt & H.P. Ramsay., *fide* Thouvenot, 2019
Macromitrium pugionifolium Müll. Hal. = *Macromitrium repandum* Müll. Hal., *fide* Vitt & Ramsay, 1985
Macromitrium pulchellum (Hornsch.) Brid. = *Macrocoma pulchella* (Hornsch.) Vitt, *fide* Vitt, 1973
Macromitrium pulchrum var. *aristatum* Thér. = *Macromitrium pulchrum* Besch., *fide* Thouvenot, 2019
Macromitrium pulchrum var. *densirete* Thér. = *Macromitrium pulchrum* Besch., *fide* Thouvenot, 2019
Macromitrium pungens Mitt. ex Bosch & Sande Lac. = *Macromitrium salakanum* Müll. Hal., *fide* Vitt et al., 1995
Macromitrium pusillum Mitt. = *Macromitrium archeri* Mitt., *fide* Vitt & Ramsay, 1985
Macromitrium pycnangium Müll. Hal. ex Broth. = *Macrocoma tenuis* subsp. *sullivantii* (Müll. Hal.) Vitt., *fide* Vitt, 1980 [1981]
Macromitrium pycnophyllum Cardot = *Groutiella chimborazensis* (Spruce ex Mitt.) Florsch. subsp. *chimborazensis.*, *fide* Vitt, 1994
**Macromitrium quadrifidum* Müll. Hal. = *Schlotheimia angulosa* (P. Beauv.) Dixon., *fide* O'Shea, 2006
Macromitrium quinquefarium Hornsch. = *Cardotiella quinquefaria* (Hornsch.) Vitt., *fide* Valente et al., 2020
Macromitrium ramentosum Thwaites & Mitt. = *Macromitrium sulcatum* (Hook.) Brid., *fide* Gangulee, 1976
Macromitrium ramosissimum Mitt. = *Leptodontium viticulosoides* var. *panamense* (A. Jaeger) R.H. Zander., *fide* Zander, 1972
Macromitrium rapaense E.B. Bartram = *Macromitrium ligulaefolium* Broth., *fide* Vitt & Ramsay, 1985
Macromitrium recurvulum Müll. Hal. = *Macrocoma tenuis* (Hook. & Grev.) Vitt., *fide* Fife, 2017
Macromitrium reflexifolium Mitt. = *Macromitrium* contextum Hampe., *fide* Vitt, 1979
**Macromitrium reflexifolium* Sande Lac. = *Macromitrium salakanum* Müll. Hal., *fide* Vitt et al., 1995
***Macromitrium rehmannii* Paris = *Macromitrium serpens* (Burch. ex Hook. & Grev.) Brid., *fide* O'Shea, 2006
Macromitrium reinwardtii Schwägr. = *Macromitrium microstomum* (Hook. & Grev.) Schwägr., *fide* Vitt & Ramsay, 1985
Macromitrium rhabdocarpum Mitt. = *Macromitrium richardii* Schwägr., *fide* Grout, 1944
Macromitrium rhizomatosum Müll. Hal. ex Besch. = *Macromitrium subtortum* (Hook. & Grev.), *fide* Schatz et al., 2023
Macromitrium rhystophylllum Müll. Hal. = *Macromitrium guatemalense* Müll. Hal., *fide* Grout, 1944
Macromitrium richardii Schwägr. = *Macromitrium viticulosum* (Raddi) Brid., *fide* Valente et al., 2020
Macromitrium rigescens Broth. & Dixon = *Macromitrium grossirete* Müll. Hal., *fide* Wijk et al., 1964
Macromitrium robinsonii R.S. Williams = *Macromitrium gymnostomum* Sull. & Lesq., *fide* Guo et al., 2007
Macromitrium rodwayi Dixon = *Macromitrium longirostre* (Hook.) Schwägr., *fide* Vitt & Ramsay, 1985
Macromitrium ruberrimum Dixon = *Macromitrium archboldii* E.B. Bartram., *fide* Vitt et al., 1995
Macromitrium rubricuspis Broth. = *Macromitrium ochraceum* (Dozy & Molk.) Müll. Hal., *fide* Wijk et al., 1964
Macromitrium rufescens Besch. = *Macromitrium sulcatum* (Hook.) Brid. subsp. *sulcatum.*, *fide* Ellis & Wilbraham, 2008
Macromitrium rugifolium (Hook.) Müll. Hal. = *Schlotheimia rugifolia* (Hook.) Schwägr., *fide* Allen, 2002
**Macromitrium rugifolium* Müll. Hal. = *Cardotiella secunda* (Müll. Hal.) Vitt., *fide* O'Shea, 2006
**Macromitrium rugifolium* Müll. Hal. ex Broth. = *Macromitrium sulcatum* (Hook.) Brid. subsp. *sulcatum, fide* O'Shea, 2006
Macromitrium runcinatella Müll. Hal. = *Macromitrium proliferum* Mitt., *fide* Wijk et al., 1964
Macromitrium rupestre Mitt. = *Macromitrium gymnostomum* Sull. & Lesq., *fide* Wijk et al., 1962
Macromitrium salakanum subsp. *celebense* (Paris) M. Fleisch. = *Macromitrium salakanum* Müll. Hal., *fide* Vitt et al., 1995
Macromitrium salakanum subsp. *pungens* (Mitt. ex Bosch & Sande Lac.) M. Fleisch. = *Macromitrium salakanum* Müll. Hal., *fide* Vitt et al., 1995
Macromitrium salakanum var. *majus* Besch. = *Macromitrium leratii* Broth. & Paris., *fide* Thouvenot, 2019
Macromitrium salakanum var. *reflexifolium* M. Fleisch. ex E.B. Bartram = *Macromitrium salakanum* Müll. Hal., *fide* Vitt et al., 1995
Macromitrium sanctae-mariae Renauld & Cardot = *Macromitrium subtortum* (Hook. & Grev.) Schwägr., *fide* Wilbraham, 2016
Macromitrium sarasinii Thér. = *Macromitrium hemitrichodes* var. *sarasinii* (Thér.) Thouvenot., *fide* Thouvenot, 2019
Macromitrium sarawaketense Nog. = *Macromitrium yuleanum* Broth. & Geh., *fide* Vitt et al., 1995

Macromitrium sarcotrichum Müll. Hal. ex Broth. = *Groutiella laxotorquata* (Müll. Hal. ex Besch.) Wijk & Margad., *fide* Schultze-Motel, 1975
Macromitrium sartorii Müll. Hal. = *Macromitrium punctatum* (Hook. & Grev.) Brid., *fide* Grout, 1944
Macromitrium saxatile Mitt. = *Macromitrium microstomum* (Hook. & Grev.) Schwägr., *fide* Robinson, 1975
Macromitrium secundum Müll. Hal. = *Cardotiella secunda* (Müll. Hal.) Vitt, *fide* Vitt, 1981
Macromitrium scabrisetum Wilson = *Macromitrium longifolium* (Hook.) Brid., *fide* Grout, 1944
Macromitrium schiffneri Broth. = *Macromitrium catharinense* Paris., *fide* Valente *et al.*, 2020
Macromitrium schimperi A. Jaeger = *Macromitrium longifolium* (Hook.) Brid., *fide* Grout, 1944
***Macromitrium schizomitrium* Besch. = *Macromitrium fasciculare* Mitt., *fide* O'Shea, 2006
Macromitrium schlotheimiiforme Paris = *Cardotiella secunda* (Müll. Hal.) Vitt., *fide* van Rooy & Wyk, 1992
Macromitrium schlumbergeri (Schimp. ex Besch.) Broth. = *Groutiella schlumbergeri* (Schimp. ex Besch.) Wijk & Margad., *fide* Bowers, 1974
Macromitrium schwaneckeanum Hampe = *Macromitrium cirrosum* (Hedw.) Brid., *fide* Grout, 1944
Macromitrium scleropelma Renauld & Cardot = *Macromitrium longifolium* (Hook.) Brid., *fide* Bowers, 1974
Macromitrium scleropodium Besch. = *Macromitrium orthostichum* Nees ex Schwägr., *fide* Willbraham, 2008
Macromitrium scottiae Müll. Hal. = *Macromitrium microstomum* (Hook. & Grev.) Schwägr., *fide* Vitt & Ramsay, 1985
Macromitrium seemannii Mitt. = *Macromitrium microstomum* (Hook. & Grev.) Schwägr., *fide* Wilbraham, 2007
Macromitrium semidiaphanum Renauld & Cardot = *Schlotheimia semidiaphana* (Renauld & Cardot) Cardot., *fide* Schatz *et al.*, 2023
Macromitrium semimarginatum Müll. Hal. = *Groutiella chimborazensis* (Spruce ex Mitt.) Florsch., *fide* Goffinet, 1993
Macromitrium seminudum Thwaites & Mitt. = *Macromitrium orthostichum* Nees ex Schwägr., *fide* Vitt *et al.*, 1995
Macromitrium semipapillosum Thér. & P. de la Varde = *Macromitrium calocalyx* Müll. Hal., *fide* Wilbraham, 2018
Macromitrium semipellucidum Dozy & Molk. = *Macromitrium fuscescens* Schwägr., *fide* Vitt *et al.*, 1995
***Macromitrium semipellucidum* Renauld & Cardot ex Broth. = *Schlotheimia semidiaphana* (Renauld & Cardot) Cardot., *fide* O'Shea, 2006
Macromitrium seriatum Paris & Broth. = *Macromitrium sulcatum* (Hook.) Brid., *fide* Wilbraham, 2016
Macromitrium serpentinum Mitt. = *Macromitrium serpens* (Burch. ex Hook. & Grev.) Brid., *fide* Wijk *et al.*, 1964
Macromitrium serrulatum Mitt. = *Macromitrium guatemalense* Müll. Hal., *fide* Vitt, 1994
**Macromitrium serrulatum* Müll. Hal. = *Macromitrium venezuelense* Paris., *fide* Wijk *et al.*, 1964
Macromitrium shankii H.A. Crum = *Calymperes rubiginosum* (Mitt.) W.D. Reese., *fide* Reese, 1993
Macromitrium sibiricum Podp. = *Herpetineuron toccoae* (Sull. & Lesq.) Cardot., *fide* Ignatov & Afonina, 1992
Macromitrium sobrinum Cardot = *Macrocoma tenuis* (Hook. & Grev.) Vitt subsp. *tenuis*, *fide* Vitt, 1980 [1981]
Macromitrium sordidevirens Müll. Hal. = *Macromitrium aurescens* Hampe., *fide* Vitt & Ramsay, 1985
Macromitrium spathulare Mitt. = *Macromitrium japonicum* Dozy & Molk., *fide* Wijk *et al.*, 1964
Macromitrium speirophyllum E.B. Bartram = *Macromitrium yuleanum* Broth. & Geh., *fide* Vitt *et al.*, 1995
Macromitrium sprengelii (Hornsch.) Müll. Hal. = *Schlotheimia sprengelii* Hornsch., *fide* Grout, 1944
Macromitrium squarrosum (Brid.) Müll. Hal. = *Schlotheimia squarrosa* Brid., fide Crosby *et al.*, 1983
Macromitrium squarrulosum Müll. Hal. = *Macrocoma tenuis* subsp. *sullivantii* (Müll. Hal.) Vitt., *fide* Vitt, 1980 [1981]
Macromitrium standleyi var. *subundulatum* E.B. Bartram = *Macromitrium standleyi* E.B. Bartram., *fide* Allen, 2002
Macromitrium stellulatum (Hornsch.) Brid. = *Macromitrium swainsonii* (Hook.) Brid., *fide* Valente *et al.*, 2020
Macromitrium stenophyllum Mitt. = *Macromitrium cirrosum* (Hedw.) Brid., *fide* Churchill & Linares, 1995
Macromitrium stolonigerum Müll. Hal. = *Macromitrium microstomum* (Hook. & Grev.) Schwägr., *fide* Allen, 2002
Macromitrium stratosum Mitt. = *Macromitrium microstomum* (Hook. & Grev.) Schwägr., *fide* Vitt & Ramsay, 1985
Macromitrium subapiculatum Broth. = *Groutiella tumidula* (Mitt.) Vitt., *fide* Valente *et al.*, 2020
Macromitrium subbrevicaule Broth. & Watts = *Macromitrium brevicaule* (Besch.) Broth., *fide* Vitt & Ramsay, 1985
Macromitrium suberosulum E.B. Bartram = *Macromitrium involutifolium* subsp. *ptychomitrioides* (Besch.) Vitt & H.P. Ramsay., *fide* Vitt & Ramsay, 1985
Macromitrium subfragile Dixon & Sainsbury = *Macromitrium brevicaule* (Besch.) Broth., *fide* Vitt & Ramsay, 1985
Macromitrium subgoniorrhynchum Broth. = *Groutiella tomentosa* (Hornsch.) Wijk & Margad., *fide* Yu & Jia, 2023
Macromitrium subleptocarpum Dixon & P. de la Varde = *Macromitrium sulcatum* var. *leptocarpum* (Broth.) J. Yu, D.D. Li, Yan Li & S.L. Guo., *fide* Yu *et al.*, 2018
Macromitrium submegalocladum Dixon = *Macromitrium megalocladon* M. Fleisch., *fide* Vitt *et al.*, 1995
Macromitrium subnitidum Müll. Hal. = *Macromitrium microstomum* (Hook. & Grev.) Schwägr., *fide* Yu *et al.*, 2018
Macromitrium subpiliferum Müll. Hal. = *Macromitrium piliferum* Schwägr., *fide* Wijk *et al.*, 1964
Macromitrium subpungens Hampe ex Müll. Hal. = *Macromitrium subtortum* (Hook. & Grev.) Schwägr., *fide* Schatz *et al.*, 2023
Macromitrium subpungens var. *madagassum* Cardot = *Macromitrium subtortum* (Hook. & Grev.) Schwägr., *fide* Schatz *et al.*, 2023
Macromitrium subpycnangium Müll. Hal. = *Macrocoma orthotrichoides* (Raddi) Wijk & Margad., *fide* Valente *et al.*, 2020
Macromitrium subreflexum Müll. Hal. = *Macromitrium guatemalense* Müll. Hal., *fide* Grout, 1944
***Macromitrium subretusum* Broth. = *Groutiella tomentosa* (Hornsch.) Wijk & Margad., *fide* Vitt & Crum, 1970
Macromitrium subsemipellucidum Broth. = *Macromitrium fuscescens* Schwägr., *fide* Vitt *et al.*, 1995
Macromitrium subsessile Broth. & Paris = *Macromitrium pilosum* Thér., *fide* Thouvenot, 2019
Macromitrium substrictifolium Müll. Hal. = *Macromitrium cirrosum* (Hedw.) Brid., *fide* Valente *et al.*, 2020
Macromitrium subtile Schwägr. = *Macromitrium incurvifolium* (Hook. & Grev.) Schwägr., *fide* Vitt & Ramsay, 1985
Macromitrium subtile subsp. *subuligerum* (Bosch & Sande Lac.) M. Fleisch. = *Macromitrium incurvifolium* (Hook. & Grev.) Schwägr., *fide* Vitt & Ramsay, 1985
Macromitrium subuligerum Bosch & Sande Lac. = *Macromitrium incurvifolium* (Hook. & Grev.) Schwägr, *fide* Vitt *et al.*, 1995
Macromitrium subvillosum Broth. & Paris = *Macromitrium tongense* Sull, *fide* Thouvenot, 2019
Macromitrium sulcatum subsp. *ceylanicum* (Mitt.) M. Fleisch. = *Macromitrium sulcatum* (Hook.) Brid., *fide* Gangulee, 1976.
Macromitrium sulcatum subsp. *neelgheriense* (Müll. Hal.) M. Fleisch. = *Macromitrium sulcatum* (Hook.) Brid., *fide* Gangulee, 1977
Macromitrium sulcatum var. *neelgheriense* (Müll. Hal.) Müll. Hal. = *Macromitrium sulcatum* (Hook.) Brid., *fide* Gangulee, 1977
Macromitrium sullivantii Müll. Hal. = *Macrocoma tenuis* subsp. *sullivantii* (Müll. Hal.) Vitt., *fide* Flora of North America Editorial Committee, 2014
Macromitrium sumichrastii Duby = *Macromitrium contextum* Hampe., *fide* Vitt, 1979
**Macromitrium tahitense* Broth. = *Macromitrium tahitisecundum* Marg., *fide* Margadant, 1972
Macromitrium tahitense Nadeaud = *Macromitrium subtile* Schwägr., *fide* Wijk *et al.*, 1964
Macromitrium tasmanicum Broth. = *Macromitrium microstomum* (Hook. & Grev.) Schwägr., *fide* Vitt & Ramsay, 1985
Macromitrium tenellum Cardot = *Macromitrium richardii* Schwägr., *fide* Grout, 1944
Macromitrium tenerum Kunze = *Ptychomitrium nigrescens* (Kunze) Wijk & Margad., *fide* Wijk *et al.*, 1964

Macromitrium tenue (Hook. & Grev.) Brid. = *Macrocoma tenuis* (Hook. & Grev.) Vitt., *fide* Vitt, 1973
***Macromitrium tenue* var. *brachypus* Müll. Hal. = *Macrocoma lycopodioides* (Schwägr.) Vitt., *fide* O'Shea, 2006
Macromitrium tenue var. *dregei* (Hornsch.) Müll. Hal. = *Macrocoma tenuis* (Hook. & Grev.) Vitt subsp. tenuis., *fide* O'Shea, 2006
Macromitrium tenue var. *lycopodioides* (Schwägr.) Müll. Hal. = *Macrocoma lycopodioides* (Schwägr.) Vitt, *fide* Vitt, 1973
Macromitrium teres (Dozy & Molk.) Müll. Hal. = *Macromitrium blumei* Nees ex Schwägr., *fide* Wijk *et al.*, 1964
Macromitrium thraustophyllum Müll. Hal. ex Broth. = *Groutiella tomentosa* (Hornsch.) Wijk & Margad., *fide* Yu & Jia, 2023
Macromitrium tomentosum Hornsch. = *Groutiella tomentosa* (Hornsch.) Wijk & Margad., *fide* Crum & Anderson, 1981
Macromitrium tonduzii Renauld & Cardot = *Macromitrium scoparium* Mitt., *fide* Churchill & Linares, 1995
Macromitrium torquatulum (Müll. Hal.) Müll. Hal. & Broth. = *Macromitrium longirostre* (Hook.) Schwägr., *fide* Vitt, 1983
Macromitrium torquatum (Sw. ex Hedw.) Müll. Hal. = *Schlotheimia torquata* (Sw. ex Hedw.) Brid., *fide* Allen, 2002
Macromitrium tortifolium Thér. = *Macromitrium sulcatum* (Hook.) Brid., *fide* O'Shea, 2006
Macromitrium tortuosum Schimp. ex Besch. = *Macromitrium guatemalense* Müll. Hal., *fide* Grout, 1944
***Macromitrium tortuosum* Wilson = *Macromitrium moorcroftii* (Hook. & Grev.) Schwägr., *fide* Gangulee, 1976
Macromitrium torulosum Mitt. = *Macromitrium sulcatum* var. *torulosum* (Mitt.) Tixier, *fide* Tixier, 1975
Macromitrium trianae Müll. Hal. = *Macromitrium guatemalense* Müll. Hal., *fide* Grout, 1944
Macromitrium trichomitrium (Schwägr.) Müll. Hal. = *Schlotheimia trichomitria* Schwägr., *fide* Valente *et al.*, 2020
Macromitrium tristratosum Dixon = *Macromitrium serpens* (Burch. ex Hook. & Grev.) Brid., *fide* van Rooy & Wyk, 1992
Macromitrium trollii Dixon = *Macromitrium sulcatum* (Hook.) Brid., *fide* Wilbraham, 2016
Macromitrium truncatum Müll. Hal. = *Macromitrium perichaetiale* (Hook. & Grev.) Müll. Hal., *fide* Grout, 1944
Macromitrium tumidulum Mitt. = *Groutiella tumidula* (Mitt.) Vitt., *fide* Crum & Anderson, 1981
Macromitrium turgidum var. *laeve* Dixon = *Macromitrium turgidum* Dixon, *fide* Guo *et al.*, 2012
Macromitrium uncinatum (Brid.) Brid. = *Macromitrium fimbriatum* (P. Beauv.) Schwägr., *fide* Crosby *et al.*, 1983
**Macromitrium uncinatum* Müll. Hal. = *Macromitrium muellerianum* Mitt., *fide* Wijk *et al.*, 1964
Macromitrium undatifolium Müll. Hal. = *Macromitrium sulcatum* (Hook.) Brid. subsp. *sulcatum.*, *fide* O'Shea, 2006
Macromitrium undosum Cardot = *Groutiella chimborazensis* (Spruce ex Mitt.) Florsch., *fide* Vitt, 1979
Macromitrium undulatum (Hook. & Grev.) Schwägr. = *Macromitrium incurvifolium* (Hook. & Grev.) Schwägr., *fide* Reese, 1997
**Macromitrium undulatum* Hampe = Leptodontium *viticulosoides* (P. Beauv.) Wijk & Margad., *fide* Churchill, 2016
Macromitrium vernicosum Schimp. = *Macromitrium dubium* Schimp. ex Müll. Hal., *fide* Grout, 1944
Macromitrium verrucosum E.B. Bartram = *Macromitrium guatemalense* Müll. Hal., *fide* Vitt, 1994
Macromitrium villosum (Besch.) Broth. = *Macromitrium tongense* Sull., *fide* Thouvenot, 2019
Macromitrium villosum var. *elongatum* Thér. = *Macromitrium tongense* Sull., *fide* Thouvenot, 2019
Macromitrium villosum var. *intermedium* Thér. = *Macromitrium tongense* Sull., *fide* Thouvenot, 2019
Macromitrium villosum var. *longisetum* Thér. = *Macromitrium tongense* Sull., *fide* Thouvenot, 2019
Macromitrium virescens Müll. Hal. = *Macrocoma abyssinica* (Müll. Hal.) Vitt., *fide* Vitt, 1980 [1981]
Macromitrium viridissimum Mitt. = *Macromitrium involutifolium* subsp. *ptychomitrioides* (Besch.) Vitt & H.P. Ramsay., *fide* Vitt & Ramsay, 1985
Macromitrium voeltzkowii Broth. = *Macromitrium pallidum* (P. Beauv.) Wijk & Margad., *fide* Wilbraham & Ellis, 2010
Macromitrium wagnerianum Müll. Hal. = *Groutiella apiculata* (Hook.) H.A. Crum & Steere., *fide* Yu & Jia, 2023
Macromitrium wattsii Broth. = *Macrocoma brevicaule* (Besch.) Broth., *fide* Vitt & Ramsay, 1985
Macromitrium wellingtonianum Vitt = *Macromitrium angulatum* Mitt., *fide* Vitt & Ramsay, 1985
Macromitrium werckleanum Thér. = *Macromitrium cirrosum* (Hedw.) Brid., *fide* Allen, 2002
Macromitrium weymouthii Broth. = *Macromitrium microstomum* (Hook. & Grev.) Schwägr., *fide* Vitt & Ramsay, 1985
Macromitrium whiteleggei Broth. & Geh. = *Macromitrium repandum* Müll. Hal., *fide* Vitt & Ramsay, 1985
***Macromitrium wilhelmii* Paris = Ulota fulva Brid., *fide* O'Shea, 2006
Macromitrium williamsii E.B. Bartram = *Macromitrium scoparium* Mitt., *fide* Churchill & Linares, 1995
Macromitrium winkleri Broth. = *Macromitrium falcatulum* Müll. Hal., *fide* Vitt *et al.*, 1995
Macromitrium woollsianum Müll. Hal. = *Macromitrium ligulaefolium* Broth., *fide* Vitt & Ramsay, 1985
Macromitrium woollsianum var. *chlorophyllosum* Müll. Hal. = *Macromitrium ligulaefolium* Broth., *fide* Vitt & Ramsay, 1985
Macromitrium xanthocarpum Hornsch. = *Macromitrium stellulatum* (Hornsch.) Brid., *fide* Wijk *et al.*, 1964
Macromitrium zikanii Herzog = *Macrocoma tenuis* (Hook. & Grev.) Vitt., *fide* Valente *et al.*, 2020
Macromitrium zippelii Bosch & Sande Lac. = *Macromitrium incurvifolium* (Hook. & Grev.) Schwägr., *fide* Vitt *et al.*, 1995
Macromitrium zollingeri Mitt. ex Bosch & Sande Lac. = *Macromitrium blumei* var. *zollingeri* (Mitt. ex Bosch & Sande Lac.) S.L. Guo, B.C. Tan & Virtanen, *fide* Guo *et al.*, 2006

36 doubtful species

1. *Macromitrium acutissimum* Müll. Hal.
2. *Macromitrium altum* Müll. Hal.
3. *Macromitrium brachycarpum* Mitt.
4. *Macromitrium braunii* Müll. Hal.
5. *Macromitrium bujongolanum* O'Shea
6. *Macromitrium clavatum* Schimp. ex Grout
7. *Macromitrium clemensiae* E.B. Bartram
8. *Macromitrium complicatulum* Müll. Hal.
9. *Macromitrium cucullatulum* Müll. Hal.
10. *Macromitrium eriomitrium* Müll. Hal.
11. *Macromitrium fragile* Hampe
12. *Macromitrium globirameum* Müll. Hal.
13. *Macromitrium kinabaluense* J. Froehl.
14. *Macromitrium liliputanum* Müll. Hal.
15. *Macromitrium lonchomitrioides* Müll. Hal.
16. *Macromitrium marginatum* Dixon
17. *Macromitrium nietneri* Müll. Hal.
18. *Macromitrium novorecurvulum* B.C. Tan & B.C. Ho
19. *Macromitrium okabei* Sakurai
20. *Macromitrium papillisetum* Dixon
21. *Macromitrium patens* Wilson
22. *Macromitrium recurvifolium* (Hook. & Grev.) Brid.
23. *Macromitrium sclerodictyon* Cardot
24. *Macromitrium sinuatum* (Hornsch.) Müll. Hal.
25. *Macromitrium stephanodictyon* J. Froehl.
26. *Macromitrium striatum* Mitt. ex Bosch & Sande Lac.
27. *Macromitrium strictfolium* Müll. Hal.
28. *Macromitrium stricticuspis* Müll. Hal.
29. *Macromitrium subdiaphanum* Renauld & Cardot
30. *Macromitrium subpaucidens* Müll. Hal.
31. *Macromitrium tahitisecundum* Margad.
32. *Macromitrium undatum* Müll. Hal.
33. *Macromitrium urceolatulum* Müll. Hal.
34. *Macromitrium venezuelense* Paris
35. *Macromitrium vitianum* E.B. Bartram
36. *Macromitrium weissioides* Müll. Hal.

List of all names in the genus Macromitrium

116 invalid or illegitimate names

**Macromitrium alatum* Müll. Hal. ex E.B. Bartram, invalid, orthographic variant
**Macromitrium anamalaiense* D. Subram., No Latin diagnosis, no type citation
**Macromitrium angusticuspis* Broth., *invalid, no description
**Macromitrium annii* Dixon, *invalid, orthographic variant
**Macromitrium armatum* Lam., *invalid, no description
**Macromitrium atrum* Müll. Hal., *invalid, no description
**Macromitrium bayleyi* Mitt., *invalid, no description
**Macromitrium borgenianum* Müll. Hal., *invalid, no description
**Macromitrium brevirete* Dixon, *invalid, no description
**Macromitrium caloblastum* Müll. Hal., *invalid, no description
**Macromitrium capillare* Mitt. ex Spruce, *invalid, no description
**Macromitrium chlororete* Müll. Hal., *invalid, no description
**Macromitrium cirrosum* var. *fuscescens* Hampe, *invalid, no description
**Macromitrium cirrosum* var. *haitense* Thér., *invalid, no Latin description
**Macromitrium clavatum* Schimp., *invalid, no description
Macromitrium commutatum Hampe, *illegitimate, type of earlier name included
**Macromitrium courtoisii* Broth., *invalid, cited as synonym
**Macromitrium damellii* Müll. Hal., *invalid, no description
**Macromitrium dentatum* Müll. Hal. ex Paris, *invalid, orthographic variant
**Macromitrium doeringii* Kindb., *invalid, orthographic variant
**Macromitrium eucalyptorum* var. *brevipedicellatum* Müll. Hal., *invalid, no description
**Macromitrium eucalyptorum* var. *gracile* Watts & Whitel., *invalid, no description
**Macromitrium falcifolium* Müll. Hal., *invalid, cited as synonym
**Macromitrium fimbriatum* Hook. f. & Wilson, *invalid, cited as synonym
**Macromitrium fragilifolium* Lindb., *invalid, no description
**Macromitrium fulgens* A. Braun, *invalid, cited as synonym
**Macromitrium glaucescens* Schimp., *invalid, cited as synonym
Macromitrium gracile (Hornsch.) Müll. Hal., illegitimate, later homonym
**Macromitrium hahnii* Müll. Hal., *invalid, no description
**Macromitrium hamulosum* Mitt. ex Spruce, *invalid, no description
**Macromitrium hartmanni* Müll. Hal., *invalid, no description
**Macromitrium hopeiense* Dixon, *invalid, no Latin description
Macromitrium immersum Müll. Hal., illegitimate, type of earlier name included & new name based on earlier epithet
**Macromitrium indistinctum* Müll. Hal., *invalid, no description
**Macromitrium insculptum* Mitt., *invalid, cited as synonym
**Macromitrium involutum* Mitt., *invalid, no description
**Macromitrium johnsoni* Hampe, *invalid, cited as synonym
**Macromitrium kliewardti* Hornsch. ex Zoll., *invalid, orthographic variant
**Macromitrium knightii* Schimp., *invalid, cited as synonym
**Macromitrium laevigatum* Schimp., *invalid, cited as synonym
**Macromitrium leucoblastum* Müll. Hal., *invalid, no description
**Macromitrium lingulare* Mitt. ex F.M. Bailey, *invalid, orthographic variant

**Macromitrium longipes* var. *acutifolium* Watts & Whitel., *invalid, no description
**Macromitrium macrocarpum* Müll. Hal., *invalid, no description
**Macromitrium macrocladulum* Broth. & Paris, *invalid, no description
**Macromitrium macrophyllum* Mitt., *invalid, no description
**Macromitrium maxwellii* Cardot & Dixon, *invalid, no description
**Macromitrium megacarpum* Müll. Hal., *invalid, no description
**Macromitrium megapterum* Müll. Hal., *invalid, no description
**Macromitrium microblastum* Broth. ex Watts, *invalid, no description
**Macromitrium micropoma* M. Fleisch., *invalid, error for *Macromitrium orthostichum* subsp. *micropoma* M. Fleisch.
**Macromitrium minutum* subsp. *micropoma* M. Fleisch. ex Broth., *invalid, error for *Macromitrium orthostichum* subsp. *micropoma* M. Fleisch.
**Macromitrium miquelianum* Mont., *invalid, cited as synonym
**Macromitrium molliculum* Broth., *invalid, no description
**Macromitrium mossmanii* Müll. Hal. ex Kindb., *invalid, orthographic variant
**Macromitrium nilghiriense* Müll. Hal. ex Mitt., *invalid, orthographic variant
**Macromitrium nitens* Müll. Hal., *invalid, orthographic variant
**Macromitrium nova-guinense* Broth., *invalid, cited as synonym
**Macromitrium osculatii* De Not. ex Kindb., *invalid, orthographic variant
**Macromitrium paccivanum* De Not., *invalid, cited as synonym
Macromitrium paraphysatum Sehnem, illegitimate, later homonym
**Macromitrium parvulum* Wilson, *invalid, no description
**Macromitrium perobtusum* Lam., *invalid, no description
**Macromitrium perpapillosum* Broth., *invalid, cited as synonym
**Macromitrium pertriste* fo. *brevifolia* Thér., *invalid, no latin description
**Macromitrium pilosum* Broth., *invalid, cited as synonym
**Macromitrium piriforme* Müll. Hal. ex Paris, *invalid, orthographic variant
**Macromitrium productinerve* Müll. Hal., *invalid, no description
**Macromitrium profusum* Müll. Hal., *invalid, cited as synonym
**Macromitrium puccioanum* De Not., invalid, no description
**Macromitrium pugionifolium* Broth., invalid, no Latin description, not C. Müller (1898)
**Macromitrium recurvatum* Müll. Hal. ex Broth., *invalid, orthographic variant
Macromitrium recurvifolium (Hornsch.) Müll. Hal., illegitimate, later homonym
Macromitrium recurvulum Cardot, illegitimate, later homonym
**Macromitrium recurvum* Lam., *invalid, no description
**Macromitrium regnellii* var. *minus* Hampe, *invalid, no description
**Macromitrium reticulatum* Müll. Hal., *invalid, no description
**Macromitrium richmondiae* Broth., *invalid, no descriptio
**Macromitrium rigidum* Schimp., *invalid, cited as synonym
**Macromitrium rotundatum* Müll. Hal., *invalid, cited as synonym

***Macromitrium ruficola* Müll. Hal., invalid, no description
***Macromitrium rupestre* var. *robustum* Broth., *invalid, no description
***Macromitrium rupicola* Müll. Hal., *invalid, no description
***Macromitrium savesii* Müll. Hal., *invalid, no description
***Macromitrium sayeria* Mitt., *invalid, no description
***Macromitrium scaberrimum* Broth., invalid, no Latin description
***Macromitrium scabrum* Broth., *invalid, cited as synonym
***Macromitrium schaalianum* Müll. Hal., invalid, no description
***Macromitrium schaalii* Müll. Hal., *invalid, no description
***Macromitrium schwaneckei* Hampe ex Kindb., *invalid, orthographic variant
***Macromitrium semihispidum* Müll. Hal., *invalid, no description
***Macromitrium serricola* Müll. Hal., invalid, no description
***Macromitrium sheareri* Broth., *invalid, no description
***Macromitrium sieberi* Schwägr., *invalid, cited as synonym
***Macromitrium spirale* Hampe, *invalid, no description
***Macromitrium spuriocrispulum* Dusén, *invalid, no description
***Macromitrium squarrosulum* Müll. Hal. ex A. Jaeger, *invalid, orthographic variant
**Macromitrium squarrosum* Müll. Hal., illegitimate, later homonym
***Macromitrium stellatum* Brid. ex E.B. Bartram, *invalid, orthographic variant
***Macromitrium stuhlmannii* Broth., *invalid, no description
***Macromitrium subhemitrichodes* Broth., *invalid, no description
***Macromitrium submicrophyllum* Hampe, *invalid, cited as synonym
***Macromitrium subpliferum* Dusén, *invalid, no description
***Macromitrium subserrulatum* Müll. Hal., *invalid, no description
***Macromitrium sulcatum* var. *lutescens* Müll. Hal. ex A. Jaeger, *invalid, no description
***Macromitrium tacaremae* Hampe, *invalid, orthographic variant
***Macromitrium tenue* var. *brevirameum* Rehmann, *invalid, no description
***Macromitrium tenue* var. *leptocladum* Müll. Hal., *invalid, no description
***Macromitrium tersum* Wilson, *invalid, cited as synonym
***Macromitrium theriotii* Cardot, *invalid, provisional name
***Macromitrium turgidum* Müll. Hal., *invalid, no description
***Macromitrium ulota* Schimp., invalid, no description
***Macromitrium viridissimum* Müll. Hal., *invalid, cited as synonym
***Macromitrium vittatum* Müll. Hal., *invalid, no description
***Macromitrium vohrai* Rajeevan, name in thesis
***Macromitrium wagneri* Müll. Hal. ex Kindb., *invalid, orthographic variant